Headline: Section 2.1.3.3. The headline of each plane-group or space-group table consists of two (somet the following information:

T0201533

① First line:

Short international (Hermann–Mauguin) symbol for the plane or space group: Sections 1.4.1.4 and 2.1.3.4.

Schoenflies symbol for the space group: Section 1.4.1.3. No Schoenflies symbols exist for the plane groups.

Crystal class: The short international (Hermann–Mauguin) symbol for the point group to which the plane or space group belongs (Section 1.4.1, *cf.* Table 2.1.1.1).

The *crystal system* to which the plane or space group belongs (Section 1.3.4.4, *cf.* Table 2.1.1.1). There exist four crystal systems in two-dimensional space (oblique, rectangular, square, hexagonal) and seven crystal systems in three-dimensional space (triclinic, monoclinic, orthorhombic, tetragonal, trigonal, hexagonal and cubic).

② Second line:

The *sequential number of the plane or space group*, as introduced in *International Tables for X-ray Crystallography* Vol. I (1952) (*cf.* Section 1.4.1.2).

The *full Hermann–Mauguin symbol* for the plane or space group: Sections 1.4.1.4 and 2.1.3.4.

Patterson symmetry: Section 2.1.3.5. The *Patterson symmetry* is a crystallographic space group denoted by its Hermann–Mauguin symbol. The Patterson symmetry has the same Bravais-lattice type as the space group itself, while the 'point-group part' of the Patterson-symmetry symbol represents the Laue class to which the plane group or space group belongs.

The third line of the headline is used, where appropriate, to indicate origin choices, settings, cell choices and coordinate axes (see Section 2.1.3.2).

③ *Space-group diagrams*, consisting of one or several orthogonal projections which show the relative location and orientation of the symmetry elements, and one illustration of a set of symmetry-equivalent points in general position. The numbers and types of the diagrams depend on the crystal system. The diagrams and their coordinate axes are described in Section 2.1.3.6, see also Section 1.4.2.5; the graphical symbols of the symmetry elements are listed in Tables 2.1.2.2 to 2.1.2.7.

④ *Origin* of the unit cell: Section 2.1.3.7. In the line *Origin*, the site symmetry of the origin is stated, if different from the identity. A further symbol indicates all symmetry elements (including glide planes and screw axes) that pass through the origin, if any. For noncentrosymmetric space groups, the origin is at a point of highest site symmetry. All centrosymmetric space groups are described with an inversion centre as origin. A further description is given if a centrosymmetric space group contains points of high site symmetry that do not coincide with a centre of symmetry. For space groups with two origin choices, for each of the two origins the location relative to the other origin is also given.

⑤ *Asymmetric unit*: Section 2.1.3.8. An asymmetric unit of a space group is a (simply connected) smallest closed part of space from which, by application of all symmetry operations of the space group, the whole of space is filled. The choice of the asymmetric unit is not unique – it may depend on its intended use. In the space-group tables the asymmetric units are chosen in such a way that Fourier summations can be performed conveniently.

⑥ *Symmetry operations*: Sections 1.4.2.1 and 2.1.3.9, and Chapter 1.2. For each point $\tilde{x}, \tilde{y}, \tilde{z}$ of the general position, the symmetry operation is listed that transforms the initial point x, y, z into the point under consideration. The symbol of the symmetry operation describes the nature of the operation, its glide or screw component (given between parentheses), if present, and the location and orientation of the corresponding geometric element. The symmetry operations are numbered in the same way as the corresponding coordinate triplets of the general position. For space groups with centred cells, several blocks of *Symmetry operations* correspond to the one *General position* block: the number of blocks equals the multiplicity of the centred cell and the numbering scheme of the general position is applied in each block.

[*Continued on inside back cover*]

EXPLANATION OF THE SPACE-GROUP DATA

International Tables for Crystallography

The definitive resource and reference work for crystallography and structural science

Available in print and online

Volume A: Space-group symmetry
Editor Mois I. Aroyo
Sixth edition 2016, 874+xxii pp., ISBN 978-0-470-97423-0

Volume A1: Symmetry relations between space groups
Editors Hans Wondratschek and Ulrich Müller
Second edition 2010, 768+xii pp., ISBN 978-0-470-66079-9

Volume B: Reciprocal space
Editor U. Shmueli
Third edition 2008, 686+xiv pp., ISBN 978-1-119-25933-6

Volume C: Mathematical, physical and chemical tables
Editor E. Prince
Corrected reprint of third edition 2011, 1000+xxxii pp., ISBN 978-0-470-71029-6

Volume D: Physical properties of crystals
Editor A. Authier
Second edition 2014, 564+xii pp., ISBN 978-1-118-76229-5

Volume E: Subperiodic groups
Editors V. Kopský and D. B. Litvin
Second edition 2010, 566+x pp., ISBN 978-0-470-68672-0

Volume F: Crystallography of biological macromolecules
Editors Eddy Arnold, Daniel M. Himmel and Michael G. Rossmann
Second edition 2012, 886+xxx pp., ISBN 978-0-470-66078-2

Volume G: Definition and exchange of crystallographic data
Editors Sydney Hall and Brian McMahon
Corrected reprint of first edition 2010, 594+xii pp., ISBN 978-0-470-68910-3

Volume H: Powder diffraction
Editors Christopher J. Gilmore, James A. Kaduk and Henk Schenk
First edition 2019, 904+xxviii pp., ISBN 978-1-118-41628-0

Volume I: X-ray absorption spectroscopy and related techniques
Editors Christopher T. Chantler, Federico Boscherini and Bruce Bunker
In preparation. Early view chapters available online, ISBN 978-1-119-43394-1

Teaching edition of International Tables for Crystallography: Crystallographic symmetry
Editor Mois I. Aroyo
Sixth edition 2021, 248 pp., ISBN 978-0-470-97422-3

it.iucr.org

INTERNATIONAL TABLES FOR CRYSTALLOGRAPHY

Teaching Edition
CRYSTALLOGRAPHIC SYMMETRY

Edited by
MOIS I. AROYO

Sixth Edition

Published for
THE INTERNATIONAL UNION OF CRYSTALLOGRAPHY
by

WILEY

2021

A C.I.P. Catalogue record for this book
is available from the Library of Congress
ISBN 978-0-470-97422-3

This edition first published 2021

First published (as the *Brief Teaching Edition of Volume A*) in 1985
Second edition 1988, third edition 1993, fourth edition 1996, fifth edition 2002
Sixth edition 2021

Published for the International Union of Crystallography by John Wiley & Sons, Ltd.
Registered offices: John Wiley & Sons, Inc., 111 River Street, Hoboken, NJ 07030, USA,
and John Wiley & Sons Ltd, The Atrium, Southern Gate, Chichester, West Sussex,
PO19 8SQ, UK.

For details of Wiley's editorial offices, customer services and more information about Wiley's products visit www.wiley.com.

Printed and bound by CPI Group (UK) Ltd, Croydon CR0 4YY

Contributing authors

H. Arnold†: Institut für Kristallographie, Rheinisch-Westfälische Technische Hochschule, Aachen, Germany. [1.5]

M. I. Aroyo: Departamento de Física de la Materia Condensada, Universidad del País Vasco (UPV/EHU), Bilbao, Spain. [1.2, 1.4, 1.5, 2.1, 2.2, 2.4]

E. F. Bertaut†: Laboratoire de Cristallographie, CNRS, Grenoble, France. [1.5]

C. P. Brock: Department of Chemistry, University of Kentucky, 505 Rose St, Lexington, KY 40506-0055, USA. [1.7]

G. Chapuis: École Polytechnique Fédérale de Lausanne, BSP/Cubotron, CH-1015 Lausanne, Switzerland. [1.4, 1.5]

H. D. Flack†: Chimie minérale, analytique et appliquée, University of Geneva, Geneva, Switzerland. [1.6, 2.1]

G de la Flor: Institute of Applied Geosciences, Karlsruhe Institute of Technology, D-76131 Karlsruhe, Germany. [2.4]

A. M. Glazer: Department of Physics, University of Oxford, Parks Road, Oxford, United Kingdom. [1.4]

Th. Hahn†: Institut für Kristallographie, RWTH Aachen University, 52062 Aachen, Germany. [2.1]

V. Kopský†: Bajkalska 1170/28, 100 00 Prague 10, Czech Republic. [1.7]

E. Kroumova: eFaber Soluciones Inteligentes, S.L., Maximo Aguirre 11, 48011 Bilbao, Spain. [2.4]

D. B. Litvin: Department of Physics, The Eberly College of Science, Penn State – Berks Campus, The Pennsylvania State University, PO Box 7009, Reading, PA 19610-6009, USA. [1.7, 2.3]

A. Looijenga-Vos†: Laboratorium voor Chemische Fysica, Rijksuniversiteit Groningen, The Netherlands. [2.1]

Ulrich Müller: Fachbereich Chemie, Philipps-Universität, D-35032 Marburg, Germany. [1.7, 2.2]

K. Momma: National Museum of Nature and Science, 4-1-1 Amakubo, Tsukuba, Ibaraki 305-0005, Japan. [2.1]

U. Shmueli: School of Chemistry, Tel Aviv University, 69978 Tel Aviv, Israel. [1.6]

B. Souvignier: Radboud University Nijmegen, Faculty of Science, Mathematics and Computing Science, Institute for Mathematics, Astrophysics and Particle Physics, Postbus 9010, 6500 GL Nijmegen, The Netherlands. [1.1, 1.3, 1.4, 1.5]

J. C. H. Spence: Department of Physics, Arizona State University, Rural Rd, Tempe, AZ 85287, USA. [1.6]

H. Wondratschek†: Laboratorium für Applikationen der Synchrotronstrahlung (LAS), Universität Karlsruhe, Germany. [1.2, 1.4, 1.5, 1.7, 2.2]

† Deceased.

Contents

PAGE

Preface (Mois I. Aroyo) x

Symbols for crystallographic items used in this book xi

PART 1. INTRODUCTION TO CRYSTALLOGRAPHIC SYMMETRY 1

1.1. A general introduction to groups (Bernd Souvignier) 2

1.1.1. Introduction 2

1.1.2. Basic properties of groups 2

1.1.3. Subgroups 4

1.1.4. Cosets 5

1.1.5. Normal subgroups, factor groups 6

1.1.6. Group actions 7

1.1.7. Conjugation, normalizers 8

1.2. Crystallographic symmetry (Hans Wondratschek and Mois I. Aroyo) 10

1.2.1. Crystallographic symmetry operations 10

1.2.2. Matrix description of symmetry operations 11

1.2.2.1. Matrix–column presentation of isometries 11

1.2.2.2. Combination of mappings and inverse mappings 12

1.2.2.3. The geometric meaning of (*W*, *w*) 12

1.2.3. Symmetry elements 14

1.3. A general introduction to space groups (Bernd Souvignier) 17

1.3.1. Introduction 17

1.3.2. Lattices 17

1.3.2.1. Basic properties of lattices 17

1.3.2.2. Metric properties 18

1.3.2.3. Unit cells 19

1.3.2.4. Primitive and centred lattices 19

1.3.2.5. Reciprocal lattice 21

1.3.3. The structure of space groups 22

1.3.3.1. Point groups of space groups 22

1.3.3.2. Coset decomposition with respect to the translation subgroup 23

1.3.3.3. Symmorphic and non-symmorphic space groups 24

1.3.4. Classification of space groups 25

1.3.4.1. Space-group types 26

1.3.4.2. Geometric crystal classes 27

1.3.4.3. Bravais types of lattices and Bravais classes 28

1.3.4.4. Other classifications of space groups 29

1.4. Space groups and their descriptions (Bernd Souvignier, Hans Wondratschek, Mois I. Aroyo, Gervais Chapuis and A. M. Glazer) 32

1.4.1. Symbols of space groups (Hans Wondratschek) 32

1.4.1.1. Introduction 32

1.4.1.2. Space-group numbers 32

1.4.1.3. Schoenflies symbols 32

1.4.1.4. Hermann–Mauguin symbols of the space groups 33

1.4.1.5. Hermann–Mauguin symbols of the plane groups 36

1.4.2. Descriptions of space-group symmetry operations (Mois I. Aroyo, Gervais Chapuis, Bernd Souvignier and A. M. Glazer) 36

1.4.2.1. Symbols for symmetry operations 36

1.4.2.2. Seitz symbols of symmetry operations 37

1.4.2.3. Symmetry operations and the general position 39

1.4.2.4. Additional symmetry operations and symmetry elements 40

1.4.2.5. Space-group diagrams 42

1.4.3. Generation of space groups (Hans Wondratschek) 44

1.4.3.1. Selected order for non-translational generators 45

1.4.4. General and special Wyckoff positions (Bernd Souvignier) 46

1.4.4.1. Crystallographic orbits 46

1.4.4.2. Wyckoff positions 47

1.4.5. Sections and projections of space groups (Bernd Souvignier) 50

1.4.5.1. Introduction 50

1.4.5.2. Sections 50

1.4.5.3. Projections 53

1.5. Transformations of coordinate systems (Hans Wondratschek, Mois I. Aroyo, Bernd Souvignier and Gervais Chapuis) 57

1.5.1. Origin shift and change of the basis (Hans Wondratschek and Mois I. Aroyo) 57

1.5.1.1. Origin shift 57

1.5.1.2. Change of the basis 58

1.5.1.3. General change of coordinate system 63

1.5.2. Transformations of crystallographic quantities under coordinate transformations (Hans Wondratschek and Mois I. Aroyo) 63

1.5.2.1. Covariant and contravariant quantities 63

1.5.2.2. Metric tensors of direct and reciprocal lattices 64

1.5.2.3. Transformation of matrix–column pairs of symmetry operations 64

1.5.2.4. Example: paraelectric-to-ferroelectric phase transition of GeTe 64

1.5.3. Transformations between different space-group descriptions (Gervais Chapuis, Hans Wondratschek and Mois I. Aroyo) 66

1.5.3.1. Space groups with more than one description in *IT* A 66

1.5.3.2. Examples 67

1.5.4. Synoptic tables of plane and space groups (Bernd Souvignier, Gervais Chapuis and Hans Wondratschek) 69

1.6. Introduction to the theory and practice of space-group determination (Uri Shmueli, Howard D. Flack and John C. H. Spence) 75

1.6.1. Overview 75

1.6.2. Symmetry determination from single-crystal studies (Uri Shmueli and Howard D. Flack) 75

1.6.2.1. Symmetry information from the diffraction pattern 75

1.6.2.2. Structure-factor statistics and crystal symmetry 76

1.6.2.3. Symmetry information from the structure solution 78

1.6.2.4. Restrictions on space groups 78

1.6.2.5. Pitfalls in space-group determination 79

1.6.3. Theoretical background of reflection conditions (Uri Shmueli) 79

1.6.3.1. Example: a determination of reflection conditions 80

1.6.4. Reflection conditions and possible space groups (Howard D. Flack and Uri Shmueli) 81

1.6.4.1. Introduction 81

1.6.4.2. Examples 82

1.6.5. Space-group determination in macromolecular crystallography (Howard D. Flack) 83

1.6.6. Space groups for nanocrystals by electron microscopy (John C. H. Spence) 83

1.6.7. Examples (Howard D. Flack) 84

1.6.7.1. Example (1), 4-chlorophenol, C_6H_5OCl 84

1.6.7.2. Example (2), $[BDTA]_2[CuCl_4]$ .. 85

1.6.7.3. Example (3), flo19, $C_{62}H_{46}N_{14}$ .. 86

1.6.7.4. Example (4), CSD refcode FOYTAO01, $C_{12}H_{20}O_6$ .. 86

1.7. Applications of crystallographic symmetry: space-group symmetry relations, subperiodic groups and magnetic symmetry (Hans Wondratschek, Ulrich Müller, Daniel B. Litvin, Vojtech Kopský and Carolyn Pratt Brock) .. 89

1.7.1. Subgroups and supergroups of space groups (Hans Wondratschek) .. 89

1.7.1.1. *Translationengleiche* (or *t*-) subgroups of space groups .. 90

1.7.1.2. *Klassengleiche* (or *k*-) subgroups of space groups .. 91

1.7.1.3. Isomorphic subgroups of space groups .. 91

1.7.1.4. Supergroups .. 91

1.7.2. Relations between Wyckoff positions for group–subgroup-related space groups (Ulrich Müller) .. 92

1.7.2.1. Symmetry relations between crystal structures .. 92

1.7.2.2. Substitution derivatives .. 92

1.7.2.3. Phase transitions .. 92

1.7.2.4. Domain structures .. 93

1.7.2.5. Presentation of the relations between the Wyckoff positions among group–subgroup-related space groups .. 93

1.7.3. Subperiodic groups .. 94

1.7.3.1. Relationships between space groups and subperiodic groups (Daniel B. Litvin and Vojtech Kopský) .. 94

1.7.3.2. Use of subperiodic groups to describe structural units (Carolyn Pratt Brock) .. 96

1.7.3.3. Applications of rod groups (Ulrich Müller) .. 98

1.7.4. Magnetic subperiodic groups and magnetic space groups (Daniel B. Litvin) .. 100

1.7.4.1. Introduction .. 100

1.7.4.2. Survey of magnetic subperiodic groups and magnetic space groups .. 101

PART 2. CRYSTALLOGRAPHIC SYMMETRY DATA .. 107

2.1. Guide to and examples of the space-group tables in *IT* A (Theo Hahn, Aafje Looijenga-Vos, Mois I. Aroyo, Howard D. Flack and Koichi Momma) .. 108

2.1.1. Conventional descriptions of plane and space groups (Theo Hahn and Aafje Looijenga-Vos) .. 108

2.1.1.1. Classification of space groups .. 108

2.1.1.2. Conventional coordinate systems and cells .. 108

2.1.2. Symbols of symmetry elements (Theo Hahn and Mois I. Aroyo) .. 110

2.1.3. Contents and arrangement of the tables (Theo Hahn, Aafje Looijenga-Vos, Mois I. Aroyo, Howard D. Flack and Koichi Momma) .. 116

2.1.3.1. General layout .. 116

2.1.3.2. Space groups with more than one description .. 116

2.1.3.3. Headline .. 117

2.1.3.4. International (Hermann–Mauguin) symbols for plane groups and space groups .. 117

2.1.3.5. Patterson symmetry .. 118

2.1.3.6. Space-group diagrams .. 119

2.1.3.7. Origin .. 123

2.1.3.8. Asymmetric unit .. 123

2.1.3.9. Symmetry operations .. 124

2.1.3.10. Generators .. 125

2.1.3.11. Positions .. 126

2.1.3.12. Oriented site-symmetry symbols .. 126

2.1.3.13. Reflection conditions .. 127

2.1.3.14. Symmetry of special projections .. 130

2.1.3.15. Crystallographic groups in one dimension .. 131

2.1.4. Examples of plane- and space-group tables .. 131

2.2. The symmetry-relations tables of *IT* A1 (Hans Wondratschek, Mois I. Aroyo and Ulrich Müller) .. 212

2.2.1. Guide to the subgroup tables (Hans Wondratschek and Mois I. Aroyo) .. 212

2.2.1.1. Contents and arrangement of the subgroup tables .. 212

2.2.1.2. I Maximal *translationengleiche* subgroups (*t*-subgroups) .. 212

2.2.1.3. II Maximal *klassengleiche* subgroups (*k*-subgroups) .. 213

2.2.1.4. Minimal supergroups .. 214

2.2.2. Examples of the subgroup tables .. 214

2.2.3. Guide to the tables of relations between Wyckoff positions (Ulrich Müller) .. 217

2.2.3.1. Guide to the use of the tables .. 217

2.2.3.2. Cell transformations .. 219

2.2.3.3. Origin shifts .. 219

2.2.3.4. Nonconventional settings of orthorhombic space groups .. 219

2.2.4. Examples of the tables of relations between Wyckoff positions .. 220

2.3. The subperiodic group tables of *IT* E (Daniel B. Litvin) .. 224

2.3.1. Guide to the subperiodic group tables .. 224

2.3.1.1. Content and arrangement of the tables .. 224

2.3.1.2. Diagrams for the symmetry elements and the general position .. 225

2.3.1.3. Symmetry operations .. 225

2.3.1.4. Subgroups and supergroups .. 225

2.3.2. Examples of subperiodic group tables .. 225

2.4. The Symmetry Database (Eli Kroumova, Gemma de la Flor and Mois I. Aroyo) .. 232

2.4.1. Space-group symmetry data .. 232

2.4.2. Symmetry relations between space groups .. 233

2.4.3. 3D Crystallographic point groups .. 233

2.4.4. Availability .. 233

Subject index .. 234

Preface

This Teaching Edition (hereafter referred to as the TE) is the successor to the *Brief Teaching Edition of Volume A* (which was last revised in 2005), although new material, substantial revisions and reorganisation of the text has resulted in a book that is quite different in structure and content to its predecessor. It focuses on the particular topic of symmetry, owing to its fundamental role in crystallography, and provides a unified and coherent introduction to the symmetry information found in three volumes of *International Tables for Crystallography*: the basic crystallographic data for the plane and space groups in Volume A (*IT* A), the symmetry relations between space groups treated in Volume A1 (*IT* A1) and the subperiodic-group data found in Volume E (*IT* E). It also introduces the Symmetry Database, which forms part of the online version of *International Tables for Crystallography* at https://it.iucr.org.

This Teaching Edition is designed for graduate (and postgraduate) students, and young researchers who have some awareness of the basics of symmetry and diffraction and who need to use crystallographic symmetry methods in their work. The TE can thus serve as an interface between elementary crystallography textbooks and the texts in *IT* A, A1 and E. Sufficient up-to-date and accessible references to further specialized sources are provided for those who need to go deeper into the subject, and to textbooks and basic crystallographic literature that could be helpful for those who need a course in basic crystallography or introduction to the mathematics that is required. The fruitful combination of tables for practical use and a didactic introduction to symmetry makes this TE a handy tool with which researchers and students can familiarize themselves with the use of crystallographic symmetry and its practical applications.

Part 1 of the TE is based on the material of the introductory part of the sixth edition of *IT* A, with additional explanations and illustrative examples to provide the reader with practical experience in the use of crystallographic symmetry data. The fundamental concepts of group theory, focusing on those properties that are of particular importance for crystallography, are introduced in the first chapter. The matrix formalism, which provides efficient instruments for the analytical description of crystallographic symmetry, is the subject of Chapter 1.2. Chapter 1.3 offers an introduction to the structure of space groups and their various classification schemes, focusing on the classifications into space-group types, geometric crystal classes and Bravais types of lattices. Chapter 1.4 deals with various crystallographic terms and the symbols used to present the symmetry data in the space-group tables of *IT* A. The purpose of Chapter 1.5 is to provide the mathematical tools needed to work out how crystallographic data are transformed if the coordinate system is changed. Chapter 1.6 offers a detailed presentation of methods for determining the symmetry of single-domain crystals from diffraction data. The tables of general reflection conditions shown in this chapter also include 'diffraction symbols' (previously known as 'extinction symbols'), and four examples of space-group determination from real intensity data are analysed. Chapter 1.7 deals with applications of crystallographic symmetry, including symmetry relations between space groups, as treated in *IT* A1, and subperiodic groups, as covered in *IT* E. Possible applications of the subperiodic groups in the description of crystal systems with rod and layer symmetry are discussed. Finally, magnetic groups, including magnetic subperiodic groups and magnetic space groups, are briefly introduced.

The material of Part 2 focuses on the presentation of the tables of crystallographic symmetry data found in Volumes A, A1 and E, including descriptions of the symbols and terms found in the tables and guides for their use. Representative sets of tables of varying complexity are included that will be useful for teaching about crystallographic symmetry. The guide to the space-group tables of Volume A is very similar to that given in Chapter 2.1 of *IT* A and is complemented by the tables for 7 of the plane groups and 35 of the space groups. In Chapter 2.2, the data for the maximal subgroups and minimal supergroups in *IT* A1 are described and illustrated using the examples of $P3_112$ (151) and $P3_121$ (152). The guide to the arrangement of the data specifying relations between Wyckoff positions in *IT* A1 uses the tables of $P6_3/mmc$ (194) and *Cmcm* (63) as examples. The subperiodic-group tables in *IT* E are described in Chapter 2.3 and are illustrated by the tables of the rod group $\not{p}mc2_1$ (No. 17), and of the layer groups $p2_1/b11$ (No. 17) and *pbam* (No. 44). Finally, Chapter 2.4 gives a brief description of the Symmetry Database, which provides online access to data for the crystallographic space and point groups that extend and enhance the symmetry information given in the printed editions of *IT* A and A1.

It is a great pleasure to express my gratitude to all authors of the TE; their contributions were indispensable for the successful completion of this project. We should dedicate this book to the memory of those authors who, to my deep regret, are not among us anymore: H. Wondratschek (KIT, Karlsruhe), Th. Hahn (RWTH, Aachen), A. Looijenga-Vos (Rijksuniversiteit Groningen), V. Kopský (Academy of Science, Prague) and H. Flack (University of Geneva).

My thanks should be extended to the International Union of Crystallography and its Editorial Office for making this Teaching Edition possible, and in particular to C. P. Brock (University of Kentucky, Lexington), P. R. Strickland and N. J. Ashcroft (IUCr Editorial Office, Chester) for providing much of the initial stimulus for this work, and for their confidence and support during it. I am particularly grateful to Nicola Ashcroft for the careful and dedicated technical editing of this book, and for her help and professionalism.

Helpful discussions were held with and invaluable comments were received from B. Souvignier (Radboud University, Nijmegen), M. Nespolo (Université de Lorraine, Nancy), U. Shmueli (Tel Aviv University), M. Glazer (Oxford University) and many others, for which I thank them most sincerely. The constant encouragement and understanding during my work on this Teaching Edition from colleagues and friends at the Universidad del País Vasco are also gratefully acknowledged. The preparation of this book was supported by projects funded by the Universidad del País Vasco, the Government of the Basque Country and the Spanish Ministry of Science and Innovation.

MOIS I. AROYO
Editor, *Teaching Edition of International Tables for Crystallography*

Symbols for crystallographic items used in this book

Direct space: points and vectors

$\mathbb{E}^3$ ($\mathbb{E}^2$)	3-dimensional (2-dimensional) Euclidean point space
$\mathbb{V}^3$ ($\mathbb{V}^2$)	3-dimensional (2-dimensional) vector space
$\mathbb{R}$, $\mathbb{Q}$, $\mathbb{Z}$	the field of real numbers, the field of rational numbers, the ring of integers
$\mathbf{L}$	lattice in $\mathbb{V}^3$
L	line in $\mathbb{E}^3$
$\mathbf{a}, \mathbf{b}, \mathbf{c}$; or $\mathbf{a}_i$	basis vectors of the lattice
a, b, c; or $\lvert\mathbf{a}\rvert, \lvert\mathbf{b}\rvert, \lvert\mathbf{c}\rvert$	lengths of basis vectors, lengths of cell edges } lattice parameters
α, β, γ; or α_j	interaxial angles $\angle(\mathbf{b}, \mathbf{c})$, $\angle(\mathbf{c}, \mathbf{a})$, $\angle(\mathbf{a}, \mathbf{b})$ } lattice parameters
$\boldsymbol{G}$, g_{ik}	fundamental matrix (metric tensor) and its coefficients
V	cell volume
X, Y, Z, P	points
$\mathbf{r}, \mathbf{d}, \mathbf{x}, \mathbf{v}, \mathbf{u}$	vectors, position vectors
r, $\lvert\mathbf{r}\rvert$	norm, length of a vector
$\mathbf{x} = x\mathbf{a} + y\mathbf{b} + z\mathbf{c}$	vector with coefficients x, y, z
x, y, z; or x_i	point coordinates expressed in units of a, b, c; coefficients of a vector
$\boldsymbol{x} = \begin{pmatrix} x \\ y \\ z \end{pmatrix} \equiv \begin{pmatrix} x_1 \\ x_2 \\ x_3 \end{pmatrix}$	column of point coordinates or vector coefficients
$\mathbf{t}$	translation vector
t_1, t_2, t_3; or t_i	coefficients of translation vector $\mathbf{t}$
$\boldsymbol{t} = \begin{pmatrix} t_1 \\ t_2 \\ t_3 \end{pmatrix}$	column of coefficients of translation vector $\mathbf{t}$
O	origin
$\mathbf{o}$	zero vector (all coefficients zero)
$\boldsymbol{o}$	(3×1) column of zero coefficients
$\mathbf{a}', \mathbf{b}', \mathbf{c}'$; or $\mathbf{a}'_i$	new basis vectors after a transformation of the coordinate system (basis transformation)
$\mathbf{r}'$; or $\mathbf{x}'$; x', y', z'; or x'_i	vector and point coordinates after a transformation of the coordinate system (basis transformation)
$\boldsymbol{x}' = \begin{pmatrix} x' \\ y' \\ z' \end{pmatrix}$	column of coordinates after a transformation of the coordinate system (basis transformation)
$\tilde{X}$	image of a point X after the action of a symmetry operation
$\tilde{x}, \tilde{y}, \tilde{z}$; or $\tilde{x}_i$	coordinates of an image point $\tilde{X}$
$\tilde{\boldsymbol{x}} = \begin{pmatrix} \tilde{x} \\ \tilde{y} \\ \tilde{z} \end{pmatrix}$	column of coordinates of an image point $\tilde{X}$

Directions and planes

$[uvw]$	indices of a lattice direction (zone axis)
$\langle uvw \rangle$	indices of a set of all symmetry-equivalent lattice directions
(hkl)	indices of a crystal face, or of a single net plane (Miller indices)
$(hkil)$	indices of a crystal face, or of a single net plane, for the hexagonal axes $\mathbf{a}_1$, $\mathbf{a}_2$, $\mathbf{a}_3$, $\mathbf{c}$ (Bravais–Miller indices)
$\{hkl\}$	indices of a set of all symmetry-equivalent crystal faces ('crystal form'), or net planes
$\{hkil\}$	indices of a set of all symmetry-equivalent crystal faces ('crystal form'), or net planes, for the hexagonal axes $\mathbf{a}_1$, $\mathbf{a}_2$, $\mathbf{a}_3$, $\mathbf{c}$
hkl	indices of the Bragg reflection (Laue indices) from the set of parallel equidistant net planes (hkl)
d_{hkl}	interplanar distance, or spacing, of neighbouring net planes (hkl)

Reciprocal space

$\mathbf{L}^*$	reciprocal lattice
$\mathbf{a}^*, \mathbf{b}^*, \mathbf{c}^*$; or $\mathbf{a}_i^*$	basis vectors of the reciprocal lattice
a^*, b^*, c^*; or $\lvert\mathbf{a}^*\rvert, \lvert\mathbf{b}^*\rvert, \lvert\mathbf{c}^*\rvert$	lengths of basis vectors of the reciprocal lattice
$\alpha^*, \beta^*, \gamma^*$; or α_j^*	interaxial angles $\angle(\mathbf{b}^*, \mathbf{c}^*)$, $\angle(\mathbf{c}^*, \mathbf{a}^*)$, $\angle(\mathbf{a}^*, \mathbf{b}^*)$ of the reciprocal lattice
$\mathbf{r}^*$, or $\mathbf{h}$	vector in reciprocal space, or vector of reciprocal lattice
r^*, or $\lvert\mathbf{r}^*\rvert$	length of a vector in reciprocal space
h, k, l; or h_i	coefficients of a reciprocal-lattice vector
$\boldsymbol{h} = (h, k, l)$	(1×3) row of coefficients of a reciprocal-lattice vector
V^*	cell volume of the reciprocal lattice
$\boldsymbol{G}^*$, g_{ik}^*	fundamental matrix (metric tensor) of the reciprocal lattice and its coefficients

Functions

$\rho(xyz)$	electron density at the point x, y, z
$P(uvw)$	Patterson function for a vector with coefficients u, v, w
$F(hkl)$, or F	structure factor (of the unit cell) corresponding to the Bragg reflection hkl
$\lvert F(hkl)\rvert$, or $\lvert F\rvert$	modulus of the structure factor $F(hkl)$

Mappings, symmetry operations and their matrix–column presentation

$\boldsymbol{A}, \boldsymbol{B}, \boldsymbol{W}$	(3 × 3) matrices describing the linear part of a mapping
A_{ik}, W_{ik}	matrix coefficients
$\boldsymbol{I}$	(3 × 3) unit matrix
$\boldsymbol{A}^{\mathrm{T}}$	matrix $\boldsymbol{A}$ transposed
$\det(\boldsymbol{A})$, $\mathrm{tr}(\boldsymbol{A})$	determinant of matrix $\boldsymbol{A}$, trace of matrix $\boldsymbol{A}$
$\boldsymbol{w} = \begin{pmatrix} w_1 \\ w_2 \\ w_3 \end{pmatrix}$	(3 × 1) column of coefficients w_i describing the translation part of a mapping
$\boldsymbol{w}_g$	intrinsic translation part of a symmetry operation
$\boldsymbol{w}_l$	location translation part of a symmetry operation
A, I, W	mappings, symmetry operations
t	translation symmetry operation
ϕ	rotation angle of a symmetry operation
$(\boldsymbol{W}, \boldsymbol{w})$	matrix–column pair of a symmetry operation given by a (3 × 3) matrix $\boldsymbol{W}$ and a (3 × 1) column $\boldsymbol{w}$
$(\boldsymbol{I}, \boldsymbol{t})$	matrix–column pair of a translation
$(\boldsymbol{I}, \boldsymbol{o})$	matrix–column pair of the identity
$\{\boldsymbol{R}\|\boldsymbol{v}\}$	Seitz symbol of a symmetry operation
$(\boldsymbol{P}, \boldsymbol{p})$	transformation of the coordinate system, described by a (3 × 3) matrix $\boldsymbol{P}$ and a (3 × 1) column $\boldsymbol{p}$
$(\boldsymbol{Q}, \boldsymbol{q})$	inverse transformation of $(\boldsymbol{P}, \boldsymbol{p})$: $(\boldsymbol{Q}, \boldsymbol{q}) = (\boldsymbol{P}, \boldsymbol{p})^{-1}$

Groups

$\mathcal{G}$	group, space group
$\mathcal{P}, \mathcal{S}, \mathcal{F}, \mathcal{D}, \mathcal{R}$	groups
a, b, g, h, m, t	group elements
e	unit element of a group
$\mathcal{H}, \mathcal{U}$	subgroups
$\mathcal{I}$	trivial group, consisting of the unit element e only
$\|\mathcal{G}\|$	order of the group $\mathcal{G}$
i, or $[i]$	index of a subgroup in a group
$\mathcal{T}$, or $\mathcal{T}_{\mathcal{G}}$	group of all translations of a space group, or of the space group $\mathcal{G}$
t	element of the translation group $\mathcal{T}$
$\mathcal{P}$, or $\mathcal{P}_{\mathcal{G}}$	point group of a space group, or of the space group $\mathcal{G}$
$\mathcal{M}$	Hermann's group
$\mathcal{E}$	group of all isometries (motions) (Euclidean group)
φ	homomorphic mapping (homomorphism)
$\mathcal{G}/\mathcal{H}$	factor group or quotient group of $\mathcal{G}$ by $\mathcal{H}$
$\mathcal{N}_{\mathcal{G}}(\mathcal{H})$	normalizer of $\mathcal{H}$ in $\mathcal{G}$
Ω, ω	set on which a group acts, object in Ω
$\mathcal{G}(\omega)$	orbit of ω under the group $\mathcal{G}$
$\mathcal{S}_{\mathcal{G}}(\omega)$, $\mathcal{S}_{\mathcal{H}}(\omega)$	stabilizer of ω in the group $\mathcal{G}$, or $\mathcal{H}$
$\mathcal{O} = \mathcal{G}(X)$	orbit of point X under the group $\mathcal{G}$
$\mathcal{S}_X = \mathcal{S}_{\mathcal{G}}(X)$	site-symmetry group of point X

1. INTRODUCTION TO CRYSTALLOGRAPHIC SYMMETRY

By M. I. Aroyo, C. P. Brock, G. Chapuis, H. D. Flack, A. M. Glazer, V. Kopský, D. B. Litvin, U. Müller, U. Shmueli, B. Souvignier, J. C. H. Spence and H. Wondratschek

1.1. A general introduction to groups 2
1.1.1. Introduction 2
1.1.2. Basic properties of groups 2
1.1.3. Subgroups 4
1.1.4. Cosets 5
1.1.5. Normal subgroups, factor groups 6
1.1.6. Group actions 7
1.1.7. Conjugation, normalizers 8
1.2. Crystallographic symmetry 10
1.2.1. Crystallographic symmetry operations 10
1.2.2. Matrix description of symmetry operations 11
1.2.3. Symmetry elements 14
1.3. A general introduction to space groups 17
1.3.1. Introduction 17
1.3.2. Lattices 17
1.3.3. The structure of space groups 22
1.3.4. Classification of space groups 25
1.4. Space groups and their descriptions 32
1.4.1. Symbols of space groups 32
1.4.2. Descriptions of space-group symmetry operations 36
1.4.3. Generation of space groups 44
1.4.4. General and special Wyckoff positions 46
1.4.5. Sections and projections of space groups 50
1.5. Transformations of coordinate systems 57
1.5.1. Origin shift and change of the basis 57
1.5.2. Transformations of crystallographic quantities under coordinate transformations 63
1.5.3. Transformations between different space-group descriptions 66
1.5.4. Synoptic tables of plane and space groups 69
1.6. Introduction to the theory and practice of space-group determination 75
1.6.1. Overview 75
1.6.2. Symmetry determination from single-crystal studies 75
1.6.3. Theoretical background of reflection conditions 79
1.6.4. Reflection conditions and possible space groups 81
1.6.5. Space-group determination in macromolecular crystallography 83
1.6.6. Space groups for nanocrystals by electron microscopy 83
1.6.7. Examples 84
1.7. Applications of crystallographic symmetry: space-group symmetry relations, subperiodic groups and magnetic symmetry 89
1.7.1. Subgroups and supergroups of space groups 89
1.7.2. Relations between Wyckoff positions for group–subgroup-related space groups 92
1.7.3. Subperiodic groups 94
1.7.4. Magnetic subperiodic groups and magnetic space groups 100

1.1. A general introduction to groups

Bernd Souvignier

In this chapter we give a general introduction to group theory, which provides the mathematical background for considering symmetry properties. Starting from basic principles, we discuss those properties of groups that are of particular interest in crystallography. To readers interested in a more elaborate treatment of the theoretical background, the standard textbooks by Armstrong (2010), Hill (1999) or Sternberg (2008) are recommended; an account from the perspective of crystallography can also be found in Müller (2013).

1.1.1. Introduction

Crystal structures may be investigated and classified according to their symmetry properties. But in a strict sense, crystal structures in nature are never perfectly symmetric, due to impurities, structural imperfections and especially their finite extent. Therefore, symmetry considerations deal with *idealized* crystal structures that are free from impurities and structural imperfections and that extend infinitely in all directions. In the mathematical model of such an idealized crystal structure, the atoms are replaced by points in a three-dimensional point space and this model will be called a *crystal pattern*.

A symmetry operation of a crystal pattern is a transformation of three-dimensional space that preserves distances and angles and that leaves the crystal pattern as a whole unchanged. The symmetry of a crystal pattern is then understood as the collection of all symmetry operations of the pattern.

The following simple statements about the symmetry operations of a crystal pattern are almost self-evident:

(*a*) If two symmetry operations are applied successively, the crystal pattern is still invariant, thus the combination of the two operations (called their *composition*) is again a symmetry operation.

(*b*) Every symmetry operation can be reversed by simply moving every point back to its original position.

These observations (together with the fact that leaving all points in their position is also a symmetry operation) show that the symmetry operations of a crystal pattern form an algebraic structure called a *group*.

1.1.2. Basic properties of groups

Although groups occur in innumerable contexts, their basic properties are very simple and are captured by the following definition.

Definition. Let $\mathcal{G}$ be a set of elements on which a binary operation is defined which assigns to each pair (g, h) of elements the composition $g \circ h \in \mathcal{G}$. Then $\mathcal{G}$, together with the binary operation $\circ$, is called a *group* if the following hold:

(i) the binary operation is associative, *i.e.* $(g \circ h) \circ k = g \circ (h \circ k)$;

(ii) there exists a *unit element* or *identity element* $e \in \mathcal{G}$ such that $g \circ e = g$ and $e \circ g = g$ for all $g \in \mathcal{G}$;

(iii) every $g \in \mathcal{G}$ has an inverse element, denoted by g^{-1}, for which $g \circ g^{-1} = g^{-1} \circ g = e$.

In most cases, the composition of group elements is regarded as a *product* and is written as $g \cdot h$ or even gh instead of $g \circ h$. An exception is groups where the composition is addition, *e.g.* a group of translations. In such a case, the composition $\mathbf{a} \circ \mathbf{b}$ is more conveniently written as $\mathbf{a} + \mathbf{b}$.

Examples

(i) The set $\mathbb{Z}$ of all integers forms a group with addition as operation. The identity element is 0, the inverse element for $a \in \mathbb{Z}$ is $-a$.

(ii) The group $3m$ of all symmetries of an equilateral triangle is a group with the composition of symmetry operations as binary operation. The group contains six elements, namely three reflections, two rotations and the identity element. It is schematically displayed in Fig. 1.1.2.2.

(iii) The set of all real $n \times n$ matrices with determinant $\neq 0$ is a group with matrix multiplication as operation. This group is called the *general linear group* and denoted by $\mathrm{GL}_n(\mathbb{R})$.

If a group $\mathcal{G}$ contains finitely many elements, it is called a *finite group* and the number of its elements is called the *order* of the group, denoted by $|\mathcal{G}|$. A group with infinitely many elements is called an *infinite group*.

For a group element g, its *order* is the smallest integer $n > 0$ such that $g^n = e$ is the identity element. If there is no such integer, then g is said to be of *infinite order*.

The group operation is not required to be *commutative*, *i.e.* in general one will have $gh \neq hg$. However, a group $\mathcal{G}$ in which $gh = hg$ for all g, h is said to be a *commutative* or *abelian group*.

The inverse of the product gh of two group elements is the product of the inverses of the two elements in reversed order, *i.e.* $(gh)^{-1} = h^{-1}g^{-1}$.

Groups of small order may be displayed by their *multiplication table*, which is a square table with rows and columns indexed by the group elements and where the intersection of the row labelled by g and of the column labelled by h is the product gh. It follows immediately from the invertibility of the group elements that each row and column of the multiplication table contains every group element precisely once.

Examples

(i) The group of rotations leaving an equilateral triangle invariant consists of the rotations by 0, 120 and 240°, which are denoted by 1, 3^+ and 3^-, respectively. Its multiplication table is

	1	3^+	3^-
1	1	3^+	3^-
3^+	3^+	3^-	1
3^-	3^-	1	3^+

Note that in this and all subsequent examples of crystallographic point groups we will use *Seitz symbols* (*cf.* Section 1.4.2.2) for the symmetry operations and

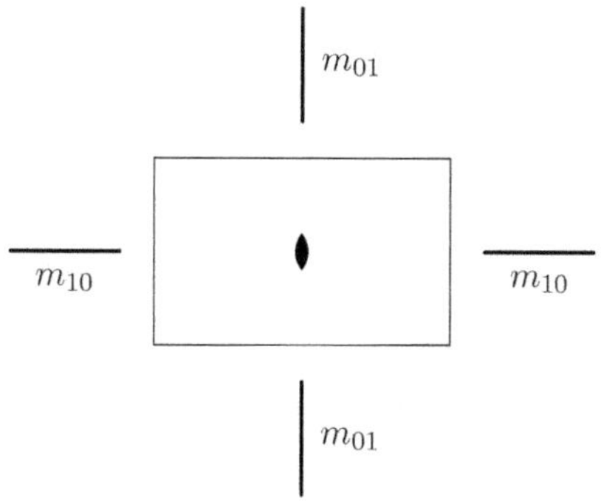

Figure 1.1.2.1
Symmetry group $2mm$ of a rectangle.

Hermann–Mauguin symbols (*cf.* Section 1.4.1) for the point groups.

(ii) The symmetry group $2mm$ of a rectangle (with unequal sides) consists of a twofold rotation 2, two reflections m_{10}, m_{01} with mirror lines along the coordinate axes and the identity element 1 (see Fig. 1.1.2.1; the small black oval in the centre represents the twofold rotation point).
The multiplication table of the group $2mm$ is

	1	2	m_{10}	m_{01}
1	1	2	m_{10}	m_{01}
2	2	1	m_{01}	m_{10}
m_{10}	m_{10}	m_{01}	1	2
m_{01}	m_{01}	m_{10}	2	1

The symmetry of the multiplication table (with respect to the main diagonal) shows that this is an abelian group.

(iii) The symmetry group $3m$ of an equilateral triangle consists (apart from the identity element 1) of the threefold rotations 3^+ and 3^- and the reflections m_{10}, m_{01}, m_{11} with mirror lines through a corner of the triangle and the centre of the opposite side (see Fig. 1.1.2.2; the small black triangle in the centre represents the threefold rotation point).
The multiplication table of the group $3m$ is

	1	3^+	3^-	m_{10}	m_{01}	m_{11}
1	1	3^+	3^-	m_{10}	m_{01}	m_{11}
3^+	3^+	3^-	1	m_{11}	m_{10}	m_{01}
3^-	3^-	1	3^+	m_{01}	m_{11}	m_{10}
m_{10}	m_{10}	m_{01}	m_{11}	1	3^+	3^-
m_{01}	m_{01}	m_{11}	m_{10}	3^-	1	3^+
m_{11}	m_{11}	m_{10}	m_{01}	3^+	3^-	1

The fact that $3^+ \cdot m_{10} = m_{11}$, but $m_{10} \cdot 3^+ = m_{01}$ shows that this group is not abelian. It is actually the smallest group (in terms of order) that is not abelian.

(iv) The symmetry group $4mm$ of the square consists of the rotations $1, 4^+, 2, 4^-$ (the different powers of the fourfold rotation 4^+) and the reflections m_{10}, m_{01}, m_{11}, $m_{1\bar{1}}$ with mirror lines along the coordinate axes and the diagonals of the square (see Fig. 1.1.2.3; the small black square in the centre represents the fourfold rotation point).
The multiplication table of the group $4mm$ is

	1	2	4^+	4^-	m_{10}	m_{01}	m_{11}	$m_{1\bar{1}}$
1	1	2	4^+	4^-	m_{10}	m_{01}	m_{11}	$m_{1\bar{1}}$
2	2	1	4^-	4^+	m_{01}	m_{10}	$m_{1\bar{1}}$	m_{11}
4^+	4^+	4^-	2	1	m_{11}	$m_{1\bar{1}}$	m_{01}	m_{10}
4^-	4^-	4^+	1	2	$m_{1\bar{1}}$	m_{11}	m_{10}	m_{01}
m_{10}	m_{10}	m_{01}	$m_{1\bar{1}}$	m_{11}	1	2	4^-	4^+
m_{01}	m_{01}	m_{10}	m_{11}	$m_{1\bar{1}}$	2	1	4^+	4^-
m_{11}	m_{11}	$m_{1\bar{1}}$	m_{10}	m_{01}	4^+	4^-	1	2
$m_{1\bar{1}}$	$m_{1\bar{1}}$	m_{11}	m_{01}	m_{10}	4^-	4^+	2	1

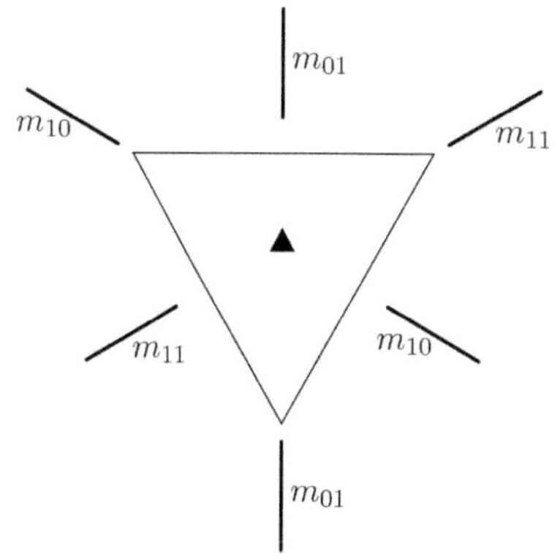

Figure 1.1.2.2
Symmetry group $3m$ of an equilateral triangle.

This group is not abelian, because for example $4^+ \cdot m_{10} = m_{11}$, but $m_{10} \cdot 4^+ = m_{1\bar{1}}$.

The groups that are considered in crystallography do not consist of abstract elements but of symmetry operations with a geometric meaning. In the figures illustrating the groups and also in the symbols used for the group elements, this geometric nature is taken into account. For example, the fourfold rotation 4^+ in the group $4mm$ is represented by the small black square placed at the rotation point and the reflection m_{10} by the line fixed by the reflection. To each crystallographic symmetry operation a *geometric element* is assigned which characterizes the type of the symmetry operation. The precise definition of the geometric elements for the different types of operations is given in Section 1.2.3. For a rotation in three-dimensional space the geometric element is the line along the rotation axis and for a reflection it is the plane fixed by the reflection. Different symmetry operations may share the same geometric element, but these operations are then closely related, such as rotations around the same line. One therefore introduces the notion of a *symmetry element*, which is a geometric element together with its associated symmetry operations. In the figures for the crystallographic groups, the symbols like the little black square or the lines actually represent these symmetry elements (and not just a symmetry operation or a geometric element).

Two groups $\mathcal{G}$ and $\mathcal{H}$ are said to be *isomorphic* (notation: $\mathcal{G} \cong \mathcal{H}$) if there is a mapping φ between the elements of $\mathcal{G}$ and those of $\mathcal{H}$ such that every element of $\mathcal{G}$ is associated with precisely one element of $\mathcal{H}$ and *vice versa*, and such that the mapping φ is *compatible* with the group operation in the two groups, *i.e.* that $\varphi(gg') = \varphi(g)\varphi(g')$ for all g, g' in $\mathcal{G}$. If for $\mathcal{G} = g_1, g_2, g_3, \ldots$ the elements of $\mathcal{H}$ are ordered accordingly as $\varphi(g_1), \varphi(g_2), \varphi(g_3), \ldots$, this means that φ transforms the multi-

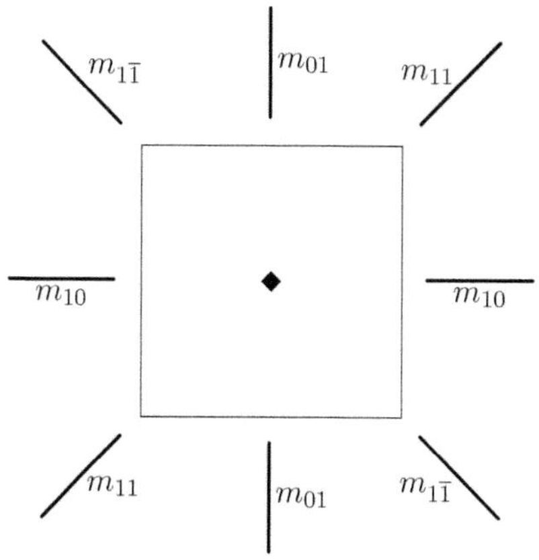

Figure 1.1.2.3
Symmetry group $4mm$ of the square.

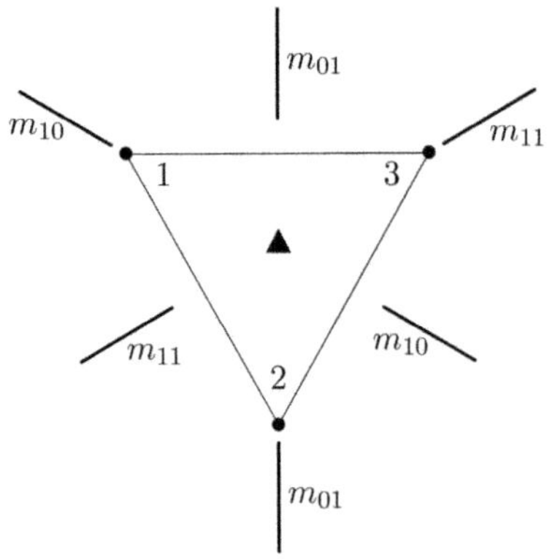

Figure 1.1.2.4
Illustration of the isomorphism between the symmetry group $3m$ and the group S_3 of permutations of $\{1, 2, 3\}$.

plication table of $\mathcal{G}$ into that of $\mathcal{H}$. It thus does not matter whether two elements are first multiplied in $\mathcal{G}$ and the result is mapped to $\mathcal{H}$ or whether the elements are first mapped to $\mathcal{H}$ and then multiplied.

Isomorphic groups may differ in the way they are realized, but they coincide in their structure. In essence, one can regard isomorphic groups as the same group with different names or labels for the group elements. For example, isomorphic groups have the same multiplication table if the elements are relabelled according to the isomorphism identifying the elements of the first group with those of the second. If one wants to stress that a certain property of a group $\mathcal{G}$ will be the same for all groups which are isomorphic to $\mathcal{G}$, one speaks of $\mathcal{G}$ as an *abstract group*.

Example
The symmetry group $3m$ of an equilateral triangle is isomorphic to the group S_3 of all permutations of $\{1, 2, 3\}$. This can be seen as follows: labelling the corners of the triangle by 1, 2, 3, each element of $3m$ gives rise to a permutation of the labels. Writing a permutation f of $\{1, 2, 3\}$ as the sequence $[f(1), f(2), f(3)]$, this gives rise to the following mapping from $3m$ to S_3 (see Fig. 1.1.2.4):

$$1 \mapsto [1, 2, 3], \quad 3^+ \mapsto [2, 3, 1], \quad 3^- \mapsto [3, 1, 2],$$
$$m_{10} \mapsto [1, 3, 2], \quad m_{01} \mapsto [3, 2, 1], \quad m_{11} \mapsto [2, 1, 3].$$

This is a one-to-one mapping and one checks that the multiplication table for S_3 (with composition as group operation) indeed agrees with that of $3m$.

It is clear that for larger groups the multiplication table becomes unwieldy to set up and use. Fortunately, for many purposes a full list of all products in the group is actually not required. A very economic alternative of describing a group is to give only a small subset of the group elements from which all other elements can be obtained by forming products.

Definition. A subset $\mathcal{X} \subseteq \mathcal{G}$ is called a set of *generators* for $\mathcal{G}$ if every element of $\mathcal{G}$ can be obtained as a finite product of elements from $\mathcal{X}$ or their inverses. If $\mathcal{X}$ is a set of generators for $\mathcal{G}$, one writes $\mathcal{G} = \langle \mathcal{X} \rangle$.

Examples
(i) A group may be generated by a single element, in which case it is called a *cyclic group*. For example, the integers $(\mathbb{Z}, +)$ form an infinite cyclic group generated by $\mathcal{X} = \{1\}$, but also by $\mathcal{X} = \{-1\}$.
(ii) The symmetry group $4mm$ of the square is generated by a fourfold rotation and any of the reflections, *e.g.* by $\{4^+, m_{10}\}$, but also by two reflections with reflection lines which are not perpendicular, *e.g.* by $\{m_{10}, m_{11}\}$.
(iii) The full symmetry group $m\bar{3}m$ of the cube consists of 48 elements. It can be generated by a fourfold rotation 4^+_{100} around the a axis, a threefold rotation 3^+_{111} around a space diagonal and the inversion $\bar{1}$. It is also possible to generate the group by only two elements, *e.g.* by the fourfold rotation 4^+_{100} and a reflection m_{110} in a plane with normal vector along one of the face diagonals of the cube.

Although one usually chooses generating sets with as few elements as possible, it is sometimes convenient to actually include some redundancy. For example, it may be useful to generate the symmetry group $4mm$ of the square by $\{2, m_{10}, m_{11}\}$. The element 2 is redundant, since $2 = (m_{10}m_{11})^2$, but this generating set explicitly shows that the symmetry group $2mm = \langle 2, m_{10} \rangle$ of a rectangle is contained in that of the square.

1.1.3. Subgroups

The group of symmetry operations of a crystal pattern may alter if the crystal undergoes a phase transition. Often, some symmetries are preserved, while others are lost, *i.e.* symmetry breaking takes place. The symmetry operations that are preserved form a subset of the original symmetry group which is itself a group. This gives rise to the concept of a subgroup.

Definition. A subset $\mathcal{H} \subseteq \mathcal{G}$ is called a *subgroup* of $\mathcal{G}$ if its elements form a group by themselves. This is denoted by $\mathcal{H} \leq \mathcal{G}$. If $\mathcal{H}$ is a subgroup of $\mathcal{G}$, then $\mathcal{G}$ is called a *supergroup* of $\mathcal{H}$. In order to be a subgroup, $\mathcal{H}$ is required to contain the identity element e of $\mathcal{G}$, to contain inverse elements and to be closed with respect to composition of elements. Thus, technically, every group is a subgroup of itself.

The subgroups $\mathcal{H}$ of $\mathcal{G}$ that are not equal to $\mathcal{G}$ are called *proper subgroups* of $\mathcal{G}$. A proper subgroup $\mathcal{H}$ of $\mathcal{G}$ is called a *maximal subgroup* if there is no intermediate group $\mathcal{H}'$ between $\mathcal{H}$ and $\mathcal{G}$ such that $\mathcal{H}$ is a proper subgroup of $\mathcal{H}'$ and $\mathcal{H}'$ is a proper subgroup of $\mathcal{G}$.

It is often convenient to specify a subgroup $\mathcal{H}$ of $\mathcal{G}$ by a set $\{h_1, \ldots, h_s\}$ of generators. This is denoted by $\mathcal{H} = \langle h_1, \ldots, h_s \rangle$. The order of $\mathcal{H}$ is not *a priori* obvious from the set of generators. For example, in the symmetry group $4mm$ of the square the pairs $\{m_{10}, m_{01}\}$ and $\{m_{11}, m_{1\bar{1}}\}$ both generate subgroups of order 4, whereas the pair $\{m_{10}, m_{11}\}$ generates the full group of order 8.

The subgroups of a group can be visualized in a *subgroup diagram*. In such a diagram the subgroups are arranged with subgroups of higher order above subgroups of lower order. Two subgroups are connected by a line if one is a maximal subgroup of the other. By following downward paths in this diagram, all group–subgroup relations in a group can be derived. Additional information is provided by connecting subgroups of the same order by a horizontal line if they are *conjugate* (see Section 1.1.5).

Examples
(i) For the group $\mathbb{Z}$ of the integers, all subgroups are cyclic and generated by some integer n, *i.e.* they are of the form $n\mathbb{Z} := \{na \mid a \in \mathbb{Z}\}$ for an integer n. Such a subgroup is maximal if n is a prime number.

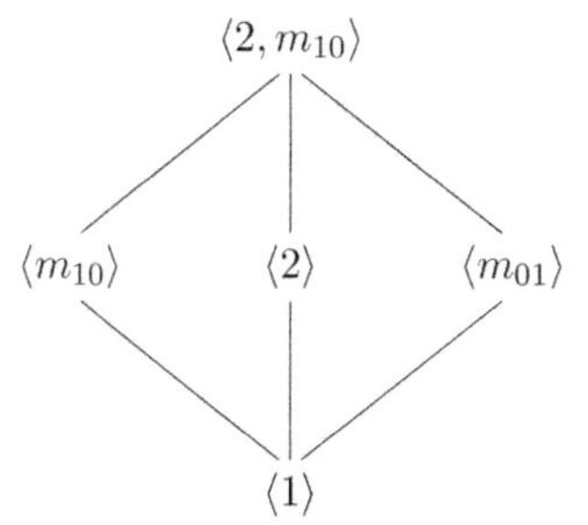

Figure 1.1.3.1
Subgroup diagram for the symmetry group $2mm$ of a rectangle.

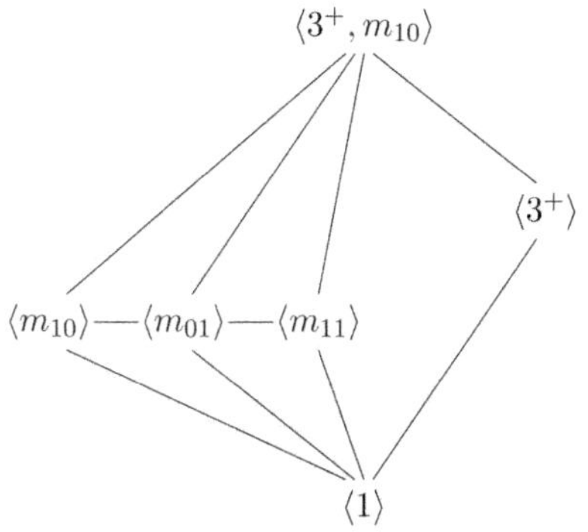

Figure 1.1.3.2
Subgroup diagram for the symmetry group $3m$ of an equilateral triangle.

(ii) In $GL_n(\mathbb{R})$ the matrices of determinant 1 form a subgroup, since the determinant of the matrix product $\boldsymbol{A} \cdot \boldsymbol{B}$ is equal to the product of the determinants of $\boldsymbol{A}$ and $\boldsymbol{B}$.

(iii) The symmetry group $2mm$ of a rectangle has three subgroups of order 2, generated by the reflection m_{10}, the twofold rotation 2 and the reflection m_{01}, respectively (see Fig. 1.1.3.1).

(iv) In the symmetry group $3m$ of an equilateral triangle the rotations form a subgroup of order 3 (see Fig. 1.1.3.2).

(v) In the symmetry group $4mm$ of the square, the reflections m_{10} and m_{01} together with their product 2 and the identity element 1 form a subgroup of order 4. This subgroup is indicated by $\langle 2, m_{10}\rangle$ in the left part of Fig. 1.1.3.3 and the subdiagram of its subgroups clearly coincides with the subgroup diagram of $2mm$ in Fig. 1.1.3.1. A different subgroup of order 4 is formed by the other pair of perpendicular reflections m_{11}, $m_{1\bar{1}}$ together with 2 and 1, and a third subgroup of order 4 is the cyclic subgroup $\langle 4^+\rangle$ generated by the fourfold rotation (see Fig. 1.1.3.3).

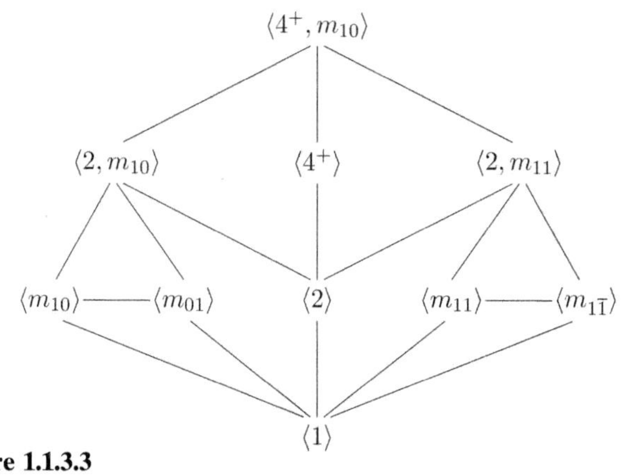

Figure 1.1.3.3
Subgroup diagram for the symmetry group $4mm$ of the square.

1.1.4. Cosets

A subgroup allows us to partition a group into disjoint subsets of the same size, called *cosets*.

Definition. Let $\mathcal{H} = \{h_1, h_2, h_3, \ldots\}$ be a subgroup of $\mathcal{G}$. Then for $g \in \mathcal{G}$ the set

$$g\mathcal{H} := \{gh_1, gh_2, gh_3, \ldots\} = \{gh \mid h \in \mathcal{H}\}$$

is called the *left coset* of $\mathcal{H}$ with *representative* g. Analogously, the *right coset* with representative g is defined as

$$\mathcal{H}g := \{h_1g, h_2g, h_3g, \ldots\} = \{hg \mid h \in \mathcal{H}\}.$$

The coset $e\mathcal{H} = \mathcal{H} = \mathcal{H}e$ is called the *trivial coset* of $\mathcal{H}$.

Remarks

(i) Since two elements gh and gh' in the same coset $g\mathcal{H}$ can only be the same if $h = h'$, the elements of $g\mathcal{H}$ are in one-to-one correspondence with the elements of $\mathcal{H}$. In particular, for a finite subgroup $\mathcal{H}$ the number of elements in each coset of $\mathcal{H}$ equals the order $|\mathcal{H}|$ of the subgroup $\mathcal{H}$.

(ii) Every element contained in $g\mathcal{H}$ may serve as representative for this coset, *i.e.* $g'\mathcal{H} = g\mathcal{H}$ for every $g' \in g\mathcal{H}$. In particular, if an element g'' is contained in the intersection $g\mathcal{H} \cap g'\mathcal{H}$ of two cosets, one has $g''\mathcal{H} = g\mathcal{H}$ and $g''\mathcal{H} = g'\mathcal{H}$. This implies that two cosets are either disjoint (*i.e.* contain no common element) or they are equal.

These two remarks have an important consequence: since an element $g \in \mathcal{G}$ is contained in the coset $g\mathcal{H}$, the cosets of $\mathcal{H}$ partition the elements of $\mathcal{G}$ into sets of the same cardinality as $\mathcal{H}$ (which is the order of $\mathcal{H}$ in the case where this is finite).

Definition. If the number of different cosets of a subgroup $\mathcal{H} \le \mathcal{G}$ is finite, this number is called the *index* of $\mathcal{H}$ in $\mathcal{G}$, denoted by $[i]$ or $[\mathcal{G} : \mathcal{H}]$. Otherwise, $\mathcal{H}$ is said to have *infinite index* in $\mathcal{G}$.

In the case of a finite group, the partitioning of the elements of $\mathcal{G}$ into the cosets of $\mathcal{H}$ (which are all of size $|\mathcal{H}|$) shows that $|\mathcal{G}| = |\mathcal{H}| \cdot [\mathcal{G} : \mathcal{H}]$, *i.e.* the order of a subgroup multiplied by its index gives the order of the full group. This is a famous result known as *Lagrange's theorem*. In particular, both the order of $\mathcal{H}$ and the index of $\mathcal{H}$ in $\mathcal{G}$ divide the order of $\mathcal{G}$.

Whether or not two cosets of a subgroup $\mathcal{H}$ are equal depends on whether the quotient of their representatives is contained in $\mathcal{H}$: for left cosets one has $g\mathcal{H} = g'\mathcal{H}$ if and only if $g^{-1}g' \in \mathcal{H}$ and for right cosets $\mathcal{H}g = \mathcal{H}g'$ if and only if $g'g^{-1} \in \mathcal{H}$.

Definition. If $\mathcal{H}$ is a subgroup of $\mathcal{G}$ and $g_1, g_2, g_3, \ldots \in \mathcal{G}$ are such that $g_i\mathcal{H} \ne g_j\mathcal{H}$ for $i \ne j$, and every $g \in \mathcal{G}$ is contained in some left coset $g_i\mathcal{H}$, then $g_1, g_2, g_3, \ldots$ is called a system of *left coset representatives* of $\mathcal{G}$ relative to $\mathcal{H}$. It is customary to choose $g_1 = e$ so that the coset $g_1\mathcal{H} = e\mathcal{H} = \mathcal{H}$ is the subgroup $\mathcal{H}$ itself. The decomposition

$$\mathcal{G} = \mathcal{H} \cup g_2\mathcal{H} \cup g_3\mathcal{H} \ldots$$

is called the *coset decomposition* of $\mathcal{G}$ into left cosets relative to $\mathcal{H}$.

Analogously, $g'_1, g'_2, g'_3, \ldots \in \mathcal{G}$ is called a system of *right coset representatives* if $\mathcal{H}g'_i \ne \mathcal{H}g'_j$ for $i \ne j$ and every $g \in \mathcal{G}$ is contained in some right coset $\mathcal{H}g_i$. Again, one usually chooses $g'_1 = e$ and the decomposition

$$\mathcal{G} = \mathcal{H} \cup \mathcal{H}g'_2 \cup \mathcal{H}g'_3 \ldots$$

is called the *coset decomposition* of $\mathcal{G}$ into right cosets relative to $\mathcal{H}$.

To obtain the coset decomposition one starts by choosing $\mathcal{H}$ as the first coset (with representative e). Next, an element $g_2 \in \mathcal{G}$ with $g_2 \notin \mathcal{H}$ is selected as representative for the second coset $g_2\mathcal{H}$. For the third coset, an element $g_3 \in \mathcal{G}$ with $g_3 \notin \mathcal{H}$ and $g_3 \notin g_2\mathcal{H}$ is required. If at a certain stage the cosets $\mathcal{H}, g_2\mathcal{H}, \ldots, g_m\mathcal{H}$ have been defined but do not yet exhaust $\mathcal{G}$, an element g_{m+1} not contained in the union $\mathcal{H} \cup g_2\mathcal{H} \cup \ldots \cup g_m\mathcal{H}$ is chosen as representative for the next coset.

Examples

(i) For any integer n, the set $n\mathbb{Z} := \{na \mid a \in \mathbb{Z}\}$ of multiples of n forms an infinite subgroup of index n in $\mathbb{Z}$. A system of coset representatives of $\mathbb{Z}$ relative to $n\mathbb{Z}$ is formed by the numbers $0, 1, 2, \ldots, n-1$. The coset with representative 0 is $\{\ldots, -n, 0, n, 2n, \ldots\}$, the coset with representative 1 is $\{\ldots, -n+1, 1, n+1, 2n+1, \ldots\}$ and an integer a belongs to the coset with representative k if and only if a gives remainder k upon division by n.

(ii) Let $\mathcal{G} = 3m$ be the symmetry group of an equilateral triangle and $\mathcal{H} = \langle 3^+ \rangle$ its subgroup containing the rotations. Since each coset must consist of three elements, the coset decomposition is necessarily of the form $\mathcal{G} = \{1, 3^+, 3^-\} \cup \{m_{10}, m_{01}, m_{11}\}$. This shows that for every reflection $m \in \mathcal{G}$ the elements $1, m$ form a system of coset representatives of $\mathcal{G}$ relative to $\mathcal{H}$.

(iii) Let $\mathcal{G} = 4mm$ be the symmetry group of the square and $\mathcal{H} = \langle m_{10} \rangle$. Then a system of coset representatives of $\mathcal{G}$ relative to $\mathcal{H}$ is $1, 4^+, 2, 4^-$, since the corresponding cosets $1\{1, m_{10}\} = \{1, m_{10}\}$, $4^+\{1, m_{10}\} = \{4^+, m_{11}\}$, $2\{1, m_{10}\} = \{2, m_{01}\}$ and $4^-\{1, m_{10}\} = \{4^-, m_{1\bar{1}}\}$ are all different. The coset decomposition of $\mathcal{G}$ with respect to $\mathcal{H}$ is thus $\{1, m_{10}\} \cup \{4^+, m_{11}\} \cup \{2, m_{01}\} \cup \{4^-, m_{1\bar{1}}\}$. Note that a system of coset representatives has to contain precisely one element from each coset, therefore $1, m_{11}, 2, m_{1\bar{1}}$ would be another possible choice for a system of coset representatives.

1.1.5. Normal subgroups, factor groups

In general, the left and right cosets of a subgroup $\mathcal{H}$ differ, for example in the symmetry group $3m$ of an equilateral triangle the left coset decomposition with respect to the subgroup $\mathcal{H} = \{1, m_{10}\}$ is

$$\begin{aligned}&\{1, m_{10}\} \cup 3^+\{1, m_{10}\} \cup 3^-\{1, m_{10}\}\\ =&\{1, m_{10}\} \cup \{3^+, m_{11}\} \cup \{3^-, m_{01}\},\end{aligned}$$

whereas the right coset decomposition is

$$\begin{aligned}&\{1, m_{10}\} \cup \{1, m_{10}\}3^+ \cup \{1, m_{10}\}3^-\\ =&\{1, m_{10}\} \cup \{3^+, m_{01}\} \cup \{3^-, m_{11}\}.\end{aligned}$$

For particular subgroups, however, it turns out that the left and right cosets coincide, *i.e.* one has $g\mathcal{H} = \mathcal{H}g$ for all $g \in \mathcal{G}$. This means that for every $h \in \mathcal{H}$ and every $g \in \mathcal{G}$ the element gh is of the form $gh = h'g$ for some $h' \in \mathcal{H}$ and thus $ghg^{-1} = h' \in \mathcal{H}$. The element $h' = ghg^{-1}$ is called the *conjugate of h by g*. Note that in the definition of the conjugate element there is a choice whether the inverse element g^{-1} is placed to the left or right of h. Depending on the applications that are envisaged and on the preferences of the author, both versions ghg^{-1} and $g^{-1}hg$ are found in the literature, but in the context of crystallographic groups it is more convenient to have the inverse g^{-1} to the right of h.

An important aspect of conjugate elements is that they share many properties, such as the order or the type of symmetry operation. As a consequence, conjugate symmetry operations have the same type of geometric elements. For example, if h is a threefold rotation in three-dimensional space, its geometric element is the line along the rotation axis. The geometric element of a conjugate element ghg^{-1} is then also a line fixed by a threefold rotation, but in general this line has a different orientation.

Definition. A subgroup $\mathcal{H}$ of $\mathcal{G}$ is called a *normal subgroup* if $ghg^{-1} \in \mathcal{H}$ for all $g \in \mathcal{G}$ and all $h \in \mathcal{H}$. This is denoted by $\mathcal{H} \trianglelefteq \mathcal{G}$. For a normal subgroup $\mathcal{H}$, the left and right cosets of $\mathcal{G}$ with respect to $\mathcal{H}$ coincide.

Remarks

(i) In abelian groups, every subgroup is a normal subgroup, because $gh = hg$ implies $ghg^{-1} = h \in \mathcal{H}$.

(ii) A subgroup $\mathcal{H}$ of index 2 in $\mathcal{G}$ is always a normal subgroup, since the coset decomposition relative to $\mathcal{H}$ consists of only two cosets and for any element $g \notin \mathcal{H}$ the left and right cosets $g\mathcal{H}$ and $\mathcal{H}g$ both consist precisely of those elements of $\mathcal{G}$ that are not contained in $\mathcal{H}$. Therefore, $g\mathcal{H} = \mathcal{H}g$ for $g \notin \mathcal{H}$ and for $h \in \mathcal{H}$ clearly $h\mathcal{H} = \mathcal{H} = \mathcal{H}h$ holds.

Examples

(i) In the symmetry group $3m$ of an equilateral triangle, the subgroup generated by the threefold rotation 3^+ is a normal subgroup because it is of index 2 in $3m$.
The subgroups of order 2 generated by the reflections m_{10}, m_{01} and m_{11} are not normal because $3^+ \cdot m_{10} \cdot 3^- = m_{01} \notin \langle m_{10} \rangle$, $3^+ \cdot m_{01} \cdot 3^- = m_{11} \notin \langle m_{01} \rangle$ and $3^+ \cdot m_{11} \cdot 3^- = m_{10} \notin \langle m_{11} \rangle$.

(ii) In the symmetry group $4mm$ of the square, the subgroups $\langle 2, m_{10} \rangle$, $\langle 4^+ \rangle$, and $\langle 2, m_{11} \rangle$ are normal subgroups because they are subgroups of index 2.
The subgroups of order 2 generated by the reflections m_{10}, m_{01}, m_{11} and $m_{1\bar{1}}$ are not normal because $4^+ \cdot m_{10} \cdot 4^- = m_{01} \notin \langle m_{10} \rangle$, $4^+ \cdot m_{01} \cdot 4^- = m_{10} \notin \langle m_{01} \rangle$, $4^+ \cdot m_{11} \cdot 4^- = m_{1\bar{1}} \notin \langle m_{11} \rangle$ and $4^+ \cdot m_{1\bar{1}} \cdot 4^- = m_{11} \notin \langle m_{1\bar{1}} \rangle$.
The subgroup of order 2 generated by the twofold rotation 2 is normal because $4^+ \cdot 2 \cdot 4^- = 2$ and $m_{10} \cdot 2 \cdot m_{10}^{-1} = 2$.

For a subgroup $\mathcal{H}$ of $\mathcal{G}$ and an element $g \in \mathcal{G}$, the conjugates ghg^{-1} form a subgroup

$$\mathcal{H}' = g\mathcal{H}g^{-1} = \{ghg^{-1} \mid h \in \mathcal{H}\}$$

because $gh_1g^{-1} \cdot gh_2g^{-1} = gh_1h_2g^{-1}$. This subgroup is called the *conjugate subgroup* of $\mathcal{H}$ by g. As already noted, conjugation does not alter the type of symmetry operations and their geometric elements, but it is possible that the orientations of the geometric elements are changed.

Using the concept of conjugate subgroups, a normal subgroup is a subgroup $\mathcal{H}$ that coincides with all its conjugate subgroups $g\mathcal{H}g^{-1}$. This means that the set of geometric elements of a normal subgroup is not changed by conjugation; the single geometric elements may, however, be permuted by the conjugating element. In the example of the symmetry group $4mm$ discussed above, the normal subgroup $\langle 2, m_{10} \rangle$ contains the reflections m_{10} and m_{01} with the lines along the coordinate axes as geometric elements.

These two lines are interchanged by the fourfold rotation 4^+, corresponding to the fact that conjugation by 4^+ interchanges m_{10} and m_{01}. The concept of conjugation will be discussed in more detail in Section 1.1.7.

One of the main motivations for studying normal subgroups is that they allow us to define a group operation on the cosets of $\mathcal{H}$ in $\mathcal{G}$. The products of any element in the coset $g\mathcal{H}$ with any element in the coset $g'\mathcal{H}$ lie in a single coset, namely in the coset $gg'\mathcal{H}$. Thus we can define the product of the two cosets $g\mathcal{H}$ and $g'\mathcal{H}$ as the coset with representative gg'.

Definition. The set $\mathcal{G}/\mathcal{H} := \{g\mathcal{H} \mid g \in \mathcal{G}\}$ together with the binary operation

$$g\mathcal{H} \circ g'\mathcal{H} := gg'\mathcal{H}$$

forms a group, called the *factor group* or *quotient group* of $\mathcal{G}$ by $\mathcal{H}$.

The identity element of the factor group $\mathcal{G}/\mathcal{H}$ is the coset $\mathcal{H}$ and the inverse element of $g\mathcal{H}$ is the coset $g^{-1}\mathcal{H}$.

A familiar example of a factor group is provided by the times on a clock. If it is 8 o'clock (in the morning) now, then we say that in nine hours it will be 5 o'clock (in the afternoon). We regard times as elements of the factor group $\mathbb{Z}/12\mathbb{Z}$ in which $(8+12\mathbb{Z})+(9+12\mathbb{Z}) = 17+12\mathbb{Z} = 5+12\mathbb{Z}$. In the factor group $\mathbb{Z}/12\mathbb{Z}$, the clock is imagined as a circle of circumference 12 around which the line of integers is wrapped so that integers with a difference of 12 are located at the same position on the circle.

Examples

(i) The clock example given above is a special case of factor groups of the integers. We have already seen that the set $n\mathbb{Z} = \{na \mid a \in \mathbb{Z}\}$ of multiples of a natural number n forms a subgroup of index n in $\mathbb{Z}$. This is a normal subgroup, since $\mathbb{Z}$ is an abelian group. The factor group $\mathbb{Z}/n\mathbb{Z}$ represents the addition of integers *modulo n*.

(ii) If we take $\mathcal{G}$ to be the symmetry group $4mm$ of the square and choose as normal subgroup the subgroup $\mathcal{H} = \langle 4^+\rangle$ generated by the fourfold rotation, we obtain a factor group $\mathcal{G}/\mathcal{H}$ with two elements, namely the cosets $\mathcal{H} = \{1, 2, 4^+, 4^-\}$ and $m_{10}\mathcal{H} = \{m_{10}, m_{01}, m_{11}, m_{1\bar{1}}\}$. The trivial coset $\mathcal{H}$ is the identity element in the factor group $\mathcal{G}/\mathcal{H}$ and contains the rotations in $4mm$. The other element $m_{10}\mathcal{H}$ in the factor group $\mathcal{G}/\mathcal{H}$ consists of the reflections in $4mm$.
In this example, the separation of the rotations and reflections in $4mm$ into the two cosets $\mathcal{H}$ and $m_{10}\mathcal{H}$ makes it easy to see that the product of two cosets is independent of the chosen representative of the coset: the product of two rotations is again a rotation, hence $\mathcal{H}\cdot\mathcal{H} = \mathcal{H}$, the product of a rotation and a reflection is a reflection, hence $\mathcal{H}\cdot m_{10}\mathcal{H} = m_{10}\mathcal{H}\cdot\mathcal{H} = m_{10}\mathcal{H}$, and finally the product of two reflections is a rotation, hence $m_{10}\mathcal{H}\cdot m_{10}\mathcal{H} = \mathcal{H}$. The multiplication table of the factor group is thus

	$\mathcal{H}$	$m_{10}\mathcal{H}$
$\mathcal{H}$	$\mathcal{H}$	$m_{10}\mathcal{H}$
$m_{10}\mathcal{H}$	$m_{10}\mathcal{H}$	$\mathcal{H}$

(iii) Take again $\mathcal{G} = 4mm$ but now consider the cosets with respect to the normal subgroup $\mathcal{H} = \langle 2\rangle = \{1, 2\}$. The coset decomposition of $\mathcal{G}$ with respect to $\mathcal{H}$ is $\mathcal{G} = \{1, 2\} \cup \{4^+, 4^-\} \cup \{m_{10}, m_{01}\} \cup \{m_{11}, m_{1\bar{1}}\}$. Looking at the multiplication table of $4mm$, one derives that the factor group $\mathcal{G}/\mathcal{H}$ has the multiplication table

	$\{1,2\}$	$\{4^+,4^-\}$	$\{m_{10},m_{01}\}$	$\{m_{11},m_{1\bar{1}}\}$
$\{1,2\}$	$\{1,2\}$	$\{4^+,4^-\}$	$\{m_{10},m_{01}\}$	$\{m_{11},m_{1\bar{1}}\}$
$\{4^+,4^-\}$	$\{4^+,4^-\}$	$\{1,2\}$	$\{m_{11},m_{1\bar{1}}\}$	$\{m_{10},m_{01}\}$
$\{m_{10},m_{01}\}$	$\{m_{10},m_{01}\}$	$\{m_{11},m_{1\bar{1}}\}$	$\{1,2\}$	$\{4^+,4^-\}$
$\{m_{11},m_{1\bar{1}}\}$	$\{m_{11},m_{1\bar{1}}\}$	$\{m_{10},m_{01}\}$	$\{4^+,4^-\}$	$\{1,2\}$

By comparing this multiplication table with that of the symmetry group $2mm = \{1', 2', m'_{10}, m'_{01}\}$ of a rectangle (where we added primes to the elements to distinguish them from the elements of $4mm$), one sees that the factor group $\mathcal{G}/\mathcal{H}$ is in fact isomorphic to $2mm$ *via* the mapping $\{1,2\}\longmapsto 1'$, $\{4^+,4^-\}\longmapsto 2'$, $\{m_{10},m_{01}\}\longmapsto m'_{10}$ and $\{m_{11},m_{1\bar{1}}\}\longmapsto m'_{01}$.

Remarks

(i) If one takes cosets with respect to a subgroup that is not normal, the products of elements from two cosets do in general not lie in a single coset. As we have seen, the left cosets of the group $3m$ of an equilateral triangle with respect to the non-normal subgroup $\mathcal{H} = \{1, m_{10}\}$ are $\{1, m_{10}\}$, $\{3^+, m_{11}\}$ and $\{3^-, m_{01}\}$. Taking products from elements of the first and second coset, we get $1\cdot 3^+ = 3^+$ and $1\cdot m_{11} = m_{11}$, which are both in the second coset, whereas $m_{10}\cdot 3^+ = m_{01}$ and $m_{10}\cdot m_{11} = 3^-$ are both in the third coset.

(ii) In the above examples we observed that the factor groups of $4mm$ are actually isomorphic to subgroups of $4mm$: $\mathcal{G}/\mathcal{H} \cong \langle 2\rangle$ for $\mathcal{H} = \langle 4^+\rangle$ and $\mathcal{G}/\mathcal{H} \cong \langle 2, m_{10}\rangle$ for $\mathcal{H} = \langle 2\rangle$. Note that this is a mere coincidence: in general factor groups of a group $\mathcal{G}$ do not have to be isomorphic to any of the subgroups of $\mathcal{G}$.

1.1.6. Group actions

The concept of a group is the essence of an abstraction process which distils the common features of various examples of groups. On the other hand, although abstract groups are important and interesting objects in their own right, they are particularly useful because the group elements *act* on something, *i.e.* they can be applied to certain objects. For example, symmetry groups act on the points in space, but they also act on lines or planes. Groups of permutations act on the symbols themselves, but also on ordered and unordered pairs. Groups of matrices act on the vectors of a vector space, but also on the subspaces. All these different actions can be described in a uniform manner and common concepts can be developed.

Definition. A *group action* of a group $\mathcal{G}$ on a set $\Omega = \{\omega \mid \omega \in \Omega\}$ assigns to each pair (g, ω) an object $\omega' = g(\omega)$ of Ω such that the following hold:

(i) applying two group elements g and g' consecutively has the same effect as applying the product $g'g$, *i.e.* $g'(g(\omega)) = (g'g)(\omega)$ (note that since the group elements act *from the left* on the objects in Ω, the elements in a product of two (or more) group elements are applied right-to-left);

(ii) applying the identity element e of $\mathcal{G}$ has no effect on ω, *i.e.* $e(\omega) = \omega$ for all ω in Ω.

One says that the object ω is *moved* to $g(\omega)$ by g.

Example

The abstract group $C_2 = \{e, g\}$ occurs as symmetry group in three-dimensional space with three different actions of g on the points of three-dimensional space:

(i) If g is a reflection, then the points fixed by g form a two-dimensional plane.

(ii) If g is a twofold rotation, then the fixed points of g form a one-dimensional line.

(iii) If g is an inversion, then only a single point is fixed by g.

Often, two objects ω and ω' are regarded as equivalent if there is a group element moving ω to ω'. *Via* this equivalence, the action of $\mathcal{G}$ partitions the objects in Ω into equivalence classes, called orbits, where the equivalence class of an object $\omega \in \Omega$ consists of all objects which are equivalent to ω.

Definition. Two objects $\omega, \omega' \in \Omega$ lie in the same *orbit* under $\mathcal{G}$ if there exists $g \in \mathcal{G}$ such that $\omega' = g(\omega)$.

The set $\mathcal{G}(\omega) := \{g(\omega) \mid g \in \mathcal{G}\}$ of all objects in the orbit of ω is called the *orbit of ω under $\mathcal{G}$*.

The set $S_\mathcal{G}(\omega) := \{g \in \mathcal{G} \mid g(\omega) = \omega\}$ of group elements that do not move the object ω is a subgroup of $\mathcal{G}$ called the *stabilizer* of ω in $\mathcal{G}$.

If the orbit of a group action is finite, the length of the orbit is equal to the index of the stabilizer and thus in particular a divisor of the group order (in the case of a finite group). Actually, the objects in an orbit are in a very explicit one-to-one correspondence with the cosets relative to the stabilizer, as is summarized in the *orbit–stabilizer theorem*.

Orbit–stabilizer theorem

For a group $\mathcal{G}$ acting on a set Ω let ω be an object in Ω and let $S_\mathcal{G}(\omega)$ be the stabilizer of ω in $\mathcal{G}$.

(i) If $g_1 S_\mathcal{G}(\omega) \cup g_2 S_\mathcal{G}(\omega) \cup \ldots \cup g_m S_\mathcal{G}(\omega)$ is the coset decomposition of $\mathcal{G}$ relative to $S_\mathcal{G}(\omega)$, then the coset $g_i S_\mathcal{G}(\omega)$ consists of precisely those elements of $\mathcal{G}$ that move ω to $g_i(\omega)$. As a consequence, the full orbit of ω is already obtained by applying only the coset representatives to ω, *i.e.* $\mathcal{G}(\omega) = \{g_1(\omega), g_2(\omega), \ldots, g_m(\omega)\}$ and the number of cosets equals the length of the orbit.

(ii) For objects in the same orbit under $\mathcal{G}$, the stabilizers are *conjugate subgroups* of $\mathcal{G}$ (*cf.* Section 1.1.5). If $\omega' = g(\omega)$, then $S_\mathcal{G}(\omega') = g S_\mathcal{G}(\omega) g^{-1}$, *i.e.* the stabilizer of ω' is obtained by conjugating the stabilizer of ω by the element g moving ω to ω'.

Example

The symmetry group $\mathcal{G} = 4mm$ of the square acts on the corners of a square as displayed in Fig. 1.1.6.1. All four points lie in a single orbit under $\mathcal{G}$ and the stabilizer of the point 1 is $\mathcal{H} = \langle m_{1\bar{1}} \rangle$, *i.e.* a subgroup of index 4, as required by the orbit–stabilizer theorem. The stabilizers of the other points are conjugate to $\mathcal{H}$: The stabilizer of corner 3 equals $\mathcal{H}$ and the stabilizer of both the corners 2 and 4 is $\langle m_{11} \rangle$, which is conjugate to $\mathcal{H}$ by the fourfold rotation 4^+ which moves corner 1 to corner 2.

For a group $\mathcal{G}$ acting on the points of $\mathbb{R}^3$, the stabilizer of a point P is called the *site-symmetry group* of P (in $\mathcal{G}$). According to the orbit–stabilizer theorem, points that are in the same orbit under $\mathcal{G}$ and which are thus symmetry equivalent have site-symmetry groups that are conjugate subgroups of $\mathcal{G}$. This gives rise to the concept of *Wyckoff positions*: points with site-symmetry groups that are conjugate subgroups of $\mathcal{G}$ belong to the same Wyckoff position. As a consequence, points in the same orbit under $\mathcal{G}$ certainly belong to the same Wyckoff position, but points may have the same site-symmetry group without being symmetry equivalent. The Wyckoff position of a point P consists of the union of the orbits of all points Q which have the same site-symmetry group as P. For a detailed discussion of the fundamental concept of Wyckoff positions we refer to Section 1.4.4.

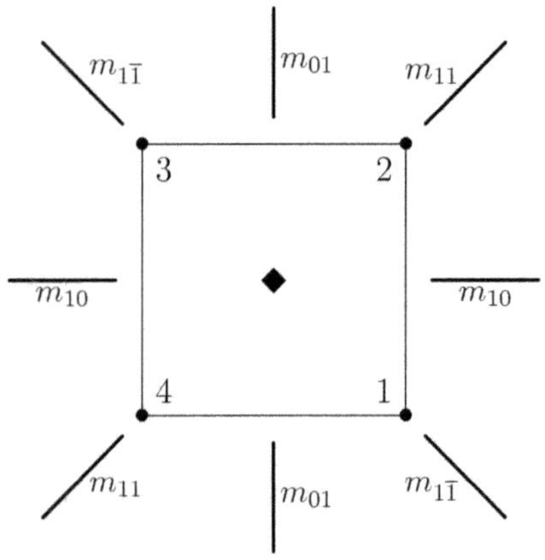

Figure 1.1.6.1
Stabilizers in the symmetry group $4mm$ of the square.

1.1.7. Conjugation, normalizers

In this section we focus on two group actions which are of particular importance for describing intrinsic properties of a group, namely the conjugation of group elements and the conjugation of subgroups. These actions were mentioned earlier in Section 1.1.5 when we introduced normal subgroups.

A group $\mathcal{G}$ acts on its elements *via* $g(h) := ghg^{-1}$, *i.e.* by conjugation. Note that the inverse element g^{-1} is required on the right-hand side of h in order to fulfil the rule $g(g'(h)) = (gg')(h)$ for a group action.

The orbits for this action are called the *conjugacy classes of elements* of $\mathcal{G}$ or simply *conjugacy classes* of $\mathcal{G}$; the conjugacy class of an element h consists of all its conjugates ghg^{-1} with g running over all elements of $\mathcal{G}$. Elements in one conjugacy class have *e.g.* the same order, and in the case of groups of symmetry operations they also share geometric properties such as being a reflection, rotation or rotoinversion. In particular, conjugate elements have the same type of geometric element.

The connection between conjugate symmetry operations and their geometric elements is even more explicit by the orbit–stabilizer theorem: If h and h' are conjugate by g, *i.e.* $h' = ghg^{-1}$, then g maps the geometric element of h to the geometric element of h'.

Example

The rotation group of a cube contains six fourfold rotations and if the cube is in standard orientation with the origin in its centre, the fourfold rotations 4^+_{100}, 4^+_{010} and 4^+_{001} and their inverses have the lines along the coordinate axes

$$\left\{\begin{pmatrix} x \\ 0 \\ 0 \end{pmatrix} \mid x \in \mathbb{R}\right\}, \left\{\begin{pmatrix} 0 \\ y \\ 0 \end{pmatrix} \mid y \in \mathbb{R}\right\} \text{ and } \left\{\begin{pmatrix} 0 \\ 0 \\ z \end{pmatrix} \mid z \in \mathbb{R}\right\}$$

as their geometric elements, respectively. The twofold rotation 2_{110} around the line

$$\left\{ \begin{pmatrix} x \\ x \\ 0 \end{pmatrix} \mid x \in \mathbb{R} \right\}$$

maps the a axis to the b axis and *vice versa*, therefore the symmetry operation 2_{110} conjugates 4^+_{100} to a fourfold rotation with the line along the b axis as geometric element. Since the positive part of the a axis is mapped to the positive part of the b axis and conjugation also preserves the sense of a rotation, 4^+_{100} is conjugated to 4^+_{010} and not to the inverse element 4^-_{010}. The line along the c axis is fixed by 2_{110}, but its direction is reversed, *i.e.* the positive and negative parts of the c axis are interchanged. Therefore, 4^+_{001} is conjugated to its inverse 4^-_{001} by 2_{110}.

A group $\mathcal{G}$ acts on its subgroups *via* $g(\mathcal{H}) := g\mathcal{H}g^{-1} = \{ghg^{-1} \mid h \in \mathcal{H}\}$, *i.e.* by conjugating all elements of the subgroup. The orbits are called *conjugacy classes of subgroups* of $\mathcal{G}$. Considering the conjugation action of $\mathcal{G}$ on its subgroups is often convenient, because conjugate subgroups are in particular isomorphic: an isomorphism from $\mathcal{H}$ to $g\mathcal{H}g^{-1}$ is provided by the mapping $h \mapsto ghg^{-1}$.

The stabilizer of a subgroup $\mathcal{H}$ of $\mathcal{G}$ under this conjugation action is called the *normalizer* $\mathcal{N}_\mathcal{G}(\mathcal{H})$ of $\mathcal{H}$ in $\mathcal{G}$. The normalizer of a subgroup $\mathcal{H}$ of $\mathcal{G}$ is the largest subgroup of $\mathcal{G}$ in which $\mathcal{H}$ is a normal subgroup, *i.e.* if $\mathcal{H} \trianglelefteq \mathcal{N}$ for some $\mathcal{N} \leq \mathcal{G}$, then $\mathcal{N} \leq \mathcal{N}_\mathcal{G}(\mathcal{H})$. In particular, a subgroup is a normal subgroup of $\mathcal{G}$ if and only if its normalizer in $\mathcal{G}$ is the full group $\mathcal{G}$.

The number of conjugate subgroups of $\mathcal{H}$ in $\mathcal{G}$ is equal to the index of $\mathcal{N}_\mathcal{G}(\mathcal{H})$ in $\mathcal{G}$. According to the orbit–stabilizer theorem, the different conjugate subgroups of $\mathcal{H}$ are obtained by conjugating $\mathcal{H}$ with coset representatives for the cosets of $\mathcal{G}$ relative to $\mathcal{N}_\mathcal{G}(\mathcal{H})$.

Examples

(i) The conjugacy classes of the symmetry group $\mathcal{G} = 3m$ of an equilateral triangle are $\{1\}$, $\{m_{10}, m_{01}, m_{11}\}$ and $\{3^+, 3^-\}$. The subgroups $\langle m_{10}\rangle$, $\langle m_{01}\rangle$ and $\langle m_{11}\rangle$ are conjugate subgroups (with conjugating elements 3^+ and 3^-). By the orbit–stabilizer theorem, the normalizer of each of these subgroups has index 3 in $\mathcal{G}$ and since these groups themselves have already index 3 in $\mathcal{G}$ they must coincide with their normalizers.

(ii) The conjugacy classes of the symmetry group $4mm$ of a square are $\{1\}$, $\{2\}$, $\{m_{10}, m_{01}\}$, $\{m_{11}, m_{1\bar{1}}\}$ and $\{4^+, 4^-\}$. The five subgroups of order 2 in $4mm$ fall into three conjugacy classes, namely the normal subgroup $\langle 2\rangle$ and the two pairs $\{\langle m_{10}\rangle, \langle m_{01}\rangle\}$ and $\{\langle m_{11}\rangle, \langle m_{1\bar{1}}\rangle\}$. The normalizer of both $\langle m_{10}\rangle$ and $\langle m_{01}\rangle$ is $\langle 2, m_{10}\rangle$, the two groups are conjugate by all elements of $4mm$ that are not contained in $\langle 2, m_{10}\rangle$, *e.g.* by the fourfold rotation 4^+. Analogously, the normalizer of both $\langle m_{11}\rangle$ and $\langle m_{1\bar{1}}\rangle$ is $\langle 2, m_{11}\rangle$ and again 4^+ can be chosen as conjugating element, because it is not contained in the normalizer $\langle 2, m_{11}\rangle$.

In the context of crystallographic groups, conjugate subgroups are not only isomorphic, but have the same types of geometric elements, possibly with different orientations. In many situations it is therefore sufficient to restrict attention to representatives of the conjugacy classes of subgroups. Furthermore, conjugation with elements from the normalizer of a group $\mathcal{H}$ permutes the geometric elements of the symmetry operations of $\mathcal{H}$. The role of the normalizer may in this situation be expressed by the phrase:

The normalizer describes the *symmetry of the symmetries*.

Thus, the normalizer reflects an intrinsic ambiguity between different but equivalent descriptions of an object by its symmetries.

Example

The subgroup $\mathcal{H} = \langle 2, m_{10}\rangle$ is a normal subgroup of the symmetry group $\mathcal{G} = 4mm$ of the square, and thus $\mathcal{G}$ is the normalizer of $\mathcal{H}$ in $\mathcal{G}$. As can be seen in the diagram in Fig. 1.1.6.1, the fourfold rotation 4^+ maps the geometric element of the reflection m_{10} to the geometric element of m_{01} and *vice versa*, and fixes the geometric element of the rotation 2. Consequently, conjugation by 4^+ fixes $\mathcal{H}$ as a set, but interchanges the reflections m_{10} and m_{01}. These two reflections are geometrically indistinguishable (crystallographically equivalent), since their geometric elements are both lines through the centres of opposite edges of the square.

Analogously, 4^+ interchanges the geometric elements of the reflections m_{11} and $m_{1\bar{1}}$ of the subgroup $\mathcal{H}' = \langle 2, m_{11}\rangle$. These are the two reflection lines through opposite corners of the square.

In contrast to that, $\mathcal{G}$ does not contain an element mapping the geometric element of m_{10} to that of m_{11}. Note that an eightfold rotation would be such an element, but this is, however, not a symmetry of the square. The reflections m_{10} and m_{11} are thus crystallographically inequivalent symmetry operations of the square.

References

Armstrong, M. A. (2010). *Groups and Symmetry*. New York: Springer.

Hill, V. E. (1999). *Groups and Characters*. Boca Raton: Chapman & Hall/CRC.

Müller, U. (2013). *Symmetry Relationships between Crystal Structures*. Oxford: IUCr/Oxford University Press.

Sternberg, S. (2008). *Group Theory and Physics*. Cambridge University Press.

1.2. Crystallographic symmetry

HANS WONDRATSCHEK AND MOIS I. AROYO

1.2.1. Crystallographic symmetry operations

Geometric mappings have the property that for each point X of the space, and thus of the object, there is a uniquely determined point $\tilde{X}$, the *image point*. If also for each image point $\tilde{X}$ there is a uniquely determined preimage or original point X, then the mapping is called *reversible*. Examples of non-reversible mappings are *projections*, *cf.* Section 1.4.5.

A mapping is called a *motion*, a *rigid motion* or an *isometry* if it leaves all distances invariant (and thus all angles, as well as the size and shape of an object). In Volume A of *International Tables for Crystallography* (2016), abbreviated as *IT* A, and in this edition, the term 'isometry' is used.

Isometries are a special kind of affine mappings. In an *affine mapping*, parallel lines are mapped onto parallel lines; lengths and angles may be distorted but distances along the same line are preserved.

A mapping is called a *symmetry operation* of an object if (i) it is an isometry, and (ii) it maps the object onto itself. Instead of 'maps the object onto itself' one frequently says 'leaves the object invariant (as a whole)'.

Real crystals are finite objects in physical space, which because of the presence of impurities and structural imperfections such as disorder, dislocations *etc.* are not perfectly symmetric. In order to describe their symmetry properties, real crystals are modelled as blocks of ideal, infinitely extended periodic structures, known as *ideal crystals* or (ideal) crystal structures. Crystal patterns are models of crystal structures in point space. In other words, while the crystal structure is an infinite periodic spatial arrangement of the atoms (ions, molecules) of which the real crystal is composed, the crystal pattern is the related model of the ideal crystal (crystal structure) consisting of a strictly three-dimensional periodic set of points in point space. If the growth of the ideal crystal is undisturbed, then it forms an *ideal macroscopic crystal* and displays its ideal shape with planar faces.

Both the symmetry operations of an ideal crystal and of a crystal pattern are called *crystallographic symmetry operations*. The symmetry operations of the ideal macroscopic crystal form the finite point group of the crystal, those of the crystal pattern form the (infinite) space group of the crystal pattern. Because of its periodicity, a crystal pattern always has translations among its symmetry operations.

The symmetry operations are divided into two main kinds depending whether they preserve or not the so-called *handedness* or *chirality* of chiral objects. *Isometries of the first kind* or *proper isometries* are those that preserve the handedness of chiral objects: *e.g.* if a right (left) glove is mapped by one of these isometries, then the image is also a right (left) glove of equal size and shape. Isometries that change the handedness, *i.e.* the image of a right glove is a left one, of a left glove is a right one, are called *isometries of the second kind* or *improper isometries*. Improper isometries cannot be performed in space physically but can nevertheless be observed as symmetries of objects.

The notion of *fixed points* is essential for the characterization of symmetry operations. A point X is a *fixed point* of a mapping if it is mapped onto itself, *i.e.* the *image point* $\tilde{X}$ is the same as the original point X: $\tilde{X} = X$. The set of all fixed points of an isometry may be the whole space, a plane in the space, a straight line, a point, or the set may be empty (no fixed point).

Crystallographic symmetry operations are also characterized by their *order*: a symmetry operation W is of *order* k if its application k times results in the identity mapping, *i.e* $\mathsf{W}^k = \mathsf{I}$, where I is the identity operation, and $k > 0$ is the smallest number for which this equation is fulfilled.

There are eight different types of isometries that may be crystallographic symmetry operations:

(1) The *identity operation* I maps each point of the space onto itself, *i.e.* the set of fixed points is the whole space. It is the only operation whose order is 1. The identity operation is a symmetry operation of the first kind. It is a symmetry operation of any object and although trivial, it is indispensable for the group properties of the set of symmetry operations of the object (*cf.* Section 1.1.2).

(2) A *translation* t is characterized by its translation vector $\mathbf{t}$. Under translation every point of space is shifted by $\mathbf{t}$, hence a translation has no fixed point. A translation is a symmetry operation of infinite order as there is no number $k \neq 0$ such that $t^k = \mathsf{I}$ with translation vector $\mathbf{o}$. It preserves the handedness of any chiral object.

(3) A *rotation* is an isometry which leaves one line fixed pointwise. This line is called the *rotation axis*. The degree of rotation about this axis is described by its rotation angle ϕ. Because of the periodicity of crystals, the rotation angles of crystallographic rotations are restricted to $\phi = k \times 2\pi/N$, where N = 2, 3, 4 or 6 and k is an integer which is relative prime to N (for details, see Section 1.3.3.1). A rotation of rotation angle $\phi = k \times 2\pi/N$ is of order N and is called an N-fold rotation. A rotation preserves the handedness of any chiral object.

The rotations are also characterized by their *sense of rotation*. The adopted convention for *positive* (*negative*) sense of rotation follows the mathematical convention for *positive* (*negative*) sense of rotation: the sense of rotation is positive (negative) if the rotation is counter-clockwise (clockwise) when viewed down the rotation axis.

(4) A *screw rotation* is a rotation coupled with a translation parallel to the rotation axis. The rotation axis is called the *screw axis*. The translation vector is called the *screw vector* or the *intrinsic translation component* $\mathbf{w}_g$ (of the screw rotation), *cf.* Section 1.2.2.3. A screw rotation has no fixed points because of its translation component. However, the screw axis is invariant pointwise under the so-called *reduced symmetry operation* of the screw rotation: it is the rotation obtained from the screw rotation by removing its intrinsic translation component.

The screw rotation is a proper symmetry operation. If $\phi = 2\pi/N$ is the smallest rotation angle of a screw rotation, then the screw rotation is called N-fold. Owing to its translation component, the order of any screw rotation is infinite. Let **u** be the shortest lattice vector in the direction of the screw axis, and $n\mathbf{u}/N$, with $n \neq 0$ and integer, be the screw vector of the screw rotation by the angle ϕ. After N screw rotations with rotation angle $\phi = 2\pi/N$ the crystal pattern has its original orientation but is shifted parallel to the screw axis by the lattice vector $n\mathbf{u}$.

(5) An N-fold *rotoinversion* $\overline{N}$ is an N-fold rotation coupled with an inversion through a point on the rotation axis. This point is called the *centre of the rotoinversion*. For $N \neq 2$ it is the only fixed point. The axis of the rotation is invariant as a whole under the rotoinversion and is called its *rotoinversion axis*. The restrictions on the angles ϕ of the rotational parts of rotoinversions are the same as for rotations. The order of an *N-fold* rotoinversion is N for even N and $2N$ for odd N. A rotoinversion changes the handedness by its inversion component: it maps any right-hand glove onto a left-hand one and *vice versa*. Special rotoinversions are those for $N = 1$ and $N = 2$ which are dealt with separately.

The rotoinversions $\overline{N}$ can be described equally as rotoreflections S_N. The N-fold rotation is now coupled with a reflection through a plane which is perpendicular to the rotation axis and cuts the axis in its centre. The following equivalences hold: $\bar{1} = S_2$, $\bar{2} = m = S_1$, $\bar{3} = S_6^{-1}$, $\bar{4} = S_4^{-1}$ and $\bar{6} = S_3^{-1}$. In *IT* A and in this edition the description by rotoinversions is chosen.

(6) The *inversion* can be considered as a onefold rotoinversion ($\bar{1}$, $N = 1$) or equally as a twofold rotoreflection S_2. The fixed point is called the *inversion centre*. The inversion is a symmetry operation of the second kind, its order is 2.

(7) A twofold rotoinversion ($N = 2$) is equivalent to a *reflection* or a *reflection through a plane* and is simultaneously a onefold rotoreflection ($\bar{2} = m = S_1$). It is an isometry which leaves the plane perpendicular to the twofold rotoinversion axis fixed pointwise. This plane is called the *reflection plane* or *mirror plane*; it intersects the rotation axis in its centre. Its orientation is described by the direction of its normal vector, *i.e.* of the rotation axis. (Note that in the space-group tables of *IT* A and Part 2 of this edition the reflection planes are specified by their locations, and not by their normal vectors, *cf.* Section 1.4.2.) The order of a reflection is 2. As for any rotoinversion, the reflection changes the handedness of a chiral object.

(8) A *glide reflection* is a reflection through a plane coupled with a translation parallel to this plane. The translation vector is called the *glide vector* (or the *intrinsic translation component* $\boldsymbol{w}_g$ of the glide reflection, *cf.* Section 1.2.2.3). A glide reflection changes the handedness and has no fixed point. The set of fixed points of the related reduced symmetry operation (*i.e.* the reflection that is obtained by removing the glide component from the glide reflection) is called the *glide plane*. The glide vector of a glide reflection is 1/2 of a lattice vector **t** (including centring translations of centred-cell lattice descriptions, *cf.* Table 2.1.1.2). Whereas twice the application of a reflection restores the original position of the crystal pattern, applying a glide reflection twice results in a translation of the crystal pattern with the translation vector $\mathbf{t} = 2\boldsymbol{w}_g$. The order of any glide reflection is infinite.

1.2.2. Matrix description of symmetry operations

1.2.2.1. Matrix–column presentation of isometries

In order to describe mappings analytically one introduces a coordinate system $\{O, \mathbf{a}, \mathbf{b}, \mathbf{c}\}$, consisting of three linearly independent (*i.e.* not coplanar) basis vectors **a**, **b**, **c** (or $\mathbf{a}_1$, $\mathbf{a}_2$, $\mathbf{a}_3$) and an origin O. Referred to this coordinate system each point X can be described by three coordinates x, y, z (or x_1, x_2, x_3). A mapping can be regarded as an instruction for how to calculate the coordinates $\tilde{x}, \tilde{y}, \tilde{z}$ of the image point $\tilde{X}$ from the coordinates x, y, z of the original point X.

The instruction for the calculation of the coordinates of $\tilde{X}$ from the coordinates of X is simple for an affine mapping and thus for an isometry. The equations are

$$\begin{aligned}\tilde{x} &= W_{11}x + W_{12}y + W_{13}z + w_1\\ \tilde{y} &= W_{21}x + W_{22}y + W_{23}z + w_2\\ \tilde{z} &= W_{31}x + W_{32}y + W_{33}z + w_3,\end{aligned} \qquad (1.2.2.1)$$

where the coefficients W_{ik} and w_j are constant. These equations can be written using the matrix formalism:

$$\begin{pmatrix}\tilde{x}\\ \tilde{y}\\ \tilde{z}\end{pmatrix} = \begin{pmatrix}W_{11} & W_{12} & W_{13}\\ W_{21} & W_{22} & W_{23}\\ W_{31} & W_{32} & W_{33}\end{pmatrix}\begin{pmatrix}x\\ y\\ z\end{pmatrix} + \begin{pmatrix}w_1\\ w_2\\ w_3\end{pmatrix}. \qquad (1.2.2.2)$$

This matrix equation is usually abbreviated by

$$\tilde{\boldsymbol{x}} = \boldsymbol{W}\boldsymbol{x} + \boldsymbol{w}, \qquad (1.2.2.3)$$

where

$$\tilde{\boldsymbol{x}} = \begin{pmatrix}\tilde{x}\\ \tilde{y}\\ \tilde{z}\end{pmatrix}, \quad \boldsymbol{x} = \begin{pmatrix}x\\ y\\ z\end{pmatrix}, \quad \boldsymbol{w} = \begin{pmatrix}w_1\\ w_2\\ w_3\end{pmatrix} \quad \text{and}$$

$$\boldsymbol{W} = \begin{pmatrix}W_{11} & W_{12} & W_{13}\\ W_{21} & W_{22} & W_{23}\\ W_{31} & W_{32} & W_{33}\end{pmatrix}.$$

The matrix $\boldsymbol{W}$ is called the *linear part* or *matrix part* and the column $\boldsymbol{w}$ is the *translation part* or *column part* of the mapping. The rotation parts $\boldsymbol{W}$ referring to conventional coordinate systems of all space-group symmetry operations are listed in Tables 1.2.2.1 and 1.2.2.2 of *IT* A as matrices for point-group symmetry operations.

Very often, equation (1.2.2.3) is written in the form

$$\tilde{\boldsymbol{x}} = (\boldsymbol{W}, \boldsymbol{w})\boldsymbol{x} \quad \text{or} \quad \tilde{\boldsymbol{x}} = \{\boldsymbol{W} \,|\, \boldsymbol{w}\}\boldsymbol{x}. \qquad (1.2.2.4)$$

The symbols $(\boldsymbol{W}, \boldsymbol{w})$ and $\{\boldsymbol{W} \,|\, \boldsymbol{w}\}$ which describe the mapping referred to the chosen coordinate system are called the *matrix–column pair* and can be considered as *Seitz symbols* (Seitz, 1935) (*cf.* Section 1.4.2.2 for an introduction to Seitz symbols of crystallographic symmetry operations).

1.2.2.1.1. Shorthand notation of matrix–column pairs

In crystallography in general, and in both *IT* A and this edition in particular, an efficient procedure is used to condense the description of symmetry operations by matrix–column pairs considerably. The so-called *shorthand notation* of the matrix–column pair $(\boldsymbol{W}, \boldsymbol{w})$ consists of a coordinate triplet $W_{11}x + W_{12}y + W_{13}z + w_1$, $W_{21}x + W_{22}y + W_{23}z + w_2$, $W_{31}x + W_{32}y + W_{33}z + w_3$. All coefficients '+1' and the terms with coefficients 0 are omitted, while coefficients '−1' are replaced by '−' and are frequently written on top of the variable:

$\bar{x}$ instead of $-x$ *etc.* The following examples illustrate the assignments of the coordinate triplets to the matrix–column pairs.

Examples

(1) The coordinate triplet of $y+1/2, \bar{x}+1/2, z+1/4$ stands for the symmetry operation with the rotation part

$$\boldsymbol{W} = \begin{pmatrix} 0 & 1 & 0 \\ \bar{1} & 0 & 0 \\ 0 & 0 & 1 \end{pmatrix}$$

and the translation part $\boldsymbol{w} = \begin{pmatrix} 1/2 \\ 1/2 \\ 1/4 \end{pmatrix}$. This symmetry operation is found under space group $P4_32_12$, No. 96 in the space-group tables of Part 2. It is the entry (4) of the first block (the so-called *General position* block) starting with 8 *b* 1 under the heading **Positions**.

(2) The matrix–column pair

$$(\boldsymbol{W}, \boldsymbol{w}) = \left[\begin{pmatrix} \bar{1} & 1 & 0 \\ 0 & 1 & 0 \\ 0 & 0 & \bar{1} \end{pmatrix}, \begin{pmatrix} 0 \\ 0 \\ 1/2 \end{pmatrix}\right]$$

is represented in shorthand notation by the coordinate triplet $\bar{x}+y, y, \bar{z}+1/2$. This is the entry (11) of the general positions of the space group $P6_3/mmc$, No. 194 (*cf.* the space-group tables of Part 2).

1.2.2.2. Combination of mappings and inverse mappings

The combination of two symmetry operations $(\boldsymbol{W}_1, \boldsymbol{w}_1)$ and $(\boldsymbol{W}_2, \boldsymbol{w}_2)$ is again a symmetry operation. The linear and translation part of the combined symmetry operation is derived from the rotation and translation parts of $(\boldsymbol{W}_1, \boldsymbol{w}_1)$ and $(\boldsymbol{W}_2, \boldsymbol{w}_2)$ in a straightforward way:

Applying first the symmetry operation $(\boldsymbol{W}_1, \boldsymbol{w}_1)$, on the one hand,

$$\tilde{\boldsymbol{x}} = \boldsymbol{W}_1\boldsymbol{x} + \boldsymbol{w}_1,$$

$$\tilde{\tilde{\boldsymbol{x}}} = \boldsymbol{W}_2\tilde{\boldsymbol{x}} + \boldsymbol{w}_2 = \boldsymbol{W}_2(\boldsymbol{W}_1\boldsymbol{x} + \boldsymbol{w}_1) + \boldsymbol{w}_2 = \boldsymbol{W}_2\boldsymbol{W}_1\boldsymbol{x} + \boldsymbol{W}_2\boldsymbol{w}_1 + \boldsymbol{w}_2. \qquad (1.2.2.5)$$

On the other hand

$$\tilde{\tilde{\boldsymbol{x}}} = (\boldsymbol{W}_2, \boldsymbol{w}_2)\tilde{\boldsymbol{x}} = (\boldsymbol{W}_2, \boldsymbol{w}_2)(\boldsymbol{W}_1, \boldsymbol{w}_1)\boldsymbol{x}. \qquad (1.2.2.6)$$

By comparing equations (1.2.2.5) and (1.2.2.6) one obtains

$$(\boldsymbol{W}_2, \boldsymbol{w}_2)(\boldsymbol{W}_1, \boldsymbol{w}_1) = (\boldsymbol{W}_2\boldsymbol{W}_1, \boldsymbol{W}_2\boldsymbol{w}_1 + \boldsymbol{w}_2). \qquad (1.2.2.7)$$

The formula for the inverse of an affine mapping follows from the equations $\tilde{\boldsymbol{x}} = (\boldsymbol{W}, \boldsymbol{w})\boldsymbol{x} = \boldsymbol{W}\boldsymbol{x} + \boldsymbol{w}$, *i.e.* $\boldsymbol{x} = \boldsymbol{W}^{-1}\tilde{\boldsymbol{x}} - \boldsymbol{W}^{-1}\boldsymbol{w}$, which compared with $\boldsymbol{x} = (\boldsymbol{W}, \boldsymbol{w})^{-1}\tilde{\boldsymbol{x}}$ gives

$$(\boldsymbol{W}, \boldsymbol{w})^{-1} = (\boldsymbol{W}^{-1}, -\boldsymbol{W}^{-1}\boldsymbol{w}). \qquad (1.2.2.8)$$

Because of the inconvenience of these relations, especially for the column parts of the isometries, it is often preferable to use so-called *augmented matrices*, by which one can describe the combination of affine mappings and the inverse mapping by equations of matrix multiplication (*cf.* Section 1.2.2.3 of *IT* A for an introduction to and examples of augmented-matrix formalism).

1.2.2.3. The geometric meaning of (*W*, *w*)

Given the matrix–column pair $(\boldsymbol{W}, \boldsymbol{w})$ of a symmetry operation W, the geometric interpretation of W, *i.e.* the type of operation, screw or glide component, location *etc.*, can be calculated provided the coordinate system to which $(\boldsymbol{W}, \boldsymbol{w})$ refers is known.

(1) Evaluation of the matrix part $\boldsymbol{W}$:

(*a*) *Type of operation*: In general the coefficients of the matrix depend on the choice of the basis; a change of basis changes the coefficients, see Section 1.5.2.3. However, there are geometric quantities that are independent of the basis.

(i) The preservation of the handedness of a chiral object, *i.e.* the question of whether the symmetry operation is a rotation or rotoinversion, is a geometric property which is deduced from the determinant of $\boldsymbol{W}$: $\det(\boldsymbol{W}) = +1$: *rotation*; $\det(\boldsymbol{W}) = -1$: *rotoinversion*.

(ii) The *angle of rotation* ϕ. This does not depend on the coordinate basis. The corresponding invariant of the matrix $\boldsymbol{W}$ is the trace and it is defined by $tr(\boldsymbol{W}) = W_{11} + W_{22} + W_{33}$. The rotation angle ϕ of the rotation or of the rotation part of a rotoinversion can be calculated from the trace by the formula

$$\pm\mathrm{tr}(\boldsymbol{W}) = 1 + 2\cos\phi \quad \text{or} \quad \cos\phi = (\pm\mathrm{tr}(\boldsymbol{W}) - 1)/2. \qquad (1.2.2.9)$$

The + sign is used for rotations, the − sign for rotoinversions.

The type of isometry: the types 1, 2, 3, 4, 6 or $\bar{1}$, $\bar{2} = m$, $\bar{3}$, $\bar{4}$, $\bar{6}$ can be uniquely specified by the matrix invariants: the determinant $\det(\boldsymbol{W})$ and the trace $\mathrm{tr}(\boldsymbol{W})$:

	$\det(\boldsymbol{W}) = +1$					$\det(\boldsymbol{W}) = -1$				
$\mathrm{tr}(\boldsymbol{W})$	3	2	1	0	−1	−3	−2	−1	0	1
Type	1	6	4	3	2	$\bar{1}$	$\bar{6}$	$\bar{4}$	$\bar{3}$	$\bar{2} = m$
Order	1	6	4	3	2	2	6	4	6	2

(*b*) *Rotation or rotoinversion axis*: All symmetry operations (except 1 and $\bar{1}$) have a characteristic axis (the *rotation* or *rotoinversion axis*). The direction $\boldsymbol{u}$ of this axis is invariant under the symmetry operation:

$$\pm\boldsymbol{W}\boldsymbol{u} = \boldsymbol{u} \quad \text{or} \quad (\pm\boldsymbol{W} - \boldsymbol{I})\boldsymbol{u} = \boldsymbol{o}. \qquad (1.2.2.10)$$

The + sign is for rotations, the − sign for rotoinversions.

In the case of a *k*-fold rotation, the direction $\boldsymbol{u}$ can be calculated by the equation

$$\boldsymbol{u} = \boldsymbol{Y}(\boldsymbol{W})\boldsymbol{v} = (\boldsymbol{W}^{k-1} + \boldsymbol{W}^{k-2} + \ldots + \boldsymbol{W} + \boldsymbol{I})\boldsymbol{v}, \qquad (1.2.2.11)$$

where $\boldsymbol{v} = \begin{pmatrix} v_1 \\ v_2 \\ v_3 \end{pmatrix}$ is an arbitrary direction. The direction $\boldsymbol{Y}(\boldsymbol{W})\boldsymbol{v}$ is invariant under the symmetry operation $\boldsymbol{W}$ as the multiplication with $\boldsymbol{W}$ just permutes the terms of $\boldsymbol{Y}$. If the application of equation (1.2.2.11) results in $\boldsymbol{u} = \boldsymbol{o}$, then the direction $\boldsymbol{v}$ is perpendicular to $\boldsymbol{u}$ and another direction $\boldsymbol{v}$ has to be selected. In the case of a rotoinversion $\boldsymbol{W}$, the direction $\boldsymbol{Y}(-\boldsymbol{W})\boldsymbol{v}$ gives the direction of the rotoinversion axis. For $\bar{2} = m$, $\boldsymbol{Y}(-\boldsymbol{W}) = -\boldsymbol{W} + \boldsymbol{I}$.

(*c*) *Sense of rotation* (for rotations or rotoinversions with $k > 2$): The sense of rotation is determined by the sign of the determinant of the matrix $\boldsymbol{Z}$, given by $\boldsymbol{Z} = [\boldsymbol{u}|\boldsymbol{x}|(\det \boldsymbol{W})\boldsymbol{W}\boldsymbol{x}]$, where $\boldsymbol{u}$ is the vector of equation (1.2.2.11) and $\boldsymbol{x}$ is a non-parallel vector of $\boldsymbol{u}$, *e.g.* one of the basis vectors.

(2) Analysis of the translation column $\boldsymbol{w}$:

(*a*) If $\boldsymbol{W}$ is the matrix of a rotation of order k or of a reflection ($k = 2$), then $\boldsymbol{W}^k = \boldsymbol{I}$, and one determines the *intrinsic translation part* (or *screw part* or *glide part*) of the symmetry operation, also called the *intrinsic translation component* of the symmetry operation, $\boldsymbol{w}_g = \boldsymbol{t}/k$ by

$$(\boldsymbol{W}, \boldsymbol{w})^k = (\boldsymbol{W}^k, \boldsymbol{W}^{k-1}\boldsymbol{w} + \boldsymbol{W}^{k-2}\boldsymbol{w} + \ldots + \boldsymbol{W}\boldsymbol{w} + \boldsymbol{w}) = (\boldsymbol{I}, \boldsymbol{t}) \qquad (1.2.2.12)$$

or

$$\boldsymbol{w}_g = \boldsymbol{t}/k = \frac{1}{k}(\boldsymbol{W}^{k-1} + \boldsymbol{W}^{k-2} + \ldots + \boldsymbol{W} + \boldsymbol{I})\boldsymbol{w} = \frac{1}{k}\boldsymbol{Y}(\boldsymbol{W})\,\boldsymbol{w}. \qquad (1.2.2.13)$$

The vector with the column of coefficients $\boldsymbol{w}_g = \boldsymbol{t}/k$ is called the *screw* or *glide vector*. This vector is invariant under the symmetry operation: $\boldsymbol{W}\boldsymbol{w}_g = \boldsymbol{w}_g$. Indeed, multiplication with $\boldsymbol{W}$ permutes only the terms on the right side of equation (1.2.2.13). Thus, the screw vector of a screw rotation is parallel to the screw axis. The glide vector of a glide reflection is left invariant for the same reason. It is parallel to the glide plane because $(-\boldsymbol{W} + \boldsymbol{I})(\boldsymbol{I} + \boldsymbol{W}) = \boldsymbol{O}$.

If in equation (1.2.2.12) $\boldsymbol{t} = \boldsymbol{o}$ holds, then $(\boldsymbol{W}, \boldsymbol{w})$ describes a *rotation* or *reflection*. For $\boldsymbol{t} \neq \boldsymbol{o}$, $(\boldsymbol{W}, \boldsymbol{w})$ describes a *screw rotation* or *glide reflection*. One forms the so-called *reduced operation* by subtracting the *intrinsic translation part* $\boldsymbol{w}_g = \boldsymbol{t}/k$ from $(\boldsymbol{W}, \boldsymbol{w})$:

$$(\boldsymbol{I}, -\boldsymbol{t}/k)(\boldsymbol{W}, \boldsymbol{w}) = (\boldsymbol{W}, \boldsymbol{w} - \boldsymbol{w}_g) = (\boldsymbol{W}, \boldsymbol{w}_l). \qquad (1.2.2.14)$$

The column $\boldsymbol{w}_l = \boldsymbol{w} - \boldsymbol{t}/k$ is called the *location part* (or the *location component* of the translation part) of the symmetry operation because it determines the position of the rotation or screw–rotation axis or of the reflection or glide–reflection plane in space.

(*b*) The set of *fixed points* of a symmetry operation is obtained by solving the equation

$$\boldsymbol{W}\boldsymbol{x}_F + \boldsymbol{w} = \boldsymbol{x}_F. \qquad (1.2.2.15)$$

Equation (1.2.2.15) has a unique solution for all rotoinversions (including $\bar{1}$, excluding $\bar{2} = m$). There is a one-dimensional set of solutions for rotations (the rotation axis) and a two-dimensional set of solutions for reflections (the mirror plane). For translations, screw rotations and glide reflections, there are no solutions: there are no fixed points. However, a solution is found for the reduced operation, *i.e.* after subtraction of the intrinsic translation part, *cf.* equation (1.2.2.14)

$$\boldsymbol{W}\boldsymbol{x}_F + \boldsymbol{w}_l = \boldsymbol{x}_F. \qquad (1.2.2.16)$$

(Note that the reduced operation of a translation is the identity, whose set of fixed points is the whole space.)

The formulae of this section enable the user to find the geometric contents of any symmetry operation. In practice, the geometric meanings for all symmetry operations which are listed in the *General position* blocks of the space-group tables of *IT* A can be found in the corresponding **Symmetry operations** blocks of the space-group tables. The explanation of the symbols for the symmetry operations is found in Sections 1.4.2 and 2.1.3.9.

The procedure for the geometric interpretation of the matrix–column pairs $(\boldsymbol{W}, \boldsymbol{w})$ of the symmetry operations is illustrated by three examples of the space group $Ia\bar{3}d$, No. 230 (*cf.* the tables of Part 2).

Examples

(1) Consider the symmetry operation $y + \frac{1}{4}, \bar{x} + \frac{1}{4}, z + \frac{3}{4}$ [symmetry operation (15) of the *General position* (0, 0, 0) block of the space group $Ia\bar{3}d$]. Its matrix–column pair is given by

$$\boldsymbol{W} = \begin{pmatrix} 0 & 1 & 0 \\ \bar{1} & 0 & 0 \\ 0 & 0 & 1 \end{pmatrix}, \boldsymbol{w} = \begin{pmatrix} 1/4 \\ 1/4 \\ 3/4 \end{pmatrix}.$$

Type of operation: the values of $\det(\boldsymbol{W}) = 1$ and $\mathrm{tr}(\boldsymbol{W}) = 1$ show that the symmetry operation is a fourfold rotation.
The direction of rotation axis $\boldsymbol{u}$: The application of equation (1.2.2.11) with the matrix

$$\boldsymbol{Y}(\boldsymbol{W}) = (\boldsymbol{W}^3 + \boldsymbol{W}^2 + \boldsymbol{W} + \boldsymbol{I}) = \left[\begin{pmatrix} 0 & \bar{1} & 0 \\ 1 & 0 & 0 \\ 0 & 0 & 1 \end{pmatrix} + \begin{pmatrix} \bar{1} & 0 & 0 \\ 0 & \bar{1} & 0 \\ 0 & 0 & 1 \end{pmatrix} + \begin{pmatrix} 0 & 1 & 0 \\ \bar{1} & 0 & 0 \\ 0 & 0 & 1 \end{pmatrix} + \begin{pmatrix} 1 & 0 & 0 \\ 0 & 1 & 0 \\ 0 & 0 & 1 \end{pmatrix}\right] = \begin{pmatrix} 0 & 0 & 0 \\ 0 & 0 & 0 \\ 0 & 0 & 4 \end{pmatrix}$$

yields the direction $\boldsymbol{u}$ = [001] of the fourfold rotation axis.
Sense of rotation: The negative sense of rotation follows from $\det(\boldsymbol{Z}) = -1$, where the matrix

$$\boldsymbol{Z} = [\boldsymbol{u}|\boldsymbol{x}|(\det \boldsymbol{W})\boldsymbol{W}\boldsymbol{x}] = \begin{pmatrix} 0 & 1 & 0 \\ 0 & 0 & \bar{1} \\ 1 & 0 & 0 \end{pmatrix}$$

(here, $\boldsymbol{x} = \begin{pmatrix} 1 \\ 0 \\ 0 \end{pmatrix}$ is taken as a vector non-parallel to $\boldsymbol{u}$).

Screw component: The intrinsic translation part (screw component) $\boldsymbol{w}_g$ of the symmetry operation is calculated from

$$\boldsymbol{w}_g = \tfrac{1}{4}\boldsymbol{Y}(\boldsymbol{W})\boldsymbol{w} = 1/4\begin{pmatrix} 0 & 0 & 0 \\ 0 & 0 & 0 \\ 0 & 0 & 4 \end{pmatrix}\begin{pmatrix} 1/4 \\ 1/4 \\ 3/4 \end{pmatrix} = \begin{pmatrix} 0 \\ 0 \\ 3/4 \end{pmatrix}.$$

Location of the symmetry operation: The location of the fourfold screw rotation is given by the fixed points of the reduced symmetry operation $(\boldsymbol{W}, \boldsymbol{w} - \boldsymbol{w}_g)$. The set of fixed points $\boldsymbol{x}_F = \begin{pmatrix} 1/4 \\ 0 \\ z \end{pmatrix}$ is obtained from the equation

$$\begin{pmatrix} 0 & 1 & 0 \\ \bar{1} & 0 & 0 \\ 0 & 0 & 1 \end{pmatrix}\begin{pmatrix} x_F \\ y_F \\ z_F \end{pmatrix} + \begin{pmatrix} 1/4 \\ 1/4 \\ 0 \end{pmatrix} = \begin{pmatrix} x_F \\ y_F \\ z_F \end{pmatrix}.$$

Following the conventions for the designation of symmetry operations adopted in *IT* A (*cf.* Section 1.4.2 and 2.1.3.9), the symbol of the symmetry operation $y + \frac{1}{4}, -x + \frac{1}{4}, z + \frac{3}{4}$ is given by $4^-(0, 0, \frac{3}{4})\ \frac{1}{4}, 0, z$.

(2) The symmetry operation $\bar{z}+\frac{1}{2}, x+\frac{1}{2}, y$ with

$$\boldsymbol{W} = \begin{pmatrix} 0 & 0 & \bar{1} \\ 1 & 0 & 0 \\ 0 & 1 & 0 \end{pmatrix}, \boldsymbol{w} = \begin{pmatrix} 1/2 \\ 1/2 \\ 0 \end{pmatrix}$$

corresponds to the entry No. (30) of the *General position* (0, 0, 0) block of the space group $Ia\bar{3}d$, No. 230.

Type of operation: the values of $\det(\boldsymbol{W}) = -1$ and $\mathrm{tr}(\boldsymbol{W}) = 0$ show that symmetry operation is a threefold rotoinversion.

The direction of rotoinversion axis $\boldsymbol{u}$:

$$\begin{aligned} \boldsymbol{Y}(-\boldsymbol{W}) &= (-\boldsymbol{W}^5 + \boldsymbol{W}^4 - \boldsymbol{W}^3 + \boldsymbol{W}^2 - \boldsymbol{W} + \boldsymbol{I}) \\ &= 2(\boldsymbol{W}^2 - \boldsymbol{W} + \boldsymbol{I}) \\ &= 2\left[\begin{pmatrix} 0 & \bar{1} & 0 \\ 0 & 0 & \bar{1} \\ 1 & 0 & 0 \end{pmatrix} + \begin{pmatrix} 0 & 0 & 1 \\ \bar{1} & 0 & 0 \\ 0 & \bar{1} & 0 \end{pmatrix} + \begin{pmatrix} 1 & 0 & 0 \\ 0 & 1 & 0 \\ 0 & 0 & 1 \end{pmatrix}\right] \\ &= 2\begin{pmatrix} 1 & \bar{1} & 1 \\ \bar{1} & 1 & \bar{1} \\ 1 & \bar{1} & 1 \end{pmatrix} \end{aligned}$$

yields the direction $\boldsymbol{u} = [\bar{1}1\bar{1}]$ from $\boldsymbol{v} = \begin{pmatrix} 0 \\ 1 \\ 0 \end{pmatrix}$.

Sense of rotation: The positive sense of rotation follows from the positive sign of the determinant of the matrix $\mathbf{Z}$, $\det(\mathbf{Z}) = 1$, where the matrix

$$\boldsymbol{Z} = [\boldsymbol{u}|\boldsymbol{x}|(\det \boldsymbol{W})\boldsymbol{W}\boldsymbol{x}] = \begin{pmatrix} \bar{1} & 0 & 1 \\ 1 & 0 & 0 \\ \bar{1} & 1 & 0 \end{pmatrix}$$

(here, $\boldsymbol{x} = \begin{pmatrix} 0 \\ 0 \\ 1 \end{pmatrix}$ is taken as a vector non-parallel to $\boldsymbol{u}$).

Location of the symmetry operation: The solution $x_F = 0$, $y_F = 1/2$, $z_F = 1/2$ of the fixed-point equation of the rotoinversion

$$\begin{pmatrix} 0 & 0 & \bar{1} \\ 1 & 0 & 0 \\ 0 & 1 & 0 \end{pmatrix}\begin{pmatrix} x_F \\ y_F \\ z_F \end{pmatrix} + \begin{pmatrix} 1/2 \\ 1/2 \\ 0 \end{pmatrix} = \begin{pmatrix} x_F \\ y_F \\ z_F \end{pmatrix}$$

gives the coordinates of the inversion centre on the rotoinversion axis. An obvious description of a line along the direction $\boldsymbol{u} = [\bar{1}1\bar{1}]$ and passing through the point $(0, 1/2, 1/2)$ is given by the parametric expression $\bar{u}, u + 1/2, \bar{u} + 1/2$. The choice of the free parameter $u = x + 1/2$ results in the description $\bar{x} - 1/2, x + 1, \bar{x}$ of the rotoinversion axis found in the **Symmetry operation** $(0, 0, 0)$ block of the space-group table of $Ia\bar{3}d$. The convention adopted in *IT* A to have zero constant at the z coordinate of the description of the $\bar{3}$ axis determines the specific choice of the free parameter.

The geometric characteristics of the symmetry operation $\bar{z}+\frac{1}{2}, x+\frac{1}{2}, y$ are reflected in its symbol $3^+\ \bar{x} - 1/2, x + 1, \bar{x};\ 0, \frac{1}{2}, \frac{1}{2}$.

(3) The matrix–column pair $(\boldsymbol{W}, \boldsymbol{w})$ of the symmetry operation (37) $\bar{y} + 3/4, \bar{x} + 1/4, z + 1/4$ of the *General position* $(1/2, 1/2, 1/2)$ block of the space group $Ia\bar{3}d$ is given by

$$\boldsymbol{W} = \begin{pmatrix} 0 & \bar{1} & 0 \\ \bar{1} & 0 & 0 \\ 0 & 0 & 1 \end{pmatrix}, \boldsymbol{w} = \begin{pmatrix} 3/4 \\ 1/4 \\ 1/4 \end{pmatrix}.$$

Type of operation: The values of the determinant $\det(\boldsymbol{W}) = -1$ and the trace $\mathrm{tr}(\boldsymbol{W}) = +1$ indicate that the symmetry operation is a reflection.

Normal $\boldsymbol{u}$ *of the reflection plane*: The orientation of the reflection plane in space is determined by its normal $\boldsymbol{u}$, which is directed along [110]. The direction of $\boldsymbol{u}$ follows from the matrix equation

$$\begin{aligned} \boldsymbol{u} &= (-\boldsymbol{W} + \boldsymbol{I})\boldsymbol{v} \\ &= \left[\begin{pmatrix} 0 & 1 & 0 \\ 1 & 0 & 0 \\ 0 & 0 & \bar{1} \end{pmatrix} + \begin{pmatrix} 1 & 0 & 0 \\ 0 & 1 & 0 \\ 0 & 0 & 1 \end{pmatrix}\right]\boldsymbol{v} = \begin{pmatrix} 1 & 1 & 0 \\ 1 & 1 & 0 \\ 0 & 0 & 0 \end{pmatrix}\boldsymbol{v}, \end{aligned}$$

where $\boldsymbol{v} = \begin{pmatrix} v_1 \\ v_2 \\ v_3 \end{pmatrix}$ is arbitrary.

Glide component: The glide component $\boldsymbol{w}_g = \begin{pmatrix} 1/4 \\ -1/4 \\ 1/4 \end{pmatrix}$, determined from the equation

$$\begin{aligned} \boldsymbol{w}_g &= \tfrac{1}{2}(\boldsymbol{W} + \boldsymbol{I})\boldsymbol{w} \\ &= \tfrac{1}{2}\left[\begin{pmatrix} 0 & \bar{1} & 0 \\ \bar{1} & 0 & 0 \\ 0 & 0 & 1 \end{pmatrix} + \begin{pmatrix} 1 & 0 & 0 \\ 0 & 1 & 0 \\ 0 & 0 & 1 \end{pmatrix}\right]\begin{pmatrix} 3/4 \\ 1/4 \\ 1/4 \end{pmatrix}, \end{aligned}$$

indicates that the symmetry operation is a d-glide reflection. As expected, the translation vector

$$\mathbf{t} = 2\boldsymbol{w}_g = \begin{pmatrix} 1/2 \\ -1/2 \\ 1/2 \end{pmatrix}$$

corresponds to a centring translation.

Location of the symmetry operation: The set of fixed points of the reduced symmetry operation

$$(\boldsymbol{W}, \boldsymbol{w} - \boldsymbol{w}_g) = \left[\begin{pmatrix} 0 & \bar{1} & 0 \\ \bar{1} & 0 & 0 \\ 0 & 0 & 1 \end{pmatrix}, \begin{pmatrix} 3/4 - 1/4 \\ 1/4 - (-1/4) \\ 1/4 - 1/4 \end{pmatrix}\right]$$

determines the location of the d-glide plane

$$\begin{pmatrix} 0 & \bar{1} & 0 \\ \bar{1} & 0 & 0 \\ 0 & 0 & 1 \end{pmatrix}\begin{pmatrix} x_F \\ y_F \\ z_F \end{pmatrix} + \begin{pmatrix} 1/2 \\ 1/2 \\ 0 \end{pmatrix} = \begin{pmatrix} x_F \\ y_F \\ z_F \end{pmatrix}.$$

Thus, the set of fixed points (the d-glide plane) can be described as $x + 1/2, \bar{x}, z$.

The symbol $d\,(1/4, -1/4, 1/4)\ x + 1/2, \bar{x}, z$ of the symmetry operation (37) $\bar{y} + 3/4, \bar{x} + 1/4, z + 1/4$, found in the **Symmetry operations** $(1/2, 1/2, 1/2)$ block of the space-group table of $Ia\bar{3}d$, comprises the essential geometric characteristics of the symmetry operation, *i.e.* its type, glide component and location. It is worth repeating that according to the conventions adopted in the space-group tables of *IT* A, the mirror planes are specified by their sets of fixed points and not by the normals to the planes (*cf.* Section 1.4.2 for more details).

1.2.3. Symmetry elements

In the 1970s, when the International Union of Crystallography (IUCr) planned a new series of *International Tables for Crystallography* to replace the series *International Tables for X-ray Crystallography* (1952), there was some confusion about the use

Table 1.2.3.1
Symmetry elements in point and space groups

Name of symmetry element	Geometric element	Defining operation (d.o.)	Operations in element set
Mirror plane	Plane p	Reflection through p	D.o. and its coplanar equivalents†
Glide plane	Plane p	Glide reflection through p; $2\mathbf{v}$ (not $\mathbf{v}$) a lattice-translation vector	D.o. and its coplanar equivalents†
Rotation axis	Line l	Rotation around l, angle $2\pi/N$, $N = 2, 3, 4$ or 6	1st ... $(N-1)$th powers of d.o. and their coaxial equivalents‡
Screw axis	Line l	Screw rotation around l, angle $2\pi/N$, $u = j/N$ times shortest lattice translation along l, right-hand screw, $N = 2, 3, 4$ or 6, $j = 1, \ldots, (N-1)$	1st ... $(N-1)$th powers of d.o. and their coaxial equivalents‡
Rotoinversion axis	Line l and point P on l	Rotoinversion: rotation around l, angle $2\pi/N$, followed by inversion through P, $N = 3, 4$ or 6	D.o. and its inverse
Centre	Point P	Inversion through P	D.o. only

† That is, all glide reflections through the same reflection plane, with glide vectors $\mathbf{v}$ differing from that of the d.o. (taken to be zero for reflections) by a lattice-translation vector. The glide planes a, b, c, n, d and e are distinguished (*cf.* Table 2.1.2.1). ‡ That is, all rotations and screw rotations around the same axis l, with the same angle and sense of rotation and the same screw vector $\mathbf{u}$ (zero for rotation) up to a lattice-translation vector.

of the term *symmetry element*. Crystallographers and mineralogists had used this term for rotation and rotoinversion axes and reflection planes, in particular for the description of the morphology of crystals, for a long time, although there had been no strict definition of 'symmetry element'. With the impact of mathematical group theory in crystallography the term *element* was introduced with another meaning, in which an element is a member of a set, in particular as a group element of a group. In crystallography these group elements, however, were the symmetry operations of the symmetry groups, not the crystallographic symmetry elements. Therefore, the IUCr Commission on Crystallographic Nomenclature appointed an *Ad-hoc* Committee on the Nomenclature of Symmetry with P. M. de Wolff as Chairman to propose definitions for terms of crystallographic symmetry and for several classifications of crystallographic space groups and point groups.

In the reports of the *Ad-hoc* Committee, de Wolff *et al.* (1989) and (1992) with *Addenda*, Flack *et al.* (2000), the results were published. To define the term *symmetry element* for any symmetry operation was more complicated than had been envisaged previously, in particular for unusual screw and glide components.

According to the proposals of the Committee the following procedure has been adopted (*cf.* also Table 1.2.3.1):

(1) No symmetry element is defined for the identity and the (lattice) translations.

(2) For any symmetry operation of point groups and space groups with the exception of the rotoinversions $\bar{3}$, $\bar{4}$ and $\bar{6}$, the *geometric element* is defined as the *set of fixed points* (the second column of Table 1.2.3.1) of the *reduced operation*, *cf.* equation (1.2.2.14). For reflections and glide reflections this is a plane; for rotations and screw rotations it is a line, for the inversion it is a point. For the rotoinversions $\bar{3}$, $\bar{4}$ and $\bar{6}$ the geometric element is a line with a point (the inversion centre) on this line.

(3) The *element set* (*cf.* the last column of Table 1.2.3.1) is defined as a set of operations that share the same geometric element. The element set can consist of symmetry operations of the same type (such as the powers of a rotation) or of different types, *e.g.* by a reflection and a glide reflection through the same plane. The *defining operation* (d.o.) may be any symmetry operation from the element set that suffices to identify the symmetry element. In most cases, the 'simplest' symmetry operation from the element set is chosen as the d.o. (*cf.* the third column of Table 1.2.3.1). For reflections and glide reflections the element set includes the defining operation and all glide reflections through the same reflection plane but with glide vectors differing by a lattice-translation vector, *i.e.* the so-called *coplanar equivalents*. For rotations and screw rotations of angle $2\pi/N$ the element set is the defining operation, its 1st ... $(N-1)$th powers and all rotations and screw rotations with screw vectors differing from that of the defining operation by a lattice-translation vector, known as *coaxial equivalents*. For a rotoinversion the element set includes the defining operation and its inverse.

(4) The combination of the geometric element and its element set is indicated by the name *symmetry element*. The names of the symmetry elements (first column of Table 1.2.3.1) are combinations of the name of the defining operation attached to the name of the corresponding geometric element. Names of symmetry elements are *mirror plane*, *glide plane*, *rotation axis*, *screw axis*, *rotoinversion axis* and *centre*.[1] This allows such statements as *this point lies on a rotation axis* or *these operations belong to a glide plane*.

Examples

(1) *Glide and mirror planes*. The element set of a glide plane with a glide vector $\mathbf{v}$ consists of infinitely many different glide reflections with glide vectors that are obtained from $\mathbf{v}$ by adding any lattice-translation vector parallel to the glide plane, including centring translations of centred cells.

(*a*) It is important to note that if among the infinitely many glide reflections of the element set of the same plane there exists one operation with zero glide vector, then this operation is taken as the *defining operation* (d.o). Consider, for example, the symmetry operation $x+1/2,\ y+1/2,\ -z+1/2$ of *Cmcm* (63) [*General position* $(1/2, 1/2, 0)$ block]. This is an n-glide reflection through the plane $x, y, 1/4$. However, the corresponding symmetry element is a mirror plane, as among the glide reflections of the element set of the plane $x, y, 1/4$ one finds the reflection $x, y, -z+1/2$ [symmetry operation (6) of the *General position* $(0, 0, 0)$ block].

(*b*) The symmetry operation $x+5/2, y-7/2, -z+3$ is a glide reflection. Its geometric element is the plane $x, y, 3/2$. Its symmetry element is a glide plane in space

[1] The proposal of the Committee to introduce the symbols for the symmetry elements *Em*, *Eg*, *En*, En_j, $E\bar{n}$ and $E\bar{1}$ was not taken up in practice. The printed and graphical symbols of symmetry elements used throughout the space-group tables of *IT* A are introduced in Section 2.1.2 and listed in Tables 2.1.2.1 to 2.1.2.7.

group *Pmmn* (59) because there is no lattice translation by which the glide vector can be changed to **o**. If, however, the same mapping is a symmetry operation of space group *Cmmm* (65), then its symmetry element is a reflection plane because the glide vector with components $5/2, -7/2, 0$ can be cancelled through a translation $(2+\frac{1}{2})\mathbf{a} + (-4+\frac{1}{2})\mathbf{b}$, which is a lattice translation in a *C* lattice. Evidently, the correct specification of the symmetry element is possible only with respect to a specific translation lattice.

(*c*) Similarly, in *Aem*2 (39) with a *b*-glide reflection $\bar{x}, y+1/2, z$, the *c*-glide reflection $\bar{x}, y, z+1/2$ also occurs. The geometric element is the plane $0, y, z$ and the symmetry element is an *e*-glide plane.

Likewise, all vectors $(u+\frac{1}{2})\mathbf{a} + v\mathbf{b} + \frac{1}{2}k(\mathbf{a}+\mathbf{b})$, u, v, k integers, are glide vectors of glide reflections through the (001) plane of a space group with a *C*-centred lattice. Among them one finds a glide reflection *b* with a glide vector $\frac{1}{2}\mathbf{b}$ related to $\frac{1}{2}\mathbf{a}$ by the centring translation; an *a*-glide reflection and a *b*-glide reflection share the same plane as a geometric element. Their symmetry element is thus an *e*-glide plane.

(*d*) In general, the *e*-glide planes are symmetry elements characterized by the existence of two glide reflections through the same plane with perpendicular glide vectors and with the additional requirement that at least one glide vector is along a crystal axis (de Wolff *et al.*, 1992). The *e*-glide designation of glide planes occurs only when a centred cell represents the choice of basis (*cf.* Table 2.1.2.1). The 'double' *e*-glide planes are indicated by special graphical symbols on the symmetry-element diagrams of the space groups (*cf.* Tables 2.1.2.2 to 2.1.2.4). For example, consider the space group *I*4*cm* (108). The symmetry operations (8) $y, x, z+1/2$ [*General position* (0, 0, 0) block] and (8) $y+1/2, x+1/2, z$ [*General position* (1/2, 1/2, 1/2) block] are glide reflections through the same x, x, z plane, and their glide vectors $\frac{1}{2}\mathbf{c}$ and $\frac{1}{2}(\mathbf{a}+\mathbf{b})$ are related by the centring $(1/2, 1/2, 1/2)$ translation. The corresponding symmetry element is an *e*-glide plane and it is easily recognized on the symmetry-element diagram of *I*4*cm* shown in the space-group tables of *IT* A.

(2) *Screw and rotation axes*. The element set of a screw axis is formed by a screw rotation of angle $2\pi/N$ with a screw vector **u**, its $(N-1)$ powers and all its co-axial equivalents, *i.e.* screw rotations around the same axis, with the same angle and sense of rotation, with screw vectors obtained by adding a lattice-translation vector parallel to **u**.

(*a*) Twofold screw axis ∥ [001] in a primitive cell: the element set is formed by all twofold screw rotations around the same axis with screw vectors of the type $(u+\frac{1}{2})\mathbf{c}$, *i.e.* screw components as $\frac{1}{2}\mathbf{c}$, $-\frac{1}{2}\mathbf{c}$, $\frac{3}{2}\mathbf{c}$ *etc.*

(*b*) The symmetry operation $4-x, -2-y, z+5/2$ is a screw rotation of space group $Pna2_1$ (33). Its geometric element is the line $2, -1, z$ and its symmetry element is a screw axis.

(*c*) The determination of the complete element set of a geometric element is important for the correct designation of the corresponding symmetry element. For example, the symmetry element of a twofold screw rotation with an axis along the line $\frac{1}{2}, 0, z$ is a twofold screw axis in the space group $Pna2_1$ but a fourfold screw axis in $P4_12_12$ (92).

(3) *Special case*. In point groups $6/m$, $6/mmm$ and space groups $P6/m$ (175), $P6/mmm$ (191) and $P6/mcc$ (192) the geometric elements of the defining operations $\bar{6}$ and $\bar{3}$ are the same. To make the element sets unique, the geometric elements should not be given just by a line and a point on it, but should be labelled by these operations. Then the element sets and thus the symmetry element are unique (Flack *et al.*, 2000).

References

Flack, H. D., Wondratschek, H., Hahn, Th. & Abrahams, S. C. (2000). *Symmetry elements in space groups and point groups. Addenda to two IUCr Reports on the Nomenclature of Symmetry. Acta Cryst.* **A56**, 96–98.

International Tables for Crystallography (2016). Volume A, *Space-Group Symmetry*, 6th ed., edited by M. I. Aroyo. Chichester: Wiley. (Abbreviated as *IT* A.)

International Tables for X-ray Crystallography (1952). Vol. I, edited by N. F. M. Henry & K. Lonsdale. Birmingham: Kynoch Press.

Seitz, F. (1935). *A matrix-algebraic development of crystallographic groups. III. Z. Kristallogr.* **71**, 336–366.

Wolff, P. M. de, Billiet, Y., Donnay, J. D. H., Fischer, W., Galiulin, R. B., Glazer, A. M., Senechal, M., Shoemaker, D. P., Wondratschek, H., Hahn, Th., Wilson, A. J. C. & Abrahams, S. C. (1989). *Definition of symmetry elements in space groups and point groups. Report of the International Union of Crystallography Ad-Hoc Committee on the Nomenclature of Symmetry. Acta Cryst.* **A45**, 494–499.

Wolff, P. M. de, Billiet, Y., Donnay, J. D. H., Fischer, W., Galiulin, R. B., Glazer, A. M., Hahn, Th., Senechal, M., Shoemaker, D. P., Wondratschek, H., Wilson, A. J. C. & Abrahams, S. C. (1992). *Symbols for symmetry elements and symmetry operations. Final Report of the International Union of Crystallography Ad-Hoc Committee on the Nomenclature of Symmetry. Acta Cryst.* **A48**, 727–732.

1.3. A general introduction to space groups

Bernd Souvignier

1.3.1. Introduction

We recall from Chapter 1.2 that an *isometry* is a mapping of the point space $\mathbb{E}^3$ which preserves distances and angles. From the mathematical viewpoint, $\mathbb{E}^3$ is an *affine space* in which two points differ by a unique vector in the underlying *vector space* $\mathbb{V}^3$. The crucial difference between these two types of spaces is that in an affine space no point is distinguished, whereas in a vector space the zero vector plays a special role, namely as the identity element for the addition of vectors. After choosing an origin O, the points of the affine space $\mathbb{E}^3$ are in one-to-one correspondence with the vectors of $\mathbb{V}^3$ by identifying a point X with the difference vector $\overrightarrow{OX}$.

A *crystallographic space-group operation* is an isometry that maps a crystal pattern (*cf.* Section 1.2.1) onto itself. Since isometries are invertible and the composition of two isometries leaves a crystal pattern invariant as a whole if the two single isometries do so, the space-group operations form a group $\mathcal{G}$, called a *crystallographic space group*.

As a mapping of points in an affine space, a space-group operation is an affine mapping and is thus composed of a linear mapping of the underlying vector space and a translation. Once a coordinate system has been chosen, space-group operations are conveniently represented as *matrix–column pairs* $(\boldsymbol{W}, \boldsymbol{w})$, where $\boldsymbol{W}$ is the *linear part* and $\boldsymbol{w}$ the *translation part* and a point with coordinates $\boldsymbol{x}$ is mapped to $\boldsymbol{W}\boldsymbol{x} + \boldsymbol{w}$ (*cf.* Section 1.2.2).

A translation is a matrix–column pair of the form $(\boldsymbol{I}, \boldsymbol{t})$, where $\boldsymbol{I}$ is the unit matrix and the coefficients of the translation part $\boldsymbol{t}$ define the corresponding translation vector. All translations taken together form the *translation subgroup* $\mathcal{T}$ of $\mathcal{G}$, which is infinite and forms an abelian normal subgroup of $\mathcal{G}$. The factor group $\mathcal{G}/\mathcal{T}$ is a finite group that can be identified with the group of linear parts of $\mathcal{G}$ *via* the mapping $(\boldsymbol{W}, \boldsymbol{w}) \mapsto \boldsymbol{W}$, which simply forgets about the translation part. The group $\mathcal{P} = \{\boldsymbol{W} \mid (\boldsymbol{W}, \boldsymbol{w}) \in \mathcal{G}\}$ of linear parts occurring in $\mathcal{G}$ is called the *point group* $\mathcal{P}$ of $\mathcal{G}$.

The representation of space-group operations as matrix–column pairs is clearly adapted to the fact that space groups can be built from these two parts, the translation subgroup and the point group. This viewpoint will be discussed in detail in Section 1.3.3. It allows one to treat space groups in many aspects analogously to finite groups, although, due to the infinite translation subgroup, they are of course infinite groups.

1.3.2. Lattices

A crystal pattern is defined to be periodic in three linearly independent directions, which means that it is invariant under translations in three linearly independent directions. This periodicity implies that the crystal pattern extends infinitely in all directions. Since the atoms of a crystal form a discrete pattern in which two different points have a certain minimal distance, the translations that fix the crystal pattern as a whole cannot have arbitrarily small lengths. If $\mathbf{v}$ is a vector such that the crystal pattern is invariant under a translation by $\mathbf{v}$, the periodicity implies that the pattern is invariant under a translation by $m\mathbf{v}$ for every integer m. Furthermore, if a crystal pattern is invariant under translations by $\mathbf{v}$ and $\mathbf{w}$, it is also invariant by the composition of these two translations, which is the translation by $\mathbf{v} + \mathbf{w}$. This shows that the set of vectors by which the translations in a space group move the crystal pattern is closed under taking integral linear combinations. This property is formalized by the mathematical concept of a *lattice* and the translation subgroups of space groups are best understood by studying their corresponding lattices. These lattices capture the periodic nature of the underlying crystal patterns and reflect their geometric properties.

1.3.2.1. Basic properties of lattices

The two-dimensional vector space $\mathbb{V}^2$ is the space of columns $\begin{pmatrix} x \\ y \end{pmatrix}$ with two real components $x, y \in \mathbb{R}$ and the three-dimensional vector space $\mathbb{V}^3$ is the space of columns $\begin{pmatrix} x \\ y \\ z \end{pmatrix}$ with three real components $x, y, z \in \mathbb{R}$.

Definition

For vectors $\mathbf{a}, \mathbf{b}, \mathbf{c}$ forming a basis of the three-dimensional vector space $\mathbb{V}^3$, the set

$$\mathbf{L} := \{l\mathbf{a} + m\mathbf{b} + n\mathbf{c} \mid l, m, n \in \mathbb{Z}\}$$

of all *integral* linear combinations of $\mathbf{a}, \mathbf{b}, \mathbf{c}$ is called a *lattice* in $\mathbb{V}^3$ and the vectors $\mathbf{a}, \mathbf{b}, \mathbf{c}$ are called a *lattice basis* of $\mathbf{L}$.

It is inherent in the definition of a crystal pattern that the translation vectors of the translations leaving the pattern invariant are closed under taking integral linear combinations. Since the crystal pattern is assumed to be discrete, it follows that a set of three translation vectors $\mathbf{a}, \mathbf{b}, \mathbf{c}$ can be found such that all translation vectors are integral linear combinations of these three vectors. This shows that the translation vectors of a crystal pattern form a lattice with lattice basis $\mathbf{a}, \mathbf{b}, \mathbf{c}$ in the sense of the definition above.

By definition, a lattice is determined by a lattice basis. Note, however, that every two- or three-dimensional lattice has infinitely many bases.

Example

The square lattice

$$\mathbf{L} = \mathbb{Z}^2 = \left\{ \begin{pmatrix} m \\ n \end{pmatrix} \mid m, n \in \mathbb{Z} \right\}$$

in $\mathbb{V}^2$ has the vectors

$$\mathbf{a} = \begin{pmatrix} 1 \\ 0 \end{pmatrix}, \quad \mathbf{b} = \begin{pmatrix} 0 \\ 1 \end{pmatrix}$$

as its standard lattice basis. But

$$\mathbf{a}' = \begin{pmatrix} 1 \\ \bar{2} \end{pmatrix}, \quad \mathbf{b}' = \begin{pmatrix} \bar{2} \\ 3 \end{pmatrix}$$

is also a lattice basis of $\mathbf{L}$: on the one hand $\mathbf{a}'$ and $\mathbf{b}'$ are integral linear combinations of $\mathbf{a}$, $\mathbf{b}$ and are thus contained in $\mathbf{L}$. On the other hand

$$-3\mathbf{a}' - 2\mathbf{b}' = \begin{pmatrix} \bar{3} \\ 6 \end{pmatrix} + \begin{pmatrix} 4 \\ \bar{6} \end{pmatrix} = \begin{pmatrix} 1 \\ 0 \end{pmatrix} = \mathbf{a}$$

and

$$-2\mathbf{a}' - \mathbf{b}' = \begin{pmatrix} \bar{2} \\ 4 \end{pmatrix} + \begin{pmatrix} 2 \\ \bar{3} \end{pmatrix} = \begin{pmatrix} 0 \\ 1 \end{pmatrix} = \mathbf{b},$$

hence $\mathbf{a}$ and $\mathbf{b}$ are also integral linear combinations of $\mathbf{a}'$, $\mathbf{b}'$ and thus the two bases $\mathbf{a}$, $\mathbf{b}$ and $\mathbf{a}'$, $\mathbf{b}'$ both span the same lattice (see Fig. 1.3.2.1).

The example indicates how the different lattice bases of a lattice $\mathbf{L}$ are related. For two lattice bases $\mathbf{a}, \mathbf{b}, \mathbf{c}$ and $\mathbf{a}', \mathbf{b}', \mathbf{c}'$, the basis transformation $\boldsymbol{P}$ such that $(\mathbf{a}', \mathbf{b}', \mathbf{c}') = (\mathbf{a}, \mathbf{b}, \mathbf{c})\boldsymbol{P}$ is an integral 3×3 matrix. Since the inverse transformation $\boldsymbol{P}^{-1}$ must also be integral, one requires that $\det \boldsymbol{P} = \pm 1$. In the example above the two basis transformations are

$$\boldsymbol{P} = \begin{pmatrix} 1 & \bar{2} \\ \bar{2} & 3 \end{pmatrix} \text{ and } \boldsymbol{P}^{-1} = \begin{pmatrix} \bar{3} & \bar{2} \\ \bar{2} & \bar{1} \end{pmatrix}.$$

In this case, $\det \boldsymbol{P} = -1$ and one can observe in Fig. 1.3.2.1 that the two coordinate systems $\mathbf{a}, \mathbf{b}$ and $\mathbf{a}', \mathbf{b}'$ have opposite handedness.

1.3.2.2. Metric properties

In the three-dimensional vector space $\mathbb{V}^3$, the *norm* or *length* of a vector $\mathbf{v} = \begin{pmatrix} v_x \\ v_y \\ v_z \end{pmatrix}$ is (due to Pythagoras' theorem) given by

$$|\mathbf{v}| = \sqrt{v_x^2 + v_y^2 + v_z^2}.$$

From this, the *scalar product*

$$\mathbf{v} \cdot \mathbf{w} = v_x w_x + v_y w_y + v_z w_z \text{ for } \mathbf{v} = \begin{pmatrix} v_x \\ v_y \\ v_z \end{pmatrix}, \mathbf{w} = \begin{pmatrix} w_x \\ w_y \\ w_z \end{pmatrix}$$

is derived, which allows one to express angles by

$$\cos \angle(\mathbf{v}, \mathbf{w}) = \frac{\mathbf{v} \cdot \mathbf{w}}{|\mathbf{v}|\,|\mathbf{w}|}.$$

The definition of a norm function for the vectors turns $\mathbb{V}^3$ into a *Euclidean space*. A lattice $\mathbf{L}$ that is contained in $\mathbb{V}^3$ inherits the metric properties of this space. But for the lattice, these properties are most conveniently expressed with respect to a lattice basis. It is customary to choose basis vectors $\mathbf{a}, \mathbf{b}, \mathbf{c}$ which define a right-handed coordinate system, *i.e.* such that the matrix with columns $\mathbf{a}, \mathbf{b}, \mathbf{c}$ has a positive determinant.

Definition

For a lattice $\mathbf{L} \subseteq \mathbb{V}^3$ with lattice basis $\mathbf{a}, \mathbf{b}, \mathbf{c}$ the *metric tensor* of $\mathbf{L}$ is the 3×3 matrix

$$\boldsymbol{G} = \begin{pmatrix} \mathbf{a}\cdot\mathbf{a} & \mathbf{a}\cdot\mathbf{b} & \mathbf{a}\cdot\mathbf{c} \\ \mathbf{b}\cdot\mathbf{a} & \mathbf{b}\cdot\mathbf{b} & \mathbf{b}\cdot\mathbf{c} \\ \mathbf{c}\cdot\mathbf{a} & \mathbf{c}\cdot\mathbf{b} & \mathbf{c}\cdot\mathbf{c} \end{pmatrix}.$$

If $\boldsymbol{A}$ is the 3×3 matrix with the vectors $\mathbf{a}, \mathbf{b}, \mathbf{c}$ as its columns, then the metric tensor is obtained as the matrix product $\boldsymbol{G} = \boldsymbol{A}^{\mathrm{T}} \cdot \boldsymbol{A}$. It follows immediately that the metric tensor is a symmetric matrix, *i.e.* $\boldsymbol{G}^{\mathrm{T}} = \boldsymbol{G}$.

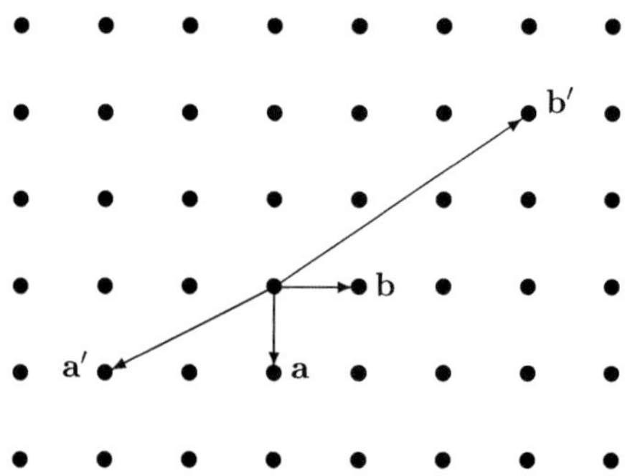

Figure 1.3.2.1
Conventional basis $\mathbf{a}$, $\mathbf{b}$ and a non-conventional basis $\mathbf{a}'$, $\mathbf{b}'$ for the square lattice.

Example

Let

$$\mathbf{a} = \begin{pmatrix} 1 \\ 1 \\ 1 \end{pmatrix}, \quad \mathbf{b} = \begin{pmatrix} 1 \\ 1 \\ 0 \end{pmatrix}, \quad \mathbf{c} = \begin{pmatrix} \bar{1} \\ 1 \\ 0 \end{pmatrix}$$

be the basis of a lattice $\mathbf{L}$. Then the metric tensor of $\mathbf{L}$ (with respect to the given basis) is

$$\boldsymbol{G} = \begin{pmatrix} 3 & 2 & 0 \\ 2 & 2 & 0 \\ 0 & 0 & 2 \end{pmatrix}.$$

With the help of the metric tensor the scalar products of arbitrary vectors, given as linear combinations of the lattice basis, can be computed from their coordinate columns as follows: If $\mathbf{v} = x_1\mathbf{a} + y_1\mathbf{b} + z_1\mathbf{c}$ and $\mathbf{w} = x_2\mathbf{a} + y_2\mathbf{b} + z_2\mathbf{c}$, then

$$\mathbf{v} \cdot \mathbf{w} = (x_1\, y_1\, z_1) \cdot \boldsymbol{G} \cdot \begin{pmatrix} x_2 \\ y_2 \\ z_2 \end{pmatrix}.$$

From this it follows how the metric tensor transforms under a basis transformation $\boldsymbol{P}$. If $(\mathbf{a}', \mathbf{b}', \mathbf{c}') = (\mathbf{a}, \mathbf{b}, \mathbf{c})\boldsymbol{P}$, then the metric tensor $\boldsymbol{G}'$ of $\mathbf{L}$ with respect to the new basis $\mathbf{a}', \mathbf{b}', \mathbf{c}'$ is given by

$$\boldsymbol{G}' = \boldsymbol{P}^{\mathrm{T}} \cdot \boldsymbol{G} \cdot \boldsymbol{P}.$$

An alternative way to specify the geometry of a lattice in $\mathbb{V}^3$ is using the *cell parameters*, which are the lengths of the lattice basis vectors and the angles between them.

Definition

For a lattice $\mathbf{L}$ in $\mathbb{V}^3$ with lattice basis $\mathbf{a}, \mathbf{b}, \mathbf{c}$ the *cell parameters* (also called *lattice parameters*, *lattice constants* or *metric parameters*) are given by the lengths

$$a = |\mathbf{a}| = \sqrt{\mathbf{a}\cdot\mathbf{a}}, \quad b = |\mathbf{b}| = \sqrt{\mathbf{b}\cdot\mathbf{b}}, \quad c = |\mathbf{c}| = \sqrt{\mathbf{c}\cdot\mathbf{c}}$$

of the basis vectors and by the interaxial angles

$$\alpha = \angle(\mathbf{b}, \mathbf{c}), \quad \beta = \angle(\mathbf{c}, \mathbf{a}), \quad \gamma = \angle(\mathbf{a}, \mathbf{b}).$$

Owing to the relation $\mathbf{v} \cdot \mathbf{w} = |\mathbf{v}|\,|\mathbf{w}| \cos \angle(\mathbf{v}, \mathbf{w})$ for the scalar product of two vectors, one can immediately write down the metric tensor in terms of the cell parameters:

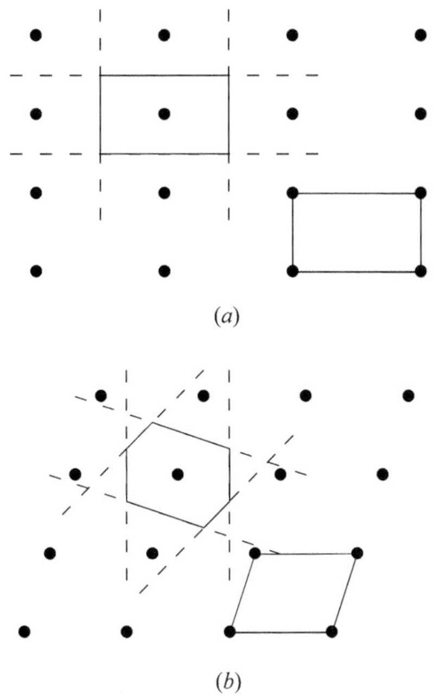

Figure 1.3.2.2
Voronoï domains and primitive unit cells for a rectangular lattice (*a*) and an oblique lattice (*b*).

$$G = \begin{pmatrix} a^2 & ab\cos\gamma & ac\cos\beta \\ ab\cos\gamma & b^2 & bc\cos\alpha \\ ac\cos\beta & bc\cos\alpha & c^2 \end{pmatrix}.$$

1.3.2.3. Unit cells

A lattice **L** can be used to subdivide $\mathbb{V}^3$ into cells of finite volume which all have the same shape. The idea is to define a suitable subset **C** of $\mathbb{V}^3$ such that the translates of **C** by the vectors in **L** cover $\mathbb{V}^3$ without overlapping. Such a subset **C** is called a *unit cell* of **L**, or, in the more mathematically inclined literature, a *fundamental domain* of $\mathbb{V}^3$ with respect to **L**. Two standard constructions for such unit cells are the *primitive unit cell* and the *Voronoï domain* (which is also known by many other names).

Definition

Let **L** be a lattice in $\mathbb{V}^3$ with lattice basis **a**, **b**, **c**.

(i) The set $\mathbf{C} := \{x\mathbf{a} + y\mathbf{b} + z\mathbf{c} \mid 0 \le x, y, z < 1\}$ is called the *primitive unit cell* of **L** with respect to the basis **a**, **b**, **c**. The primitive unit cell is the parallelepiped spanned by the vectors of the given basis.

(ii) The set $\mathbf{C} := \{\mathbf{w} \in \mathbb{V}^3 \mid |\mathbf{w}| \le |\mathbf{w} - \mathbf{v}| \text{ for all } \mathbf{v} \in \mathbf{L}\}$ is called the *Voronoï domain* or *Dirichlet domain* or *Wigner–Seitz cell* or *Wirkungsbereich* or *first Brillouin zone* (for the case of reciprocal lattices in dual space, see Section 1.3.2.5) of **L** (around the origin).

The Voronoï domain consists of those points of $\mathbb{V}^3$ that are closer to the origin than to any other lattice point of **L**.

See Fig. 1.3.2.2 for examples of these two types of unit cells in two-dimensional space.

It should be noted that the attribute 'primitive' for a unit cell is often omitted. The term 'unit cell' then either denotes a primitive unit cell in the sense of the definition above or a slight generalization of this, namely a cell spanned by vectors **a**, **b**, **c** which are not necessarily a lattice basis. This will be discussed in detail in the next section.

The construction of the Voronoï domain is independent of the basis of **L**, as the Voronoï domain is bounded by planes bisecting the line segment between the origin and a lattice point and perpendicular to this segment. The boundaries of the Voronoï domain and its translates overlap, thus in order to get a proper fundamental domain, part of the boundary has to be excluded from the Voronoï domain.

The volume V of the unit cell can be expressed both *via* the metric tensor and *via* the cell parameters. One has

$$\begin{aligned} V^2 &= \det G \\ &= a^2b^2c^2(1 - \cos^2\alpha - \cos^2\beta - \cos^2\gamma + 2\cos\alpha\cos\beta\cos\gamma) \end{aligned}$$

and thus

$$V = abc\sqrt{1 - \cos^2\alpha - \cos^2\beta - \cos^2\gamma + 2\cos\alpha\cos\beta\cos\gamma}.$$

Although the cell parameters depend on the chosen lattice basis, the volume of the unit cell is not affected by a transition to a different lattice basis $\mathbf{a}', \mathbf{b}', \mathbf{c}'$. As remarked in Section 1.3.2.1, two lattice bases are related by an integral basis transformation $\boldsymbol{P}$ of determinant ±1 and therefore $\det G' = \det(\boldsymbol{P}^{\rm T} \cdot \boldsymbol{G} \cdot \boldsymbol{P}) = \det \boldsymbol{G}$, *i.e.* the determinant of the metric tensor is the same for all lattice bases.

Assuming that the vectors **a**, **b**, **c** form a *right-handed* system, the volume can also be obtained *via*

$$V = \mathbf{a} \cdot (\mathbf{b} \times \mathbf{c}) = \mathbf{b} \cdot (\mathbf{c} \times \mathbf{a}) = \mathbf{c} \cdot (\mathbf{a} \times \mathbf{b}).$$

1.3.2.4. Primitive and centred lattices

The definition of a lattice as given in Section 1.3.2.1 states that a lattice consists precisely of the integral linear combinations of the vectors in a lattice basis. However, in crystallographic applications it has turned out to be convenient to work with bases that have particularly nice metric properties. For example, many calculations are simplified if the basis vectors are perpendicular to each other, *i.e.* if the metric tensor has all non-diagonal entries equal to zero. Moreover, it is preferable that the basis vectors reflect the symmetry properties of the lattice. By a case-by-case analysis of the different types of lattices a set of rules for convenient bases has been identified and bases conforming with these rules are called *conventional bases*. The conventional bases are chosen such that in all cases the integral linear combinations of the basis vectors are lattice vectors, but it is admitted that not all lattice vectors are obtained as integral linear combinations.

To emphasize that a basis has the property that the vectors of a lattice are precisely the integral linear combinations of the basis vectors, such a basis is called a *primitive basis* for this lattice.

If the conventional basis of a lattice is not a primitive basis for this lattice, the price to be paid for the transition to the conventional basis is that in addition to the integral linear combinations of the basis vectors one requires one or more *centring vectors* in order to obtain all lattice vectors. These centring vectors have non-integral (but rational) coordinates with respect to the conventional basis. The name *centring* vectors reflects the fact that the additional vectors are usually the centres of the unit cell or of faces of the unit cell spanned by the conventional basis.

Definition

Let **a**, **b**, **c** be linearly independent vectors in $\mathbb{V}^3$.

(i) A lattice **L** is called a *primitive lattice* with respect to a basis **a**, **b**, **c** if **L** consists precisely of all integral linear combinations of **a**, **b**, **c**, *i.e.* if $\mathbf{L} = \mathbf{L}_P = \{l\mathbf{a} + m\mathbf{b} + n\mathbf{c} \mid l, m, n \in \mathbb{Z}\}$.

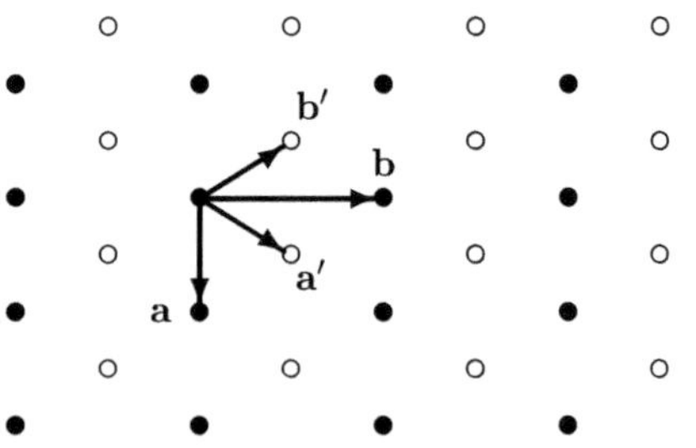

Figure 1.3.2.3
Primitive rectangular lattice (only the filled nodes) and centred rectangular lattice (filled and open nodes).

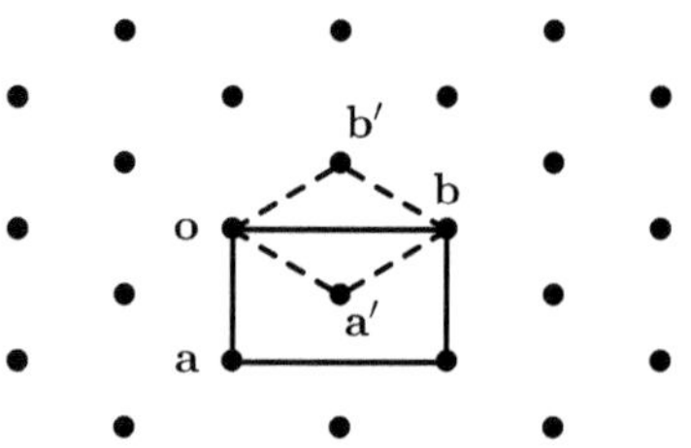

Figure 1.3.2.4
Primitive cell (dashed line) and centred cell (solid lines) for the centred rectangular lattice.

(ii) A lattice $\mathbf{L}$ is called a *centred lattice* with respect to a basis $\mathbf{a}, \mathbf{b}, \mathbf{c}$ if the integral linear combinations $\mathbf{L}_P = \{l\mathbf{a} + m\mathbf{b} + n\mathbf{c} \mid l, m, n \in \mathbb{Z}\}$ form a proper sublattice of $\mathbf{L}$ such that $\mathbf{L}$ is the union of $\mathbf{L}_P$ with the translates of $\mathbf{L}_P$ by centring vectors $\mathbf{v}_1, \ldots, \mathbf{v}_s$, *i.e.* $\mathbf{L} = \mathbf{L}_P \cup (\mathbf{v}_1 + \mathbf{L}_P) \cup \ldots \cup (\mathbf{v}_s + \mathbf{L}_P)$.

Typically, the basis $\mathbf{a}, \mathbf{b}, \mathbf{c}$ is a conventional basis and in this case one often briefly says that a lattice $\mathbf{L}$ is a *primitive lattice* or a *centred lattice* without explicitly mentioning the conventional basis.

Example

A rectangular lattice has as conventional basis a vector $\mathbf{a}$ of minimal length and a vector $\mathbf{b}$ of minimal length amongst the vectors perpendicular to $\mathbf{a}$. The resulting primitive lattice $\mathbf{L}_P$ is indicated by the filled nodes in Fig. 1.3.2.3. Now consider the lattice $\mathbf{L}$ having both the filled and the open nodes in Fig. 1.3.2.3 as its lattice nodes. One sees that $\mathbf{a}' = \frac{1}{2}\mathbf{a} + \frac{1}{2}\mathbf{b}$, $\mathbf{b}' = -\frac{1}{2}\mathbf{a} + \frac{1}{2}\mathbf{b}$ is a primitive basis for $\mathbf{L}$, but it is more convenient to regard $\mathbf{L}$ as a centred lattice with respect to the basis $\mathbf{a}, \mathbf{b}$ with centring vector $\mathbf{v} = \frac{1}{2}\mathbf{a} + \frac{1}{2}\mathbf{b}$. The filled nodes then show the sublattice $\mathbf{L}_P$ of $\mathbf{L}$, the open nodes are the translate $\mathbf{v} + \mathbf{L}_P$ and $\mathbf{L}$ is the union $\mathbf{L}_P \cup (\mathbf{v} + \mathbf{L}_P)$.

The concepts of primitive and centred lattices suggest corresponding notions of primitive and centred unit cells. If $\mathbf{a}, \mathbf{b}, \mathbf{c}$ is a primitive basis for the lattice $\mathbf{L}$, then the parallelepiped spanned by $\mathbf{a}, \mathbf{b}, \mathbf{c}$ is called a *primitive unit cell* (or primitive cell); if $\mathbf{a}, \mathbf{b}, \mathbf{c}$ spans a proper sublattice $\mathbf{L}_P$ of index $[i]$ in $\mathbf{L}$, then the parallelepiped spanned by $\mathbf{a}, \mathbf{b}, \mathbf{c}$ is called a *centred unit cell* (or centred cell). Because the centred cell contains $[i]$ lattice vectors of the centred lattice, it is also called a *multiple cell*. Its volume is $[i]$ times as large as that of a primitive cell for $\mathbf{L}$.

For a conventional basis $\mathbf{a}, \mathbf{b}, \mathbf{c}$ of the lattice $\mathbf{L}$, the parallelepiped spanned by $\mathbf{a}, \mathbf{b}, \mathbf{c}$ is called a *conventional unit cell* (or conventional cell) of $\mathbf{L}$. Depending on whether the conventional basis is a primitive basis or not, *i.e.* whether the lattice is primitive or centred, the conventional cell is a primitive or a centred cell.

Remark: It is important to note that the cell parameters given in the description of a crystallographic structure almost always refer to a conventional cell. When in the crystallographic literature the term 'unit cell' is used without further attributes, in most cases a conventional unit cell (as specified by the cell parameters) is meant, which is a primitive or centred (multiple) cell depending on whether the lattice is primitive or centred.

Example (continued)

In the example of a centred rectangular lattice, the conventional basis $\mathbf{a}, \mathbf{b}$ spans the centred unit cell indicated by solid lines in Fig. 1.3.2.4, whereas the primitive basis $\mathbf{a}' = \frac{1}{2}\mathbf{a} + \frac{1}{2}\mathbf{b}$, $\mathbf{b}' = -\frac{1}{2}\mathbf{a} + \frac{1}{2}\mathbf{b}$ spans the primitive unit cell indicated by dashed lines. One observes that the centred cell contains two lattice vectors, $\mathbf{o}$ and $\mathbf{a}'$, whereas the primitive cell only contains the zero vector $\mathbf{o}$ (note that due to the condition $0 \le x, y < 1$ for the points in the unit cell the other vertices $\mathbf{a}', \mathbf{b}', \mathbf{b}$ of the cell are excluded). The volume of the centred cell is clearly twice as large as that of the primitive cell.

Figures displaying the different primitive and centred unit cells as well as tables describing the metric properties of the different primitive and centred lattices are given in Section 3.1.2 of *International Tables for Crystallography* Volume A (2016) (abbreviated as *IT* A).

Examples

(i) The conventional basis for a *primitive cubic lattice* (*cP*) is a basis $\mathbf{a}, \mathbf{b}, \mathbf{c}$ of vectors of equal length which are pairwise perpendicular, *i.e.* with $|\mathbf{a}| = |\mathbf{b}| = |\mathbf{c}|$ and $\mathbf{a} \cdot \mathbf{b} = \mathbf{b} \cdot \mathbf{c} = \mathbf{c} \cdot \mathbf{a} = 0$. As the name indicates, this basis is a primitive basis.

(ii) A *body-centred cubic lattice* (*cI*) has as its conventional basis the conventional basis $\mathbf{a}, \mathbf{b}, \mathbf{c}$ of a primitive cubic lattice, but the lattice also contains the centring vector $\mathbf{v} = \frac{1}{2}\mathbf{a} + \frac{1}{2}\mathbf{b} + \frac{1}{2}\mathbf{c}$ which points to the centre of the conventional cell. If we denote the primitive cubic lattice by $\mathbf{L}_P$, then the body-centred cubic lattice $\mathbf{L}_I$ is the union of $\mathbf{L}_P$ and the translate $\mathbf{v} + \mathbf{L}_P = \{\mathbf{v} + \mathbf{w} \mid \mathbf{w} \in \mathbf{L}_P\}$. Since $\mathbf{L}_P$ is a sublattice of index 2 in $\mathbf{L}_I$, the ratio of the volumes of the centred and the primitive cell of the body-centred cubic lattice is 2.

A possible primitive basis for $\mathbf{L}_I$ is $\mathbf{a}' = \mathbf{a}$, $\mathbf{b}' = \mathbf{b}$, $\mathbf{c}' = \frac{1}{2}(\mathbf{a} + \mathbf{b} + \mathbf{c})$. With respect to this basis, the metric tensor of $\mathbf{L}_I$ is

$$a^2 \cdot \begin{pmatrix} 1 & 0 & \frac{1}{2} \\ 0 & 1 & \frac{1}{2} \\ \frac{1}{2} & \frac{1}{2} & \frac{3}{4} \end{pmatrix}$$

(where $a = \mathbf{a} \cdot \mathbf{a}$). However, it is more common to use a primitive basis with vectors of the same length and equal interaxial angles. Such a basis is $\mathbf{a}'' = \frac{1}{2}(-\mathbf{a} + \mathbf{b} + \mathbf{c})$, $\mathbf{b}'' = \frac{1}{2}(\mathbf{a} - \mathbf{b} + \mathbf{c})$, $\mathbf{c}'' = \frac{1}{2}(\mathbf{a} + \mathbf{b} - \mathbf{c})$ (*cf.* Fig. 1.5.1.3), and with respect to this basis the metric tensor of $\mathbf{L}_I$ is

$$\frac{a^2}{4} \cdot \begin{pmatrix} 3 & \bar{1} & \bar{1} \\ \bar{1} & 3 & \bar{1} \\ \bar{1} & \bar{1} & 3 \end{pmatrix}.$$

(iii) The conventional basis for a *face-centred cubic lattice* (*cF*) is again the conventional basis $\mathbf{a}, \mathbf{b}, \mathbf{c}$ of a primitive cubic lattice, but the lattice also contains the three centring

vectors $\mathbf{v}_1 = \frac{1}{2}\mathbf{b} + \frac{1}{2}\mathbf{c}$, $\mathbf{v}_2 = \frac{1}{2}\mathbf{a} + \frac{1}{2}\mathbf{c}$, $\mathbf{v}_3 = \frac{1}{2}\mathbf{a} + \frac{1}{2}\mathbf{b}$ which point to the centres of faces of the conventional cell.
The face-centred cubic lattice $\mathbf{L}_F$ is the union of the primitive cubic lattice $\mathbf{L}_P$ with its translates $\mathbf{v}_i + \mathbf{L}_P$ by the three centring vectors. The ratio of the volumes of the centred and the primitive cell of the face-centred cubic lattice is 4. In this case, the centring vectors actually form a primitive basis of $\mathbf{L}_F$. With respect to the basis $\mathbf{a}' = \frac{1}{2}(\mathbf{b} + \mathbf{c})$, $\mathbf{b}' = \frac{1}{2}(\mathbf{a} + \mathbf{c})$, $\mathbf{c}' = \frac{1}{2}(\mathbf{a} + \mathbf{b})$ (*cf.* Fig. 1.5.1.4) the metric tensor of $\mathbf{L}_F$ is

$$\frac{a^2}{4} \cdot \begin{pmatrix} 2 & 1 & 1 \\ 1 & 2 & 1 \\ 1 & 1 & 2 \end{pmatrix}.$$

(iv) In the conventional basis of a primitive hexagonal lattice, the basis vector $\mathbf{c}$ is chosen as a shortest vector along a sixfold axis. The vectors $\mathbf{a}$ and $\mathbf{b}$ then are shortest vectors along twofold axes in a plane perpendicular to $\mathbf{c}$ and such that they enclose an angle of 120°. The corresponding metric tensor has the form

$$\begin{pmatrix} a^2 & -\dfrac{a^2}{2} & 0 \\ -\dfrac{a^2}{2} & a^2 & 0 \\ 0 & 0 & c^2 \end{pmatrix}.$$

(v) In the unit cell of the primitive hexagonal lattice $\mathbf{L}_P$, a point with coordinates $\frac{2}{3}, \frac{1}{3}, z$ is mapped to the points $-\frac{1}{3}, \frac{1}{3}, z$ and $-\frac{1}{3}, -\frac{2}{3}, z$ under the threefold rotation around the c axis. Both of these points are translates of $\frac{2}{3}, \frac{1}{3}, z$ by lattice vectors of $\mathbf{L}_P$. This means that a centring vector of the form $\frac{2}{3}\mathbf{a} + \frac{1}{3}\mathbf{b} + z\mathbf{c}$ will result in a lattice which is invariant under the threefold rotation. Choosing $\mathbf{v}_1 = \frac{1}{3}(2\mathbf{a} + \mathbf{b} + \mathbf{c})$ as centring vector, the lattice generated by $\mathbf{L}_P$ and $\mathbf{v}_1$ contains $\mathbf{L}_P$ as a sublattice of index 3 with coset representatives $\mathbf{0}$, $\mathbf{v}_1$ and $2\mathbf{v}_1 = \frac{1}{3}(4\mathbf{a} + 2\mathbf{b} + 2\mathbf{c})$. The coset representative $2\mathbf{v}_1$ is commonly replaced by $\mathbf{v}_2 = \frac{1}{3}(\mathbf{a} + 2\mathbf{b} + 2\mathbf{c})$ and the centred lattice $\mathbf{L}_R$ with centring vectors $\mathbf{v}_1$ and $\mathbf{v}_2$ so obtained is called the *rhombohedrally centred lattice* (*hR*). The ratio of the volumes of the centred and the primitive cell of the rhombohedrally centred lattice is 3.
For this lattice, the primitive basis of $\mathbf{L}_R$ consisting of three shortest non-coplanar vectors which are permuted by the threefold rotation is also regarded as a conventional basis. With respect to the above lattice basis of the primitive hexagonal lattice, this basis can be chosen as $\mathbf{a}' = \frac{1}{3}(2\mathbf{a} + \mathbf{b} + \mathbf{c})$, $\mathbf{b}' = \frac{1}{3}(-\mathbf{a} + \mathbf{b} + \mathbf{c})$, $\mathbf{c}' = \frac{1}{3}(-\mathbf{a} - 2\mathbf{b} + \mathbf{c})$. The metric tensor with respect to this basis is

$$\frac{1}{9} \cdot \begin{pmatrix} 3a^2 + c^2 & -\dfrac{3}{2}a^2 + c^2 & -\dfrac{3}{2}a^2 + c^2 \\ -\dfrac{3}{2}a^2 + c^2 & 3a^2 + c^2 & -\dfrac{3}{2}a^2 + c^2 \\ -\dfrac{3}{2}a^2 + c^2 & -\dfrac{3}{2}a^2 + c^2 & 3a^2 + c^2 \end{pmatrix}.$$

Details about the transformations between hexagonal and rhombohedral cells are given in Section 1.5.3.1 and Table 1.5.1.1 (see also Fig. 1.5.1.6).

1.3.2.5. Reciprocal lattice

For crystallographic applications, a lattice $\mathbf{L}^*$ related to $\mathbf{L}$ is of utmost importance. If the atoms are placed at the nodes of a lattice $\mathbf{L}$, then the diffraction pattern will have sharp Bragg peaks at the nodes of the *reciprocal lattice* $\mathbf{L}^*$. More generally, if the crystal pattern is invariant under translations from $\mathbf{L}$, then the locations of the Bragg peaks in the diffraction pattern will be invariant under translations from $\mathbf{L}^*$.

Definition

Let $\mathbf{L} \subset \mathbb{V}^3$ be a lattice with lattice basis $\mathbf{a}, \mathbf{b}, \mathbf{c}$. Then the *reciprocal basis* $\mathbf{a}^*, \mathbf{b}^*, \mathbf{c}^*$ is defined by the properties

$$\mathbf{a} \cdot \mathbf{a}^* = \mathbf{b} \cdot \mathbf{b}^* = \mathbf{c} \cdot \mathbf{c}^* = 1$$

and

$$\mathbf{b} \cdot \mathbf{a}^* = \mathbf{c} \cdot \mathbf{a}^* = \mathbf{c} \cdot \mathbf{b}^* = \mathbf{a} \cdot \mathbf{b}^* = \mathbf{a} \cdot \mathbf{c}^* = \mathbf{b} \cdot \mathbf{c}^* = 0,$$

which can conveniently be written as the matrix equation

$$\begin{pmatrix} \mathbf{a} \cdot \mathbf{a}^* & \mathbf{a} \cdot \mathbf{b}^* & \mathbf{a} \cdot \mathbf{c}^* \\ \mathbf{b} \cdot \mathbf{a}^* & \mathbf{b} \cdot \mathbf{b}^* & \mathbf{b} \cdot \mathbf{c}^* \\ \mathbf{c} \cdot \mathbf{a}^* & \mathbf{c} \cdot \mathbf{b}^* & \mathbf{c} \cdot \mathbf{c}^* \end{pmatrix} = \begin{pmatrix} 1 & 0 & 0 \\ 0 & 1 & 0 \\ 0 & 0 & 1 \end{pmatrix} = \boldsymbol{I}_3.$$

This means that $\mathbf{a}^*$ is perpendicular to the plane spanned by $\mathbf{b}$ and $\mathbf{c}$ and its projection to the line along $\mathbf{a}$ has length $1/|\mathbf{a}|$. Analogous properties hold for $\mathbf{b}^*$ and $\mathbf{c}^*$.
The *reciprocal lattice* $\mathbf{L}^*$ of $\mathbf{L}$ is defined to be the lattice with lattice basis $\mathbf{a}^*, \mathbf{b}^*, \mathbf{c}^*$.

In three-dimensional space $\mathbb{V}^3$, the reciprocal basis can be determined *via* the vector product. Assuming that $\mathbf{a}, \mathbf{b}, \mathbf{c}$ form a right-handed system that spans a unit cell of volume V, the relation $\mathbf{a} \cdot (\mathbf{b} \times \mathbf{c}) = V$ and the defining conditions $\mathbf{a} \cdot \mathbf{a}^* = 1$, $\mathbf{b} \cdot \mathbf{a}^* = \mathbf{c} \cdot \mathbf{a}^* = 0$ imply that $\mathbf{a}^* = \frac{1}{V}(\mathbf{b} \times \mathbf{c})$. Analogously, one has $\mathbf{b}^* = \frac{1}{V}(\mathbf{c} \times \mathbf{a})$ and $\mathbf{c}^* = \frac{1}{V}(\mathbf{a} \times \mathbf{b})$.
The reciprocal lattice can also be defined independently of a lattice basis by stating that the vectors of the reciprocal lattice have integral scalar products with all vectors of the lattice:

$$\mathbf{L}^* = \{\mathbf{w}^* \in \mathbb{V}^3 \mid \mathbf{v} \cdot \mathbf{w}^* \in \mathbb{Z} \text{ for all } \mathbf{v} \in \mathbf{L}\}.$$

Remark: In parts of the literature, especially in physics, the reciprocal lattice is defined slightly differently. The condition there is that $\mathbf{a}_i \cdot \mathbf{a}_j^* = 2\pi$ if $i = j$ and 0 otherwise and thus the reciprocal lattice is scaled by the factor 2π as compared to the above definition. By this variation the exponential function $\exp(-2\pi i \, \mathbf{v} \cdot \mathbf{w})$ is changed to $\exp(-i \, \mathbf{v} \cdot \mathbf{w})$, which simplifies the formulas for the Fourier transform.

Example

Let $\mathbf{a}, \mathbf{b}, \mathbf{c}$ be the lattice basis of a primitive cubic lattice with lattice parameter a. Then the body-centred cubic lattice $\mathbf{L}_I$ with lattice parameter a spanned by $\mathbf{a}, \mathbf{b}, \mathbf{c}$ and the centring vector $\frac{1}{2}(\mathbf{a} + \mathbf{b} + \mathbf{c})$ has as its reciprocal lattice the rescaled face-centred cubic lattice with lattice parameter $2/a$, spanned by $(2/a^2)\mathbf{a}$, $(2/a^2)\mathbf{b}$, $(2/a^2)\mathbf{c}$ and the centring vectors $(1/a^2)(\mathbf{b} + \mathbf{c})$, $(1/a^2)(\mathbf{a} + \mathbf{c})$, $(1/a^2)(\mathbf{a} + \mathbf{b})$. Conversely, the face-centred cubic lattice with lattice parameter a has as its reciprocal lattice the rescaled body-centred cubic lattice with lattice parameter $2/a$.

This example illustrates that a lattice and its reciprocal lattice need not have the same type. The reciprocal lattice of a body-centred cubic lattice is a face-centred cubic lattice and *vice versa*. However, the conventional bases are chosen such that for a primitive lattice with a conventional basis as lattice basis, the reciprocal lattice is a primitive lattice of the same type. Therefore the reciprocal lattice of a centred lattice is always a centred lattice for the same type of primitive lattice.

The reciprocal basis can be read off the inverse matrix of the metric tensor $\boldsymbol{G}$: We denote by $\boldsymbol{P}^*$ the matrix containing the coordinate columns of $\mathbf{a}^*, \mathbf{b}^*, \mathbf{c}^*$ with respect to the basis $\mathbf{a}, \mathbf{b}, \mathbf{c}$, so that $\mathbf{a}^* = P^*_{11}\mathbf{a} + P^*_{21}\mathbf{b} + P^*_{31}\mathbf{c}$ *etc.* Computing the scalar products between the basis $\mathbf{a}, \mathbf{b}, \mathbf{c}$ and the reciprocal basis $\mathbf{a}^*, \mathbf{b}^*, \mathbf{c}^*$ by multiplying the metric tensor $\boldsymbol{G}$ from the left and right with coordinate columns results in the matrix equation $\boldsymbol{I}_3 \cdot \boldsymbol{G} \cdot \boldsymbol{P}^* = \boldsymbol{I}_3$, from which one concludes that $\boldsymbol{P}^* = \boldsymbol{G}^{-1}$. The coordinate columns of $\mathbf{a}^*, \mathbf{b}^*, \mathbf{c}^*$ with respect to the basis $\mathbf{a}, \mathbf{b}, \mathbf{c}$ are thus precisely the columns of the inverse matrix $\boldsymbol{G}^{-1}$ of the metric tensor $\boldsymbol{G}$.

From $\boldsymbol{P}^* = \boldsymbol{G}^{-1}$ one also derives that the metric tensor $\boldsymbol{G}^*$ of the reciprocal basis is

$$\boldsymbol{G}^* = \boldsymbol{P}^{*\mathrm{T}} \cdot \boldsymbol{G} \cdot \boldsymbol{P}^* = \boldsymbol{G}^{-1} \cdot \boldsymbol{G} \cdot \boldsymbol{G}^{-1} = \boldsymbol{G}^{-1}.$$

This means that the metric tensors of a basis and its reciprocal basis are inverse matrices of each other. As a further consequence, the volume V^* of the unit cell spanned by the reciprocal basis is $V^* = V^{-1}$, *i.e.* the inverse of the volume of the unit cell spanned by $\mathbf{a}, \mathbf{b}, \mathbf{c}$.

Of course, the reciprocal basis can also be computed from the basis vectors $\mathbf{a}_i$ directly. If $\boldsymbol{B}$ and $\boldsymbol{B}^*$ are the matrices containing as ith column the vectors $\mathbf{a}_i$ and $\mathbf{a}_i^*$, respectively, then the relation defining the reciprocal basis reads as $\boldsymbol{B}^{\mathrm{T}} \cdot \boldsymbol{B}^* = \boldsymbol{I}_3$, *i.e.* $\boldsymbol{B}^* = (\boldsymbol{B}^{-1})^{\mathrm{T}}$. Thus, the reciprocal basis vector $\mathbf{a}_i^*$ is the ith column of the transposed matrix of $\boldsymbol{B}^{-1}$ and thus the ith *row* of the inverse of the matrix $\boldsymbol{B}$ containing the $\mathbf{a}_i$ as columns.

The cell parameters of the reciprocal lattice can either be obtained from the vector-product expressions for $\mathbf{a}^*, \mathbf{b}^*, \mathbf{c}^*$ or by explicitly inverting the metric tensor $\boldsymbol{G}$ (*e.g.* using Cramer's rule). Either way, one finds

$$\begin{aligned}
a^* &= \frac{bc\sin\alpha}{V}, & b^* &= \frac{ca\sin\beta}{V}, & c^* &= \frac{ab\sin\gamma}{V},\\
\sin\alpha^* &= \frac{V}{abc\sin\beta\sin\gamma}, & \cos\alpha^* &= \frac{\cos\beta\cos\gamma - \cos\alpha}{\sin\beta\sin\gamma}, &&\\
\sin\beta^* &= \frac{V}{abc\sin\gamma\sin\alpha}, & \cos\beta^* &= \frac{\cos\gamma\cos\alpha - \cos\beta}{\sin\gamma\sin\alpha}, &&\\
\sin\gamma^* &= \frac{V}{abc\sin\alpha\sin\beta}, & \cos\gamma^* &= \frac{\cos\alpha\cos\beta - \cos\gamma}{\sin\alpha\sin\beta}. &&
\end{aligned}$$

Example

The body-centred cubic lattice $\mathbf{L}$ with lattice parameter a has the vectors

$$\mathbf{a} = \frac{a}{2}\begin{pmatrix}\bar{1}\\1\\1\end{pmatrix}, \quad \mathbf{b} = \frac{a}{2}\begin{pmatrix}1\\\bar{1}\\1\end{pmatrix}, \quad \mathbf{c} = \frac{a}{2}\begin{pmatrix}1\\1\\\bar{1}\end{pmatrix}$$

as primitive basis.

The matrix

$$\boldsymbol{B} = \frac{a}{2}\begin{pmatrix}\bar{1} & 1 & 1\\ 1 & \bar{1} & 1\\ 1 & 1 & \bar{1}\end{pmatrix}$$

with the basis vectors $\mathbf{a}, \mathbf{b}, \mathbf{c}$ as columns has as its inverse the matrix

$$\boldsymbol{B}^{-1} = \frac{1}{a}\begin{pmatrix}0 & 1 & 1\\ 1 & 0 & 1\\ 1 & 1 & 0\end{pmatrix}.$$

The rows of $\boldsymbol{B}^{-1}$, giving the reciprocal basis, are the vectors

$$\mathbf{a}^* = \frac{1}{a}\begin{pmatrix}0\\1\\1\end{pmatrix}, \quad \mathbf{b}^* = \frac{1}{a}\begin{pmatrix}1\\0\\1\end{pmatrix}, \quad \mathbf{c}^* = \frac{1}{a}\begin{pmatrix}1\\1\\0\end{pmatrix},$$

showing that the reciprocal lattice of a body-centred cubic lattice with lattice parameter a is a face-centred cubic lattice with lattice parameter $2/a$.

The metric tensors $\boldsymbol{G}$ and $\boldsymbol{G}^* = \boldsymbol{G}^{-1}$ of $\mathbf{L}$ and its reciprocal lattice $\mathbf{L}^*$ are

$$\boldsymbol{G} = \frac{a^2}{4}\begin{pmatrix}3 & \bar{1} & \bar{1}\\ \bar{1} & 3 & \bar{1}\\ \bar{1} & \bar{1} & 3\end{pmatrix} \text{ and } \boldsymbol{G}^* = \frac{1}{a^2}\begin{pmatrix}2 & 1 & 1\\ 1 & 2 & 1\\ 1 & 1 & 2\end{pmatrix},$$

and interpreting the columns of $\boldsymbol{G}^*$ as coordinate columns with respect to the original basis shows that

$$\mathbf{a}^* = \frac{1}{a^2}(2\mathbf{a} + \mathbf{b} + \mathbf{c}), \quad \mathbf{b}^* = \frac{1}{a^2}(\mathbf{a} + 2\mathbf{b} + \mathbf{c}),$$
$$\mathbf{c}^* = \frac{1}{a^2}(\mathbf{a} + \mathbf{b} + 2\mathbf{c}),$$

which results in the same reciprocal basis as computed above.

1.3.3. The structure of space groups

1.3.3.1. Point groups of space groups

The multiplication rule for symmetry operations

$$(\boldsymbol{W}_2, \boldsymbol{w}_2)(\boldsymbol{W}_1, \boldsymbol{w}_1) = (\boldsymbol{W}_2\boldsymbol{W}_1, \boldsymbol{W}_2\boldsymbol{w}_1 + \boldsymbol{w}_2)$$

shows that the linear parts of a space group form a group themselves, called the point group of $\mathcal{G}$, because the first component of the combined operation is simply the product of the linear parts of the two operations.

The translation subgroup $\mathcal{T}$ consisting of the operations $(\boldsymbol{I}, \boldsymbol{t})$ having linear part $\boldsymbol{I}$ forms a normal subgroup of $\mathcal{G}$, as can be concluded from the following explicit computation. Let $t = (\boldsymbol{I}, \boldsymbol{t})$ be a translation in $\mathcal{T}$ and $W = (\boldsymbol{W}, \boldsymbol{w})$ an arbitrary operation in $\mathcal{G}$, then one has

$$\begin{aligned}
WtW^{-1} &= (\boldsymbol{W}, \boldsymbol{w})(\boldsymbol{I}, \boldsymbol{t})(\boldsymbol{W}^{-1}, -\boldsymbol{W}^{-1}\boldsymbol{w})\\
&= (\boldsymbol{W}, \boldsymbol{W}\boldsymbol{t} + \boldsymbol{w})(\boldsymbol{W}^{-1}, -\boldsymbol{W}^{-1}\boldsymbol{w})\\
&= (\boldsymbol{I}, -\boldsymbol{w} + \boldsymbol{W}\boldsymbol{t} + \boldsymbol{w}) = (\boldsymbol{I}, \boldsymbol{W}\boldsymbol{t}),
\end{aligned}$$

which is again a translation in $\mathcal{G}$, namely by $\boldsymbol{W}\boldsymbol{t}$.

Definition

The *point group* $\mathcal{P}$ of a space group $\mathcal{G}$ is the group of linear parts of operations occurring in $\mathcal{G}$. It is isomorphic to the factor group $\mathcal{G}/\mathcal{T}$ of $\mathcal{G}$ by the translation subgroup $\mathcal{T}$.

When $\mathcal{G}$ is considered with respect to a coordinate system, the operations of $\mathcal{P}$ are simply 3×3 matrices.

The point group plays an important role in the analysis of the macroscopic properties of crystals: it describes the symmetry of the set of face normals and can thus be directly observed. It is usually obtained from the *diffraction record* of the crystal, where adding the information about the translation subgroup explains the sharpness of the Bragg peaks in the diffraction pattern.

Table 1.3.3.1
Automorphism groups (Bravais groups) of two-dimensional primitive lattices

Lattice	Metric tensor	Bravais group	
		Hermann–Mauguin symbol	Generators
Oblique	$\begin{pmatrix} g_{11} & g_{12} \\ & g_{22} \end{pmatrix}$	2	2: $\bar{x}, \bar{y}$
Rectangular	$\begin{pmatrix} g_{11} & 0 \\ & g_{22} \end{pmatrix}$	$2mm$	2: $\bar{x}, \bar{y}$ m_{10}: $\bar{x}, y$
Square	$\begin{pmatrix} g_{11} & 0 \\ & g_{11} \end{pmatrix}$	$4mm$	4^+: $\bar{y}, x$ m_{10}: $\bar{x}, y$
Hexagonal	$\begin{pmatrix} g_{11} & -\frac{1}{2}g_{11} \\ & g_{11} \end{pmatrix}$	$6mm$	6^+: $x-y, x$ m_{21}: $\bar{x}, \bar{x}+y$

The above computation that $(\boldsymbol{W}, \boldsymbol{w})(\boldsymbol{I}, \boldsymbol{t})(\boldsymbol{W}, \boldsymbol{w})^{-1} = (\boldsymbol{I}, \boldsymbol{Wt})$ not only demonstrates that the translations in a space group form a normal subgroup, but also shows an important property of the translation subgroup with respect to the point group, namely that every vector from the translation lattice is mapped again to a lattice vector by each operation of the point group of $\mathcal{G}$.

Proposition. Let $\mathcal{G}$ be a space group with point group $\mathcal{P}$ and translation subgroup $\mathcal{T}$ and let $\mathbf{L} = \{\boldsymbol{t} \mid (\boldsymbol{I}, \boldsymbol{t}) \in \mathcal{T}\}$ be the lattice of translations in $\mathcal{T}$. Then $\mathcal{P}$ *acts* on the lattice $\mathbf{L}$, *i.e.* for every $\boldsymbol{W} \in \mathcal{P}$ and $\boldsymbol{t} \in \mathbf{L}$ one has $\boldsymbol{Wt} \in \mathbf{L}$.

A point group that acts on a lattice is a subgroup of the full group of symmetries of the lattice, obtained as the group of orthogonal mappings that map the lattice to itself. With respect to a primitive basis, the group of symmetries of a lattice consists of all integral basis transformations that fix the metric tensor of the lattice.

Definition

Let $\mathbf{L}$ be a three-dimensional lattice with metric tensor $\boldsymbol{G}$ with respect to a primitive basis **a**, **b**, **c**.

(i) An *automorphism* of $\mathbf{L}$ is an isometry mapping $\mathbf{L}$ to itself. Written with respect to the basis **a**, **b**, **c**, an automorphism of $\mathbf{L}$ is an integral basis transformation fixing the metric tensor of $\mathbf{L}$, *i.e.* it is an integral matrix $\boldsymbol{W} \in \mathrm{GL}_3(\mathbb{Z})$ with $\boldsymbol{W}^{\mathrm{T}} \cdot \boldsymbol{G} \cdot \boldsymbol{W} = \boldsymbol{G}$.

(ii) The group

$$\mathcal{B} := Aut(\mathbf{L}) = \{\boldsymbol{W} \in \mathrm{GL}_3(\mathbb{Z}) \mid \boldsymbol{W}^{\mathrm{T}} \cdot \boldsymbol{G} \cdot \boldsymbol{W} = \boldsymbol{G}\}$$

of all automorphisms of $\mathbf{L}$ is called the *automorphism group* or *Bravais group* of $\mathbf{L}$. Note that $Aut(\mathbf{L})$ acts on the coordinate columns of $\mathbf{L}$, which are simply columns with integral coordinates.

Since the isometries in the Bravais group of a lattice preserve distances, the possible images of the vectors in a basis are vectors of the same lengths as the basis vectors. But due to its discreteness, a lattice contains only finitely many lattice vectors up to a given length. This means that a lattice automorphism can only permute the finitely many vectors up to the maximum length of a basis vector. Thus, there can only be finitely many automorphisms of a lattice. This argument proves the following important fact:

Theorem. The Bravais group of a lattice is finite. As a consequence, point groups of space groups are finite groups.

As subgroups of the Bravais group of a lattice, point groups can be realized as integral matrix groups when written with respect to a primitive basis. For a centred lattice, it is possible that the Bravais group of a lattice contains non-integral matrices, because the centring vector is a column with non-integral entries. However, in dimensions two and three the conventional bases are chosen such that the Bravais groups of all lattices are integral when written with respect to a conventional basis.

Information on the Bravais groups of the primitive lattices in two- and three-dimensional space is displayed in Tables 1.3.3.1 and 1.3.3.2. The columns of the tables contain the names of the lattices, the metric tensor with respect to the conventional basis (with only the upper half given, the lower half following by the symmetry of the metric tensor), the Hermann–Mauguin symbol for the type of the Bravais group and generators of the Bravais group (given in the shorthand notation introduced in Section 1.2.2.1 and the corresponding Seitz symbols discussed in Section 1.4.2.2).

The finiteness and integrality of the point groups have important consequences. For example, they imply the *crystallographic restriction* that rotations in space groups of two- and three-dimensional space can only have orders 1, 2, 3, 4 or 6: with respect to a suitable basis, the matrix of a k-fold rotation is of the form

$$\begin{pmatrix} \cos(2\pi/k) & -\sin(2\pi/k) & 0 \\ \sin(2\pi/k) & \cos(2\pi/k) & 0 \\ 0 & 0 & 1 \end{pmatrix}.$$

Since the trace of a matrix (*i.e.*, the sum of its diagonal entries, equalling the sum of its eigenvalues) is preserved under a basis transformation, it must be an integer for a rotation acting on a lattice and thus represented by an integral matrix. But $2\cos(2\pi/k)$ is only an integer for $k = 1, 2, 3, 4, 6$.

1.3.3.2. Coset decomposition with respect to the translation subgroup

The translation subgroup $\mathcal{T}$ of a space group $\mathcal{G}$ can be used to distribute the operations of $\mathcal{G}$ into different classes by grouping together all operations that differ only by a translation. This results in the decomposition of $\mathcal{G}$ into cosets with respect to $\mathcal{T}$ (see Section 1.1.4 for details of cosets).

Definition

Let $\mathcal{G}$ be a space group with translation subgroup $\mathcal{T}$.

(i) The *right coset* $\mathcal{T}W$ of an operation $W \in \mathcal{G}$ with respect to $\mathcal{T}$ is the set $\{tW \mid t \in \mathcal{T}\}$.
Analogously, the set $W\mathcal{T} = \{Wt \mid t \in \mathcal{T}\}$ is called the *left coset* of W with respect to $\mathcal{T}$.

(ii) A set $\{W_1, \ldots, W_m\}$ of operations in $\mathcal{G}$ is called a system of *coset representatives* relative to $\mathcal{T}$ if every operation W in $\mathcal{G}$ is contained in exactly one coset $\mathcal{T}W_i$.

(iii) Writing $\mathcal{G}$ as the disjoint union

$$\mathcal{G} = \mathcal{T}W_1 \cup \ldots \cup \mathcal{T}W_m$$

is called the *coset decomposition of* $\mathcal{G}$ *relative to* $\mathcal{T}$.

If the translation subgroup $\mathcal{T}$ is a subgroup of index $[i]$ in $\mathcal{G}$, a set of coset representatives for $\mathcal{G}$ relative to $\mathcal{T}$ consists of $[i]$

Table 1.3.3.2
Automorphism groups (Bravais groups) of three-dimensional primitive lattices

Lattice	Metric tensor	Bravais group	
		Hermann–Mauguin symbol	Generators
Triclinic	$\begin{pmatrix} g_{11} & g_{12} & g_{13} \\ & g_{22} & g_{23} \\ & & g_{33} \end{pmatrix}$	$\bar{1}$	$\bar{1}$: $\bar{x}, \bar{y}, \bar{z}$
Monoclinic	$\begin{pmatrix} g_{11} & 0 & g_{13} \\ & g_{22} & 0 \\ & & g_{33} \end{pmatrix}$	$2/m$	2_{010}: $\bar{x}, y, \bar{z}$ m_{010}: $x, \bar{y}, z$
Orthorhombic	$\begin{pmatrix} g_{11} & 0 & 0 \\ & g_{22} & 0 \\ & & g_{33} \end{pmatrix}$	mmm	m_{100}: $\bar{x}, y, z$ m_{010}: $x, \bar{y}, z$ m_{001}: $x, y, \bar{z}$
Tetragonal	$\begin{pmatrix} g_{11} & 0 & 0 \\ & g_{11} & 0 \\ & & g_{33} \end{pmatrix}$	$4/mmm$	4_{001}: $\bar{y}, x, z$ m_{001}: $x, y, \bar{z}$ m_{100}: $\bar{x}, y, z$
Hexagonal	$\begin{pmatrix} g_{11} & -\frac{1}{2}g_{11} & 0 \\ & g_{11} & 0 \\ & & g_{33} \end{pmatrix}$	$6/mmm$	6_{001}: $x-y, x, z$ m_{001}: $x, y, \bar{z}$ m_{100}: $\bar{x}+y, y, z$
Rhombohedral	$\begin{pmatrix} g_{11} & g_{12} & g_{12} \\ & g_{11} & g_{12} \\ & & g_{11} \end{pmatrix}$	$\bar{3}m$	$\bar{3}_{111}$: $\bar{z}, \bar{x}, \bar{y}$ $m_{1\bar{1}0}$: y, x, z
Cubic	$\begin{pmatrix} g_{11} & 0 & 0 \\ & g_{11} & 0 \\ & & g_{11} \end{pmatrix}$	$m\bar{3}m$	m_{001}: $x, y, \bar{z}$ $\bar{3}_{111}$: $\bar{z}, \bar{x}, \bar{y}$ m_{110}: $\bar{y}, \bar{x}, z$

Table 1.3.3.3
Right-coset decomposition of $\mathcal{G}$ relative to $\mathcal{T}$

$W_1 = e$	W_2	W_3	$\ldots$	$W_{[i]}$
t_1	$t_1 W_2$	$t_1 W_3$	$\ldots$	$t_1 W_{[i]}$
t_2	$t_2 W_2$	$t_2 W_3$	$\ldots$	$t_2 W_{[i]}$
t_3	$t_3 W_2$	$t_3 W_3$	$\ldots$	$t_3 W_{[i]}$
t_4	$t_4 W_2$	$t_4 W_3$	$\ldots$	$t_4 W_{[i]}$
$\vdots$	$\vdots$	$\vdots$		$\vdots$

operations $W_1, W_2, \ldots, W_{[i]}$, where W_1 is assumed to be the identity element e of $\mathcal{G}$. The cosets of $\mathcal{G}$ relative to $\mathcal{T}$ can be imagined as columns of an infinite array with $[i]$ columns, labelled by the coset representatives, as displayed in Table 1.3.3.3.

Writing out the matrix–column pairs, the coset $\mathcal{T}(\boldsymbol{W}, \boldsymbol{w})$ consists of the operations of the form $(\boldsymbol{I}, \boldsymbol{t})(\boldsymbol{W}, \boldsymbol{w}) = (\boldsymbol{W}, \boldsymbol{w} + \boldsymbol{t})$ with $\boldsymbol{t}$ running over the lattice translations of $\mathcal{T}$. This means that the operations of a coset with respect to the translation subgroup all have the same linear part, which is also evident from the listing of the cosets as columns of an infinite array, as given in Table 1.3.3.3.

Proposition

Let $W = (\boldsymbol{W}, \boldsymbol{w})$ and $W' = (\boldsymbol{W}', \boldsymbol{w}')$ be two operations of a space group $\mathcal{G}$ with translation subgroup $\mathcal{T}$.

(1) If $\boldsymbol{W} \neq \boldsymbol{W}'$, then the cosets $\mathcal{T}W$ and $\mathcal{T}W'$ are disjoint, *i.e.* their intersection is empty.

(2) If $\boldsymbol{W} = \boldsymbol{W}'$, then the cosets $\mathcal{T}W$ and $\mathcal{T}W'$ are equal, because WW'^{-1} has linear part $\boldsymbol{I}$ and is thus an operation contained in $\mathcal{T}$.

The one-to-one correspondence between the point-group operations and the cosets relative to $\mathcal{T}$ explicitly displays the isomorphism between the point group $\mathcal{P}$ of $\mathcal{G}$ and the factor group $\mathcal{G}/\mathcal{T}$. This correspondence is also exploited in the listing of the general-position coordinates. What is given there are the coordinate triplets for coset representatives of $\mathcal{G}$ relative to $\mathcal{T}$, which correspond to the first row of the array in Table 1.3.3.3. As just explained, the other operations in $\mathcal{G}$ can be obtained from these coset representatives by adding a lattice translation to the translational part.

Furthermore, the correspondence between the point group and the coset decomposition relative to $\mathcal{T}$ makes it easy to find a system of coset representatives $\{W_1, \ldots, W_m\}$ of $\mathcal{G}$ relative to $\mathcal{T}$. What is required is that the linear parts of the W_i are precisely the operations in the point group of $\mathcal{G}$. If $\boldsymbol{W}_1, \ldots, \boldsymbol{W}_m$ are the different operations in the point group $\mathcal{P}$ of $\mathcal{G}$, then a system of coset representatives is obtained by choosing for every linear part $\boldsymbol{W}_i$ a translation part $\boldsymbol{w}_i$ such that $W_i = (\boldsymbol{W}_i, \boldsymbol{w}_i)$ is an operation in $\mathcal{G}$.

It is customary to choose the translation parts $\boldsymbol{w}_i$ of the coset representatives such that their coordinates lie between 0 and 1, excluding 1. In particular, if the translation part of a coset representative is a lattice vector, it is usually chosen as the zero vector $\boldsymbol{o}$.

More details on the coset decomposition with respect to the translation subgroup and examples of its application can be found in Section 1.4.2.

1.3.3.3. Symmorphic and non-symmorphic space groups

If a coset with respect to the translation subgroup contains an operation of the form $(\boldsymbol{W}, \boldsymbol{w})$ with $\boldsymbol{w}$ a vector in the translation lattice, it is clear that the same coset also contains the operation $(\boldsymbol{W}, \boldsymbol{o})$ with trivial translation part. On the other hand, if a coset does not contain an operation of the form $(\boldsymbol{W}, \boldsymbol{o})$, this may be caused by an inappropriate choice of origin. For example, the operation $\{\bar{1} \mid \frac{1}{2}, \frac{1}{2}, \frac{1}{2}\}$ is turned into the inversion $\{\bar{1} \mid 0\}$ by moving the origin to $\frac{1}{4}, \frac{1}{4}, \frac{1}{4}$ (*cf.* Section 1.5.1.1 for a detailed treatment of origin-shift transformations).

Depending on the actual space group $\mathcal{G}$, it may or may not be possible to choose the origin such that every coset with respect to $\mathcal{T}$ contains an operation of the form $(\boldsymbol{W}, \boldsymbol{o})$.

Definition

Let $\mathcal{G}$ be a space group with translation subgroup $\mathcal{T}$. If it is possible to choose the coordinate system such that every coset of $\mathcal{G}$ with respect to $\mathcal{T}$ contains an operation $(\boldsymbol{W}, \boldsymbol{o})$ with trivial translation part, $\mathcal{G}$ is called a *symmorphic* space group, otherwise $\mathcal{G}$ is called a *non-symmorphic* space group.

One sees that the operations with trivial translation part form a subgroup of $\mathcal{G}$ which is isomorphic to a subgroup of the point group $\mathcal{P}$. This subgroup is the group of operations in $\mathcal{G}$ that fix the origin and is called the *site-symmetry group* of the origin (site-symmetry groups are discussed in detail in Section 1.4.4). It is the distinctive property of symmorphic space groups that they contain a subgroup which is isomorphic to the full point group. This may in fact be seen as an alternative definition for symmorphic space groups.

Proposition. A space group $\mathcal{G}$ with point group $\mathcal{P}$ is symmorphic if and only if it contains a subgroup isomorphic to $\mathcal{P}$. For a non-symmorphic space group $\mathcal{G}$, every finite subgroup of $\mathcal{G}$ is isomorphic to a proper subgroup of the point group.

Note that every finite subgroup of a space group is a subgroup of the site-symmetry group for some point, because finite groups cannot contain translations. Therefore, a symmorphic space group is characterized by the fact that it contains a site-symmetry group isomorphic to its point group, whereas in non-symmorphic space groups all site-symmetry groups have orders strictly smaller than the order of the point group.

Symmorphic space groups can easily be constructed by choosing a lattice $\mathbf{L}$ and a point group $\mathcal{P}$ which acts on $\mathbf{L}$. Then $\mathcal{G} = \{(\boldsymbol{W}, \boldsymbol{w}) \mid \boldsymbol{W} \in \mathcal{P}, \boldsymbol{w} \in \mathbf{L}\}$ is a space group in which the coset representatives can be chosen as $(\boldsymbol{W}, \boldsymbol{o})$.

Non-symmorphic space groups can also be constructed from a lattice $\mathbf{L}$ and a point group $\mathcal{P}$. What is required is a system of coset representatives with respect to $\mathcal{T}$ and these are obtained by choosing for each operation $\boldsymbol{W} \in \mathcal{P}$ a translation part $\boldsymbol{w}$. Owing to the translations, it is sufficient to consider vectors $\boldsymbol{w}$ with components between 0 and 1. However, the translation parts cannot be chosen arbitrarily, because for three coset representatives $(\boldsymbol{W}_1, \boldsymbol{w}_1), (\boldsymbol{W}_2, \boldsymbol{w}_2), (\boldsymbol{W}_3, \boldsymbol{w}_3)$ with $\boldsymbol{W}_2\boldsymbol{W}_1 = \boldsymbol{W}_3$ the composition rule

$$(\boldsymbol{W}_2, \boldsymbol{w}_2)(\boldsymbol{W}_1, \boldsymbol{w}_1) = (\boldsymbol{W}_2\boldsymbol{W}_1, \boldsymbol{W}_2\boldsymbol{w}_1 + \boldsymbol{w}_2)$$

imposes the restriction that

$$\boldsymbol{W}_2\boldsymbol{w}_1 + \boldsymbol{w}_2 - \boldsymbol{w}_3 \in \mathbf{L}.$$

Once translation parts $\boldsymbol{w}$ are found that fulfil all these restrictions, one finally has to check whether the space group obtained this way is (by accident) symmorphic, but written with respect to an inappropriate origin. A change of origin by $\boldsymbol{p}$ is realized by conjugating the matrix–column pair $(\boldsymbol{W}, \boldsymbol{w})$ by the translation $(\boldsymbol{I}, -\boldsymbol{p})$ (*cf.* Section 1.5.2; for an introduction to coordinate transformations, see Section 1.5.1) which gives

$$(\boldsymbol{I}, -\boldsymbol{p})(\boldsymbol{W}, \boldsymbol{w})(\boldsymbol{I}, \boldsymbol{p}) = (\boldsymbol{W}, \boldsymbol{W}\boldsymbol{p} + \boldsymbol{w} - \boldsymbol{p}) = (\boldsymbol{W}, \boldsymbol{w} + (\boldsymbol{W} - \boldsymbol{I})\boldsymbol{p}).$$

Thus, the space group just constructed is symmorphic if there is a vector $\boldsymbol{p}$ such that $(\boldsymbol{W} - \boldsymbol{I})\boldsymbol{p} + \boldsymbol{w} \in \mathbf{L}$ for each of the coset representatives $(\boldsymbol{W}, \boldsymbol{w})$.

The above considerations also show how every space group can be assigned to a symmorphic space group in a canonical way, namely by setting the translation parts of coset representatives with respect to $\mathcal{T}$ to $\boldsymbol{o}$. This has the effect that screw rotations are turned into rotations and glide reflections into reflections. The Hermann–Mauguin symbol (for a discussion of Hermann–Mauguin symbols, see Section 1.4.1) of the symmorphic space group to which an arbitrary space group is assigned is simply obtained by replacing any screw rotation symbol N_m by the corresponding rotation symbol N and every glide reflection symbol a, b, c, d, e, n by the symbol m for a reflection. A space group is found to be symmorphic if no such replacement is required, *i.e.* if the Hermann–Mauguin symbol only contains the symbols 1, 2, 3, 4, 6 for rotations, $\bar{1}, \bar{3}, \bar{4}, \bar{6}$ for rotoinversions and m for reflections.

Example

The space groups with Hermann-Mauguin symbols $P4mm$, $P4bm$, $P4_2cm$, $P4_2nm$, $P4cc$, $P4nc$, $P4_2mc$, $P4_2bc$ are all assigned to the symmorphic space group with Hermann–Mauguin symbol $P4mm$.

The assignment of arbitrary space groups to symmorphic space groups, given by the corresponding Hermann–Mauguin symbols, is displayed in Table 2.1.3.2.

1.3.4. Classification of space groups

In this section we will consider various ways in which space groups may be grouped together. For the space groups themselves, the natural notion of equivalence is the classification into *space-group types*, but the point groups and lattices from which the space groups are built also have their own classification schemes into *geometric crystal classes* and *Bravais types of lattices*, respectively.

Some other types of classifications are relevant for certain applications, and these will also be considered. The hierarchy of the different classification levels and the numbers of classes on the different levels in dimension 3 are displayed in Fig. 1.3.4.1.

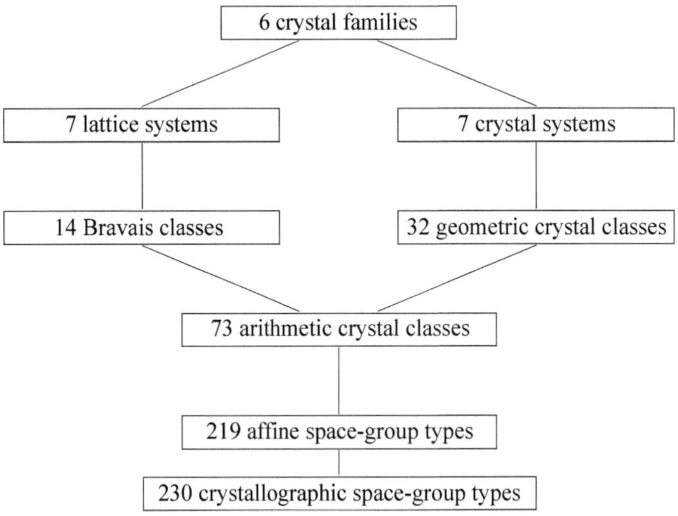

Figure 1.3.4.1
Classification levels for three-dimensional space groups.

1.3.4.1. Space-group types

The main motivation behind studying space groups is that they allow the classification of crystal structures according to their symmetry properties. Since many properties of a structure can be derived from its group of symmetries alone, this allows the investigation of the properties of many structures simultaneously.

On the other hand, even for the same crystal structure the corresponding space group may look different, depending on the chosen coordinate system (see Chapter 1.5 for a detailed discussion of transformations to different coordinate systems). Because it is natural to regard two realizations of a group of symmetry operations with respect to two different coordinate systems as equivalent, the following notion of equivalence between space groups is introduced.

Definition

Two space groups $\mathcal{G}$ and $\mathcal{G}'$ are called *affinely equivalent* if $\mathcal{G}'$ can be obtained from $\mathcal{G}$ by a change of the coordinate system. In terms of matrix–column pairs this means that there must exist a matrix–column pair $(\boldsymbol{P}, \boldsymbol{p})$ such that

$$\mathcal{G}' = \{(\boldsymbol{P}, \boldsymbol{p})^{-1}(\boldsymbol{W}, \boldsymbol{w})(\boldsymbol{P}, \boldsymbol{p}) \mid (\boldsymbol{W}, \boldsymbol{w}) \in \mathcal{G}\}.$$

The collection of space groups that are affinely equivalent with $\mathcal{G}$ forms the *affine type* of $\mathcal{G}$.

In dimension 2 there are 17 affine types of plane groups and in dimension 3 there are 219 affine space-group types. Note that in order to avoid misunderstandings we refrain from calling the space-group types *affine classes*, since the term classes is usually associated with *geometric crystal classes* (see below).

Grouping together space groups according to their space-group type serves different purposes. On the one hand, it is sometimes convenient to consider the same crystal structure and thus also its space group with respect to different coordinate systems, usually called space-group *settings*, *e.g.* when the origin can be chosen in different natural ways or when a phase transition to a higher- or lower-symmetry phase with a different conventional cell is described. On the other hand, different crystal structures may give rise to the same space group once suitable coordinate systems have been chosen for both. We illustrate both of these perspectives by an example.

Examples

(i) The space group $\mathcal{G}$ of type *Pban* (50) has a subgroup $\mathcal{H}$ of index 2 for which the coset representatives relative to the translation subgroup are the identity e: x, y, z, the twofold rotation g: $-x, y, -z$, the n glide h: $x+\frac{1}{2}, y+\frac{1}{2}, -z$ and the b glide k: $-x+\frac{1}{2}, y+\frac{1}{2}, z$. This subgroup is of type *Pb*2*n*, which is a non-conventional setting for *Pnc*2 (30). In the conventional setting, the coset representatives of *Pnc*2 are given by g': $-x, -y, z$, h': $-x, y+\frac{1}{2}, z+\frac{1}{2}$ and k': $x, -y+\frac{1}{2}, z+\frac{1}{2}$, *i.e.* with the z axis as rotation axis for the twofold rotation. The subgroup $\mathcal{H}$ can be transformed to its conventional setting by the basis transformation $\mathbf{a}' = \mathbf{c}$, $\mathbf{b}' = \mathbf{a}$, $\mathbf{c}' = \mathbf{b}$. Depending on whether the perspective of the full group $\mathcal{G}$ or the subgroup $\mathcal{H}$ is more important for a crystal structure, the groups $\mathcal{G}$ and $\mathcal{H}$ will be considered either with respect to the basis $\mathbf{a}, \mathbf{b}, \mathbf{c}$ (conventional for $\mathcal{G}$) or to the basis $\mathbf{a}', \mathbf{b}', \mathbf{c}'$ (conventional for $\mathcal{H}$).

(ii) The elements carbon, silicon and germanium all crystallize in the *diamond structure*, which has a face-centred cubic unit cell with two atoms shifted by 1/4 along the space diagonal of the conventional cubic cell. The space group is in all cases of type $Fd\bar{3}m$ (227), but the cell parameters differ: $a_{\rm C}$ = 3.5668 Å for carbon, $a_{\rm Si}$ = 5.4310 Å for silicon and $a_{\rm Ge}$ = 5.6579 Å for germanium (measured at 298 K). In order to scale the conventional cell of carbon to that of silicon, the coordinate system has to be transformed by the diagonal matrix

$$a_{\rm Si}/a_{\rm C} \cdot \boldsymbol{I}_3 \approx \begin{pmatrix} 1.523 & 0 & 0 \\ 0 & 1.523 & 0 \\ 0 & 0 & 1.523 \end{pmatrix}.$$

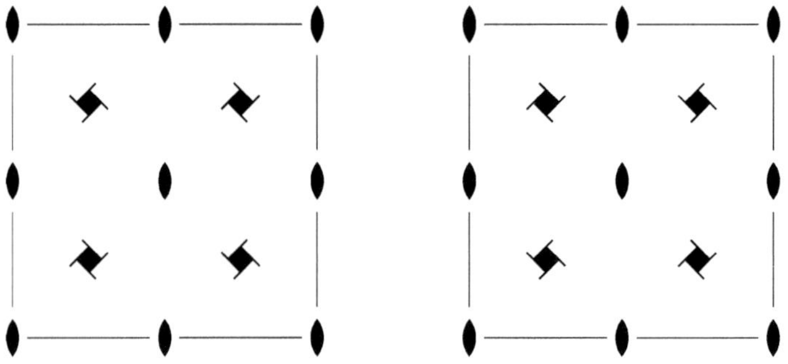

Figure 1.3.4.2
Space-group diagram of $I4_1$ (left) and its reflection in the plane $z = 0$ (right).

In crystallography, a notion of equivalence slightly stronger than affine equivalence is usually used. Since crystals occur in physical space and physical space can only be transformed by orientation-preserving mappings, space groups are only regarded as equivalent if they are conjugate by an *orientation-preserving* coordinate transformation, *i.e.* by an affine mapping that has a linear part with positive determinant.

Definition

Two space groups $\mathcal{G}$ and $\mathcal{G}'$ are said to belong to the same *space-group type* if $\mathcal{G}'$ can be obtained from $\mathcal{G}$ by an orientation-preserving coordinate transformation, *i.e.* by conjugation with a matrix–column pair $(\boldsymbol{P}, \boldsymbol{p})$ with $\det \boldsymbol{P} > 0$. In order to distinguish the space-group types explicitly from the affine space-group types, they are often called *crystallographic space-group types*.

The (crystallographic) space-group type collects together the infinitely many space groups that are obtained by expressing a single space group with respect to all possible right-handed coordinate systems for the point space.

Example

We consider the space group $\mathcal{G}$ of type $I4_1$ (80) which is generated by the right-handed fourfold screw rotation g: $-y, x+1/2, z+1/4$ (located at $-1/4, 1/4, z$), the centring translation t: $x+1/2, y+1/2, z+1/2$ and the integral translations of a primitive tetragonal lattice. Conjugating the group $\mathcal{G}$ to $\mathcal{G}' = m\mathcal{G}m^{-1}$ by the reflection m in the plane $z = 0$ turns the right-handed screw rotation g into the left-handed screw rotation g': $-y, x+1/2, z-1/4$, and one might suspect that $\mathcal{G}'$ is a space group of the same affine type but of a different crystallographic space-group type as $\mathcal{G}$. However, this is not the case because conjugating $\mathcal{G}$ by the translation $n = t(0, 1/2, 0)$ conjugates g to $g'' = ngn^{-1}$: $-y+1/2, x+1, z+1/4$. One sees that g'' is the composition of g' with the centring translation t and hence g'' belongs to $\mathcal{G}'$. This shows that conjugating

$\mathcal{G}$ by either the reflection m or the translation n both result in the same group $\mathcal{G}'$. This can also be concluded directly from the space-group diagrams in Fig. 1.3.4.2. Reflecting in the plane $z = 0$ turns the diagram on the left into the diagram on the right, but the same effect is obtained when the left diagram is shifted by $\frac{1}{2}$ along either **a** or **b**.

The groups $\mathcal{G}$ and $\mathcal{G}'$ thus belong to the same crystallographic space-group type because $\mathcal{G}$ is transformed to $\mathcal{G}'$ by a shift of the origin by $\frac{1}{2}\mathbf{b}$, which is clearly an orientation-preserving coordinate transformation.

Enantiomorphism

The 219 affine space-group types in dimension 3 result in 230 crystallographic space-group types. Since an affine type either forms a single space-group type (in the case where the group obtained by an orientation-reversing coordinate transformation can also be obtained by an orientation-preserving transformation) or splits into two space-group types, this means that there are 11 affine space-group types such that an orientation-reversing coordinate transformation cannot be compensated by an orientation-preserving transformation.

Groups that differ only by their handedness are closely related to each other and share many properties. One addresses this phenomenon by the concept of *enantiomorphism*.

Definition

Two space groups $\mathcal{G}$ and $\mathcal{G}'$ are said to form an *enantiomorphic pair* if they are conjugate under an affine mapping, but not under an orientation-preserving affine mapping.

If $\mathcal{G}$ is the group of isometries of some crystal pattern, then its enantiomorphic counterpart $\mathcal{G}'$ is the group of isometries of the mirror image of this crystal pattern.

The splitting of affine space-group types of three-dimensional space groups into pairs of crystallographic space-group types gives rise to the following 11 enantiomorphic pairs of space-group types: $P4_1/P4_3$ (76/78), $P4_122/P4_322$ (91/95), $P4_12_12/P4_32_12$ (92/96), $P3_1/P3_2$ (144/145), $P3_112/P3_212$ (151/153), $P3_121/P3_221$ (152/154), $P6_1/P6_5$ (169/173), $P6_2/P6_4$ (170/172), $P6_122/P6_522$ (178/179), $P6_222/P6_422$ (180/181), $P4_332/P4_132$ (212/213).

Example

A well known example of a crystal that occurs in forms whose symmetry is described by enantiomorphic pairs of space groups is quartz. For low-temperature α-quartz there exists a left-handed and a right-handed form with space groups $P3_121$ (152) and $P3_221$ (154), respectively. The two individuals of opposite chirality occur together in the so-called Brazil twin of quartz. At higher temperatures, a phase transition leads to the higher-symmetry β-quartz forms, with space groups $P6_422$ (181) and $P6_222$ (180), which still form an enantiomorphic pair.

1.3.4.2. Geometric crystal classes

We recall that the point group of a space group is the group of linear parts occurring in the space group. Once a basis for the underlying vector space is chosen, such a point group is a group of 3×3 matrices. A point group is characterized by the relative positions between the rotation and rotoinversion lines and the reflection planes of its operations, and in this sense a point group is independent of the chosen basis. However, a suitable choice of basis is useful to highlight the geometric properties of a point group.

Example

A point group of type $3m$ is generated by a threefold rotation and a reflection in a plane with normal vector perpendicular to the rotation axis. Choosing a basis **a**, **b**, **c** such that **c** is along the rotation axis, **a** is perpendicular to the reflection plane and **b** is the image of **a** under the threefold rotation (*i.e.* **b** lies in the plane perpendicular to the rotation axis and makes an angle of 120° with **a**), the matrices of the threefold rotation and the reflection with respect to this basis are

$$\begin{pmatrix} 0 & \bar{1} & 0 \\ 1 & \bar{1} & 0 \\ 0 & 0 & 1 \end{pmatrix} \text{ and } \begin{pmatrix} \bar{1} & 1 & 0 \\ 0 & 1 & 0 \\ 0 & 0 & 1 \end{pmatrix}.$$

A different useful basis is obtained by choosing a vector $\mathbf{a}'$ in the reflection plane but neither along the rotation axis nor perpendicular to it and taking $\mathbf{b}'$ and $\mathbf{c}'$ to be the images of $\mathbf{a}'$ under the threefold rotation and its square. Then the matrices of the threefold rotation and the reflection with respect to the basis $\mathbf{a}', \mathbf{b}', \mathbf{c}'$ are

$$\begin{pmatrix} 0 & 0 & 1 \\ 1 & 0 & 0 \\ 0 & 1 & 0 \end{pmatrix} \text{ and } \begin{pmatrix} 1 & 0 & 0 \\ 0 & 0 & 1 \\ 0 & 1 & 0 \end{pmatrix}.$$

Different choices of a basis for a point group in general result in different matrix groups, and it is natural to consider two point groups as equivalent if they are transformed into each other by a basis transformation. This is entirely analogous to the situation of space groups, where space groups that only differ by the choice of coordinate system are regarded as equivalent. This notion of equivalence is applied at both the level of space groups and point groups.

Definition

Two space groups $\mathcal{G}$ and $\mathcal{G}'$ with point groups $\mathcal{P}$ and $\mathcal{P}'$, respectively, are said to belong to the same *geometric crystal class* if $\mathcal{P}$ and $\mathcal{P}'$ become the same matrix group once suitable bases for the three-dimensional space are chosen.

Equivalently, $\mathcal{G}$ and $\mathcal{G}'$ belong to the same geometric crystal class if the point group $\mathcal{P}'$ can be obtained from $\mathcal{P}$ by a basis transformation of the underlying vector space $\mathbb{V}^3$, *i.e.* if there is an invertible 3×3 matrix $\boldsymbol{P}$ such that

$$\mathcal{P}' = \{\boldsymbol{P}^{-1}\boldsymbol{W}\boldsymbol{P} \mid \boldsymbol{W} \in \mathcal{P}\}.$$

Also, two matrix groups $\mathcal{P}$ and $\mathcal{P}'$ are said to belong to the same geometric crystal class if they are conjugate by an invertible 3×3 matrix $\boldsymbol{P}$.

Historically, the geometric crystal classes in dimension 3 were determined much earlier than the space groups. They were obtained as the symmetry groups for the set of normal vectors of crystal faces which describe the morphological symmetry of crystals.

Remark: One often speaks of the geometric crystal classes as the *types of point groups*. This emphasizes the point of view in which a point group is regarded as the group of linear parts of a space group, written with respect to an *arbitrary basis* of $\mathbb{R}^3$ (not necessarily a lattice basis).

It is also common to state that *there are 32 point groups in three-dimensional space*. This is just as imprecise as saying that *there are 230 space groups*, since there are in fact infinitely many point groups and space groups.

What is meant when we say that two space groups have *the same point group* is usually that their point groups are of the same type (*i.e.* lie in the same geometric crystal class) and can thus be *made to coincide* by a suitable basis transformation.

Example

In the space group $P3$ the threefold rotation generating the point group is given by the matrix

$$W = \begin{pmatrix} 0 & \bar{1} & 0 \\ 1 & \bar{1} & 0 \\ 0 & 0 & 1 \end{pmatrix},$$

whereas in the space group $R3$ (in the rhombohedral setting) the threefold rotation is given by the matrix

$$W' = \begin{pmatrix} 0 & 0 & 1 \\ 1 & 0 & 0 \\ 0 & 1 & 0 \end{pmatrix}.$$

These two matrices are conjugate by the basis transformation

$$P = \frac{1}{3}\begin{pmatrix} 2 & \bar{1} & \bar{1} \\ 1 & 1 & \bar{2} \\ 1 & 1 & 1 \end{pmatrix},$$

which transforms the basis of the hexagonal setting into that of the rhombohedral setting. This shows that the space groups $P3$ and $R3$ belong to the same geometric crystal class.

The example is typical in the sense that different groups in the same geometric crystal class usually describe the same group of linear parts acting on different lattices, *e.g.* primitive and centred. Writing the action of the linear parts with respect to primitive bases of different lattices gives rise to different matrix groups.

1.3.4.3. Bravais types of lattices and Bravais classes

In the classification of space groups into geometric crystal classes, only the point-group part is considered and the translation lattice is ignored. It is natural that the converse point of view is also adopted, where space groups are grouped together according to their translation lattices, irrespective of what the point groups are.

We have already seen that a lattice can be characterized by its metric tensor, containing the scalar products of a primitive basis. If a point group $\mathcal{P}$ acts on a lattice $\mathbf{L}$, it fixes the metric tensor G of $\mathbf{L}$, *i.e.* $W^{\rm T} \cdot G \cdot W = G$ for all W in $\mathcal{P}$ and is thus a subgroup of the Bravais group $Aut(\mathbf{L})$ of $\mathbf{L}$. Also, a matrix group $\mathcal{B}$ is called a *Bravais group* if it is the Bravais group $Aut(\mathbf{L})$ for some lattice $\mathbf{L}$. The Bravais groups govern the classification of lattices.

Definition

Two lattices $\mathbf{L}$ and $\mathbf{L}'$ belong to the same *Bravais type of lattices* if their Bravais groups $Aut(\mathbf{L})$ and $Aut(\mathbf{L}')$ are the same matrix group when written with respect to suitable primitive bases of $\mathbf{L}$ and $\mathbf{L}'$.

The different Bravais types of lattices, their cell parameters and metric tensors are displayed in Tables 3.1.2.1 (dimension 2) and 3.1.2.2 (dimension 3) of *IT* A: in dimension 2 there are 5 Bravais types and in dimension 3 there are 14 Bravais types of lattices.

It is crucial for the classification of lattices *via* their Bravais groups that one works with primitive bases, because a primitive and a body-centred cubic lattice have the same automorphisms when written with respect to the conventional cubic basis, but are clearly different types of lattices.

It is important to note that in order to have the same Bravais group, the metric tensors of the two lattices $\mathbf{L}$ and $\mathbf{L}'$ do not have to be the same or scalings of each other. However, if they are written with respect to suitable bases they are found to have the same structure, differing only in the specific values for certain free parameters.

Definition

Let $\mathbf{L}$ be a lattice with metric tensor G with respect to a primitive basis and let

$$\mathcal{B} = Aut(\mathbf{L}) = \{W \in {\rm GL}_3(\mathbb{Z}) \mid W^{\rm T} \cdot G \cdot W = G\}$$

be the Bravais group of $\mathbf{L}$. Then

$$\mathbf{M}(\mathcal{B}) := \{G' \text{ symmetric } 3 \times 3 \text{ matrix} \mid W^{\rm T} \cdot G' \cdot W = G' \text{ for all } W \in \mathcal{B}\}$$

is called the *space of metric tensors* of $\mathcal{B}$. The dimension of $\mathbf{M}(\mathcal{B})$ is called the *number of free parameters* of the lattice $\mathbf{L}$. Analogously, for an arbitrary integral matrix group $\mathcal{P}$,

$$\mathbf{M}(\mathcal{P}) := \{G' \text{ symmetric } 3 \times 3 \text{ matrix} \mid W^{\rm T} \cdot G' \cdot W = G' \text{ for all } W \in \mathcal{P}\}$$

is called the *space of metric tensors* of $\mathcal{P}$. If $\dim \mathbf{M}(\mathcal{P}') = \dim \mathbf{M}(\mathcal{P})$ for a subgroup $\mathcal{P}'$ of $\mathcal{P}$, the spaces of metric tensors are the same for both groups and one says that $\mathcal{P}'$ *does not act on a more general lattice* than $\mathcal{P}$ does.

The space of metric tensors obtained from a lattice can be interpreted as an expression of the metric tensor with general entries, *i.e.* as a generic metric tensor describing the different lattices within the same Bravais type. It is precisely these generic metric tensors that are given in Tables 3.1.2.1 and 3.1.2.2 of *IT* A to describe the Bravais types of lattices.

Example

The mineral rutile (TiO_2) has a space group of type $P4_2/mnm$ (136) with a primitive tetragonal cell with cell parameters $a = b = 4.594$ Å and $c = 2.959$ Å. The metric tensor of the translation lattice $\mathbf{L}$ is therefore

$$G = \begin{pmatrix} 4.594^2 & 0 & 0 \\ 0 & 4.594^2 & 0 \\ 0 & 0 & 2.959^2 \end{pmatrix}.$$

The silicate mineral cristobalite also has (at low temperatures) a primitive tetragonal cell with $a = b = 4.971$ Å and $c = 6.928$ Å, and the space-group type is $P4_12_12$ (92). In this case, the metric tensor of the translation lattice $\mathbf{L}'$ is

$$G' = \begin{pmatrix} 4.971^2 & 0 & 0 \\ 0 & 4.971^2 & 0 \\ 0 & 0 & 6.928^2 \end{pmatrix}.$$

One checks that the point groups $\mathcal{P}$ and $\mathcal{P}'$ for rutile and cristobalite have the same space of metric tensors

$$\mathbf{M}(\mathcal{P}) = \mathbf{M}(\mathcal{P}') = \left\{ \begin{pmatrix} g_{11} & 0 & 0 \\ 0 & g_{11} & 0 \\ 0 & 0 & g_{33} \end{pmatrix} \mid g_{11}, g_{33} \in \mathbb{R} \right\}$$

of dimension 2, corresponding to a primitive tetragonal lattice. The metric tensor for the lattice of rutile is obtained by specializing $g_{11} = 4.595$ and $g_{33} = 2.959$, whereas for cristobalite one has $g_{11} = 4.971$ and $g_{33} = 6.928$. Since both lattices

$\mathbf{L}$ and $\mathbf{L}'$ have the same Bravais group (generated by the fourfold rotation 4^+_{001}: $\bar{y}, x, z$ around the z axis, the reflection m_{001}: $x, y, \bar{z}$ in the plane $z = 0$ and the reflection m_{100}: $\bar{x}, y, z$ in the plane $x = 0$), the translation lattices $\mathbf{L}$ for rutile and $\mathbf{L}'$ for cristobalite belong to the same Bravais type of lattices.

Special choices for the entries in the generic metric tensor may lead to lattices with accidental higher symmetry, which is in fact a common phenomenon in phase transitions caused by changes of temperature or pressure. One says that the translation lattice $\mathbf{L}$ of a space group $\mathcal{G}$ with point group $\mathcal{P}$ has a *specialized metric* if the dimension of the space of metric tensors of $\mathcal{B} = Aut(\mathbf{L})$ is smaller than the dimension of the space of metric tensors of $\mathcal{P}$. Viewed from a slightly different angle, a specialized metric occurs if the location of the atoms within the unit cell reduces the symmetry of the translation lattice to that of a different lattice type.

Example

A space group $\mathcal{G}$ of type $P2/m$ (10) with cell parameters $a = 4.4$, $b = 5.5$, $c = 6.6$ Å, $\alpha = \beta = \gamma = 90°$ has a specialized metric, because the point group $\mathcal{P}$ of type $2/m$ is generated by

$$\boldsymbol{W} = \begin{pmatrix} \bar{1} & 0 & 0 \\ 0 & 1 & 0 \\ 0 & 0 & \bar{1} \end{pmatrix}$$

and $-\boldsymbol{I}$, and has

$$\mathbf{M}(\mathcal{P}) = \left\{ \begin{pmatrix} g_{11} & 0 & g_{13} \\ 0 & g_{22} & 0 \\ g_{13} & 0 & g_{33} \end{pmatrix} \mid g_{11}, g_{22}, g_{33}, g_{13} \in \mathbb{R} \right\}$$

as its space of metric tensors, which is of dimension 4. The lattice $\mathbf{L}$ with the given cell parameters, however, is orthorhombic, since the free parameter g_{13} is specialized to $g_{13} = 0$. The automorphism group $Aut(\mathbf{L})$ is of type mmm and has a space of metric tensors of dimension 3, namely

$$\left\{ \begin{pmatrix} g_{11} & 0 & 0 \\ 0 & g_{22} & 0 \\ 0 & 0 & g_{33} \end{pmatrix} \mid g_{11}, g_{22}, g_{33} \in \mathbb{R} \right\}.$$

The higher symmetry of the translation lattice would, for example, be destroyed by an atomic configuration compatible with the lattice and represented by only two atoms in the unit cell located at 0.17, 1/2, 0.42 and 0.83, 1/2, 0.58. The two atoms are related by a twofold rotation around the b axis, which indicates the invariance of the configuration under twofold rotations with axes parallel to $\mathbf{b}$, but in contrast to the lattice $\mathbf{L}$, the atomic configuration is not compatible with rotations around the a or the c axes.

By looking at the spaces of metric tensors, space groups can be classified according to the Bravais types of their translation lattices, without suffering from complications due to specialized metrics.

Definition

Let $\mathbf{L}$ be a lattice with Bravais group $\mathcal{B} = Aut(\mathbf{L})$ and let $\mathbf{M}(\mathcal{B})$ be the space of metric tensors of $\mathcal{B}$ represented by the generic metric tensor for the lattice type of $\mathbf{L}$. Then those space groups $\mathcal{G}$ form the *Bravais class* corresponding to the Bravais type of $\mathbf{L}$ for which the generic metric tensor of the point group $\mathcal{P}$ of $\mathcal{G}$ is equal to that of $\mathcal{B}$ when $\mathcal{P}$ is written with respect to a suitable primitive basis of the translation lattice of $\mathcal{G}$. The names for the Bravais classes are the same as those for the corresponding Bravais types of lattices.

The Bravais groups of lattices provide a link between lattices and point groups, the two building blocks of space groups. However, although the Bravais group of a lattice is simply a matrix group, the fact that it is expressed with respect to a primitive basis and fixes the metric tensor of the lattice preserves the necessary information about the lattice. When the Bravais group is regarded as a point group, the information about the lattice is lost, since point groups can be written with respect to an arbitrary basis. In order to distinguish Bravais groups of lattices at the level of point groups and geometric crystal classes, the concept of a holohedry is introduced.

Definition

The geometric crystal class of a point group $\mathcal{P}$ is called a *holohedry* (or *lattice point group*, *cf.* Chapters 3.1 and 3.3 of *IT* A) if $\mathcal{P}$ is the Bravais group of some lattice $\mathbf{L}$.

Example

Let $\mathcal{P}$ be the point group of type $\bar{3}m$ generated by the threefold rotoinversion

$$\boldsymbol{W}_1 = \begin{pmatrix} 0 & 1 & 0 \\ \bar{1} & 1 & 0 \\ 0 & 0 & \bar{1} \end{pmatrix}$$

around the z axis and the twofold rotation

$$\boldsymbol{W}_2 = \begin{pmatrix} 1 & \bar{1} & 0 \\ 0 & \bar{1} & 0 \\ 0 & 0 & \bar{1} \end{pmatrix},$$

expressed with respect to the conventional basis $\mathbf{a}, \mathbf{b}, \mathbf{c}$ of a hexagonal lattice. The group $\mathcal{P}$ is not the Bravais group of the lattice $\mathbf{L}$ spanned by $\mathbf{a}, \mathbf{b}, \mathbf{c}$ because this lattice also allows a sixfold rotation around the z axis, which is not contained in $\mathcal{P}$. But $\mathcal{P}$ also acts on the rhombohedrally centred lattice $\mathbf{L}'$ with primitive basis $\mathbf{a}' = \frac{1}{3}(2\mathbf{a} + \mathbf{b} + \mathbf{c})$, $\mathbf{b}' = \frac{1}{3}(-\mathbf{a} + \mathbf{b} + \mathbf{c})$, $\mathbf{c}' = \frac{1}{3}(-\mathbf{a} - 2\mathbf{b} + \mathbf{c})$. With respect to the basis $\mathbf{a}', \mathbf{b}', \mathbf{c}'$ the rotoinversion and twofold rotation are transformed to

$$\boldsymbol{W}'_1 = \begin{pmatrix} 0 & 0 & \bar{1} \\ \bar{1} & 0 & 0 \\ 0 & \bar{1} & 0 \end{pmatrix} \text{ and } \boldsymbol{W}'_2 = \begin{pmatrix} 0 & \bar{1} & 0 \\ \bar{1} & 0 & 0 \\ 0 & 0 & \bar{1} \end{pmatrix},$$

and these matrices indeed generate the Bravais group of $\mathbf{L}'$. The geometric crystal class with symbol $\bar{3}m$ is therefore a holohedry.

Note that in dimension 3 the above is actually the only example of a geometric crystal class in which the point groups are Bravais groups for some but not for all the lattices on which they act. In all other cases, each matrix group $\mathcal{P}$ corresponding to a holohedry is actually the Bravais group of the lattice spanned by the basis with respect to which $\mathcal{P}$ is written.

1.3.4.4. Other classifications of space groups

In this section we summarize a number of other classification schemes which are perhaps of slightly lower significance than those of space-group types, geometric crystal classes and Bravais types of lattices, but also play an important role for certain applications.

1.3.4.4.1. Arithmetic crystal classes

We have already seen that every space group can be assigned to a symmorphic space group in a natural way by setting the translation parts of coset representatives with respect to the translation subgroup to $\boldsymbol{o}$. The groups assigned to a symmorphic space group in this way all have the same translation lattice and the same point group but the different possibilities for the interplay between these two parts are ignored.

If we want to collect together all space groups that correspond to symmorphic space groups of the same type, we arrive at the classification into *arithmetic crystal classes*. This can also be seen as a classification of the symmorphic space-group types. The distribution of the space groups into arithmetic classes, represented by the corresponding symmorphic space-group types, is given in Table 2.1.3.2.

The crucial observation for characterizing this classification is that space groups that correspond to the same symmorphic space group all have translation lattices of the same Bravais type. This means that the freedom in the choice of a basis transformation of the underlying vector space is restricted, because a primitive basis has to be mapped again to a primitive basis. Assuming that the point groups are written with respect to primitive bases, this means that the basis transformation is an integral matrix with determinant ±1.

Definition

Two space groups $\mathcal{G}$ and $\mathcal{G}'$ with point groups $\mathcal{P}$ and $\mathcal{P}'$, respectively, both written with respect to primitive bases of their translation lattices, are said to lie in the same *arithmetic crystal class* if $\mathcal{P}'$ can be obtained from $\mathcal{P}$ by an integral basis transformation of determinant ±1, *i.e.* if there is an integral 3×3 matrix $\boldsymbol{P}$ with $\det \boldsymbol{P} = \pm 1$ such that

$$\mathcal{P}' = \{\boldsymbol{P}^{-1}\boldsymbol{W}\boldsymbol{P} \mid \boldsymbol{W} \in \mathcal{P}\}.$$

Also, two integral matrix groups $\mathcal{P}$ and $\mathcal{P}'$ are said to belong to the same arithmetic crystal class if they are conjugate by an integral 3×3 matrix $\boldsymbol{P}$ with $\det \boldsymbol{P} = \pm 1$.

Example

Let

$$\boldsymbol{M}_1 = \begin{pmatrix} \bar{1} & 0 & 0 \\ 0 & 1 & 0 \\ 0 & 0 & 1 \end{pmatrix}, \quad \boldsymbol{M}_2 = \begin{pmatrix} 1 & 0 & 0 \\ 0 & \bar{1} & 0 \\ 0 & 0 & 1 \end{pmatrix}$$
$$\text{and } \boldsymbol{M}_3 = \begin{pmatrix} 0 & 1 & 0 \\ 1 & 0 & 0 \\ 0 & 0 & 1 \end{pmatrix}$$

be reflections in the planes $x = 0$, $y = 0$ and $x = y$, respectively, and let $\mathcal{P}_1 = \langle \boldsymbol{M}_1 \rangle$, $\mathcal{P}_2 = \langle \boldsymbol{M}_2 \rangle$ and $\mathcal{P}_3 = \langle \boldsymbol{M}_3 \rangle$ be the integral matrix groups generated by these reflections. Then $\mathcal{P}_1$ and $\mathcal{P}_2$ belong to the same arithmetic crystal class because they are transformed into each other by the basis transformation

$$\boldsymbol{P} = \begin{pmatrix} 0 & 1 & 0 \\ 1 & 0 & 0 \\ 0 & 0 & 1 \end{pmatrix}$$

interchanging the x and y axes. But $\mathcal{P}_3$ belongs to a different arithmetic crystal class, because $\boldsymbol{M}_3$ is not conjugate to $\boldsymbol{M}_1$ by an integral matrix $\boldsymbol{P}$ of determinant ±1. The two groups $\mathcal{P}_1$ and $\mathcal{P}_3$ belong, however, to the same geometric crystal class, because $\boldsymbol{M}_1$ and $\boldsymbol{M}_3$ are transformed into each other by the basis transformation

$$\boldsymbol{P} = \begin{pmatrix} \frac{1}{2} & -\frac{1}{2} & 0 \\ \frac{1}{2} & \frac{1}{2} & 0 \\ 0 & 0 & 1 \end{pmatrix},$$

which has determinant $\frac{1}{2}$. This basis transformation shows that $\boldsymbol{M}_1$ and $\boldsymbol{M}_3$ can be interpreted as the action of the same reflection on a primitive lattice and on a *C*-centred lattice.

As explained above, the number of arithmetic crystal classes is equal to the number of symmorphic space-group types: in dimension 2 there are 13 such classes, in dimension 3 there are 73 arithmetic crystal classes.

1.3.4.4.2. Lattice systems

It is sometimes convenient to group together those space groups for which the Bravais groups of their translation lattices belong to the same holohedry.

Definition

Two space groups $\mathcal{G}$ and $\mathcal{G}'$ belong to the same *lattice system* if the Bravais groups of their translation lattices lie in the same geometric crystal class (which is thus a holohedry). Also, two lattices are said to belong to the same lattice system if their Bravais groups belong to the same holohedry.

Remark: The lattice systems were called *Bravais systems* in earlier editions of *IT* A and this Teaching Edition.

Example

The primitive cubic, face-centred cubic and body-centred cubic lattices all belong to the same lattice system, because their Bravais groups all belong to the holohedry with symbol $m\bar{3}m$.

On the other hand, the hexagonal and the rhombohedral lattices belong to different lattice systems, because their Bravais groups are not even of the same order and lie in different holohedries (with symbols $6/mmm$ and $\bar{3}m$, respectively).

Every lattice system contains the lattices of precisely one holohedry and a holohedry determines a unique lattice system, containing the lattices of the arithmetic crystal classes in the holohedry. Therefore, there is a one-to-one correspondence between holohedries and lattice systems. There are four lattice systems in dimension 2 and seven lattice systems in dimension 3. The lattice systems in three-dimensional space are displayed in Table 1.3.4.1. Along with the name of each lattice system, the Bravais types of lattices contained in it and the corresponding holohedry are given.

Table 1.3.4.1
Lattice systems in three-dimensional space

Lattice system	Bravais types of lattices	Holohedry
Triclinic (anorthic)	*aP*	$\bar{1}$
Monoclinic	*mP*, *mS*	$2/m$
Orthorhombic	*oP*, *oS*, *oF*, *oI*	mmm
Tetragonal	*tP*, *tI*	$4/mmm$
Hexagonal	*hP*	$6/mmm$
Rhombohedral	*hR*	$\bar{3}m$
Cubic	*cP*, *cF*, *cI*	$m\bar{3}m$

Table 1.3.4.2
Crystal systems in three-dimensional space

Crystal system	Point-group types
Triclinic	$\bar{1}$, 1
Monoclinic	2/*m*, *m*, 2
Orthorhombic	*mmm*, *mm*2, 222
Tetragonal	4/*mmm*, $\bar{4}$2*m*, 4*mm*, 422, 4/*m*, $\bar{4}$, 4
Hexagonal	6/*mmm*, $\bar{6}$2*m*, 6*mm*, 622, 6/*m*, $\bar{6}$, 6
Trigonal	$\bar{3}$*m*, 3*m*, 32, $\bar{3}$, 3
Cubic	*m*$\bar{3}$*m*, $\bar{4}$3*m*, 432, *m*$\bar{3}$, 23

1.3.4.4.3. Crystal systems

The action on different types of lattices is exploited for further classification of point groups by joining those geometric crystal classes that act on the same Bravais types of lattices. For example, the holohedry *m*$\bar{3}$*m* acts on primitive, face-centred and body-centred cubic lattices. The other geometric crystal classes that act on these three types of lattices are 23, *m*$\bar{3}$, 432 and $\bar{4}$3*m*.

Definition

Two space groups $\mathcal{G}$ and $\mathcal{G}'$ with point groups $\mathcal{P}$ and $\mathcal{P}'$, respectively, belong to the same *crystal system* if the sets of Bravais types of lattices on which $\mathcal{P}$ and $\mathcal{P}'$ act coincide. Since point groups in the same geometric crystal class act on the same types of lattices, crystal systems consist of full geometric crystal classes and the point groups $\mathcal{P}$ and $\mathcal{P}'$ are also said to belong to the same crystal system.

Remark: In the literature there are many different notions of crystal systems. In *IT* A, only the one defined above is used.

The typical situation in three-dimensional space is that a crystal class consists of a holohedry together with the geometric crystal classes of those subgroups of a point group in the holohedry that do not act on lattices with a larger number of free parameters. For example, the crystal system containing the holohedry of type 4/*mmm* consists of the geometric classes of types 4, $\bar{4}$, 4/*m*, 422, 4*mm*, $\bar{4}$2*m* and 4/*mmm*, since these are precisely the point groups acting on tetragonal and body-centred tetragonal lattices.

The only exceptions from this situation occur for the hexagonal and rhombohedral lattices.

Example

A point group containing a threefold rotation but no sixfold rotation or rotoinversion acts both on a hexagonal lattice and on a rhombohedral lattice. On the other hand, point groups containing a sixfold rotation only act on a hexagonal but not on a rhombohedral lattice. The geometric crystal classes of point groups containing a threefold rotation or rotoinversion but not a sixfold rotation or rotoinversion form a crystal system which is called the *trigonal crystal system*. The geometric crystal classes of point groups containing a sixfold rotation or rotoinversion form a different crystal system, which is called the *hexagonal crystal system*.

The classification of the point-group types into crystal systems is summarized in Table 1.3.4.2.

1.3.4.4.4. Crystal families

The classification into crystal systems has many important applications, but it has the disadvantage that it is not compatible with the classification into lattice systems. Space groups that belong to the hexagonal lattice system are distributed over the trigonal and the hexagonal crystal system. Conversely, space groups in the trigonal crystal system belong to either the rhombohedral or the hexagonal lattice system. It is therefore desirable to define a further classification level in which the classes consist of full crystal systems and of full lattice systems, or, equivalently, of full geometric crystal classes and full Bravais classes. Since crystal systems already contain only geometric crystal classes with spaces of metric tensors of the same dimension, this can be achieved by the following definition.

Definition

For a space group $\mathcal{G}$ with point group $\mathcal{P}$ the *crystal family* of $\mathcal{G}$ is the union of all geometric crystal classes that contain a space group $\mathcal{G}'$ that has the same Bravais type of lattices as $\mathcal{G}$.

The crystal family of $\mathcal{G}$ thus consists of those geometric crystal classes that contain a point group $\mathcal{P}'$ such that $\mathcal{P}$ and $\mathcal{P}'$ are contained in a common supergroup $\mathcal{B}$ (which is a Bravais group) and such that $\mathcal{P}$, $\mathcal{P}'$ and $\mathcal{B}$ all act on lattices with the same number of free parameters.

In two-dimensional space, the crystal families coincide with the crystal systems and in three-dimensional space only the trigonal and hexagonal crystal system are merged into a single crystal family, whereas all other crystal systems again form a crystal family on their own.

Example

The trigonal and hexagonal crystal systems belong to a single crystal family, called the *hexagonal crystal family*, because for both crystal systems the number of free parameters of the corresponding lattices is 2 and a point group of type $\bar{3}$*m* in the trigonal crystal system is a subgroup of a point group of type 6/*mmm* in the hexagonal crystal system.

For the space groups within one crystal family the same coordinate system is usually used, which is called the *conventional coordinate system* (for this crystal family). However, depending on the application it may be useful to work with a different coordinate system. To avoid confusion, it is recommended to state explicitly when a coordinate system differing from the conventional coordinate system is used.

References

International Tables for Crystallography (2016). Volume A, *Space-Group Symmetry*, 6th ed., edited by M. I. Aroyo. Chichester: Wiley. (Abbreviated as *IT* A.)

1.4. Space groups and their descriptions

Bernd Souvignier, Hans Wondratschek, Mois I. Aroyo, Gervais Chapuis and A. M. Glazer

1.4.1. Symbols of space groups

By Hans Wondratschek

1.4.1.1. Introduction

Space groups describe the symmetries of crystal patterns; the point group of the space group is the symmetry of the ideal macroscopic crystal, *cf.* Section 1.2.1. Both kinds of symmetry are characterized by symbols of which there are different kinds. In this section the space-group numbers as well as the Schoenflies symbols and the Hermann–Mauguin symbols of the space groups and point groups will be dealt with and compared, because these are used throughout Volume A of *International Tables for Crystallography* (2016), hereafter referred to as *IT* A, and in this edition. They are rather different in their aims. For the Fedorov symbols, mainly used in Russian crystallographic literature, *cf.* Chapter 3.3 of *IT* A. For computer-adapted symbols of space groups implemented in crystallographic software, such as *Hall symbols* (Hall, 1981*a,b*) or *explicit symbols* (Shmueli, 1984), the reader is referred to Chapter 1.4 of *International Tables for Crystallography*, Volume B (2008).

For the definition of space groups and plane groups, *cf.* Chapter 1.3. The plane groups characterize the symmetries of two-dimensional periodic arrangements, realized in sections and projections of crystal structures or by periodic wallpapers or tilings of planes. They are described individually and in detail in Chapter 2.2 of IT A. Groups of one- and two-dimensional periodic arrangements embedded in two-dimensional and three-dimensional space are called subperiodic groups. They are listed in Vol. E of International Tables for Crystallography (2010) (referred to as IT E) with symbols similar to the Hermann–Mauguin symbols of plane groups and space groups, and are related to these groups as their subgroups. The space groups sensu stricto are the symmetries of periodic arrangements in three-dimensional space, e.g. of normal crystals, see also Chapter 1.3. They are described individually and in detail in the space-group tables of Chapter 2.3 of IT A. In the following, if not specified separately, both space groups and plane groups are covered by the term space group.

The description of each space group in the tables of Chapter 2.3 of *IT* A (see also the examples of space-group tables in Part 2) starts with two headlines in which the different symbols of the space group are listed. All these names are explained in this section with the exception of the data for *Patterson symmetry* (*cf.* Chapter 1.6 and Section 2.1.3.5 for explanations of Patterson symmetry).

1.4.1.2. Space-group numbers

The space-group numbers were introduced in *International Tables for X-ray Crystallography* (1952) [referred to as *IT* (1952)] for plane groups (Nos. 1–17) and space groups (Nos. 1–230). They provide a short way of specifying the type of a space group uniquely, albeit without reference to its symmetries. They are particularly convenient for use with computers and have been in use since their introduction.

There are no numbers for the point groups.

1.4.1.3. Schoenflies symbols

The Schoenflies symbols were introduced by Schoenflies (1891, 1923). They describe the point-group type, also known as the geometric crystal class or (for short) crystal class (*cf.* Section 1.3.4.2), of the space group geometrically. The different space-group types within the same crystal class are denoted by a superscript index appended to the point-group symbol.

1.4.1.3.1. Schoenflies symbols of the crystal classes

Schoenflies derived the point groups as groups of crystallographic symmetry operations, but described these crystallographic point groups geometrically by their representation through lines (axes) of rotation or rotoreflection and planes of reflection (also called mirror planes), *i.e.* by *geometric elements*; for geometric elements of symmetry elements, *cf.* Section 1.2.3, de Wolff *et al.* (1989, 1992) and Flack *et al.* (2000). Rotation axes dominate the description and planes of reflection are added when necessary. Rotoreflection axes are also indicated when necessary. The orientation of a reflection plane, whether *horizontal*, *vertical* or *diagonal*, refers to the plane itself, not to its normal.

A coordinate basis may be chosen by the user: the basis vectors start at the origin which is placed in front of the user. The basis vector **c** points vertically upwards, the basis vectors **a** and **b** lie more or less horizontal; the basis vector **a** pointing at the user, **b** pointing to the user's right-hand side, *i.e.* the basis vectors **a**, **b** and **c** form a *right-handed* set. Such a basis will be called a *conventional crystallographic basis* in this chapter. (In the usual basis of mathematics and physics the basis vector **a** points to the right-hand side and **b** points away from the user.) The lengths of the basis vectors, the inclination of the **ab** plane relative to the **c** axis and the angles between the basis vectors are determined by the symmetry of the point group and the specific values of the lattice parameters of the crystal structure.

The letter C is used for *cyclic groups* of rotations around a rotation axis which is conventionally **c**. The order n of the rotation is appended as a subscript index: C_n; Fig. 1.4.1.1(*a*) represents C_2. The values of n that are possible in the rotation symmetry of a crystal are 1, 2, 3, 4 and 6 (*cf.* Section 1.3.3.1 for a discussion of this basic result). An n-fold rotoreflection, *i.e.* an n-fold rotation followed or preceded by a reflection through a plane perpendicular to the rotation axis (such that neither the rotation nor the reflection is in general a symmetry operation) is designated by S_n, see Fig. 1.4.1.1(*b*) for S_4.

The following types of point groups exist:

(1) Cyclic groups

(*a*) of rotations (C):

$$C_1,\ C_2,\ C_3,\ C_4,\ C_6;$$

(*b*) of rotoreflections [S, for the names in parentheses see item (7) below]:

$$S_1\ (= C_{1h} = C_s),\ S_2\ (= C_i),\ S_3\ (= C_{3h}),\ S_4,\ S_6\ (= C_{3i}).$$

(2) In dihedral groups D_n an n-fold (vertical) rotation axis is accompanied by n symmetry-equivalent horizontal twofold rotation axes. The symbols are D_2 [in older literature, as in

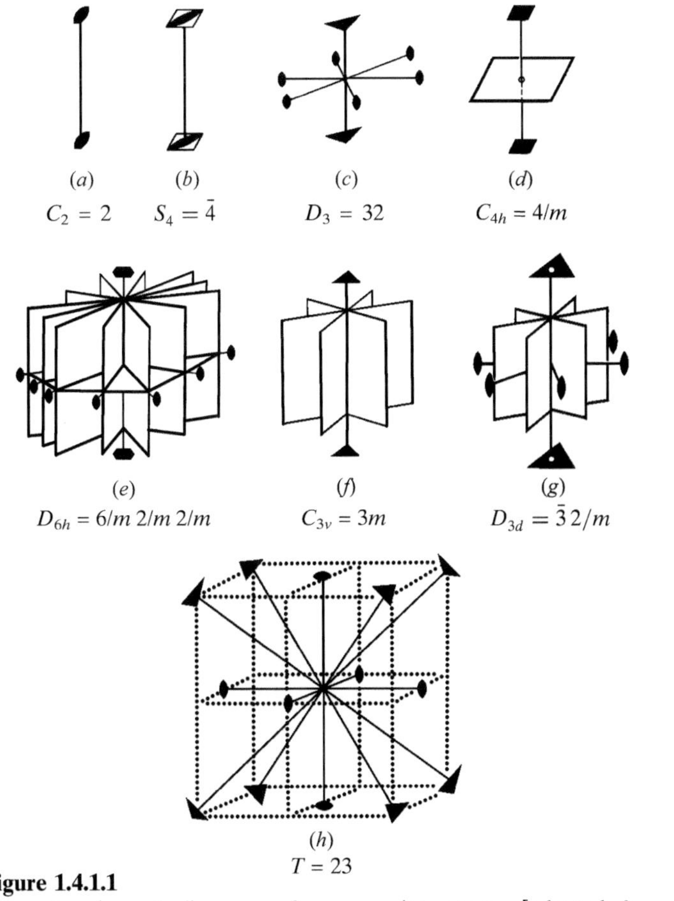

Figure 1.4.1.1
Symmetry-element diagrams of some point groups [adapted from Vainshtein (1994)]. The point groups are specified by their Schoenflies and Hermann–Mauguin symbols. [The cubic frame in part (*h*) has no crystallographic meaning: it has been included to aid visualization of the orientation of the symmetry elements.]

IT (1952), one also finds V instead of D_2, taken from the *Vierergruppe* of Klein (1884)], D_3, D_4, D_6; D_3 is visualized in Fig. 1.4.1.1(*c*).

(3) Other crystallographic point groups can be constructed by a C_n rotation or a D_n combination of rotations with a reflection through a horizontal symmetry plane, leading to symbols C_{nh} or D_{nh}:

$$C_{2h}, C_{3h}, C_{4h}, C_{6h}, D_{2h}, D_{3h}, D_{4h}, D_{6h}.$$

The point groups C_{4h} and D_{6h} are represented by Figs. 1.4.1.1(*d*) and 1.4.1.1(*e*).

(4) Vertical rotation axes C_n can be combined with a vertical reflection plane, leading to n symmetry-equivalent vertical reflection planes (denoted v) which all contain the rotation axis:

$$C_{2v}, C_{3v}, C_{4v}, C_{6v}$$

with Fig. 1.4.1.1(*f*) for C_{3v}.

(5) Combinations D_n of rotation axes may be combined with vertical reflection planes which bisect the angles between the horizontal twofold axes, such that the vertical planes (designated by the index d for 'diagonal') alternate with the horizontal twofold axes:

$$D_{2d} \text{ with } n = 2 \text{ or } D_{3d} \text{ with } n = 3;$$

see Fig. 1.4.1.1(*g*) for D_{3d}. In both point groups rotoreflections S_{2n}, *i.e.* S_4 or S_6, occur. Note that the classification of crystal classes into crystal systems follows the order of rotoinversions $\overline{N}$, not that of rotoreflections S_n (*cf.* Section 1.2.1 for the definition of rotoinversions). Therefore, D_{2d} is tetragonal ($S_4 \sim \bar{4}$) and D_{3d} is trigonal because of $S_6^5 = \bar{3}$). Analogously, C_{3h} and D_{3h} are hexagonal because they contain $S_3 \sim \bar{6}$. The point groups D_{4d} and D_{6d} are not crystallographic as they contain noncrystallographic eightfold or 12-fold rotoreflections S_8 or S_{12}.

(6) In all these groups the directions of the vectors $\pm\mathbf{c}$ are not equivalent to any other directions. There are, however, also cubic point groups and thus cubic space groups in which the basis vector **c** is symmetry-equivalent to both basis vectors **a** and **b**. T, T_h and T_d can be derived from the rotation group T of the tetrahedron, see Fig. 1.4.1.1(*h*). O and O_h can be derived from the rotation group O of the octahedron. The indices h and d have the same meaning as before.

(7) Some of these symbols are no longer used but are replaced by more visual ones. S_1 describes a reflection through a horizontal plane, it is replaced now by C_{1h} or by C_s; S_2 describes an inversion in a centre, it is replaced by C_i. The symbol S_3 describes the same arrangement as C_{3h} and is thus not used. S_6 contains an inversion centre combined with a threefold rotation axis and is replaced by C_{3i}.

The description of crystal classes using Schoenflies symbols is intuitive and much more graphic than that by Hermann–Mauguin symbols. It is useful for morphological studies investigating the symmetry of the ideal shape of crystals. Schoenflies symbols of crystal classes are also still used traditionally by physicists and chemists, in particular in spectroscopy and quantum chemistry.

1.4.1.3.2. Schoenflies symbols of the space groups

Different space groups of the same crystal class are distinguished by their superscript index, for example C_1^1; $D_{2h}^1, D_{2h}^2, \ldots, D_{2h}^{28}$ or $O_h^1, \ldots, O_h^{10}$.

Schoenflies symbols display the space-group symmetry only partly. Therefore, they are nowadays rarely used for the description of the symmetry of crystal structures. In comparison with the Schoenflies symbols, the Hermann–Mauguin symbols are more indicative of the space-group symmetry and that of the crystal structures.

1.4.1.4. Hermann–Mauguin symbols of the space groups

1.4.1.4.1. Introduction

The Hermann–Mauguin symbols, abbreviated as HM symbols in the following sections, were proposed by Hermann (1928, 1931) and Mauguin (1931), and introduced to the *Internationale Tabellen zur Bestimmung von Kristallstrukturen* (1935) according to the decision of the corresponding Programme Committee (Ewald, 1930). There are different kinds of HM symbols of a space group. One distinguishes *short HM symbols*, *full HM symbols* and *extended HM symbols*. The *full HM symbols* will be the basis of this description. They form the most transparent kind of HM symbols and their use will minimize confusion, especially for those who are new to crystallography.

As the name suggests, the *short HM symbols* are mostly shortened versions of the full HM symbols: some symmetry information of the full HM symbols is omitted such that these symbols are more convenient in daily use. The full HM symbol can be reconstructed from the short symbol. In the *extended HM symbols* the symmetry of the space group is listed in a more complete fashion (*cf.* Section 1.5.4). They are rarely used in crystallographic practice.

In the next section general features of the HM symbols will be discussed. (For a detailed discussion of the HM symbols and their dependence on the crystal system to which the space group belongs, see Section 1.4.1.4 of *IT* A.)

1.4.1.4.2. General aspects

The Hermann–Mauguin symbol for a space group consists of a sequence of letters and numbers, here called the *constituents of the HM symbol*. The first constituent is always a symbol for the conventional cell of the translation lattice of the space group (*cf.* Section 1.3.2.1 for the definition of the translation lattice); the following constituents, namely rotations, screw rotations, rotoinversions, reflections and glide reflections, are marked by conventional symbols, *cf.* Table 2.1.2.1.[1] Together with the generating translations of the lattice, the set of these symmetry operations forms a *set of generating symmetry operations* (not necessarily minimal) of the space group. The space group can thus be generated from its HM symbol.

The symmetry operations of the constituents are referred to the basis that is used conventionally for the crystal system of the space group. The kind of symmetry operation can be read from its symbol; the orientation of its geometric element, *cf.* de Wolff *et al.* (1989, 1992), *i.e.* its invariant axis or plane normal, can be concluded from the position of the corresponding constituent in the HM symbol. The origin is not specified. It is chosen by the user, who selects it in such a way that the matrices of the symmetry operations appear in the most convenient form. This is often, but not necessarily, the conventional origin chosen in the space-group tables of *IT* A (and of this Teaching Edition). The choice of a different origin may make other tasks, *e.g.* the derivation of the space group from its generators, particularly easy and transparent.

The first constituent (the lattice symbol) characterizes the lattice of the space group referred to the conventional coordinate system. (Each lattice can be referred to a lattice basis, also called a *primitive basis*: the lattice vectors have only integer coefficients and the lattice is called a *primitive lattice*.) Lattice vectors with non-integer coefficients can occur if the lattice is referred to a non-primitive basis. In this way similarities and relations between different space-group types are emphasized.

The lattice symbol of a primitive basis consists of an upper-case letter P (**p**rimitive). Lattices with conventional non-primitive bases are called *centred lattices*, *cf.* Section 1.3.2.4 and Table 2.1.1.2. For these other letters are used: if the **ab** plane of the unit cell is centred with a lattice vector $\frac{1}{2}(\mathbf{a}+\mathbf{b})$, the letter is C; for **ca** centring [$\frac{1}{2}(\mathbf{c}+\mathbf{a})$ as additional *centring vector*] the letter is B, and A is the letter for centring the **bc** plane of the unit cell by $\frac{1}{2}(\mathbf{b}+\mathbf{c})$. The letter is F for centring all side faces of the cell with centring vectors $\frac{1}{2}(\mathbf{a}+\mathbf{b})$, $\frac{1}{2}(\mathbf{c}+\mathbf{a})$ and $\frac{1}{2}(\mathbf{b}+\mathbf{c})$. It is I (German: ***i**nnenzentriert*) for body centring by the vector $\frac{1}{2}(\mathbf{a}+\mathbf{b}+\mathbf{c})$ and R for the rhombohedral centring of the hexagonal cell by the vectors $\frac{1}{3}(2\mathbf{a}+\mathbf{b}+\mathbf{c})$ and $\frac{1}{3}(\mathbf{a}+2\mathbf{b}+2\mathbf{c})$. In 1985, the letter S was introduced as a setting-independent 'centring symbol' for monoclinic and orthorhombic Bravais lattices (*cf.* de Wolff *et al.*, 1985).

To describe the structure of the HM symbols the introduction of the term *symmetry direction* is useful.

Definition

A direction is called a *symmetry direction* of a crystal structure if it is parallel to an axis of rotation, screw rotation or rotoinversion or if it is parallel to the normal of a reflection or glide-reflection plane. A symmetry direction is thus the direction of the geometric element of a symmetry operation when the normal of a symmetry plane is used for the description of its orientation.

The corresponding symmetry operations [the *element set* of de Wolff *et al.* (1989 & 1992)] specify the type of the symmetry direction. The symmetry direction is always a lattice direction of the space group; the shortest lattice vector in the symmetry direction will be called **q**.

If **q** represents both a rotation or screw rotation and a reflection or glide reflection, then their symbols are connected in the HM symbol by a slash '/', *e.g.* $2/m$ or $4_1/a$ *etc.*

The symmetry directions of a space group form *sets of equivalent symmetry directions* under the symmetry of the space group. For example, in a cubic space group the **a**, **b** and **c** axes are equivalent and form the set of six directions $\langle 100\rangle$: [100], $[\bar{1}00]$, [010] *etc.* Another set of equivalent directions is formed by the eight space diagonals $\langle 111\rangle$: [111], $[1\bar{1}\bar{1}]$, ... If there are twofold rotations around the twelve face diagonals $\langle 110\rangle$, as in the space group of the crystal structure of NaCl, $\langle 110\rangle$ forms a third set of 12 symmetry directions.[2]

Instead of listing the symmetry operations (element set) for each symmetry direction of a set of symmetry directions, it is sufficient to choose one *representative direction of the set*. In the HM symbol, generators for the element set of each representative direction are listed.

It can be shown that there are zero (triclinic space groups), one (monoclinic), up to two (trigonal and rhombohedral) or up to three (most other space groups) sets of symmetry directions in each space group and thus zero, one, two or three representative symmetry directions.

The non-translation generators of a symmetry direction may include only one kind of symmetry operation, *e.g.* for twofold rotations 2 in space group $P121$, but they may also include several symmetry operations, *e.g.* 2, 2_1, m and a in space group $C12/m1$. To search for such directions it is helpful simply to look at the space-group diagrams to find out whether more than one kind of symmetry operation belongs to the generators of a symmetry direction. In general, only the simplest symbols are listed (*simplest-operation rule*): if we use '>' to mean '*has priority*', then pure rotations > screw rotations; pure rotations > rotoinversions; reflection $m > a, b, c > n$.[3] The space group mentioned above is conventionally called $C12/m1$ and not $C12_1/m1$ or $C12/a1$ or $C12_1/a1$.

There are two exceptions to the 'simplest symmetry operation' rule. If the I centring is added to the P space groups of the crystal

[1] According to the recommendations of the International Union of Crystallography Ad Hoc Committee on the Nomenclature of Symmetry (de Wolff *et al.*, 1992), the characters appearing after the lattice letter in the HM symbol of a space group should represent symmetry elements, which is reflected, for example, in the introduction of the '*e*-glide' notation in the HM space-group symbols. To avoid misunderstandings, it is worth noting that in the following discussion of the HM symbolism, the author preferred to keep strictly to the original idea according to which the characters of the HM symbols were meant to represent (generating) symmetry operations of the space group, and not symmetry elements.

[2] The numbers listed are those for bipolar directions, for which direction and opposite direction are equivalent. For the corresponding polar directions in cubic space groups only the four equivalent polar directions $\langle 111\rangle$ or $\langle\bar{1}\bar{1}\bar{1}\rangle$ of the tetrahedron occur.

[3] The 'symmetry-element' interpretation of the constituents of the HM symbols (*cf.* footnote 1) results in the following modification of the 'simplest-operation' rule [known as the 'priority rule', *cf.* Section 4.1.2.3 of *International Tables for Crystallography*, Volume A (2002) (referred to as *IT* A5)]: When more than one kind of symmetry element exists in a given direction, the choice of the corresponding symbols in the space-group symbol is made in order of descending priority $m > e > a, b, c > n$, and rotation axes before screw axes.

Table 1.4.1.1
The structure of the Hermann–Mauguin symbols for the space groups

The positions of the representative symmetry directions for the different crystal systems are given. The description of the non-translational part of the HM symbol is always preceded by the lattice symbol, which in conventional settings is *P*, *A*, *B*, *C*, *F*, *I* or *R*. For monoclinic **b** setting and monoclinic **c** setting, *cf.* Section 1.5.4 (see Sections 1.4.1.4.4 and 1.5.4.3 of *IT* A for more details); the primitive hexagonal lattice is called *H* in this table.

Crystal system	First position	Second position	Third position
Triclinic (anorthic)	1 or $\bar{1}$	—	—
Monoclinic **b** setting Monoclinic **c** setting	1 1	**b** 1	1 **c**
Orthorhombic	**a**	**b**	**c**
Tetragonal	**c**	**a**	**a** − **b**
Trigonal *H* lattice	**c** **c**	**a** or 1	1 **a** − **b**
Trigonal, *R* lattice, hexagonal coordinates Trigonal, *R* lattice, rhombohedral coordinates	$\mathbf{c}_H$ $\mathbf{a}_R + \mathbf{b}_R + \mathbf{c}_R$	$\mathbf{a}_H$ or $\mathbf{a}_R - \mathbf{b}_R$	— —
Hexagonal	**c**	**a**	**a** − **b**
Cubic	**c**	**a** + **b** + **c**	**a** − **b**

class 222, one obtains two different space groups with an *I* lattice, each has 2 and 2_1 operations in each of the symmetry directions. One space group is derived by adding the *I* centring to the space group *P*222, the other is obtained by adding the *I* centring to a space group $P2_12_12_1$. In the first case the twofold axes intersect, in the second they do not. According to the rules both should get the HM symbol *I*222, but only the space group generated from *P*222 is named *I*222, whereas the space group generated from $P2_12_12_1$ is called $I2_12_12_1$.

The second exception to the 'simplest symmetry operation' rule occurs for the pair of cubic groups *I*23 and $I2_13$. In both space groups twofold rotations and screw rotations around **a**, **b** and **c** occur simultaneously. In *I*23 the rotation axes intersect, in $I2_13$ they do not. The first space group can be generated by adding the *I*-centring to the space group *P*23, the second is obtained by adding the *I*-centring to the space group $P2_13$.

The position of a plane is fixed by one parameter if its orientation is known. On the other hand, fixing an axis of known direction needs two parameters. Glide components also show two-dimensional variability, whereas there is only one parameter of a screw component. Therefore, reflections and glide reflections can better express the geometric relations between the symmetry operations than can rotations and screw rotations; reflections and glide reflections are more important for HM symbols than are rotations and screw rotations. The latter are frequently omitted to form short HM symbols from the full ones.

The second part of the *full HM symbol* of a space group consists of one position for each of up to three representative symmetry directions. To each position belong the generating symmetry operations of their representative symmetry direction. The position is thus occupied either by a rotation, screw rotation or rotoinversion and/or by a reflection or glide reflection.

The representative symmetry directions are different in the different crystal systems. For example, the directions of the basis vectors **a**, **b** and **c** are symmetry independent in orthorhombic crystals and are thus all representative, whereas **a** and **b** are symmetry equivalent and thus dependent in tetragonal crystals. All three directions are symmetry equivalent in cubic crystals; they belong to the same set and are represented by one of the directions. Therefore, the symmetry directions and their sequence in the HM symbols depend on the crystal system to which the crystal and thus its space group belongs.

Table 1.4.1.1 gives the positions of the representative lattice-symmetry directions in the HM symbols for the different crystal systems.

Examples of full HM symbols are (from triclinic to cubic) $P\bar{1}$, $P12/c1$, $A112/m$, $F2/d\,2/d\,2/d$, $I4_1/a$, $P4/m\,2_1/n\,2/c$, $P\bar{3}$, $P3m1$, $P3_112$, $R\bar{3}2/c$, $P6_3/m$, $P6_122$ and $F4_132$.

Monoclinic space groups have exactly one symmetry direction, often called the monoclinic axis. The *b* axis is the symmetry direction of the (most frequently used) conventional setting, called the *b*-axis setting. Another conventional setting has *c* as its symmetry direction (*c*-axis setting).

There are crystal systems, for example tetragonal, for which the high-symmetry space groups display symmetry in all symmetry directions whereas lower-symmetry space groups display symmetry in only some of them. In such cases, the symmetry of the 'empty' symmetry direction is denoted by the constituent 1 or it is simply omitted. For example, instead of three symmetry directions in *P*4*mm*, there is only one in $I4_1/a11$, for which the HM symbol is usually written $I4_1/a$. However, in some trigonal space groups the designation of a symmetry direction by '1' ($P3_112$) is necessary to maintain the uniqueness of the HM symbols.[4]

The HM symbols can not only describe the space groups in their conventional settings but they can also indicate the setting of the space group relative to the conventional coordinate system mentioned in Section 1.4.1.3.1. For example, the orthorhombic space group $P2/m\,2/n\,2_1/a$ may appear as $P2/n\,2/m\,2_1/b$ or $P2/n\,2_1/c\,2/m$ or $P2_1/c\,2/n\,2/m$ or $P2_1/b\,2/m\,2/n$ or $P2/m\,2_1/a\,2/n$ depending on its orientation relative to the conventional coordinate basis. On the one hand this is an advantage, because the HM symbols include some indication of the orientation of the space group and form a more powerful tool than being just a space-group nomenclature. On the other hand, it is sometimes not easy to recognize the space-group type that is described by an unconventional HM symbol.

[4] In the original HM symbols the constituent '1' was avoided by the use of different centred cells.

The full HM symbols describe the symmetry of a space group in a transparent way, but they are redundant. They can be shortened to the *short HM symbols* such that the set of generators is reduced to a necessary set. The *conventional short HM symbols* still provide a unique description and enable the generation of the space group. For the monoclinic space groups with their many conventional settings they are not variable and are taken as standard for their space-group types. Monoclinic short HM symbols may look quite different from the full HM symbol, *e.g.* Cc instead of $A1n1$ or $I1a1$ or $B11n$ or $I11b$.

The *extended HM symbols* display the additional symmetry that is often generated by lattice centrings. The full HM symbol denotes only the simplest symmetry operations for each symmetry direction, by the 'simplest symmetry operation' rule; the other operations can be found in the extended symbols, which are discussed in Section 1.5.4; see also Table 1.5.4.1, where the extended symbols for the monoclinic and orthorhombic space groups are listed. (For the extended symbols for all the plane and space groups, see Tables 1.5.4.3 and 1.5.4.4 of *IT* A.)

From the HM symbol of the space group, the full or short *HM symbol for a crystal class* of a space group is obtained easily: one omits the lattice symbol, cancels all screw components such that only the symbol for the rotation is left and replaces any letter for a glide reflection by the letter m for a reflection. Examples are $P2_1/b2_1/a2/m \rightarrow 2/m2/m2/m$ and $I4_1/a11 \rightarrow 4/m$.

1.4.1.5. Hermann–Mauguin symbols of the plane groups

The principles of the HM symbols for space groups are retained in the HM symbols for plane groups (also known as *wallpaper groups*). The rotation axes along **c** of three dimensions are replaced by *rotation points* in the **ab** plane; the possible orders of rotations are the same as in three-dimensional space: 2, 3, 4 and 6. The lattice (sometimes called *net*) of a plane group is spanned by the two basis vectors **a** and **b**, and is designated by a lower-case letter. The choice of a lattice basis, *i.e.* of a minimal cell, leads to a primitive lattice p, in addition a c-centred lattice is conventionally used. The nets are listed in Table 3.1.2.1 of *IT* A. The reflections and glide reflections through planes of the space groups are replaced by *reflections and glide reflections through lines*. Glide reflections are called g independent of the direction of the glide line. The arrangement of the constituents in the HM symbol is displayed in Table 1.4.1.2.

Short HM symbols are used only if there is at most one symmetry direction, *e.g.* $p411$ is replaced by $p4$ (no symmetry direction), $p1m1$ is replaced by pm (one symmetry direction) *etc.*

There are four crystal systems of plane groups, *cf.* Table 2.1.1.1. The analogue of the triclinic crystal system is called *oblique*, the analogues of the monoclinic and orthorhombic crystal systems are *rectangular*. Both have rotations of order 2 at most. The presence of reflection or glide reflection lines in the rectangular crystal system allows one to choose a rectangular basis with one basis vector perpendicular to a symmetry line and one basis vector parallel to it. The *square* crystal system is analogous to the tetragonal crystal system for space groups by the occurrence of fourfold rotation points and a square net. Plane groups with threefold and sixfold rotation points are united in the *hexagonal* crystal system with a hexagonal net.

Plane groups occur as sections and projections of the space groups, *cf.* Section 1.4.5. In order to maintain the relations to the space groups, the symmetry directions of the symmetry lines are determined by their normals, not by the directions of the lines themselves. This is important because the normal of the line, not the direction of the line itself, determines the position in the HM symbol.

Table 1.4.1.2
The structure of the Hermann–Mauguin symbols for the plane groups

The positions of the representative symmetry directions for the crystal systems are given. The lattice symbol and the maximal order of rotations around a point are followed by two positions for symmetry directions.

Crystal system	Lattice(s)	First position	Second position	Third position
Oblique	p	1 or 2	—	—
Rectangular	p, c	1 or 2	**a**	**b**
Tetragonal	p	4	**a**	**a** − **b**
Hexagonal	p	3	**a** or	1
		3	1	**a** − **b**
		6	**a**	**a** − **b**

1.4.2. Descriptions of space-group symmetry operations

BY MOIS I. AROYO, GERVAIS CHAPUIS, BERND SOUVIGNIER AND A. M. GLAZER

One of the aims of the space-group tables of *IT* A is to represent the symmetry operations of each of the 17 plane groups and 230 space groups. The following sections offer a short description of the symbols of the symmetry operations, their listings and their graphical representations as found in the space-group tables of *IT* A (see also the space-group tables of Part 2). For a detailed discussion of crystallographic symmetry operations and their matrix–column presentation $(\boldsymbol{W}, \boldsymbol{w})$ the reader is referred to Chapter 1.2.

1.4.2.1. Symbols for symmetry operations

Given the analytical description of the symmetry operations by matrix–column pairs $(\boldsymbol{W}, \boldsymbol{w})$, their geometric meaning can be determined following the procedure discussed in Section 1.2.2.3. The notation scheme of the symmetry operations applied in the space-group tables was designed by W. Fischer and E. Koch, and the following description of the symbols partly reproduces the explanations by the authors given in Section 11.1.2 of *IT* A5. Further explanations of the symbolism and examples are presented in Section 2.1.3.9.

The symbol of a symmetry operation indicates the type of the operation, its screw or glide component (if relevant) and the location of the corresponding geometric element (*cf.* Section 1.2.3 and Table 1.2.3.1 for a discussion of geometric elements). The symbols of the symmetry operations explained below are based on the Hermann–Mauguin symbols (*cf.* Section 1.4.1.4), modified and supplemented where necessary.

The symbol for the *identity* mapping is 1.

A *translation* is symbolized by the letter t followed by the components of the translation vector between parentheses. *Example*: $t(\frac{1}{2}, \frac{1}{2}, 0)$ represents a translation by a vector $\frac{1}{2}\mathbf{a} + \frac{1}{2}\mathbf{b}$, *i.e.* a C centring.

A *rotation* is symbolized by a number $n = 2, 3, 4$ or 6 (according to the rotation angle $360°/n$) and a superscript + or −, which specifies the sense of rotation ($n > 2$). The symbol of rotation is followed by the location of the rotation axis. *Example*: $4^+\ 0, y, 0$

Positions

Multiplicity,
Wyckoff letter,
Site symmetry

Coordinates

4	*e*	1	(1) x,y,z	(2) $\bar{x},y+\frac{1}{2},\bar{z}+\frac{1}{2}$	(3) $\bar{x},\bar{y},\bar{z}$	(4) $x,\bar{y}+\frac{1}{2},z+\frac{1}{2}$

Symmetry operations

(1) 1	(2) $2(0,\frac{1}{2},0)$ $0,y,\frac{1}{4}$	(3) $\bar{1}$ $0,0,0$	(4) c $x,\frac{1}{4},z$

Figure 1.4.2.1

General-position and symmetry-operations blocks for the space group $P2_1/c$, No. 14 (unique axis b, cell choice 1). The coordinate triplets of the general position, numbered from (1) to (4), correspond to the four coset representatives of the decomposition of $P2_1/c$ with respect to its translation subgroup, *cf.* Table 1.4.2.2. The entries of the symmetry-operations block numbered from (1) to (4) describe geometrically the symmetry operations represented by the four coordinate triplets of the general-position block.

indicates a rotation of 90° about the line $0, y, 0$ that brings point $0, 0, 1$ onto point $1, 0, 0$, *i.e.* a counter-clockwise rotation (or rotation in the mathematically *positive sense*) if viewed from point $0, 1, 0$ to point $0, 0, 0$.

A *screw rotation* is symbolized in the same way as a pure rotation, but with the screw part added between parentheses. *Example*: $3^{-}(0, 0, \frac{1}{3})$ $\frac{2}{3}, \frac{1}{3}, z$ indicates a clockwise rotation of 120° around the line $\frac{2}{3}, \frac{1}{3}, z$ (or rotation in the mathematically *negative sense*) if viewed from the point $\frac{2}{3}, \frac{1}{3}, 1$ towards $\frac{2}{3}, \frac{1}{3}, 0$, combined with a translation of $\frac{1}{3}\mathbf{c}$.

A *reflection* is symbolized by the letter m, followed by the location of the mirror plane.

A *glide reflection* in general is symbolized by the letter g, with the glide part given between parentheses, followed by the location of the glide plane. These specifications characterize every glide reflection uniquely. Exceptions are the traditional symbols a, b, c, n and d that are used instead of g. In the case of a glide plane a, b or c, the explicit statement of the glide vector is omitted if it is $\frac{1}{2}\mathbf{a}$, $\frac{1}{2}\mathbf{b}$ or $\frac{1}{2}\mathbf{c}$, respectively. *Examples*: a $x, y, \frac{1}{4}$ means a glide reflection with glide vector $\frac{1}{2}\mathbf{a}$ and through a plane $x, y, \frac{1}{4}$; $d(\frac{1}{4}, \frac{1}{4}, \frac{3}{4})$ $x, x-\frac{1}{4}, z$ denotes a glide reflection with glide part $(\frac{1}{4}, \frac{1}{4}, \frac{3}{4})$ and the glide plane d at $x, x-\frac{1}{4}, z$.

An *inversion* is symbolized by $\bar{1}$ followed by the location of the inversion centre.

A *rotoinversion* is symbolized, in analogy with a rotation, by $\bar{3}$, $\bar{4}$ or $\bar{6}$ and the superscript + or −, again followed by the location of the (rotoinversion) axis. Note that angle and sense of rotation refer to the pure rotation and not to the combination of rotation and inversion. In addition, the location of the inversion point is given by the appropriate coordinate triplet after a semicolon. *Example*: $\bar{4}^{+}$ $0, \frac{1}{2}, z; 0, \frac{1}{2}, \frac{1}{4}$ means a 90° rotoinversion with axis at $0, \frac{1}{2}, z$ and inversion point at $0, \frac{1}{2}, \frac{1}{4}$. The rotation is performed in the mathematically positive sense when viewed from $0, \frac{1}{2}, 1$ towards $0, \frac{1}{2}, 0$. Therefore, the rotoinversion maps point $0, 0, 0$ onto point $-\frac{1}{2}, \frac{1}{2}, \frac{1}{2}$.

The notation scheme is extensively applied in the symmetry-operations blocks of the space-group descriptions in the tables of Chapter 2.3 of *IT* A (see also the space-group tables of Part 2). The numbering of the entries of the symmetry-operations block corresponds to that of the coordinate triplets of the general position, and in space groups with primitive cells the two lists contain the same number of entries. As an example consider the symmetry-operations block of the space group $P2_1/c$ shown in Fig. 1.4.2.1. The four entries correspond to the four coordinate triplets of the general-position block of the group and provide the geometric description of the symmetry operations chosen as coset representatives of the decomposition of $P2_1/c$ with respect to its translation subgroup.

For space groups with conventional *centred* cells, there are several (2, 3 or 4) blocks of symmetry operations: one block for each of the translations listed below the subheading 'Coordinates'. Consider, for example, the four symmetry-operations blocks of the space group *Fmm*2 (42) reproduced in Fig. 1.4.2.2. They correspond to the four sets of coordinate triplets of the general position obtained by the translations $t(0, 0, 0)$, $t(0, \frac{1}{2}, \frac{1}{2})$, $t(\frac{1}{2}, 0, \frac{1}{2})$ and $t(\frac{1}{2}, \frac{1}{2}, 0)$, *cf.* Fig. 1.4.2.2. The numbering scheme of the entries in the different symmetry-operations blocks follows that of the general position. For example, the geometric description of entry (4) in the symmetry-operations block under the heading 'For $(\frac{1}{2}, \frac{1}{2}, 0)+$ set' of *Fmm*2 corresponds to the coordinate triplet $\bar{x}+\frac{1}{2}, y+\frac{1}{2}, z$, which is obtained by adding $t(\frac{1}{2}, \frac{1}{2}, 0)$ to the translation part of the printed coordinate triplet (4) $\bar{x}, y, z$ (*cf.* Fig. 1.4.2.2).

1.4.2.2. Seitz symbols of symmetry operations

Apart from the notation for the geometric interpretation of the matrix–column representation of symmetry operations $(\boldsymbol{W}, \boldsymbol{w})$ discussed in detail in the previous section, there is another

Positions

Multiplicity,
Wyckoff letter,
Site symmetry

Coordinates

$(0,0,0)+$ $(0,\frac{1}{2},\frac{1}{2})+$ $(\frac{1}{2},0,\frac{1}{2})+$ $(\frac{1}{2},\frac{1}{2},0)+$

16	*e*	1	(1) x,y,z	(2) $\bar{x},\bar{y},z$	(3) $x,\bar{y},z$	(4) $\bar{x},y,z$

Symmetry operations

For $(0,0,0)+$ set			
(1) 1	(2) 2 $0,0,z$	(3) m $x,0,z$	(4) m $0,y,z$
For $(0,\frac{1}{2},\frac{1}{2})+$ set			
(1) $t(0,\frac{1}{2},\frac{1}{2})$	(2) $2(0,0,\frac{1}{2})$ $0,\frac{1}{4},z$	(3) c $x,\frac{1}{4},z$	(4) $n(0,\frac{1}{2},\frac{1}{2})$ $0,y,z$
For $(\frac{1}{2},0,\frac{1}{2})+$ set			
(1) $t(\frac{1}{2},0,\frac{1}{2})$	(2) $2(0,0,\frac{1}{2})$ $\frac{1}{4},0,z$	(3) $n(\frac{1}{2},0,\frac{1}{2})$ $x,0,z$	(4) c $\frac{1}{4},y,z$
For $(\frac{1}{2},\frac{1}{2},0)+$ set			
(1) $t(\frac{1}{2},\frac{1}{2},0)$	(2) 2 $\frac{1}{4},\frac{1}{4},z$	(3) a $x,\frac{1}{4},z$	(4) b $\frac{1}{4},y,z$

Figure 1.4.2.2

General-position and symmetry-operations blocks as given in the space-group tables for space group *Fmm*2 (42). The numbering scheme of the entries in the different symmetry-operations blocks follows that of the general position.

notation which has been adopted and is widely used by solid-state physicists and chemists. This is the so-called Seitz notation $\{\boldsymbol{R}|\boldsymbol{v}\}$ introduced by Seitz in a series of papers on the matrix-algebraic development of crystallographic groups (Seitz, 1935).

Seitz symbols $\{\boldsymbol{R}|\boldsymbol{v}\}$ reflect the fact that space-group operations are affine mappings and are essentially shorthand descriptions of the matrix–column representations of the symmetry operations of the space groups. They consist of two parts: a rotation (or linear) part $\boldsymbol{R}$ and a translation part $\boldsymbol{v}$. The Seitz symbol is specified between braces and the rotational and the translational parts are separated by a vertical line. The translation parts $\boldsymbol{v}$ correspond exactly to the columns $\boldsymbol{w}$ of the coordinate triplets of the general-position blocks of the space-group tables. The rotation parts $\boldsymbol{R}$ consist of symbols that specify (i) the type and the order of the symmetry operation, and (ii) the orientation of the corresponding symmetry element with respect to the basis. The orientation is denoted by the direction of the axis for rotations or rotoinversions, or the direction of the normal to reflection planes. (Note that in the latter case this is different from the way the orientation of reflection planes is given in the symmetry-operations block.)

The linear parts of Seitz symbols are denoted in many different ways in the literature (Litvin & Kopsky, 2011). According to the conventions approved by the Commission of Crystallographic Nomenclature of the International Union of Crystallography (Glazer *et al.*, 2014) the symbol $\boldsymbol{R}$ is 1 and $\bar{1}$ for the identity and the inversion, m for reflections, the symbols 2, 3, 4 and 6 are used for rotations and $\bar{3}$, $\bar{4}$ and $\bar{6}$ for rotoinversions. For rotations and rotoinversions of order higher than 2, a superscript + or − is used to indicate the sense of the rotation. Subscripts of the symbols $\boldsymbol{R}$ denote the characteristic direction of the operation: for example, the subscripts 100, 010 and $1\bar{1}0$ refer to the directions [100], [010] and $[1\bar{1}0]$, respectively.

Examples

(*a*) Consider the coordinate triplets of the general positions of $P2_12_12$ (18):

$$(1)\ x,y,z\quad (2)\ \bar{x},\bar{y},z\quad (3)\ \bar{x}+\tfrac{1}{2},y+\tfrac{1}{2},\bar{z}\quad (4)\ x+\tfrac{1}{2},\bar{y}+\tfrac{1}{2},\bar{z}$$

The corresponding geometric interpretations of the symmetry operations are given by

$$(1)\ 1\quad (2)\ 2\ 0,0,z\quad (3)\ 2(0,\tfrac{1}{2},0)\ \tfrac{1}{4},y,0\quad (4)\ 2(\tfrac{1}{2},0,0)\ x,\tfrac{1}{4},0$$

In Seitz notation the symmetry operations are denoted by

$$(1)\ \{1|0\}\quad (2)\ \{2_{001}|0\}\quad (3)\ \{2_{010}|\tfrac{1}{2},\tfrac{1}{2},0\}\quad (4)\ \{2_{100}|\tfrac{1}{2},\tfrac{1}{2},0\}$$

(*b*) Similarly, the symmetry operations corresponding to the general-position coordinate triplets of $P2_1/c$ (14), *cf.* Fig. 1.4.2.1, in Seitz notation are given as

$$(1)\ \{1|0\}\quad (2)\ \{2_{010}|0,\tfrac{1}{2},\tfrac{1}{2}\}\quad (3)\ \{\bar{1}|0\}\quad (4)\ \{m_{010}|0,\tfrac{1}{2},\tfrac{1}{2}\}$$

The linear parts $\boldsymbol{R}$ of the Seitz symbols of the space-group symmetry operations of tetragonal, orthorhombic, monoclinic and triclinic systems are shown in Table 1.4.2.1 (for the symmetry operations of the rest of the crystal systems, see Tables 1.4.2.1–1.4.2.3 of *IT* A). Each symbol $\boldsymbol{R}$ is specified by the shorthand notation of its (3 × 3) matrix representation (also known as the *Jones' faithful representation symbol*, *cf.* Bradley & Cracknell, 1972), the type of symmetry operation and its orientation as described in the corresponding symmetry-operations block of the space-group tables of Part 2 (and of *IT* A). The sequence of $\boldsymbol{R}$ symbols in Table 1.4.2.1 corresponds to the numbering scheme of the general-position coordinate triplets of the space groups of the crystal class $4/mmm$.

Table 1.4.2.1

Linear parts $\boldsymbol{R}$ of the Seitz symbols $\{\boldsymbol{R}|\boldsymbol{v}\}$ for space-group symmetry operations of tetragonal, orthorhombic, monoclinic and triclinic crystal systems

Each symmetry operation is specified by the shorthand description of the rotation part of its matrix–column presentation, the type of symmetry operation and its characteristic direction.

IT A description				Seitz symbol
No.	Coordinate triplet	Type	Orientation	
1	x,y,z	1		1
2	$\bar{x},\bar{y},z$	2	$0,0,z$	2_{001}
3	$\bar{y},x,z$	4^+	$0,0,z$	4^+_{001}
4	$y,\bar{x},z$	4^-	$0,0,z$	4^-_{001}
5	$\bar{x},y,\bar{z}$	2	$0,y,0$	2_{010}
6	$x,\bar{y},\bar{z}$	2	$x,0,0$	2_{100}
7	$y,x,\bar{z}$	2	$x,x,0$	2_{110}
8	$\bar{y},\bar{x},\bar{z}$	2	$x,\bar{x},0$	$2_{1\bar{1}0}$
9	$\bar{x},\bar{y},\bar{z}$	$\bar{1}$		$\bar{1}$
10	$x,y,\bar{z}$	m	$x,y,0$	m_{001}
11	$y,\bar{x},\bar{z}$	$\bar{4}^+$	$0,0,z$	$\bar{4}^+_{001}$
12	$\bar{y},x,\bar{z}$	$\bar{4}^-$	$0,0,z$	$\bar{4}^-_{001}$
13	$x,\bar{y},z$	m	$x,0,z$	m_{010}
14	$\bar{x},y,z$	m	$0,y,z$	m_{100}
15	$\bar{y},\bar{x},z$	m	$x,\bar{x},z$	m_{110}
16	y,x,z	m	x,x,z	$m_{1\bar{1}0}$

The same symbols $\boldsymbol{R}$ can be used for the construction of Seitz symbols for the symmetry operations of subperiodic layer and rod groups (Litvin & Kopsky, 2014), and magnetic groups, or for the designation of the symmetry operations of the point groups of space groups. [One should note that the Seitz symbols applied in the first and second editions of *IT* E and in the IUCr e-book on magnetic groups (Litvin, 2012) differ from the standard symbols adopted by the Commission of Crystallographic Nomenclature.]

The Seitz symbols for plane groups are constructed following similar rules to those for space groups. The rotation part $\boldsymbol{R}$ is 1 for the identity, m for reflections, and 2, 3, 4 and 6 are used for rotations. The orientation of a reflection line is specified by a subscript indicating the direction of its 'normal'. Obviously, the direction indicators are of no relevance for the rotation points. The linear parts $\boldsymbol{R}$ of the Seitz symbols of the plane-group symmetry operations can be found in Tables 1.4.2.4 and 1.4.2.5 of *IT* A. The same symbols $\boldsymbol{R}$ can be used for the construction of Seitz symbols for the symmetry operations of subperiodic frieze groups (Litvin & Kopsky, 2014).

As illustrated in the examples above, zero translations are normally specified by a single zero in the Seitz symbols, but in cases where it is unclear whether the symbol refers to a space- or a plane-group symmetry operation, an explicit indication of the components of the translation vector is recommended.

From the description given above, it is clear that Seitz symbols can be considered as shorthand modifications of the matrix–column presentation $(\boldsymbol{W},\boldsymbol{w})$ of symmetry operations discussed in detail in Chapter 1.2: the translation parts of $\{\boldsymbol{R}|\boldsymbol{v}\}$ and $(\boldsymbol{W},\boldsymbol{w})$ coincide, while the different (3 × 3) matrices $\boldsymbol{W}$ are represented by the symbols $\boldsymbol{R}$. As a result, the expressions for the product and the inverse of symmetry operations in Seitz notation are rather similar to those of the matrix–column pairs $(\boldsymbol{W}, \boldsymbol{w})$ discussed in detail in Chapter 1.2:

(*a*) product of symmetry operations:

$$\{\boldsymbol{R}_1|\boldsymbol{v}_1\}\{\boldsymbol{R}_2|\boldsymbol{v}_2\} = \{\boldsymbol{R}_1\boldsymbol{R}_2|\boldsymbol{R}_1\boldsymbol{v}_2 + \boldsymbol{v}_1\};$$

(*b*) inverse of a symmetry operation:

$$\{\boldsymbol{R}|\boldsymbol{v}\}^{-1} = \{\boldsymbol{R}^{-1}| - \boldsymbol{R}^{-1}\boldsymbol{v}\}.$$

Similarly, the action of a symmetry operation $\{\boldsymbol{R}|\boldsymbol{v}\}$ on the column of point coordinates $\boldsymbol{x}$ is given by $\{\boldsymbol{R}|\boldsymbol{v}\}\boldsymbol{x} = \boldsymbol{R}\boldsymbol{x} + \boldsymbol{v}$ [*cf.* Chapter 1.2, equation (1.2.2.4)].

The rotation parts of the Seitz symbols partly resemble the geometric-description symbols of symmetry operations described in Section 1.4.2.1 and listed in the symmetry-operation blocks of the space-group tables of Part 2 (and of *IT* A): $\boldsymbol{R}$ contains the information on the type and order of the symmetry operation, and its characteristic direction. The Seitz symbols do not *directly* indicate the location of the symmetry operation, nor its glide or screw component, if any.

1.4.2.3. Symmetry operations and the general position

The classifications of space groups introduced in Chapter 1.3 allow one to reduce the practically unlimited number of possible space groups to a finite number of space-group types. However, each individual space-group type still consists of an infinite number of symmetry operations generated by the set of all translations of the space group. A practical way to represent the symmetry operations of space groups is based on the coset decomposition of a space group with respect to its translation subgroup, which was introduced and discussed in Section 1.3.3.2. For our further considerations, it is important to note that the listings of the general position in the space-group tables can be interpreted in two ways:

(i) Each of the numbered entries lists the coordinate triplets of an image point of a starting point with coordinates x, y, z under a symmetry operation of the space group. This feature of the general position will be discussed in detail in Section 1.4.4.

(ii) Each of the numbered entries of the general position lists a symmetry operation of the space group by the shorthand notation of its matrix–column pair $(\boldsymbol{W}, \boldsymbol{w})$ (*cf.* Section 1.2.2.1). This fact is not as obvious as the more 'crystallographic' aspect described under (i), but its importance becomes evident from the following discussion, where it is shown how to extract the full analytical symmetry information of space groups from the general-position data in the space-group tables of *IT* A (and of Part 2).

With reference to a conventional coordinate system, the set of symmetry operations $\{W\}$ of a space group $\mathcal{G}$ is described by the set of matrix–column pairs $\{(\boldsymbol{W}, \boldsymbol{w})\}$. The set $\mathcal{T}_{\mathcal{G}} = \{(\boldsymbol{I}, \boldsymbol{t})\}$ of all translations forms the *translation subgroup* $\mathcal{T}_{\mathcal{G}} \lhd \mathcal{G}$, which is a normal subgroup of $\mathcal{G}$ of finite index $[i]$. If $(\boldsymbol{W}, \boldsymbol{w})$ is a fixed symmetry operation, then all the products $\mathcal{T}_{\mathcal{G}}(\boldsymbol{W}, \boldsymbol{w}) = \{(\boldsymbol{I}, \boldsymbol{t})(\boldsymbol{W}, \boldsymbol{w})\} = \{(\boldsymbol{W}, \boldsymbol{w} + \boldsymbol{t})\}$ of translations with $(\boldsymbol{W}, \boldsymbol{w})$ have the same rotation part $\boldsymbol{W}$. Conversely, every symmetry operation W of $\mathcal{G}$ with the same matrix part $\boldsymbol{W}$ is represented in the set $\mathcal{T}_{\mathcal{G}}(\boldsymbol{W}, \boldsymbol{w})$. The infinite set of symmetry operations $\mathcal{T}_{\mathcal{G}}(\boldsymbol{W}, \boldsymbol{w})$ is called a coset of the right coset decomposition of $\mathcal{G}$ with respect to $\mathcal{T}_{\mathcal{G}}$, and $(\boldsymbol{W}, \boldsymbol{w})$ its coset representative. In this way, the symmetry operations of $\mathcal{G}$ can be distributed into a finite set of infinite cosets, the elements of which are obtained by the combination of a coset representative $(\boldsymbol{W}_j, \boldsymbol{w}_j)$ and the infinite set $\mathcal{T}_{\mathcal{G}} = \{(\boldsymbol{I}, \boldsymbol{t})\}$ of translations (*cf.* Section 1.3.3.2):

$$\mathcal{G} = \mathcal{T}_{\mathcal{G}} \cup \mathcal{T}_{\mathcal{G}}(\boldsymbol{W}_2, \boldsymbol{w}_2) \cup \cdots \cup \mathcal{T}_{\mathcal{G}}(\boldsymbol{W}_m, \boldsymbol{w}_m) \cup \cdots \cup \mathcal{T}_{\mathcal{G}}(\boldsymbol{W}_i, \boldsymbol{w}_i), \tag{1.4.2.1}$$

where $(\boldsymbol{W}_1, \boldsymbol{w}_1) = (\boldsymbol{I}, \boldsymbol{o})$ is omitted. Obviously, the coset representatives $(\boldsymbol{W}_j, \boldsymbol{w}_j)$ of the decomposition $(\mathcal{G} : \mathcal{T}_{\mathcal{G}})$ represent in a clear and compact way the infinite number of symmetry operations of the space group $\mathcal{G}$. Each coset in the decomposition $(\mathcal{G} : \mathcal{T}_{\mathcal{G}})$ is characterized by its linear part $\boldsymbol{W}_j$ and its entries differ only by lattice translations. The translations $(\boldsymbol{I}, \boldsymbol{t}) \in \mathcal{T}_{\mathcal{G}}$ form the first coset with the identity $(\boldsymbol{I}, \boldsymbol{o})$ as a coset representative. The symmetry operations with rotation part $\boldsymbol{W}_2$ form the second coset *etc.* The number of cosets equals the number of different matrices $\boldsymbol{W}_j$ of the symmetry operations of the space group. This number $[i]$ is always finite and is equal to the order of the point group $\mathcal{P}_{\mathcal{G}}$ of the space group (*cf.* Section 1.3.3.2).

For each space group, a set of coset representatives $\{(\boldsymbol{W}_j, \boldsymbol{w}_j), 1 \le j \le [i]\}$ of the decomposition $(\mathcal{G} : \mathcal{T}_{\mathcal{G}})$ is listed under the general-position block of the space-group tables. In general, any element of a coset may be chosen as a coset representative. For convenience, the representatives listed in the space-group tables are always chosen such that the components $w_{j,k}$, $k = 1, 2, 3$, of the translation parts $\boldsymbol{w}_j$ fulfil $0 \le w_{j,k} < 1$ (by subtracting integers). To save space, each matrix–column pair $(\boldsymbol{W}_j, \boldsymbol{w}_j)$ is represented by the corresponding *coordinate triplet* (*cf.* Section 1.2.2.1 for the shorthand notation of matrix–column pairs).

Example

The right coset decomposition of $P2_1/c$, No. 14 (unique axis b, cell choice 1) with respect to its translation subgroup is shown in Table 1.4.2.2. All possible symmetry operations of $P2_1/c$ are distributed into four cosets:

The first column represents the infinitely many translations $t = (\boldsymbol{I}, \boldsymbol{t}) = x + u_1, y + u_2, z + u_3 = \{1|u_1, u_2, u_3\}$ of the translation subgroup $\mathcal{T}$ of $P2_1/c$. The numbers u_1, u_2 and u_3 are positive or negative integers. The identity operation $(\boldsymbol{I}, \boldsymbol{o})$ is usually chosen as a coset representative.

The third coset of the decomposition $(\mathcal{G} : \mathcal{T}_{\mathcal{G}})$ represents the infinite set of inversions $(-\boldsymbol{I}, \boldsymbol{t}) = \bar{x} + u_1, \bar{y} + u_2, \bar{z} + u_3 = \{\bar{1}|u_1, u_2, u_3\}$ of the space group $P2_1/c$ with inversion centres located at $u_1/2, u_2/2, u_3/2$ (*cf.* Section 1.2.2.3 for the determination of the location of the inversion centres). The inversion in the origin, *i.e.* $\bar{x}, \bar{y}, \bar{z} = \{\bar{1}|0\}$, is taken as a coset representative.

The coset representative of the second coset is the twofold screw rotation $\{2_{010}|0, \frac{1}{2}, \frac{1}{2}\}$ around the line $0, y, \frac{1}{4}$, followed by its infinite combinations with all lattice translations: $\bar{x} + u_1, y + \frac{1}{2} + u_2, \bar{z} + \frac{1}{2} + u_3 = \{2_{010}|u_1, \frac{1}{2} + u_2, \frac{1}{2} + u_3\}$. These are twofold screw rotations around the lines $u_1/2, y, u_3/2 + \frac{1}{4}$ with screw components $\begin{pmatrix} 0 \\ \frac{1}{2} + u_2 \\ 0 \end{pmatrix}$.

The symmetry operations of the fourth column represented by $x + u_1$, $\bar{y} + \frac{1}{2} + u_2$, $z + \frac{1}{2} + u_3 = \{m_{010}|u_1, \frac{1}{2} + u_2, \frac{1}{2} + u_3\}$ correspond to glide reflections with glide components $\begin{pmatrix} u_1 \\ 0 \\ \frac{1}{2} + u_3 \end{pmatrix}$ through the (infinite) set of glide planes at $x, \frac{1}{4}, z$; $x, \frac{3}{4}, z$; $x, \frac{5}{4}, z$; ...; $x, (2u_2 + 1)/4, z$. As usual, the symmetry operation with $u_1 = u_2 = u_3 = 0$, *i.e.* $x, \bar{y} + \frac{1}{2}, z + \frac{1}{2} = \{m_{010}|0, \frac{1}{2}, \frac{1}{2}\}$, is taken as a coset representative of the coset of glide reflections.

Table 1.4.2.2
Right coset decomposition of space group $P2_1/c$, No. 14 (unique axis b, cell choice 1) with respect to the normal subgroup of translations $\mathcal{T}$

The numbers u_1, u_2 and u_3 are positive or negative integers.

x	y	z	$\bar{x}$	$y+\frac{1}{2}$	$\bar{z}+\frac{1}{2}$	$\bar{x}$	$\bar{y}$	$\bar{z}$	x	$\bar{y}+\frac{1}{2}$	$z+\frac{1}{2}$
$x+1$	y	z	$\bar{x}+1$	$y+\frac{1}{2}$	$\bar{z}+\frac{1}{2}$	$\bar{x}+1$	$\bar{y}$	$\bar{z}$	$x+1$	$\bar{y}+\frac{1}{2}$	$z+\frac{1}{2}$
$x+2$	y	z	$\bar{x}+2$	$y+\frac{1}{2}$	$\bar{z}+\frac{1}{2}$	$\bar{x}+2$	$\bar{y}$	$\bar{z}$	$x+2$	$\bar{y}+\frac{1}{2}$	$z+\frac{1}{2}$
	⋮			⋮			⋮			⋮	
x	$y+1$	z	$\bar{x}$	$y+\frac{3}{2}$	$\bar{z}+\frac{1}{2}$	$\bar{x}$	$\bar{y}+1$	$\bar{z}$	x	$\bar{y}+\frac{3}{2}$	$z+\frac{1}{2}$
$x+1$	$y+1$	z	$\bar{x}+1$	$y+\frac{3}{2}$	$\bar{z}+\frac{1}{2}$	$\bar{x}+1$	$\bar{y}+1$	$\bar{z}$	$x+1$	$\bar{y}+\frac{3}{2}$	$z+\frac{1}{2}$
$x+2$	$y+1$	z	$\bar{x}+2$	$y+\frac{3}{2}$	$\bar{z}+\frac{1}{2}$	$\bar{x}+2$	$\bar{y}+1$	$\bar{z}$	$x+2$	$\bar{y}+\frac{3}{2}$	$z+\frac{1}{2}$
	⋮			⋮			⋮			⋮	
x	$y+2$	z	$\bar{x}$	$y+\frac{5}{2}$	$\bar{z}+\frac{1}{2}$	$\bar{x}$	$\bar{y}+2$	$\bar{z}$	x	$\bar{y}+\frac{5}{2}$	$z+\frac{1}{2}$
$x+1$	$y+2$	z	$\bar{x}+1$	$y+\frac{5}{2}$	$\bar{z}+\frac{1}{2}$	$\bar{x}+1$	$\bar{y}+2$	$\bar{z}$	$x+1$	$\bar{y}+\frac{5}{2}$	$z+\frac{1}{2}$
$x+2$	$y+2$	z	$\bar{x}+2$	$y+\frac{5}{2}$	$\bar{z}+\frac{1}{2}$	$\bar{x}+2$	$\bar{y}+2$	$\bar{z}$	$x+2$	$\bar{y}+\frac{5}{2}$	$z+\frac{1}{2}$
	⋮			⋮			⋮			⋮	
x	y	$z+1$	$\bar{x}$	$y+\frac{1}{2}$	$\bar{z}+\frac{3}{2}$	$\bar{x}$	$\bar{y}$	$\bar{z}+1$	x	$\bar{y}+\frac{1}{2}$	$z+\frac{3}{2}$
$x+1$	y	$z+1$	$\bar{x}+1$	$y+\frac{1}{2}$	$\bar{z}+\frac{3}{2}$	$\bar{x}+1$	$\bar{y}$	$\bar{z}+1$	$x+1$	$\bar{y}+\frac{1}{2}$	$z+\frac{3}{2}$
$x+2$	y	$z+1$	$\bar{x}+2$	$y+\frac{1}{2}$	$\bar{z}+\frac{3}{2}$	$\bar{x}+2$	$\bar{y}$	$\bar{z}+1$	$x+2$	$\bar{y}+\frac{1}{2}$	$z+\frac{3}{2}$
	⋮			⋮			⋮			⋮	
x	$y+1$	$z+1$	$\bar{x}$	$y+\frac{3}{2}$	$\bar{z}+\frac{3}{2}$	$\bar{x}$	$\bar{y}+1$	$\bar{z}+1$	x	$\bar{y}+\frac{3}{2}$	$z+\frac{3}{2}$
$x+1$	$y+1$	$z+1$	$\bar{x}+1$	$y+\frac{3}{2}$	$\bar{z}+\frac{3}{2}$	$\bar{x}+1$	$\bar{y}+1$	$\bar{z}+1$	$x+1$	$\bar{y}+\frac{3}{2}$	$z+\frac{3}{2}$
$x+2$	$y+1$	$z+1$	$\bar{x}+2$	$y+\frac{3}{2}$	$\bar{z}+\frac{3}{2}$	$\bar{x}+2$	$\bar{y}+1$	$\bar{z}+1$	$x+2$	$\bar{y}+\frac{3}{2}$	$z+\frac{3}{2}$
	⋮			⋮			⋮			⋮	
$x+u_1$	$y+u_2$	$z+u_3$	$\bar{x}+u_1$	$y+u_2+\frac{1}{2}$	$\bar{z}+u_3+\frac{1}{2}$	$\bar{x}+u_1$	$\bar{y}+u_2$	$\bar{z}+u_3$	$x+u_1$	$\bar{y}+u_2+\frac{1}{2}$	$z+u_3+\frac{1}{2}$
	⋮			⋮			⋮			⋮	

The coordinate triplets of the general-position block of $P2_1/c$ (unique axis b, cell choice 1) (*cf.* Fig. 1.4.2.1) correspond to the coset representatives of the decomposition of the group listed in the first line of Table 1.4.2.2.

When the space group is referred to a primitive basis (which is always done for 'P' space groups), each coordinate triplet of the general-position block corresponds to one coset of $(\mathcal{G} : \mathcal{T}_{\mathcal{G}})$, *i.e.* the *multiplicity* of the general position and the number of cosets is the same. If, however, the space group is referred to a centred cell, then the complete set of general-position coordinate triplets is obtained by the combinations of the listed coordinate triplets with the centring translations. In this way, the total number of coordinate triplets per conventional unit cell, *i.e.* the multiplicity of the general position, is given by the product $[i] \times [p]$, where $[i]$ is the index of $\mathcal{T}_{\mathcal{G}}$ in $\mathcal{G}$ and $[p]$ is the index of the group of integer translations in the group $\mathcal{T}_{\mathcal{G}}$ of all (integer and centring) translations.

Example

The listing of the general position for the space–group type $Fmm2$ (42) of the space-group tables is reproduced in Fig. 1.4.2.2. The four entries, numbered (1) to (4), are to be taken as they are printed [indicated by (0, 0, 0)+]. The additional 12 more entries are obtained by adding the centring translations $(0, \frac{1}{2}, \frac{1}{2})$, $(\frac{1}{2}, 0, \frac{1}{2})$, $(\frac{1}{2}, \frac{1}{2}, 0)$ to the translation parts of the printed entries [indicated by $(0, \frac{1}{2}, \frac{1}{2})+$, $(\frac{1}{2}, 0, \frac{1}{2})+$ and $(\frac{1}{2}, \frac{1}{2}, 0)+$, respectively]. Altogether there are 16 entries, which is announced by the multiplicity of the general position, *i.e.* by the first number in the row. (The additional information specified on the left of the general-position block, namely the Wyckoff letter and the site symmetry, will be dealt with in Section 1.4.4.)

1.4.2.4. Additional symmetry operations and symmetry elements

The symmetry operations of a space group are conveniently partitioned into the cosets with respect to the translation subgroup. All operations which belong to the same coset have the same linear part and, if a single operation from a coset is given, all other operations in this coset are obtained by composition with a translation. However, not all symmetry operations in a coset with respect to the translation subgroup are operations of the same type and, furthermore, they may belong to element sets of different symmetry elements. In general, one can distinguish the following cases:

(i) The composition $W' = tW$ of a symmetry operation W with a translation t is an operation of the same type as W, with the same or a different type of symmetry element.

(ii) The composition $W' = tW$ is an operation of a different type to W with the same or a different type of symmetry element.

In order to distinguish the different cases, a closer analysis of the type of a symmetry operation and its symmetry element is required. These types, however, might be obscured by two obstacles:

(1) The origin in the chosen coordinate system might not lie on the geometric element of the symmetry operation. For example, the symmetry operation represented by the coordinate triplet $\bar{x}+1, \bar{y}+1, \bar{z}$ (*cf.* Section 1.4.2.3) is in fact an inversion through the point $1/2, 1/2, 0$ and thus of the same type as the inversion $\{\bar{1}|0\}$ through the origin.

(2) The screw or glide part might not be reduced to a vector within the unit cell. For example, the symmetry operation $\bar{x}, \bar{y}, z+1$, which is a twofold screw rotation 2 (0, 0, 1) 0, 0, z along the c axis, is the composition of the twofold rotation $\bar{x}, \bar{y}, z$ with the lattice translation $t(0, 0, 1)$ along the screw axis. Although the two operations $\bar{x}, \bar{y}, z$ and $\bar{x}, \bar{y}, z+1$ are of different types, they are coaxial equivalents and belong to the element set of the same symmetry element (*cf.* Section 1.2.3).

These issues can be overcome by decomposing the translation part $\boldsymbol{w}$ of a symmetry operation $W = (\boldsymbol{W}, \boldsymbol{w})$ into an intrinsic translation part $\boldsymbol{w}_g$ which is fixed by the linear part $\boldsymbol{W}$ of W and thus parallel to the geometric element of W, and a location part $\boldsymbol{w}_l$, which is perpendicular to the intrinsic translation part. Note that the subspace of vectors fixed by $\boldsymbol{W}$ and the subspace perpendicular to this space of fixed vectors are complementary subspaces, *i.e.* their dimensions add up to 3, therefore this decomposition is always possible.

The procedure for determining the intrinsic translation part of a symmetry operation is described in Section 1.2.2.3, and is based on the fact that the kth power of a symmetry operation $W = (\boldsymbol{W}, \boldsymbol{w})$ with linear part $\boldsymbol{W}$ of order k must be a pure translation, *i.e.* $W^k = (\boldsymbol{I}, \boldsymbol{t})$ for some lattice translation $\boldsymbol{t}$. The *intrinsic translation part* of W is then defined as $\boldsymbol{w}_g = \frac{1}{k}\boldsymbol{t}$.

The difference $\boldsymbol{w}_l = \boldsymbol{w} - \boldsymbol{w}_g$ is perpendicular to $\boldsymbol{w}_g$ and it is called the *location part* of $\boldsymbol{w}$. This terminology is justified by the fact that the location part can be reduced to $\boldsymbol{o}$ by an origin shift, *i.e.* the location part indicates whether the origin of the chosen coordinate system lies on the geometric element of W.

The transformation of point coordinates and matrix–column pairs under an origin shift is explained in detail in Sections 1.5.1.1 and 1.5.2.3, and the complete procedure for determining the additional symmetry operations is discussed in the context of the synoptic tables in Section 1.5.4 of *IT* A. In this section we will restrict ourselves to a detailed discussion of two examples which illustrate typical phenomena.

Example 1

Consider a space group of type *Fmm*2 (42). The information on the general position and on the symmetry operations given in the space-group tables are reproduced in Fig. 1.4.2.2. From this information one deduces that coset representatives with respect to the translation subgroup are the identity element $W_1 = x, y, z$, a rotation $W_2 = \bar{x}, \bar{y}, z$ with the c axis as geometric element, a reflection $W_3 = x, \bar{y}, z$ with the plane $x, 0, z$ as geometric element and a reflection $W_4 = \bar{x}, y, z$ with the plane $0, y, z$ as geometric element (with the indices following the numbering in the table).

Composing these coset representatives with the centring translations $t(0, \frac{1}{2}, \frac{1}{2})$, $t(\frac{1}{2}, 0, \frac{1}{2})$ and $t(\frac{1}{2}, \frac{1}{2}, 0)$ gives rise to elements in the same cosets, but with different types of symmetry operations and symmetry elements in several cases.

(i) $(0, \frac{1}{2}, \frac{1}{2})$: The composition of the rotation W_2 with $t(0, \frac{1}{2}, \frac{1}{2})$ results in the symmetry operation $\bar{x}, \bar{y} + \frac{1}{2}, z + \frac{1}{2}$, which is a twofold screw rotation with screw axis $0, \frac{1}{4}, z$. This means that both the type of the symmetry operation and the location of the geometric element are changed. Composing the reflection W_3 with $t(0, \frac{1}{2}, \frac{1}{2})$ gives the symmetry operation $x, \bar{y} + \frac{1}{2}, z + \frac{1}{2}$, which is a c glide with the plane $x, \frac{1}{4}, z$ as geometric element, *i.e.* shifted by $\frac{1}{4}$ along the b axis relative to the geometric element of W_3. In the composition of W_4 with $t(0, \frac{1}{2}, \frac{1}{2})$, the translation lies in the plane forming the geometric element of W_4. The geometric element of the resulting symmetry operation $\bar{x}, y + \frac{1}{2}, z + \frac{1}{2}$ is still the plane $0, y, z$, but the symmetry operation is now an n glide, *i.e.* a glide reflection with diagonal glide vector.

(ii) $(\frac{1}{2}, 0, \frac{1}{2})$: Analogous to the first centring translation, the composition of W_2 with $t(\frac{1}{2}, 0, \frac{1}{2})$ results in a twofold screw rotation with screw axis $\frac{1}{4}, 0, z$ as geometric element. The roles of the reflections W_3 and W_4 are interchanged, because the translation vector now lies in the plane forming the geometric element of W_3. Therefore, the composition of W_3 with $t(\frac{1}{2}, 0, \frac{1}{2})$ is an n glide with the plane $x, 0, z$ as geometric element, whereas the composition of W_4 with $t(\frac{1}{2}, 0, \frac{1}{2})$ is a c glide with the plane $\frac{1}{4}, y, z$ as geometric element.

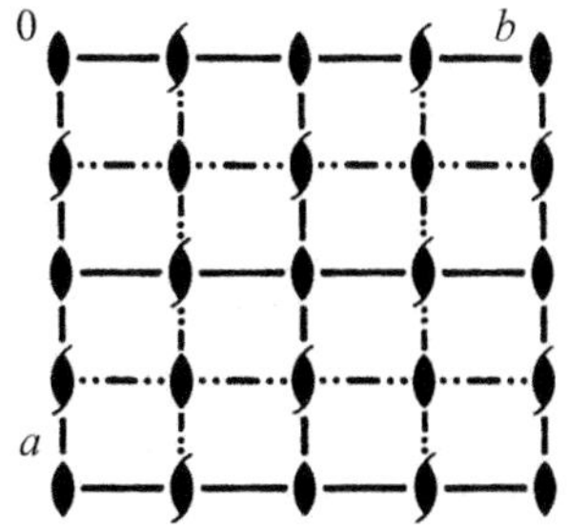

Figure 1.4.2.3
Symmetry-element diagram for space group *Fmm*2 (42) (orthogonal projection along [001]).

(iii) $(\frac{1}{2}, \frac{1}{2}, 0)$: Because this translation vector lies in the plane perpendicular to the rotation axis of W_2, the composition of W_2 with $t(\frac{1}{2}, \frac{1}{2}, 0)$ is still a twofold rotation, *i.e.* a symmetry operation of the same type, but the rotation axis is shifted by $\frac{1}{4}, \frac{1}{4}, 0$ in the xy plane to become the axis $\frac{1}{4}, \frac{1}{4}, z$. The composition of W_3 with $t(\frac{1}{2}, \frac{1}{2}, 0)$ results in the symmetry operation $x + \frac{1}{2}, \bar{y} + \frac{1}{2}, z$, which is an a glide with the plane $x, \frac{1}{4}, z$ as geometric element, *i.e.* shifted by $\frac{1}{4}$ along the b axis relative to the geometric element of W_3. Similarly, the composition of W_4 with $t(\frac{1}{2}, \frac{1}{2}, 0)$ is a b glide with the plane $\frac{1}{4}, y, z$ as geometric element.

In this example, all additional symmetry operations are listed in the symmetry-operations block of the space-group tables of *Fmm*2 because they are due to compositions of the coset representatives with centring translations.

The additional symmetry operations can easily be recognized in the symmetry-element diagrams (*cf.* Section 1.4.2.5). Fig. 1.4.2.3 shows the symmetry-element diagram of *Fmm*2 for the projection along the c axis. One sees that twofold rotation axes alternate with twofold screw axes and that mirror planes alternate with 'double' or e-glide planes, *i.e.* glide planes with two glide vectors. For example, the dot–dashed lines at $x = \frac{1}{4}$ and $x = \frac{3}{4}$ in Fig. 1.4.2.3 represent the b and c glides with normal vector along the a axis [for a discussion of e-glide notation, see Sections 1.2.3 and 2.1.2, and de Wolff *et al.*, 1992].

Example 2

In a space group of type *P4mm* (99), representatives of the space group with respect to the translation subgroup are the powers of a fourfold rotation and reflections with normal vectors along the a and the b axis and along the diagonals [110] and [$1\bar{1}0$] (*cf.* Fig. 1.4.2.4).

In this case, additional symmetry operations occur although there are no centring translations. Consider for example the reflection W_8 with the plane x, x, z as geometric element.

Positions

Multiplicity, Wyckoff letter, Site symmetry			Coordinates			
8	g	1	(1) x,y,z	(2) $\bar{x},\bar{y},z$	(3) $\bar{y},x,z$	(4) $y,\bar{x},z$
			(5) $x,\bar{y},z$	(6) $\bar{x},y,z$	(7) $\bar{y},\bar{x},z$	(8) y,x,z

Symmetry operations

(1) 1	(2) 2 $0,0,z$	(3) 4^+ $0,0,z$	(4) 4^- $0,0,z$
(5) m $x,0,z$	(6) m $0,y,z$	(7) m $x,\bar{x},z$	(8) m x,x,z

Figure 1.4.2.4
General-position and symmetry-operations blocks as given in the space-group tables for space group *P4mm* (99).

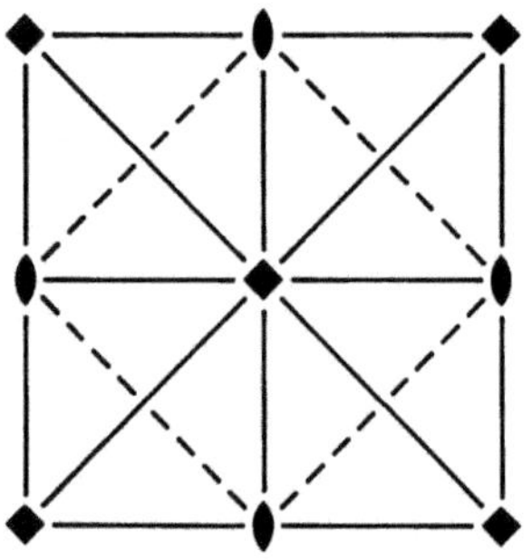

Figure 1.4.2.5
Symmetry-element diagram for space group *P4mm* (99) (orthogonal projection along [001]).

Composing this reflection with the translation $t(1, 0, 0)$ gives rise to the symmetry operation represented by $y+1, x, z$. This operation maps a point with coordinates $x+\frac{1}{2}, x, z$ to $x+1, x+\frac{1}{2}, z$ and is thus a glide reflection with the plane $x+\frac{1}{2}, x, z$ as geometric element and $(\frac{1}{2}, \frac{1}{2}, 0)$ as glide vector. In a similar way, composing the other diagonal reflection with translations yields further glide reflections.
These glide reflections are symmetry operations which are not listed in the symmetry-operations block, although they are clearly of a different type to the operations given there. However, in the symmetry-element diagram as shown in Fig. 1.4.2.5, the corresponding symmetry elements are displayed as diagonal dashed lines which alternate with the solid diagonal lines representing the diagonal reflections.

1.4.2.5. Space-group diagrams

In the space-group tables of *IT* A (see also Part 2), for each space group there are at least two diagrams displaying the symmetry (there are more diagrams for space groups of low symmetry). The *symmetry-element* diagram displays the location and orientation of the symmetry elements of the space group. The *general-position* diagrams show the arrangement of a set of symmetry-equivalent points of the general position. Because of the periodicity of the arrangements, the presentation of the contents of one unit cell is sufficient. Both types of diagrams are orthogonal projections of the space-group unit cell onto the plane of projection along a basis vector of the conventional crystallographic coordinate system. The symmetry elements of triclinic, monoclinic and orthorhombic groups are shown in three different projections along the basis vectors. The thin lines outlining the projection are the traces of the side planes of the unit cell.

Detailed explanations of the diagrams of space groups are found in Section 2.1.3.6. In this section, after a very brief introduction to the diagrams, we will focus mainly on certain important but very often overlooked features of the diagrams.

Symmetry-element diagram

The graphical symbols of the symmetry elements used in the diagrams are explained in Section 2.1.2. The heights along the projection direction above the plane of the diagram are indicated for rotation or screw axes and mirror or glide planes parallel to the projection plane, for rotoinversion axes and inversion centres. The heights (if different from zero) are given as fractions of the shortest translation vector along the projection direction. In Fig. 1.4.2.6 (left) the symmetry elements of $P2_1/c$ (unique axis *b*, cell choice 1) are represented graphically in a projection of the unit cell along the monoclinic axis **b**. The directions of the basis vectors **c** and **a** can be read directly from the figure. The origin (upper left corner of the unit cell) lies on a centre of inversion indicated by a small open circle. The black lenticular symbols with tails represent the twofold screw axes parallel to **b**. The *c*-glide plane at height $\frac{1}{4}$ along **b** is shown as a bent arrow with the arrowhead pointing along **c**.

The crystallographic symmetry operations are visualized geometrically by the related symmetry elements. Whereas the symmetry element of a symmetry operation is uniquely defined, more than one symmetry operation may belong to the same symmetry element (*cf.* Section 1.2.3). The following examples illustrate some important features of the diagrams related to the fact that the symmetry-element symbols that are displayed visualize all symmetry operations that belong to the element sets of the symmetry elements.

Examples

(1) *Visualization of the twofold screw rotations of* $P2_1/c$ (Fig. 1.4.2.6). The second coset of the decomposition of $P2_1/c$ with respect to its translation subgroup shown in Table 1.4.2.2 is formed by the infinite set of twofold screw rotations represented by the coordinate triplets $\bar{x}+u_1, y+\frac{1}{2}+u_2, \bar{z}+\frac{1}{2}+u_3$ (where u_1, u_2, u_3 are integers). To analyse how these symmetry operations are visualized, it is convenient to consider two special cases:

(i) $u_2 = 0$, *i.e.* $\bar{x}+u_1, y+\frac{1}{2}, \bar{z}+\frac{1}{2}+u_3 = \{2_{010}|u_1, \frac{1}{2}, \frac{1}{2}+u_3\}$; these operations correspond to twofold screw rotations around the infinitely many screw axes parallel to the line $0, y, \frac{1}{4}$, *i.e.* around the lines $u_1/2, y, u_3/2+\frac{1}{4}$. The symbols of the symmetry elements (*i.e.* of the twofold screw axes) located in the unit cell at $0, y, \frac{1}{4}$; $0, y, \frac{3}{4}$; $\frac{1}{2}, y, \frac{1}{4}$; $\frac{1}{2}, y, \frac{3}{4}$ (and the translationally equivalent $1, y, \frac{1}{4}$ and $1, y, \frac{3}{4}$) are shown in the symmetry-element diagram (Fig. 1.4.2.6);

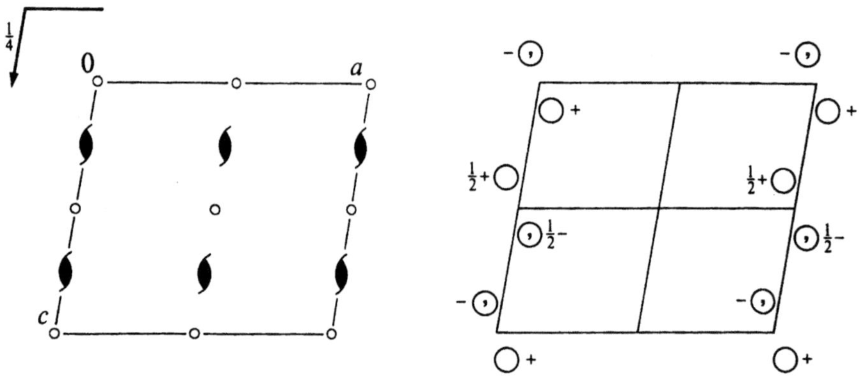

Figure 1.4.2.6
Symmetry-element diagram (left) and general-position diagram (right) for the space group $P2_1/c$, No. 14 (unique axis *b*, cell choice 1).

(ii) $u_1 = u_3 = 0$, *i.e.* $\bar{x}, y+\frac{1}{2}+u_2, \bar{z}+\frac{1}{2} = \{2_{010}|0, \frac{1}{2}+u_2, \frac{1}{2}\}$; these symmetry operations correspond to screw rotations around the line $0, y, \frac{1}{4}$ with screw components $\frac{1}{2}, \frac{3}{2}, \frac{5}{2}, \ldots, -\frac{1}{2}, -\frac{3}{2}, -\frac{5}{2}, \ldots$, *i.e.* with a screw component $\frac{1}{2}$ to which all lattice translations parallel to the screw axis are added. These operations, infinite in number, share the same geometric element, *i.e.* they form the element set of the same symmetry element, and geometrically they are represented just by one graphical symbol on the symmetry-element diagrams located exactly at $0, y, \frac{1}{4}$.

(iii) The rest of the symmetry operations in the coset, *i.e.* those with the translation parts $\begin{pmatrix} u_1 \\ \frac{1}{2}+u_2 \\ \frac{1}{2}+u_3 \end{pmatrix}$, are combinations of the two special cases above.

(2) *Inversion centres of* $P2_1/c$ (Fig. 1.4.2.6). The element set of an inversion centre consists of only one symmetry operation, *viz.* the inversion through the point located at the centre. In other words, to each inversion centre displayed on a symmetry-element diagram there corresponds one symmetry operation of inversion. The infinitely many inversions $(-\boldsymbol{I}, \boldsymbol{t}) = \bar{x}+u_1, \bar{y}+u_2, \bar{z}+u_3 = \{\bar{1}|u_1, u_2, u_3\}$ of $P2_1/c$ are located at points $u_1/2$, $u_2/2$, $u_3/2$. Apart from translational equivalence, there are eight centres located in the unit cell: four at $y = 0$, namely at $0, 0, 0$; $\frac{1}{2}, 0, 0$; $0, 0, \frac{1}{2}$; $\frac{1}{2}, 0, \frac{1}{2}$ and four at height $\frac{1}{2}$ of b. It is important to note that only inversion centres at $y = 0$ are indicated on the diagram.

A similar rule is applied to all pairs of symmetry elements of the same type (such as *e.g.* twofold rotation axes, planes *etc.*) whose heights differ by $\frac{1}{2}$ of the shortest lattice direction along the projection direction. For example, the *c*-glide plane symbol in Fig. 1.4.2.6 with the fraction $\frac{1}{4}$ next to it represents not only the *c*-glide plane located at height $\frac{1}{4}$ but also the one at height $\frac{3}{4}$.

(3) *Glide reflections visualized by mirror planes*. As discussed in Section 1.2.3, the element set of a mirror or glide plane consists of a defining operation and all its coplanar equivalents (*cf.* Table 1.2.3.1). The corresponding symmetry element is a mirror plane if among the infinite set of the coplanar glide reflections there is one with zero glide vector. Thus, the symmetry element is a mirror plane and the graphical symbol for a mirror plane is used for its representation on the symmetry-element diagrams of the space groups. For example, the mirror plane $0, y, z$ shown on the symmetry-element diagram of *Fmm*2 (42), *cf.* Fig. 1.4.2.3, represents all glide reflections of the element set of the defining operation $0, y, z$ [symmetry operation (4) of the general-position $(0, 0, 0)+$ set, *cf.* Fig. 1.4.2.2], including the *n*-glide reflection $\bar{x}, y+\frac{1}{2}, z+\frac{1}{2}$ [entry (4) of the general-position $(0, \frac{1}{2}, \frac{1}{2})+$ set]. In a similar way, the graphical symbols of the mirror planes $x, 0, z$ also represent the *n*-glide reflections $x+\frac{1}{2}, \bar{y}, z+\frac{1}{2}$ [entry (3) of the general-position $(\frac{1}{2}, 0, \frac{1}{2})+$ set] of *Fmm*2.

General-position diagram

The graphical presentations of the space-group symmetries provided by the general-position diagrams consist of a set of general-position points which are symmetry equivalent under the symmetry operations of the space group. Starting with a point in the upper left corner of the unit cell, indicated by an open circle with a sign '+', all the displayed points inside and near the unit cell are images of the starting point under some symmetry operation of the space group. Because of the one-to-one correspondence between the image points and the symmetry operations, the number of general-position points in the unit cell (excluding the points that are equivalent by integer translations) equals the multiplicity of the general position. The coordinates of the points in the projection plane can be read directly from the diagram. For all systems except cubic, only one parameter is necessary to describe the height along the projection direction. For example, if the height of the starting point above the projection plane is indicated by a '+' sign, then signs '+', '−' or their combinations with fractions (*e.g.* $\frac{1}{2}$+, $\frac{1}{2}$− *etc.*) are used to specify the heights of the image points. A circle divided by a vertical line represents two points with different coordinates along the projection direction but identical coordinates in the projection plane. A comma ',' in the circle indicates an image point obtained by a symmetry operation $W = (\boldsymbol{W}, \boldsymbol{w})$ of the second kind [*i.e.* with $\det(\boldsymbol{W}) = -1$, *cf.* Section 1.2.2.3].

Example

The general-position diagram of $P2_1/c$ (unique axis *b*, cell choice 1) is shown in Fig. 1.4.2.6 (right). The open circles indicate the location of the four symmetry-equivalent points of the space group within the unit cell along with additional eight translation-equivalent points to complete the presentation. The circles with a comma inside indicate the image points generated by operations of the second kind – inversions and glide planes in the present case. The fractions and signs close to the circles indicate their heights in units of *b* of the symmetry-equivalent points along the monoclinic axis. For example, $\frac{1}{2}$− is a shorthand notation for $\frac{1}{2} - y$.

Notes:

(1) The close relation between the symmetry-element and the general-position diagrams is obvious. For example, the points shown on the general-position diagram are images of a general-position point under the action of the space-group symmetry operations displayed by the corresponding symmetry elements on the symmetry-element diagram. With some practice each of the diagrams can be generated from the other. In a number of texts, the two diagrams are considered as completely equivalent descriptions of the same space group. This statement is true for most of the space groups. However, there are a number of space groups for which the point configuration displayed on the general-position diagram has higher symmetry than the generating space group (Suescun & Nespolo, 2012; Müller, 2012). For example, consider the diagrams of the space group *P*2, No. 3 (unique axis *b*, cell choice 1) shown in Fig. 1.4.2.7. It is easy to recognise that, apart from the twofold rotations, the point configuration shown in the general-position diagram is symmetric with respect to a reflection through a plane containing the general-position points, and as a result the space group of the general-position configuration is of *P*2/*m* type, and not of *P*2. There are a number of space groups for which the general-position diagram displays higher space-group symmetry, for example: *P*1, $P2_1$, *P*4*mm*, *P*6 *etc.* The analysis of the eigensymmetry groups of the general-position orbits results in a systematic procedure for the determination of such space groups: the general-position diagrams do not reflect the space-group symmetry correctly if the general-

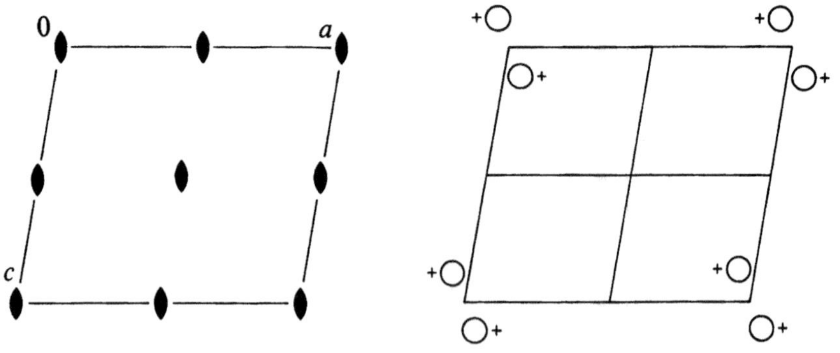

Figure 1.4.2.7
Symmetry-element diagram (left) and general-position diagram (right) for the space group $P2$, No. 3 (unique axis b, cell choice 1).

position orbits are *non-characteristic*, *i.e.* their eigensymmetry groups are supergroups of the space groups. (An introduction to terms like eigensymmetry groups, characteristic and non-characteristic orbits, and further discussion of space groups with non-characteristic general-position orbits are given in Section 1.4.4.4 of *IT* A.)

(2) The graphical presentation of the general-position points of cubic groups is more difficult: three different parameters are required to specify the height of the points along the projection direction. To make the presentation clearer, the general-position points are grouped around points of higher site symmetry and represented in the form of polyhedra. For the general-position diagrams of *IT* A for most of the space groups the initial general point is taken as 0.048, 0.12, 0.089, and the polyhedra are centred at 0, 0, 0 (and its equivalent points). Additional general-position diagrams are shown for space groups with special sites different from 0, 0, 0 that have site-symmetry groups of equal or higher order. Consider, for example, the two general-position diagrams of the space group $I4_132$ (214) shown in Fig. 1.4.2.8. The polyhedra of the left-hand diagram are centred at special points of highest site-symmetry, namely, at $\frac{1}{8}, \frac{1}{8}, \frac{1}{8}$ and its equivalent points in the unit cell. The site-symmetry groups are of the type 32 leading to polyhedra in the form of *twisted trigonal antiprisms* (*cf.* Table 3.2.3.2 of *IT* A). The polyhedra (sphenoids) of the right-hand diagram are attached to the origin 0, 0, 0 and its equivalent points in the unit cell, site-symmetry group of the type 3. The fractions attached to the polyhedra indicate the heights of the high-symmetry points along the projection direction (*cf.* Section 2.1.3.6 for further explanations of the diagrams).

1.4.3. Generation of space groups

By Hans Wondratschek

In group theory, a *set of generators* of a group is a set of group elements such that each group element may be obtained as a finite ordered product of the generators. For space groups of one, two and three dimensions, generators may always be chosen and ordered in such a way that each symmetry operation W can be written as the product of powers of h generators g_j ($j = 1, 2, \ldots, h$). Thus,

$$W = g_h^{k_h} \cdot g_{h-1}^{k_{h-1}} \cdot \ldots \cdot g_p^{k_p} \cdot \ldots \cdot g_3^{k_3} \cdot g_2^{k_2} \cdot g_1,$$

where the powers k_j are positive or negative integers (including zero). The description of a group by means of generators has the advantage of compactness. For instance, the 48 symmetry operations in point group $m\bar{3}m$ can be described by two generators. Different choices of generators are possible. For the

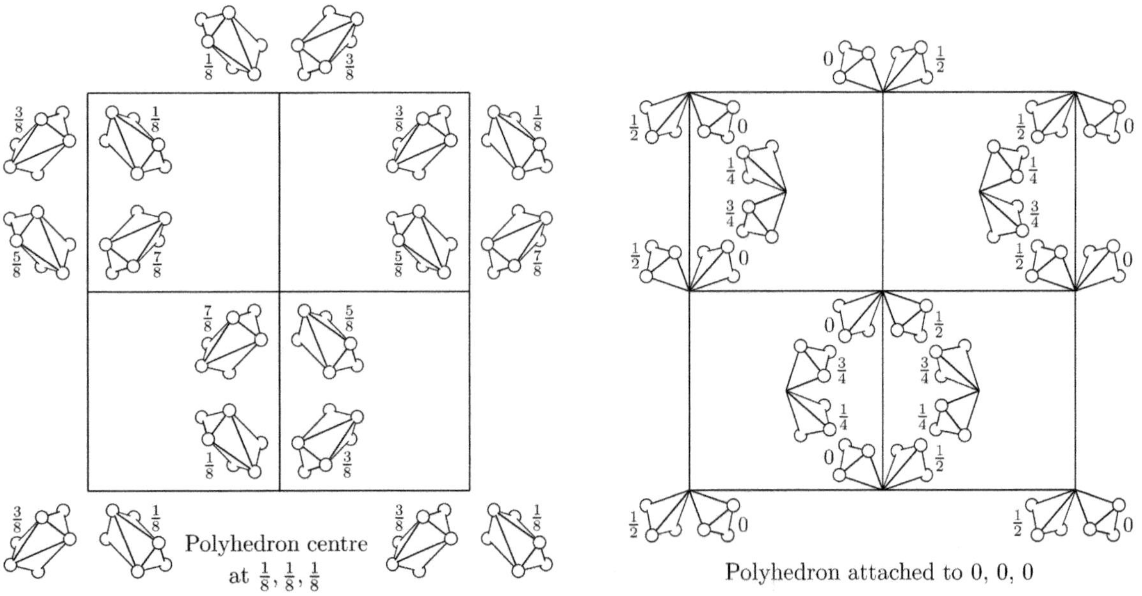

Figure 1.4.2.8
General-position diagrams for the space group $I4_132$ (214). Left: polyhedra (twisted trigonal antiprisms) with centres at $\frac{1}{8}, \frac{1}{8}, \frac{1}{8}$ and its equivalent points (site-symmetry group .32). Right: polyhedra (sphenoids) attached to 0, 0, 0 and its equivalent points (site-symmetry group .3.).

Table 1.4.3.1
Sequence of generators for the crystal classes

The space-group generators differ from those listed here by their glide or screw components. The generator 1 is omitted, except for crystal class 1. The generators are represented by the corresponding Seitz symbols (*cf.* Section 1.4.2.2). Following the conventions, the subscript of a symbol denotes the characteristic direction of that operation, where necessary. For example, the subscripts 001, 010, 110 *etc.* refer to the directions [001], [010], [110] *etc.* For mirror reflections *m*, the 'direction of *m*' refers to the normal of the mirror plane.

Hermann–Mauguin symbol of crystal class	Generators g_i (sequence left to right)
1	1
$\bar{1}$	$\bar{1}$
2	2
m	*m*
2/*m*	$2, \bar{1}$
222	$2_{001}, 2_{010}$
*mm*2	$2_{001}, m_{010}$
mmm	$2_{001}, 2_{010}, \bar{1}$
4	$2_{001}, 4^+_{001}$
$\bar{4}$	$2_{001}, \bar{4}^+_{001}$
4/*m*	$2_{001}, 4^+_{001}, \bar{1}$
422	$2_{001}, 4^+_{001}, 2_{010}$
4*mm*	$2_{001}, 4^+_{001}, m_{010}$
$\bar{4}2m$	$2_{001}, \bar{4}^+_{001}, 2_{010}$
$\bar{4}m2$	$2_{001}, \bar{4}^+_{001}, m_{010}$
4/*mmm*	$2_{001}, 4^+_{001}, 2_{010}, \bar{1}$
3	3^+_{001}
(rhombohedral coordinates	3^+_{111})
$\bar{3}$	$3^+_{001}, \bar{1}$
(rhombohedral coordinates	$3^+_{111}, \bar{1}$
321	$3^+_{001}, 2_{110}$
(rhombohedral coordinates	$3^+_{111}, 2_{\bar{1}01}$)
312	$3^+_{001}, 2_{1\bar{1}0}$
3*m*1	$3^+_{001}, m_{110}$
(rhombohedral coordinates	$3^+_{111}, m_{\bar{1}01}$)
31*m*	$3^+_{001}, m_{1\bar{1}0}$
$\bar{3}m1$	$3^+_{001}, 2_{110}, \bar{1}$
(rhombohedral coordinates	$3^+_{111}, 2_{\bar{1}01}, \bar{1}$)
$\bar{3}1m$	$3^+_{001}, 2_{1\bar{1}0}, \bar{1}$
6	$3^+_{001}, 2_{001}$
$\bar{6}$	$3^+_{001}, m_{001}$
6/*m*	$3^+_{001}, 2_{001}, \bar{1}$
622	$3^+_{001}, 2_{001}, 2_{110}$
6*mm*	$3^+_{001}, 2_{001}, m_{110}$
$\bar{6}m2$	$3^+_{001}, m_{001}, m_{110}$
$\bar{6}2m$	$3^+_{001}, m_{001}, 2_{110}$
6/*mmm*	$3^+_{001}, 2_{001}, 2_{110}, \bar{1}$
23	$2_{001}, 2_{010}, 3^+_{111}$
$m\bar{3}$	$2_{001}, 2_{010}, 3^+_{111}, \bar{1}$
432	$2_{001}, 2_{010}, 3^+_{111}, 2_{110}$
$\bar{4}3m$	$2_{001}, 2_{010}, 3^+_{111}, m_{1\bar{1}0}$
$m\bar{3}m$	$2_{001}, 2_{010}, 3^+_{111}, 2_{110}, \bar{1}$

space-group tables of *IT* A, generators and generating procedures have been chosen such as to make the entries in the blocks 'General position' (*cf.* Section 2.1.3.11) and 'Symmetry operations' (*cf.* Section 2.1.3.9) as transparent as possible. Space groups of the same crystal class are generated in the same way (see Table 1.4.3.1 for the sequences that have been chosen), and the aim has been to accentuate important subgroups of space groups as much as possible. Accordingly, a process of generation in the form of a *composition series* has been adopted, see Ledermann (1976). The generator g_1 is defined as the identity operation, represented by (1) x, y, z. The generators g_2, g_3, and g_4 are the translations with translation vectors **a**, **b** and **c**, respectively. Thus, the coefficients k_2, k_3 and k_4 may have any integral value. If centring translations exist, they are generated by translations g_5 (and g_6 in the case of an *F* lattice) with translation vectors **d** (and **e**). For a *C* lattice, for example, **d** is given by $\mathbf{d} = \frac{1}{2}(\mathbf{a}+\mathbf{b})$. The exponents k_5 (and k_6) are restricted to the following values:

Lattice letter *A*, *B*, *C*, *I*: $k_5 = 0$ or 1.
Lattice letter *R* (hexagonal axes): $k_5 = 0$, 1 or 2.
Lattice letter *F*: $k_5 = 0$ or 1; $k_6 = 0$ or 1.

As a consequence, any translation *t* of $\mathcal{G}$ with translation vector

$$\mathbf{t} = k_2\mathbf{a} + k_3\mathbf{b} + k_4\mathbf{c}\,(+k_5\mathbf{d} + k_6\mathbf{e})$$

can be obtained as a product

$$t = (g_6^{k_6}).(g_5^{k_5}).g_4^{k_4}.g_3^{k_3}.g_2^{k_2}.g_1,$$

where $k_2, \ldots, k_6$ are integers determined by *t*. The generators g_6 and g_5 are enclosed between parentheses because they are effective only in centred lattices.

The remaining generators generate those symmetry operations that are not translations. They are chosen in such a way that only terms g_j or g_j^2 occur. For further specific rules, see below.

The process of generating the entries of the space-group tables may be demonstrated by the example in Table 1.4.3.2, where $\mathcal{G}_j$ denotes the group generated by $g_1, g_2, \ldots, g_j$. For $j \geq 5$, the next generator g_{j+1} is introduced when $g_j^{k_j} \in \mathcal{G}_{j-1}$, because in this case no new symmetry operation would be generated by $g_j^{k_j}$. The generating process is terminated when there is no further generator. In the present example, g_7 completes the generation: $\mathcal{G}_7 \equiv P6_122$ (178).

1.4.3.1. Selected order for non-translational generators

For the non-translational generators, the following sequence has been adopted:

(*a*) In all centrosymmetric space groups, an inversion (if possible at the origin *O*) has been selected as the last generator.

(*b*) Rotations precede symmetry operations of the second kind. In crystal classes $\bar{4}2m$ and $\bar{4}m2$ and $\bar{6}2m$ and $\bar{6}m2$, as an exception, $\bar{4}$ and $\bar{6}$ are generated first in order to take into account the conventional choice of origin in the fixed points of $\bar{4}$ and $\bar{6}$.

(*c*) The non-translational generators of space groups with *C*, *A*, *B*, *F*, *I* or *R* symbols are those of the corresponding space group with a *P* symbol, if possible. For instance, the generators of $I2_12_12_1$ (24) are those of $P2_12_12_1$ (19) and the generators of *Ibca* (73) are those of *Pbca* (61), apart from the centring translations.

(*d*) In some cases, rule (*c*) could not be followed without breaking rule (*a*), *e.g.* in *Cmme* (67). In such cases, the generators are chosen to correspond to the Hermann–Mauguin symbol as far as possible. For instance, the generators (apart from centring) of *Cmme* and *Imma* (74) are those of *Pmmb*, which is a non-standard setting of *Pmma* (51). (A combination of the generators of *Pmma* with the *C*- or *I*-centring translation results in non-standard settings of *Cmme* and *Imma*.)

For the space groups with lattice symbol *P*, the generation procedure has given the same triplets (except for their sequence) as in *IT* (1952). In non-*P* space groups, the triplets listed sometimes differ from those of *IT* (1952) by a centring translation.

Table 1.4.3.2
Generation of the space group $P6_122 \equiv D_6^2$ (178)

The entries in the second column designated by the numbers (1)–(12) correspond to the coordinate triplets of the general position of $P6_122$.

	Coordinate triplets	Symmetry operations
g_1	(1) x, y, z;	Identity I
g_2	$t(100)$	Generating translations
g_3	$t(010)$	
g_4	$t(001)$	
	The group $\mathcal{G}_4 \equiv \mathcal{T}$ of all translations of $P6_122$ has been generated	
g_5	(2) $\bar{y}, x-y, z+\frac{1}{3}$;	Threefold screw rotation
g_5^2	(3) $\bar{x}+y, \bar{x}, z+\frac{2}{3}$;	Threefold screw rotation
$g_5^3 = t(001)$:	Now the space group $\mathcal{G}_5 \equiv P3_1$ has been generated	
g_6	(4) $\bar{x}, \bar{y}, z+\frac{1}{2}$;	Twofold screw rotation
$g_6 * g_5$	(5) $y, \bar{x}+y, z+\frac{5}{6}$;	Sixfold screw rotation
$g_6 * g_5^2$	$x-y, x, z+\frac{7}{6} \sim$ (6) $x-y, x, z+\frac{1}{6}$;	Sixfold screw rotation
$g_6^2 = t(001)$:	Now the space group $\mathcal{G}_6 \equiv P6_1$ has been generated	
g_7	(7) $y, x, \bar{z}+\frac{1}{3}$;	Twofold rotation, direction of axis [110]
$g_7 * g_5$	(8) $x-y, \bar{y}, \bar{z}$;	Twofold rotation, axis [100]
$g_7 * g_5^2$	$\bar{x}, \bar{x}+y, \bar{z}-\frac{1}{3} \sim$ (9) $\bar{x}, \bar{x}+y, \bar{z}+\frac{2}{3}$;	Twofold rotation, axis [010]
$g_7 * g_6$	$\bar{y}, \bar{x}, \bar{z}-\frac{1}{6} \sim$ (10) $\bar{y}, \bar{x}, \bar{z}+\frac{5}{6}$;	Twofold rotation, axis $[1\bar{1}0]$
$g_7 * g_6 * g_5$	$\bar{x}+y, y, \bar{z}-\frac{1}{2} \sim$ (11) $\bar{x}+y, y, \bar{z}+\frac{1}{2}$;	Twofold rotation, axis [120]
$g_7 * g_6 * g_5^2$	$x, x-y, \bar{z}-\frac{5}{6} \sim$ (12) $x, x-y, \bar{z}+\frac{1}{6}$;	Twofold rotation, axis [210]
$g_7^2 = I$	$\mathcal{G}_7 \sim P6_122$	

1.4.4. General and special Wyckoff positions

By Bernd Souvignier

One of the first tasks in the analysis of crystal patterns is to determine the actual positions of the atoms. Since the full crystal pattern can be reconstructed from a single unit cell or even an asymmetric unit, it is clearly sufficient to focus on the atoms inside such a restricted volume. What one observes is that the atoms typically do not occupy arbitrary positions in the unit cell, but that they often lie on geometric elements, *e.g.* reflection planes or lines along rotation axes. It is therefore very useful to analyse the symmetry properties of the points in a unit cell in order to predict likely positions of atoms.

We note that in this chapter all statements and definitions refer to the usual three-dimensional space $\mathbb{E}^3$, but also can be formulated, *mutatis mutandis*, for plane groups acting on $\mathbb{E}^2$ and for higher-dimensional groups acting on n-dimensional space $\mathbb{E}^n$.

1.4.4.1. Crystallographic orbits

Since the operations of a space group provide symmetries of a crystal pattern, two points X and Y that are mapped onto each other by a space-group operation are regarded as being *geometrically equivalent*. Starting from a point $X \in \mathbb{E}^3$, infinitely many points Y equivalent to X are obtained by applying all space-group operations $g = (\boldsymbol{W}, \boldsymbol{w})$ to X: $Y = g(X) = (\boldsymbol{W}, \boldsymbol{w})X = (\boldsymbol{W}X + \boldsymbol{w})$.

Definition

For a space group $\mathcal{G}$ acting on the three-dimensional space $\mathbb{E}^3$, the (infinite) set

$$\mathcal{O} = \mathcal{G}(X) := \{g(X) | g \in \mathcal{G}\}$$

is called the *orbit of X under $\mathcal{G}$*.
The orbit of X is the smallest subset of $\mathbb{E}^3$ that contains X and is closed under the action of $\mathcal{G}$. It is also called a *crystallographic orbit*.

Every point in direct space $\mathbb{E}^3$ belongs to precisely one orbit under $\mathcal{G}$ and thus the orbits of $\mathcal{G}$ partition the direct space into disjoint subsets. It is clear that an orbit is completely determined by its points in the unit cell, since translating the unit cell by the translation subgroup $\mathcal{T}$ of $\mathcal{G}$ entirely covers $\mathbb{E}^3$.

It may happen that two different symmetry operations g and h in $\mathcal{G}$ map X to the same point. Since $g(X) = h(X)$ implies that $h^{-1}g(X) = X$, the point X is fixed by the nontrivial operation $h^{-1}g$ in $\mathcal{G}$.

Definition

The subgroup $\mathcal{S}_X = \mathcal{S}_{\mathcal{G}}(X) := \{g \in \mathcal{G} | g(X) = X\}$ of symmetry operations from $\mathcal{G}$ that fix X is called the *site-symmetry group of X in $\mathcal{G}$*.

Since translations, glide reflections and screw rotations fix no point in $\mathbb{E}^3$, a site-symmetry group $\mathcal{S}_X$ never contains operations of these types and thus consists only of reflections, rotations, inversions and rotoinversions. Because of the absence of translations, $\mathcal{S}_X$ contains at most one operation from a coset $\mathcal{T}g$ relative to the translation subgroup $\mathcal{T}$ of $\mathcal{G}$, since otherwise the quotient of two such operations tg and $t'g$ would be the nontrivial translation $tgg^{-1}t'^{-1} = tt'^{-1}$ (see Chapter 1.3 for a discussion of coset decompositions). In particular, the operations in $\mathcal{S}_X$ all have different linear parts and because these linear parts form a subgroup of the point group $\mathcal{P}$ of $\mathcal{G}$, the order of the site-symmetry group $\mathcal{S}_X$ is a divisor of the order of the point group of $\mathcal{G}$.

The site-symmetry group of a point X is thus a finite subgroup of the space group $\mathcal{G}$, a subgroup which is isomorphic to a subgroup of the point group $\mathcal{P}$ of $\mathcal{G}$.

Example

For a space group $\mathcal{G}$ of type $P\bar{1}$, the site-symmetry group of the origin $X = \begin{pmatrix} 0 \\ 0 \\ 0 \end{pmatrix}$ is clearly generated by the inversion in the origin: $\{\bar{1}|0\}(X) = X$. On the other hand, the point $Y = \begin{pmatrix} \frac{1}{2} \\ 0 \\ \frac{1}{2} \end{pmatrix}$

is fixed by the inversion in Y, *i.e.*

$$\{\bar{1}|1,0,1\}(Y) = \begin{pmatrix} \bar{1} & 0 & 0 \\ 0 & \bar{1} & 0 \\ 0 & 0 & \bar{1} \end{pmatrix} \begin{pmatrix} \frac{1}{2} \\ 0 \\ \frac{1}{2} \end{pmatrix} + \begin{pmatrix} 1 \\ 0 \\ 1 \end{pmatrix} = \begin{pmatrix} \frac{1}{2} \\ 0 \\ \frac{1}{2} \end{pmatrix} = Y.$$

The symmetry operation $\{\bar{1}|1,0,1\}$ also belongs to $\mathcal{G}$ and generates the site-symmetry group of Y. The site-symmetry groups $\mathcal{S}_X = \{\{1|0\}, \{\bar{1}|0\}\}$ of X and $\mathcal{S}_Y = \{\{1|0\}, \{\bar{1}|1,0,1\}\}$ of Y are thus different subgroups of order 2 of $\mathcal{G}$ which are isomorphic to the point group of $\mathcal{G}$ (which is generated by $\bar{1}$).

The order $|\mathcal{S}_X|$ of the site-symmetry group $\mathcal{S}_X$ is closely related to the number of points in the orbit of X that lie in the unit cell. An application of the orbit–stabilizer theorem (see Section 1.1.6) yields the crucial observation that each point $Y = g(X)$ in the orbit of X under $\mathcal{G}$ is obtained precisely $|\mathcal{S}_X|$ times as an orbit point: for each $h \in \mathcal{S}_X$ one has $gh(X) = g(X) = Y$ and conversely $g'(X) = g(X)$ implies that $g^{-1}g' = h \in \mathcal{S}_X$ and thus $g' = gh$ for an operation h in $\mathcal{S}_X$.

Assuming first that we are dealing with a space group $\mathcal{G}$ described by a *primitive* lattice, each coset of $\mathcal{G}$ relative to the translation subgroup $\mathcal{T}$ contains precisely one operation g such that $g(X)$ lies in the primitive unit cell. Since the number of cosets equals the order $|\mathcal{P}|$ of the point group $\mathcal{P}$ of $\mathcal{G}$ and since each orbit point is obtained $|\mathcal{S}_X|$ times, it follows that the number of orbit points in the unit cell is $|\mathcal{P}|/|\mathcal{S}_X|$.

If we deal with a space group with a centred unit cell, the above result has to be modified slightly. If there are $k - 1$ centring vectors, the lattice spanned by the conventional basis is a sublattice of index k in the full translation lattice. The conventional cell therefore is built up from k primitive unit cells (spanned by a primitive lattice basis) and thus in particular contains k times as many points as the primitive cell (see Chapter 1.3 for a detailed discussion of conventional and primitive bases and cells).

Proposition

Let $\mathcal{G}$ be a space group with point group $\mathcal{P}$ and let $\mathcal{S}_X$ be the site-symmetry group of a point X in $\mathbb{E}^3$. Then the number of orbit points of the orbit of X which lie in a conventional cell for $\mathcal{G}$ is equal to the product $k \times |\mathcal{P}|/|\mathcal{S}_X|$, where k is the volume of the conventional cell divided by the volume of a primitive unit cell.

1.4.4.2. Wyckoff positions

As already mentioned, one of the first issues in the analysis of crystal structures is the determination of the actual atom positions. Energetically favourable configurations in inorganic compounds are often achieved when the atoms occupy positions that have a nontrivial site-symmetry group. This suggests that one should classify the points in $\mathbb{E}^3$ into equivalence classes according to their site-symmetry groups.

Definition

A point $X \in \mathbb{E}^3$ is called a point in a *general position* for the space group $\mathcal{G}$ if its site-symmetry group contains only the identity element of $\mathcal{G}$. Otherwise, X is called a point in a *special position*.

The distinctive feature of a point in a general position is that the points in its orbit are in one-to-one correspondence with the symmetry operations of the group $\mathcal{G}$ by associating the orbit point $g(X)$ with the group operation g. For different group elements g and g', the orbit points $g(X)$ and $g'(X)$ must be different, since otherwise $g^{-1}g'$ would be a non-trivial operation in the site-symmetry group of X. Therefore, the entries listed in the space-group tables for the general positions can not only be interpreted as a shorthand notation for the symmetry operations in $\mathcal{G}$ (as seen in Section 1.4.2.3), but also as coordinates of the points in the orbit of a point X in a general position with coordinates x, y, z (up to translations).

Whereas points in general positions exist for every space group, not every space group has points in a special position. Such groups are called *fixed-point-free space groups* or *Bieberbach groups* and are precisely those groups that may contain glide reflections or screw rotations, but no proper reflections, rotations, inversions and rotoinversions.

Example

The group $\mathcal{G}$ of type $Pna2_1$ (33) has a point group of order 4 and representatives for the non-trivial cosets relative to the translation subgroup are the twofold screw rotation $\bar{x}, \bar{y}, z + \frac{1}{2}$, the a glide $x + \frac{1}{2}, \bar{y} + \frac{1}{2}, z$ and the n glide $\bar{x} + \frac{1}{2}, y + \frac{1}{2}, z + \frac{1}{2}$. No operation in the coset of the twofold screw rotation can have a fixed point, since such an operation maps the z component to $z + \frac{1}{2} + t_z$ for an integer t_z, and this is never equal to z. The same argument applies to the x component of the a glide and to the y component of the n glide, hence this group contains no operation with a fixed point (apart from the identity element) and is thus a fixed-point-free space group.

The distinction into general and special positions is of course very coarse. In a finer classification, it is certainly desirable that two points in the same orbit under the space group belong to the same class, since they are symmetry equivalent. Such points have *conjugate* site-symmetry groups (*cf.* the orbit–stabilizer theorem in Section 1.1.6).

Lemma

Let X and Y be points in the same orbit of a space group $\mathcal{G}$ and let $g \in \mathcal{G}$ such that $g(X) = Y$. Then the site-symmetry groups of X and Y are conjugate by the operation mapping X to Y, *i.e.* one has $\mathcal{S}_Y = g \cdot \mathcal{S}_X \cdot g^{-1}$.

The classification motivated by the conjugacy relation between the site-symmetry groups of points in the same orbit is the classification into *Wyckoff positions*.

Definition

Two points X and Y in $\mathbb{E}^3$ belong to the same *Wyckoff position* with respect to $\mathcal{G}$ if their site-symmetry groups $\mathcal{S}_X$ and $\mathcal{S}_Y$ are conjugate subgroups of $\mathcal{G}$.

In particular, the Wyckoff position containing a point X also contains the full orbit $\mathcal{G}(X)$ of X under $\mathcal{G}$.

Remark: It is built into the definition of Wyckoff positions that points that are related by a symmetry operation of $\mathcal{G}$ belong to the same Wyckoff position. However, a single site-symmetry group may have more than one fixed point, *e.g.* points on the same rotation axis or in the same reflection plane. These points are in general not symmetry related but, having identical site-symmetry groups, clearly belong to the same Wyckoff position. This situation can be analyzed more explicitly:

Let $\mathcal{S}_X$ be the site-symmetry group of the point X and assume that Y is another point with the same site-symmetry group

$\mathcal{S}_Y = \mathcal{S}_X$. Choosing a coordinate system with origin X, the operations in $\mathcal{S}_X$ all have translational part equal to zero and are thus matrix–column pairs of the form $(\boldsymbol{W}, \boldsymbol{o})$. In particular, these operations are *linear* operations, and since both points X and Y are fixed by all operations in $\mathcal{S}_X$, the vector $\mathbf{v} = Y - X$ is also fixed by the linear operations $(\boldsymbol{W}, \boldsymbol{o})$ in $\mathcal{S}_X$. But with the vector $\mathbf{v}$ each scaling $c \cdot \mathbf{v}$ of $\mathbf{v}$ is fixed as well, and therefore all the points on the line through X and Y are fixed by the operations in $\mathcal{S}_X$. This shows that the Wyckoff position of X is a union of infinitely many orbits if $\mathcal{S}_X$ has more than one fixed point.

Lemma

Let $\mathcal{S}_X$ be the site-symmetry group of X in $\mathcal{G}$:

(i) The points belonging to the same Wyckoff position as X are precisely the points in the orbit of X under $\mathcal{G}$ if and only if X is the only point fixed by all operations in $\mathcal{S}_X$. In this case the coordinates of a point belonging to this Wyckoff position have fixed values not depending on a parameter.

(ii) If Y is a further point fixed by all operations in $\mathcal{S}_X$ but there is no fixed point of $\mathcal{S}_X$ outside the line through X and Y, then all the points on the line through X and Y are fixed by $\mathcal{S}_X$. The Wyckoff position of X is then the union of the orbits of points on this line (with the exception of a possibly empty discrete subset of points which have a larger site-symmetry group). In this case the coordinates of a point belonging to this Wyckoff position have values depending on a single variable parameter.

(iii) If Y and Z are points fixed by all operations in $\mathcal{S}_X$ such that X, Y, Z do not lie on a line, then all the points on the plane through X, Y and Z are fixed by $\mathcal{S}_X$. The Wyckoff position of X is then the union of the orbits of points in this plane with the exception of a (possibly empty) discrete subset of lines or points which have a larger site-symmetry group. In this case the coordinates of a point belonging to this Wyckoff position have values depending on two variable parameters.

(iv) Only the points belonging to the general position depend on three variable parameters.

The space-group tables of *IT* A (see also the space-group tables of Part 2) contain the following information about the Wyckoff positions of a space group $\mathcal{G}$:

Multiplicity: The Wyckoff multiplicity is the number of points in an orbit for this Wyckoff position which lie in the conventional cell. For a group with a primitive unit cell, the multiplicity for the general position equals the order of the point group of $\mathcal{G}$, while for a centred cell this is multiplied by the quotient of the volumes of the conventional cell and a primitive unit cell.

The quotient of the multiplicity for the general position by that of a special position gives the order of the site-symmetry group of the special position.

Wyckoff letter: Each Wyckoff position is labelled by a letter in alphabetical order, starting with '*a*' for a position with site-symmetry group of maximal order and ending with the highest letter (corresponding to the number of different Wyckoff positions) for the general position.

It is common to specify a Wyckoff position by its multiplicity and Wyckoff letter, *e.g.* by 4*a* for a position with multiplicity 4 and letter *a*.

Site symmetry: The point group isomorphic to the site-symmetry group is indicated by an *oriented symbol*, which is a variation of the Hermann–Mauguin point-group symbol that provides information about the orientation of the symmetry elements. The constituents of the oriented symbol are ordered according to the symmetry directions of the corresponding crystal lattice (primary, secondary and tertiary). A symmetry operation in the site-symmetry group gives rise to a symbol in the position corresponding to the direction of its geometric element. Directions for which no symmetry operation contributes to the site-symmetry group are represented by a dot in the oriented symbol.

Coordinates: Under this heading, the coordinates of the points in an orbit belonging to the Wyckoff position are given, possibly depending on one or two variable parameters (three for the general position). The points given represent the orbit up to translations from the full translational subgroup. For a space group with a centred lattice, centring vectors which are coset representatives for the translation lattice relative to the lattice spanned by the conventional basis are given at the top of the table. To obtain representatives of the orbit up to translations from the lattice spanned by the conventional basis, these centring vectors have to be added to each of the given points.

As already mentioned, the coordinates given for the general position can also be interpreted as a compact notation for the symmetry operations, specified up to translations.

The entries in the last column, the *reflection conditions*, are discussed in detail in Chapter 1.6. This column lists the conditions for the reflection indices *hkl* for which the corresponding structure factor is not systematically zero.

Examples

(1) Let $\mathcal{G}$ be the space group of type *Pbca* (61) generated by the twofold screw rotations $\{2_{001}|\frac{1}{2},0,\frac{1}{2}\}$: $x+\frac{1}{2}, \bar{y}, z+\frac{1}{2}$ and $\{2_{010}|0,\frac{1}{2},\frac{1}{2}\}$: $\bar{x}, y+\frac{1}{2}, \bar{z}+\frac{1}{2}$, the inversion $\{\bar{1}|0\}$: $\bar{x},\bar{y},\bar{z}$ and the translations $t(1,0,0)$, $t(0,1,0)$, $t(0,0,1)$.

Applying the eight coset representatives of $\mathcal{G}$ with respect to the translation subgroup, the points in the orbit of the origin $X_1 = \begin{pmatrix}0\\0\\0\end{pmatrix}$ that lie in the unit cell are found to be X_1, $X_2 = \begin{pmatrix}\frac{1}{2}\\0\\\frac{1}{2}\end{pmatrix}$, $X_3 = \begin{pmatrix}0\\\frac{1}{2}\\\frac{1}{2}\end{pmatrix}$ and $X_4 = \begin{pmatrix}\frac{1}{2}\\\frac{1}{2}\\0\end{pmatrix}$, and the Wyckoff position to which X_1 belongs has multiplicity 4 and is labelled 4*a*.

Since the point group $\mathcal{P}$ of $\mathcal{G}$ has order 8, the site-symmetry group $\mathcal{S}_{X_1}$ has order 8/4 = 2. The inversion in the origin X_1 obviously fixes X_1, hence $\mathcal{S}_{X_1} = \{\{1|0\}, \{\bar{1}|0\}\}$. The oriented symbol for the site symmetry is $\bar{1}$, indicating that the site-symmetry group is generated by an inversion.

The points X_2, X_3 and X_4 belong to the same Wyckoff position as X_1, since they lie in the orbit of X_1 and thus have conjugate site-symmetry groups.

The point $Y_1 = \begin{pmatrix}0\\0\\\frac{1}{2}\end{pmatrix}$ also has an orbit with 4 points in the unit cell, namely Y_1, $Y_2 = \begin{pmatrix}\frac{1}{2}\\0\\0\end{pmatrix}$, $Y_3 = \begin{pmatrix}0\\\frac{1}{2}\\0\end{pmatrix}$ and $Y_4 = \begin{pmatrix}\frac{1}{2}\\\frac{1}{2}\\\frac{1}{2}\end{pmatrix}$. These points therefore belong to a common Wyckoff position, namely position 4*b*. Moreover, the site-symmetry group of Y_1 is also generated by an inversion, namely the inversion $\{\bar{1}|0,0,1\}$: $\bar{x},\bar{y},\bar{z}+1$ located at Y_1 and is thus denoted by the oriented symbol $\bar{1}$.

Positions

Multiplicity, Wyckoff letter, Site symmetry	Coordinates			
8 *d* 1	(1) x,y,z	(2) $\bar{x},\bar{y},z$	(3) $\bar{y},x,z$	(4) $y,\bar{x},z$
	(5) $x+\frac{1}{2},\bar{y}+\frac{1}{2},z$	(6) $\bar{x}+\frac{1}{2},y+\frac{1}{2},z$	(7) $\bar{y}+\frac{1}{2},\bar{x}+\frac{1}{2},z$	(8) $y+\frac{1}{2},x+\frac{1}{2},z$

Figure 1.4.4.1
General-position block as given in the space-group tables for space group *P*4*bm* (100).

The points X_1 and Y_1 do not belong to the same Wyckoff position, because an operation $(\boldsymbol{W}, \boldsymbol{w})$ in $\mathcal{G}$ conjugates the inversion $\{\bar{1}|0, 0, 0\}$ in the origin to an inversion in $\boldsymbol{w}$. Since the translational parts of the operations in $\mathcal{G}$ are (up to integers) $(0, 0, 0)$, $(\frac{1}{2}, \frac{1}{2}, 0)$, $(\frac{1}{2}, 0, \frac{1}{2})$ and $(0, \frac{1}{2}, \frac{1}{2})$, an inversion in $Y_1 = \begin{pmatrix} 0 \\ 0 \\ \frac{1}{2} \end{pmatrix}$ can not be obtained by conjugation with operations from $\mathcal{G}$.

(2) Let $\mathcal{G}$ be the space group of type *P*4*bm* (100) generated by the fourfold rotation $\{4^+|0\}$: $\bar{y}, x, z$, the glide reflection (of *b* type) $\{m_{100}|\frac{1}{2}, \frac{1}{2}, 0\}$: $\bar{x}+\frac{1}{2}, y+\frac{1}{2}, z$ and the translations $t(1, 0, 0)$, $t(0, 1, 0)$, $t(0, 0, 1)$. The general-position coordinate triplets are shown in Fig. 1.4.4.1
From this information, the coordinates for the orbit of a specific point X in a special position can be derived by simply inserting the coordinates of X into the general-position coordinates, normalizing to values between 0 and 1 (by adding ± 1 if required) and eliminating duplicates.
For example, for the point $X = \begin{pmatrix} \frac{1}{2} \\ 0 \\ \frac{1}{4} \end{pmatrix}$ in Wyckoff position 2*b* one obtains X and $Y = \begin{pmatrix} 0 \\ \frac{1}{2} \\ \frac{1}{4} \end{pmatrix}$ as the points in the orbit of X that lie in the unit cell. Since the point group $\mathcal{P}$ of $\mathcal{G}$ has order 8, the site-symmetry group $\mathcal{S}_X$ is a group of order 8/2 = 4. Its four operations are

Coordinate triplet	Description
x, y, z	Identity operation
$\bar{x}+1, \bar{y}, z$	Twofold rotation with axis $\frac{1}{2}, 0, z$
$\bar{y}+\frac{1}{2}, \bar{x}+\frac{1}{2}, z$	Reflection with plane $x+\frac{1}{2}, -x, z$
$y+\frac{1}{2}, x-\frac{1}{2}, z$	Reflection with plane $x+\frac{1}{2}, x, z$

The corresponding oriented symbol for the site-symmetry is 2.*mm*, indicating that the site-symmetry group contains a twofold rotation along a primary lattice direction, no symmetry operations along the secondary directions and two reflections along tertiary directions.
Since X and Y lie in the same orbit, they clearly belong to the same Wyckoff position. But every point $X' = \begin{pmatrix} \frac{1}{2} \\ 0 \\ z \end{pmatrix}$ with $0 \le z < 1$ has the same site-symmetry group as X and therefore also belongs to the same Wyckoff position as X. Inserting the coordinates of X' in the general-position coordinates, one obtains $Y' = \begin{pmatrix} 0 \\ \frac{1}{2} \\ z \end{pmatrix}$ as the only other point in the orbit of X' that lies in the unit cell. Clearly, Y' has the same site-symmetry group as Y. The Wyckoff position 2*b* to which X belongs therefore consists of the union of the orbits of the points $X' = \begin{pmatrix} \frac{1}{2} \\ 0 \\ z \end{pmatrix}$ with $0 \le z < 1$.
In the space-group diagram in Fig. 1.4.4.2, the points belonging to Wyckoff position 2*b* can be identified as the points on the intersection of a twofold rotation axis directed along [001] and two reflection planes normal to the square diagonals and crossing the centres of the sides bordering the unit cell. It is clear that for every value of z, the four intersection points in the unit cell lie in one orbit under the fourfold rotation located in the centre of the displayed cell.
Applying the same procedure to a point $X = \begin{pmatrix} 0 \\ 0 \\ z \end{pmatrix}$ in Wyckoff position 2*a*, the points in the orbit that lie in the unit cell are seen to be X and $Y = \begin{pmatrix} \frac{1}{2} \\ \frac{1}{2} \\ z \end{pmatrix}$. The site-symmetry group $\mathcal{S}_X$ is again of order 4 and since the fourfold rotation $\{4^+|0\}$ fixes X, $\mathcal{S}_X$ is the cyclic group of order 4 generated by this fourfold rotation. The oriented symbol for this site-symmetry group is 4.. and the corresponding points can easily be identified in the space-group diagram in Fig. 1.4.4.2 by the symbol for a fourfold rotation.
Since a point in a special position has to lie on the geometric element of a reflection, rotation or inversion, the special positions can in principle be read off from the space-group diagrams. In the present example, we have dealt with the positions fixed by twofold or fourfold rotations, and from the diagram in Fig. 1.4.4.2 one sees that the only remaining case is that of points on reflection planes, indicated by the solid lines. A point on such a reflection plane is $X = \begin{pmatrix} x \\ x+\frac{1}{2} \\ z \end{pmatrix}$ and by inserting these coordinates

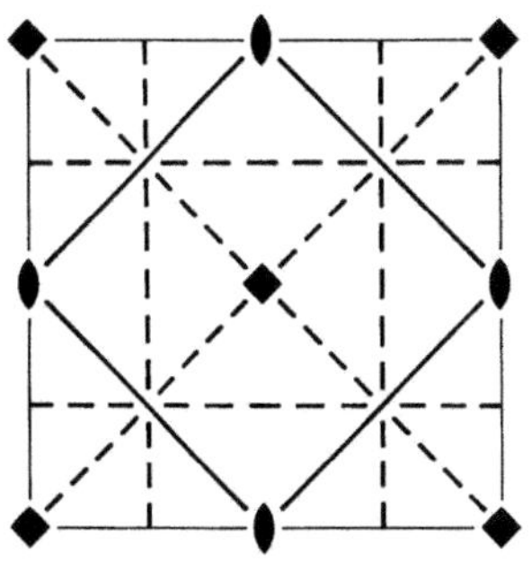

Figure 1.4.4.2
Symmetry-element diagram for the space group *P*4*bm* (100) for the orthogonal projection along [001].

into the general-position coordinates one obtains the points $\bar{x}, \bar{x}+\frac{1}{2}, z$, $\bar{x}+\frac{1}{2}, x, z$ and $x+\frac{1}{2}, \bar{x}, z$ as the other points in the orbit of X (up to translations). Here, the site-symmetry group $\mathcal{S}_X$ is of order 2, it is generated by the reflection $\{m_{1\bar{1}0}|-\frac{1}{2},\frac{1}{2},0\}$: $y-\frac{1}{2}, x+\frac{1}{2}, z$ having the plane $x, x+\frac{1}{2}, z$ as geometric element. The oriented symbol of $\mathcal{S}_X$ is $..m$, since the reflection is along a tertiary direction.

1.4.5. Sections and projections of space groups

By Bernd Souvignier

In crystallography, two-dimensional sections and projections of crystal structures play an important role, *e.g.* in structure determination by Fourier and Patterson methods or in the treatment of twin boundaries and domain walls. Planar sections of three-dimensional scattering density functions are used for finding approximate locations of atoms in a crystal structure. They are indispensable for the location of Patterson peaks corresponding to vectors between equivalent atoms in different asymmetric units (the Harker vectors).

1.4.5.1. Introduction

A two-dimensional section of a crystal pattern takes out a slice of a crystal pattern. In the mathematical idealization, this slice is regarded as a two-dimensional plane, allowing one, however, to distinguish its upper and lower side. Depending on how the slice is oriented with respect to the crystal lattice, the slice will be invariant by translations of the crystal pattern along zero, one or two linearly independent directions. A section resulting in a slice with two-dimensional translational symmetry is called a *rational section* and is by far the most important case for crystallography.

Because the slice is regarded as a two-sided plane, the symmetries of the full crystal pattern that leave the slice invariant fall into two types:

(i) If a symmetry operation of the slice maps its upper side to the upper side, a vector normal to the slice is fixed.

(ii) If a symmetry operation of the slice maps the upper side to the lower side, a vector normal to the slice is mapped to its opposite and the slice is turned upside down.

Therefore, the symmetries of two-dimensional rational sections are described by *layer groups*, *i.e.* subgroups of space groups with a two-dimensional translation lattice. Layer groups are *subperiodic groups* and for their elaborate discussion we refer to Section 1.7.3 and *IT* E (2010).

Analogous to two-dimensional sections of a crystal pattern, one can also consider the penetration of crystal patterns by a straight line, which is the idealization of a one-dimensional section taking out a rod of the crystal pattern. If the penetration line is along the direction of a translational symmetry of the crystal pattern, the rod has one-dimensional translational symmetry and its group of symmetries is a *rod group*, *i.e.* a subgroup of a space group with a one-dimensional translation lattice. Rod groups are also subperiodic groups, *cf.* *IT* E for their detailed treatment and listing.

A projection along a direction **d** into a plane maps a point of a crystal pattern to the intersection of the plane with the line along **d** through the point. If the projection direction is not along a rational lattice direction, the projection of the crystal pattern will contain points with arbitrarily small distances and additional restrictions are required to obtain a discrete pattern (*e.g.* the cut-and-project method used in the context of quasicrystals). We avoid any such complication by assuming that **d** is along a rational lattice direction. Furthermore, one is usually only interested in *orthogonal* projections in which the projection direction is perpendicular to the projection plane. This has the effect that spheres in three-dimensional space are mapped to circles in the projection plane.

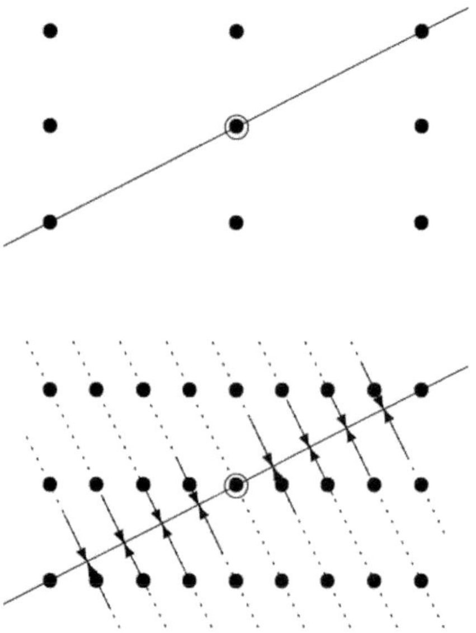

Figure 1.4.5.1
Duality between section and projection.

Although it is also possible to regard the projection plane as a two-sided plane by taking into account from which side of the plane a point is projected into it, this is usually not done. Therefore, the symmetries of projections are described by ordinary plane groups.

Sections and projections are related by the *projection–slice theorem* (Bracewell, 2003) of Fourier theory: A section in reciprocal space containing the origin (the so-called zero layer) corresponds to a projection in direct space and *vice versa*. The projection direction in the one space is normal to the slice in the other space. This correspondence is illustrated schematically in Fig. 1.4.5.1. The top part shows a rectangular lattice with $b/a = 2$ and a slice along the line defined by $2x + y = 0$. Normalizing $a = 1$, the distance between two neighbouring lattice points in the slice is $\sqrt{5}$. If the pattern is restricted to this slice, the points of the corresponding diffraction pattern in reciprocal space must have distance $1/\sqrt{5}$ and this is precisely obtained by projecting the lattice points of the reciprocal lattice onto the slice.

The different, but related, viewpoints of sections and projections can be stated in a simple way as follows: For a section perpendicular to the c axis, only those points of a crystal pattern are considered which have z coordinate equal to a fixed value z_0 or in a small interval around z_0. For a projection along the c axis, all points of the crystal pattern are considered, but their z coordinate is simply ignored. This means that all points of the crystal pattern that differ only by their z coordinate are regarded as the same point.

1.4.5.2. Sections

For a space group $\mathcal{G}$ and a point X in the three-dimensional point space $\mathbb{E}^3$, the site-symmetry group of X is the subgroup of operations of $\mathcal{G}$ that fix X. Analogously, one can also look at the subgroup of operations fixing a one-dimensional line or a two-dimensional plane. If the line is along a rational direction, it will be fixed at least by the translations of $\mathcal{G}$ along that direction. However, it may also be fixed by a symmetry operation that reverses the direction of the line. The resulting subgroup of $\mathcal{G}$ that fixes the line is a *rod group*.

Similarly, a plane having a normal vector along a rational direction is fixed by translations of $\mathcal{G}$ corresponding to a two-dimensional lattice. Again, the plane may also be fixed by additional symmetry operations, *e.g.* by a twofold rotation around an axis lying in the plane, by a rotation around an axis normal to the plane or by a reflection in the plane.

Definition

A *rational planar section* of a crystal pattern is the intersection of the crystal pattern with a plane containing two linearly independent translation vectors of the crystal pattern. The intersecting plane is called the *section plane*.

A *rational linear section* of a crystal pattern is the intersection of the crystal pattern with a line containing a translation vector of the crystal pattern. The intersecting line is called the *penetration line*.

A planar section is determined by a vector **d** which is perpendicular to the section plane and a continuous parameter s, called the *height*, which gives the position of the plane on the line along **d**.

A linear section is specified by a vector **d** parallel to the penetration line and a point in a plane perpendicular to **d** giving the intersection of the line with that plane.

Definition

(i) The symmetry group of a planar section of a crystal pattern is the subgroup of the space group $\mathcal{G}$ of the crystal pattern that leaves the section plane invariant as a whole.

If the section is a rational section, this symmetry group is a *layer group*, *i.e.* a subgroup of a space group which contains translations only in a two-dimensional plane.

(ii) The symmetry group of a linear section of a crystal pattern is the subgroup of the space group $\mathcal{G}$ of the crystal pattern that leaves the penetration line invariant as a whole.

If the section is a rational section, this symmetry group is a *rod group*, *i.e.* a subgroup of a space group which contains translations only along a one-dimensional line.

From now on we will only consider rational sections and omit this attribute. Moreover, we will concentrate on the case of planar sections, since this is by far the most relevant case for crystallographic applications. The treatment of one-dimensional sections is analogous, but in general much easier.

Let **d** be a vector perpendicular to the section plane. In most cases, **d** is chosen as the shortest lattice vector perpendicular to the section plane. However, in the triclinic and monoclinic crystal family this may not be possible, since the translations of the crystal pattern may not contain a vector perpendicular to the section plane. In that case, we assume that **d** captures the periodicity of the crystal pattern perpendicular to the section plane. This is achieved by choosing **d** as the shortest non-zero projection of a lattice vector to the line through the origin which is perpendicular to the section plane. Because of the periodicity of the crystal pattern along **d**, it is enough to consider heights s with $0 \le s < 1$, since for an integer m the sectional layer groups at heights s and $s + m$ are conjugate subgroups of $\mathcal{G}$. This is a consequence of the orbit–stabilizer theorem in Section 1.1.6, applied to the group $\mathcal{G}$ acting on the planes in $\mathbb{E}^3$. The layer at height s is mapped to the layer at height $s + m$ by the translation through $m\mathbf{d}$. Thus, the two layers lie in the same orbit under $\mathcal{G}$. According to the orbit–stabilizer theorem, the corresponding stabilizers, being just the layer groups at heights s and $s + m$, are then conjugate by the translation through $m\mathbf{d}$.

Since we assume a rational section, the sectional layer group will always contain translations along two independent directions $\mathbf{a}'$, $\mathbf{b}'$ which, we assume, form a crystallographic basis for the lattice of translations fixing the section plane. The points in the section plane at height s are then given by $x\mathbf{a}' + y\mathbf{b}' + s\mathbf{d}$. In order to determine whether the sectional layer group contains additional symmetry operations which are not translations, the following simple remark is crucial:

Let g be an operation of a sectional layer group. Then the rotational part of g maps **d** either to $+\mathbf{d}$ or to $-\mathbf{d}$. In the former case, g is side-preserving, in the latter case it is side-reversing. Moreover, since the section plane remains fixed under g, the vectors $\mathbf{a}'$ and $\mathbf{b}'$ are mapped to linear combinations of $\mathbf{a}'$ and $\mathbf{b}'$ by the rotational part of g. Therefore, with respect to the (usually non-conventional) basis $\mathbf{a}'$, $\mathbf{b}'$, **d** of three-dimensional space and some choice of origin, the operation g is represented by the matrix–column pair

$$\left[\begin{pmatrix} r_{11} & r_{12} & 0 \\ r_{21} & r_{22} & 0 \\ 0 & 0 & r_{33} \end{pmatrix}, \begin{pmatrix} t_1 \\ t_2 \\ t_3 \end{pmatrix}\right].$$

Here, $r_{33} = \pm 1$. Moreover, if $r_{33} = 1$, *i.e.* g is side-preserving, then t_3 is necessarily zero, since otherwise the plane is shifted along **d**. On the other hand, if $r_{33} = -1$, *i.e.* g is side-reversing, then a plane situated at height s along **d** is only fixed if $t_3 = 2s$.

From these considerations it is straightforward to determine the conditions under which a space-group operation belongs to a certain sectional layer group (excluding translations):

The side-preserving operations will belong to the sectional layer groups for all planes perpendicular to **d**, independent of the height s:

(i) rotations with axis parallel to **d**;
(ii) reflections with normal vector perpendicular to **d**;
(iii) glide reflections with normal vector and glide vector perpendicular to **d**.

Side-reversing operations will only occur in the sectional layer groups for planes at special heights along **d**:

(i) inversion with inversion point in the section plane;
(ii) twofold rotations or twofold screw rotations with rotation axis in the section plane;
(iii) reflections or glide reflections through the section plane with glide vector perpendicular to **d**;
(iv) rotoinversions with axis parallel to **d** and inversion point in the section plane.

Note that, because of the periodicity along **d**, a side-reversing operation that occurs at height s gives rise to a side-reversing operation of the same type occurring at height $s + \frac{1}{2}$: if g is a side-reversing symmetry operation fixing a layer at height s, then g maps a point in the layer at height $s + \frac{1}{2}$ with coordinates $x, y, s + \frac{1}{2}$ (with respect to the layer-adapted basis $\mathbf{a}'$, $\mathbf{b}'$, **d**) to a point with coordinates $x', y', s - \frac{1}{2}$ and hence the composition $t_{\mathbf{d}}g$ of g with the translation by **d** maps $x, y, s + \frac{1}{2}$ to $x', y', s + \frac{1}{2}$, *i.e.* it fixes the layer at height $s + \frac{1}{2}$. This shows that the composition with the translation by **d** provides a one-to-one correspondence between the side-reversing symmetry operations in the layer group at height s with those at height $s + \frac{1}{2}$.

Table 1.4.5.1
Coset representatives of $Pmn2_1$ (31) relative to its translation subgroup

Seitz symbol	Coordinate triplet	Description
$\{1\|0\}$	x, y, z	Identity
$\{2_{001}\|\frac{1}{2}, 0, \frac{1}{2}\}$	$\bar{x}+\frac{1}{2}, \bar{y}, z+\frac{1}{2}$	Twofold screw rotation with axis along [001]
$\{m_{010}\|\frac{1}{2}, 0, \frac{1}{2}\}$	$x+\frac{1}{2}, \bar{y}, z+\frac{1}{2}$	*n*-glide reflection with normal vector along [010]
$\{m_{100}\|0\}$	$\bar{x}, y, z$	Reflection with normal vector along [100]

If a section allows any side-reversing symmetry at all, then the side-preserving symmetries of the section form a subgroup of index 2 in the sectional layer group. Since the side-preserving symmetries exist independently of the height parameter s, the full sectional layer group is always generated by the side-preserving subgroup and either none or a single side-reversing symmetry.

Summarizing, one can conclude that for a given space group the interesting sections are those for which the perpendicular vector **d** is parallel or perpendicular to a symmetry direction of the group, *e.g.* an axis of a rotation or rotoinversion or the normal vector of a reflection or glide reflection.

Example

Consider the space group $\mathcal{G}$ of type $Pmn2_1$ (31). In its standard setting, the cosets of $\mathcal{G}$ relative to the translation subgroup are represented by the operations given in Table 1.4.5.1.

Since this is an orthorhombic group, it is natural to consider sections along the coordinate axes. The space-group diagrams displayed in Fig. 1.4.5.2, which show the orthogonal projections of the symmetry elements along these directions, are very helpful.

d *along* [100]: A point x, y, z in a plane perpendicular to the coordinate axis along [100] is mapped to a point x', y', z' in the same plane if $x' = x$, *i.e.* if $x' - x = 0$.

A general operation from the coset of $\{2_{001}|\frac{1}{2}, 0, \frac{1}{2}\}$ maps a point with coordinates x, y, z to a point with coordinates $x' = \bar{x} + \frac{1}{2} + u_1, y' = \bar{y} + u_2, z' = z + \frac{1}{2} + u_3$ for integers u_1, u_2, u_3. One has $x' - x = -2x + \frac{1}{2} + u_1$ which becomes zero for $x = \frac{1}{4}$ (and $u_1 = 0$) and $x = \frac{3}{4}$ (and $u_1 = 1$), thus operations from the coset of $\{2_{001}|\frac{1}{2}, 0, \frac{1}{2}\}$ fix planes at heights $s = \frac{1}{4}$ and $\frac{3}{4}$. In the left-hand diagram in Fig. 1.4.5.2, the symmetry elements to which these operations belong are indicated by the half-arrows, the label $\frac{1}{4}$ indicating that they are at level $x = \frac{1}{4}$ and $x = \frac{3}{4}$.

An operation from the coset of $\{m_{010}|\frac{1}{2}, 0, \frac{1}{2}\}$ maps x, y, z to $x' = x + \frac{1}{2} + u_1, y' = \bar{y} + u_2, z' = z + \frac{1}{2} + u_3$ and one has $x' - x = \frac{1}{2} + u_1$. Since this is never zero, no operation from this coset fixes a plane perpendicular to [100].

Finally, an operation from the coset of $\{m_{100}|0\}$ maps x, y, z to $x' = \bar{x} + u_1, y' = y + u_2, z' = z + u_3$ and one has $x' - x = -2x + u_1$, which becomes zero for $x = 0$ (and $u_1 = 0$) and $x = \frac{1}{2}$ (and $u_1 = 1$). Thus, operations from the coset of $\{m_{100}|0\}$ fix planes at heights $s = 0$ and $\frac{1}{2}$. The symmetry elements of these reflections with mirror plane parallel to the projection plane are indicated by the right-angle symbol in the upper left corner of the left-hand diagram in Fig. 1.4.5.2.

The sectional layer groups are thus layer groups of type $pm11$ (layer group No. 4 with symbol $p11m$ in a non-standard setting) for $s = 0$ and $s = \frac{1}{2}$, of type $p112_1$ (layer group No. 9 with symbol $p2_111$ in a non-standard setting) for $s = \frac{1}{4}$ and $s = \frac{3}{4}$ and of type $p1$ (layer group No. 1) for all other s between 0 and 1. The side-preserving operations are in all cases just the translations.

It is worthwhile noting that in many cases most of the information about the sectional layer groups can be read off the space-group diagrams. In the present example, the left-hand diagram in Fig. 1.4.5.2 displays the twofold screw rotation at height $s = \frac{1}{4}$ (and thus also at $s = \frac{3}{4}$) and the reflection at height $s = 0$ (and thus also at $s = \frac{1}{2}$). On the other hand, the n glide, indicated by the dashed-dotted lines in the diagram, does not give rise to an element of the sectional layer group, because its glide vector has a component along the [100] direction and can thus not fix any layer along this direction.

d *along* [010]: A point x, y, z in a plane perpendicular to the coordinate axis along [010] is mapped to a point x', y', z' in the same plane if $y' = y$, *i.e.* if $y' - y = 0$.

From the calculations above one sees that for operations in the coset of $\{m_{100}|0\}$ one has $y' - y = u_2$, hence operations in this coset fix the plane for any value of s and are side-preserving operations. In the middle diagram in Fig. 1.4.5.2 the symmetry elements for these reflections are indicated by the horizontal solid lines.

For the operations in the coset of $\{2_{001}|\frac{1}{2}, 0, \frac{1}{2}\}$ one has $y' - y = -2y + u_2$, and so these operations fix planes only for $s = 0$ and $s = \frac{1}{2}$. The same is true for the operations in the coset of $\{m_{010}|\frac{1}{2}, 0, \frac{1}{2}\}$, because here one also has $y' - y = -2y + u_2$. The symmetry elements to which the screw rotations belong

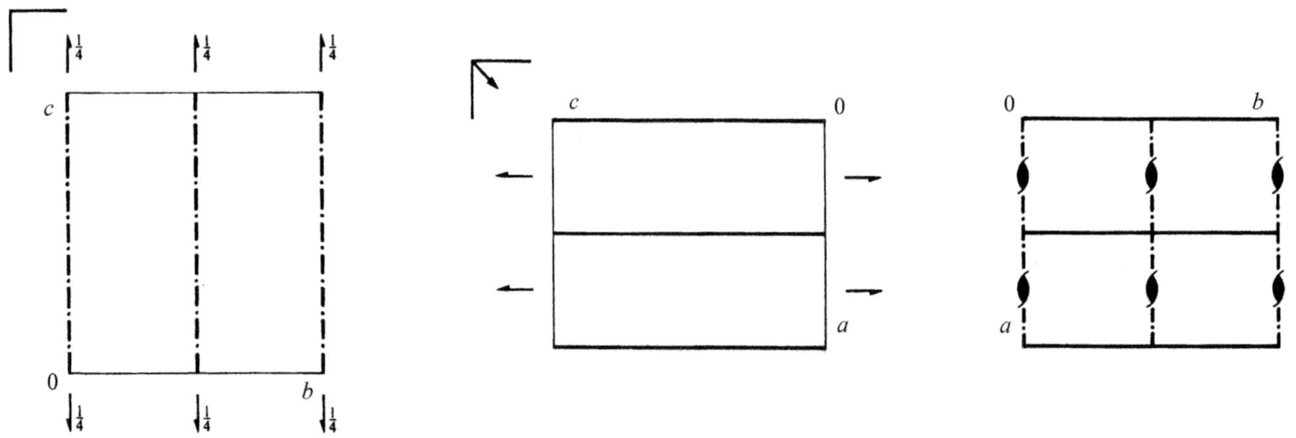

Figure 1.4.5.2
Symmetry-element diagrams for the space group $Pmn2_1$ (31) for orthogonal projections along [100] (left), [010] (middle) and [001] (right).

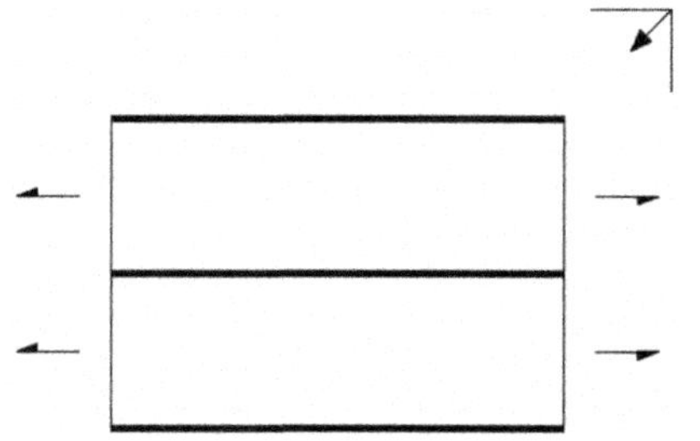

Figure 1.4.5.3
Symmetry-element diagram for the layer group $pm2_1n$ (32).

are indicated by the half arrows in the middle diagram of Fig. 1.4.5.2, and the symmetry elements for the glide reflections are symbolized by the right angle with diagonal arrow in the upper left corner, indicating that the geometric element is a diagonal glide plane.

The sectional layer groups are thus of type $pmn2_1$ (layer group No. 32 with symbol $pm2_1n$ in a non-standard setting) for $s = 0, \frac{1}{2}$ and of type $pm11$ (layer group No. 11) for all other s. The group of side-preserving operations is in all cases of type $pm11$.

In Fig. 1.4.5.3 the diagram of the symmetry elements for the layer group $pm2_1n$ (layer group No. 32) is displayed. It coincides with the middle diagram in Fig. 1.4.5.2 (up to the placement of the symbol for the diagonal glide plane), showing that in this case the sectional layer groups can also be read off directly from the space-group diagrams.

d *along* [001]: A point x, y, z in a plane perpendicular to the coordinate axis along [001] is mapped to a point x', y', z' in the same plane if $z' = z$, *i.e.* if $z' - z = 0$.

As in the case of **d** along [010], operations in the coset of $\{m_{100}|0\}$ fix such a plane for any value of s, since $z' - z = u_3$. Again, these are side-preserving operations. The symmetry elements to which these reflections belong are indicated by the horizontal solid lines in the right-hand diagram in Fig. 1.4.5.2. For the operations in the cosets of $\{2_{001}|\frac{1}{2}, 0, \frac{1}{2}\}$ and $\{m_{010}|\frac{1}{2}, 0, \frac{1}{2}\}$ one has $z' - z = \frac{1}{2} + u_3$, which is never zero (for an integer u_3), and so operations in these cosets never fix a plane perpendicular to [001].

Thus, for any value of s the sectional layer group is of type $pm11$ (layer group No. 11) and contains only side-preserving operations.

1.4.5.3. Projections

As we have seen, a section of a crystal pattern is determined by a vector **d** and a height s along this vector. Choosing two vectors $\mathbf{a}'$ and $\mathbf{b}'$ perpendicular to **d**, the points of the section plane at height s are precisely given by the vectors $x\mathbf{a}' + y\mathbf{b}' + s\mathbf{d}$. In contrast to that, a *projection* of a crystal pattern along **d** is obtained by mapping an arbitrary point $x\mathbf{a}' + y\mathbf{b}' + z\mathbf{d}$ to the point $x\mathbf{a}' + y\mathbf{b}'$ of the plane spanned by $\mathbf{a}'$ and $\mathbf{b}'$, thereby ignoring the coordinate along the **d** direction.

Definition

In a *projection* of a crystal pattern along the *projection direction* **d**, a point X of the crystal pattern is mapped to the intersection of the line through X along **d** with a fixed plane perpendicular to **d**.

One may think of the projection plane as the plane perpendicular to **d** and containing the origin, but every plane perpendicular to **d** will give the same result.

Let L be the line along **d**. If a symmetry operation g of a space group $\mathcal{G}$ maps L to a line parallel to L, then g maps every plane perpendicular to **d** again to a plane perpendicular to **d**. This means that points that are projected to a single point (*i.e.* points on a line parallel to L) are mapped by g to points that are again projected to a single point and thus the operation g gives rise to a symmetry of the projection of the crystal pattern. Conversely, an operation g that maps L to a line that is inclined to L does not result in a symmetry of the projection, since the points on L are projected to a single point, whereas the image points under g are projected to a line. In summary, the operations of $\mathcal{G}$ that map L to a line parallel to L give rise to symmetries of the projection forming a *plane group*, sometimes called a wallpaper group.

Let $\mathcal{H}$ be the subgroup of $\mathcal{G}$ consisting of those $g \in \mathcal{G}$ mapping the line L to a line parallel to L, then $\mathcal{H}$ is called the *scanning group along* **d**. The scanning group $\mathcal{H}$ can be read off a coset decomposition $\mathcal{G} = g_1\mathcal{T} \cup \cdots \cup g_s\mathcal{T}$ relative to the translation subgroup $\mathcal{T}$ of $\mathcal{G}$. Since translations map lines to parallel lines, one only has to check whether a coset representative g_i maps L to a line parallel to L. This is precisely the case if the linear part of g_i maps **d** to **d** or to $-\mathbf{d}$. Therefore, $\mathcal{H}$ is the union of those cosets $g_i\mathcal{T}$ relative to $\mathcal{T}$ for which the linear part of g_i maps **d** to **d** or to $-\mathbf{d}$.

If the operations of a space group $\mathcal{G}$ are represented by matrix–column pairs with respect to a (usually non-conventional) basis $\mathbf{a}'$, $\mathbf{b}'$, **d** such that $\mathbf{a}'$ and $\mathbf{b}'$ are perpendicular to **d**, then an operation g of the scanning group $\mathcal{H}$ is of the form

$$g = \left[\begin{pmatrix} r_{11} & r_{12} & 0 \\ r_{21} & r_{22} & 0 \\ 0 & 0 & r_{33} \end{pmatrix}, \begin{pmatrix} t_1 \\ t_2 \\ t_3 \end{pmatrix}\right]$$

with $r_{33} = \pm 1$ (just as for planar sections). Then the action of g on the projection along **d** is obtained by ignoring the z coordinate, *i.e.* by cutting out the upper 2×2 block of the linear part and the first two components of the translation part. This gives rise to the plane-group operation

$$g' = \left[\begin{pmatrix} r_{11} & r_{12} \\ r_{21} & r_{22} \end{pmatrix}, \begin{pmatrix} t_1 \\ t_2 \end{pmatrix}\right].$$

The mapping that assigns to each operation g of the scanning group its action g' on the projection is in fact a homomorphism from $\mathcal{H}$ to a plane group and the kernel $\mathcal{K}$ of this homomorphism are the operations of the form

$$\left[\begin{pmatrix} 1 & 0 & 0 \\ 0 & 1 & 0 \\ 0 & 0 & r_{33} \end{pmatrix}, \begin{pmatrix} 0 \\ 0 \\ t_3 \end{pmatrix}\right],$$

i.e. translations along **d** and reflections with normal vector parallel to **d**.

Definition

The symmetry group of the projection along the projection direction **d** is the plane group of actions on the projection of those operations of $\mathcal{G}$ that map the line L along **d** to a line parallel to L.

This group is isomorphic to the quotient group of the scanning group $\mathcal{H}$ along **d** by the group $\mathcal{K}$ of translations along **d** and reflections with normal vector parallel to **d**.

Example

We consider again the space group $\mathcal{G}$ of type $Pmn2_1$ (31) for which the matrix–column pairs of the coset representatives with respect to the translation subgroup (in the standard setting) are given by

$$\{1|0\} = \left[\begin{pmatrix} 1 & 0 & 0 \\ 0 & 1 & 0 \\ 0 & 0 & 1 \end{pmatrix}, \begin{pmatrix} 0 \\ 0 \\ 0 \end{pmatrix}\right],$$

$$\{2_{001}|\tfrac{1}{2}, 0, \tfrac{1}{2}\} = \left[\begin{pmatrix} \bar{1} & 0 & 0 \\ 0 & \bar{1} & 0 \\ 0 & 0 & 1 \end{pmatrix}, \begin{pmatrix} \frac{1}{2} \\ 0 \\ \frac{1}{2} \end{pmatrix}\right]$$

$$\{m_{010}|\tfrac{1}{2}, 0, \tfrac{1}{2}\} = \left[\begin{pmatrix} 1 & 0 & 0 \\ 0 & \bar{1} & 0 \\ 0 & 0 & 1 \end{pmatrix}, \begin{pmatrix} \frac{1}{2} \\ 0 \\ \frac{1}{2} \end{pmatrix}\right],$$

$$\{m_{100}|0\} = \left[\begin{pmatrix} \bar{1} & 0 & 0 \\ 0 & 1 & 0 \\ 0 & 0 & 1 \end{pmatrix}, \begin{pmatrix} 0 \\ 0 \\ 0 \end{pmatrix}\right].$$

Since the linear parts of all four matrices are diagonal matrices, the scanning group for projections along the coordinate axes is always the full group $\mathcal{G}$.

For the projection along the direction [100], one has to cut out the lower 2×2 part of the linear parts and the second and third component of the translation part, thus choosing $\mathbf{a}' = \mathbf{b}, \mathbf{b}' = \mathbf{c}$ as a basis for the projection plane. This gives as matrices for the projected operations

$$\left[\begin{pmatrix} 1 & 0 \\ 0 & 1 \end{pmatrix}, \begin{pmatrix} 0 \\ 0 \end{pmatrix}\right] \quad \left[\begin{pmatrix} \bar{1} & 0 \\ 0 & 1 \end{pmatrix}, \begin{pmatrix} 0 \\ \frac{1}{2} \end{pmatrix}\right]$$

$$\left[\begin{pmatrix} \bar{1} & 0 \\ 0 & 1 \end{pmatrix}, \begin{pmatrix} 0 \\ \frac{1}{2} \end{pmatrix}\right] \quad \left[\begin{pmatrix} 1 & 0 \\ 0 & 1 \end{pmatrix}, \begin{pmatrix} 0 \\ 0 \end{pmatrix}\right],$$

in which the third and fourth operations are clearly redundant and which is thus a plane group of type $p1g1$ (plane group No. 4 with short symbol pg).

The projection along the direction [010] gives for the basis $\mathbf{a}' = \mathbf{a}, \mathbf{b}' = \mathbf{c}$ of the projection plane (thus picking out the first and third rows and columns) the matrices

$$\left[\begin{pmatrix} 1 & 0 \\ 0 & 1 \end{pmatrix}, \begin{pmatrix} 0 \\ 0 \end{pmatrix}\right] \quad \left[\begin{pmatrix} \bar{1} & 0 \\ 0 & 1 \end{pmatrix}, \begin{pmatrix} \frac{1}{2} \\ \frac{1}{2} \end{pmatrix}\right]$$

$$\left[\begin{pmatrix} 1 & 0 \\ 0 & 1 \end{pmatrix}, \begin{pmatrix} \frac{1}{2} \\ \frac{1}{2} \end{pmatrix}\right], \quad \left[\begin{pmatrix} \bar{1} & 0 \\ 0 & 1 \end{pmatrix}, \begin{pmatrix} 0 \\ 0 \end{pmatrix}\right],$$

where the second matrix is the product of the third and fourth. The third operation is a centring translation, the fourth a reflection, thus the resulting plane group is of type $c1m1$ (plane group No. 5 with short symbol cm).

Finally, the projection along the direction [001] results for the basis $\mathbf{a}' = \mathbf{a}, \mathbf{b}' = \mathbf{b}$ of the projection plane in the matrices

$$\left[\begin{pmatrix} 1 & 0 \\ 0 & 1 \end{pmatrix}, \begin{pmatrix} 0 \\ 0 \end{pmatrix}\right], \quad \left[\begin{pmatrix} \bar{1} & 0 \\ 0 & \bar{1} \end{pmatrix}, \begin{pmatrix} \frac{1}{2} \\ 0 \end{pmatrix}\right],$$

$$\left[\begin{pmatrix} 1 & 0 \\ 0 & \bar{1} \end{pmatrix}, \begin{pmatrix} \frac{1}{2} \\ 0 \end{pmatrix}\right], \quad \left[\begin{pmatrix} \bar{1} & 0 \\ 0 & 1 \end{pmatrix}, \begin{pmatrix} 0 \\ 0 \end{pmatrix}\right],$$

where again the second matrix is the product of two others. The third operation is a glide reflection and the fourth is a reflection, thus the corresponding plane group is of type $p2mg$ (plane group No. 7). Note that in order to obtain the plane group $p2mg$ in its standard setting, the origin has to be shifted to $\frac{1}{4}, 0$ (with respect to the plane basis $\mathbf{a}', \mathbf{b}'$).

As for the sectional layer groups, the typical projection directions considered are symmetry directions of the space group $\mathcal{G}$, *i.e.* directions along rotation or screw axes or normal to reflection or glide planes. In order to relate the coordinate system of the plane group to that of the space group, not only the basis vectors $\mathbf{a}', \mathbf{b}'$ perpendicular to the projection direction $\mathbf{d}$ have to be given, but also the origin for the plane group. This is done by specifying a line parallel to the projection direction which is projected to the origin of the plane group in its conventional setting. The space-group tables list the plane groups for the projections along symmetry directions of the group in the block 'Symmetry of special projections'.

It is not hard to determine the corresponding types of plane-group operations for the different types of space-group operations, as is shown by the following list of simple rules:

(i) a translation becomes a translation (possibly the identity);

(ii) an inversion becomes a twofold rotation;

(iii) a k-fold rotation or screw rotation with axis parallel to $\mathbf{d}$ becomes a k-fold rotation;

(iv) a three-, four- or sixfold rotoinversion with axis parallel to $\mathbf{d}$ becomes a six-, four- or threefold rotation, respectively;

(v) a reflection or glide reflection with normal vector parallel to $\mathbf{d}$ becomes a translation (possibly the identity);

(vi) a twofold rotation and a screw rotation with axis perpendicular to $\mathbf{d}$ become a reflection and glide reflection, respectively;

(vii) a reflection or a glide reflection with normal vector perpendicular to $\mathbf{d}$ becomes a reflection or glide reflection depending on whether there is a glide component perpendicular to $\mathbf{d}$ or not.

The relationship between the symmetry operations in three-dimensional space and the corresponding symmetry operations of a projection as listed above can be seen directly in the diagrams of the corresponding groups. In Fig. 1.4.5.4, the top diagram shows the orthogonal projection of the symmetry-element diagram of $Pmn2_1$ along the [001] direction and the bottom diagram shows the diagram for the plane group $p2mg$, which is precisely the symmetry group of the projection of $Pmn2_1$ along [001]. Firstly, one sees immediately that in order to match the two diagrams, the origin in the projection plane has to be shifted to $\frac{1}{4}, 0$ (as already noted in the example above). Secondly, keeping in mind that the projection direction $\mathbf{d}$ is perpendicular to the drawing plane, one sees the correspondence between the twofold screw rotations in $Pmn2_1$ with the twofold rotations in $p2mg$ [rule (iii)], the correspondence between the reflections with normal vector perpendicular to $\mathbf{d}$ in $Pmn2_1$ and the reflections in $p2mg$ [rule (vii)] and the correspondence between the diagonal glide reflections in $Pmn2_1$ (indicated by the

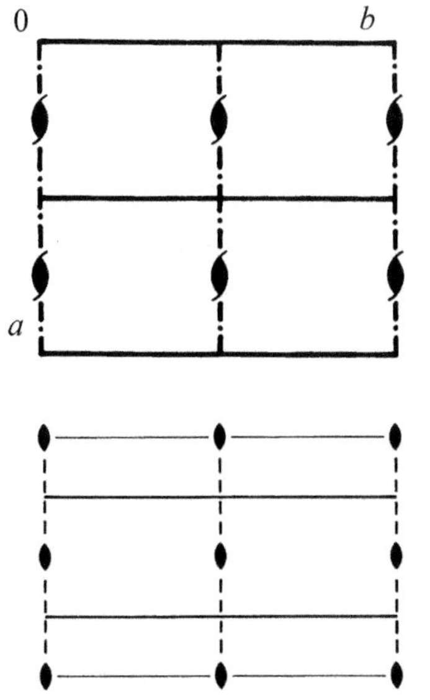

Figure 1.4.5.4
Orthogonal projection along [001] of the symmetry-element diagram for *Pmn*2_1 (31) (top) and the diagram for plane group *p2mg* (7) (bottom).

dot-dash lines) and the glide reflections in *p2mg* {rule (vii); note that the diagonal glide vector has a component perpendicular to the projection direction [001]}.

Example

Let $\mathcal{G}$ be a space group of type $P\bar{4}b2$ (117), then the interesting projection directions (*i.e.* symmetry directions) are [100], [010], [001], [110] and [$1\bar{1}0$]. However, the directions [100] and [010] are symmetry-related by the fourfold rotoinversion and thus result in the same projection. The same holds for the directions [110] and [$\bar{1}10$]. The three remaining directions are genuinely different and the projections along these directions will be discussed in detail below. The corresponding information given in the space-group tables under the heading 'Symmetry of special projections' is reproduced in Fig. 1.4.5.5 for $P\bar{4}b2$.

Coset representatives of $\mathcal{G}$ relative to its translation subgroup can be extracted from the general-positions block in the space-group tables of $P\bar{4}b2$ and are given in Table 1.4.5.2.

d *along* [001]: The linear parts of all coset representatives map [001] to $\pm$[001], and therefore the scanning group $\mathcal{H}$ is the full group $\mathcal{G}$. A conventional basis for the translations of the projection is $\mathbf{a}' = \mathbf{a}$ and $\mathbf{b}' = \mathbf{b}$. The operation g_3 acts as a fourfold rotation, g_5 acts as a glide reflection with normal vector $\mathbf{b}'$ and g_8 as a reflection with normal vector $\mathbf{a}' + \mathbf{b}'$. Thus, the resulting plane group has type *p4gm* (plane group No. 12). The line parallel to the projection direction [001] which is projected to the origin of *p4gm* in its conventional setting is the line 0, 0, *z*.

Again, it is instructive to look at the symmetry-element diagrams for the respective space and plane groups, as displayed in Fig. 1.4.5.6. The twofold rotations and fourfold rotoinversions with axis along [001] are turned into twofold rotations and fourfold rotations, respectively [rules (iii) and

Table 1.4.5.2
Coset representatives of $P\bar{4}b2$ (117) relative to its translation subgroup

	Coordinate triplet	Description
g_1:	x, y, z	Identity
g_2:	$\bar{x}, \bar{y}, z$	Twofold rotation with axis along [001]
g_3:	$y, \bar{x}, \bar{z}$	Fourfold rotoinversion with axis along [001]
g_4:	$\bar{y}, x, \bar{z}$	Fourfold rotoinversion with axis along [001]
g_5:	$x+\frac{1}{2}, \bar{y}+\frac{1}{2}, z$	Glide reflection with normal vector [010] and glide component along [100]
g_6:	$\bar{x}+\frac{1}{2}, y+\frac{1}{2}, z$	Glide reflection with normal vector [100] and glide component along [010]
g_7:	$y+\frac{1}{2}, x+\frac{1}{2}, \bar{z}$	Twofold screw rotation with axis parallel to [110]
g_8:	$\bar{y}+\frac{1}{2}, \bar{x}+\frac{1}{2}, \bar{z}$	Twofold rotation with axis parallel to [$1\bar{1}0$]

(iv)]. The glide reflections with both normal vector and glide vector perpendicular to [001] (dashed lines) result in glide reflections [rule (vii)]. The twofold rotations (full arrows) and screw rotations (half arrows) with rotation axis perpendicular to [001] give reflections and glide reflections, respectively [rule (vi)]. Note that the two diagrams can be matched directly, because the line 0, 0, *z* which is projected to the origin of *p4gm* runs through the origin of $P\bar{4}b2$.

d *along* [100]: Only the linear parts of the coset representatives g_1, g_2, g_5 and g_6 map [100] to $\pm$[100], thus these four cosets form the scanning group $\mathcal{H}$ (which is of index 2 in $\mathcal{G}$). The operation g_6 acts as a translation by $\frac{1}{2}\mathbf{b}$, thus a conventional basis for the translations of the projection is $\mathbf{a}' = \frac{1}{2}\mathbf{b}$ and $\mathbf{b}' = \mathbf{c}$. The operation g_2 acts as a reflection with normal vector $\mathbf{a}'$ and g_5 acts as the same reflection composed with the translation $\mathbf{a}'$. The resulting plane group is thus of type *p*1*m*1 (plane group No. 3 with short symbol *pm*). The line which is mapped to the origin of *p*1*m*1 in its conventional setting is *x*, 0, 0.

d *along* [110]: Only the linear parts of the coset representatives g_1, g_2, g_7 and g_8 map [110] to $\pm$[110], thus these four cosets form the scanning group $\mathcal{H}$ (of index 2 in $\mathcal{G}$). The translation by $\mathbf{b}$ is projected to a translation by $\frac{1}{2}(-\mathbf{a}+\mathbf{b})$, thus a conventional basis for the translations of the projection is $\mathbf{a}' = \frac{1}{2}(-\mathbf{a}+\mathbf{b})$ and $\mathbf{b}' = \mathbf{c}$. The operation g_2 acts as a reflection with normal

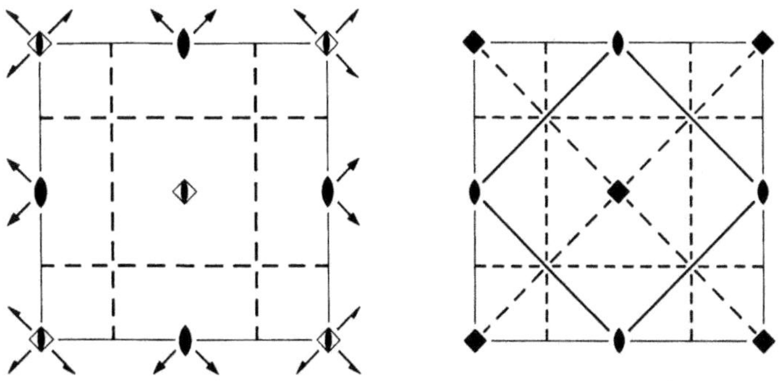

Figure 1.4.5.6
Orthogonal projection along [001] of the symmetry-element diagram for $P\bar{4}b2$ (117) (left) and the diagram for plane group *p4gm* (12) (right).

Symmetry of special projections

Along [001] *p*4*gm*	Along [100] *p*1*m*1	Along [110] *p*2*mm*
$\mathbf{a}' = \mathbf{a}$ $\quad \mathbf{b}' = \mathbf{b}$	$\mathbf{a}' = \frac{1}{2}\mathbf{b}$ $\quad \mathbf{b}' = \mathbf{c}$	$\mathbf{a}' = \frac{1}{2}(-\mathbf{a}+\mathbf{b})$ $\quad \mathbf{b}' = \mathbf{c}$
Origin at 0, 0, *z*	Origin at *x*, 0, 0	Origin at *x*, *x*, 0

Figure 1.4.5.5
'Symmetry of special projections' block of $P\bar{4}b2$ (117) as given in the space-group tables of *IT* A.

vector $\mathbf{a}'$, g_7 acts as a twofold rotation and g_8 acts as a reflection with normal vector $\mathbf{b}'$. The resulting plane group is thus of type $p2mm$ (plane group No. 6). The line parallel to the projection direction [110] that is mapped to the origin of $p2mm$ (in its conventional setting) is $x, x, 0$.

Note that for directions different from those considered above, additional non-trivial plane groups may be obtained. For example, for the projection direction $\mathbf{d} = [1\bar{1}1]$, the scanning group consists of the cosets of g_1 and g_7. The operation g_7 acts as a glide reflection and the resulting plane group is of type $c1m1$ (plane group No. 5).

References

Bracewell, R. N. (2003). *Fourier Analysis and Imaging*. New York: Springer Science+Business Media.

Bradley, C. J. & Cracknell, A. P. (1972). *The Mathematical Theory of Symmetry in Solids*. Oxford University Press.

Ewald, P. P. (1930). *Tagung des erweiterten Tabellenkomitees in Zürich, 28. - 31. Juli 1930. Z. Kristallogr.* **75**, 159–160.

Flack, H. D., Wondratschek, H., Hahn, Th. & Abrahams, S. C. (2000). *Symmetry elements in space groups and point groups. Addenda to two IUCr reports on the nomenclature of symmetry. Acta Cryst.* A**56**, 96–98.

Glazer, A. M., Aroyo, M. I. & Authier, A. (2014). *Seitz symbols for crystallographic symmetry operations. Acta Cryst.* A**70**, 300–302.

Hall, S. R. (1981*a*). *Space-group notation with an explicit origin. Acta Cryst.* A**37**, 517–525.

Hall, S. R. (1981*b*). *Space-group notation with an explicit origin; erratum. Acta Cryst.* A**37**, 921.

Hermann, C. (1928). *Zur systematischen Strukturtheorie. I. Eine neue Raumgruppensymbolik. Z. Kristallogr.* **68**, 257–287.

Hermann, C. (1931). *Bemerkung zu der vorstehenden Arbeit von Ch. Mauguin. Z. Kristallogr.* **76**, 559–561.

International Tables for Crystallography (2002). Vol. A, *Space-Group Symmetry*, 5th ed., edited by Th. Hahn. Dordrecht: Kluwer Academic Publishers. [Abbreviated as *IT* A5.]

International Tables for Crystallography (2016). Volume A, *Space-Group Symmetry*, 6th ed., edited by M. I. Aroyo. Chichester: Wiley. [Abbreviated as *IT* A.]

International Tables for Crystallography (2008). Vol. B, *Reciprocal Space*, 3rd ed., edited by U. Shmueli. Heidelberg: Springer.

International Tables for Crystallography (2010). Vol. E, *Subperiodic Groups*, 2nd ed., edited by V. Kopský & D. B. Litvin. Chichester: John Wiley. [Abbreviated as *IT* E.]

International Tables for X-ray Crystallography (1952). Vol. I, *Symmetry Groups*, edited by N. F. M. Henry & K. Lonsdale. Birmingham: Kynoch Press. [Abbreviated as *IT* (1952).]

Internationale Tabellen zur Bestimmung von Kristallstrukturen (1935). 1. Band, *Gruppentheoretische Tafeln*, edited by C. Hermann. Berlin: Gebrüder Borntraeger.

Klein, F. (1884). *Vorlesungen über das Ikosaeder*. Leipzig: Teubner.

Ledermann, W. (1976). *Introduction to Group Theory*. London: Longman.

Litvin, D. B. (2012). *Magnetic Group Tables*. IUCr e-book. http://www.iucr.org/publ/978-0-9553602-2-0.

Litvin, D. B. & Kopský, V. (2011). *Seitz notation for symmetry operations of space groups. Acta Cryst.* A**67**, 415–418.

Litvin, D. B. & Kopsky, V. (2014). *Seitz symbols for symmetry operations of subperiodic groups. Acta Cryst.* A**70**, 677–678.

Mauguin, M. (1931). *Sur le symbolisme des groupes de répétition ou de symétrie des assemblages cristallins. Z. Kristallogr.* **76**, 542–558.

Müller, U. (2012). Personal communication.

Schoenflies, A. (1891). *Krystallsysteme und Krystallstruktur*. Leipzig: B. G. Teubner. [Reprint (1984). Berlin: Springer-Verlag.]

Schoenflies, A. (1923). *Theorie der Kristallstruktur*. Berlin: Gebrüder Borntraeger.

Seitz, F. (1935). *A matrix-algebraic development of the crystallographic groups. III. Z. Kristallogr.* **91**, 336–366.

Shmueli, U. (1984). *Space-group algorithms. I. The space group and its symmetry elements. Acta Cryst.* A**40**, 559–567.

Suescun, L. & Nespolo, M. (2012). *From patterns to space groups and the eigensymmetry of crystallographic orbits: a reinterpretation of some symmetry diagrams in IUCr Teaching Pamphlet No. 14. J. Appl. Cryst.* **45**, 834–837.

Vainshtein, B. K. (1994). *Fundamentals of Crystals. Symmetry, and Methods of Structural Crystallography*. Berlin, Heidelberg, New York: Springer.

Wolff, P. M. de, Belov, N. V., Bertaut, E. F., Buerger, M. J., Donnay, J. D. H., Fischer, W., Hahn, Th., Koptsik, V. A., Mackay, A. L., Wondratschek, H., Wilson, A. J. C. & Abrahams, S. C. (1985). *Nomenclature for crystal families, Bravais-lattice types and arithmetic classes. Report of the International Union of Crystallography Ad-Hoc Committee on the Nomenclature of Symmetry. Acta Cryst.* A**41**, 278–280.

Wolff, P. M. de, Billiet, Y., Donnay, J. D. H., Fischer, W., Galiulin, R. B., Glazer, A. M., Hahn, Th., Senechal, M., Shoemaker, D. P., Wondratschek, H., Wilson, A. J. C. & Abrahams, S. C. (1992). *Symbols for symmetry elements and symmetry operations. Final report of the International Union of Crystallography Ad-Hoc Committee on the Nomenclature of Symmetry. Acta Cryst.* A**48**, 727–732.

Wolff, P. M. de, Billiet, Y., Donnay, J. D. H., Fischer, W., Galiulin, R. B., Glazer, A. M., Senechal, M., Shoemaker, D. P., Wondratschek, H., Hahn, Th., Wilson, A. J. C. & Abrahams, S. C. (1989). *Definition of symmetry elements in space groups and point groups. Report of the International Union of Crystallography Ad-Hoc Committee on the Nomenclature of Symmetry. Acta Cryst.* A**45**, 494–499.

1.5. Transformations of coordinate systems

Hans Wondratschek, Mois I. Aroyo, Bernd Souvignier and Gervais Chapuis

It is in general advantageous to describe crystallographic objects and their symmetries using the most appropriate coordinate system. The best coordinate system may be different for different steps of the calculations and for different objects which have to be considered simultaneously. Therefore, a change of the origin and/or the basis is frequently necessary when treating crystallographic problems, for example in the study of phase-transition phenomena, or in the comparison of crystal structures described with respect to different coordinate systems.

1.5.1. Origin shift and change of the basis[1]

By Hans Wondratschek and Mois I. Aroyo

1.5.1.1. Origin shift

Let a coordinate system be given with a basis **a**, **b**, **c** and an origin O. Referred to this coordinate system, the column of coordinates of a point X is $\boldsymbol{x} = \begin{pmatrix} x_1 \\ x_2 \\ x_3 \end{pmatrix}$ and the corresponding vector is $\mathbf{x} = x_1\mathbf{a} + x_2\mathbf{b} + x_3\mathbf{c}$. Referred to a new coordinate system, specified by the basis $\mathbf{a}', \mathbf{b}', \mathbf{c}'$ and the origin O', the column of coordinates of the point X is $\boldsymbol{x}' = \begin{pmatrix} x_1' \\ x_2' \\ x_3' \end{pmatrix}$. Let $\boldsymbol{p} = \overrightarrow{OO'} = \begin{pmatrix} p_1 \\ p_2 \\ p_3 \end{pmatrix}$ be the column of coefficients for the vector **p** from the old origin O to the new origin O', see Fig. 1.5.1.1.

Then the columns are related by $\boldsymbol{p} + \boldsymbol{x}' = \boldsymbol{x}$, *i.e.*

$$\boldsymbol{x}' = \boldsymbol{x} - \boldsymbol{p} \quad \text{or} \quad \begin{pmatrix} x_1' \\ x_2' \\ x_3' \end{pmatrix} = \begin{pmatrix} x_1 \\ x_2 \\ x_3 \end{pmatrix} - \begin{pmatrix} p_1 \\ p_2 \\ p_3 \end{pmatrix} = \begin{pmatrix} x_1 - p_1 \\ x_2 - p_2 \\ x_3 - p_3 \end{pmatrix}. \tag{1.5.1.1}$$

This can be written in the formalism of matrix–column pairs (*cf.* Section 1.2.2 for details of the matrix–column formalism) as

$$\boldsymbol{x}' = (\boldsymbol{I}, -\boldsymbol{p})\boldsymbol{x} \quad \text{or} \quad \boldsymbol{x}' = (\boldsymbol{I}, \boldsymbol{p})^{-1}\boldsymbol{x}, \tag{1.5.1.2}$$

where $(\boldsymbol{I}, \boldsymbol{p})$ represents the translation corresponding to the vector **p** of the origin shift.

The vector **r** from the point X to the point Y (also known as a 'distance vector') for which $\mathbf{x} + \mathbf{r} = \mathbf{y}$ (*cf.* Fig. 1.5.1.1), and which thus has coefficients

$$\boldsymbol{r} = \boldsymbol{y} - \boldsymbol{x} = \begin{pmatrix} y_1 - x_1 \\ y_2 - x_2 \\ y_3 - x_3 \end{pmatrix},$$

shows a different transformation behaviour under the origin shift. From the diagram one can see that $\mathbf{x} + \mathbf{r} = \mathbf{y}$ and $\mathbf{x}' + \mathbf{r} = \mathbf{y}'$, and thus

$$\mathbf{r} = \mathbf{y}' - \mathbf{x}' = \mathbf{y} - \mathbf{x}, \tag{1.5.1.3}$$

i.e. the vector coefficients of **r** are not affected by the origin shift.

Example

The description of a crystal structure is closely related to its space-group symmetry: different descriptions of the underlying space group, in general, result in different descriptions of the crystal structure. This example illustrates the comparison of two structure descriptions corresponding to different origin choices of the space group.

To compare the two structures it is not only necessary to apply the origin-shift transformation but also to adjust the selection of the representative atoms of the two descriptions.

In the Inorganic Crystal Structure Database (2012) (abbreviated as ICSD) one finds the following two descriptions of the mineral zircon $ZrSiO_4$:

(*a*) Wyckoff & Hendricks (1927), ICSD No. 31101, space group $I4_1/amd = D_{4h}^{19}$, No. 141, cell parameters a = 6.61 Å, c = 5.98 Å.

The coordinates of the atoms in the unit cell are (normalized so that $0 \le x_i < 1$):

Zr:	$4a$	$0, 0, 0;\ 0, \frac{1}{2}, \frac{1}{4}$ [and the same with $(\frac{1}{2}, \frac{1}{2}, \frac{1}{2})+$]
Si:	$4b$	$0, 0, \frac{1}{2};\ 0, \frac{1}{2}, \frac{3}{4}$ [and the same with $(\frac{1}{2}, \frac{1}{2}, \frac{1}{2})+$]
O:	$16h$	0, 0.2, 0.34; 0.5, 0.3, 0.84; 0.8, 0.5, 0.59;
		0.7, 0, 0.09; 0.5, 0.2, 0.41; 0, 0.3, 0.91;
		0.7, 0.5, 0.16; 0.8, 0, 0.66
		[and the same with $(\frac{1}{2}, \frac{1}{2}, \frac{1}{2})+$].

The coordinates of Zr and Si atoms together with the fact that they occupy Wyckoff positions $4a$ and $4b$ of maximal symmetry indicate that the space-group setting corresponds to the origin choice 1 description of $I4_1/amd$ given in *International Tables for Crystallography*, Vol. A (2016) (hereafter abbreviated as *IT* A), *i.e.* origin at $\bar{4}m2$ (*cf.* the space-group tables for $I4_1/amd$ in Part 2).

(*b*) Krstanovic (1958), ICSD No. 45520, space group $I4_1/amd = D_{4h}^{19}$, No. 141, cell parameters a = 6.6164 (5) Å, c = 6.0150 (5) Å.

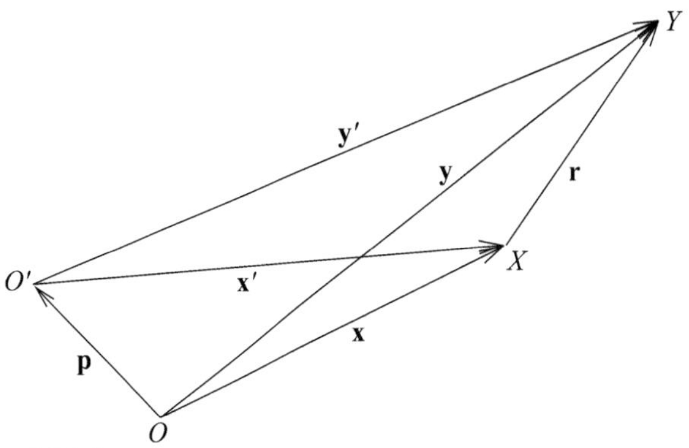

Figure 1.5.1.1

The coordinates of the points X (or Y) with respect to the old origin O are $\boldsymbol{x}$ ($\boldsymbol{y}$), and with respect to the new origin O' they are $\boldsymbol{x}'$ ($\boldsymbol{y}'$). From the diagram one can see that $\mathbf{p} + \mathbf{x}' = \mathbf{x}$ and $\mathbf{p} + \mathbf{y}' = \mathbf{y}$.

[1] With Table 1.5.1.1 and Figs. 1.5.1.2 and 1.5.1.5–1.5.1.7 by H. Arnold.

The coordinates of the atoms in the unit cell are (normalized so that $0 \le x_i < 1$):

Zr: 4*a* $0, \frac{3}{4}, \frac{1}{8}$; $\frac{1}{2}, \frac{3}{4}, \frac{3}{8}$ [and the same with $(\frac{1}{2}, \frac{1}{2}, \frac{1}{2})+$]
Si: 4*b* $0, \frac{1}{4}, \frac{3}{8}$; $0, \frac{3}{4}, \frac{5}{8}$ [and the same with $(\frac{1}{2}, \frac{1}{2}, \frac{1}{2})+$]
O: 16*h* 0, 0.067, 0.198; 0.5, 0.933, 0.698;
0.183, 0.75, 0.448; 0.317, 0.25, 0.948;
0.5, 0.067, 0.302; 0, 0.933, 0.802;
0.317, 0.75, 0.052; 0.183, 0.25, 0.552
[and the same with $(\frac{1}{2}, \frac{1}{2}, \frac{1}{2})+$].

In this case, the structure is described with respect to the origin choice 2 setting of $I4_1/amd$, *i.e.* the origin is located at an inversion centre, and it is specified as 'Origin at centre (2/*m*) at $0, -\frac{1}{4}, \frac{1}{8}$ from $\bar{4}m2$' (*cf.* the space-group tables for $I4_1/amd$ in Part 2).

In order to compare the different structure descriptions, the atomic coordinates of the origin choice 1 description are to be transformed to 'Origin at centre 2/*m*', *i.e.* origin choice 2.
Origin choice 2 has coordinates $0, -\frac{1}{4}, \frac{1}{8}$ referred to origin choice 1. Therefore, the change of coordinates consists of *subtracting* $\boldsymbol{p} = \begin{pmatrix} 0 \\ -\frac{1}{4} \\ \frac{1}{8} \end{pmatrix}$ from the origin choice 1 values, *i.e.* leave the x coordinate unchanged, add $\frac{1}{4} = 0.25$ to the y coordinate and subtract $\frac{1}{8} = 0.125$ from the z coordinate [*cf.* equation (1.5.1.1)].
The transformed and normalized coordinates (so that $0 \le x_i < 1$) are

(i) Zr: 4*a* $0, \frac{1}{4}, \frac{7}{8}$; $0, \frac{3}{4}, \frac{1}{8}$; $\frac{1}{2}, \frac{1}{4}, \frac{5}{8}$; $\frac{1}{2}, \frac{3}{4}, \frac{3}{8}$;
(ii) Si: 4*b* $0, \frac{1}{4}, \frac{3}{8}$; $0, \frac{3}{4}, \frac{5}{8}$; $\frac{1}{2}, \frac{1}{4}, \frac{1}{8}$; $\frac{1}{2}, \frac{3}{4}, \frac{7}{8}$;
(iii) O: 16*h* 0, 0.20 + 0.25, 0.34 − 0.125 = 0, 0.45, 0.215.
While the transformed coordinates of Zr and Si atoms coincide with those of the origin-2 description, the oxygen atom obviously does not correspond to the representative 0, 0.067, 0.198 given by Krstanovic (1958). However, it is seen to correspond to the second position with coordinates 0.5, 0.933, 0.698 after adding the centring vector $(\frac{1}{2}, \frac{1}{2}, \frac{1}{2})$ to its coordinates 0, 0.45, 0.215. The transformed (and normalized) coordinates of the rest of the oxygen atoms in the unit cell are:

0.5, 0.55, 0.715; 0.8, 0.75, 0.465; 0.7, 0.25, 0.965;
0.5, 0.45, 0.285; 0, 0.55, 0.785; 0.7, 0.75, 0.035;
0.8, 0.25, 0.535;

all also with $(\frac{1}{2}, \frac{1}{2}, \frac{1}{2})+$. The difference in the oxygen-atom coordinates of the two descriptions could be explained by the differences in the accuracy of the two refinements (note also the differing cell parameters).

1.5.1.2. Change of the basis

A change of the basis is described by a (3 × 3) matrix:

$$\boldsymbol{P} = \begin{pmatrix} P_{11} & P_{12} & P_{13} \\ P_{21} & P_{22} & P_{23} \\ P_{31} & P_{32} & P_{33} \end{pmatrix}.$$

The matrix $\boldsymbol{P}$ relates the new basis $\mathbf{a}', \mathbf{b}', \mathbf{c}'$ to the old basis $\mathbf{a}, \mathbf{b}, \mathbf{c}$ according to

$$(\mathbf{a}', \mathbf{b}', \mathbf{c}') = (\mathbf{a}, \mathbf{b}, \mathbf{c})\boldsymbol{P} = (\mathbf{a}, \mathbf{b}, \mathbf{c})\begin{pmatrix} P_{11} & P_{12} & P_{13} \\ P_{21} & P_{22} & P_{23} \\ P_{31} & P_{32} & P_{33} \end{pmatrix}$$
$$= (\mathbf{a}P_{11} + \mathbf{b}P_{21} + \mathbf{c}P_{31}, \mathbf{a}P_{12} + \mathbf{b}P_{22} + \mathbf{c}P_{32}, \mathbf{a}P_{13} + \mathbf{b}P_{23} + \mathbf{c}P_{33}). \quad (1.5.1.4)$$

The matrix $\boldsymbol{P}$ is often referred to as the *linear part* of the coordinate transformation and it describes a change of direction and/or length of the basis vectors. It is preferable to choose the matrix $\boldsymbol{P}$ in such a way that its determinant is positive: a negative determinant of $\boldsymbol{P}$ implies a change from a right-handed coordinate system to a left-handed coordinate system or *vice versa*. If $\det(\boldsymbol{P}) = 0$, then the new vectors $\mathbf{a}', \mathbf{b}', \mathbf{c}'$ are linearly dependent, *i.e.* they do not form a complete set of basis vectors.

For a point X (*cf.* Fig. 1.5.1.1), the vector $\overrightarrow{OX} = \mathbf{x}$ is

$$\mathbf{x} = \mathbf{a}x_1 + \mathbf{b}x_2 + \mathbf{c}x_3 = \mathbf{a}'x'_1 + \mathbf{b}'x'_2 + \mathbf{c}'x'_3 \text{ or}$$
$$\mathbf{x} = (\mathbf{a}, \mathbf{b}, \mathbf{c})\begin{pmatrix} x_1 \\ x_2 \\ x_3 \end{pmatrix} = (\mathbf{a}', \mathbf{b}', \mathbf{c}')\begin{pmatrix} x'_1 \\ x'_2 \\ x'_3 \end{pmatrix}.$$

By inserting equation (1.5.1.4) one obtains

$$\mathbf{x} = (\mathbf{a}', \mathbf{b}', \mathbf{c}')\begin{pmatrix} x'_1 \\ x'_2 \\ x'_3 \end{pmatrix} = (\mathbf{a}, \mathbf{b}, \mathbf{c})\begin{pmatrix} P_{11} & P_{12} & P_{13} \\ P_{21} & P_{22} & P_{23} \\ P_{31} & P_{32} & P_{33} \end{pmatrix}\begin{pmatrix} x'_1 \\ x'_2 \\ x'_3 \end{pmatrix}$$

or

$$\begin{pmatrix} x_1 \\ x_2 \\ x_3 \end{pmatrix} = \begin{pmatrix} P_{11} & P_{12} & P_{13} \\ P_{21} & P_{22} & P_{23} \\ P_{31} & P_{32} & P_{33} \end{pmatrix}\begin{pmatrix} x'_1 \\ x'_2 \\ x'_3 \end{pmatrix},$$

i.e. $\boldsymbol{x} = \boldsymbol{P}\boldsymbol{x}'$ or $\boldsymbol{x}' = \boldsymbol{P}^{-1}\boldsymbol{x} = (\boldsymbol{P}, \boldsymbol{o})^{-1}\boldsymbol{x}$, which is often written as

$$\begin{pmatrix} x'_1 \\ x'_2 \\ x'_3 \end{pmatrix} = \boldsymbol{x}' = \boldsymbol{Q}\boldsymbol{x} = (\boldsymbol{Q}, \boldsymbol{o})\boldsymbol{x} = \begin{pmatrix} Q_{11}x_1 + Q_{12}x_2 + Q_{13}x_3 \\ Q_{21}x_1 + Q_{22}x_2 + Q_{23}x_3 \\ Q_{31}x_1 + Q_{32}x_2 + Q_{33}x_3 \end{pmatrix}. \quad (1.5.1.5)$$

Here the inverse matrix $\boldsymbol{P}^{-1}$ is designated by $\boldsymbol{Q}$, while $\boldsymbol{o}$ is the (3 × 1) column with zero coefficients. [Note that in equation (1.5.1.4) the sum is over the row (first) index of $\boldsymbol{P}$, while in equation (1.5.1.5), the sum is over the column (second) index of $\boldsymbol{Q}$.]

A selected set of transformation matrices $\boldsymbol{P}$ and their inverses $\boldsymbol{P}^{-1} = \boldsymbol{Q}$ that are frequently used in crystallographic calculations are listed in Table 1.5.1.1 and illustrated in Figs. 1.5.1.2 to 1.5.1.7 (*cf.* Table 1.5.1.1 of *IT* A for a larger set of transformation matrices).

Example

Consider an *F*-centred cell with conventional basis $\mathbf{a}_F, \mathbf{b}_F, \mathbf{c}_F$ and a corresponding primitive cell with basis $\mathbf{a}_P, \mathbf{b}_P, \mathbf{c}_P$, *cf.* Fig. 1.5.1.4. The transformation matrix $\boldsymbol{P}$ from the conventional basis to a primitive basis can either be deduced from Fig. 1.5.1.4 or can be found directly in Table 1.5.1.1: $\mathbf{a}_P = \frac{1}{2}(\mathbf{b}_F + \mathbf{c}_F)$, $\mathbf{b}_P = \frac{1}{2}(\mathbf{a}_F + \mathbf{c}_F)$, $\mathbf{c}_P = \frac{1}{2}(\mathbf{a}_F + \mathbf{b}_F)$, which in matrix notation is

$$(\mathbf{a}_P, \mathbf{b}_P, \mathbf{c}_P) = (\mathbf{a}_F, \mathbf{b}_F, \mathbf{c}_F)\boldsymbol{P} = (\mathbf{a}_F, \mathbf{b}_F, \mathbf{c}_F)\begin{pmatrix} 0 & \frac{1}{2} & \frac{1}{2} \\ \frac{1}{2} & 0 & \frac{1}{2} \\ \frac{1}{2} & \frac{1}{2} & 0 \end{pmatrix}.$$

(Continued on page 63)

Table 1.5.1.1
Selected 3 × 3 transformation matrices $\boldsymbol{P}$ and $\boldsymbol{Q} = \boldsymbol{P}^{-1}$

For opposite transformations (with arrows reversed) replace $\boldsymbol{P}$ by $\boldsymbol{Q}$ and *vice versa.*

Transformation	$\boldsymbol{P}$	$\boldsymbol{Q} = \boldsymbol{P}^{-1}$	Crystal system
Cell choice 1 → cell choice 2: $\begin{cases} P \to P \\ C \to A \end{cases}$ Cell choice 2 → cell choice 3: $\begin{cases} P \to P \\ A \to I \end{cases}$ Unique axis **b** invariant Cell choice 3 → cell choice 1: $\begin{cases} P \to P \\ I \to C \end{cases}$ (Fig. 1.5.1.2*a*)	$\begin{pmatrix} \bar{1} & 0 & 1 \\ 0 & 1 & 0 \\ \bar{1} & 0 & 0 \end{pmatrix}$	$\begin{pmatrix} 0 & 0 & \bar{1} \\ 0 & 1 & 0 \\ 1 & 0 & \bar{1} \end{pmatrix}$	Monoclinic (*cf.* Sections 1.5.3 and 1.5.4)
Cell choice 1 → cell choice 2: $\begin{cases} P \to P \\ A \to B \end{cases}$ Cell choice 2 → cell choice 3: $\begin{cases} P \to P \\ B \to I \end{cases}$ Unique axis **c** invariant Cell choice 3 → cell choice 1: $\begin{cases} P \to P \\ I \to A \end{cases}$ (Fig. 1.5.1.2*b*)	$\begin{pmatrix} 0 & \bar{1} & 0 \\ 1 & \bar{1} & 0 \\ 0 & 0 & 1 \end{pmatrix}$	$\begin{pmatrix} \bar{1} & 1 & 0 \\ \bar{1} & 0 & 0 \\ 0 & 0 & 1 \end{pmatrix}$	Monoclinic (*cf.* Sections 1.5.3 and 1.5.4)
Cell choice 1 → cell choice 2: $\begin{cases} P \to P \\ B \to C \end{cases}$ Cell choice 2 → cell choice 3: $\begin{cases} P \to P \\ C \to I \end{cases}$ Unique axis **a** invariant Cell choice 3 → cell choice 1: $\begin{cases} P \to P \\ I \to B \end{cases}$ (Fig. 1.5.1.2*c*)	$\begin{pmatrix} 1 & 0 & 0 \\ 0 & 0 & \bar{1} \\ 0 & 1 & \bar{1} \end{pmatrix}$	$\begin{pmatrix} 1 & 0 & 0 \\ 0 & \bar{1} & 1 \\ 0 & \bar{1} & 0 \end{pmatrix}$	Monoclinic (*cf.* Sections 1.5.3 and 1.5.4)
Unique axis **b** → unique axis **c** Cell choice 1: $\begin{cases} P \to P \\ C \to A \end{cases}$ Cell choice 2: $\begin{cases} P \to P \\ A \to B \end{cases}$ Cell choice invariant Cell choice 3: $\begin{cases} P \to P \\ I \to I \end{cases}$	$\begin{pmatrix} 0 & 1 & 0 \\ 0 & 0 & 1 \\ 1 & 0 & 0 \end{pmatrix}$	$\begin{pmatrix} 0 & 0 & 1 \\ 1 & 0 & 0 \\ 0 & 1 & 0 \end{pmatrix}$	Monoclinic (*cf.* Sections 1.5.3 and 1.5.4)
Unique axis **b** → unique axis **a** Cell choice 1: $\begin{cases} P \to P \\ C \to B \end{cases}$ Cell choice 2: $\begin{cases} P \to P \\ A \to C \end{cases}$ Cell choice invariant Cell choice 3: $\begin{cases} P \to P \\ I \to I \end{cases}$	$\begin{pmatrix} 0 & 0 & 1 \\ 1 & 0 & 0 \\ 0 & 1 & 0 \end{pmatrix}$	$\begin{pmatrix} 0 & 1 & 0 \\ 0 & 0 & 1 \\ 1 & 0 & 0 \end{pmatrix}$	Monoclinic (*cf.* Sections 1.5.3 and 1.5.4)
Unique axis **c** → unique axis **a** Cell choice 1: $\begin{cases} P \to P \\ A \to B \end{cases}$ Cell choice 2: $\begin{cases} P \to P \\ B \to C \end{cases}$ Cell choice invariant Cell choice 3: $\begin{cases} P \to P \\ I \to I \end{cases}$	$\begin{pmatrix} 0 & 1 & 0 \\ 0 & 0 & 1 \\ 1 & 0 & 0 \end{pmatrix}$	$\begin{pmatrix} 0 & 0 & 1 \\ 1 & 0 & 0 \\ 0 & 1 & 0 \end{pmatrix}$	Monoclinic (*cf.* Sections 1.5.3 and 1.5.4)
$I \to P$ (Fig. 1.5.1.3)	$\begin{pmatrix} \frac{\bar{1}}{2} & \frac{1}{2} & \frac{1}{2} \\ \frac{1}{2} & \frac{\bar{1}}{2} & \frac{1}{2} \\ \frac{1}{2} & \frac{1}{2} & \frac{\bar{1}}{2} \end{pmatrix}$	$\begin{pmatrix} 0 & 1 & 1 \\ 1 & 0 & 1 \\ 1 & 1 & 0 \end{pmatrix}$	Orthorhombic Tetragonal Cubic
$F \to P$ (Fig. 1.5.1.4)	$\begin{pmatrix} 0 & \frac{1}{2} & \frac{1}{2} \\ \frac{1}{2} & 0 & \frac{1}{2} \\ \frac{1}{2} & \frac{1}{2} & 0 \end{pmatrix}$	$\begin{pmatrix} \bar{1} & 1 & 1 \\ 1 & \bar{1} & 1 \\ 1 & 1 & \bar{1} \end{pmatrix}$	Orthorhombic Tetragonal Cubic

Table 1.5.1.1 (continued)

Transformation	$\boldsymbol{P}$	$\boldsymbol{Q} = \boldsymbol{P}^{-1}$	Crystal system
$(\mathbf{b}, \mathbf{a}, \bar{\mathbf{c}}) \rightarrow (\mathbf{a}, \mathbf{b}, \mathbf{c})$	$\begin{pmatrix} 0 & 1 & 0 \\ 1 & 0 & 0 \\ 0 & 0 & \bar{1} \end{pmatrix}$	$\begin{pmatrix} 0 & 1 & 0 \\ 1 & 0 & 0 \\ 0 & 0 & \bar{1} \end{pmatrix}$	Unconventional orthorhombic setting (*cf.* Section 1.5.4)
$(\mathbf{c}, \mathbf{a}, \mathbf{b}) \rightarrow (\mathbf{a}, \mathbf{b}, \mathbf{c})$	$\begin{pmatrix} 0 & 0 & 1 \\ 1 & 0 & 0 \\ 0 & 1 & 0 \end{pmatrix}$	$\begin{pmatrix} 0 & 1 & 0 \\ 0 & 0 & 1 \\ 1 & 0 & 0 \end{pmatrix}$	Unconventional orthorhombic setting (*cf.* Section 1.5.4)
$(\bar{\mathbf{c}}, \mathbf{b}, \mathbf{a}) \rightarrow (\mathbf{a}, \mathbf{b}, \mathbf{c})$	$\begin{pmatrix} 0 & 0 & \bar{1} \\ 0 & 1 & 0 \\ 1 & 0 & 0 \end{pmatrix}$	$\begin{pmatrix} 0 & 0 & 1 \\ 0 & 1 & 0 \\ \bar{1} & 0 & 0 \end{pmatrix}$	Unconventional orthorhombic setting (*cf.* Section 1.5.4)
$(\mathbf{b}, \mathbf{c}, \mathbf{a}) \rightarrow (\mathbf{a}, \mathbf{b}, \mathbf{c})$	$\begin{pmatrix} 0 & 1 & 0 \\ 0 & 0 & 1 \\ 1 & 0 & 0 \end{pmatrix}$	$\begin{pmatrix} 0 & 0 & 1 \\ 1 & 0 & 0 \\ 0 & 1 & 0 \end{pmatrix}$	Unconventional orthorhombic setting (*cf.* Section 1.5.4)
$(\mathbf{a}, \bar{\mathbf{c}}, \mathbf{b}) \rightarrow (\mathbf{a}, \mathbf{b}, \mathbf{c})$	$\begin{pmatrix} 1 & 0 & 0 \\ 0 & 0 & \bar{1} \\ 0 & 1 & 0 \end{pmatrix}$	$\begin{pmatrix} 1 & 0 & 0 \\ 0 & 0 & 1 \\ 0 & \bar{1} & 0 \end{pmatrix}$	Unconventional orthorhombic setting (*cf.* Section 1.5.4)
$\left.\begin{matrix} P \rightarrow C_1 \\ I \rightarrow F_1 \end{matrix}\right\}$ (Fig. 1.5.1.5), **c** axis invariant	$\begin{pmatrix} 1 & 1 & 0 \\ \bar{1} & 1 & 0 \\ 0 & 0 & 1 \end{pmatrix}$	$\begin{pmatrix} \frac{1}{2} & \frac{\bar{1}}{2} & 0 \\ \frac{1}{2} & \frac{1}{2} & 0 \\ 0 & 0 & 1 \end{pmatrix}$	Tetragonal
$\left.\begin{matrix} P \rightarrow C_2 \\ I \rightarrow F_2 \end{matrix}\right\}$ (Fig. 1.5.1.5), **c** axis invariant	$\begin{pmatrix} 1 & \bar{1} & 0 \\ 1 & 1 & 0 \\ 0 & 0 & 1 \end{pmatrix}$	$\begin{pmatrix} \frac{1}{2} & \frac{1}{2} & 0 \\ \frac{\bar{1}}{2} & \frac{1}{2} & 0 \\ 0 & 0 & 1 \end{pmatrix}$	Tetragonal
Primitive rhombohedral cell → triple hexagonal cell R_1, obverse setting (Fig. 1.5.1.6*a*,*c*)	$\begin{pmatrix} 1 & 0 & 1 \\ \bar{1} & 1 & 1 \\ 0 & \bar{1} & 1 \end{pmatrix}$	$\begin{pmatrix} \frac{2}{3} & \frac{\bar{1}}{3} & \frac{\bar{1}}{3} \\ \frac{1}{3} & \frac{1}{3} & \frac{\bar{2}}{3} \\ \frac{1}{3} & \frac{1}{3} & \frac{1}{3} \end{pmatrix}$	Rhombohedral space groups (*cf.* Section 1.5.3)
Primitive rhombohedral cell → triple hexagonal cell R_2, obverse setting (Fig. 1.5.1.6*c*)	$\begin{pmatrix} 0 & \bar{1} & 1 \\ 1 & 0 & 1 \\ \bar{1} & 1 & 1 \end{pmatrix}$	$\begin{pmatrix} \frac{\bar{1}}{3} & \frac{2}{3} & \frac{\bar{1}}{3} \\ \frac{\bar{2}}{3} & \frac{1}{3} & \frac{1}{3} \\ \frac{1}{3} & \frac{1}{3} & \frac{1}{3} \end{pmatrix}$	Rhombohedral space groups (*cf.* Section 1.5.3)
Primitive rhombohedral cell → triple hexagonal cell R_3, obverse setting (Fig. 1.5.1.6*c*)	$\begin{pmatrix} \bar{1} & 1 & 1 \\ 0 & \bar{1} & 1 \\ 1 & 0 & 1 \end{pmatrix}$	$\begin{pmatrix} \frac{\bar{1}}{3} & \frac{\bar{1}}{3} & \frac{2}{3} \\ \frac{1}{3} & \frac{\bar{2}}{3} & \frac{1}{3} \\ \frac{1}{3} & \frac{1}{3} & \frac{1}{3} \end{pmatrix}$	Rhombohedral space groups (*cf.* Section 1.5.3)
Primitive rhombohedral cell → triple hexagonal cell R_1, reverse setting (Fig. 1.5.1.6*d*)	$\begin{pmatrix} \bar{1} & 0 & 1 \\ 1 & \bar{1} & 1 \\ 0 & 1 & 1 \end{pmatrix}$	$\begin{pmatrix} \frac{\bar{2}}{3} & \frac{1}{3} & \frac{1}{3} \\ \frac{\bar{1}}{3} & \frac{\bar{1}}{3} & \frac{2}{3} \\ \frac{1}{3} & \frac{1}{3} & \frac{1}{3} \end{pmatrix}$	Rhombohedral space groups (*cf.* Section 1.5.3)
Primitive rhombohedral cell → triple hexagonal cell R_2, reverse setting (Fig. 1.5.1.6*b*,*d*)	$\begin{pmatrix} 0 & 1 & 1 \\ \bar{1} & 0 & 1 \\ 1 & \bar{1} & 1 \end{pmatrix}$	$\begin{pmatrix} \frac{1}{3} & \frac{\bar{2}}{3} & \frac{1}{3} \\ \frac{2}{3} & \frac{\bar{1}}{3} & \frac{\bar{1}}{3} \\ \frac{1}{3} & \frac{1}{3} & \frac{1}{3} \end{pmatrix}$	Rhombohedral space groups (*cf.* Section 1.5.3)
Primitive rhombohedral cell → triple hexagonal cell R_3, reverse setting (Fig. 1.5.1.6*d*)	$\begin{pmatrix} 1 & \bar{1} & 1 \\ 0 & 1 & 1 \\ \bar{1} & 0 & 1 \end{pmatrix}$	$\begin{pmatrix} \frac{1}{3} & \frac{1}{3} & \frac{\bar{2}}{3} \\ \frac{\bar{1}}{3} & \frac{2}{3} & \frac{\bar{1}}{3} \\ \frac{1}{3} & \frac{1}{3} & \frac{1}{3} \end{pmatrix}$	Rhombohedral space groups (*cf.* Section 1.5.3)
Hexagonal cell P → orthohexagonal centred cell C_1 (Fig. 1.5.1.7)	$\begin{pmatrix} 1 & 1 & 0 \\ 0 & 2 & 0 \\ 0 & 0 & 1 \end{pmatrix}$	$\begin{pmatrix} 1 & \frac{\bar{1}}{2} & 0 \\ 0 & \frac{1}{2} & 0 \\ 0 & 0 & 1 \end{pmatrix}$	Trigonal Hexagonal

Table 1.5.1.1 (continued)

Transformation	P	$Q = P^{-1}$	Crystal system
Hexagonal cell $P \rightarrow$ orthohexagonal centred cell C_2 (Fig. 1.5.1.7)	$\begin{pmatrix} 1 & \bar{1} & 0 \\ 1 & 1 & 0 \\ 0 & 0 & 1 \end{pmatrix}$	$\begin{pmatrix} \frac{1}{2} & \frac{1}{2} & 0 \\ \bar{\frac{1}{2}} & \frac{1}{2} & 0 \\ 0 & 0 & 1 \end{pmatrix}$	Trigonal Hexagonal
Hexagonal cell $P \rightarrow$ orthohexagonal centred cell C_3 (Fig. 1.5.1.7)	$\begin{pmatrix} 0 & \bar{2} & 0 \\ 1 & \bar{1} & 0 \\ 0 & 0 & 1 \end{pmatrix}$	$\begin{pmatrix} \bar{\frac{1}{2}} & 1 & 0 \\ \bar{\frac{1}{2}} & 0 & 0 \\ 0 & 0 & 1 \end{pmatrix}$	Trigonal Hexagonal

(*a*) Unique axis *b*:
Cell choice 1: *C*-centred cell a_1, b, c_1.
Cell choice 2: *A*-centred cell a_2, b, c_2.
Cell choice 3: *I*-centred cell a_3, b, c_3.

(*b*) Unique axis *c*:
Cell choice 1: *A*-centred cell a_1, b_1, c.
Cell choice 2: *B*-centred cell a_2, b_2, c.
Cell choice 3: *I*-centred cell a_3, b_3, c.

(*c*) Unique axis *a*:
Cell choice 1: *B*-centred cell a, b_1, c_1.
Cell choice 2: *C*-centred cell a, b_2, c_2.
Cell choice 3: *I*-centred cell a, b_3, c_3.

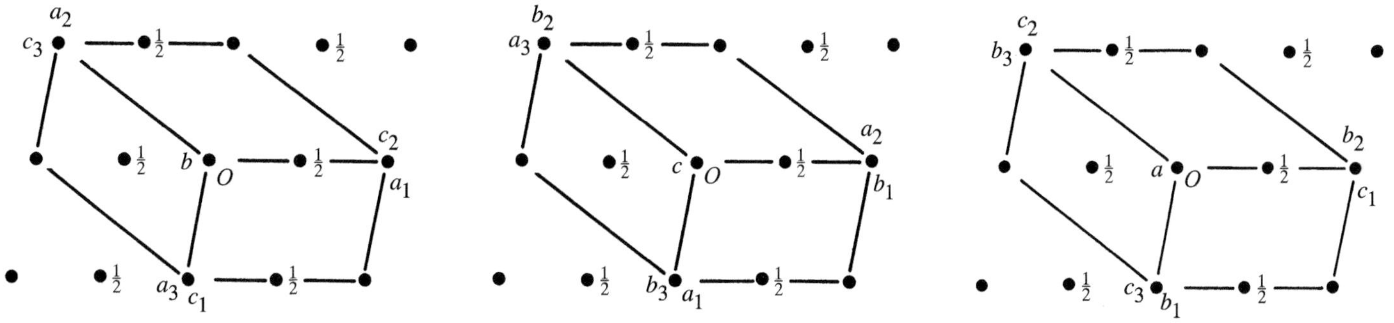

Figure 1.5.1.2
Monoclinic centred lattice, projected along the unique axis. The origin for all the cells is the same. The fractions $\frac{1}{2}$ indicate the height of the lattice points along the axis of projection.

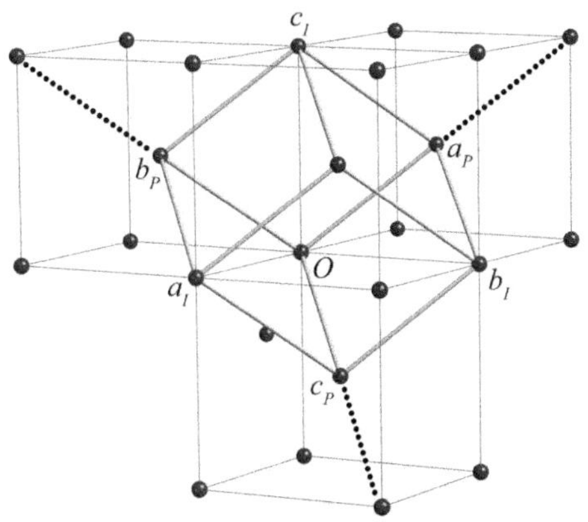

Figure 1.5.1.3
Body-centred cell *I* with a_I, b_I, c_I and a corresponding primitive cell *P* with a_P, b_P, c_P. The origin for both cells is *O*. A cubic *I* cell with lattice constant a_c can be considered as a primitive rhombohedral cell with $a_{\rm rh} = a_c \frac{1}{2}\sqrt{3}$ and $\alpha = 109.47°$ (rhombohedral axes) or a triple hexagonal cell with $a_{\rm hex} = a_c\sqrt{2}$ and $c_{\rm hex} = a_c \frac{1}{2}\sqrt{3}$ (hexagonal axes).

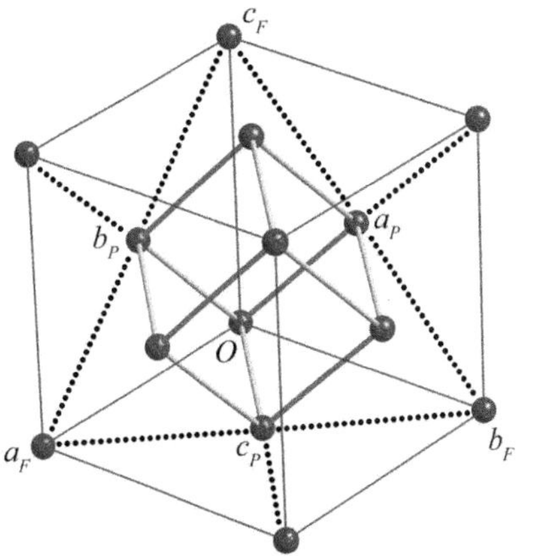

Figure 1.5.1.4
Face-centred cell *F* with a_F, b_F, c_F and a corresponding primitive cell *P* with a_P, b_P, c_P. The origin for both cells is *O*. A cubic *F* cell with lattice constant a_c can be considered as a primitive rhombohedral cell with $a_{\rm rh} = a_c \frac{1}{2}\sqrt{2}$ and $\alpha = 60°$ (rhombohedral axes) or a triple hexagonal cell with $a_{\rm hex} = a_c \frac{1}{2}\sqrt{2}$ and $c_{\rm hex} = a_c\sqrt{3}$ (hexagonal axes).

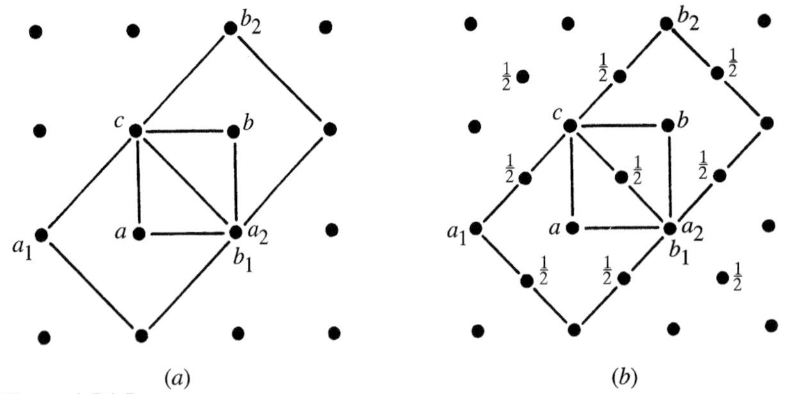

Figure 1.5.1.5
Tetragonal lattices, projected along $[00\bar{1}]$. (a) Primitive cell P with a, b, c and the C-centred cells C_1 with a_1, b_1, c and C_2 with a_2, b_2, c. The location of the origin for all three cells is the same, and coincides with the label c. (b) Body-centred cell I with a, b, c and the F-centred cells F_1 with a_1, b_1, c and F_2 with a_2, b_2, c. The origin for all three cells is the same, and coincides with the label c. The fractions $\frac{1}{2}$ indicate the height of the lattice points along the axis of projection.

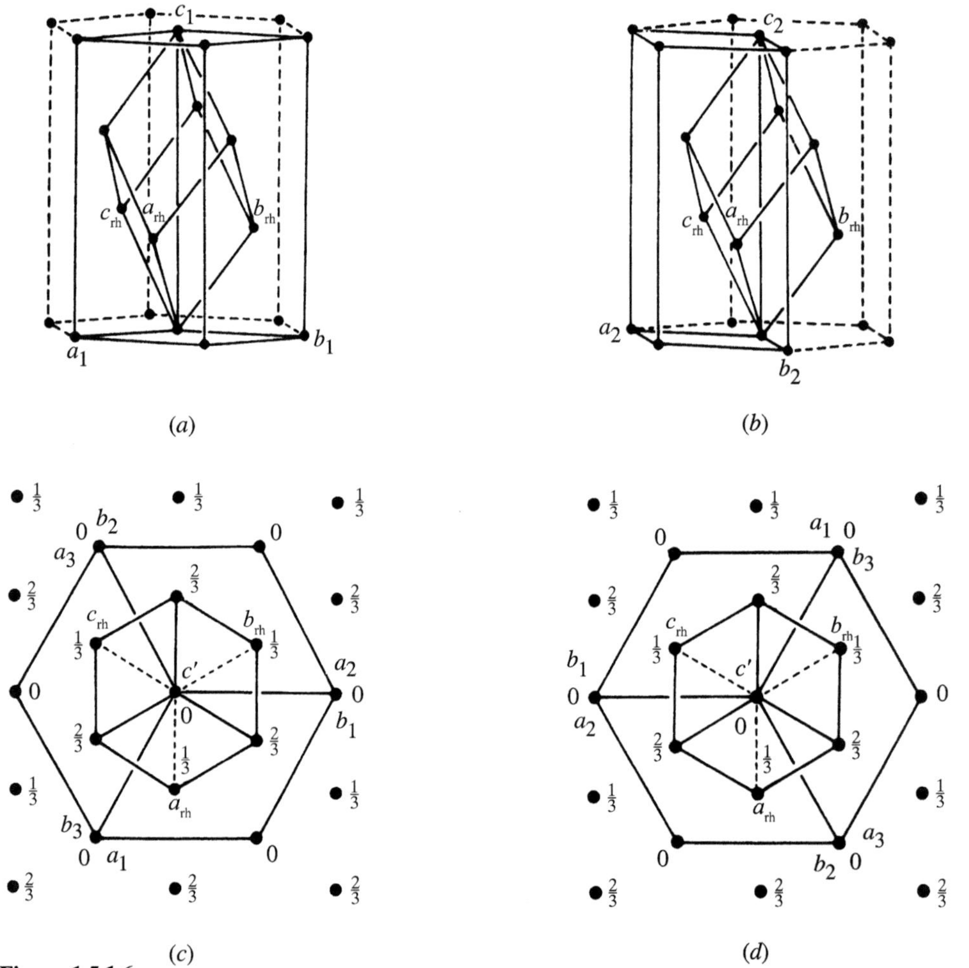

Figure 1.5.1.6
Unit cells in the rhombohedral lattice: same origin for all cells. The basis of the rhombohedral cell is labelled $a_{\rm rh}$, $b_{\rm rh}$, $c_{\rm rh}$. Two settings of the triple hexagonal cell are possible with respect to a primitive rhombohedral cell, related by a 180° rotation: The *obverse setting* with the lattice points 0, 0, 0; $\frac{2}{3}, \frac{1}{3}, \frac{1}{3}$; $\frac{1}{3}, \frac{2}{3}, \frac{2}{3}$ has been used in *International Tables* since 1952. Its general reflection condition is $-h + k + l = 3n$. The *reverse setting* with lattice points 0, 0, 0; $\frac{1}{3}, \frac{2}{3}, \frac{1}{3}$; $\frac{2}{3}, \frac{1}{3}, \frac{2}{3}$ was used in the 1935 edition. Its general reflection condition is $h - k + l = 3n$. The fractions indicate the height of the lattice points along the axis of projection. (a) Obverse setting of triple hexagonal cell a_1, b_1, c_1 in relation to the primitive rhombohedral cell $a_{\rm rh}$, $b_{\rm rh}$, $c_{\rm rh}$. (b) Reverse setting of triple hexagonal cell a_2, b_2, c_2 in relation to the primitive rhombohedral cell $a_{\rm rh}$, $b_{\rm rh}$, $c_{\rm rh}$. (c) Primitive rhombohedral cell (- - - lower edges), $a_{\rm rh}$, $b_{\rm rh}$, $c_{\rm rh}$ in relation to the three triple hexagonal cells in obverse setting a_1, b_1, c'; a_2, b_2, c'; a_3, b_3, c'. Projection along c'. (d) Primitive rhombohedral cell (- - - lower edges), $a_{\rm rh}$, $b_{\rm rh}$, $c_{\rm rh}$ in relation to the three triple hexagonal cells in reverse setting a_1, b_1, c'; a_2, b_2, c'; a_3, b_3, c'. Projection along c'.

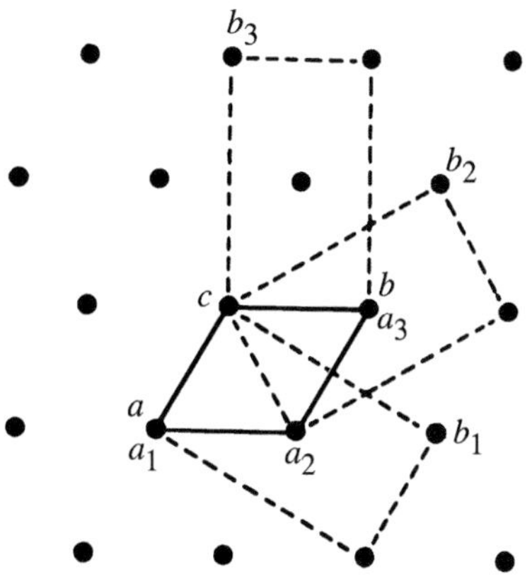

Figure 1.5.1.7
Hexagonal lattice projected along $[00\bar{1}]$. Primitive hexagonal cell P with a, b, c and the three C-centred (orthohexagonal) cells a_1, b_1, c; a_2, b_2, c; a_3, b_3, c. The location of the origin for all three cells is the same, and coincides with the label c.

The inverse matrix $\boldsymbol{P}^{-1} = \boldsymbol{Q}$ is also listed in Table 1.5.1.1 or can be deduced from Fig. 1.5.1.4. It is the matrix that describes the conventional basis vectors $\mathbf{a}_F, \mathbf{b}_F, \mathbf{c}_F$ as linear combinations of $\mathbf{a}_P, \mathbf{b}_P, \mathbf{c}_P$:

$$\mathbf{a}_F = -\mathbf{a}_P + \mathbf{b}_P + \mathbf{c}_P, \quad \mathbf{b}_F = \mathbf{a}_P - \mathbf{b}_P + \mathbf{c}_P, \quad \mathbf{c}_F = \mathbf{a}_P + \mathbf{b}_P - \mathbf{c}_P,$$

or

$$(\mathbf{a}_F, \mathbf{b}_F, \mathbf{c}_F) = (\mathbf{a}_P, \mathbf{b}_P, \mathbf{c}_P)\boldsymbol{P}^{-1} = (\mathbf{a}_P, \mathbf{b}_P, \mathbf{c}_P)\begin{pmatrix} \bar{1} & 1 & 1 \\ 1 & \bar{1} & 1 \\ 1 & 1 & \bar{1} \end{pmatrix}.$$

Correspondingly, the point coordinates transform as

$$\begin{pmatrix} x_P \\ y_P \\ z_P \end{pmatrix} = \boldsymbol{P}^{-1}\begin{pmatrix} x_F \\ y_F \\ z_F \end{pmatrix} = \begin{pmatrix} \bar{1} & 1 & 1 \\ 1 & \bar{1} & 1 \\ 1 & 1 & \bar{1} \end{pmatrix}\begin{pmatrix} x_F \\ y_F \\ z_F \end{pmatrix}$$
$$= \begin{pmatrix} -x_F + y_F + z_F \\ x_F - y_F + z_F \\ x_F + y_F - z_F \end{pmatrix}.$$

For example, the coordinates $\begin{pmatrix} 1 \\ 0 \\ 0 \end{pmatrix}_F$ of the end point of $\mathbf{a}_F$ with respect to the conventional basis become $\begin{pmatrix} \bar{1} \\ 1 \\ 1 \end{pmatrix}_P$ in the primitive basis, the centring point $\begin{pmatrix} \frac{1}{2} \\ \frac{1}{2} \\ 0 \end{pmatrix}_F$ of the $\mathbf{a}_F, \mathbf{b}_F$ plane becomes the end point $\begin{pmatrix} 0 \\ 0 \\ 1 \end{pmatrix}_P$ of $\mathbf{c}_P$ *etc.*

1.5.1.3. General change of coordinate system

A general change of the coordinate system involves both an origin shift and a change of the basis. Such a transformation of the coordinate system is described by the matrix–column pair $(\boldsymbol{P}, \boldsymbol{p})$, where the (3×3) matrix $\boldsymbol{P}$ relates the new basis $\mathbf{a}', \mathbf{b}', \mathbf{c}'$ to the old one $\mathbf{a}, \mathbf{b}, \mathbf{c}$ according to equation (1.5.1.4). The origin shift is described by the *shift vector* $\mathbf{p} = p_1\mathbf{a} + p_2\mathbf{b} + p_3\mathbf{c}$. The coordinates of the new origin O' with respect to the old coordinate system $\mathbf{a}, \mathbf{b}, \mathbf{c}$ are given by the (3×1) column $\boldsymbol{p} = \begin{pmatrix} p_1 \\ p_2 \\ p_3 \end{pmatrix}$.

The general coordinate transformation can be performed in two consecutive steps. Because the origin shift $\boldsymbol{p}$ refers to the old basis $\mathbf{a}, \mathbf{b}, \mathbf{c}$, it has to be applied first (as described in Section 1.5.1.1), followed by the change of the basis (*cf.* Section 1.5.1.2):

$$\boldsymbol{x}' = (\boldsymbol{P}, \boldsymbol{o})^{-1}(\boldsymbol{I}, \boldsymbol{p})^{-1}\boldsymbol{x} = ((\boldsymbol{I}, \boldsymbol{p})(\boldsymbol{P}, \boldsymbol{o}))^{-1}\boldsymbol{x} = (\boldsymbol{P}, \boldsymbol{p})^{-1}\boldsymbol{x}. \quad (1.5.1.6)$$

Here, $\boldsymbol{I}$ is the three-dimensional unit matrix and $\boldsymbol{o}$ is the (3×1) column containing only zeros as coefficients.

The formulae for the change of the *point coordinates* from $\boldsymbol{x}$ to $\boldsymbol{x}'$ uses $(\boldsymbol{Q},\ \boldsymbol{q}) = (\boldsymbol{P}, \boldsymbol{p})^{-1} = (\boldsymbol{P}^{-1},\ -\boldsymbol{P}^{-1}\boldsymbol{p})$, *i.e.*

$$\begin{pmatrix} x'_1 \\ x'_2 \\ x'_3 \end{pmatrix} = \begin{pmatrix} Q_{11} & Q_{12} & Q_{13} \\ Q_{21} & Q_{22} & Q_{23} \\ Q_{31} & Q_{32} & Q_{33} \end{pmatrix}\begin{pmatrix} x_1 \\ x_2 \\ x_3 \end{pmatrix} + \begin{pmatrix} q_1 \\ q_2 \\ q_3 \end{pmatrix}$$
$$\text{with } \boldsymbol{Q} = \boldsymbol{P}^{-1} \text{ and } \boldsymbol{q} = -\boldsymbol{P}^{-1}\boldsymbol{p},$$
$$\text{thus } \boldsymbol{x}' = \boldsymbol{P}^{-1}\boldsymbol{x} - \boldsymbol{P}^{-1}\boldsymbol{p} = \boldsymbol{P}^{-1}(\boldsymbol{x} - \boldsymbol{p}). \quad (1.5.1.7)$$

The effect of a general change of the coordinate system $(\boldsymbol{P}, \boldsymbol{p})$ on the coefficients of a vector $\mathbf{r}$ is reduced to the linear transformation described by $\boldsymbol{P}$, as the vector coefficients are not affected by the origin shift [*cf.* equation (1.5.1.3)].

Hereafter, the data for the matrix–column pair

$$(\boldsymbol{P}, \boldsymbol{p}) = \left[\begin{pmatrix} P_{11} & P_{12} & P_{13} \\ P_{21} & P_{22} & P_{23} \\ P_{31} & P_{32} & P_{33} \end{pmatrix}, \begin{pmatrix} p_1 \\ p_2 \\ p_3 \end{pmatrix}\right]$$

are often written in the following concise form:

$$P_{11}\mathbf{a} + P_{21}\mathbf{b} + P_{31}\mathbf{c},\ P_{12}\mathbf{a} + P_{22}\mathbf{b} + P_{32}\mathbf{c},\ P_{13}\mathbf{a} + P_{23}\mathbf{b} + P_{33}\mathbf{c};$$
$$p_1, p_2, p_3. \quad (1.5.1.8)$$

The concise notation of the transformation matrices is widely used in the tables of maximal subgroups of space groups in *International Tables for Crystallography* Volume A1 (2010), where $(\boldsymbol{P}, \boldsymbol{p})$ describes the relation between the conventional bases of a group and its maximal subgroups. For example, the expression $(\boldsymbol{P}, \boldsymbol{p}) = (-\mathbf{a} - \mathbf{b}, \mathbf{a} - \mathbf{b}, \mathbf{c}; 0, 0, \frac{1}{3})$ (*cf.* the table of maximal subgroups of $P3_112$, No. 151, in Part 2) stands for

$$\boldsymbol{P} = \begin{pmatrix} \bar{1} & 1 & 0 \\ \bar{1} & \bar{1} & 0 \\ 0 & 0 & 1 \end{pmatrix} \text{ and } \boldsymbol{p} = \begin{pmatrix} 0 \\ 0 \\ \frac{1}{3} \end{pmatrix}.$$

Note that the matrix elements of $\boldsymbol{P}$ in equation (1.5.1.8) are read by *columns* since they act on (1×3) *rows* of basis vectors, whereas in the shorthand notation of symmetry operations they are read by *rows*, since these are applied to (3×1) *columns* of coordinates (*cf.* Section 1.2.2.1).

1.5.2. Transformations of crystallographic quantities under coordinate transformations

BY HANS WONDRATSCHEK AND MOIS I. AROYO

1.5.2.1. Covariant and contravariant quantities

If the *direct* or *crystal* basis is transformed by the transformation matrix $\boldsymbol{P}$: $(\mathbf{a}', \mathbf{b}', \mathbf{c}') = (\mathbf{a}, \mathbf{b}, \mathbf{c})\boldsymbol{P}$, the corresponding basis vectors of the *reciprocal (or dual) basis* transform as (*cf.* Section 1.3.2.5)

$$\begin{pmatrix} \mathbf{a}^{*\prime} \\ \mathbf{b}^{*\prime} \\ \mathbf{c}^{*\prime} \end{pmatrix} = \boldsymbol{Q}\begin{pmatrix} \mathbf{a}^{*} \\ \mathbf{b}^{*} \\ \mathbf{c}^{*} \end{pmatrix} = \begin{pmatrix} Q_{11} & Q_{12} & Q_{13} \\ Q_{21} & Q_{22} & Q_{23} \\ Q_{31} & Q_{32} & Q_{33} \end{pmatrix}\begin{pmatrix} \mathbf{a}^{*} \\ \mathbf{b}^{*} \\ \mathbf{c}^{*} \end{pmatrix}, \quad (1.5.2.1)$$

where the notation $\boldsymbol{Q} = \boldsymbol{P}^{-1}$ is applied (*cf.* Section 1.5.1.2).

The quantities that transform in the same way as the basis vectors **a**, **b**, **c** are called *covariant* with respect to the basis **a**, **b**, **c** and *contravariant* with respect to the reciprocal basis $\begin{pmatrix}\mathbf{a}^*\\ \mathbf{b}^*\\ \mathbf{c}^*\end{pmatrix}$. Such quantities are the *Miller indices* (*hkl*) of a plane (or a set of planes) in direct space and the vector coefficients (*h*, *k*, *l*) of the vector perpendicular to those planes, referred to the reciprocal basis $\begin{pmatrix}\mathbf{a}^*\\ \mathbf{b}^*\\ \mathbf{c}^*\end{pmatrix}$:

$$(h', k', l') = (h, k, l)\boldsymbol{P}. \qquad (1.5.2.2)$$

Quantities like the vector coefficients of any vector $\boldsymbol{u} = \begin{pmatrix}u\\ v\\ w\end{pmatrix}$ in direct space (or the *indices of a direction* in direct space) are covariant with respect to the reciprocal basis vectors $\begin{pmatrix}\mathbf{a}^*\\ \mathbf{b}^*\\ \mathbf{c}^*\end{pmatrix}$ and contravariant with respect to **a**, **b**, **c**:

$$\begin{pmatrix}u'\\ v'\\ w'\end{pmatrix} = \boldsymbol{Q}\begin{pmatrix}u\\ v\\ w\end{pmatrix}. \qquad (1.5.2.3)$$

1.5.2.2. Metric tensors of direct and reciprocal lattices

The metric tensor of a crystal lattice with a basis **a**, **b**, **c** is the (3×3) matrix

$$\boldsymbol{G} = \begin{pmatrix}\mathbf{a}\cdot\mathbf{a} & \mathbf{a}\cdot\mathbf{b} & \mathbf{a}\cdot\mathbf{c}\\ \mathbf{b}\cdot\mathbf{a} & \mathbf{b}\cdot\mathbf{b} & \mathbf{b}\cdot\mathbf{c}\\ \mathbf{c}\cdot\mathbf{a} & \mathbf{c}\cdot\mathbf{b} & \mathbf{c}\cdot\mathbf{c}\end{pmatrix},$$

which can be written formally as

$$\boldsymbol{G} = (\mathbf{a}, \mathbf{b}, \mathbf{c})^T \cdot (\mathbf{a}, \mathbf{b}, \mathbf{c}) = \begin{pmatrix}\mathbf{a}\\ \mathbf{b}\\ \mathbf{c}\end{pmatrix} \cdot (\mathbf{a}, \mathbf{b}, \mathbf{c})$$

(*cf.* Section 1.3.2.2). The transformation of the metric tensor under the coordinate transformation $(\boldsymbol{P}, \boldsymbol{p})$ follows directly from its definition:

$$\begin{aligned}\boldsymbol{G}' &= (\mathbf{a}', \mathbf{b}', \mathbf{c}')^T \cdot (\mathbf{a}', \mathbf{b}', \mathbf{c}') = [(\mathbf{a}, \mathbf{b}, \mathbf{c})\boldsymbol{P}]^T \cdot (\mathbf{a}, \mathbf{b}, \mathbf{c})\boldsymbol{P}\\ &= \boldsymbol{P}^T(\mathbf{a}, \mathbf{b}, \mathbf{c})^T \cdot (\mathbf{a}, \mathbf{b}, \mathbf{c})\boldsymbol{P} = \boldsymbol{P}^T\boldsymbol{G}\boldsymbol{P}, \qquad (1.5.2.4)\end{aligned}$$

where $\boldsymbol{P}^T$ is the transposed matrix of $\boldsymbol{P}$. The transformation behaviour of $\boldsymbol{G}$ under $(\boldsymbol{P}, \boldsymbol{p})$ is thus determined by the linear part $\boldsymbol{P}$ of the coordinate transformation alone and is not affected by an origin shift $\boldsymbol{p}$.

The volume V of the unit cell defined by the basis vectors **a**, **b**, **c** can be obtained from the determinant of the metric tensor, $V^2 = \det(\boldsymbol{G})$. The transformation behaviour of V under a coordinate transformation follows from the transformation behaviour of the metric tensor [note that $\det(\boldsymbol{P}) = \det(\boldsymbol{P}^T)$]: $(V')^2 = \det(\boldsymbol{G}') = \det(\boldsymbol{P}^T\boldsymbol{G}\boldsymbol{P}) = \det(\boldsymbol{P})\det(\boldsymbol{P}^T)\det(\boldsymbol{G}) = \det(\boldsymbol{P})^2V^2$, *i.e.*

$$V' = |\det(\boldsymbol{P})|V, \qquad (1.5.2.5)$$

which is reduced to $V' = \det(\boldsymbol{P})V$ if $\det(\boldsymbol{P}) > 0$.

Similarly, the metric tensor $\boldsymbol{G}^*$ of the reciprocal lattice and the volume V^* of the unit cell defined by the basis vectors $\mathbf{a}^*, \mathbf{b}^*, \mathbf{c}^*$ transform as

$$\boldsymbol{G}^{*\prime} = \boldsymbol{Q}\boldsymbol{G}^*\boldsymbol{Q}^T, \qquad (1.5.2.6)$$

$$V^{*\prime} = |\det(\boldsymbol{Q})|V^* \text{ or } V^{*\prime} = \det(\boldsymbol{Q})V^* = [1/\det(\boldsymbol{P})]V^* \text{ if } \det(\boldsymbol{Q}) > 0. \qquad (1.5.2.7)$$

Again, it is only the linear part $\boldsymbol{Q} = \boldsymbol{P}^{-1}$ that determines the transformation behaviour of $\boldsymbol{G}^*$ and V^* under coordinate transformations.

1.5.2.3. Transformation of matrix–column pairs of symmetry operations

The *matrix–column pairs for the symmetry operations* are changed by a change of the coordinate system (see Section 1.2.2 for details of the matrix description of symmetry operations). A symmetry operation $\mathcal{W}$ that maps a point X to an image point $\tilde{X}$ is described in the 'old' (unprimed) coordinate system by the system of equations

$$\begin{aligned}\tilde{x}_1 &= W_{11}x_1 + W_{12}x_2 + W_{13}x_3 + w_1\\ \tilde{x}_2 &= W_{21}x_1 + W_{22}x_2 + W_{23}x_3 + w_2 \qquad (1.5.2.8)\\ \tilde{x}_3 &= W_{31}x_1 + W_{32}x_2 + W_{33}x_3 + w_3,\end{aligned}$$

i.e. by the matrix–column pair $(\boldsymbol{W}, \boldsymbol{w})$:

$$\tilde{\boldsymbol{x}} = \boldsymbol{W}\boldsymbol{x} + \boldsymbol{w} = (\boldsymbol{W}, \boldsymbol{w})\boldsymbol{x}. \qquad (1.5.2.9)$$

In the new (primed) coordinate system, the symmetry operation $\mathcal{W}$ is described by the pair $(\boldsymbol{W}', \boldsymbol{w}')$:

$$\tilde{\boldsymbol{x}}' = (\boldsymbol{W}', \boldsymbol{w}')\boldsymbol{x}' = \boldsymbol{W}'\boldsymbol{x}' + \boldsymbol{w}'. \qquad (1.5.2.10)$$

The relation between $(\boldsymbol{W}, \boldsymbol{w})$ and $(\boldsymbol{W}', \boldsymbol{w}')$ is derived *via* the transformation matrix–column pair $(\boldsymbol{P}, \boldsymbol{p})$, which specifies the change of the coordinate system. The successive application of equations (1.5.1.6), (1.5.2.9) and again (1.5.1.6) results in $\tilde{\boldsymbol{x}}' = (\boldsymbol{P}, \boldsymbol{p})^{-1}\tilde{\boldsymbol{x}} = (\boldsymbol{P}, \boldsymbol{p})^{-1}(\boldsymbol{W}, \boldsymbol{w})\boldsymbol{x} = (\boldsymbol{P}, \boldsymbol{p})^{-1}(\boldsymbol{W}, \boldsymbol{w})(\boldsymbol{P}, \boldsymbol{p})\boldsymbol{x}'$, which compared with equation (1.5.2.10) gives

$$(\boldsymbol{W}', \boldsymbol{w}') = (\boldsymbol{P}, \boldsymbol{p})^{-1}(\boldsymbol{W}, \boldsymbol{w})(\boldsymbol{P}, \boldsymbol{p}). \qquad (1.5.2.11)$$

The result indicates that the change of the matrix–column pairs of symmetry operations $(\boldsymbol{W}, \boldsymbol{w})$ under a coordinate transformation described by the matrix–column pair $(\boldsymbol{P}, \boldsymbol{p})$ is realized by the conjugation of $(\boldsymbol{W}, \boldsymbol{w})$ by $(\boldsymbol{P}, \boldsymbol{p})$.

1.5.2.4. Example: paraelectric-to-ferroelectric phase transition of GeTe

Coordinate transformations are essential in the study of structural relationships between crystal structures. Consider as an example two phases **A** (*basic* or *parent* structure) and **B** (*derivative* structure) of the same compound. Let the space group $\mathcal{H}$ of **B** be a proper subgroup of the space group $\mathcal{G}$ of **A**, $\mathcal{H} < \mathcal{G}$. The relationship between the two structures is characterized by a global distortion that, in general, can be decomposed into a homogeneous strain describing the distortion of the lattice of **B** relative to that of **A** and an atomic displacement field representing the displacements of the atoms of **B** from their positions in **A**. In order to facilitate the determination of the global distortion, the description of structure **A** is transformed by an appropriate transformation $(\boldsymbol{P}, \boldsymbol{p})$ to an equivalent description that is most similar to that of **B**. This new description of **A** is called the *reference description* of structure **A** relative to structure **B**. Now, the metric tensors $\boldsymbol{G}_{\mathbf{A}}$ of the reference description of **A** and $\boldsymbol{G}_{\mathbf{B}}$ are of the same type and are distinguished only by the values of their parameters. The comparison of the (reference) description of **A** and the description of **B** is performed in two steps. In the first step the parameter values of $\boldsymbol{G}_{\mathbf{A}}$ are adapted to those of $G_{\mathbf{B}}$ by an affine transformation which determines the metric deformation (spontaneous strain) of structure **B** relative to structure **A**. The result is a hypothetical structure which still differs from structure **B** by atomic displacements. In the second step these displacements are balanced out by shifting the individual atoms to those of structure **B**.

In summary, if **a**, **b**, **c** represents the basis of the parent phase, then its image under the transformation $(\mathbf{a}', \mathbf{b}', \mathbf{c}') = (\mathbf{a}, \mathbf{b}, \mathbf{c})\boldsymbol{P}$ should be similar to the basis of the derivative phase $\mathbf{a}_{\mathcal{H}}, \mathbf{b}_{\mathcal{H}}, \mathbf{c}_{\mathcal{H}}$. The difference between $\mathbf{a}', \mathbf{b}', \mathbf{c}'$ and $\mathbf{a}_{\mathcal{H}}, \mathbf{b}_{\mathcal{H}}, \mathbf{c}_{\mathcal{H}}$ determines the metric deformation (spontaneous strain) accompanying the transition between the two phases. Similarly, the differences between the images X' of the atomic positions X of the basic structure under the transformation $(\boldsymbol{P}, \boldsymbol{p})$ and the atomic positions $X_{\mathcal{H}}$ of the derivative structure give the atomic displacements that occur during the phase transition.

As an example we will consider the structural phase transition of GeTe, which is of displacive type, *i.e.* the phase transition is accomplished through small atomic displacements. The room-temperature ferroelectric phase belongs to the rhombohedral space group $R3m$ (160). At about 720 K a structural phase transition takes place to a high-symmetry paraelectric cubic phase of the NaCl type. The following descriptions of the two phases of GeTe are taken from the ICSD:

(*a*) Wiedemeier & Siemers (1989), ICSD No. 56037. The symmetry of the high-temperature phase is described by the space group $Fm\bar{3}m$ (225) with cell parameters a_c = 6.009 Å and atomic coordinates listed as

Ge: $4a$ 0, 0, 0

Te: $4b$ $\frac{1}{2}, \frac{1}{2}, \frac{1}{2}$

(*b*) Chattopadhyay *et al.* (1987), ICSD No. 56038. The structure is described with respect to the hexagonal-axes setting of $R3m$ (160) (*cf.* Section 1.5.3.1) with cell parameters $a_{\rm hex}$ = 4.164 (2) Å, $c_{\rm hex}$ = 10.69 (4) Å. The coordinates of the atoms in the asymmetric unit are given as

Ge: $3a$ 0, 0, 0.2376

Te: $3a$ 0, 0, 0.7624

The relation between the basis $\mathbf{a}_c, \mathbf{b}_c, \mathbf{c}_c$ of the F-centred cubic lattice and the basis $\mathbf{a}'_c, \mathbf{b}'_c, \mathbf{c}'_c$ of the reference description can be obtained by inspection. The $\mathbf{c}'_c$ axis of the reference hexagonal basis must be one of the cubic threefold axes, say [111]. The axes $\mathbf{a}'_c$ and $\mathbf{b}'_c$ must be lattice vectors of the F-centred lattice, perpendicular to the rhombohedral axis. They must have equal length, form an angle of 120°, and together with $\mathbf{c}'_c$ define a right-handed basis. For example, the vectors $\mathbf{a}'_c = \frac{1}{2}(-\mathbf{a}_c + \mathbf{b}_c)$, $\mathbf{b}'_c = \frac{1}{2}(-\mathbf{b}_c + \mathbf{c}_c)$ fulfil these conditions.

The transformation matrix $\boldsymbol{P}$ between the bases $\mathbf{a}_c, \mathbf{b}_c, \mathbf{c}_c$ and $\mathbf{a}'_c, \mathbf{b}'_c, \mathbf{c}'_c$ can also be derived from the data listed in Table 1.5.1.1 in two steps:

(i) A cubic F cell can be considered as a primitive rhombohedral cell with $a_p = a_c\frac{1}{2}\sqrt{2}$ and α = 60°. The relation between the two cells is described by the transformation matrix $\boldsymbol{P}_1$ (*cf.* Table 1.5.1.1 and Fig. 1.5.1.4):

$$(\mathbf{a}_p, \mathbf{b}_p, \mathbf{c}_p) = (\mathbf{a}_c, \mathbf{b}_c, \mathbf{c}_c)\boldsymbol{P}_1 = (\mathbf{a}_c, \mathbf{b}_c, \mathbf{c}_c)\begin{pmatrix} 0 & \frac{1}{2} & \frac{1}{2} \\ \frac{1}{2} & 0 & \frac{1}{2} \\ \frac{1}{2} & \frac{1}{2} & 0 \end{pmatrix}. \qquad (1.5.2.12)$$

(ii) The transformation matrix $\boldsymbol{P}_2$ between the rhombohedral primitive cell and the triple hexagonal cell (obverse setting) of the reference description is obtained from Table 1.5.1.1 (*cf.* Fig. 1.5.1.6):

$$(\mathbf{a}'_c, \mathbf{b}'_c, \mathbf{c}'_c) = (\mathbf{a}_p, \mathbf{b}_p, \mathbf{c}_p)\boldsymbol{P}_2 = (\mathbf{a}_p, \mathbf{b}_p, \mathbf{c}_p)\begin{pmatrix} 1 & 0 & 1 \\ \bar{1} & 1 & 1 \\ 0 & \bar{1} & 1 \end{pmatrix}. \qquad (1.5.2.13)$$

Combining equations (1.5.2.12) and (1.5.2.13) gives the orientational relationship between the F-centred cubic cell and the rhombohedrally centred hexagonal cell $(\mathbf{a}'_c, \mathbf{b}'_c, \mathbf{c}'_c) = (\mathbf{a}_c, \mathbf{b}_c, \mathbf{c}_c)\boldsymbol{P}$, where

$$\boldsymbol{P} = \boldsymbol{P}_1\boldsymbol{P}_2 = \begin{pmatrix} 0 & \frac{1}{2} & \frac{1}{2} \\ \frac{1}{2} & 0 & \frac{1}{2} \\ \frac{1}{2} & \frac{1}{2} & 0 \end{pmatrix}\begin{pmatrix} 1 & 0 & 1 \\ \bar{1} & 1 & 1 \\ 0 & \bar{1} & 1 \end{pmatrix} = \begin{pmatrix} -\frac{1}{2} & 0 & 1 \\ \frac{1}{2} & -\frac{1}{2} & 1 \\ 0 & \frac{1}{2} & 1 \end{pmatrix}. \qquad (1.5.2.14)$$

Formally, the lattice parameters of the reference unit cell can be extracted from the metric tensor $\boldsymbol{G}'_c$ obtained from the metric tensor $\boldsymbol{G}_c$ transformed by $\boldsymbol{P}$, *cf.* equation (1.5.2.4):

$$\begin{aligned} \boldsymbol{G}'_c &= \boldsymbol{P}^T\boldsymbol{G}_c\boldsymbol{P} \\ &= \begin{pmatrix} -\frac{1}{2} & \frac{1}{2} & 0 \\ 0 & -\frac{1}{2} & \frac{1}{2} \\ 1 & 1 & 1 \end{pmatrix}\begin{pmatrix} a_c^2 & 0 & 0 \\ 0 & a_c^2 & 0 \\ 0 & 0 & a_c^2 \end{pmatrix}\begin{pmatrix} -\frac{1}{2} & 0 & 1 \\ \frac{1}{2} & -\frac{1}{2} & 1 \\ 0 & \frac{1}{2} & 1 \end{pmatrix} \\ &= a_c^2\begin{pmatrix} \frac{1}{2} & -\frac{1}{4} & 0 \\ -\frac{1}{4} & \frac{1}{2} & 0 \\ 0 & 0 & 3 \end{pmatrix}, \end{aligned} \qquad (1.5.2.15)$$

which gives $a'_c = a_c\frac{1}{2}\sqrt{2}$ = 4.249 Å and $c'_c = a_c\sqrt{3}$ = 10.408 Å. The comparison of these values with the experimentally determined lattice parameters of the low-symmetry phase [$a_{\rm hex}$ = 4.164 (2) Å, $c_{\rm hex}$ = 10.69 (4) Å (Chattopadhyay *et al.*, 1987)] determines the lattice deformation accompanying the displacive phase transition, which basically consists of expanding the cubic unit cell along the [111] direction. (In fact, the elongation along [111] is accompanied by a contraction in the *ab* plane that leads to an overall volume reduction of about 1.3%.)

Owing to the polar character of $R3m$, the symmetry conditions following from the group–subgroup relation $Fm\bar{3}m > R3m$ [*cf.* equation (1.5.2.11)] are not sufficient to determine the origin shift of the transformation between the high- and the low-symmetry space groups. The origin shift of $\boldsymbol{p} = (-\frac{1}{4}, -\frac{1}{4}, -\frac{1}{4})$ in this specific case is chosen in such a way that the relative displacements of Ge and Te are equal in size but in opposite direction along [111].

The inverse transformation matrix–column pair $(\boldsymbol{Q}, \boldsymbol{q}) = (\boldsymbol{P}, \boldsymbol{p})^{-1} = (\boldsymbol{P}^{-1}, -\boldsymbol{P}^{-1}\boldsymbol{p})$ is necessary for the calculation of the atomic coordinates of the reference description X'_c. Given the matrix $\boldsymbol{P}$, its inverse $\boldsymbol{P}^{-1}$ can be calculated either directly (*i.e.* applying the algebraic procedure for inversion of a matrix) or using the inverse matrices $\boldsymbol{Q}_1 = \boldsymbol{P}_1^{-1}$ and $\boldsymbol{Q}_2 = \boldsymbol{P}_2^{-1}$ listed in Table 1.5.1.1:

$$\begin{aligned} \boldsymbol{Q} &= \boldsymbol{P}^{-1} = (\boldsymbol{P}_1\boldsymbol{P}_2)^{-1} = \boldsymbol{P}_2^{-1}\boldsymbol{P}_1^{-1} \\ &= \begin{pmatrix} \frac{2}{3} & -\frac{1}{3} & -\frac{1}{3} \\ \frac{1}{3} & \frac{1}{3} & -\frac{2}{3} \\ \frac{1}{3} & \frac{1}{3} & \frac{1}{3} \end{pmatrix}\begin{pmatrix} \bar{1} & 1 & 1 \\ 1 & \bar{1} & 1 \\ 1 & 1 & \bar{1} \end{pmatrix} = \begin{pmatrix} -\frac{4}{3} & \frac{2}{3} & \frac{2}{3} \\ -\frac{2}{3} & -\frac{2}{3} & \frac{4}{3} \\ \frac{1}{3} & \frac{1}{3} & \frac{1}{3} \end{pmatrix}. \end{aligned} \qquad (1.5.2.16)$$

(Note the change in the order of multiplication of the matrices $\boldsymbol{P}_1^{-1}$ and $\boldsymbol{P}_2^{-1}$ in $\boldsymbol{Q}$.) The corresponding origin shift $\boldsymbol{q}$ is given by

$$q = -P^{-1}p = -\begin{pmatrix} -\frac{4}{3} & \frac{2}{3} & \frac{2}{3} \\ -\frac{2}{3} & -\frac{2}{3} & \frac{4}{3} \\ \frac{1}{3} & \frac{1}{3} & \frac{1}{3} \end{pmatrix} \begin{pmatrix} -\frac{1}{4} \\ -\frac{1}{4} \\ -\frac{1}{4} \end{pmatrix} = \begin{pmatrix} 0 \\ 0 \\ \frac{1}{4} \end{pmatrix}. \quad (1.5.2.17)$$

The atomic positions of the reference description become

$$\begin{pmatrix} x'_c \\ y'_c \\ z'_c \end{pmatrix} = \begin{pmatrix} -\frac{4}{3} & \frac{2}{3} & \frac{2}{3} \\ -\frac{2}{3} & -\frac{2}{3} & \frac{4}{3} \\ \frac{1}{3} & \frac{1}{3} & \frac{1}{3} \end{pmatrix} \begin{pmatrix} x_c \\ y_c \\ z_c \end{pmatrix} + \begin{pmatrix} 0 \\ 0 \\ \frac{1}{4} \end{pmatrix}.$$

The coordinates of the representative Ge atom occupying position $4a$ 0, 0, 0 in $Fm\bar{3}m$ are transformed to $0, 0, \frac{1}{4}$, while those of Te are transformed from $4b$ $\frac{1}{2}, \frac{1}{2}, \frac{1}{2}$ in $Fm\bar{3}m$ to $0, 0, \frac{3}{4}$. The comparison of these values with the experimentally determined atomic coordinates of Ge 0, 0, 0.2376 and Te 0, 0, 0.7624 reveals the corresponding atomic displacements associated with the displacive phase transition. The low-symmetry phase is a result of relative atomic displacements of the Ge and Te atoms along the polar (rhombohedral) [111] direction, giving rise to non-zero polarization along the same direction, *i.e.* the phase transition is a *paraelectric-to-ferroelectric* one.

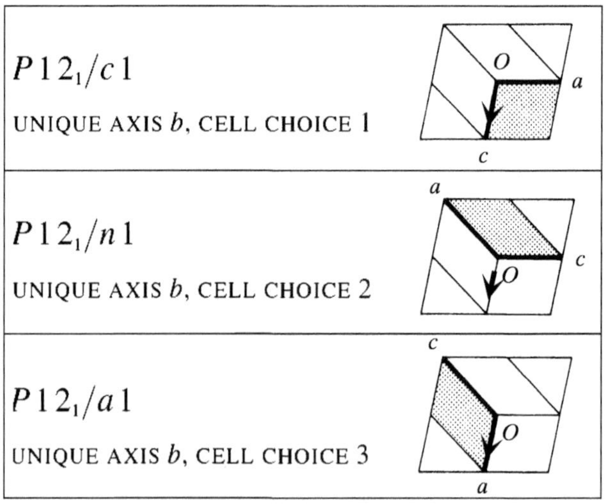

Figure 1.5.3.1
Three possible cell choices for the monoclinic space group $P2_1/c$ (14) with unique axis b. Note the corresponding changes in the full Hermann–Mauguin symbols. The glide vector is indicated by an arrow.

1.5.3. Transformations between different space-group descriptions

By Gervais Chapuis, Hans Wondratschek and Mois I. Aroyo

1.5.3.1. Space groups with more than one description in *IT* A

In the description of the space-group symbols presented in Section 1.4.1, we have already seen that in the conventional, *unique axis b* description of monoclinic space groups, the unique symmetry direction is chosen as **b**; it is normal to **c** and **a**, which form the angle β. However, it is often the case that this standard direction is not the most appropriate choice and that another choice would be more convenient. An example of this is a phase transition from an orthorhombic parent phase to a monoclinic phase. Here, it is often preferable to keep the same orientation of the axes even if the resulting monoclinic setting is not standard.

In some of the space groups, and especially in the monoclinic ones, the space-group tables of *IT* A (see also the space-group tables in Part 2) provide a selection of possible alternative settings. For example, in space group $P2_1/c$, two possible orientations of the unit-cell axes are provided, namely with unique axis b and c. This is reflected in the corresponding full Hermann–Mauguin symbols by the explicit specification of the unique-axis position (dummy indices '1' indicate 'empty' symmetry directions), and by the corresponding change in the direction of the glide plane: $P12_1/c1$ or $P112_1/a$ (*cf.* Section 1.4.1 for a treatment of Hermann–Mauguin symbols of space groups).

It is not just the unique monoclinic axis that can be varied: the choice of the other axes can vary as well. There are cases where the selection of the conventional setting leads to an inconvenient monoclinic angle that deviates greatly from 90°. If another cell choice minimizes the deviation from 90°, it is preferred. Fig. 1.5.3.1 illustrates three cell choices for the monoclinic axis b setting of $P2_1/c$.

In centrosymmetric space groups the origin of the unit cell is located at an inversion centre ('origin choice 2'). If, however, another point has higher site symmetry $\mathcal{S}$, a second diagram is displayed with the origin at a point with site symmetry $\mathcal{S}$ ('origin choice 1'). Fig. 1.5.3.2 illustrates the space group *Pban* (50) with two possible origins. The origin of the first choice is located on a point with site symmetry 222, whereas the origin for the second choice is located on an inversion centre. Among the 230 space groups, *IT* A lists 24 centrosymmetric space groups with an additional alternative origin.

Finally, the seven rhombohedral space-group types (*i.e.* space groups with a rhombohedral lattice) also have alternative descriptions included in the space-group tables of *IT* A. The rhombohedral lattice is first presented with an R-centred hexa-

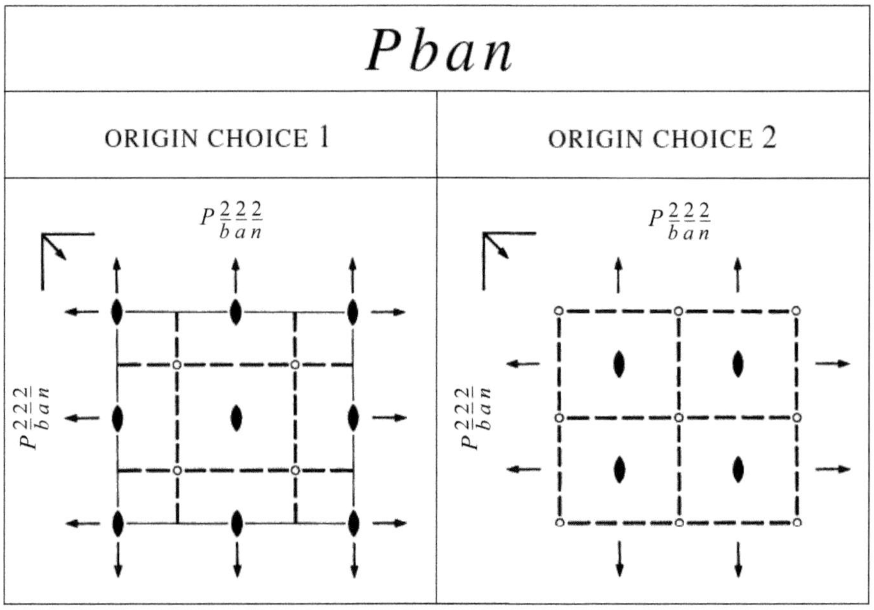

Figure 1.5.3.2
Two possible origin choices for the orthorhombic space group *Pban* (50). Origin choice 1 is on 222, whereas origin choice 2 is on $\bar{1}$.

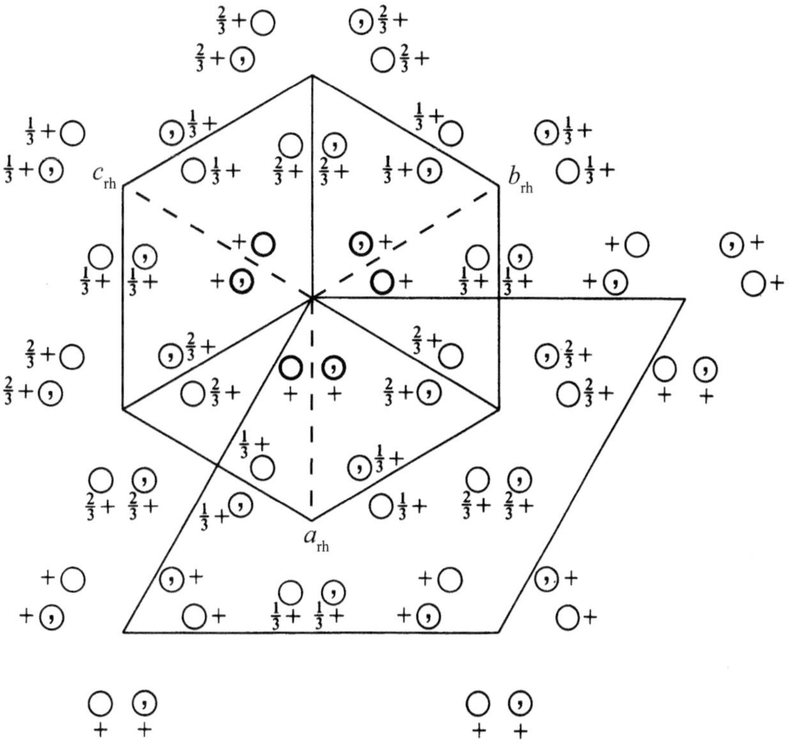

Figure 1.5.3.3
General-position diagram of the space group $R3m$ (160) showing the relation between the hexagonal and rhombohedral axes in the obverse setting: $\mathbf{a}_{rh} = \frac{1}{3}(2\mathbf{a}_{hex} + \mathbf{b}_{hex} + \mathbf{c}_{hex})$, $\mathbf{b}_{rh} = \frac{1}{3}(-\mathbf{a}_{hex} + \mathbf{b}_{hex} + \mathbf{c}_{hex})$, $\mathbf{c}_{rh} = \frac{1}{3}(-\mathbf{a}_{hex} - 2\mathbf{b}_{hex} + \mathbf{c}_{hex})$.

gonal cell ($|\mathbf{a}_{hex}| = |\mathbf{b}_{hex}|$; $\mathbf{c}_{hex} \perp \mathbf{a}_{hex}, \mathbf{b}_{hex}$; $\gamma = 120°$) with a volume three times larger than that of the primitive rhombohedral cell. The second presentation is given with a primitive rhombohedral cell with $a_{rh} = b_{rh} = c_{rh}$ and $\alpha_{rh} = \beta_{rh} = \gamma_{rh}$. The relation between the two types of cell is illustrated in Fig. 1.5.3.3 for the space group $R3m$ (160). In the hexagonal cell, the coordinates of the special position with site symmetry $3m$ are $0, 0, z$, whereas in the rhombohedral cell the same special position has coordinates x, x, x. If we refer to the transformations of the primitive rhombohedral cell cited in Table 1.5.1.1, we observe two different centrings with three possible orientations R_1, R_2 and R_3 which are related by $\pm 120°$ to each other. The two kinds of centrings, called *obverse* and *reverse*, are illustrated in Fig. 1.5.1.6. A rotation of 180° around the threefold axis relates the obverse and reverse descriptions of the rhombohedral lattice. The obverse triple R cells have lattice points at $0, 0, 0$; $\frac{2}{3}, \frac{1}{3}, \frac{1}{3}$; $\frac{1}{3}, \frac{2}{3}, \frac{2}{3}$, whereas the reverse R cells have lattice points at $0, 0, 0$; $\frac{1}{3}, \frac{2}{3}, \frac{1}{3}$; $\frac{2}{3}, \frac{1}{3}, \frac{2}{3}$. The triple hexagonal cell R_1 of the obverse setting (*i.e.* $\mathbf{a}_{hex} = \mathbf{a}_{rh} - \mathbf{b}_{rh}$, $\mathbf{b}_{hex} = \mathbf{b}_{rh} - \mathbf{c}_{rh}$, $\mathbf{c}_{hex} = \mathbf{a}_{rh} + \mathbf{b}_{rh} + \mathbf{b}_{rh}$) has been used in the description of the rhombohedral space groups in *IT* A (*cf.* Table 1.5.1.1 and Fig. 1.5.3.3).

1.5.3.2. Examples

Example 1: Transformations between different settings of $P2_1/c$

In the space-group tables of Part 2 (see also the space-group tables of *IT* A), the monoclinic space group $P2_1/c$ (14) is described in six different settings: for each of the 'unique axis b' and 'unique axis c' settings there are three descriptions specified by different cell choices (*cf.* Section 1.5.4 and Table 1.5.4.1; for more details see Section 2.1.3.15 of *IT* A). The different settings are identified by the appropriate full Hermann–Mauguin symbols. The basis transformations $(\boldsymbol{P}, \boldsymbol{p})$ between the different settings are completely specified by the linear part of the transformation, the 3×3 matrix $\boldsymbol{P}$ [*cf.* equation (1.5.1.4)], as all settings of $P2_1/c$ refer to the same origin, *i.e.* $\boldsymbol{p} = \boldsymbol{o}$. The transformation matrices $\boldsymbol{P}$ necessary for switching between the different descriptions of $P2_1/c$ can either be read off directly or constructed from the transformation-matrix data listed in Table 1.5.1.1.

The transformation from $P12_1/c1$ (unique axis b, cell choice 1) to $P112_1/a$ (unique axis c, cell choice 1) is illustrated by the following example. The change of the direction of the screw axis 2_1 indicates that the unique direction $\mathbf{b}$ transforms to the unique direction $\mathbf{c}$, while the glide vector along $\mathbf{c}$ transforms to a glide vector along $\mathbf{a}$. These changes are reflected in the transformation matrix $\boldsymbol{P}$ between the basis $\mathbf{a}_b, \mathbf{b}_b, \mathbf{c}_b$ of $P12_1/c1$ and $\mathbf{a}_c, \mathbf{b}_c, \mathbf{c}_c$ of $P112_1/a$, which can be obtained directly from Table 1.5.1.1:

$$(\mathbf{a}_c, \mathbf{b}_c, \mathbf{c}_c) = (\mathbf{a}_b, \mathbf{b}_b, \mathbf{c}_b)\boldsymbol{P} = (\mathbf{a}_b, \mathbf{b}_b, \mathbf{c}_b)\begin{pmatrix} 0 & 1 & 0 \\ 0 & 0 & 1 \\ 1 & 0 & 0 \end{pmatrix}$$
$$= (\mathbf{c}_b, \mathbf{a}_b, \mathbf{b}_b).$$

The matrix for the inverse transformation is

$$\boldsymbol{Q} = \boldsymbol{P}^{-1} = \begin{pmatrix} 0 & 0 & 1 \\ 1 & 0 & 0 \\ 0 & 1 & 0 \end{pmatrix}.$$

(i) *Transformation of point coordinates.* From $\boldsymbol{x}' = \boldsymbol{P}^{-1}\boldsymbol{x}$, *cf.* equation (1.5.1.5), it follows that

$$\begin{pmatrix} x_c \\ y_c \\ z_c \end{pmatrix} = \begin{pmatrix} 0 & 0 & 1 \\ 1 & 0 & 0 \\ 0 & 1 & 0 \end{pmatrix}\begin{pmatrix} x_b \\ y_b \\ z_b \end{pmatrix} = \begin{pmatrix} z_b \\ x_b \\ y_b \end{pmatrix}.$$

For example, the representative coordinate triplets of the special Wyckoff position $2d\ \bar{1}$ of $P12_1/c1$ transform exactly to the representative coordinate triplets of the special

Table 1.5.3.1
Transformation of reflection-condition data for $P12_1/c1$ to $P112_1/a$

	$P12_1/c1$ $h_bk_bl_b$	$P112_1/a$ $h_ck_cl_c$
General conditions	$h0l$: $l = 2n$ $0k0$: $k = 2n$ $00l$: $l = 2n$	$hk0$: $h = 2n$ $00l$: $l = 2n$ $h00$: $h = 2n$
Special conditions for the inversion centres	hkl: $k + l = 2n$	hkl: $h + l = 2n$

Wyckoff position $2d\ \bar{1}$ of $P112_1/a$: $\begin{pmatrix}\frac{1}{2}\\0\\\frac{1}{2}\end{pmatrix}$ and $\begin{pmatrix}\frac{1}{2}\\\frac{1}{2}\\0\end{pmatrix}$ transform to $\begin{pmatrix}\frac{1}{2}\\\frac{1}{2}\\0\end{pmatrix}$ and $\begin{pmatrix}0\\\frac{1}{2}\\\frac{1}{2}\end{pmatrix}$.

(ii) *Transformation of the indices in the 'Reflection conditions' block*. Under a coordinate transformation specified by a matrix $\boldsymbol{P}$, the indices of the reflection conditions (Miller indices) transform according to $(h'k'l') = (hkl)\boldsymbol{P}$, *cf.* equation (1.5.2.2). The transformation under

$$\boldsymbol{P} = \begin{pmatrix}0&1&0\\0&0&1\\1&0&0\end{pmatrix}$$

of the set of general or special reflection conditions $h_bk_bl_b$ for $P12_1/c1$ should result in the set of general or special reflection conditions $h_ck_cl_c$ of $P112_1/a$:

$$(h_ck_cl_c) = (h_bk_bl_b)\begin{pmatrix}0&1&0\\0&0&1\\1&0&0\end{pmatrix} = (l_bh_bk_b),$$

i.e. $h_c = l_b$, $k_c = h_b$, $l_c = k_b$ (see Table 1.5.3.1).

(iii) *Transformation of the matrix–column pairs* $(\boldsymbol{W}, \boldsymbol{w})$ *of the symmetry operations*. The matrix–column pairs $(\boldsymbol{W}, \boldsymbol{w})_b$ of the representatives of the symmetry operations of $P12_1/c1$ can be constructed from the coordinate triplets listed in the *General position* block of the group:

$$(1)\, x, y, z \quad (2)\, \bar{x}, y+\tfrac{1}{2}, \bar{z}+\tfrac{1}{2} \quad (3)\, \bar{x}, \bar{y}, \bar{z} \quad (4)\, x, \bar{y}+\tfrac{1}{2}, z+\tfrac{1}{2}$$

According to equation (1.5.2.11), the matrix–column pairs $(\boldsymbol{W}, \boldsymbol{w})_c$ of the symmetry operations of $P112_1/a$ are related to the matrices $(\boldsymbol{W}, \boldsymbol{w})_b$ of $P12_1/c1$ by the equation $(\boldsymbol{W}, \boldsymbol{w})_c = (\boldsymbol{Q}, \boldsymbol{q})(\boldsymbol{W}, \boldsymbol{w})_b(\boldsymbol{P}, \boldsymbol{p})$, where $\boldsymbol{p} = \boldsymbol{q} = \boldsymbol{o}$, since both settings refer to the same origin.

The unit matrix representing the identity operation (1) is invariant under any basis transformation, *i.e.* x, y, z transforms to x, y, z. Similarly, the matrix of inversion $\bar{1}$ (3) (the linear part of which is a multiple of the unit matrix) is also invariant under any basis transformation, *i.e.* $\bar{x}, \bar{y}, \bar{z}$ transforms to $\bar{x}, \bar{y}, \bar{z}$. The symmetry operation $(2)\, \bar{x}, y+\frac{1}{2}, \bar{z}+\frac{1}{2}$, represented by the matrix–column pair

$$\left[\begin{pmatrix}\bar{1}&0&0\\0&1&0\\0&0&\bar{1}\end{pmatrix}, \begin{pmatrix}0\\\frac{1}{2}\\\frac{1}{2}\end{pmatrix}\right],$$

transforms to

$$\left[\begin{pmatrix}0&0&1\\1&0&0\\0&1&0\end{pmatrix}, \begin{pmatrix}0\\0\\0\end{pmatrix}\right]\left[\begin{pmatrix}\bar{1}&0&0\\0&1&0\\0&0&\bar{1}\end{pmatrix}, \begin{pmatrix}0\\\frac{1}{2}\\\frac{1}{2}\end{pmatrix}\right]\left[\begin{pmatrix}0&1&0\\0&0&1\\1&0&0\end{pmatrix}, \begin{pmatrix}0\\0\\0\end{pmatrix}\right]$$
$$= \left[\begin{pmatrix}\bar{1}&0&0\\0&\bar{1}&0\\0&0&1\end{pmatrix}, \begin{pmatrix}\frac{1}{2}\\0\\\frac{1}{2}\end{pmatrix}\right],$$

which corresponds to $\bar{x}+\frac{1}{2}, \bar{y}, z+\frac{1}{2}$.

Finally, the symmetry operation $(4)\, x, \bar{y}+\frac{1}{2}, z+\frac{1}{2}$ transforms to the coordinate triplet $x+\frac{1}{2}, y, \bar{z}+\frac{1}{2}$, since (4) is the composition of (2) with the inversion $(3)\, \bar{x}, \bar{y}, \bar{z}$, which remains invariant under the basis transformation.

The coordinate triplets of the transformed symmetry operations correspond to the entries of the *general-position* block of $P112_1/a$ (*cf.* the space-group tables of $P2_1/c$ of this volume).

Example 2: Transformation between the two origin-choice settings of $I4_1/amd$

The zircon example of Section 1.5.1.1 illustrates how the atomic coordinates change under an origin-choice transformation. Here, the case of the two origin-choice descriptions of the same space group $I4_1/amd$ (141) will be used to demonstrate how the rest of the crystallographic quantities are affected by an origin shift.

The two descriptions of $I4_1/amd$ in the space-group tables of Part 2 are distinguished by the origin choices of the reference coordinate systems: the origin statement of the origin choice 1 setting indicates that its origin O_1 is taken at a point of $\bar{4}m2$ symmetry, which is located at $0, \frac{1}{4}, -\frac{1}{8}$ with respect to the origin O_2 of origin choice 2, taken at a centre $(2/m)$. Conversely, the origin O_2 is taken at a centre $(2/m)$ at $0, -\frac{1}{4}, \frac{1}{8}$ from the origin O_1. These origin descriptions in fact specify explicitly the origin-shift vector $\boldsymbol{p}$ necessary for the transformation between the two settings. For example, the shift vector listed for origin choice 2 expresses the origin O_2 with respect to O_1, *i.e.* the corresponding transformation matrix

$$(\boldsymbol{P}, \boldsymbol{p}) = (\boldsymbol{I}, \boldsymbol{p}) = \left[\begin{pmatrix}1&0&0\\0&1&0\\0&0&1\end{pmatrix}, \begin{pmatrix}0\\-\frac{1}{4}\\\frac{1}{8}\end{pmatrix}\right]$$

transforms the crystallographic data from the origin choice 1 setting to the origin choice 2 setting.

(i) *Transformation of point coordinates*. In accordance with the discussion of Section 1.5.1.1 [*cf.* equation (1.5.1.2)], the transformation of point coordinates $\boldsymbol{x}_1 = \begin{pmatrix}x_1\\y_1\\z_1\end{pmatrix}$ of the origin choice 1 setting of $I4_1/amd$ to $\boldsymbol{x}_2 = \begin{pmatrix}x_2\\y_2\\z_2\end{pmatrix}$ of the origin choice 2 setting is given by

$$\boldsymbol{x}_2 = \begin{pmatrix}x_2\\y_2\\z_2\end{pmatrix} = (\boldsymbol{P}, \boldsymbol{p})^{-1}\boldsymbol{x}_1 = (\boldsymbol{I}, -\boldsymbol{p})\boldsymbol{x}_1 = \begin{pmatrix}x_1\\y_1+\frac{1}{4}\\z_1-\frac{1}{8}\end{pmatrix}. \tag{1.5.3.1}$$

(ii) *Metric tensors and the data for the reflection conditions*. The metric tensors and the data for the reflection conditions are not affected by an origin shift as $\boldsymbol{P} = \boldsymbol{I}$, *cf.* equations (1.5.2.4) and (1.5.2.2).

(iii) *Transformation of the matrix–column pairs* $(\boldsymbol{W}, \boldsymbol{w})$ *of the symmetry operations*. The origin-shift transformation $(\boldsymbol{I}, \boldsymbol{p})$ relates the matrix–column pairs $(\boldsymbol{W}_1, \boldsymbol{w}_1)$ of the symmetry operations of the origin choice 1 setting of $I4_1/amd$ to $(\boldsymbol{W}_2, \boldsymbol{w}_2)$ of the origin choice 2 setting [*cf.* equation (1.5.2.11)]:

$$(\boldsymbol{W}_2, \boldsymbol{w}_2) = (\boldsymbol{I}, -\boldsymbol{p})(\boldsymbol{W}_1, \boldsymbol{w}_1)(\boldsymbol{I}, \boldsymbol{p}) = (\boldsymbol{W}_1, \boldsymbol{w}_1 + [\boldsymbol{W}_1 - \boldsymbol{I}]\boldsymbol{p}). \qquad (1.5.3.2)$$

The rotation part of the symmetry operation is not affected by the origin shift, but the translation part is affected, *i.e.* $\boldsymbol{W}_2 = \boldsymbol{W}_1$ and $\boldsymbol{w}_2 = \boldsymbol{w}_1 + (\boldsymbol{W}_1 - \boldsymbol{I})\boldsymbol{p}$. For example, the translation and unit element generators of $I4_1/amd$ are not changed under the origin-shift transformation, as $\boldsymbol{W}_1 = \boldsymbol{I}$. For the first non-translation generator given by the coordinate triplet $\bar{y}, x+\frac{1}{2}, z+\frac{1}{4}$ and represented by the matrix

$$\left[\begin{pmatrix} 0 & \bar{1} & 0 \\ 1 & 0 & 0 \\ 0 & 0 & 1 \end{pmatrix}, \begin{pmatrix} 0 \\ \frac{1}{2} \\ \frac{1}{4} \end{pmatrix}\right],$$

the translation part is changed by

$$(\boldsymbol{W}_1 - \boldsymbol{I})\boldsymbol{p} = \begin{pmatrix} \bar{1} & \bar{1} & 0 \\ 1 & \bar{1} & 0 \\ 0 & 0 & 0 \end{pmatrix}\begin{pmatrix} 0 \\ -\frac{1}{4} \\ \frac{1}{8} \end{pmatrix} = \begin{pmatrix} \frac{1}{4} \\ \frac{1}{4} \\ 0 \end{pmatrix}$$

and thus the matrix transforms to

$$\left[\begin{pmatrix} 0 & \bar{1} & 0 \\ 1 & 0 & 0 \\ 0 & 0 & 1 \end{pmatrix}, \begin{pmatrix} \frac{1}{4} \\ \frac{3}{4} \\ \frac{1}{4} \end{pmatrix}\right],$$

which corresponds to the coordinate triplet $\bar{y}+\frac{1}{4}, x+\frac{3}{4}, z+\frac{1}{4}$.

The second non-translation generator $\bar{x}, \bar{y}+\frac{1}{2}, \bar{z}+\frac{1}{4}$, represented by the matrix

$$\left[\begin{pmatrix} \bar{1} & 0 & 0 \\ 0 & \bar{1} & 0 \\ 0 & 0 & \bar{1} \end{pmatrix}, \begin{pmatrix} 0 \\ \frac{1}{2} \\ \frac{1}{4} \end{pmatrix}\right] \text{ transforms to } \left[\begin{pmatrix} \bar{1} & 0 & 0 \\ 0 & \bar{1} & 0 \\ 0 & 0 & \bar{1} \end{pmatrix}, \begin{pmatrix} 0 \\ 1 \\ 0 \end{pmatrix}\right],$$

which under the normalization $0 \leq w_i < 1$ is written as the coordinate triplet $\bar{x}, \bar{y}, \bar{z}$ (confirming that the origin is now located at an inversion centre). The coordinate triplets of the transformed symmetry operations are the entries of the corresponding generators of the origin choice 2 setting of $I4_1/amd$ (*cf.* the space-group tables of $I4_1/amd$ in Part 2).

1.5.4. Synoptic tables of plane and space groups[2]

By Bernd Souvignier, Gervais Chapuis and Hans Wondratschek

It is already clear from Section 1.5.3.1 that the Hermann–Mauguin symbol of a space group depends on the choice of the basis vectors. The purpose of this section is to give an overview of a large selection of possible alternative settings of the monoclinic and orthorhombic space groups and their Hermann–Mauguin symbols covering most practical cases (*cf.* Section 1.5.4 of *IT* A for the synoptic tables of all plane and space groups). In particular, the synoptic tables include two main types of information:

(i) Space-group symbols for various settings and choices of the basis. The axis transformations involve permutations of axes conserving the shape of the cell and also transformations leading to different cell shapes and multiple cells.

(ii) *Extended* Hermann–Mauguin space-group symbols in addition to the short and full symbols. The three types of symbols, short, full and extended, provide different levels of information about the symmetry elements and the related symmetry operations of the space group (*cf.* Section 1.2.3 for definitions and discussion of the concepts of symmetry element, geometric element, element set and defining symmetry operation). The short and full Hermann–Mauguin symbols only display information about a chosen set of generators for a space group from which all the elements of a space group can in principle be deduced (for an introduction to short and full Hermann–Mauguin symbols, see Section 1.4.1.4). The multiplicity of the general position in each space group gives the number of symmetry operations *modulo* the lattice translations. As already discussed in Section 1.4.2.4, the combinations of this representative set of symmetry operations with lattice translations give rise to *additional symmetry operations* and *additional symmetry elements*, displayed in the symmetry-element diagrams. The additional symmetry operations are also reflected in the so-called *extended Hermann–Mauguin symbols*, which were introduced in *International Tables for X-ray Crystallography* Volume I (1952). They were systematically developed and tabulated by Bertaut for the first edition of Volume A of *International Tables for Crystallography*, published in 1983.

An extended Hermann–Mauguin symbol is a complex multi-line symbol: (i) the first line contains those symmetry operations for which the coordinate triplets are explicitly printed under 'Positions' in the space-group tables in *IT* A; (ii) the entries of the lines below indicate the additional symmetry operations generated by the compositions of the symmetry operations of the first line with lattice translations. For example, for *A*-, *B*-, *C*- and *I*-centred space groups, the entries of the second line of the two-line extended symbol denote the symmetry operations generated by combinations with the corresponding centring translations.[3]

The monoclinic and orthorhombic systems present the largest number of alternative Hermann–Mauguin symbols owing to various settings and cell choices. A comprehensive listing of the possible Hermann–Mauguin symbols for different settings and unit-cell choices for the monoclinic and orthorhombic space groups is given in Table 1.5.4.1 (the corresponding data for the rest of the space groups can be found in Table 1.5.4.4 of *IT* A).

In the monoclinic system there are three choices for the unique axis, namely *b*, *c* and *a*. In each case, two permutations of the other axes are possible, thus yielding six possible settings given in terms of three pairs, namely $\mathbf{a\underline{b}c}$ and $\mathbf{c\underline{\bar{b}}a}$, $\mathbf{ab\underline{c}}$ and $\mathbf{ba\underline{\bar{c}}}$, $\mathbf{\underline{a}bc}$ and $\mathbf{\underline{\bar{a}}cb}$. The unique axes are underlined and the negative sign, placed over the letter, maintains the correct handedness of the reference system. The three possible cell choices indicated in Fig. 1.5.3.1

[2] With Table 1.5.4.1 by E. F. Bertaut.

[3] After the introduction of the *e*-glide convention and the symmetry-element interpretation of the characters of the Hermann–Mauguin symbols (de Wolff *et al.*, 1992), the tabulated data for the extended symbols were partially modified by introducing the *e*-glide notation in the symbols of only some of the groups [*cf.* Table 4.3.2.1 of the fifth edition of *IT* A (2002)]. In contrast to the fifth edition, in Table 1.5.4.1 (see also Table 1.5.4.4 of *IT* A) extended symbols similar to those that can be found in the first four editions of *IT* A have been reinstated.

Table 1.5.4.1
List of space-group symbols for various settings and cells of monoclinic and orthorhombic space groups

MONOCLINIC SYSTEM

No. of space group	Schoenflies symbol	Standard short Hermann–Mauguin symbol	Extended Hermann–Mauguin symbols for various settings and cell choices						
			$\mathbf{a\underline{b}c}$	$\mathbf{c\underline{b}\bar{a}}$	$\mathbf{ab\underline{c}}$	$\mathbf{ba\underline{\bar{c}}}$	$\mathbf{\underline{a}bc}$	$\mathbf{\underline{\bar{a}}cb}$	Unique axis b Unique axis c Unique axis a
3	C_2^1	$P2$	$P121$	$P121$	$P112$	$P112$	$P211$	$P211$	
4	C_2^2	$P2_1$	$P12_11$	$P12_11$	$P112_1$	$P112_1$	$P2_111$	$P2_111$	
5	C_2^3	$C2$	$C121$	$A121$	$A112$	$B112$	$B211$	$C211$	Cell choice 1
			2_1	2_1	2_1	2_1	2_1	2_1	
			$A121$	$C121$	$B112$	$A112$	$C211$	$B211$	Cell choice 2
			2_1	2_1	2_1	2_1	2_1	2_1	
			$I121$	$I121$	$I112$	$I112$	$I211$	$I211$	Cell choice 3
			2_1	2_1	2_1	2_1	2_1	2_1	
6	C_s^1	Pm	$P1m1$	$P1m1$	$P11m$	$P11m$	$Pm11$	$Pm11$	
7	C_s^2	Pc	$P1c1$	$P1a1$	$P11a$	$P11b$	$Pb11$	$Pc11$	Cell choice 1
			$P1n1$	$P1n1$	$P11n$	$P11n$	$Pn11$	$Pn11$	Cell choice 2
			$P1a1$	$P1c1$	$P11b$	$P11a$	$Pc11$	$Pb11$	Cell choice 3
8	C_s^3	Cm	$C1m1$	$A1m1$	$A11m$	$B11m$	$Bm11$	$Cm11$	Cell choice 1
			a	c	b	a	c	b	
			$A1m1$	$C1m1$	$B11m$	$A11m$	$Cm11$	$Bm11$	Cell choice 2
			c	a	a	b	b	c	
			$I1m1$	$I1m1$	$I11m$	$I11m$	$Im11$	$Im11$	Cell choice 3
			n	n	n	n	n	n	
9	C_s^4	Cc	$C1c1$	$A1a1$	$A11a$	$B11b$	$Bb11$	$Cc11$	Cell choice 1
			n	n	n	n	n	n	
			$A1n1$	$C1n1$	$B11n$	$A11n$	$Cn11$	$Bn11$	Cell choice 2
			a	c	b	a	c	b	
			$I1a1$	$I1c1$	$I11b$	$I11a$	$Ic11$	$Ib11$	Cell choice 3
			c	a	a	b	b	c	
10	C_{2h}^1	$P2/m$	$P1\frac{2}{m}1$	$P1\frac{2}{m}1$	$P11\frac{2}{m}$	$P11\frac{2}{m}$	$P\frac{2}{m}11$	$P\frac{2}{m}11$	
11	C_{2h}^2	$P2_1/m$	$P1\frac{2_1}{m}1$	$P1\frac{2_1}{m}1$	$P11\frac{2_1}{m}$	$P11\frac{2_1}{m}$	$P\frac{2_1}{m}11$	$P\frac{2_1}{m}11$	
12	C_{2h}^3	$C2/m$	$C1\frac{2}{m}1$	$A1\frac{2}{m}1$	$A11\frac{2}{m}$	$B11\frac{2}{m}$	$B\frac{2}{m}11$	$C\frac{2}{m}11$	Cell choice 1
			$\frac{2_1}{a}$	$\frac{2_1}{c}$	$\frac{2_1}{b}$	$\frac{2_1}{a}$	$\frac{2_1}{c}$	$\frac{2_1}{b}$	
			$A1\frac{2}{m}1$	$C1\frac{2}{m}1$	$B11\frac{2}{m}$	$A11\frac{2}{m}$	$C\frac{2}{m}11$	$B\frac{2}{m}11$	Cell choice 2
			$\frac{2_1}{c}$	$\frac{2_1}{a}$	$\frac{2_1}{a}$	$\frac{2_1}{b}$	$\frac{2_1}{b}$	$\frac{2_1}{c}$	
			$I1\frac{2}{m}1$	$I1\frac{2}{m}1$	$I11\frac{2}{m}$	$I11\frac{2}{m}$	$I\frac{2}{m}11$	$I\frac{2}{m}11$	Cell choice 3
			$\frac{2_1}{n}$	$\frac{2_1}{n}$	$\frac{2_1}{n}$	$\frac{2_1}{n}$	$\frac{2_1}{n}$	$\frac{2_1}{n}$	
13	C_{2h}^4	$P2/c$	$P1\frac{2}{c}1$	$P1\frac{2}{a}1$	$P11\frac{2}{a}$	$P11\frac{2}{b}$	$P\frac{2}{b}11$	$P\frac{2}{c}11$	Cell choice 1
			$P1\frac{2}{n}1$	$P1\frac{2}{n}1$	$P11\frac{2}{n}$	$P11\frac{2}{n}$	$P\frac{2}{n}11$	$P\frac{2}{n}11$	Cell choice 2
			$P1\frac{2}{a}1$	$P1\frac{2}{c}1$	$P11\frac{2}{b}$	$P11\frac{2}{a}$	$P\frac{2}{c}11$	$P\frac{2}{b}11$	Cell choice 3
14	C_{2h}^5	$P2_1/c$	$P1\frac{2_1}{c}1$	$P1\frac{2_1}{a}1$	$P11\frac{2_1}{a}$	$P11\frac{2_1}{b}$	$P\frac{2_1}{b}11$	$P\frac{2_1}{c}11$	Cell choice 1
			$P1\frac{2_1}{n}1$	$P1\frac{2_1}{n}1$	$P11\frac{2_1}{n}$	$P11\frac{2_1}{n}$	$P\frac{2_1}{n}11$	$P\frac{2_1}{n}11$	Cell choice 2
			$P1\frac{2_1}{a}1$	$P1\frac{2_1}{c}1$	$P11\frac{2_1}{b}$	$P11\frac{2_1}{a}$	$P\frac{2_1}{c}11$	$P\frac{2_1}{b}11$	Cell choice 3

Table 1.5.4.1 (continued)

No. of space group	Schoenflies symbol	Standard short Hermann–Mauguin symbol	Extended Hermann–Mauguin symbols for various settings and cell choices						
			$a\underline{b}c$	$\bar{c}\underline{b}a$	$ab\underline{c}$	$ba\underline{\bar{c}}$	$\underline{a}bc$	$\underline{\bar{a}}cb$	Unique axis *b* Unique axis *c* Unique axis *a*
15	C_{2h}^6	$C2/c$	$C1\frac{2}{c}1$ $\frac{2_1}{n}$	$A1\frac{2}{a}1$ $\frac{2_1}{n}$	$A11\frac{2}{a}$ $\frac{2_1}{n}$	$B11\frac{2}{b}$ $\frac{2_1}{n}$	$B\frac{2}{b}11$ $\frac{2_1}{n}$	$C\frac{2}{c}11$ $\frac{2_1}{n}$	Cell choice 1
			$A1\frac{2}{n}1$ $\frac{2_1}{a}$	$C1\frac{2}{n}1$ $\frac{2_1}{c}$	$B11\frac{2}{n}$ $\frac{2_1}{b}$	$A11\frac{2}{n}$ $\frac{2_1}{a}$	$C\frac{2}{n}11$ $\frac{2_1}{c}$	$B\frac{2}{n}11$ $\frac{2_1}{b}$	Cell choice 2
			$I1\frac{2}{a}1$ $\frac{2_1}{c}$	$I1\frac{2}{c}1$ $\frac{2_1}{a}$	$I11\frac{2}{b}$ $\frac{2_1}{a}$	$I11\frac{2}{a}$ $\frac{2_1}{b}$	$I\frac{2}{c}11$ $\frac{2_1}{b}$	$I\frac{2}{b}11$ $\frac{2_1}{c}$	Cell choice 3

ORTHORHOMBIC SYSTEM

No. of space group	Schoen-flies symbol	Standard full Hermann–Mauguin symbol **abc**	Extended Hermann–Mauguin symbols for the six settings of the same unit cell					
			abc (standard)	$\mathbf{ba\bar{c}}$	**cab**	$\mathbf{\bar{c}ba}$	**bca**	$\mathbf{a\bar{c}b}$
16	D_2^1	$P222$	$P222$	$P222$	$P222$	$P222$	$P222$	$P222$
17	D_2^2	$P222_1$	$P222_1$	$P222_1$	$P2_122$	$P2_122$	$P22_12$	$P22_12$
18	D_2^3	$P2_12_12$	$P2_12_12$	$P2_12_12$	$P22_12_1$	$P22_12_1$	$P2_122_1$	$P2_122_1$
19	D_2^4	$P2_12_12_1$	$P2_12_12_1$	$P2_12_12_1$	$P2_12_12_1$	$P2_12_12_1$	$P2_12_12_1$	$P2_12_12_1$
20	D_2^5	$C222_1$	$C222_1$ $2_12_12_1$	$C222_1$ $2_12_12_1$	$A2_122$ $2_12_12_1$	$A2_122$ $2_12_12_1$	$B22_12$ $2_12_12_1$	$B22_12$ $2_12_12_1$
21	D_2^6	$C222$	$C222$ 2_12_12	$C222$ 2_12_12	$A222$ 22_12_1	$A222$ 22_12_1	$B222$ 2_122_1	$B222$ 2_122_1
22	D_2^7	$F222$	$F222$ 2_12_12 22_12_1 2_122_1	$F222$ 2_12_12 2_122_1 22_12_1	$F222$ 22_12_1 2_122_1 2_12_12	$F222$ 22_12_1 2_12_12 2_122_1	$F222$ 2_122_1 2_12_12 22_12_1	$F222$ 2_122_1 22_12_1 2_12_12
23	D_2^8	$I222$	$I222$ $2_12_12_1$	$I222$ $2_12_12_1$	$I222$ $2_12_12_1$	$I222$ $2_12_12_1$	$I222$ $2_12_12_1$	$I222$ $2_12_12_1$
24	D_2^9	$I2_12_12_1$	$I2_12_12_1$ 222	$I2_12_12_1$ 222	$I2_12_12_1$ 222	$I2_12_12_1$ 222	$I2_12_12_1$ 222	$I2_12_12_1$ 222
25	C_{2v}^1	$Pmm2$	$Pmm2$	$Pmm2$	$P2mm$	$P2mm$	$Pm2m$	$Pm2m$
26	C_{2v}^2	$Pmc2_1$	$Pmc2_1$	$Pcm2_1$	$P2_1ma$	$P2_1am$	$Pb2_1m$	$Pm2_1b$
27	C_{2v}^3	$Pcc2$	$Pcc2$	$Pcc2$	$P2aa$	$P2aa$	$Pb2b$	$Pb2b$
28	C_{2v}^4	$Pma2$	$Pma2$	$Pbm2$	$P2mb$	$P2cm$	$Pc2m$	$Pm2a$
29	C_{2v}^5	$Pca2_1$	$Pca2_1$	$Pbc2_1$	$P2_1ab$	$P2_1ca$	$Pc2_1b$	$Pb2_1a$
30	C_{2v}^6	$Pnc2$	$Pnc2$	$Pcn2$	$P2na$	$P2an$	$Pb2n$	$Pn2b$
31	C_{2v}^7	$Pmn2_1$	$Pmn2_1$	$Pnm2_1$	$P2_1mn$	$P2_1nm$	$Pn2_1m$	$Pm2_1n$
32	C_{2v}^8	$Pba2$	$Pba2$	$Pba2$	$P2cb$	$P2cb$	$Pc2a$	$Pc2a$
33	C_{2v}^9	$Pna2_1$	$Pna2_1$	$Pbn2_1$	$P2_1nb$	$P2_1cn$	$Pc2_1n$	$Pn2_1a$
34	C_{2v}^{10}	$Pnn2$	$Pnn2$	$Pnn2$	$P2nn$	$P2nn$	$Pn2n$	$Pn2n$
35	C_{2v}^{11}	$Cmm2$	$Cmm2$ $ba2$	$Cmm2$ $ba2$	$A2mm$ $2cb$	$A2mm$ $2cb$	$Bm2m$ $c2a$	$Bm2m$ $c2a$
36	C_{2v}^{12}	$Cmc2_1$	$Cmc2_1$ $bn2_1$	$Ccm2_1$ $na2_1$	$A2_1ma$ 2_1cn	$A2_1am$ 2_1nb	$Bb2_1m$ $n2_1a$	$Bm2_1b$ $c2_1n$
37	C_{2v}^{13}	$Ccc2$	$Ccc2$ $nn2$	$Ccc2$ $nn2$	$A2aa$ $2nn$	$A2aa$ $2nn$	$Bb2b$ $n2n$	$Bb2b$ $n2n$
38	C_{2v}^{14}	$Amm2$	$Amm2$ $nc2_1$	$Bmm2$ $cn2_1$	$B2mm$ 2_1na	$C2mm$ 2_1an	$Cm2m$ $b2_1n$	$Am2m$ $n2_1b$
39†	C_{2v}^{15}	$Aem2$	$Abm2$ ($Aem2$) $cc2_1$	$Bma2$ ($Bme2$) $cc2_1$	$B2cm$ ($B2em$) 2_1aa	$C2mb$ ($C2me$) 2_1aa	$Cm2a$ ($Cm2e$) $b2_1b$	$Ac2m$ ($Ae2m$) $b2_1b$

Table 1.5.4.1 (continued)

No. of space group	Schoen-flies symbol	Standard full Hermann–Mauguin symbol **abc**	Extended Hermann–Mauguin symbols for the six settings of the same unit cell					
			abc (standard)	$\mathbf{ba\bar{c}}$	**cab**	$\mathbf{\bar{c}ba}$	**bca**	$\mathbf{a\bar{c}b}$
40	C_{2v}^{16}	*Ama*2	*Ama*2 $nn2_1$	*Bbm*2 $nn2_1$	*B*2*mb* 2_1nn	*C*2*cm* 2_1nn	*Cc*2*m* $n2_1n$	*Am*2*a* $n2_1n$
41†	C_{2v}^{17}	*Aea*2	*Aba*2 (*Aea*2) $cn2_1$	*Bba*2 (*Bbe*2) $nc2_1$	*B*2*cb* (*B*2*eb*) 2_1an	*C*2*cb* (*C*2*ce*) 2_1na	*Cc*2*a* (*Cc*2*e*) $n2_1b$	*Ac*2*a* (*Ae*2*a*) $b2_1n$
42	C_{2v}^{18}	*Fmm*2	*Fmm*2 *ba*2 $nc2_1$ $cn2_1$	*Fmm*2 *ba*2 $cn2_1$ $nc2_1$	*F*2*mm* 2*cb* 2_1na 2_1an	*F*2*mm* 2*cb* 2_1an 2_1na	*Fm*2*m* *c*2*a* $b2_1n$ $n2_1b$	*Fm*2*m* *c*2*a* $n2_1b$ $b2_1n$
43	C_{2v}^{19}	*Fdd*2	*Fdd*2 $dd2_1$	*Fdd*2 $dd2_1$	*F*2*dd* 2_1dd	*F*2*dd* 2_1dd	*Fd*2*d* $d2_1d$	*Fd*2*d* $d2_1d$
44	C_{2v}^{20}	*Imm*2	*Imm*2 $nn2_1$	*Imm*2 $nn2_1$	*I*2*mm* 2_1nn	*I*2*mm* 2_1nn	*Im*2*m* $n2_1n$	*Im*2*m* $n2_1n$
45	C_{2v}^{21}	*Iba*2	*Iba*2 $cc2_1$	*Iba*2 $cc2_1$	*I*2*cb* 2_1aa	*I*2*cb* 2_1aa	*Ic*2*a* $b2_1b$	*Ic*2*a* $b2_1b$
46	C_{2v}^{22}	*Ima*2	*Ima*2 $nc2_1$	*Ibm*2 $cn2_1$	*I*2*mb* 2_1na	*I*2*cm* 2_1an	*Ic*2*m* $b2_1n$	*Im*2*a* $n2_1b$
47	D_{2h}^{1}	$P\frac{2}{m}\frac{2}{m}\frac{2}{m}$	*Pmmm*	*Pmmm*	*Pmmm*	*Pmmm*	*Pmmm*	*Pmmm*
48	D_{2h}^{2}	$P\frac{2}{n}\frac{2}{n}\frac{2}{n}$	*Pnnn*	*Pnnn*	*Pnnn*	*Pnnn*	*Pnnn*	*Pnnn*
49	D_{2h}^{3}	$P\frac{2}{c}\frac{2}{c}\frac{2}{m}$	*Pccm*	*Pccm*	*Pmaa*	*Pmaa*	*Pbmb*	*Pbmb*
50	D_{2h}^{4}	$P\frac{2}{b}\frac{2}{a}\frac{2}{n}$	*Pban*	*Pban*	*Pncb*	*Pncb*	*Pcna*	*Pcna*
51	D_{2h}^{5}	$P\frac{2_1}{m}\frac{2}{m}\frac{2}{a}$	*Pmma*	*Pmmb*	*Pbmm*	*Pcmm*	*Pmcm*	*Pmam*
52	D_{2h}^{6}	$P\frac{2}{n}\frac{2_1}{n}\frac{2}{a}$	*Pnna*	*Pnnb*	*Pbnn*	*Pcnn*	*Pncn*	*Pnan*
53	D_{2h}^{7}	$P\frac{2}{m}\frac{2}{n}\frac{2_1}{a}$	*Pmna*	*Pnmb*	*Pbmn*	*Pcnm*	*Pncm*	*Pman*
54	D_{2h}^{8}	$P\frac{2_1}{c}\frac{2}{c}\frac{2}{a}$	*Pcca*	*Pccb*	*Pbaa*	*Pcaa*	*Pbcb*	*Pbab*
55	D_{2h}^{9}	$P\frac{2_1}{b}\frac{2_1}{a}\frac{2}{m}$	*Pbam*	*Pbam*	*Pmcb*	*Pmcb*	*Pcma*	*Pcma*
56	D_{2h}^{10}	$P\frac{2_1}{c}\frac{2_1}{c}\frac{2}{n}$	*Pccn*	*Pccn*	*Pnaa*	*Pnaa*	*Pbnb*	*Pbnb*
57	D_{2h}^{11}	$P\frac{2}{b}\frac{2_1}{c}\frac{2_1}{m}$	*Pbcm*	*Pcam*	*Pmca*	*Pmab*	*Pbma*	*Pcmb*
58	D_{2h}^{12}	$P\frac{2_1}{n}\frac{2_1}{n}\frac{2}{m}$	*Pnnm*	*Pnnm*	*Pmnn*	*Pmnn*	*Pnmn*	*Pnmn*
59	D_{2h}^{13}	$P\frac{2_1}{m}\frac{2_1}{m}\frac{2}{n}$	*Pmmn*	*Pmmn*	*Pnmm*	*Pnmm*	*Pmnm*	*Pmnm*
60	D_{2h}^{14}	$P\frac{2_1}{b}\frac{2}{c}\frac{2_1}{n}$	*Pbcn*	*Pcan*	*Pnca*	*Pnab*	*Pbna*	*Pcnb*
61	D_{2h}^{15}	$P\frac{2_1}{b}\frac{2_1}{c}\frac{2_1}{a}$	*Pbca*	*Pcab*	*Pbca*	*Pcab*	*Pbca*	*Pcab*
62	D_{2h}^{16}	$P\frac{2_1}{n}\frac{2_1}{m}\frac{2_1}{a}$	*Pnma*	*Pmnb*	*Pbnm*	*Pcmn*	*Pmcn*	*Pnam*
63	D_{2h}^{17}	$C\frac{2}{m}\frac{2}{c}\frac{2_1}{m}$	*Cmcm* *bnn*	*Ccmm* *nan*	*Amma* *ncn*	*Amam* *nnb*	*Bbmm* *nna*	*Bmmb* *cnn*
64†	D_{2h}^{18}	$C\frac{2}{m}\frac{2}{c}\frac{2_1}{e}$	*Cmca* (*Cmce*) *bnb*	*Ccmb* (*Ccme*) *naa*	*Abma* (*Aema*) *ccn*	*Acam* (*Aeam*) *bnb*	*Bbcm* (*Bbem*) *naa*	*Bmab* (*Bmeb*) *cnn*
65	D_{2h}^{19}	$C\frac{2}{m}\frac{2}{m}\frac{2}{m}$	*Cmmm* *ban*	*Cmmm* *ban*	*Ammm* *ncb*	*Ammm* *ncb*	*Bmmm* *cna*	*Bmmm* *cna*
66	D_{2h}^{20}	$C\frac{2}{c}\frac{2}{c}\frac{2}{m}$	*Cccm* *nnn*	*Cccm* *nnn*	*Amaa* *nnn*	*Amaa* *nnn*	*Bbmb* *nnn*	*Bbmb* *nnn*
67†	D_{2h}^{21}	$C\frac{2}{m}\frac{2}{m}\frac{2}{e}$	*Cmma* (*Cmme*) *bab*	*Cmmb* (*Cmme*) *baa*	*Abmm* (*Aemm*) *ccb*	*Acmm* (*Aemm*) *bcb*	*Bmcm* (*Bmem*) *caa*	*Bmam* (*Bmem*) *cca*

Table 1.5.4.1 (continued)

No. of space group	Schoen-flies symbol	Standard full Hermann–Mauguin symbol **abc**	Extended Hermann–Mauguin symbols for the six settings of the same unit cell					
			abc (standard)	**bac̄**	**cab**	**c̄ba**	**bca**	**ac̄b**
68†	D_{2h}^{22}	$C\frac{2}{c}\frac{2}{c}\frac{2}{e}$	*Ccca* (*Ccce*) *nnb*	*Cccb* (*Ccce*) *nna*	*Abaa* (*Aeaa*) *cnn*	*Acaa* (*Aeaa*) *bnn*	*Bbcb* (*Bbeb*) *nan*	*Bbab* (*Bbeb*) *ncn*
69	D_{2h}^{23}	$F\frac{2}{m}\frac{2}{m}\frac{2}{m}$	*Fmmm* *ban* *ncb* *cna*	*Fmmm* *ban* *cna* *ncb*	*Fmmm* *ncb* *cna* *ban*	*Fmmm* *ncb* *ban* *cna*	*Fmmm* *cna* *ban* *ncb*	*Fmmm* *cna* *ncb* *ban*
70	D_{2h}^{24}	$F\frac{2}{d}\frac{2}{d}\frac{2}{d}$	*Fddd*	*Fddd*	*Fddd*	*Fddd*	*Fddd*	*Fddd*
71	D_{2h}^{25}	$I\frac{2}{m}\frac{2}{m}\frac{2}{m}$	*I mmm* *nnn*	*I mmm* *nnn*	*I mmm* *nnn*	*I mmm* *nnn*	*I mmm* *nnn*	*I mmm* *nnn*
72	D_{2h}^{26}	$I\frac{2}{b}\frac{2}{a}\frac{2}{m}$	*I bam* *ccn*	*I bam* *ccn*	*I mcb* *naa*	*I mcb* *naa*	*I cma* *bnb*	*I cma* *bnb*
73	D_{2h}^{27}	$I\frac{2_1}{b}\frac{2_1}{c}\frac{2_1}{a}$	*I bca* *cab*	*I cab* *bca*	*I bca* *cab*	*I cab* *bca*	*I bca* *cab*	*I cab* *bca*
74	D_{2h}^{28}	$I\frac{2_1}{m}\frac{2_1}{m}\frac{2_1}{a}$	*I mma* *nnb*	*I mmb* *nna*	*I bmm* *cnn*	*I cmm* *bnn*	*I mcm* *nan*	*I mam* *ncn*

† For the five space groups *Aem*2 (39), *Aea*2 (41), *Cmce* (64), *Cmme* (67) and *Ccce* (68), the 'new' space-group symbols, containing the symbol '*e*' for the 'double' glide plane, are given for all settings. These symbols were first introduced in the fourth edition of Vol. A of *International Tables for Crystallography* (1995). For further explanations, see Sections 1.2.3 and 2.1.2, and de Wolff *et al.* (1992).

increase the number of possible symbols by a factor of three, thus yielding 18 different cases for each monoclinic space group, except for five cases, namely *P*2 (3), $P2_1$ (4), *Pm* (6), *P*2/*m* (10) and $P2_1/m$ (11) with only six variants.

In monoclinic *P* lattices, the symmetry operations along the symmetry direction are always unique. Here the cell centrings give rise to additional entries in the extended Hermann–Mauguin symbols. Consider, for example, the data for monoclinic *P*12/*m*1 (10), *C*12/*m*1 (12) and *C*12/*c*1 (15) in Table 1.5.4.1. For *P*12/*m*1 and its various settings there is only one line, which corresponds to the full Hermann–Mauguin symbols; these contain only rotations 2 and reflections *m*. The first line for *C*12/*m*1 is followed by a second line describing the composition of the generating symmetry operations from the full symbol with the centring translation. The first entry in the second line is the symbol $2_1/a$, since composing 2 with the *C*-centring translation results in a 2_1 screw rotation and composing *m* with the centring translation gives an *a*-glide reflection. Similarly, in *C*12/*c*1, composing rotations 2 and *c*-glide reflections with the *C*-centring translation gives screw rotations 2_1 and *n*-glide reflections, and thus one finds the entry $2_1/n$ under the full symbol *C*12/*c*1.

In Table 1.5.4.1 the Hermann–Mauguin symbols of the orthorhombic space groups are listed in six different settings: the *standard setting* **abc**, and the settings **bac̄**, **cab**, **c̄ba**, **bca** and **ac̄b**. These six settings result from the possible permutations of the three axes. Let us compare for a few space groups the standard setting **abc** with the **cab** setting. For *Pmm*2 (25) the permutation yields the new setting *P*2*mm*, reflecting the fact that the twofold axes parallel to the *c* direction change to the *a* direction. The mirrors normal to **a** and **b** become normal to **b** and **c**, respectively.

The case of *Cmm*2 (35) is slightly more complex due to the centring. As a result of the permutation the *C* centring becomes an *A* centring. The changes in the twofold axes and mirrors are similar to those of the previous example and result in the *A*2*mm* setting of *Cmm*2.

The extended Hermann–Mauguin symbol of the centred space group *Aem*2 (39) reveals the nature of the *e*-glide plane (also called the 'double' glide plane): among the set of glide reflections through the same (100) plane, there exist two glide reflections with glide components $\frac{1}{2}$**b** and $\frac{1}{2}$**c** (for details of the *e*-glide notation the reader is referred to Section 1.2.3, see also de Wolff *et al.*, 1992). In the **cab** setting, the *A* centring changes to a *B* centring and the double glide plane is now normal to **b** and the glide reflections have glide components $\frac{1}{2}$**a** and $\frac{1}{2}$**c**. The corresponding symbol is thus *B*2*em*. Note that in the cases of the five orthorhombic space groups whose Hermann–Mauguin symbols contain the *e*-glide symbol, namely *Aem*2 (39), *Aea*2 (41), *Cmce* (64), *Cmme* (67) and *Ccce* (68), the characters in the first lines of the extended symbols differ from the short symbols because the characters in the extended symbol represent symmetry operations, whereas those in the short and full symbol represent symmetry elements (*cf.* footnote 3). In all these cases, the extended symbols listed in Table 1.5.4.1 are complemented by the short symbols, given in brackets.

The discussion in Section 1.4.2.4 about the additional symmetry operations that occur as a result of combinations with lattice translations provides some rules for the construction of the extended Hermann–Mauguin symbols in the orthorhombic crystal system (*cf.* Section 1.5.4.1 of *IT* A for a detailed discussion of the additional symmetry operations). In orthorhombic space groups with primitive lattices, the symmetry operations of any symmetry direction are always unique: either 2 or 2_1, either *m* or *a* or *b* or *c* or *n*. In *C*-centred lattices, owing to the possible combination of the original symmetry operations with the centring translations, the axes 2 along [100] and [010] alternate with axes 2_1. However, parallel to **c** there are either 2 or 2_1 axes because the combination of a rotation or screw rotation with a centring translation results in another operation of the same kind. Similarly, m_{100} alternates with b_{100}, m_{010} with a_{010}, c_{100} with n_{100} *etc.* The m_{001} reflection plane is simultaneously an n_{001} glide plane and an a_{001} glide plane is simultaneously a b_{001} glide plane. This latter plane with its double role is the *e*-glide plane, as found for example in the full symbol of *C*2/*m* 2/*m* 2/*e* (67) and the corresponding short symbol *Cmme*. As another example, consider the space group *C*2/*m* 2/*c* 2_1/*m* (63). In Table 1.5.4.1, in the line of various settings for this space group the short Hermann–Mauguin symbols are listed, and the rotations or screw rotations do not appear. The m_{100}, c_{010} and m_{001} reflections and glide reflections occur alternating with b_{100}, n_{010} and n_{001} glide reflections, respectively. The entry under *Cmcm* is thus *bnn*.

F and *I* centrings cause alternating symmetry operations for all three coordinate axes *a*, *b* and *c*. For these centrings, the permutation of the axes does not affect the symbol *F* or *I* of the centring type. However, the number of symmetry operations increases by a factor of four for *F* centrings and by a factor of two for *I* centrings when compared to those of a space group with a primitive lattice. In *Fmm*2 (42) for example, three additional lines appear in the extended symbol, namely *ba*2, *nc*2_1 and *cn*2_1. These operations are obtained by combining successively the centring translations $t(\frac{1}{2},\frac{1}{2},0)$, $t(0,\frac{1}{2},\frac{1}{2})$ and $t(\frac{1}{2},0,\frac{1}{2})$ with the symmetry operations of *Pmm*2. However, in space groups *Fdd*2 (43) and *Fddd* (70) the nature of the *d* planes is not altered by the translations of the *F*-centred lattice; for this reason, in Table 1.5.4.1 a two-line symbol for *Fdd*2 and a one-line symbol for *Fddd* are sufficient.

References

Chattopadhyay, T. K., Boucherle, J. X. & von Schnering, H. G. (1987). *Neutron diffraction study on the structural phase transition in GeTe. J. Phys. C*, **20**, 1431–1440.

Inorganic Crystal Structure Database (2012). Release 2012/2. Fachinformationszentrum Karlsruhe and National Institute of Standards and Technology. http://www.fiz-karlsruhe.de/icsd.html. (Abbreviated as ICSD.)

International Tables for Crystallography (1983). Vol. A, *Space-Group Symmetry*, edited by Th. Hahn. Dordrecht: D. Reidel Publishing Company.

International Tables for Crystallography (1995). Vol. A, *Space-Group Symmetry*, 4th revised ed., edited by Th. Hahn. Dordrecht: Kluwer Academic Publishers.

International Tables for Crystallography (2002). Vol. A, *Space-Group Symmetry*, 5th ed., edited by Th. Hahn. Dordrecht: Kluwer Academic Publishers.

International Tables for Crystallography (2010). Vol. A1, *Symmetry Relations between Space Groups*, 2nd ed., edited by H. Wondratschek & U. Müller. Chichester: John Wiley & Sons.

International Tables for Crystallography (2016). Vol. A, *Space-Group Symmetry*, 6th ed., edited by M. I. Aroyo. Chichester: Wiley. (Abbreviated as *IT* A.)

International Tables for X-ray Crystallography (1952). Vol. I, *Symmetry Groups*, edited by N. F. M. Henry & K. Lonsdale. Birmingham: Kynoch Press.

Krstanovic, I. R. (1958). *Redetermination of the oxygen parameters in zircon* ($ZrSiO_4$). *Acta Cryst.* **11**, 896–897.

Wiedemeier, H. & Siemers, P. A. (1989). *The thermal expansion of GeS and GeTe. J. Less Common Met.* **146**, 279–298.

Wolff, P. M. de, Billiet, Y., Donnay, J. D. H., Fischer, W., Galiulin, R. B., Glazer, A. M., Hahn, Th., Senechal, M., Shoemaker, D. P., Wondratschek, H., Wilson, A. J. C. & Abrahams, S. C. (1992). *Symbols for symmetry elements and symmetry operations. Final Report of the International Union of Crystallography Ad-hoc Committee on the Nomenclature of Symmetry. Acta Cryst.* A**48**, 727–732.

Wyckoff, R. W. G. & Hendricks, S. B. (1927). *Die Kristallstruktur von Zirkon und die Kriterien fuer spezielle Lagen in tetragonalen Raumgruppen. Z. Kristallogr. Kristallgeom. Kristallphys. Kristallchem.* **66**, 73–102.

1.6. Introduction to the theory and practice of space-group determination

Uri Shmueli, Howard D. Flack and John C. H. Spence

1.6.1. Overview

This chapter describes and discusses several methods of symmetry determination of single-domain crystals. A detailed presentation of symmetry determination from diffraction data is followed by a brief discussion of intensity statistics, ideal as well as non-ideal, with an application of the latter to real intensity data from a $P\bar{1}$ crystal structure. Several methods of retrieving symmetry information from a solved crystal structure are then discussed. This is followed by a discussion of chemical and physical restrictions on space-group symmetry, including some aids in symmetry determination, and by a brief section on pitfalls in space-group determination.

The following two sections deal with reflection conditions. The first presents briefly the theoretical background of conditions for possible general reflections and their corresponding derivation. A relevant example follows. The second presents some examples of tabulations of general reflection conditions and possible space groups.

These are followed by a concise section on space-group determination in macromolecular crystallography.

Next, a description and illustration of symmetry determination based on electron-diffraction methods are presented. These are principally focused on convergent-beam electron diffraction.

The chapter concludes with four examples of space-group determination from real intensity data, employing methods described in the previous sections of this chapter.

This chapter deals only with single crystals. A supplement (Flack, 2015) deals with twinned crystals and those displaying a specialized metric.

1.6.2. Symmetry determination from single-crystal studies

By Uri Shmueli and Howard D. Flack

1.6.2.1. Symmetry information from the diffraction pattern

The extraction of symmetry information from the diffraction pattern takes place in three stages.

In the first stage, the unit-cell dimensions are determined and analysed in order to establish to which Bravais lattice the crystal belongs. The Bravais lattice symbol consists of two characters. The first is the first letter of the name of a crystal family and the second is the centring mode of a conventional unit cell. For example, the Bravais-lattice types *tP* and *tI* belong to the tetragonal family and have conventional unit-cell parameters $a = b, c$, $\alpha = \beta = \gamma = 90°$.

Knowledge of the Bravais lattice leads to a conventional choice of the unit cell, primitive or centred. The determination of the Bravais lattice of the crystal is achieved by the process of cell reduction, in which the lattice is first described by a primitive unit cell, and then linear combinations of the unit-cell vectors are taken to reduce the metric tensor (and the cell dimensions) to a standard form. From the relationships amongst the elements of the metric tensor, one obtains the Bravais lattice, together with a conventional choice of the unit cell, with the aid of standard tables [*e.g.* those in *International Tables for Crystallography* Volume A (2016), hereafter abbreviated as *IT* A]. A detailed description of cell reduction is given in Chapter 3.1 of *IT* A and in Part 9 of earlier editions (*e.g.* Burzlaff *et al.*, 2002). Formally, cell reduction leads to a conventional lattice basis for the crystal, primitive or centred, depending on its Bravais-lattice type. An alternative approach (Le Page, 1982) seeks the Bravais lattice directly from the cell dimensions by searching for all the twofold axes present. All these operations are automated in software. Regardless of the technique employed, at the end of the process one obtains an indication of the Bravais lattice and a unit cell in a conventional setting for the crystal system. These are usually good indications which, however, must be confirmed by an examination of the distribution of diffracted intensities as outlined below.

In the second stage, it is the point-group symmetry of the intensities of the Bragg reflections which is determined. We recall that the average reduced intensity of a pair of Friedel opposites (hkl and $\overline{hkl}$) is given by

$$\begin{aligned}|F_{\rm av}(\mathbf{h})|^2 &= \tfrac{1}{2}[|F(\mathbf{h})|^2 + |F(\bar{\mathbf{h}})|^2] \\ &= \sum_{i,j}[(f_i + f_i')(f_j + f_j') + f_i''f_j'']\cos[2\pi\mathbf{h}(\boldsymbol{r}_i - \boldsymbol{r}_j)] \equiv A(\mathbf{h}),\end{aligned} \quad (1.6.2.1)$$

where the atomic scattering factor of atom j, taking into account resonant scattering (sometimes called anomalous scattering or, in old literature, anomalous dispersion), is given by

$$\mathbf{f}_j = f_j + f_j' + if_j'',$$

the wavelength-dependent components f_j' and f_j'' being the real and imaginary parts, respectively, of the contribution of atom j to the resonant scattering, $\mathbf{h}$ contains in the (row) matrix (1×3) the diffraction orders (hkl) and $\boldsymbol{r}_j$ contains in the (column) matrix (3×1) the coordinates (x_j, y_j, z_j) of atom j. The components of the $\mathbf{f}_j$ are assumed to contain implicitly the displacement parameters. Equation (1.6.2.1) can be found *e.g.* in Okaya & Pepinsky (1955), Rossmann & Arnold (2001) and Flack & Shmueli (2007). It follows from (1.6.2.1) that

$$|F_{\rm av}(\mathbf{h})|^2 = |F_{\rm av}(\bar{\mathbf{h}})|^2 \text{ or } A(\mathbf{h}) = A(\bar{\mathbf{h}}),$$

regardless of the contribution of resonant scattering. Hence the averaging introduces a centre of symmetry in the (averaged) diffraction pattern. We must thus mention the well known Friedel's law, which states that $|F(\mathbf{h})|^2 = |F(\bar{\mathbf{h}})|^2$ and which is only a reasonable approximation for noncentrosymmetric crystals if resonant scattering is negligibly small. This law holds well for centrosymmetric crystals, independently of the resonant-scattering contribution. In fact, working with the average of Friedel opposites, one may determine the Laue group of the diffraction pattern by comparing the intensities of reflections which should be symmetry equivalent under each of the Laue groups. These are the 11 centrosymmetric point groups: $\bar{1}$, $2/m$, mmm, $4/m$, $4/mmm$, $\bar{3}$, $\bar{3}m$, $6/m$, $6/mmm$, $m\bar{3}$ and $m\bar{3}m$. For example, the reflections of which the intensities are to be compared for the Laue group $\bar{3}$ are: *hkl*, *kil*, *ihl*, $\overline{hkl}$, $\overline{kil}$ and $\overline{ihl}$, where $i = -h - k$. An extensive listing of the indices of symmetry-related reflections in all the point groups, including of course the Laue groups, is given in Appendix 1.4.4 of *International Tables for Crystallography* Volume B (Shmueli, 2008). (The tables in Appendix 1.4.4 actually deal with space groups in reciprocal space; however, the left part of any entry is just the indices of a reflection generated by the point-group operation corresponding to this entry.) In the past, one used to inspect the diffraction images to see which classes of reflections are symmetry equivalent within experimental and other uncertainty. Nowadays, the whole intensity data set is analysed by software. The intensities are merged and averaged under each of the 11 Laue groups in various settings (*e.g.* 2/*m* unique axis *b* and unique axis *c*) and orientations (*e.g.* $\bar{3}m1$ and $\bar{3}1m$). For each choice of Laue group and its variant, an $R_{\rm merge}$ factor is calculated as follows:

$$R_{{\rm merge},i} = \frac{\sum_{\bf h}\sum_{s=1}^{|G|_i} |\langle|F_{\rm av}({\bf h})|^2\rangle_i - |F_{\rm av}({\bf h}{\bf W}_{si})|^2|}{|G|_i\sum_{\bf h}\langle|F_{\rm av}({\bf h})|^2\rangle_i}, \qquad (1.6.2.2)$$

where $\mathbf{W}_{si}$ is the *s*th symmetry operation of the *i*th Laue group, $|G|_i$ is the order of that group, the average in the first term in the numerator and in the denominator ranges over the intensities of the trial Laue group and the outer summations $\sum_{\bf h}$ range over the *hkl* reflections. Choices with low $R_{{\rm merge},i}$ display the chosen symmetry, whereas for those with high $R_{{\rm merge},i}$ the symmetry is inappropriate. The Laue group of highest symmetry with a low $R_{{\rm merge},i}$ is considered the best indication of the Laue group. Several variants of the above procedure exist in the available software. Whichever of them is used, it is important for the discrimination of the averaging process to choose a strategy of data collection such that the intensities of the greatest possible number of Bragg reflections are measured. In practice, validation of symmetry can often be carried out with a few initial images and the data-collection strategy may be based on this assignment.

In the third stage, the intensities of the Bragg reflections are studied to identify the conditions for systematic absences. Some space groups give rise to zero intensity for certain classes of reflections. These 'zeros' occur in a systematic manner and are commonly called systematic absences (*e.g.* in the *h*0*l* class of reflections, if all rows with *l* odd are absent, then the corresponding reflection condition is *h*0*l*: $l = 2n$). In practice, as implemented in software, statistics are produced on the intensity observations of all possible sets of 'reflection conditions' as given in *IT* A, Chapter 2.3 (*e.g.* in the example above, *h*0*l* reflections are separated into sets with $l = 2n$ and those with $l = 2n + 1$). In one approach, the number of observations in each set having an intensity (I) greater than n standard uncertainties [$u(I)$] [*i.e.* $I/u(I) > n$] is displayed for various values of n. Clearly, if a trial condition for systematic absence has observations with strong or medium intensity [*i.e.* $I/u(I) > 3$], the systematic-absence condition is not fulfilled (*i.e.* the reflections are not systematically absent). If there are no such observations, the condition for systematic absence may be valid and the statistics for smaller values of n need then to be examined. These are more problematic to evaluate, as the set of reflections under examination may have many weak reflections due to structural effects of the crystal or to perturbations of the measurements by other systematic effects. An alternative approach to examining numbers of observations is to compare the mean value, $\langle I/u(I)\rangle$, taken over reflections obeying or not a trial reflection condition. For a valid reflection condition, one expects the former value to be considerably larger than the latter. In Section 3.1 of Palatinus & van der Lee (2008), real examples of marginal cases are described.

The third stage continues by noting that the systematic absences are characteristic of the space group of the crystal, although some sets of space groups have identical reflection conditions. In Chapter 2.3 of *IT* A, one finds all the reflection conditions listed individually for all the settings of the 230 space groups that are tabulated there. It is recalled that a reflection condition is a condition limiting possible reflections. That is, if a reflection condition is fulfilled the reflection *may* appear, and if it is not fulfilled the reflection *must* be absent (*cf.* the example in Section 1.6.3.1). A classification of reflection conditions into general and special is introduced in Section 2.1.3.13. Only general reflection conditions, *i.e.* those related to the presence of lattice centring, glide planes and/or screw axes, are dealt with in the present chapter. For practical use in space-group determination, tables have been set up that present a list of all those space groups that are characterized by a given set of reflection conditions. The tables for all the Bravais lattices and Laue groups are given in Section 1.6.4 of *IT* A. So, once the reflection conditions have been determined, all compatible space groups can be identified from these tables. For example, 85 space groups may be unequivocally determined by the procedures defined in this section. For other sets of reflection conditions, there are a larger number of compatible space groups, attaining the value of 6 in one case.

It should be pointed out in connection with this third stage that a possible weakness of the analysis of systematic absences for crystals with small unit-cell dimensions is that there may be a small number of axial reflections capable of being systematically absent.

It goes without saying that the selected space groups must be compatible with the Bravais lattice determined in stage 1, with the Laue class determined in stage 2 and with the set of space-group absences determined in stage 3.

1.6.2.2. Structure-factor statistics and crystal symmetry

Most structure-solving software packages contain a section dedicated to several probabilistic methods based on the Wilson (1949) paper on the probability distribution of structure-factor magnitudes. These statistics sometimes correctly indicate whether the intensity data set was collected from a centrosymmetric or noncentrosymmetric crystal. However, not infrequently these indications are erroneous. The reasons for this may be many, but outstandingly important are (i) the presence of a few very heavy atoms amongst a host of lighter ones, and (ii) a very small number of nearly equal atoms. Omission of weak reflections from the data set also contributes to failures of Wilson (1949) statistics. These erroneous indications are also rather strongly space-group dependent.

The well known probability density functions (hereafter p.d.f.'s) of the magnitude of the normalized structure factor E, also known as ideal p.d.f.'s, are

$$p(|E|) = \begin{cases} \sqrt{2/\pi}\exp(-|E|^2/2) & \text{for } P\bar{1} \\ 2|E|\exp(-|E|^2) & \text{for } P1 \end{cases}, \qquad (1.6.2.3)$$

where it is assumed that all the atoms are of the same chemical element. Let us see their graphical representations.

It is seen from Fig. 1.6.2.1 that the two p.d.f.'s are significantly different, but usually they are not presented as such by the software. What is usually shown are the cumulative distributions of $|E|^2$, the moments: $\langle |E|^n \rangle$ for $n = 1, 2, 3, 4, 5, 6$, and the averages of low powers of $|E^2 - 1|$ for ideal centric and acentric distributions, based on equation (1.6.2.3). Table 1.6.2.1 shows the numerical values of several low-order moments of $|E|$ and that of the lowest power of $|E^2 - 1|$. The higher the value of n the greater is the difference between their values for centric and acentric cases. However, it is most important to remember that the influence of measurement uncertainties also increases with n and therefore the higher the moment the less reliable it tends to be.

There are several ideal indicators of the status of centrosymmetry of a crystal structure. The most frequently used are: (i) the $N(z)$ test (Howells *et al.*, 1950), a cumulative distribution of $z = |E|^2$, based on equation (1.6.2.3), and (ii) the low-order moments of $|E|$, also based on equation (1.6.2.3). Equation (1.6.2.3), however, is very seldom used as an indicator of the status of centrosymmetry of a crystal stucture.

Let us now briefly consider p.d.f.'s that are valid for any atomic composition as well as any space-group symmetry, and exemplify their performance by comparing a histogram derived from observed intensities from a $P\bar{1}$ structure with theoretical p.d.f.'s for the space groups $P1$ and $P\bar{1}$. The p.d.f.'s considered presume that all the atoms are in general positions and that the reflections considered are general (see, *e.g.*, Section 1.6.3). A general treatment of the problem is given in the literature and summarized in the book *Introduction to Crystallographic Statistics* (Shmueli & Weiss, 1995).

The basics of the exact p.d.f.'s are conveniently illustrated in the following. The normalized structure factor for the space group $P\bar{1}$, assuming that all the atoms occupy general positions and resonant scattering is neglected, is given by

$$E(\mathbf{h}) = 2 \sum_{j=1}^{N/2} n_j \cos(2\pi \mathbf{h} \mathbf{r}_j),$$

where n_j is the normalized scattering factor. The maximum possible value of E is $E_{\max} = \sum_{j=1}^{N} n_j$ and the minimum possible value of E is $-E_{\max}$. Therefore, $E(\mathbf{h})$ must be confined to the $(-E_{\max}, E_{\max})$ range. The probability of finding E outside this range is of course zero. Such a probability density function can be expanded in a Fourier series within this range (*cf.* Shmueli *et al.*, 1984). This is the basis of the derivation, the details of which are well documented (*e.g.* Shmueli *et al.*, 1984; Shmueli & Weiss, 1995; Shmueli, 2007). Exact p.d.f.'s for any centrosymmetric space group have the form

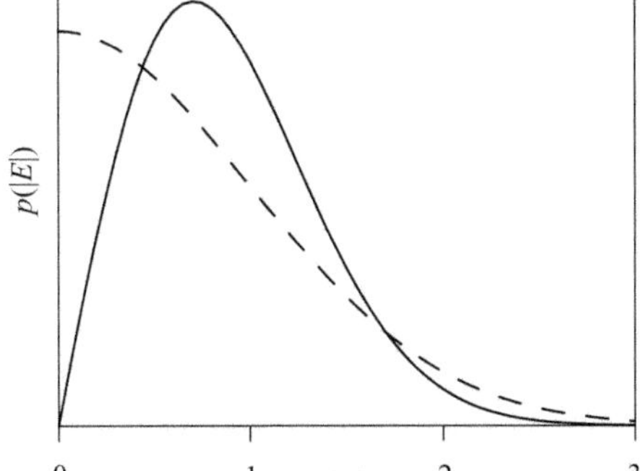

Figure 1.6.2.1
Ideal p.d.f.'s for the equal-atom case. The dashed line is the centric, and the solid line the acentric ideal p.d.f.

Table 1.6.2.1
The numerical values of several low-order moments of $|E|$ based on equation (1.6.2.3), and the first six moments of the ideal bicentric p.d.f. derived by Rogers & Wilson (1953)

The values observed from the intensity data of Examples (1) and (2) of Section 1.6.7 are also shown.

	Theoretical values			Observed values	
Moment	$P\bar{1}$	$P1$	Bicentric	Example (1)	Example (2)
$\langle \lvert E\rvert \rangle$	0.798	0.886	0.718	0.805	0.791
$\langle \lvert E\rvert^2 \rangle$	1.000	1.000	1.000	1.000	1.000
$\langle \lvert E\rvert^3 \rangle$	1.596	1.329	1.910	1.601	1.778
$\langle \lvert E\rvert^4 \rangle$	3.000	2.000	4.500	3.075	4.037
$\langle \lvert E\rvert^5 \rangle$	6.383	3.323	12.260	6.844	10.917
$\langle \lvert E\rvert^6 \rangle$	15.000	6.000	37.500	17.313	33.425
$\langle \lvert E^2 - 1\rvert \rangle$	0.968	0.736		0.945	1.026

$$p(|E|) = \alpha \left\{ 1 + 2 \sum_{m=1}^{\infty} C_m \cos(\pi m |E| \alpha) \right\}, \quad (1.6.2.4)$$

where $\alpha = 1/E_{\max}$, and exact p.d.f.'s for any noncentrosymmetric space group can be computed as the double Fourier series

$$p(|E|) = \tfrac{1}{2} \pi \alpha^2 |E| \sum_{m=1}^{\infty} \sum_{n=1}^{\infty} C_{mn} J_0[\pi \alpha |E| (m^2 + n^2)^{1/2}], \quad (1.6.2.5)$$

where $J_0(X)$ is a Bessel function of the first kind and of order zero. Expressions for the coefficients C_m and C_{mn} are given by Rabinovich *et al.* (1991) and by Shmueli & Wilson (2008) for all the space groups up to and including $Fd\bar{3}$.

The following example deals with a very high sensitivity to atomic heterogeneity. Consider the crystal structure of [(*Z*)-ethyl *N*-isopropylthiocarbamato-κS](tricyclohexylphosphine-κP)gold(I), published as $P\bar{1}$ with $Z = 2$, the content of its asymmetric unit being $AuSPONC_{24}H_{45}$ (Tadbuppa & Tiekink, 2010). Let us construct a histogram from the $|E|$ data computed from all the observed reflections with non-negative reduced intensities and compare the histogram with the p.d.f.'s for the space groups $P1$ and $P\bar{1}$, computed from equations (1.6.2.5) and (1.6.2.4), respectively. The histogram and the p.d.f.'s were put on the same scale. The result is shown in Fig. 1.6.2.2.

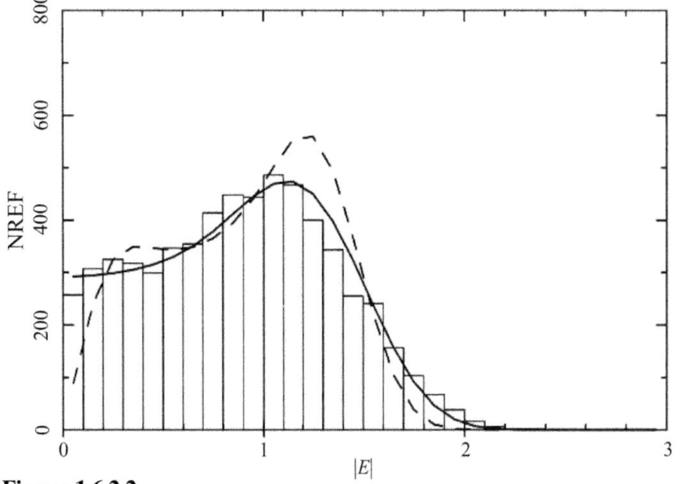

Figure 1.6.2.2
Exact p.d.f.'s. for a crystal in the triclinic system (see Section 1.6.2.2). Solid curve: $P\bar{1}$, computed from (1.6.2.4); dashed curve: $P1$, computed from (1.6.2.5); histogram based on the $|E|$ data computed from all the reflections with non-negative reduced intensities. The height of each bin corresponds to the number of reflections (NREF) in its range of $|E|$ values.

A visual comparison strongly indicates that the space-group assignment as $P\bar{1}$ was correct, since the recalculated histogram agrees rather well with the p.d.f. (1.6.2.4) and much less with (1.6.2.5). The ideal Wilson-type statistics incorrectly indicated that this crystal is noncentrosymmetric. It is seen that the ideal p.d.f. breaks down in the presence of strong atomic heterogeneity (gold among many lighter atoms) in the space group $P\bar{1}$. Other space groups behave differently, as shown in the literature (*e.g.* Rabinovich *et al.*, 1991; Shmueli & Weiss, 1995).

Additional examples of applications of structure-factor statistics and some relevant computing considerations can be found in Shmueli (2012) and a corresponding computer program is described in Shmueli (2013).

1.6.2.3. Symmetry information from the structure solution

It is also possible to obtain information on the symmetry of the crystal after structure solution. The latter is obtained either in space group $P1$ (*i.e.* no symmetry assumed) or in some other candidate space group. The analysis may take place either on the electron-density map, or on its interpretation in terms of atomic coordinates and atomic types (*i.e.* chemical elements). The analysis of the electron-density map has become increasingly popular with the advent of dual-space methods, as introduced in the charge-flipping method of Oszlányi & Sütő (2004), which solve structures in $P1$ by default. The analysis of the atomic coordinates and atomic types obtained from least-squares refinement in a candidate space group is used extensively in structure validation. Symmetry operations present in the structure solution but not in the candidate space group are sought.

An exhaustive search for symmetry operations is undertaken. However, those to be investigated may be very efficently limited by making use of knowledge of the highest point-group symmetry of the lattice compatible with the known cell dimensions of the crystal. It is well established that the point-group symmetry of any lattice is one of the following seven centrosymmetric point groups: $\bar{1}$, $2/m$, mmm, $4/mmm$, $\bar{3}m$, $6/mmm$, $m\bar{3}m$. This point group is known as the holohedry of the lattice. The relationship between the symmetry operations of the space group and its holohedry is rather simple. A rotation or screw axis of symmetry in the crystal has as its counterpart a corresponding rotation axis of symmetry of the lattice and a mirror or glide plane in the crystal has as its counterpart a corresponding mirror plane in the lattice. The holohedry may be equal to or higher than the point group of the crystal. Hence, at least the rotational part of any space-group operation should have its counterpart in the symmetry of the lattice. If and when this rotational part is found by a systematic comparison either of the electron density or of the positions of the independent atoms of the solved structure, the location and intrinsic parts of the translation parts of the space-group operation can be easily completed.

Palatinus and van der Lee (2008) describe their procedure in detail with useful examples. It uses the structure solution both in the form of an electron-density map and a set of phased structure factors obtained by Fourier transformation. No interpretation of the electron-density map in the form of atomic coordinates and chemical-element type is required. The algorithm of the procedure proceeds in the following steps:

(1) The lattice centring is determined by a search for strong peaks in the autocorrelation (self-convolution, Patterson) function of the electron density and the potential centring vectors are evaluated through a reciprocal-space R value.

(2) A complete list of possible symmetry operations compatible with the lattice is generated by searching for the invariance of the direct-space metric under potential symmetry operations.

(3) A figure of merit is then assigned to each symmetry operation evaluated from the convolution of the symmetry-transformed electron density with that of the structure solution. Those symmetry operations that have a good figure of merit are selected as belonging to the space group of the crystal structure.

(4) The space group is completed by group multiplication of the selected operations and then validated.

(5) The positions of the symmetry elements are shifted to those of a conventional setting for the space group.

Palatinus & van der Lee (2008) report a very high success rate in the use of this algorithm. It is also a powerful technique to apply in structure validation.

Le Page's (1987) pioneering software *MISSYM* for the detection of 'missed' symmetry operations uses refined atomic coordinates, unit-cell dimensions and space group assigned from the crystal-structure solution. The algorithm follows all the principles described above in this section. In *MISSYM*, the metric symmetry is established as described in the first stage of Section 1.6.2.1. The 'missed' symmetry operations are those that are present in the arrangement of the atoms but are not part of the space group used for the structure refinement. Indeed, this procedure has its main applications in structure validation. The algorithm used in Le Page's software is also implemented in *ADDSYM* (Spek, 2003). There are numerous reports of successful applications of this software in the literature.

1.6.2.4. Restrictions on space groups

The values of certain chemical and physical properties of a bulk compound, or its crystals, have implications for the assignment of the space group of a crystal structure. In the chemical domain, notably in proteins and small-molecule natural products, information concerning the enantiomeric purity of the bulk compound or of its individual crystals is most useful. Further, all physical properties of a crystal are limited by the point group of the crystal structure in ways that depend on the individual nature of the physical property.

It is very well established that the crystal structure of an enantiomerically pure compound will be chiral (see Flack, 2003). By an enantiomerically pure compound one means a compound whose molecules are all chiral and all these molecules possess the same chirality. The space group of a chiral crystal structure will only contain the following types of symmetry operation: translations, pure rotations and screw rotations. Inversion in a point, mirror reflection or rotoinversion do not occur in the space group of a chiral crystal structure. Taking all this together means that the crystal structure of an enantiomerically pure compound will show one of 65 space groups (known as the Sohncke space groups), all noncentrosymmetric, containing only translations, rotations and screw rotations. As a consequence, the point group of a chiral crystal structure is limited to the 11 point groups containing only pure rotations (*i.e.* 1, 2, 222, 4, 422, 3, 32, 6, 622, 23 and 432). Particular attention must be paid as to whether a measurement of enantiomeric purity of a compound applies to the bulk material or to the single crystal used for the diffraction experiment. Clearly, a compound whose bulk is enantiomerically pure will produce crystals which are enantiomerically pure. The converse is not necessarily true (*i.e.* enantiomerically pure crystals do not necessarily come from an enantiomerically pure bulk).

For example, a bulk compound which is a racemate (*i.e.* an enantiomeric mixture containing 50% each of the opposite enantiomers) may produce either (*a*) crystals of the racemic compound (*i.e.* crystals containing 50% each of the opposite enantiomers) or (*b*) a racemic conglomerate (*i.e.* a mixture of enantiomerically pure crystals in a proportion of 50% of each pure enantiomer) or (*c*) some other rarer crystallization modes. Consequently, as part of a single-crystal structure analysis, it is highly recommended to make a measurement of the enantiomeric purity of the single crystal used for the diffraction experiment.

Much information on methods of establishing the enantiomeric purity of a compound can be found in a special issue of *Chirality* devoted to the determination of absolute configuration (Allenmark *et al.*, 2007). Measurements in the fluid state of optical activity, optical rotatory dispersion, circular dichroism (CD) and enantioselective chromatography are of prime importance. Many of these are sufficiently sensitive to be applicable not only to the bulk compound but also to the single crystal used for the diffraction experiment taken into solution. CD may also be applied in the solid state.

Many physical properties of a crystalline solid are anisotropic and the symmetry of a physical property of a crystal is limited both by the point-group symmetry of the crystal and by symmetries inherent to the physical property under study. For further information on this topic see Part 1 of Volume D (Authier *et al.*, 2014). Unfortunately, many of these physical properties are intrinsically centrosymmetric, so few of them are of use in distinguishing between the subgroups of a Laue group, a common problem in space-group determination. In Chapter 3.2 of *IT* A, Hahn & Klapper (2016) show to which point groups a crystal must belong to be capable of displaying some of the principal physical properties of crystals (see Table 3.2.2.1 in that chapter). Measurement of morphology, pyroelectricity, piezoelectricity, second harmonic generation and optical activity of a crystalline sample can be of use.

1.6.2.5. Pitfalls in space-group determination

The methods described in Section 1.6.2 rely on the crystal measured being a single-domain crystal, *i.e.* it should not be twinned. Nevertheless, some types of twin are easily identified at the measurement stage as they give rise to split reflections. Powerful data-reduction techniques may be applied to data from such crystals to produce a reasonably complete single-domain intensity data set. Consequently, the multi-domain twinned crystals that give rise to difficulties in space-group determination are those for which the reciprocal lattices of the individual domains overlap exactly without generating any splitting of the Bragg reflections. A study of the intensity data from such a crystal may display two anomalies. Firstly, the intensity distribution, as described and analysed in Section 1.6.2.2, will be broader than that of the monodomain crystal. Secondly, one may obtain a set of conditions for reflections that does not correspond to any entry in Section 1.6.4 of *IT* A. In this chapter we give no further information on the determination of the space group for such twinned crystals. For further information on this topic see Part 3 of *International Tables for Crystallography* Volume D (Boček *et al.*, 2014) and Chapter 1.3 on twinning in *International Tables for Crystallography* Volume C (Koch, 2006). A supplement (Flack, 2015) to the current section deals with the determination of the space group from twinned crystals and those displaying a specialized metric. However, it is apposite to note that the existence of twins with overlapping reciprocal lattices can be identified by recording atomic resolution transmission electron-microscope images.

In order to obtain reliable results from space-group determination, the coverage of the reciprocal space by the intensity measurements should be as complete as possible. One should attempt to attain full-sphere data coverage, *i.e.* a complete set of intensity measurements in the point group 1. All Friedel opposites should be measured. The validity and reliability of the intensity statistics described in Section 1.6.2.2 rest on a full coverage of reciprocal lattice. Any systematic omission by resolution, azimuth and declination, intensity *etc.* of part of the asymmetric region of the reciprocal lattice has an adverse effect. In particular, reflections of weak intensity should not be omitted or deleted.

There are a few other common difficulties in space-group determination due either to the nature of the crystal or the experimental setup:

(*a*) The crystal may display a pseudo-periodicity leading to systematic series of weak or very weak reflections that can be mistaken for systematic absences.

(*b*) The physical effect of multiple reflections can lead to diffraction intensity appearing at the place of systematic absences. However, the shape of these multiple-reflection intensities is usually much sharper than a normal Bragg reflection.

(*c*) Contamination of the incident radiation by a $\lambda/2$ component may also cause intensity due to the $2h\,2k\,2l$ reflection to appear at the place of the hkl one. Kirschbaum *et al.* (1997) and Macchi *et al.* (1998) have studied this probem and describe ways of circumventing it.

1.6.3. Theoretical background of reflection conditions

By Uri Shmueli

Following the notation of the previous chapters, where

$$\mathbf{W} = \begin{pmatrix} W_{11} & W_{12} & W_{13} \\ W_{21} & W_{22} & W_{23} \\ W_{31} & W_{32} & W_{33} \end{pmatrix}$$

is the rotation or rotoinversion part of a space-group operation, $\mathbf{w} = \begin{pmatrix} w_1 \\ w_2 \\ w_3 \end{pmatrix}$ is the intrinsic translation part and/or location part of a space-group operation, $\mathbf{h} = (h\ k\ l)$ are the components of the diffraction vector $\mathbf{h}$ and $\mathbf{x} = \begin{pmatrix} x \\ y \\ z \end{pmatrix}$ are the fractional coordinates of a point in the crystal, we shall now examine the effect of the space-group symmetry on the structure-factor function, $F(\mathbf{h})$. These effects are of importance in the determination of crystal symmetry. If $(\mathbf{W}, \mathbf{w})$ is a representative symmetry operation of the space group of the crystal, then, by definition

$$\rho(\mathbf{x}) = \rho(\mathbf{W}\mathbf{x} + \mathbf{w}), \tag{1.6.3.1}$$

where $\rho(\mathbf{x})$ is the value of the electron-density function at the point $\mathbf{x}$. It is known that the electron-density function at the point $\mathbf{x}$ is given by

$$\rho(\mathbf{x}) = \frac{1}{V}\sum_{\mathbf{h}} F(\mathbf{h}) \exp(-2\pi i \mathbf{h}\mathbf{x}), \tag{1.6.3.2}$$

where, as indicated above, **h** is the row matrix $(h\,k\,l)$ and $\boldsymbol{x}$ is a column matrix containing x, y and z in the first, second and third rows, respectively. Of course, $\mathbf{h}\boldsymbol{x}$ is simply equivalent to $hx + ky + lz$. If we substitute (1.6.3.2) with $\boldsymbol{x}$ replaced by $(\boldsymbol{W}\boldsymbol{x} + \boldsymbol{w})$ in (1.6.3.1) we obtain

$$\rho(\mathbf{x}) = \frac{1}{V}\sum_{\mathbf{h}} F(\mathbf{h})\exp\{-2\pi i[\mathbf{h}(\boldsymbol{W}\boldsymbol{x} + \boldsymbol{w})]\}$$
$$= \frac{1}{V}\sum_{\mathbf{h}}[F(\mathbf{h})\exp(-2\pi i\mathbf{h}\boldsymbol{w})]\exp(-2\pi i\mathbf{h}\boldsymbol{W}\boldsymbol{x}). \quad (1.6.3.3)$$

Since $\boldsymbol{W}$ is a point-group operation, the vectors $\mathbf{h}\boldsymbol{W}$ must range over all the reciprocal lattice. A comparison of the equations (1.6.3.2) and (1.6.3.3) shows that the coefficients of the exponential $\exp(-2\pi i\mathbf{h}\boldsymbol{W}\boldsymbol{x})$ in (1.6.3.3) must be the structure factors at the points $\mathbf{h}\boldsymbol{W}$ of the reciprocal lattice. Therefore

$$F(\mathbf{h}\boldsymbol{W}) = F(\mathbf{h})\exp(-2\pi i\mathbf{h}\boldsymbol{w}). \quad (1.6.3.4)$$

Equation (1.6.3.4) is the fundamental relation between symmetry-related reflections [*e.g.* Waser, 1955; Wells, 1965; and Section 1.4.2 in *International Tables for Crystallography* Volume B (Shmueli, 2008)] and is represented in component form as follows:

$$F\left[(h\ k\ l)\begin{pmatrix} W_{11} & W_{12} & W_{13} \\ W_{21} & W_{22} & W_{23} \\ W_{31} & W_{32} & W_{33} \end{pmatrix}\right] = F(h\ k\ l)\exp\left[-2\pi i(h\ k\ l)\begin{pmatrix} w_1 \\ w_2 \\ w_3 \end{pmatrix}\right].$$

We can now approach the problem of systematically absent reflections, which are alternatively called the conditions for possible reflections.

The reflection **h** is *general* if its indices remain unchanged *only* under the identity operation of the point group of the diffraction pattern. *I.e.*, if $\boldsymbol{W}$ is the identity operation of the point group, the relation $\mathbf{h}\boldsymbol{W} = \mathbf{h}$ holds true. So, if the reflection **h** is general, we must have $\boldsymbol{W} \equiv \boldsymbol{I}$, where $\boldsymbol{I}$ is the identity operation and, obviously, $\mathbf{h}\boldsymbol{I} = \mathbf{h}$. The operation $(\boldsymbol{I}, \boldsymbol{w})$ can be a space-group symmetry operation only if $\boldsymbol{w}$ is a lattice vector. Let us denote it by $\boldsymbol{w}_L$. Equation (1.6.3.4) then reduces to

$$F(\mathbf{h}) = F(\mathbf{h})\exp(-2\pi i\mathbf{h}\boldsymbol{w}_L) \quad (1.6.3.5)$$

and $F(\mathbf{h})$ can be nonzero only if $\exp(-2\pi i\mathbf{h}\boldsymbol{w}_L) = 1$. This, in turn, is possible only if $\mathbf{h}\boldsymbol{w}_L$ is an integer and leads to conditions depending on the lattice type. For example, if the components of $\boldsymbol{w}_L$ are all integers, which is the case for a *P*-type lattice, the above condition is fulfilled for all **h** – the lattice type does not impose any restrictions. If the lattice is of type *I*, there are two lattice points in the unit cell, at say 0, 0, 0 and 1/2, 1/2, 1/2. The first of these does not lead to any restrictions on possible reflections. The second, however, requires that $\exp[-\pi i(h + k + l)]$ be equal to unity. Since $\exp(\pi in) = (-1)^n$, where n is an integer, the possible reflections from a crystal with an *I*-type lattice must have indices such that their sum is an even integer; if the sum of the indices is an odd integer, the reflection is *systematically absent*. Effects of all the lattice types on conditions for possible reflections (or systematic absences) are shown in Table 2.1.3.3, see also Table 1.6.3.1 of *IT* A.

The reflection **h** is *special* if its indices remain unchanged under at least one operation (in addition to the identity operation) of the point group of the diffraction pattern. *I.e.*, the relation $\mathbf{h}\boldsymbol{W} = \mathbf{h}$ holds true for at least one operation in addition to the identity operation of the point group. We shall now assume that the reflection **h** is special. By definition, this reflection remains invariant under more than one operation $\boldsymbol{W}$ of the point group of the diffraction pattern. These operations form a subgroup of the point group of the diffraction pattern, known as the stabilizer of the reflection **h**, and we denote it by the symbol $\mathcal{S}_{\mathbf{h}}$. For each space-group symmetry operation $(\boldsymbol{W}, \boldsymbol{w})$ where $\boldsymbol{W}$ belongs to $\mathcal{S}_{\mathbf{h}}$, we must therefore have $\mathbf{h}\boldsymbol{W} = \mathbf{h}$. Equation (1.6.3.4) now reduces to

$$F(\mathbf{h}) = F(\mathbf{h})\exp(-2\pi i\mathbf{h}\boldsymbol{w}). \quad (1.6.3.6)$$

Of course, if $\boldsymbol{W}$ is the identity operation, $\boldsymbol{w}$ must be a lattice vector and the discussion of lattice absences above applies. We therefore require that $\boldsymbol{W}$ be an element of $\mathcal{S}_{\mathbf{h}}$ other than the identity. $F(\mathbf{h})$ can be nonzero only if the exponential factor in (1.6.3.6) equals unity. This, in turn, is possible only if $\mathbf{h}\boldsymbol{w}$ is an integer.

Equation (1.6.3.6) is applicable to glide reflections and to screw rotations. The effects of some glide reflections and all screw rotations are listed in Table 2.1.3.4, see also *IT* A, Section 1.6.3, Tables 1.6.3.2 and 1.6.3.3, respectively.

1.6.3.1. Example: a determination of reflection conditions

Let us apply the above results to the determination of the general reflection conditions for the space group $Pna2_1$ (No. 33). According to the conventions for symbols of orthorhombic space groups, the *n*-glide plane is normal to **a**, the *a*-glide plane is normal to **b** and the 2_1 screw axis is parallel to **c**. The general equivalent positions for this space group are listed in *IT* A as (1) x, y, z; (2) $\bar{x}, \bar{y}, z + 1/2$; (3) $x + 1/2, \bar{y} + 1/2, z$; (4) $\bar{x} + 1/2, y + 1/2, z + 1/2$ and their corresponding $(\boldsymbol{W}_i, \boldsymbol{w}_i)$ (matrix, column) representations are

$$(\boldsymbol{W}_1, \boldsymbol{w}_1)\colon \left[\begin{pmatrix} 1 & 0 & 0 \\ 0 & 1 & 0 \\ 0 & 0 & 1 \end{pmatrix}, \begin{pmatrix} 0 \\ 0 \\ 0 \end{pmatrix}\right],$$
$$(\boldsymbol{W}_2, \boldsymbol{w}_2)\colon \left[\begin{pmatrix} \bar{1} & 0 & 0 \\ 0 & \bar{1} & 0 \\ 0 & 0 & 1 \end{pmatrix}, \begin{pmatrix} 0 \\ 0 \\ 1/2 \end{pmatrix}\right],$$
$$(\boldsymbol{W}_3, \boldsymbol{w}_3)\colon \left[\begin{pmatrix} 1 & 0 & 0 \\ 0 & \bar{1} & 0 \\ 0 & 0 & 1 \end{pmatrix}, \begin{pmatrix} 1/2 \\ 1/2 \\ 0 \end{pmatrix}\right] \text{ and}$$
$$(\boldsymbol{W}_4, \boldsymbol{w}_4)\colon \left[\begin{pmatrix} \bar{1} & 0 & 0 \\ 0 & 1 & 0 \\ 0 & 0 & 1 \end{pmatrix}, \begin{pmatrix} 1/2 \\ 1/2 \\ 1/2 \end{pmatrix}\right].$$

It follows from the orientation of the geometric elements specified above that the *n* glide corresponds to $(\boldsymbol{W}_4, \boldsymbol{w}_4)$. The solution of $\mathbf{h}\boldsymbol{W}_4 = \mathbf{h}$ is $\mathbf{h} = 0kl$ and then $\mathbf{h}\boldsymbol{w}_4 = (k + l)/2$. It follows that a condition that $0kl$ be a possible reflection is $0kl$: $k + l = 2n$, where n is an integer. If the condition is not fulfilled, that is if $k + l = 2n + 1$, the reflection $0kl$ is *systematically absent*.

It also follows from the above that the *a* glide corresponds to $(\boldsymbol{W}_3, \boldsymbol{w}_3)$. The solution of $\mathbf{h}\boldsymbol{W}_3 = \mathbf{h}$ is $\mathbf{h} = h0l$ and then $\mathbf{h}\boldsymbol{w}_3 = h/2$. Therefore the condition that $h0l$ be a possible reflection is

$h0l$: $h = 2n$. If for an $h0l$ reflection $h = 2n + 1$, this $h0l$ reflection is *systematically absent*.

Finally, the 2_1 screw rotation corresponds to $(\mathbf{W}_2, \mathbf{w}_2)$. The solution of $\mathbf{h}\mathbf{W}_2 = \mathbf{h}$ is $\mathbf{h} = 00l$ and then $\mathbf{h}\mathbf{w}_2 = l/2$. Hence the condition that $00l$ be a possible reflection is $00l$: $l = 2n$. $00l$ reflections with $l = 2n + 1$ are *systematically absent*. However, this condition is included in the condition for the n glide.

There are no conditions on general hkl reflections, since the lattice is of type P and, as shown above, such a lattice does not impose any restrictions on the reflections.

It is interesting to note that the same space-group absences characterize space group $Pnam$ ($Pnma$, No. 62, in the setting $\mathbf{a\bar{c}b}$) which, unlike $Pna2_1$ (No. 33) is centrosymmetric. So, after the above absences are reliably determined, there are two possible space groups. This is an instance of a space-group ambiguity which may often be resolved with the aid of intensity statistics (see Section 1.6.2.2).

We have seen that during the derivation some components of $\mathbf{w}_i$ disappeared while others remained. For example,

$$\mathbf{w}_4 = \begin{pmatrix} w_{1,4} \\ w_{2,4} \\ w_{3,4} \end{pmatrix} = \begin{pmatrix} 1/2 \\ 1/2 \\ 1/2 \end{pmatrix},$$

but only $\begin{pmatrix} 0 \\ 1/2 \\ 1/2 \end{pmatrix}$ was required for the condition that $0kl$ be a possible reflection. In order to explain this we recall that in general the translation parts of the glide-reflection and screw-rotation operations consist of an intrinsic part (parallel to the geometric element), which gives rise to systematic absences, and a location part from which the actual location of the geometric element can be computed (*cf.* Section 1.2.2). The location parts are nonzero in n-glide and a-glide operations of the space group $Pna2_1$ (No. 33) but equal to zero for the screw-rotation operations of this space group. The relation between the location parts and the actual location of the geometric element is discussed in detail in Section 1.2.2.3 (see also *e.g.* Shmueli, 1984).

1.6.4. Reflection conditions and possible space groups

By Howard D. Flack and Uri Shmueli

1.6.4.1. Introduction

The tables for Bravais lattices type mP, oI and tP in the Laue classes $2/m$, mmm and $4/m$, respectively, are presented here as providing examples of the full set of tables which appear in *IT* A, Section 1.6.4. The primary order of presentation of these tables of reflection conditions (*cf.* Section 1.6.3) of space groups is the Bravais lattice. This order has been chosen as cell reduction on unit-cell dimensions leads to the Bravais lattice as described as stage 1 in Section 1.6.2.1.

As an aid in the study of naturally occurring macromolecules and compounds made by enantioselective synthesis, the space groups of enantiomerically pure compounds (Sohncke space groups) are typeset in bold.

The tables show, on the left, sets of reflection conditions and, on the right, those space groups that are compatible with the given set of reflection conditions. To the right of the block of reflection conditions and to the left of the block of possible space groups there is a column headed 'Diffraction symbol'. These symbols are explained below. The reflection conditions, *e.g.* h or $k + l$, are to be understood as $h = 2n$ or $k + l = 2n$, respectively. All of the space groups in each table correspond to the same Patterson symmetry, which is indicated in the table header. This makes for easy comparison with the entries for the individual space groups in Chapter 2.3 of *IT* A, in which the Patterson symmetry is also very clearly shown.

The diffraction symbols listed here have the same structure as the symbols in the column headed 'extinction symbols' in, for example, Looijenga-Vos & Buerger (2002) when based on the same reflection conditions. However, here we use the term 'diffraction symbols', as extinction is a dynamical attenuation effect and is not applicable to the interference effects that give rise to systematic absences. The notation of the diffraction symbols is the same as that summarized in Looijenga-Vos & Buerger (2002) and is described briefly below.

The dependence of a diffraction symbol on the reflection conditions to its left is similar but not identical to that of the possible space-group symbols to its right. The diffraction symbol is composed of a letter indicating the centring type and three characters, one for each of the representative symmetry directions. The integral reflection condition (involving h, k and l all different from zero) determines the centring type of the diffraction symbol, while the non-dash entries represent zonal and serial reflection conditions corresponding to glide reflections and screw rotations, respectively, along the symmetry directions. The dashes (−) correspond to symmetry directions that are associated with pure rotations, rotoinversions and mirror reflections. These operations are not associated in any way with reflection conditions and will be called condition-free operations.

For example, the diffraction symbol corresponding to the space group $Pmm2$ is $P---$ and has an empty line on its left (no conditions). If a direction is associated with both a screw rotation and a glide reflection, the corresponding entry looks the same as the possible space group (*e.g.* the diffraction symbol corresponding to the space group $P12_1/c1$ is just $P12_1/c1$; such features can be found in some monoclinic and tetragonal space groups, *cf.* Tables 1.6.4.1 and 1.6.4.3). If one or more directions are associated with screw-rotation systematic absences and the remaining directions are condition-free, the symbol(s) of the screw rotation(s) appear in the diffraction symbol (*e.g.* the diffraction symbol corresponding to space group $P222_1$ is $P--2_1$). If a screw rotation and one or two glide reflections appear in the same space-group symbol but are associated with different directions, the symbol(s) of the glide reflection(s) appear in the diffraction symbol(s) while the screw rotation appears to be regarded as condition-free (*e.g.* the diffraction symbol to the left of the space group $P2_1ca$ is $P-ca$; this apparently odd feature appears in some space groups related to the point group $mm2$ and is due to the fact that the reflection condition for the 2_1 axis is included in the reflection conditions for the glide planes c and a in the space group $P2_1ca$).

We finally wish to mention the double-glide symbols which are used in the diffraction symbols of some centred orthorhombic space groups (see Table 1.6.4.2 and the second example of Section 1.6.4.2). These symbols indicate two different glide reflections which are associated either with the same plane or with different planes. The latter are perpendicular to the same symmetry direction. In the diffraction symbols, both glide reflections are shown in parentheses at the location of the common symmetry direction. If the different glide reflections are associated with the same plane, then they belong to the same symmetry element denoted as an e-glide plane in some space-group symbols.

Table 1.6.4.1
Reflection conditions and possible space groups with Bravais lattice *mP* and Laue class 2/*m*; (monoclinic, unique axis *b*); Patterson symmetry $P12/m1$

Reflection conditions						Diffraction symbol	Space group	No.	Space group	No.	Space group	No.
$h0l$	$0kl$	$hk0$	$0k0$	$h00$	$00l$							
						$P1-1$	$\mathbf{P121}$	3	$P1m1$	6	$P12/m1$	10
			k			$P12_11$	$\mathbf{P12_11}$	4	$P12_1/m1$	11		
h				h		$P1a1$	$P1a1$	7	$P12/a1$	13		
h			k	h		$P12_1/a1$	$P12_1/a1$	14				
l					l	$P1c1$	$P1c1$	7	$P12/c1$	13		
l			k		l	$P12_1/c1$	$P12_1/c1$	14				
$h+l$				h	l	$P1n1$	$P1n1$	7	$P12/n1$	13		
$h+l$			k	h	l	$P12_1/n1$	$P12_1/n1$	14				

Table 1.6.4.2
Reflection conditions and possible space groups with Bravais lattice *oI* and Laue class *mmm*; Patterson symmetry *Immm*

Reflection conditions							Diffraction symbol	Space group	No.	Space group	No.	Space group	No.
hkl	$0kl$	$h0l$	$hk0$	$h00$	$0k0$	$00l$							
$h+k+l$	$k+l$	$h+l$	$h+k$	h	k	l	$I---$	$\mathbf{I222}$ $Im2m$	23 44	$\mathbf{I2_12_12_1}$ $I2mm$	24 44	$Imm2$ $Immm$	44 71
$h+k+l$	$k+l$	$h+l$	h, k	h	k	l	$I--(ab)$	$Im2a$ $Immb$	46 74	$I2mb$	46	$Imma$	74
$h+k+l$	$k+l$	h, l	$h+k$	h	k	l	$I-(ac)-$	$Ima2$ $Imcm$	46 74	$I2cm$	46	$Imam$	74
$h+k+l$	$k+l$	h, l	h, k	h	k	l	$I-cb$	$I2cb$	45	$Imcb$	72		
$h+k+l$	k, l	$h+l$	$h+k$	h	k	l	$I(bc)--$	$Ibm2$ $Icmm$	46 74	$Ic2m$	46	$Ibmm$	74
$h+k+l$	k, l	$h+l$	h, k	h	k	l	$Ic-a$	$Ic2a$	45	$Icma$	72		
$h+k+l$	k, l	h, l	$h+k$	h	k	l	$Iba-$	$Iba2$	45	$Ibam$	72		
$h+k+l$	k, l	h, l	h, k	h	k	l	$Icab$	$Ibca$	73	$Icab$	73		

Table 1.6.4.3
Reflection conditions and possible space groups with Bravais lattice *tP* and Laue class 4/*m*; *hk* are permutable; Patterson symmetry *P*4/*m*

Reflection conditions					Diffraction symbol	Space group	No.	Space group	No.	Space group	No.
$hk0$	$0kl$	$h\pm hl$	$00l$	$h00$							
					$P---$	$\mathbf{P4}$	75	$P\bar{4}$	81	$P4/m$	83
			l		$P4_2--$	$\mathbf{P4_2}$	77	$P4_2/m$	84		
			$l=4n$		$P4_1--$	$\mathbf{P4_1}$	76	$\mathbf{P4_3}$	78		
$h+k$				h	$Pn--$	$P4/n$	85				
$h+k$			l	h	$P4_2/n$	$P4_2/n$	86				

Although the diffraction symbols do not provide any essentially new information, they are introduced here because of their conciseness and as a service to crystallographers who may value their usefulness in research.

1.6.4.2. Examples

(1) If the Bravais lattice of a crystal is *mP*, the corresponding Laue class is 2/*m* and the observed reflection conditions are $h0l$: $h = 2n$, $h00$: $h = 2n$, then by inspection of Table 1.6.4.1 we see that the possible space groups of the crystal are $P1a1$ (No. 7) or $P12/a1$ (No. 13). If we determine by the methods of Section 1.6.2.2 that the crystal has a centre of symmetry, the only possible space group is $P12/a1$ (No. 13).

(2) If the Bravais lattice is *oI* and the Laue class is *mmm*, Table 1.6.4.1 in *IT* A directs us to Table 1.6.4.11 in *IT* A (which is reproduced here as Table 1.6.4.2). Given the observed reflection conditions

hkl: $h + k + l = 2n$, $0kl$: $k = 2n, l = 2n$, $h0l$: $h + l = 2n$,

$hk0$: $h + k = 2n$, $h00$: $h = 2n$, $0k0$: $k = 2n$, $00l$: $l = 2n$,

it is seen from Table 1.6.4.2 that the possible settings of the space groups are: *Ibm*2 (46), *Ic*2*m* (46), *Ibmm* (74) and *Icmm* (74).

The diffraction symbol based on the same observations, also appearing in Table 1.6.4.2, is given by *I*(*bc*) − −. This is an example of the case when the two glide reflections (given in parenthesis) are associated with different planes: either the *b*- or *c*-glide symbol is used to specify the settings of the space groups Nos. 46 and 74, compatible with the appropriate reflection conditions.

1.6.5. Space-group determination in macromolecular crystallography

By Howard D. Flack

For macromolecular crystallography, succinct descriptions of space-group determination have been given by Kabsch (2010*a*,*b*, 2012) and Evans (2006, 2011). Two characteristics of macromolecular crystals give rise to variations on the small-molecule procedures described above.

The first characteristic is the large size of the unit cell of macromolecular crystals and the variation of the cell dimensions from one crystal to another. This makes the determination of the Bravais lattice by cell reduction problematic, as small changes of cell dimensions give rise to differences in the assignment. Kabsch (2010*a*,*b*, 2012) uses a 'quality index' from each of Niggli's 44 lattice characters to come to a best choice. Grosse-Kunstleve *et al.* (2004) and Sauter *et al.* (2004) have found that some commonly used methods to determine the Bravais lattice are susceptible to numerical instability, making it possible for high-symmetry Bravais lattice types to be improperly identified. Sauter *et al.* (2004, 2006) find from practical experience that a deviation δ as high as 1.4° from perfect alignment of direct and reciprocal lattice rows must be allowed to construct the highest-symmetry Bravais-lattice type consistent with the data. Evans (2006) uses a value of 3.0°. The large unit-cell size also gives rise to a large number of reflections in the asymmetric region of reciprocal space and, taken with the tendency of macromolecular crystals to decompose in the X-ray beam, full-sphere data sets are uncommon. This means that confirmation of the Laue class by means of values of $R_{\rm merge}$ are rarer than with small-molecule crystallography, although Kabsch (2010*b*) does use a 'redundancy-independent *R* factor'. Evans (2006, 2011) describes methods very similar to those given as the second stage in Section 1.6.2.1. The conclusion of Sauter *et al.* (2006) and Evans (2006) is that $R_{\rm merge}$ values as high as 25% must be permitted in order to assemble an optimal set of operations to describe the diffraction symmetry. Another interesting procedure, accompanied by experimental proof, has been devised by Sauter *et al.* (2006). They show that it is clearer to calculate $R_{\rm merge}$ values individually for each potential symmetry operation of a target point group rather than comparing $R_{\rm merge}$ values for target point groups globally. According to Sauter *et al.* (2006) the reason for this improvement lies in the lack of intensity data relating some target symmetry operations.

The second characteristic of macromolecular crystals is that the compound is known, or presumed, to be chiral and enantiomerically pure, so that the crystal structure is chiral. This limits the choice of space group to the 65 Sohncke space groups containing only translations, pure rotations or screw rotations. For ease of use, these have been typeset in bold in Tables 1.6.4.1–1.6.4.3.

For the evaluation of protein structures, Poon *et al.* (2010) apply similar techniques to those described in Section 1.6.2.3. The major tactical objective is to identify pairs of α-helices that have been declared to be symmetry-independent in the structure solution but which may well be related by a rotational symmetry of the crystal structure. Poon *et al.* (2010) have been careful to test their methodology against generated structural data before proceeding to tests on real data. Their results indicate that some 2% of X-ray structures in the Protein Data Bank potentially fit in a higher-symmetry space group. Zwart *et al.* (2008) have studied the problems of under-assigned translational symmetry operations, suspected incorrect symmetry and twinned data with ambiguous space-group choices, and give illustrations of the uses of group–subgroup relations.

1.6.6. Space groups for nanocrystals by electron microscopy

By John C. H. Spence

The determination of crystal space groups may be achieved by the method of convergent-beam electron microdiffraction (CBED) using a modern transmission electron microscope (TEM). A detailed description of the CBED technique is given by Tanaka (2008) in Section 2.5.3 of *International Tables for Crystallography* Volume B; here we give a brief overview of the capabilities of the method for space-group determination, for completeness. A TEM beam focused to nanometre dimensions allows study of nanocrystals, while identification of noncentrosymmetric crystals is straightforward, as a result of the strong multiple scattering normally present in electron diffraction. (Unlike single scattering, this does not impose inversion symmetry on diffraction patterns, but preserves the symmetry of the sample and its boundaries.) CBED patterns also allow direct determination of screw and glide space-group elements, which produce characteristic absences, despite the presence of multiple scattering, in certain orientations. These absences, which remain for all sample thicknesses and beam energies, may be shown to occur as a result of an elegant cancellation theorem along symmetry-related multiple-scattering paths (Gjønnes & Moodie, 1965). Using all of the above information, most of the 230 space groups can be distinguished by CBED. The remaining more difficult cases (such as space groups that differ only in the location of their symmetry elements) are discussed in Spence & Lynch (1982), Eades (1988), and Saitoh *et al.* (2001). Enantiomorphic pairs require detailed atomistic simulations based on a model, as in the case of quartz (Goodman & Secomb, 1977). Multiple scattering renders Bragg intensities sensitive to structure-factor phases in noncentrosymmetric structures, allowing these to be measured with a tenth of a degree accuracy (Zuo *et al.*, 1993). Unlike X-ray diffraction, electron diffraction is very sensitive to ionicity and bonding effects, especially at low angles, allowing extinction-free charge-density mapping with high accuracy (Zuo, 2004; Zuo *et al.*, 1999). Because of its sensitivity to strain, CBED may also be used to map out local phase transformations which cause space-group changes on the nanoscale (Zuo, 1993; Zhang *et al.*, 2006).

In simplest terms, a CBED pattern is formed by enlarging the incident beam divergence in the transmission diffraction geometry, as first demonstrated G. Mollenstedt in 1937 (Kossel &

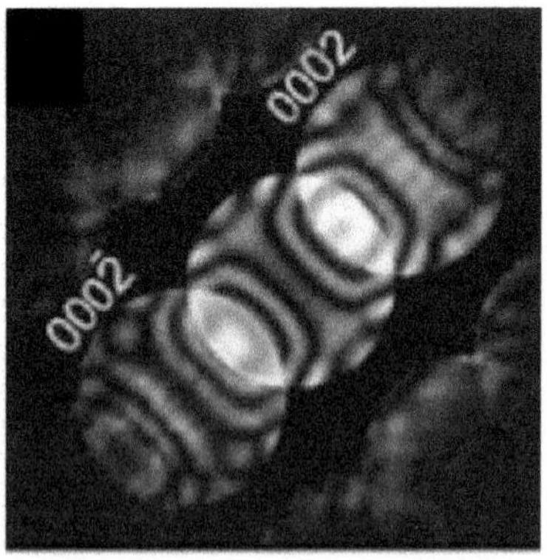

Figure 1.6.6.1
Polarity determination by convergent-beam electron diffraction. A CBED pattern from ZnO with the beam normal to the c axis is shown. The intensity distribution along **c** does not have inversion symmetry, reflecting the noncentrocentrosymmetric nature of the structure. Reproduced with permission from Wang *et al.* (2003). Copyright (2003) by The American Physical Society.

Mollenstedt, 1942). Bragg spots are then enlarged into discs, and the intensity variation within these discs is studied, in addition to that of the entire pattern, in the CBED method. The intensity variation within a disc displays a complete rocking curve in each of the many diffracted orders, which are simultaneously excited and recorded. The entire pattern thus consists of many independent 'point' diffraction patterns (each for a slightly different incident beam direction) laid beside each other. Fig. 1.6.6.1 shows a CBED pattern from the wurtzite structure of ZnO, with the beam normal to the c axis (Wang *et al.*, 2003). The intensity variation along a line running through the centres of these discs (along the c axis) is not an even function, strongly violating Friedel's law for this elastic scattering. At higher scattering angles, curvature of the Ewald sphere allows three-dimensional symmetry elements to be determined by taking account of 'out-of-zone' intensities in the outer higher-order Laue zone (HOLZ) rings near the edge of the detector. Since sub-ångstrom-diameter electron probes and nanometre X-ray laser probes (Spence *et al.*, 2012) are now being used, the effect of the inevitable coherent interference between overlapping convergent-beam orders on space-group determination must be considered (Spence & Zuo, 1992).

A systematic approach to space-group determination by CBED has been developed by several groups. In general, one would determine the symmetry of the projection diffraction group first (ignoring diffraction components along the beam direction z), then add the z-dependent information seen in HOLZ lines, allowing one to finally identify the point group from tables, by combining all this information. After indexing the pattern, in order to determine a unit cell the Bravais lattice is next determined. The form of the three-dimensional reciprocal lattice and its centring can usually be determined by noting the registry of Bragg spots in a HOLZ ring against those in the zero-order (ZOLZ) ring. Finally, by setting up certain special orientations, tests are applied for the presence of screw and glide elements, which are revealed by a characteristic dark line or cross within the CBED discs. Tables can again then be used to combine these translational symmetry elements with the previously determined point group, to find the space group. As a general experimental strategy, one first seeks mirror lines (perhaps seen in Kikuchi patterns), then follows these around using the two-axis goniometer fitted to modern TEM instruments in a systematic search for other symmetry elements. Reviews of the CBED method can be found in Steeds & Vincent (1983), in Goodman (1975), and in the texts by Tanaka *et al.* (1988). A textbook-level worked example of space-group determination by CBED can be found in Spence & Zuo (1992) and in the chapter by A. Eades in Williams & Carter (2009).

1.6.7. Examples

By Howard D. Flack

1.6.7.1. Example (1), 4-chlorophenol, C_6H_5OCl

This example was provided by Professor S. Parsons of the University of Edinburgh, UK. Data for the organic molecule 4-chlorophenol, C_6H_5OCl (Oswald *et al.*, 2005) were collected at 150 K. Cell reduction to the Niggli reduced cell found an mP Bravais lattice with an angular deviation of 0° [a = 8.7086 (11), b = 15.4523 (19), c = 8.7414 (11) Å, $\alpha = \gamma = 90°$, β = 93.954 (2)°]. Three other cells, all of angular deviation 0.22°, belong to the mS and oS Bravais-lattice types. As a consequence it was taken that the crystal has an mP Bravais-lattice type.

In the second stage, values of $R_{\rm merge}$, as defined in equation (1.6.2.2), for various point groups were evaluated. Also, for the purpose of comparison, it is useful to calculate the value of $R_{\rm int}$. The latter measures the agreement between reflections of identical reflection indices h, k and l before any merging, and hence represents the intrinsic value of the data set independent of any choice of symmetry. $R_{\rm int}$ is defined as

$$R_{\rm int} = \frac{\sum_{\bf h} \sum_{i=1}^{N({\bf h})} |\langle |F({\bf h})|^2\rangle - |F({\bf h})|_i^2|}{\sum_{\bf h} N({\bf h})\langle |F({\bf h})|^2\rangle},$$

where $N(\mathbf{h})$ is the number of observations of reflection **h**. The average in the first term in the numerator and in the denominator ranges over intensities with identical reflection indices and the other summation, $\sum_{\bf h}$, ranges over the hkl reflections. Analysis of the intensity data finds $R_{\rm int}$ = 1.93% with $\langle I/u(I)\rangle = 11.68$ for all 7471 reflections. Table 1.6.7.1 displays the $R_{\rm merge}$ values. A full set of four equivalent reflections (hkl, $\overline{hkl}$, $\bar{h}k\bar{l}$ and $h\bar{k}l$) was available for only 86 unique reflections. The latter four reflections are symmetry-equivalent in the Laue group $2/m$ and are general reflections. A full description of the use of the $R_{\rm merge}$ values, R_A and R_D, on the average (A) and difference (D) of Friedel opposites is described in *IT* A, Section 1.6.5. Values for various point groups have been calculated. From Table 1.6.7.1 one sees that all four point groups ($2/m$, m, 2 and $\bar{1}$) are reasonable choices for the diffraction symmetry of these data when compared to $R_{\rm int}$. Although there is a slight preference for point group 2, the number of complete sets of $2/m$-symmetry-equivalent reflections is rather small.

The values of the low-order moments of $|E|$ for Example (1) are given in Table 1.6.2.1. Comparison with the theoretical values for the centric and acentric distributions, given by (1.6.2.3), shows

Table 1.6.7.1
$R_{\rm merge}$ (%) values for the 86 complete sets of four general $2/m$-symmetry-equivalent reflections of Example (1) in Section 1.6.7.1

	Point group			
$R_{\rm merge}$	$2/m$	m	2	$\bar{1}$
$R_{\|F\|^2}$	2.23	2.07	1.61	1.86
R_A	1.07	1.07	1.07	0.00
R_D	100.0	155.5	64.3	100.0

Table 1.6.7.2
Reflection conditions for Example (1) in Section 1.6.7.1

The column t/f gives the ratio of columns 3 and 4. 'True' indicates that the specified condition is obeyed and 'False' indicates that the specified condition is not obeyed. Reflection conditions that appear to be in operation are highlighted in bold.

		$\langle I/u(I)\rangle$			No. of reflections	
Class	Condition	True	False	t/f	True	False
hkl	$h + l = 2n$	11.55	11.81	0.98	3711	3760
hkl	$k + l = 2n$	11.93	11.44	1.04	3734	3737
hkl	$h + k + l = 2n$	11.79	11.58	1.02	3731	3740
*h*0*l*	$h = 2n$	7.87	9.23	0.85	134	138
h*0*l	$\mathbf{l = 2n}$	**16.59**	**1.10**	**15.0**	**131**	**141**
*h*0*l*	$h + l = 2n$	8.15	8.94	0.91	131	141
0*kl*	$l = 2n$	10.99	13.05	0.84	237	244
0*kl*	$k + l = 2n$	15.24	8.90	1.71	238	243
*hk*0	$h = 2n$	15.70	14.27	1.10	281	293
*hk*0	$k = 2n$	15.22	14.73	1.03	285	289
*hk*0	$h = 2n, k = 2n$	16.17	14.59	1.10	139	435
*h*00	$h = 2n$	22.67	19.37	1.17	8	10
0*k*0	$\mathbf{k = 2n}$	**13.16**	**1.11**	**11.9**	**17**	**20**
00*l*	$\mathbf{l = 2n}$	**23.33**	**2.08**	**11.2**	**3**	**6**

Table 1.6.7.4
Reflection conditions for Example (2) in Section 1.6.7.2

The column t/f gives the ratio of columns 3 and 4. 'True' indicates that the specified condition is obeyed and 'False' indicates that the specified condition is not obeyed. Reflection conditions that appear to be in operation are highlighted in bold.

		$\langle I/u(I)\rangle$			No. of reflections	
Class	Condition	True	False	t/f	True	False
hkl	$h + k + l = 2n$	11.81	6.50	1.81	15748	15796
*hk*0	$h + k = 2n$	14.24	8.75	1.62	382	385
*hk*0	$h = 2n, k = 2n$	24.98	7.04	3.54	190	577
0*kl*	$k = 2n$	12.06	13.28	0.91	2311	2399
0*kl*	$l = 2n$	13.33	12.03	1.11	2356	2354
0*kl*	$k + l = 2n$	16.98	8.38	2.02	2355	2355
0*kl*	$k = 2n, l = 2n$	17.09	11.25	1.52	1156	3554
hhl	$l = 2n$	13.01	6.98	1.86	842	844
hhl	$2h + l = 4n$	22.66	5.76	3.93	422	1264
00*l*	$\mathbf{l = 4n}$	**34.89**	**0.53**	**66.1**	**38**	**119**
00*l*	$l = 2n$	17.02	0.77	22.0	78	79
00*l*	$l = 4n + 2$	0.04	0.77	0.05	40	79
***h*00**	$\mathbf{h = 2n}$	**34.23**	**0.47**	**73.1**	**48**	**51**
$h\bar{h}0$	$h = 2n$	40.60	2.73	14.9	16	19

that there is a strong indication that the space group is centrosymmetric.

The analysis of the average intensity of selected zones and lines of reflections is presented in Table 1.6.7.2. From this table, one sees that there are three reflection conditions in operation: (i) *h*0*l*: $l = 2n$; (ii) 0*k*0: $k = 2n$; (iii) 00*l*: $l = 2n$. These conditions are indicated in bold in Table 1.6.7.2. The lack of any reflection condition on the general *hkl* reflections confirms that the Bravais-lattice type is *mP* and not *mS*. Reference to Table 1.6.4.1 for reflection conditions and possible space groups with Bravais lattice *mP* shows that only one single space group presents these reflection conditions. It is space group $P2_1/c$, (No. 14).

Least-squares refinement in space group $P2_1/c$ led to a conventional *R* factor of 3.96%. Applying *PLATON*'s ADDSYM option (Spek, 2003) to the refined structure did not produce any alert concerning symmetry operations in the crystal-structure solution that are not part of the chosen space group.

1.6.7.2. Example (2), $[BDTA]_2[CuCl_4]$

This example was also provided by Professor S. Parsons of the University of Edinburgh, UK. Data for a crystal of the organometallic compound $[BDTA]_2[CuCl_4]$ (BDTA = benzo-1,3,2-dithiazolyl) (Staniland *et al.*, 2006) were measured at 150 K giving 31 544 reflections with $\langle I/u(I)\rangle = 9.15$, $R_{int} = 3.26\%$ and R_{merge} (for point group 2 with a twofold rotation axis parallel to **c**) = 2.40%. Cell reduction to the Niggli reduced cell finds a *tP* Bravais lattice with an angular deviation of 0° [$a = b = 8.2295$ (3), $c = 25.681$ (2) Å, $\alpha = \beta = \gamma = 90°$]. There were no other solutions with an angular deviation less than 3°. As a consequence it was taken that the crystal had a *tP* Bravais-lattice type. Table 1.6.7.3 displays the R_{merge} values calculated from the 647 sets of eight reflections ($hkl, \overline{hkl}, \bar{k}hl, k\bar{h}l, \bar{h}k\bar{l}, h\bar{k}\bar{l}, kh\bar{l}$ and $\overline{khl}$) available in the data after merging and averaging in point group 2 with a twofold rotation axis parallel to **c**. These eight reflections are symmetry-equivalent in the Laue group 4/*mmm* and are general reflections. The twofold axis parallel to **c** is common to all of the groups in Table 1.6.7.3. A full description of the use of the R_{merge} values, R_A and R_D, on the average (*A*) and difference (*D*) of Friedel opposites is described in *IT* A, Section 1.6.5. Values for the point groups 4/*mmm*, $\bar{4}2m$, $\bar{4}m2$, 4*mm*, 422 and 4/*m* were calculated. From Table 1.6.7.3, one sees that the R_{merge} value on $|F|^2$ of point group 4/*m* is only marginally lower than that of 4/*mmm*. It was concluded that the Laue group of Example (2) is 4/*mmm*. Within the Laue class 4/*mmm*, the R_{merge} value on $|F|^2$ for point group 422 is the lowest, although it is only slightly smaller that the values for the other point groups. However, the R_{merge} value on *D* (described in Section 1.6.5 of *IT* A) of point group 422 is noticeably the smallest of those of the noncentrosymmetric point groups, although larger than that of the centrosymmetric point group. On this analysis, first choices for the crystal point group are 4/*mmm* and 422.

Table 1.6.7.3
R_{merge} (%) values for the 647 complete sets of eight general 4/*mmm*-symmetry-equivalent reflections of Example (2) in Section 1.6.7.2

	Point group					
R_{merge}	4/*mmm*	$\bar{4}2m$	$\bar{4}m2$	4*mm*	422	4/*m*
$R_{\lvert F\rvert^2}$	3.60	3.36	3.37	3.33	3.15	3.23
R_A	2.47	2.47	2.47	2.47	2.47	1.98
R_D	100.0	216.8	207.1	174.8	125.8	100.0

The values of the low-order moments of $|E|$ for Example (2) are given in Table 1.6.2.1. Comparison with the theoretical values for centric and acentric distributions shows that there is a strong indication that the space group is centrosymmetric. However, it is seen from Table 1.6.2.1 that there is also a tendency to a bicentric distribution (Rogers & Wilson, 1953).

The analysis of the average intensity of selected zones and lines of reflections is presented in Table 1.6.7.4. In that table, reflection conditions that appear to be in operation are highlighted in bold. Inspection of this table shows that there are two reflection conditions in operation: (i) 00*l*: $l = 4n$; (ii) *h*00: $h = 2n$. Although it seems at first view from the values of $\langle I/u(I)\rangle$ that the condition 00*l*: $l = 2n$ might be in operation, the corresponding values for the condition 00*l*: $l = 4n + 2$ makes it clear that the reflection condition 00*l*: $l = 2n$ arises only as a consequence of the 00*l*: $l = 4n$ condition. The 00*l*: $l = 4n + 2$ reflections are very weak. The lack of any reflection condition on the general *hkl* reflections confirms that the Bravais-lattice type is *tP* and not *tI*. Reference to Table 1.6.4.3 for reflection conditions and possible space groups with Bravais lattice *tP* and Laue class 4/*mmm* shows that the two space groups $P4_12_12$ (No. 92) and $P4_32_12$ (No. 96) satisfy the observed reflection conditions. These two space groups form an enantiomorphic pair which can only be distinguished once a crystal-structure solution has been obtained.

Least-squares refinement of the crystal structure in space group $P4_32_12$ (No. 96) led to a conventional R factor of 2.70%, yielding a value of the Flack (1983) parameter of 0.05 (3). The molecules in $[BDTA]_2[CuCl_4]$ are achiral but, with space group $P4_32_12$ (No. 96), the crystal structure is chiral, and the Flack parameter indicates that the crystal is not twinned by inversion. As a consequence, the chirality sense of each crystal occurs at random during crystallization, and the batch of crystals from which this one was taken most likely contains an equal number of the opposite enantiomorphs, some in space group $P4_32_12$ (No. 96) and the opposites in space group $P4_12_12$ (No. 92). No extra symmetry in the crystal structure was detected by *PLATON*'s ADDSYM option (Spek, 2003). One recalls that there is a tendency for the intensity statistics of this crystal to follow the bicentric distribution, although the crystal structure was solved as noncentrosymmetric. As both molecules in this compound display some molecular symmetry, it is considered that the unusual intensity statistics are due to molecular rather than crystal symmetry.

1.6.7.3. Example (3), flo19, $C_{62}H_{46}N_{14}$

This example is adapted from Section 3.1 of Palatinus & van der Lee (2008). The organic molecule $C_{62}H_{46}N_{14}$ crystallizes with an oP Bravais lattice, $Z = 1$, Laue class mmm [$R_{\rm merge}(mmm)$ = 0.06, $\langle I/u(I)\rangle = 6.81$ for all data], with two short axes and one very long axis [cell parameters: $a = 3.91$, $b = 6.17$, $c = 51.22$ Å, $\alpha = \beta = \gamma = 90°$, $V = 1236.4$ Å^3]. The analysis of the relevant reflection conditions is summarized in Table 1.6.7.5. The following reflection conditions have t/f values that are noticeably large and are indicated in bold in Table 1.6.7.5: $0k0$: $k = 2n$; $00l$: $l = 2n$; $hk0$: $h + k = 2n$, but there is no space group in Table 1.6.4.7 of *IT* A that corresponds to these three conditions. Taking only two of these conditions at a time, one finds that $P22_12_1$ (No. 18), $Pm2_1n$ (No. 31), $P2_1mn$ (No. 31) and $Pmmn$ (No. 59) are possible space groups on consulting Table 1.6.4.7 of *IT* A. It appeared, however, to be impossible to solve the structure with the direct-methods software *SIR2004* (Burla *et al.*, 2005) or *SHELXS* or *SHELXD* (Sheldrick, 2008) in these space groups. The structure solution with *SUPERFLIP* (Palatinus & Chapuis, 2007) proceeded smoothly, since the structure was solved in $P1$ (No. 1); intensities were averaged according to Laue symmetry mmm and subsequently expanded to $P1$ (No. 1). No *a priori* assumptions were made concerning systematic absences, *i.e.* all reflections were included in the data set. The subsequent symmetry analysis, as described in Section 1.6.2.3, showed that the correct space group is actually $P2_122_1$ (No. 18), which was confirmed by the structural refinement that followed. A control using *PLATON*'s ADDSYM option (Spek, 2003) did not show any additional symmetry. Interestingly, the n glide perpendicular to **c**, which clearly shows up in the list of reflection conditions, is absent in the final structure with a symmetry agreement factor of only 0.72. The symmetry agreement factor $\varphi_{\rm sym}$ is defined in equation (7) of Palatinus & van der Lee (2008). A completely random electron density gives a value of 1 and a density perfectly obeying the symmetry operation being tested results in a value of 0. Palatinus & van der Lee (2008) found that for correct symmetry operations $\varphi_{\rm sym}$ is below 0.1 and almost always below 0.2. Wrong symmetry operations have values of $\varphi_{\rm sym}$ greater than 0.5.

An inspection of the refined structure shows that in the projection of the structure onto the ab plane, a large number of the atoms are related by the centring vector (1/2, 1/2) which, in combination with the generally low intensities, leads to the pseudo-reflection-condition effect in the $hk0$ plane.

Table 1.6.7.5
Reflection conditions for Example (3) in Section 1.6.7.3

The column t/f gives the ratio of columns 3 and 4. 'True' indicates that the specified condition is obeyed and 'False' indicates that the specified condition is not obeyed. Reflection conditions with noticeably large t/f are highlighted in bold.

		$\langle I/u(I)\rangle$			No. of reflections	
Class	Condition	True	False	t/f	True	False
$h00$	$h = 2n$	1.84	0.35	5.23	3	5
$0k0$	**$k = 2n$**	**52.20**	**5.69**	**9.18**	**2**	**2**
$00l$	**$l = 2n$**	**8.93**	**1.04**	**8.61**	**41**	**43**
$0kl$	$l = 2n$	8.04	5.73	1.40	428	428
$0kl$	$k = 2n$	6.19	7.68	0.81	458	398
$0kl$	$k + l = 2n$	7.20	6.57	1.10	426	430
$h0l$	$l = 2n$	8.92	8.15	1.09	230	231
$h0l$	$h = 2n$	6.05	10.50	0.58	204	257
$h0l$	$h + l = 2n$	9.27	7.81	1.19	229	232
$hk0$	$k = 2n$	4.72	16.24	0.29	41	38
$hk0$	$h = 2n$	5.43	13.91	0.39	34	45
$hk0$	**$h + k = 2n$**	**19.49**	**1.71**	**11.43**	**38**	**41**

Weak data often thwart the determination of the space group based on the analysis of reflection conditions, since the distinction between reflections with observable intensity [with *e.g.* $I > 3u(I)$] and those systematically absent [and thus necessarily $I < 3u(I)$] becomes less clear. It has been observed that problems of this nature start to arise when the mean value of the ratio of the intensity and its standard uncertainty for a given resolution, $\langle I/u(I)\rangle$, drops below 10.

1.6.7.4. Example (4), CSD refcode FOYTAO01, $C_{12}H_{20}O_6$

This example is adapted from Section 3.4 of Palatinus & van der Lee (2008). The compound has Bravais-lattice type tP with Laue class $4/m$. There are no reflection conditions and Table 1.6.4.3 shows that there are three possible space groups: $P4$ (No. 76), $P\bar{4}$ (No. 81) and $P4/m$ (No. 83). $\langle|E^2 - 1|\rangle = 0.823$, slightly favouring the two noncentrosymmetric space groups ($P4$ and $P\bar{4}$). In fact, the structure was reported in space group $P4$ with $Z = 8$, *i.e.* two independent molecules in the asymmetric unit. Various space-group-determination software modules select $P\bar{4}$ as the most probable space group, based mainly on the much higher frequency of $P\bar{4}$ (No. 81) than $P4$ (No. 76) and $P4/m$ (No. 83) in crystal-structure databases. *SUPERFLIP* (Palatinus & Chapuis, 2007) solved the structure smoothly and gave final symmetry agreement factors of 0.069, 0.848 and 0.523 for the presence of the 4 axis, the $\bar{4}$ axis and the inversion centre, respectively, which does not leave any doubt that $P4$ is the correct space group. In this crystal structure, the molecule is chiral and both molecules in the asymmetric unit are of the same chirality. No information was available concerning the enantiomeric purity of the bulk. It thus appears likely that either the bulk was a racemate which crystallized as a racemic conglomerate or the bulk was enantiomerically pure.

There are numerous cases for which the classical space-group determination procedure described in Section 1.6.2.1 leads to a choice of several probable space groups. In the present example, for which the choice of possible space groups is $P4$ (No. 76), $P\bar{4}$ (No. 81) and $P4/m$ (No. 83), the crystal structures are noncentrosymmetric–chiral, noncentrosymmetric–achiral and centrosymmetric, respectively. The distinction between these space groups can be attempted in several ways. Intensity statistics can prove very useful, as do more advanced diffraction methods not described here, and finally non-diffraction methods should not be forgotten. Nowadays space-group-determination software usually makes the distinction in these cases on the basis of the value of $\langle|E^2 - 1|\rangle$ and of the space-group frequency found in the Cambridge Structural Database (CSD) or Inorganic Crystal Structure Database (ICSD); they will therefore favour $P\bar{4}$ (No. 81) when $\langle|E^2 - 1|\rangle$ tends to the noncentrosymmetric theoretical value, since its frequency of occurrence is an order of magnitude higher than that of $P4$ (No. 76).

References

Allenmark, S., Gawronski, J. & Berova, N. (2007). Editors. *Chirality*, **20**, 605–759.

Authier, A., Borovik-Romanov, A. S., Boulanger, B., Cox, K. G., Dmitrienko, V. E., Ephraïm, M., Glazer, A. M., Grimmer, H., Janner, A., Jannsen, T., Kenzelmann, M., Kirfel, A., Kuhs, W. F., Küppers, H., Mahan, G. D., Ovchinnikova, E. N., Thiers, A., Zarembowitch, A. & Zyss, J. (2014). *International Tables for Crystallography*, Volume D, *Physical Properties of Crystals*, 2nd ed., edited by A. Authier, Part 1. Chichester: Wiley.

Boček, P., Hahn, Th., Janovec, V., Klapper, H., Kopský, V., Přivratska, J., Scott, J. F. & Tolédano, J.-C. (2014). *International Tables for Crystallography*, Volume D, *Physical Properties of Crystals*, 2nd ed., edited by A. Authier, Part 3. Chichester: Wiley.

Burla, M. C., Caliandro, R., Camalli, M., Carrozzini, B., Cascarano, G. L., De Caro, L., Giacovazzo, C., Polidori, G. & Spagna, R. (2005). *SIR2004: an improved tool for crystal structure determination and refinement. J. Appl. Cryst.* **38**, 381–388.

Burzlaff, H., Zimmermann, H. & de Wolff, P. M. (2002). *Crystal lattices.* In *International Tables for Crystallography*, Volume A, *Space-Group Symmetry*, 5th ed., edited by Th. Hahn, Part 9. Dordrecht: Kluwer Academic Publishers.

Eades, J. A. (1988). *Glide planes and screw axes in CBED.* In *Microbeam Analysis 1988*, edited by D. Newberry, pp. 75–78. San Francisco Press.

Evans, P. (2006). *Scaling and assessment of data quality. Acta Cryst.* **D62**, 72–82.

Evans, P. R. (2011). *An introduction to data reduction: space-group determination, scaling and intensity statistics. Acta Cryst.* **D67**, 282–292.

Flack, H. D. (1983). *On enantiomorph-polarity estimation. Acta Cryst.* A**39**, 876–881.

Flack, H. D. (2003). *Chiral and achiral crystal structures. Helv. Chim. Acta*, **86**, 905–921.

Flack, H. D. (2015). *Methods of space-group determination – a supplement dealing with twinned crystals and metric specialization. Acta Cryst.* C**71**, 916–920.

Flack, H. D. & Shmueli, U. (2007). *The mean-square Friedel intensity difference in P1 with a centrosymmetric substructure. Acta Cryst.* **A63**, 257–265.

Gjønnes, J. & Moodie, A. F. (1965). *Extinction conditions in the dynamic theory of electron diffraction. Acta Cryst.* **19**, 65–69.

Goodman, P. (1975). *A practical method of three-dimensional space-group analysis using convergent-beam electron diffraction. Acta Cryst.* A**31**, 804–810.

Goodman, P. & Secomb, T. W. (1977). *Identification of enantiomorphously related space groups by electron diffraction. Acta Cryst.* **A33**, 126–133.

Grosse-Kunstleve, R. W., Sauter, N. K. & Adams, P. D. (2004). *Numerically stable algorithms for the computation of reduced unit cells. Acta Cryst.* A**60**, 1–6.

Hahn, Th. & Klapper, H. (2016). *International Tables for Crystallography*, Volume A, *Space-Group Symmetry*, 6th ed., edited by M. I. Aroyo, ch. 3.2. Chichester: Wiley.

Howells, E. R., Phillips, D. C. & Rogers, D. (1950). *The probability distribution of X-ray intensities. II. Experimental investigation and the X-ray detection of centres of symmetry. Acta Cryst.* **3**, 210–214.

International Tables for Crystallography (2016). Vol. A, *Space-Group Symmetry*, 6th ed., edited by M. I. Aroyo. Chichester: Wiley. [Abbreviated as *IT* A.]

Kabsch, W. (2010*a*). *XDS. Acta Cryst.* **D66**, 125–132.

Kabsch, W. (2010*b*). *Integration, scaling, space-group assignment and post-refinement. Acta Cryst.* **D66**, 133–144.

Kabsch, W. (2012). *Space-group assignment.* In *International Tables for Crystallography*, Volume F, *Crystallography of Biological Macromolecules*, edited by E. Arnold, D. M. Himmel & M. G. Rossmann, Section 11.3.6. Chichester: Wiley.

Kirschbaum, K., Martin, A. & Pinkerton, A. A. (1997). *λ/2 Contamination in charge-coupled-device area-detector data. J. Appl. Cryst.* **30**, 514–516.

Koch, E. (2006). *Twinning.* In *International Tables for Crystallography*, Volume C, *Mathematical, Physical and Chemical Tables*, 1st online edition, edited by E. Prince, ch. 1.3. Chester: International Union of Crystallography.

Kossel, W. & Mollenstedt, G. (1942). *Electron interference in a convergent beam. Ann. Phys.* **42**, 287–296.

Le Page, Y. (1982). *The derivation of the axes of the conventional unit cell from the dimensions of the Buerger-reduced cell. J. Appl. Cryst.* **15**, 255–259.

Le Page, Y. (1987). *Computer derivation of the symmetry elements implied in a structure description. J. Appl. Cryst.* **20**, 264–269.

Looijenga-Vos, A. & Buerger, M. J. (2002). *Space-group determination and diffraction symbols.* In *International Tables for Crystallography*, Volume A, *Space-Group Symmetry*, 5th ed., edited by Th. Hahn, Section 3.1.3. Dordrecht: Kluwer Academic Publishers.

Macchi, P., Proserpio, D. M., Sironi, A., Soave, R. & Destro, R. (1998). *A test of the suitability of CCD area detectors for accurate electron-density studies. J. Appl. Cryst.* **31**, 583–588.

Okaya, Y. & Pepinsky, R. (1955). *Computing Methods and the Phase Problem in X-ray Crystal Analysis*, p. 276. Oxford: Pergamon Press.

Oswald, I. D. H., Allan, D. R., Motherwell, W. D. S. & Parsons, S. (2005). *Structures of the monofluoro- and monochlorophenols at low temperature and high pressure. Acta Cryst.* **B61**, 69–79.

Oszlányi, G. & Sütő, A. (2004). *Ab initio structure solution by charge flipping. Acta Cryst.* A**60**, 134–141.

Palatinus, L. & Chapuis, G. (2007). *SUPERFLIP – a computer program for the solution of crystal structures by charge flipping in arbitrary dimensions. J. Appl. Cryst.* **40**, 786–790.

Palatinus, L. & van der Lee, A. (2008). *Symmetry determination following structure solution in P1. J. Appl. Cryst.* **41**, 975–984.

Poon, B. K., Grosse-Kunstleve, R. W., Zwart, P. H. & Sauter, N. K. (2010). *Detection and correction of underassigned rotational symmetry prior to structure deposition. Acta Cryst.* **D66**, 503–513.

Rabinovich, S., Shmueli, U., Stein, Z., Shashua, R. & Weiss, G. H. (1991). *Exact random-walk models in crystallographic statistics. VI. P.d.f.'s of E for all plane groups and most space groups. Acta Cryst.* A**47**, 328–335.

Rogers, D. & Wilson, A. J. C. (1953). *The probability distribution of X-ray intensities. V. A note on some hypersymmetric distributions. Acta Cryst.* **6**, 439–449.

Rossmann, M. G. & Arnold, E. (2001). *Patterson and molecular replacement techniques.* In *International Tables for Crystallography*, Volume B, *Reciprocal Space*, edited by U. Shmueli, ch. 2.3, pp. 235–263. Dordrecht: Kluwer Academic Publishers.

Saitoh, K., Tsuda, K., Terauchi, M. & Tanaka, M. (2001). *Distinction between space groups having principal rotation and screw axes, which are combined with twofold rotation axes, using the coherent convergent-beam electron diffraction method. Acta Cryst.* A**57**, 219–230.

Sauter, N. K., Grosse-Kunstleve, R. W. & Adams, P. D. (2004). *Robust indexing for automatic data collection. J. Appl. Cryst.* **37**, 399–409.

Sauter, N. K., Grosse-Kunstleve, R. W. & Adams, P. D. (2006). *Improved statistics for determining the Patterson symmetry from unmerged diffraction intensities. J. Appl. Cryst.* **39**, 158–168.

Sheldrick, G. M. (2008). *A short history of SHELX. Acta Cryst.* A**64**, 112–122.

Shmueli, U. (1984). *Space-group algorithms. I. The space group and its symmetry elements. Acta Cryst.* A**40**, 559–567.

Shmueli, U. (2007). *Theories and Techniques of Crystal Structure Determination.* Oxford University Press.

Shmueli, U. (2008). *Symmetry in reciprocal space.* In *International Tables for Crystallography*, Volume B, *Reciprocal Space*, 3rd. ed., edited by U. Shmueli, ch. 1.4, Appendix A1.4.4. Dordrecht: Springer.

Shmueli, U. (2012). *Structure-factor statistics and crystal symmetry. J. Appl. Cryst.* **45**, 389–392.

Shmueli, U. (2013). *INSTAT: a program for computing non-ideal probability density functions of |E|. J. Appl. Cryst.* **46**, 1521–1522.

Shmueli, U. & Weiss, G. H. (1995). *Introduction to Crystallographic Statistics*. Oxford University Press.

Shmueli, U., Weiss, G. H., Kiefer, J. E. & Wilson, A. J. C. (1984). *Exact random-walk models in crystallographic statistics. I. Space groups $P\bar{1}$ and P1. Acta Cryst.* A**40**, 651–660.

Shmueli, U. & Wilson, A. J. C. (2008). *Statistical properties of the weighted reciprocal lattice*. In *International Tables for Crystallography*, Volume B, *Reciprocal Space*, 3rd ed., edited by U. Shmueli, ch. 2.1. Dordrecht: Springer.

Spek, A. L. (2003). *Single-crystal structure validation with the program PLATON. J. Appl. Cryst.* **36**, 7–13.

Spence, J. C. H. & Lynch, J. (1982). *Stem microanalysis by transmission electron energy loss spectroscopy in crystals. Ultramicroscopy*, **9**, 267–276.

Spence, J. C. H., Weierstall, U. & Chapman, H. (2012). *X-ray lasers for structural biology. Rep. Prog. Phys.* **75**, 102601.

Spence, J. & Zuo, J. M. (1992). *Electron Microdiffraction*. New York: Plenum.

Staniland, S. S., Harrison, A., Robertson, N., Kamenev, K. V. & Parsons, S. (2006). *Structural and magnetic properties of $[BDTA]_2[MCl_4]$ [M = Cu(1), Co(2), Mn(3)] revealing an S = 1/2 square lattice antiferromagnet with weak magnetic exchange. Inorg. Chem.* **45**, 5767–5773.

Steeds, J. W. & Vincent, R. (1983). *Use of high-symmetry zone axes in electron diffraction in determining crystal point and space groups. J. Appl. Cryst.* **16**, 317–325.

Tadbuppa, P. P. & Tiekink, E. R. T. (2010). *[(Z)-Ethyl N-isopropylthiocarbamato-κS](tricyclohexylphosphine-κP)gold(I). Acta Cryst.* E**66**, m615.

Tanaka, M. (2008). *Point-group and space-group determination by convergent-beam electron diffraction*. In *International Tables for Crystallography*, Volume B, *Reciprocal Space*, 3rd ed., edited by U. Shmueli, Section 2.5.3. Springer.

Tanaka, M., Terauchi, M. & Kaneyama, T. (1988). *Convergent Beam Electron Diffraction* (and subsequent volumes in the same series). Tokyo: JEOL Ltd.

Wang, Z. L., Kong, X. Y. & Zuo, J. M. (2003). *Induced growth of asymmetric nanocantilever on polar surfaces. Phys. Rev. Lett.* **91**, 185502.

Waser, J. (1955). *Symmetry relations between structure factors. Acta Cryst.* **8**, 595.

Wells, M. (1965). *Computational aspects of space-group symmetry. Acta Cryst.* **19**, 173–179.

Williams, D. & Carter, C. B. (2009). *Transmission Electron Microscopy*, ch. 6. New York: Springer.

Wilson, A. J. C. (1949). *The probability distribution of X-ray intensities. Acta Cryst.* **2**, 318–321.

Zhang, P., Kisielowski, C., Istratov, A., He, H., Nelson, C., Mardinly, J., Weber, E. & Spence, J. C. H. (2006). *Direct strain measurement in a 65 nm node strained silicon transistor by convergent-beam electron diffraction. Appl. Phys. Lett.* **89**, 161907.

Zuo, J. M. (1993). *New method of Bravais lattice determination. Ultramicroscopy*, **52**, 459–464.

Zuo, J. M. (2004). *Measurements of electron densities in solids: a real-space view of electronic structure and bonding in inorganic crystals. Rep. Prog. Phys.* **67**, 2053–2129.

Zuo, J. M., Kim, M., O'Keeffe, M. & Spence, J. C. H. (1999). *Observation of d holes and Cu-Cu bonding in cuprite. Nature (London)*, **401**, 49–52.

Zuo, J. M., Spence, J. C. H., Downs, J. & Mayer, J. (1993). *Measurement of individual structure-factor phases with tenth-degree accuracy: the 00.2 reflection in BeO studied by electron and X-ray diffraction. Acta Cryst.* A**49**, 422–429.

Zwart, P. H., Grosse-Kunstleve, R. W., Lebedev, A. A., Murshudov, G. N. & Adams, P. D. (2008). *Surprises and pitfalls from (pseudo)symmetry. Acta Cryst.* D**64**, 99–107.

1.7. Applications of crystallographic symmetry: space-group symmetry relations, subperiodic groups and magnetic symmetry

HANS WONDRATSCHEK, ULRICH MÜLLER, DANIEL B. LITVIN, VOJTECH KOPSKÝ AND CAROLYN PRATT BROCK

1.7.1. Subgroups and supergroups of space groups

BY HANS WONDRATSCHEK

Relations between crystal structures play an important role for the comparison and classification of crystal structures, the analysis of phase transitions in the solid state, the understanding of topotactic reactions, and other applications. The relations can often be expressed by group–subgroup relations between the corresponding space groups. Such relations may be recognized from relations between the lattices and between the point groups[1] of the crystal structures.

In the first five editions of Volume A of *International Tables for Crystallography*, subgroups and those supergroups of space groups that are space groups were listed for every space group. However, the listing was incomplete and it lacked additional information, such as, for example, possible unit-cell transformations and/or origin shifts involved. It became apparent that complete lists and more detailed data were necessary. Therefore, a supplementary volume of *International Tables for Crystallography* to Volume A was published: Volume A1, *Symmetry Relations between Space Groups* (2004; second edition 2010; abbreviated as *IT* A1). The listing of the subgroups and supergroups was thus discontinued in the sixth edition of Volume A (2016) (abbreviated as *IT* A).

Volume A1 consists of three parts. Part 1 covers the theory of space groups and their subgroups, space-group relations between crystal structures and the corresponding Wyckoff positions, and the Bilbao Crystallographic Server (http://www.cryst.ehu.es/). This server is freely accessible and offers access to computer programs that display the subgroups and supergroups of the space groups and other relevant data. Part 2 of Volume A1 contains complete lists of the maximal subgroups of the plane groups and space groups, including unit-cell transformations and origin shifts, if applicable. An overview of the group–subgroup relations is also displayed in diagrams. Part 3 contains tables of relations between the Wyckoff positions of group–subgroup-related space groups and a guide to their use. Chapter 2.2 of this Teaching Edition includes a guide to and several illustrative examples of the symmetry-relations tables of Volume A1.

Example

The crystal structures of silicon, Si, and sphalerite, ZnS, belong to space-group types $Fd\bar{3}m$ (O_h^7; No. 227) and $F\bar{4}3m$ (T_d^2; No. 216) with lattice parameters $a_{\rm Si}$ = 5.43 Å and $a_{\rm ZnS}$ = 5.41 Å. The structure of sphalerite (zinc blende) is obtained from that of silicon by replacing alternately half of the Si atoms by Zn and half by S, and by adjusting the lattice parameter (for details of the procedure, see Fig. 1.7.2.1, where the crystal-structure relation between diamond and zinc blende is illustrated). The strong connection between the two crystal structures is reflected in the relation between their space groups: the point group (crystal class) and the space group of sphalerite is a subgroup (of index 2) of that of silicon (ignoring the small difference in lattice parameters).

Data on subgroups and supergroups of the space groups are useful for the discussion of structural relations and phase transitions. It must be kept in mind, however, that group–subgroup relations only constitute symmetry relations. It is important, therefore, to ascertain that the consequential relations between the lattice parameters and between the atomic coordinates of the particles of the crystal structures also hold before a structural relation can be deduced from a symmetry relation.

Examples

NaCl and CaF_2 belong to the same space-group type, $Fm\bar{3}m$ (O_h^5; No. 225), and have lattice parameters $a_{\rm NaCl}$ = 5.64 Å and $a_{\rm CaF_2}$ = 5.46 Å. The ions, however, occupy unrelated positions and so the symmetry relation does not express a structural relation.

Pyrite, FeS_2, and solid carbon dioxide, CO_2, belong to the same space-group type, $Pa\bar{3}$ (T_h^6; No. 205). They have lattice parameters $a_{\rm FeS_2}$ = 5.42 Å and $a_{\rm CO_2}$ = 5.55 Å, and the particles occupy analogous Wyckoff positions. Nevertheless, the structures of these compounds are not related, because the positional parameters x = 0.386 of S in FeS_2 and x = 0.118 of O in CO_2 differ so much that the coordinations of the corresponding atoms are dissimilar.

To formulate group–subgroup relations some definitions are necessary. Subgroups and their distribution into conjugacy classes, normal subgroups, supergroups, maximal subgroups, minimal supergroups, proper subgroups, proper supergroups and index are defined for groups in general in Chapter 1.1. These definitions are used also for crystallographic groups like space groups. In the present section, the subgroup data of *IT* A1 are explained through many examples in order to enable the reader to use these data.

Examples

Maximal subgroups $\mathcal{H}$ of a space group $P1$ with basis vectors **a**, **b**, **c** are, among others, subgroups $P1$ for which $\mathbf{a}'' = p\mathbf{a}$, $\mathbf{b}'' = \mathbf{b}$, $\mathbf{c}'' = \mathbf{c}$, p prime. If p is not a prime number but a product of two integers $p = q \times r$, the subgroup $\mathcal{H}$ is not maximal because a proper subgroup $\mathcal{Z}$ of index q exists such that $\mathbf{a}' = q\mathbf{a}$, $\mathbf{b}' = \mathbf{b}$, $\mathbf{c}' = \mathbf{c}$. $\mathcal{Z}$ again has $\mathcal{H}$ as a proper subgroup of index r with $\mathcal{G} > \mathcal{Z} > \mathcal{H}$.

$P12_1/c1$ has maximal subgroups $P12_11$, $P1c1$ and $P\bar{1}$ with the same unit cell, whereas $P1$ is not a maximal subgroup of $P12_1/c1$: $P12_1/c1 > P12_11 > P1$; $P12_1/c1 > P1c1 > P1$; $P12_1/c1 > P\bar{1} > P1$. These are all possible chains of maximal subgroups for $P12_1/c1$ if the original translations are retained

[1] The point group determines both the symmetry of the physical properties of the macroscopic crystal and the symmetry of its ideal shape. Each space group belongs to a point group.

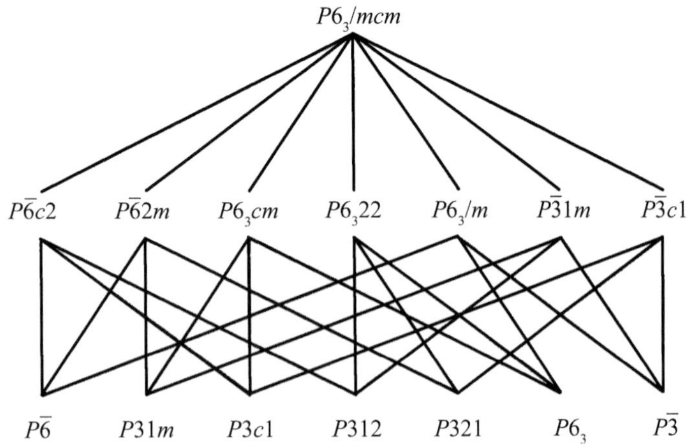

Figure 1.7.1.1
Space group $P6_3/mcm$ with *t*-subgroups of index 2 and 4. All 21 possible subgroup chains are displayed by lines.

completely. Correspondingly, the seven subgroups of index 4 with the same translations as the original space group $P6_3/mcm$ are obtained *via* the 21 different chains of Fig. 1.7.1.1.

While all group–subgroup relations considered here are relations between individual space groups, they are valid for all space groups of a space-group type, as the following example shows.

Example
A particular space group $P121$ has a subgroup $P1$ which is obtained from $P121$ by retaining all translations but eliminating all rotations and combinations of rotations with translations. For every space group of space-group type $P121$ such a subgroup $P1$ exists.

From this example it follows that the relationship exists, in an extended sense, for the two space-group types involved. One can, therefore, list these relationships by means of the symbols of the space-group types.

A three-dimensional space group may have subgroups with no translations (*i.e.* site-symmetry groups; *cf.* Section 1.4.4), or with one- or two-dimensional lattices of translations (*i.e.* line groups, frieze groups, rod groups, plane groups and layer groups), *cf.* Volume E of *International Tables for Crystallography* (2010) (abbreviated as *IT* E), or with a three-dimensional lattice of translations (space groups).

The number of subgroups of a space group is always infinite. Not only the number of all subgroups but even the number of all maximal subgroups of a given space group is infinite.

In this section, only those subgroups of a space group that are also space groups will be considered. All *maximal* subgroups of space groups are themselves space groups. To simplify the discussion, let us suppose that we know all maximal subgroups of a space group $\mathcal{G}$. In this case, *any* subgroup $\mathcal{H}$ of $\mathcal{G}$ may be obtained *via* a chain of maximal subgroups $\mathcal{H}_1, \mathcal{H}_2, \ldots, \mathcal{H}_{r-1}, \mathcal{H}_r$ such that $\mathcal{G}\ (= \mathcal{H}_0) > \mathcal{H}_1 > \mathcal{H}_2 > \ldots > \mathcal{H}_{r-1} > \mathcal{H}_r\ (= \mathcal{H})$, where $\mathcal{H}_j$ is a maximal subgroup of $\mathcal{H}_{j-1}$ of index $[i_j]$, with $j = 1, \ldots, r$. There may be many such chains between $\mathcal{G}$ and $\mathcal{H}$. On the other hand, all subgroups of $\mathcal{G}$ of a given index $[i]$ are obtained if all chains are constructed for which $[i_1] \times [i_2] \times \ldots \times [i_r] = [i]$ holds.

The index $[i]$ of a subgroup has a geometric significance. It determines the 'dilution' of symmetry operations of $\mathcal{H}$ compared with those of $\mathcal{G}$. The number of symmetry operations of $\mathcal{H}$ is $1/i$ times the number of symmetry operations of $\mathcal{G}$; since space groups are infinite groups, this is to be understood in the same way as 'the number of even numbers is one half of the number of all integer numbers'.

The infinite number of subgroups only occurs for a certain kind of subgroup and can be reduced as described below. It is thus useful to consider the different kinds of subgroups of a space group in the way introduced by Hermann (1929):

(1) By reducing the order of the point group, *i.e.* by eliminating all symmetry operations of some kind. The example $P12_11 \longrightarrow P1$ mentioned above is of this type;
(2) By loss of translations, *i.e.* by 'thinning out' the lattice of translations. For the space group $P121$ mentioned above this may happen in different ways:
   (*a*) by suppressing all translations of the kind $(2u+1)\mathbf{a} + v\mathbf{b} + w\mathbf{c}$, where u, v and w are integers. The new basis is normally written $\mathbf{a}' = 2\mathbf{a}$, $\mathbf{b}' = \mathbf{b}$, $\mathbf{c}' = \mathbf{c}$ and, hence, half of the twofold axes have been eliminated; or
   (*b*) by $\mathbf{a}' = \mathbf{a}$, $\mathbf{b}' = 2\mathbf{b}$, $\mathbf{c}' = \mathbf{c}$, *i.e.* by thinning out the translations parallel to the twofold axes; or
   (*c*) again by $\mathbf{b}' = 2\mathbf{b}$ but replacing the twofold rotation axes by twofold screw axes.
(3) By combination of (1) and (2), *e.g.* by reducing the order of the point group and by thinning out the lattice of translations.

Subgroups of the first kind, (1), are called *translationengleiche* (or *t*-) subgroups because the set $\mathcal{T}$ of all (pure) translations is retained. In case (2), the point group $\mathcal{P}$ and thus the crystal class of the space group is unchanged. These subgroups are called *klassengleiche* or *k*-subgroups. In the general case (3), both the translation subgroup $\mathcal{T}$ of $\mathcal{G}$ and the point group $\mathcal{P}$ are reduced; the subgroup has lost translations *and* belongs to a crystal class of lower order: these are *general* subgroups.

Obviously, the general subgroups are more difficult to survey than kinds (1) and (2). Fortunately, a theorem of Hermann (1929) states that if $\mathcal{H}$ is a proper subgroup of $\mathcal{G}$, then there always exists an intermediate group $\mathcal{M}$ such that $\mathcal{G} > \mathcal{M} > \mathcal{H}$, where $\mathcal{M}$ is a *t*-subgroup of $\mathcal{G}$ and $\mathcal{H}$ is a *k*-subgroup of $\mathcal{M}$. If $\mathcal{H} < \mathcal{G}$ is maximal, then either $\mathcal{M} = \mathcal{G}$ and $\mathcal{H}$ is a *k*-subgroup of $\mathcal{G}$ or $\mathcal{M} = \mathcal{H}$ and $\mathcal{H}$ is a *t*-subgroup of $\mathcal{G}$. It follows that a maximal subgroup of a space group $\mathcal{G}$ is either a *t*-subgroup or a *k*-subgroup of $\mathcal{G}$. According to this theorem, general subgroups can never occur among the maximal subgroups. They can, however, be derived by a stepwise process of linking maximal *t*-subgroups and maximal *k*-subgroups by the chains discussed above.

1.7.1.1. *Translationengleiche* (or *t*-) subgroups of space groups

The 'point group' $\mathcal{P}$ of a given space group $\mathcal{G}$ is a finite group, *cf.* Section 1.3.3. Hence, the number of subgroups and consequently the number of maximal subgroups of $\mathcal{P}$ is finite. There exist, therefore, only a finite number of maximal *t*-subgroups of $\mathcal{G}$. The possible *t*-subgroups were first listed in *Internationale Tabellen zur Bestimmung von Kristallstrukturen*, Band 1 (1935); corrections have been reported by Ascher *et al.* (1969). All maximal *t*-subgroups are listed individually for each space group $\mathcal{G}$ in *IT* A1 with the index, the (unconventional) Hermann–Mauguin symbol referred to the coordinate system of $\mathcal{G}$, the space-group number and conventional Hermann–Mauguin symbol, their general position and the transformation to the conventional coordinate system of $\mathcal{H}$. This may involve a change of basis and an origin shift from the coordinate system of $\mathcal{G}$.

1.7.1.2. *Klassengleiche* (or *k*-) subgroups of space groups

Every space group $\mathcal{G}$ has an infinite number of maximal *k*-subgroups. For dimensions 1, 2 and 3, however, it can be shown that the number of maximal *k*-subgroups is finite if subgroups belonging to the same affine space-group type as $\mathcal{G}$ are excluded. The number of maximal subgroups of $\mathcal{G}$ belonging to the same affine space-group type as $\mathcal{G}$ is always infinite; these subgroups are called maximal *isomorphic* subgroups. Maximal *non-isomorphic klassengleiche* subgroups of plane groups and space groups always have index 2, 3 or 4. They are listed individually in *IT* A1 together with the isomorphic subgroups of the same index. For practical reasons, the *k*-subgroups are distributed into two lists headed 'Loss of centring translations' and 'Enlarged (conventional) unit cell'. The data consist of the index of the subgroup $\mathcal{H}$, the lattice relation between the lattices of $\mathcal{H}$ and $\mathcal{G}$, the characterization of the space group $\mathcal{H}$, the general position or a set of generators of $\mathcal{H}$ and the transformation from the coordinate system of $\mathcal{G}$ to that of $\mathcal{H}$.

1.7.1.3. Isomorphic subgroups of space groups

The existence of isomorphic subgroups is of special interest. There can be no proper isomorphic subgroups $\mathcal{H} < \mathcal{G}$ of finite groups $\mathcal{G}$ because the difference of the orders $|\mathcal{H}| < |\mathcal{G}|$ does not allow isomorphism. The point group $\mathcal{P}$ of a space group $\mathcal{G}$ is finite and its order cannot be reduced if $\mathcal{H}$ is to be isomorphic to $\mathcal{G}$. Therefore, isomorphic subgroups are necessarily *k*-subgroups.

The number of isomorphic maximal subgroups and thus the number of all isomorphic subgroups of any space group is infinite. It can be shown that maximal subgroups of space groups of index $i > 4$ are necessarily isomorphic. Depending on the crystallographic equivalence of the coordinate axes, the index of the subgroup is p, p^2 or p^3, where p is a prime. The isomorphic subgroups cannot be listed individually because of their number, but they can be listed as members of a few series. The series are mostly determined by the index p; the members may be normal subgroups of $\mathcal{G}$ or they form conjugacy classes the size of which is either p, p^2 or p^3. The individual members of a conjugacy class are determined by the locations of their origins. The size of the conjugacy class, a basis for the lattice of the subgroup, the generators of the individual isomorphic subgroups and the coordinate transformation from the coordinate system of $\mathcal{G}$ to that of $\mathcal{H}$ are listed in *IT* A1 for all space-group types.

Examples

Isomorphic subgroups of $P1$: the space group $P1$ is an abelian space group, all of its subgroups are isomorphic and are normal subgroups. The index may be any prime p.

Isomorphic subgroups of $P\bar{1}$: the space group $P\bar{1}$ is not abelian and subgroups exist of types $P1$ and $P\bar{1}$. The latter are isomorphic. Those of index 2 are normal subgroups; for higher index $p > 2$ they form conjugacy classes of prime size p.

Enantiomorphic space groups have an infinite number of maximal isomorphic subgroups of the same type and an infinite number of maximal isomorphic subgroups of the enantiomorphic type.

Example

All *k*-subgroups $\mathcal{H}$ of a given space group $\mathcal{G} = P3_1$ with basis vectors $\mathbf{a}' = \mathbf{a}$, $\mathbf{b}' = \mathbf{b}$, $\mathbf{c}' = p\mathbf{c}$, where p is any prime number other than 3, are maximal isomorphic subgroups. They belong to space-group type $P3_1$ if $p = 1 \bmod 3$. They belong to the enantiomorphic space-group type $P3_2$ if $p = 2 \bmod 3$.

In principle there is no difference in importance between *t*-, non-isomorphic *k*- and isomorphic *k*-subgroups. Roughly speaking, a group–subgroup relation is 'strong' if the index $[i]$ of the subgroup is low. All maximal *t*- and maximal non-isomorphic *k*-subgroups have indices less than four in $\mathbb{E}^2$ and less than five in $\mathbb{E}^3$, index four already being rather exceptional. Maximal isomorphic *k*-subgroups of arbitrarily high index exist for every space group.

1.7.1.4. Supergroups

Sometimes a space group $\mathcal{H}$ is known and the possible space groups $\mathcal{G}$, of which $\mathcal{H}$ is a subgroup, are of interest. A space group $\mathcal{R}$ is called a *minimal supergroup* of a space group $\mathcal{G}$ if $\mathcal{G}$ is a maximal subgroup of $\mathcal{R}$.

Examples of minimal supergroups

In Fig. 1.7.1.1, the space group $P6_3/mcm$ is a minimal supergroup of $P\bar{6}c2, \ldots, P\bar{3}c1$; $P\bar{6}c2$ is a minimal supergroup of $P\bar{6}$, $P3c1$ and $P312$; *etc.*

If $\mathcal{G}$ is a maximal *t*-subgroup of $\mathcal{R}$, then $\mathcal{R}$ is a minimal *t*-supergroup of $\mathcal{G}$. If $\mathcal{G}$ is a maximal *k*-subgroup of $\mathcal{R}$, then $\mathcal{R}$ is a minimal *k*-supergroup of $\mathcal{G}$. Finally, if $\mathcal{G}$ is a maximal isomorphic subgroup of $\mathcal{R}$, then $\mathcal{R}$ is a minimal isomorphic supergroup of $\mathcal{G}$. Data for minimal *t*- and minimal non-isomorphic *k*-supergroups are listed in *IT* A1, although in a less explicit way than that in which the subgroups are listed. The data essentially make the detailed subgroup data usable for the search for supergroups of space groups. Data on minimal isomorphic supergroups are not listed because they can be derived from the corresponding subgroup relations.

The search for supergroups $\mathcal{R} > \mathcal{G}$ of a space group $\mathcal{G}$ differs from the search for subgroups $\mathcal{H} < \mathcal{G}$ in one essential point: when looking for subgroups one knows the available group elements, namely the elements $g \in \mathcal{G}$; when looking for supergroups, any isometry $f \in \mathcal{E}$ may be a possible element of $\mathcal{R}$, $f \in \mathcal{R}$, where $\mathcal{E}$ is the Euclidean group of all isometries.

As we are mainly interested in the symmetries of crystal structures, it is reasonable only to look for groups $\mathcal{R}$ that are themselves space groups. In this way the search for supergroups of space groups is a reversal of the search for subgroups. Nevertheless, even then there are new phenomena; only two of these shall be mentioned here.

Example

For a given space group $P\bar{1}$, there is only one *t*-subgroup $P1$. However, for a space group $P1$, there is a continuously infinite number of *t*-supergroups $P\bar{1}$. Referred to the unit cell of $P1$, an additional centre of inversion can be placed in the range $0 \le x < \frac{1}{2}$, $0 \le y < \frac{1}{2}$, $0 \le z < \frac{1}{2}$. The centre in each of these locations leads to a new supergroup resulting in a continuous set of *t*-supergroups.

If $\mathcal{R}$ is a *t*-supergroup of $\mathcal{G}$ belonging to a crystal system with higher symmetry than that of $\mathcal{G}$, then the metric of $\mathcal{G}$ has to fulfil the conditions of the metric of $\mathcal{R}$. For example, if a tetragonal space group $\mathcal{G}$ has a cubic *t*-supergroup $\mathcal{R}$, then the lattice of $\mathcal{G}$ also has to have cubic symmetry.

In practice, small differences in the lattice parameters of $\mathcal{G}$ and $\mathcal{R}$ will occur, because lattice deviations can accompany a structural relationship.

1.7.2. Relations between Wyckoff positions for group–subgroup-related space groups

BY ULRICH MÜLLER

1.7.2.1. Symmetry relations between crystal structures

The crystal structures of two compounds are *isotypic* if their atoms are arranged in the same way and if they have the same or the enantiomorphic space group. The absolute values of the lattice parameters and interatomic distances may differ and small deviations are permitted for non-fixed coordinates of corresponding atoms. The axial ratios and interaxial angles must be similar. Two structures are *homeotypic* if the conditions for isotypism are relaxed because (Lima-de-Faria *et al.*, 1990): (1) their space groups differ, allowing for a group–subgroup relation; (2) the geometric conditions differ (axial ratios, interaxial angles, atomic coordinates); or (3) an atomic position in one structure is occupied in an ordered way by various atomic species in the other structure (substitution derivatives or after a misorder–order phase transition).[2]

Group–subgroup relations between the space groups of homeotypic crystal structures are particularly suited to disclosing the relationship. A standardized procedure to set forth such relations was developed by Bärnighausen (1980). The concept is to start from a simple, highly symmetrical crystal structure and to derive more complicated structures by distortions and/or substitutions of atoms. A tree of group–subgroup relations between the space groups involved, now called a *Bärnighausen tree*, serves as the main guideline. The highly symmetrical starting structure is called the *aristotype* after Megaw (1973) or *basic structure* after Buerger (1947, 1951) or, in the literature on phase transitions in physics, *prototype* or *parent structure*. The derived structures are the *hettotypes* or *derivative structures* or, in phase-transition physics, *distorted structures* or *daughter phases*. In Megaw's terminology, the structures mentioned in the tree form a *family of structures*.

Detailed instructions on how to form a Bärnighausen tree, the information that can be drawn from it and some possible pitfalls are given in Chapter 1.6 of the second edition of *IT* A1 (2010) and in the book by Müller (2013). In any case, setting up group–subgroup relations requires a thorough monitoring of how the Wyckoff positions develop from a group to a subgroup for every position occupied. The following examples give a concise impression of such relations.

1.7.2.2. Substitution derivatives

As an example, Fig. 1.7.2.1 shows the simple relation between diamond and zinc blende. This is an example of a substitution

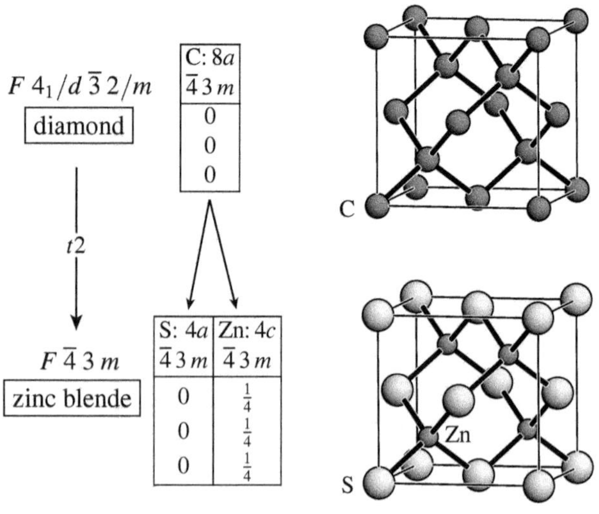

Figure 1.7.2.1
Group–subgroup relation from the aristotype diamond to its hettotype zinc blende. The numerical values in the boxes are atomic coordinates.

derivative. The reduction of the space-group symmetry from diamond to zinc blende is depicted by an arrow which points from the higher-symmetry space group of diamond to the lower-symmetry space group of zinc blende. The subgroup is *translationengleiche* of index 2, marked by *t*2 in the middle of the arrow. *Translationengleiche* means that the subgroup has the same translational lattice (the same size and dimensions of the primitive unit cell) but its crystal class is of reduced symmetry. The index [*i*] is the factor by which the total number of symmetry operations has been reduced, *i.e.* the subgroup has $1/i$ as many symmetry operations; as mentioned in Section 1.7.1, this is to be understood in the same way as 'the number of even numbers is half as many as the number of all integer numbers'.

The consequences of the symmetry reduction on the positions occupied by the atoms are important. As shown in the boxes next to the space-group symbols in Fig. 1.7.2.1, the carbon atoms in diamond occupy the Wyckoff position 8*a* of the space group $F4_1/d\bar{3}2/m$. Upon transition to zinc blende, this position splits into two independent Wyckoff positions, 4*a* and 4*c*, of the subgroup $F\bar{4}3m$, rendering possible occupation by atoms of the two different species zinc and sulfur. The site symmetry $\bar{4}3m$ remains unchanged for all atoms.

Further substitutions of atoms require additional symmetry reductions. For example, in chalcopyrite, $CuFeS_2$, the zinc atoms of zinc blende have been substituted by copper and iron atoms. This implies a symmetry reduction from $F\bar{4}3m$ to its subgroup $I\bar{4}2d$; this requires one *translationengleiche* and two steps of *klassengleiche* group–subgroup relations, including a doubling of the unit cell.

1.7.2.3. Phase transitions

Fig. 1.7.2.2 shows derivatives of the cubic ReO_3 structure type that result from distortions of this high-symmetry structure. WO_3 itself does not adopt this structure, only several distorted variants. The first step of symmetry reduction involves a tetragonal distortion of the cubic ReO_3 structure resulting in the space group *P*4/*mmm*; no example with this symmetry is yet known. The second step leads to a *klassengleiche* subgroup of index 2 (marked *k*2 in the arrow), resulting in the structure of high-temperature WO_3, which is the most symmetrical known modification of WO_3. *Klassengleiche* means that the subgroup belongs

[2] In the strict sense, two isotypic compounds do not have the same space group if their translation lattices (lattice parameters) differ. However, such a strict treatment would render it impossible to apply group-theoretical methods in crystal chemistry and crystal physics. Therefore, we treat isotypic and homeotypic structures as if their translation lattices were the same or related by an integral enlargement factor. For more details see the second edition of *IT* A1 (2010), Sections 1.2.7 and 1.6.4.1. We prefer the term 'misorder' instead of the usual 'disorder' because there still is order in the 'disordered' structure, although it is a reduced order.

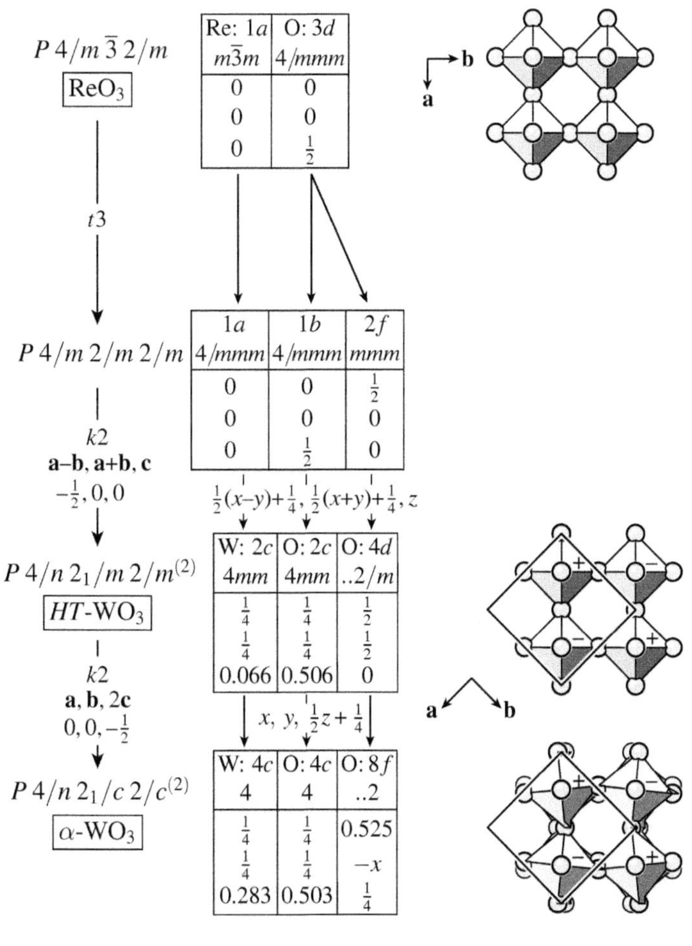

Figure 1.7.2.2

Group–subgroup relations (Bärnighausen tree) from the ReO_3 type to two polymorphic forms of WO_3. The superscript (2) after the space-group symbols states the origin choice. + and − in the images of high-temperature WO_3 and α-WO_3 indicate the direction of the z shifts of the W atoms from the octahedron centres. Structural data for WO_3 are taken from Locherer *et al.* (1999).

to the same crystal class, but it has lost translational symmetry (its primitive unit cell has been enlarged). In this case this is a doubling of the size of the unit cell (**a** − **b**, **a** + **b**, **c**) combined with an origin shift of $-\frac{1}{2}, 0, 0$ (in the coordinate system of $P4/mmm$). This cell transformation and origin shift cause a change of the atomic coordinates of the metal atom from 0, 0, 0 to $\frac{1}{4}, \frac{1}{4}, \sim0.0$ (the decimal value indicates that the coordinate is not fixed by symmetry). Simultaneously, the site symmetry of the metal atom is reduced from $4/mmm$ to $4mm$ and the z coordinate becomes independent. In fact, the W atom is shifted from $z = 0$ to $z = 0.066$, *i.e.* it is not situated in the centre of the octahedron of the surrounding O atoms. This shift is the cause of the symmetry reduction. There is no splitting of the Wyckoff positions in this step of symmetry reduction, but a decrease of the site symmetries of all atoms.

When cooled, at 1170 K HT-WO_3 is transformed to α-WO_3. This involves mutual rotations of the coordination octahedra along **c** and requires another step of symmetry reduction. Again, the Wyckoff positions do not split in this step of symmetry reduction, but the site symmetries of all atoms are further decreased.

Upon further cooling, WO_3 undergoes several other phase transitions that involve additional distortions and, in each case, an additional symmetry reduction to another subgroup (not shown in Fig. 1.7.2.2). For more details see Müller (2013), Section 11.6, and references therein.

1.7.2.4. Domain structures

In the case of phase transitions and of topotactic reactions[3] that involve a symmetry reduction, the kind of group–subgroup relation determines how many kinds of domains and what domain states can be formed. If the lower-symmetry product results from a *translationengleiche* group–subgroup relation, twinned crystals are to be expected. A *klassengleiche* group–subgroup relation will cause antiphase domains. The number of different kinds of twin or antiphase domains corresponds to the index of the symmetry reduction. For example, the phase transition from HT-WO_3 to α-WO_3 involves a *klassengleiche* group–subgroup relation of index 2 ($k2$ in Fig. 1.7.2.2); no twins will be formed, but two kinds of antiphase domains can be expected.

1.7.2.5. Presentation of the relations between the Wyckoff positions among group–subgroup-related space groups

Group–subgroup relations as outlined in the preceding sections can only be correct if all atomic positions of the hettotypes result directly from those of the aristotype. A group–subgroup tree that contains only space-group symbols and basis transformations is absolutely insufficient. If the atomic coordinates are not included in the tree as in Figs. 1.7.2.1 and 1.7.2.2, they must be supplied in a separate table. They should be presented in such a way that their relations become become clearly visible; all atoms of the asymmetric units should exhibit strict correspondence. If changes of basis vectors and/or origin shifts are necessary, the corresponding coordinate transformations should be stated (in Figs. 1.7.2.1 and 1.7.2.2 they are placed between the boxes with the atomic coordinates).

Every group–subgroup relation between space groups entails specific relations between their Wyckoff positions. The laws governing these relations are considered in Chapter 1.5 of the second edition of *IT* A1 (2010). In short, they are:

(1) Between the points of an orbit of a space group and the corresponding points in a subgroup there exists a one-to-one relation.
(2) Between the Wyckoff positions of a space group and those of its subgroups there exist unique relations. Depending on the settings of the space group and the subgroup and the relative positions of their origins, for a given Wyckoff position of the space group, the corresponding Wyckoff labels of the subgroup may differ; but they always belong to the same Wyckoff set, *i.e.* they have the same site symmetries.
(3) If the index of the symmetry reduction from a group to a subgroup is 2, a Wyckoff position either splits into two symmetry-independent positions that keep the site symmetry, or its site symmetry is reduced. If the index is 3 or higher, a Wyckoff position either splits, or its site symmetry is reduced, or both happen.
(4) Coordinates fixed by symmetry may become independent.
(5) The multiplicity of Wyckoff positions of the space group shows up in the sum of the multiplicities of the corresponding positions of the subgroup. If the unit cell selected to describe the subgroup does not change in size, then the sum of the multiplicities of the positions of the subgroup must be equal to the multiplicity of the position of the initial group. For example, from a position with a multiplicity of 6, a position with multiplicity of 6 can result, or it can split into two positions of multiplicity of 3, or into two with multiplicities of

[3] A topotactic reaction is a chemical reaction in the solid state where the orientation of the product crystal is determined by the orientation of the educt crystal.

2 and 4, or into three with multiplicity of 2 *etc.* If the unit cell of the subgroup is enlarged or reduced by a factor *f*, then the sum of the multiplicities must also be multiplied or divided by this factor *f*.

Part 3 of *IT* A1, *Relations between the Wyckoff positions*, contains tables for all space groups. For every one of them, all maximal subgroups are listed, including the corresponding coordinate transformations. For all Wyckoff positions of a space group the relations to the Wyckoff positions of the subgroups are given. This includes the infinitely many maximal isomorphic subgroups, for which general formulae are given. Isomorphic subgroups are a special kind of *klassengleiche* subgroup that belong to the same or the enantiomorphic space-group type, *i.e.* group and subgroup have the same or the enantiomorphic space-group symbol; the unit cell of the subgroup is increased by some integral factor, which is p, p^2 or p^3 (p = prime number) in the case of maximal isomorphic subgroups.

1.7.3. Subperiodic groups

1.7.3.1. Relationships between space groups and subperiodic groups

By Daniel B. Litvin and Vojtech Kopský

IT A, *Space-Group Symmetry*, treats one-, two- and three-dimensional space groups. *IT* E, *Subperiodic Groups* (2010), treats two- and three-dimensional subperiodic groups: frieze groups (groups in two-dimensional space with translations in a one-dimensional subspace), rod groups (groups in three-dimensional space with translations in a one-dimensional subspace) and layer groups (groups in three-dimensional space with translations in a two-dimensional subspace). In the same way in which three-dimensional space groups are used to classify the atomic structure of three-dimensional crystals, the subperiodic groups are used to classify the atomic structure of other crystalline structures, such as liquid crystals, domain interfaces, twins and thin films. A brief description of the content and arrangement of the subperiodic-group tables of *IT* E, and several illustrative examples are given in Chapter 2.3.

In *IT* A, the relationship between the space group of a crystal and the point-group symmetry of individual points in the crystal is given by site symmetries, the point-group subgroups of the space group that leave the points invariant. In *IT* E, an analogous relationship is given between the space group of a crystal and the subperiodic-group symmetry of planes that transect the crystal. *IT* E contains *scanning tables* (with supplementary tables in Kopský & Litvin, 2004) from which the layer-group subgroups of the space group (called *sectional layer groups*) that leave the transecting planes invariant can be determined. The first attempts to derive sectional layer groups were made by Wondratschek (1971) and by using software written by Guigas (1971). Davies & Dirl (1993*a*,*b*) developed software for finding subgroups of space groups which was modified to find sectional layer groups. The use and determination of sectional layer groups have also been discussed by Janovec *et al.* (1988), Kopský & Litvin (1989) and Fuksa *et al.* (1993).

In Fig. 1.7.3.1, part of the scanning table for the space group $P\bar{3}m1$ (164) is given. From this one can determine the layer-group subgroups of $P\bar{3}m1$ that are symmetries of planes of orientation $(hkil)$ = (0001). Vectors $\mathbf{a}'$ and $\mathbf{b}'$ are basic vectors of the translational subgroup of the layer-group symmetry of planes of this orientation. The vector **d** defines the *scanning direction* and is used to define the position of the plane within the crystal. The *linear orbit* is the set of all parallel planes obtained by applying all elements of the space group to any one plane. The *sectional layer group* is the layer subgroup of the space group that leaves the plane invariant.

Orientation orbit $(hkil)$	Conventional basis of the scanning group $\mathbf{a}'$	$\mathbf{b}'$	$\mathbf{d}$	Scanning group $\mathcal{H}$	Linear orbit $s\mathbf{d}$	Sectional layer group $\mathcal{L}(s\mathbf{d})$	
(0001)	**a**	**b**	**c**	$P\bar{3}m1$	$0\mathbf{d}$, $\frac{1}{2}\mathbf{d}$	$p\bar{3}m1$	L72
					$[s\mathbf{d}, -s\mathbf{d}]$	$p3m1$	L69

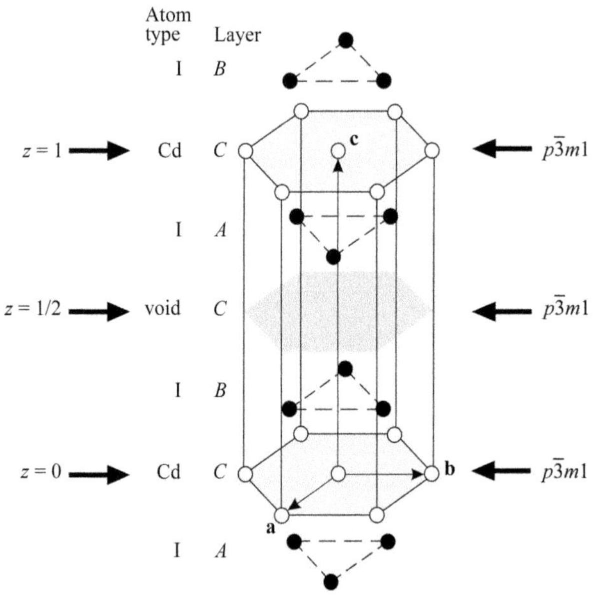

Figure 1.7.3.1
The scanning table for the space-group type $P\bar{3}m1$ (164) and orientation orbit (0001), and the structure of cadmium iodide, CdI_2. Cadmium and iodine ions are denoted by open and filled circles, respectively.

Sectional layer groups were introduced by Holser (1958*a*,*b*) in connection with the consideration of domain walls and twin boundaries as symmetry groups of planes bisecting a crystal. The mutual orientation of the two domains separated by a domain wall or twin boundary is not arbitrary, but has crystallographic restrictions. The group-theoretical basis for an analysis of domain pairs is given by Janovec (1972), and the structure of domain walls and twin boundaries is considered by Janovec (1981) and Zikmund (1984) [see also Janovec & Přívratská (2014)].

Layer symmetries have been used in *bicrystallography*. The term *bicrystal* was introduced by Pond & Bollmann (1979) in the study of grain boundaries [see also Pond & Vlachavas (1983) and Vlachavas (1985)]. A bicrystal is in general an edifice where two crystals, usually of the same structure but of different, possibly arbitrary, orientations, meet at a common boundary. The sectional layer groups describe the symmetries of such a boundary [see *IT* E (2010), Section 5.2.5.2].

An example of the application of the scanning tables to determine the layer-group symmetry of planes in a crystal is given in Section 1.7.3.1.1. In Section 1.7.3.1.2 the derivation of the layer-group symmetry of a domain wall is described.

Whereas a sectional layer group corresponds to the symmetry of a layer transecting a space group, a penetration rod group is the symmetry of a rod traversing a space group. It consist of the symmetry operations of the space group that leave a traversing straight line invariant. By analogy to the sectional layer groups, penetration rod groups can be derived by scanning the space groups. However, in *IT* E (2010) no scanning tables for penetration rod groups are provided.

1.7.3.1.1. Layer symmetries in three-dimensional crystal structures

Fig. 1.7.3.1 shows the crystal structure of cadmium iodide, CdI_2. The space group of this crystal is of type $P\bar{3}m1$ (164). The anions form a hexagonal close packing of spheres and the cations occupy half of the octahedral holes, filling one of the alternate layers. In close-packing notation, the CdI_2 structure is

A	*C*	*B*	*C*
I	Cd	I	void

From the scanning tables, we obtain for planes with the (0001) orientation and at heights $z = 0$ or $z = \frac{1}{2}$ a sectional layer-group symmetry type $p\bar{3}m1$ (layer group No. 72, or L72 for short), and for planes of this orientation at any other height a sectional layer-group symmetry type $p3m1$ (L69).

The $x, y, 0$ plane contains cadmium ions. This plane is a constituent of the orbit of planes of orientation (0001) passing through the points with coordinates $0, 0, u$, where u is an integer. All these planes contain cadmium ions in the same arrangement (*C* layer filled with Cd).

The plane at height $z = \frac{1}{2}$ is a constituent of the orbit of planes of orientation (0001) passing through the points with coordinates $0, 0, u + \frac{1}{2}$. All these planes contain only voids and lie midway

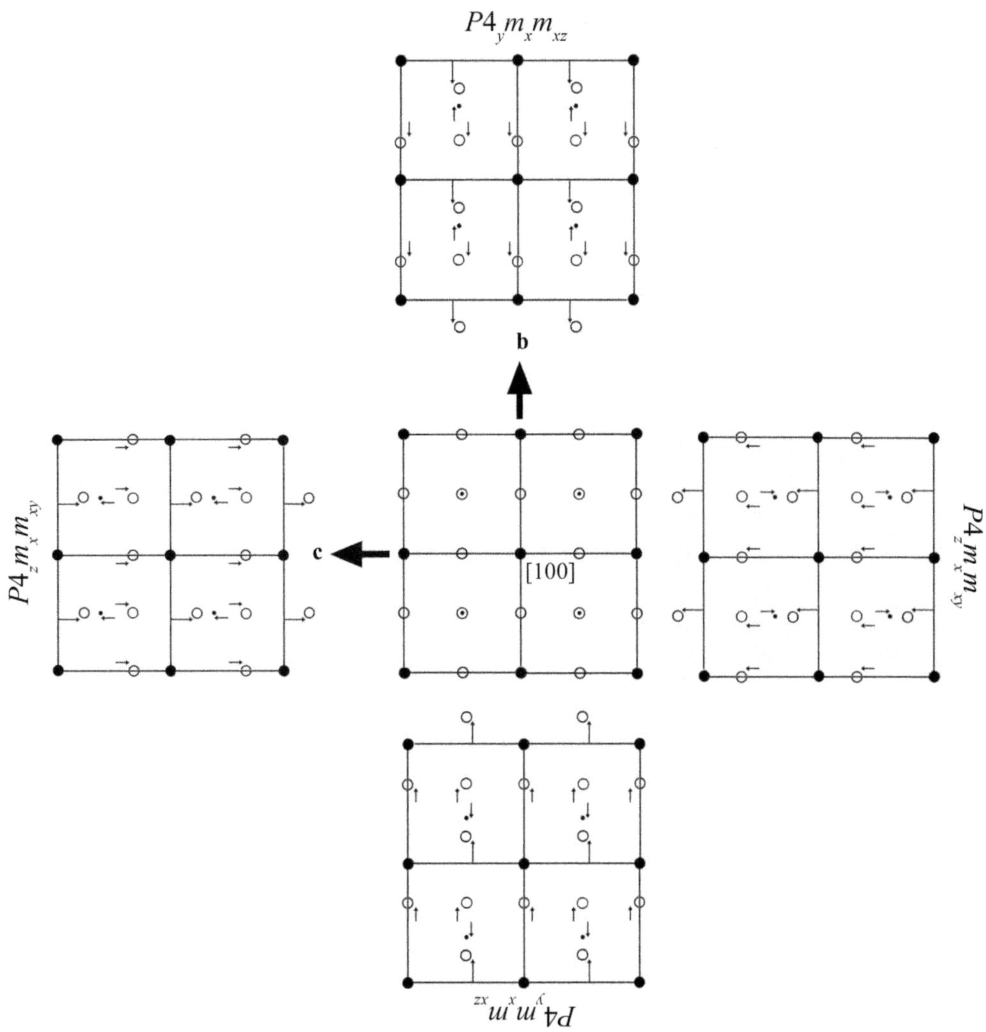

Figure 1.7.3.2

At the centre is the structure of the cubic phase of barium titanate, $BaTiO_3$, of symmetry type $Pm\bar{3}m$, surrounded by the structures of four of the six single-domain states of the tetragonal phase of symmetry type *P4mm*. All the diagrams are projections along the [100] direction. Arrows depict the atomic displacement amplitudes from their cubic-phase positions.

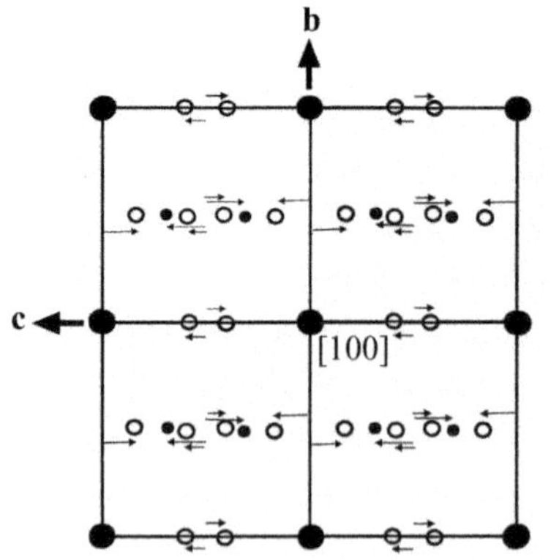

Figure 1.7.3.3
The *domain pair* of symmetry $P4_z/m_zm_xm_{xy}$ consisting of the superposition of those two single-domain states of tetragonal symmetry $P4_zm_xm_{xy}$ shown in Fig. 1.7.3.2. The diagram is a projection along the [100] direction.

between *A* and *B* layers of iodine ions with the *B* layer below and the *A* layer above the plane.

The planes at levels $z = \frac{1}{4}$ and $z = \frac{3}{4}$ contain *B* and *A* layers of iodine ions, respectively. These planes and all planes related to them by translations $t(0, 0, u)$ belong to the same orbit because the operations $\bar{3}$ exchange the *A* and *B* layers.

1.7.3.1.2. The symmetry of domain walls

The cubic phase of barium titanate $BaTiO_3$, of symmetry type $Pm\bar{3}m$, undergoes a phase transition to a tetragonal phase of symmetry type $P4mm$ which can give rise to six distinct single-domain states (Janovec *et al.*, 2004). This is represented in Fig. 1.7.3.2, where at the centre are four unit cells of the cubic phase with barium and titanium atoms represented by large and small filled circles, respectively, and oxygen atoms, which are located at the centre of each unit-cell face, as open circles. A cubic-to-tetragonal phase transition gives rise to atomic displacements represented by arrows, and to six single-domain states, four of which are depicted in the figure. The polar tetragonal symmetry of each of these tetragonal domain states is also shown.

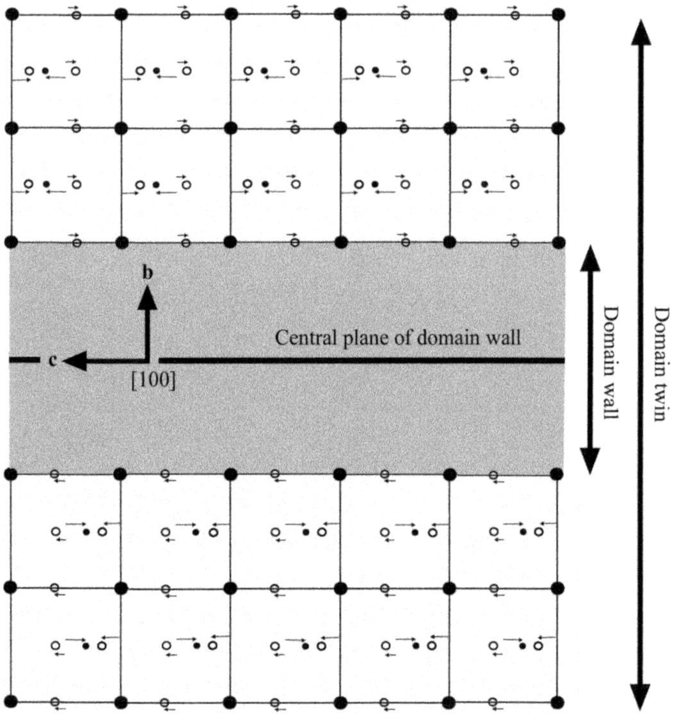

Figure 1.7.3.4
The domain twin consisting of the two domain states of tetragonal symmetry $P4_zm_xm_{xy}$ and a domain wall of orientation (010) passing through the origin of the domain twin.

In determining the symmetry of a domain wall, we first construct a *domain twin* (Janovec & Přívratská, 2014): we choose two single-domain states, for this example the two in Fig. 1.7.3.2 with symmetry $P4_zm_xm_{xy}$, and construct a *domain pair* consisting of the superposition of these two single-domain states, see Fig. 1.7.3.3. The domain twin we choose to construct is obtained by passing a plane of orientation (010) through this domain pair at the origin and deleting from one side of the plane the atoms of one of the single-domain states, and the atoms of the second single-domain state from the other side of the plane, see Fig. 1.7.3.4. The plane is referred to as the *central plane of the domain wall*, and the atoms in and near this plane as the *domain wall*.

The symmetry of the central plane of the domain wall is determined from the symmetry of the domain pair and the scanning tables: The symmetry of the domain pair is the group of operations that either leaves both single-domain states invariant or simultaneously switches the two domain states. $P4_zm_xm_{xy}$ leaves both single-domain states invariant and the symmetry operation of spatial inversion switches the two single-domain states, see Fig. 1.7.3.3. Consequently, the domain-pair symmetry is $P4_z/m_zm_xm_{xy} = P4_zm_xm_{xy} \cup \{\bar{1}|0\}P4_zm_xm_{xy}$. The symmetry of the central plane is determined from the scanning table for the space group $P4_z/m_zm_xm_{xy}$, the orientation orbit (010), the orientation of the domain wall and the linear orbit 0**d**, since the central plane of the wall passes through the origin (*IT* E, 2010). The symmetry of the central plane is the sectional layer group $pm_xm_zm_y$, where *p* denotes the lattice of translations in the *x*, 0, *z* plane.

Let **n** denote a unit vector perpendicular to the central plane of the domain wall; in this example **n** is in the [010] direction. The symmetry of the domain wall consists of:

(1) all elements of the symmetry group of the central plane that leave **n** and both domain states invariant, *i.e.* in this example, all translations of *p*, 1 and m_x; and

(2) all elements of the symmetry group of the central plane that invert **n** and switch the domain states, *i.e.* in this example, 2_x and $\bar{1}$.

The symmetry of the domain wall is then $p2_x/m_x$.

1.7.3.2. Use of subperiodic groups to describe structural units

By Carolyn Pratt Brock

An important use of the subperiodic groups is for describing structural units that are periodic in only one or two dimensions, *i.e.* structural rods or layers in a three-dimensional crystal or in a less ordered aggregate (*e.g.* a liquid crystal, a mono- or bilayer, or a disordered collection of fibrils). Scientists sometimes disagree on how a rod or layer should be identified, but if there is bonding within a subperiodic unit that is clearly stronger than the contacts between the units, then the identification is appropriate.

Graphite [Fig. 1.7.3.5(*a*)] may be the best known example of a layered structure. The overall symmetry of hexagonal graphite (*i.e.* α- or 2*H*-graphite) is $P6_3/mmc$ (space-group No. 194) with C atoms at $z = \pm\frac{1}{4}$ (Trucano & Chen, 1975). The scanning table for space group $P6_3/mmc$ in *IT* E shows that a layer of arbitrary thickness centred at $z = \pm\frac{1}{4}$ has symmetry $p\bar{6}m2$ (L78); the rotation axis is $\bar{6}$ rather than 6 because the atomic positions in adjacent atomic layers must be considered. The idealized layer symmetry of an isolated graphene sheet, however, has higher symmetry [*p*6/*mmm* (L80); Fig. 1.7.3.5(*b*)] because there are no adjacent layers to consider. Since a graphene layer is atomically thin, it is sometimes described using the plane group *p*6*mm* (P17), but the layer group is a better choice because the layer is not infinitesimally thin.

The structure of (chair-like) graphane [fully hydrogenated graphene; all C atoms sp^3; Fig. 1.7.3.5(*c*)] further illustrates the need to use layer groups to describe three-dimensional objects that are periodic in only two dimensions. In $(CH)_n$ there is a threefold axis at each C atom and a $\bar{3}$ axis at the centre of each C_6 ring. The layer group is the centrosymmetric $p\bar{3}m1$ (L72). A possible plane-group label might be *p*3*m*1 (P14), but it does not describe the relationship between the two sides of the sheet. The plane-group label for the [001] projection of a graphane layer, *p*6*mm* (P17), makes the symmetry appear higher than it is because the projection conceals the differences in the directions of the C—H bonds.

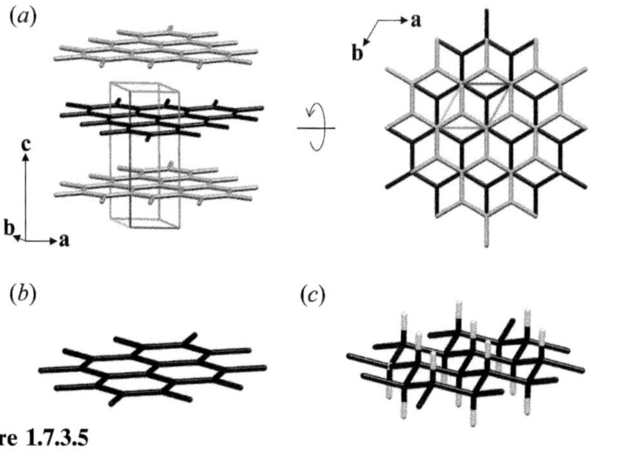

Figure 1.7.3.5
(*a*) Fragments of three layers of α- (or 2*H*-)graphite. Note the 6_3 axis along **c** that results in adjacent layers being offset in the *ab* plane. (*b*) A fragment of graphene (*i.e.* of a single graphite layer). (*c*) A fragment of (chair-like) graphane, $(CH)_n$. Note that the H atoms on adjacent C atoms point in different directions.

1.7.3.2.1. *Differences between layer and plane groups*

Plane groups cannot have twofold axes that lie within the plane because no operation that involves a third dimension is allowed (see Wood, 1964). Plane groups cannot have rotoinversion axes either. A twofold rotation point cannot be distinguished from an inversion point because both operations change *x*, *y* to −*x*, −*y*.[4] There are therefore considerably fewer plane groups (17) than layer groups (80).

A layer group reduces to a plane group in the limit of zero layer thickness. In *IT* E, **c** is the layer normal so that the projection along [001] listed for a layer group is its corresponding plane group. The correspondences between layer and plane groups can also be worked out knowing that

(1) an inversion centre in a layer group becomes a twofold rotation point in a plane group;
(2) an *n*-fold rotation axis perpendicular to the layer becomes an *n*-fold rotation point;
(3) three-, four-, or sixfold rotoinversion axes become, respectively, six-, four- or threefold rotation points;
(4) a twofold axis within the layer becomes a mirror line, while a 2_1 axis becomes a glide line;
(5) a mirror or glide with its normal in the layer is unchanged;
(6) a glide plane in the layer (*i.e.*, with its normal parallel to the layer normal) becomes a translation; and
(7) a mirror plane in the layer no longer has any effect, *i.e.*, it becomes the identity operation.

1.7.3.2.2. *Examples*

A considerable number of the molecular structures archived in the Cambridge Structural Database (Groom *et al.*, 2016) are composed of well separated molecular layers that quite often have higher approximate symmetry than does the structure as a whole (Brock, 2020). An example is the triclinic structure with refcode POVYEG (Preindl *et al.*, 2014; Fig. 1.7.3.6); layers (100) are centred at $x = \pm 0.249$ and are separated by H···H contacts at $x = 0$ and $\frac{1}{2}$. The triclinic structure has eight independent molecules ($Z' = 8$) and two angles ($\beta = 102.4°$ and $\gamma = 104.6°$) that are neither similar nor close to any special value. The axes *b* and *c*, however, are approximately the same length and nearly perpendicular. Individual layers contain two independent hydrogen-bonded tetramers that have approximate fourfold symmetry when viewed in projection but actually have approximate $\bar{4}$ symmetry because molecules lie alternately above and below the middle of the layer. The two independent tetramers are

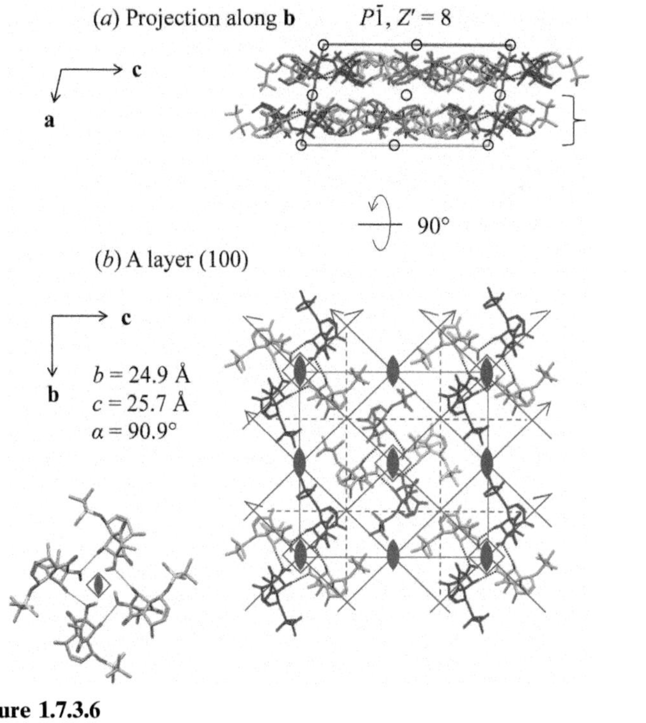

Figure 1.7.3.6
(*a*) The layered structure of the tricyclic compound $C_{20}H_{32}O_5Si$ (CSD refcode POVYEG; Preindl *et al.*, 2014) viewed along **b**. The layers are separated by H···H contacts and are related by crystallographic inversion centres. (*b*) One of the two layers shown in part (*a*) after rotation by 90°. Within each layer there are tetrameric units in which there are —OH···OMe bonds (H atoms not shown). The only crystallographic symmetry in the layer is translational but the two independent tetramers have approximate $\bar{4}$ symmetry and are related by approximate glide planes, twofold axes and 2_1 axes. (Approximate symmetry elements are shown in blue in this and the following two figures.) The lengths of the two axes in the layer differ by only 3% and the axes are nearly perpendicular. The approximate layer symmetry is $p\bar{4}b2$ (L60, labels of the *a* and *c* axes interchanged relative to the standard setting). [Figs. 1.7.3.6 to 1.7.3.8 were drawn using the program *Mercury* (Macrae *et al.*, 2020).]

[4] An inversion operation reverses the signs of all coordinates; it can be represented by a diagonal matrix having the determinant −1 in three-dimensional space and +1 in two-dimensional space. Inversion is therefore an improper operation in three dimensions but a proper operation in two dimensions. The determinant of a mirror operation (reflection), however, is −1 in both three dimensions and two dimensions.

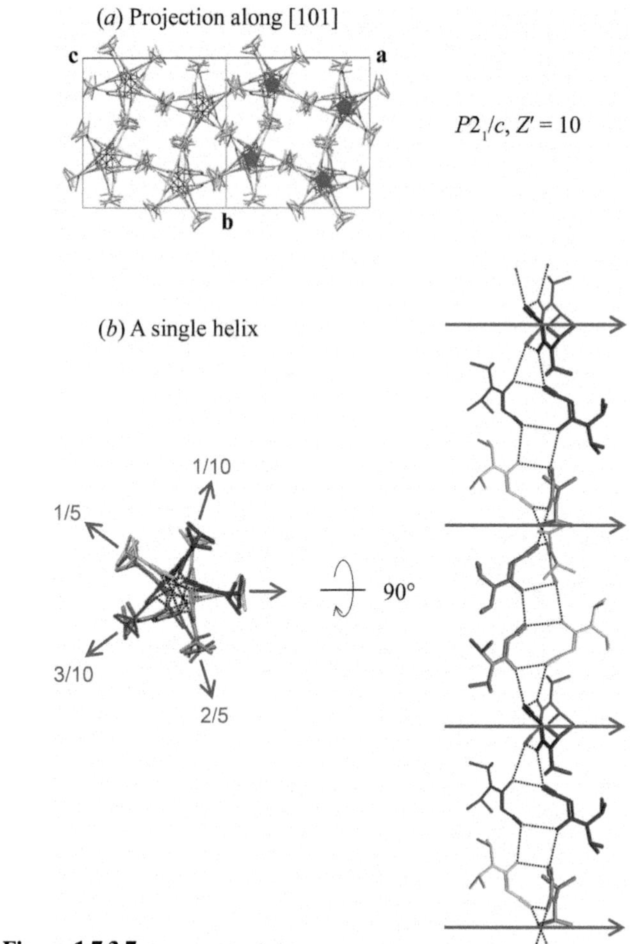

Figure 1.7.3.7
(*a*) The structure projected along [101] of *N*″-cyano-*N*,*N*-diisopropylguanidine (CSD refcode FEXCAN; Hao *et al.*, 2005; $P2_1/c$, $Z' = 10$). The crystal is composed of fivefold helices; the $N(i\text{-Pr})_2$ groups of adjacent helices are interleaved. (*b*) Two views of the ten independent molecules in a helix that are linked by NH···NC bonds (H atoms not shown). Each helix has an approximate 5_1 (or 5_4) axis and approximate perpendicular twofold axes, not all of which are shown. The approximate rod group is $\not{p}5_112$ (or $\not{p}5_412$). Rod groups having fivefold axes do not appear in *IT* E, which only illustrates groups having *n*-fold axes, $n = 2, 3, 4, 6$, but these two $n = 5$ groups are analogous to the groups $\not{p}3_112$ (R47) or $\not{p}3_212$ (R48).

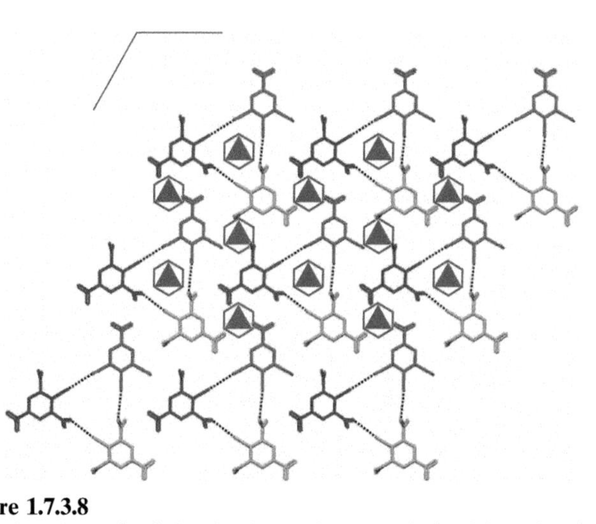

Figure 1.7.3.8
A typical layer (001) in the five polytypes of picryl bromide (CSD refcodes ZZZVXQ0*n*, $n = 2$–6; Parrish *et al.*, 2008). The space groups are $P1$ ($Z' = 18$), $P\bar{1}$ ($Z' = 6$ and 12), $P3_1$ ($Z' = 12$) and $P6_5$ ($Z' = 3$). The number of independent layers is $Z'/3$, so in the set of polytypes there are 17 independent layers, all of which have approximate symmetry $p\bar{6}$ (L74). The layer angles (γ) for the three triclinic structures are 119.9, 60.2 and 119.9°. The molecules appear to be associated into trimers by Br···O interactions, which are marked.

related by approximate glide planes, twofold axes and 2_1 screw axes. The resulting layer group is $p\bar{4}b2$ (L60), but there are clear deviations from that ideal symmetry; variations in the conformations of the $OSiMe_2(t\text{-Bu})$ ring substituent are the most obvious. Those deviations almost certainly arise from interactions between adjacent layers and so might be absent in very small molecular aggregates. In any event, a description using the tetragonal layer group is much more informative than a report of the actual symmetry of an isolated layer, which is only $p1$ (L1).

In the case of a structural rod the approximate symmetry can include, because there is no requirement of periodicity, an *n*-fold rotation parallel to the rod axis that is not allowed crystallographically; in a one-dimensional rod *n* can be any positive integer (for a more detailed discussion of rod groups, *cf.* Section 1.7.3.3). Consider the structure with refcode FEXCAN (Hao *et al.*, 2005; Fig. 1.7.3.7). In this monoclinic crystal ten independent molecules form fivefold hydrogen-bonded helices that run along [101]; adjacent helices are in contact in the region of the dimethylamino groups. Half of the helices have approximate $\not{p}5_112$ symmetry; the other half have approximate $\not{p}5_412$ symmetry. *IT* E does not include these two groups because only rod groups with axes allowed in layer and space groups are listed, but all the important features of $\not{p}5_112$ and $\not{p}5_412$ are illustrated by the groups $\not{p}3_112$ (R47) and $\not{p}3_212$ (R48).[5]

In polytypes a single layer type is stacked in different ways to generate a series of polymorphs. While the best known polytypes are inorganic [*e.g.* SiC (Ortiz *et al.*, 2013) and some silicates (Guinier *et al.*, 1984)], molecular polytypes have also been found (*e.g.* 2,4,6-trinitrobromobenzene, ZZZVXQ0*n*, $n = 2$–6; Parrish *et al.*, 2008). All of the 17 independent layers found in the five ZZZVXQ polytypes are very similar and have approximate layer symmetry $p\bar{6}$ (L74; Fig. 1.7.3.8), although there are deviations from that symmetry, which are most obvious in the rotations of the nitro groups. The symbol $p\bar{6}$ for the approximate layer group is a simple way of summarizing this series of structures.

The examples listed above are all molecular, but subperiodic structural units are perhaps even more common in inorganic structures (see *e.g.* Guinier *et al.*, 1984). Two monolayers of current interest are BN (a graphene analogue) and MoS_2 (which contains a central, planar layer of Mo atoms with trigonal prismatic coordination by S atoms). Both monolayers are described by layer group $p\bar{6}m2$ (L78) with both atoms in BN, and the Mo atoms in MoS_2, on $\bar{6}m2$ (or D_{3h}) sites, and the S atoms of MoS_2 on $3m$ sites (Li, Rao *et al.*, 2013).

1.7.3.3. Applications of rod groups

By Ulrich Müller

When we describe the symmetry of an isolated non-polymeric molecule, we mentally detach it from its surroundings. We even do this when we refer to the molecular structure in a crystal. For example, we say that a molecule of uranocene, $U(\eta^8\text{-}C_8H_8)_2$, has the point symmetry $8/mmm$ in the crystalline state, even though the molecule actually has only the $\bar{1}$ symmetry required by its site

[5] *International Tables for Crystallography* defines a screw operation N_q as a counterclockwise rotation of $2\pi/N$ coupled with a fractional translation along the screw axis of $+q/N$ (see Section 1.2.1). The screw axis itself is then right-handed. All pairs of axes N_q and N_{N-q} (*e.g.* 5_1 and 5_4) are enantiomeric. If the screw axis is compatible with space-group symmetry (*e.g.* 3_1, 4_3, 6_2) then the helix is right-handed if $q < N/2$. Note that the handedness of a chemical structure (*i.e.* of a sequence of covalent bonds) may differ from the handedness of the helix generated by the screw axis itself (Müller, 2017); *cf.* Section 1.7.3.3 and Example 3 in that section.

symmetry in the space group $P2_1/n$ (Zalkin & Raymond, 1969). From the chemical point of view and for spectroscopic studies this inaccuracy can be neglected in most cases. Apart from possible changes in conformation angles, the shape of a molecule in a crystal, as a rule, deviates only marginally from that of the free molecule.

When polymeric chain molecules crystallize, they adopt a symmetric conformation and the idealized symmetry of an individual molecule can be described by a rod group. Hermann–Mauguin symbols for rod groups correspond to those of space groups, but they begin with a script letter $\mathscr{p}$. The primary direction is usually labelled **c**; that is the direction along which the chain runs and that has the translational symmetry of the rod group. In a rod group, the order of a rotation or screw axis running along the primary direction can have any integral value. Rod groups can also be used to describe any linear aggregate.

The actual symmetry of a molecule (or aggregate) in a crystal can be no higher than the symmetry of the crystal and is usually lower. For a non-polymeric molecule this site symmetry is determined by the space group and the location of the molecule in the crystal. In *IT* E, the analogue to the site symmetry of a discrete molecule for a chain is called a *penetration rod group*. A penetration rod group is the set of all symmetry operations of the space group that leave a traversing rod invariant. It depends on the symmetry of the crystal and on the position and direction of the traversing rod.

Just as the site symmetry of a non-polymeric molecule must be a subgroup common to the space group and the point group of the idealized molecule, the penetration rod group of a polymeric chain molecule must be a subgroup common to the space group and the molecular rod group of the idealized chain. However, a higher molecular rod-group symmetry is frequently a good description even within the crystal if the arrangement of surrounding molecules is neglected. Penetration rod groups can only be crystallographic rod groups, *i.e.* rod groups with axes of the orders 1, 2, 3, 4 and 6; these rod groups are listed in *IT* E, together with their subgroups.

Penetration rod groups of the space groups are not listed in any volume of *International Tables for Crystalography*. If needed, they can be derived by comparing the images of the symmetry elements of the space groups in *IT* A with the images of the symmetry elements of the rod groups in *IT* E. The analogue for layers, called sectional layer groups, are listed in the *Scanning tables* of *IT* E (*cf.* Section 1.7.3.1).

Example 1. Polyethylene crystallizes in the space group $P2_1/n2_1/a2_1/m$ (*Pnam*, non-conventional setting of *Pnma*, No. 62) with the zigzag-shaped polymeric chains running along **c**. The molecular rod group is $\mathscr{p}2/m2/c2_1/m$ ($\mathscr{p}mcm$, No. 22); its subgroup $\mathscr{p}112_1/m$ (No. 12) corresponds to the penetration rod group located at $0, 0, z$ of *Pnam*. The CH_2 groups are on the mirror planes at $z = \frac{1}{4}$ and $z = \frac{3}{4}$ (Fig. 1.7.3.9). In the molecular rod group $\mathscr{p}mcm$ the hydrogen atoms are symmetry equivalent; in the penetration rod group $\mathscr{p}112_1/m$ in the crystal they are not; $\mathscr{p}112_1/m$ is a subgroup of $\mathscr{p}mcm$ (Avitabile *et al.*, 1975; Bunn, 1939).

Example 2. An ordered variety of the polymorphic δ-form of poly(vinylidene fluoride), $(CH_2CF_2)_\infty$, crystallizes in the space group $Pc2_1n$ (No. 33; non-conventional setting of $Pna2_1$). The chains run along **c** and have the alternating conformation *anti–gauche*. In this case, the molecular rod group and the penetration rod group are the same, $\mathscr{p}c11$ (Fig. 1.7.3.10; Bachmann *et al.*, 1980; Li, Wondergem *et al.*, 2013).

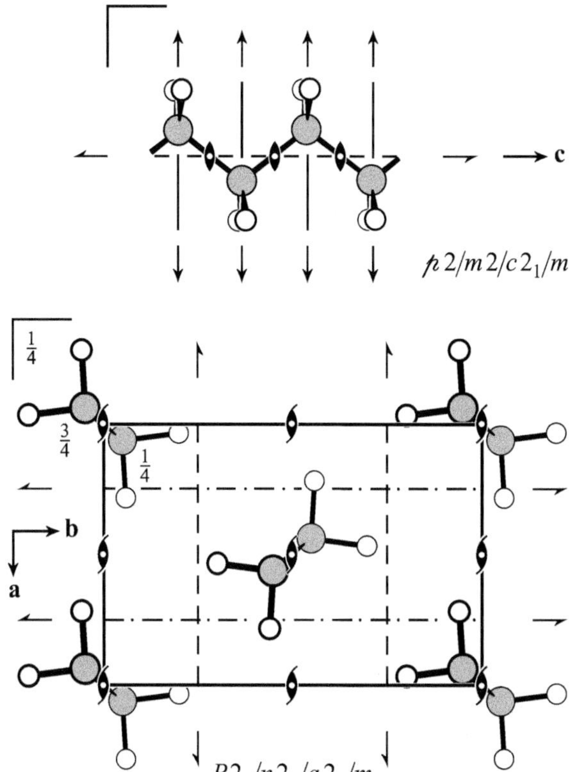

Figure 1.7.3.9
Structure of polyethylene.

Describing the symmetry of helical polymeric molecules is a little more complicated. In polymer chemistry, the structure of a helical polymeric molecule is designated as an *N*/*r* helix. *N* is the number of chemical repeating units per translation period and *r* is the number of corresponding coil turns; *N* and *r* are integers without a common divisor. Additional letters *P* or *M* indicate whether the helix is right-handed or left-handed, respectively. This nomenclature refers to the chemical structure, based on the chemical bonding, *i.e.* the turns of the (hypothetical and continuous) coil are taken to run along the main chain of covalent bonds of the polymer.

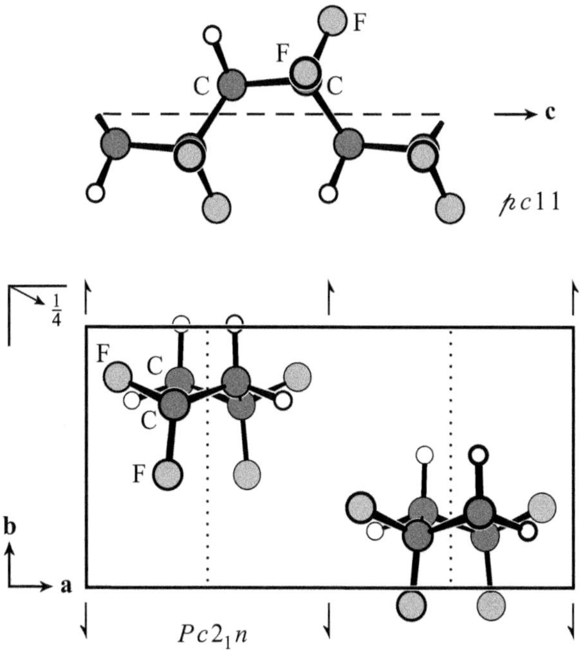

Figure 1.7.3.10
Structure of poly(vinylidene fluoride).

A chemical N/r P or M symbol can be converted to the corresponding Hermann–Mauguin screw axis symbol N_q by the following formula (Alexander, 1969; Spruiell & Clark, 1980):

$$nN \pm 1 = rq,$$

where $n = 0, 1, 2, \ldots$ and $0 < q < N$. The plus sign applies to P helices, the minus sign to M helices. The order N is the same in both symbols. It is not possible to derive the N/r symbol from the Hermann–Mauguin symbol N_q, a unique calculation is not possible (just as the point group of a molecule can be deduced from the molecular structure, but the reverse is impossible). For more details on the symmetry of helical chain molecules see Müller (2017).

If the order of the primary axis of the molecular rod group is $N = 5$ or $N \geq 7$, the order of the primary axis of the penetration rod group can be $N' = 1, 2, 3, 4$ or 6, where N' is a divisor of N. If the primary axis of the molecular rod group is a screw axis N_q, the subscript q' of the screw axis of the penetration group $N'_{q'}$ is the remainder of the integer division $q \div N'$ (Müller, 2017; the axis is a rotation axis if $q' = 0$).

Example 3. The structure of tin iodide phosphide, SnIP, is shown in Fig. 1.7.3.11 (Pfister *et al.*, 2016). It contains polyphosphide ions $(P^-)_\infty$ that form 7/2 helices. From the formula above we calculate that the helix has a 7_4 screw axis if the helix is chemically right-handed (7/2-*P*) and 7_3 if it is chemically left-handed (7/2-*M*). A second 7/2 helix $(SnI^+)_\infty$ winds around every one of the $(P^-)_\infty$ helices. The two helices (and also the ensemble of the two helices) have the same screw-axis symmetry. The non-crystallographic molecular rod groups $\not{p}7_421$ and $\not{p}7_321$ describe the symmetries in a nearly perfect way in the racemic crystal (which contains equal numbers of right- and left-handed helices).

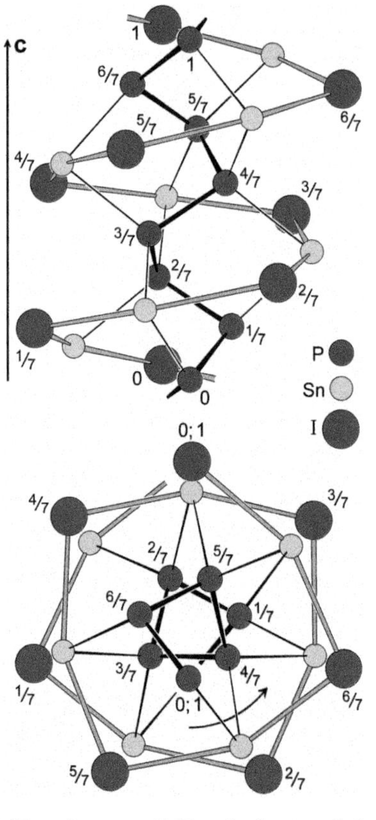

Figure 1.7.3.11
7/2-*P* helix in SnIP, rod group $\not{p}7_421$. **c** is the translation vector of the molecule. Numbers indicate the heights of the atoms (z coordinates). The curved arrow in the bottom image indicates the pitch of the 7_4 screw axis.

The actual molecular symmetry in the crystal is the penetration rod group $\not{p}121$, which is a subgroup of $\not{p}7_421$ and $\not{p}7_321$ and of the space group $P12/c1$ (No. 13). The crystallographic twofold axes are perpendicular to the helix axis and run through every seventh Sn, I and P atom (at the heights $0, 3.5/7 = \frac{1}{2}$ and 1 in Fig. 1.7.3.11). Whether the connecting Sn–I lines drawn in Fig. 1.7.3.11 are considered to be covalent bonds or not is irrelevant as far as the symmetry is concerned.

1.7.4. Magnetic subperiodic groups and magnetic space groups

By Daniel B. Litvin

1.7.4.1. Introduction

The *magnetic subperiodic groups* in the title of this section refer to generalizations of the crystallographic subperiodic groups, *i.e.* frieze groups (two-dimensional groups with one-dimensional translations), crystallographic rod groups (three-dimensional groups with one-dimensional translations) and layer groups (three-dimensional groups with two-dimensional translations). There are seven frieze-group types, 75 rod-group types and 80 layer-group types, see *IT* E, *Subperiodic Groups* (2010). The magnetic *space groups* refer to generalizations of the one-, two- and three-dimensional crystallographic space groups, n-dimensional groups with n-dimensional translations. There are two one-dimensional space-group types, 17 two-dimensional space-group types and 230 three-dimensional space-group types, see Part 2 of *IT* A.

Generalizations of the crystallographic groups began with the introduction of an operation of 'change in colour' and the 'two-colour' (black and white, antisymmetry) crystallographic point groups (Heesch, 1930; Shubnikov, 1945; Shubnikov *et al.*, 1964). Subperiodic groups and space groups were also extended into two-colour groups. Two-colour subperiodic groups consist of 31 two-colour frieze-group types (Belov, 1956*a*,*b*), 394 two-colour rod-group types (Shubnikov, 1959*a*,*b*; Neronova & Belov, 1961*a*,*b*; Galyarskii & Zamorzaev, 1965*a*,*b*) and 528 two-colour layer-group types (Neronova & Belov, 1961*a*,*b*; Palistrant & Zamorzaev, 1964*a*,*b*). Of the two-colour space groups, there are seven two-colour one-dimensional space-group types (Neronova & Belov, 1961*a*,*b*), 80 two-colour two-dimensional space-group types (Heesch, 1929; Cochran, 1952) and 1651 two-colour three-dimensional space-group types (Zamorzaev, 1953, 1957*a*,*b*; Belov *et al.*, 1957). See also Zamorzaev (1976), Shubnikov & Koptsik (1974), Koptsik (1966, 1967, 1968), and Zamorzaev & Palistrant (1980). [Extensive listings of references on colour symmetry, magnetic symmetry and related topics can be found in the books by Shubnikov *et al.* (1964), Shubnikov & Koptsik (1974), and Opechowski (1986).]

The so-called *magnetic groups*, groups to describe the symmetry of spin arrangements, were introduced by Landau & Lifschitz (1951, 1957) by re-interpreting the operation of 'change in colour' in two-colour crystallographic groups as 'time inversion'. This section introduces the structure, properties and symbols of *magnetic* subperiodic groups and *magnetic* space groups as given in the extensive tables by Litvin (2013), which are an extension of the classic tables of properties of the two- and three-dimensional subperiodic groups found in *IT* E and the one-, two- and three-dimensional space groups found in *IT* A. A survey of magnetic group types is also presented in Litvin

(2013), listing the elements of one representative group in each *reduced superfamily* of the two- and three-dimensional magnetic subperiodic groups and one-, two- and three-dimensional magnetic space groups. Two notations for magnetic groups, the Opechowski–Guccione notation (OG notation) (Guccione, 1963*a*,*b*; Opechowski & Guccione, 1965; Opechowski, 1986) and the Belov–Neronova–Smirnova notation (BNS notation) (Belov *et al.*, 1957) are compared, and the maximal subgroups of index $\leq$ 4 of the magnetic subperiodic groups and magnetic space groups are also given in Litvin (2013).

1.7.4.2. Survey of magnetic subperiodic groups and magnetic space groups

We review the concept of a reduced magnetic superfamily (Opechowski, 1986) to provide a classification scheme for magnetic groups. This is used to obtain the survey of the two- and three-dimensional magnetic subperiodic group types and the one-, two- and three-dimensional magnetic space groups given in Litvin (2013). In that survey a specification of a single representative group from each group type is provided.

1.7.4.2.1. Reduced magnetic superfamilies of magnetic groups

Let $\mathcal{F}$ denote a crystallographic group. The magnetic superfamily of $\mathcal{F}$ consists of the following set of groups:

(1) The group $\mathcal{F}$.
(2) The group $\mathcal{F}1' \equiv \mathcal{F} \times 1'$, the direct product of the group $\mathcal{F}$ and the time-inversion group $1'$, the latter consisting of the identity 1 and time inversion $1'$.
(3) All groups $\mathcal{F}(\mathcal{D}) \equiv \mathcal{D} \cup (\mathcal{F} - \mathcal{D})1' \equiv \mathcal{F} \underline{\times} 1'$, subdirect products of the groups $\mathcal{F}$ and $1'$. $\mathcal{D}$ is a subgroup of index 2 of $\mathcal{F}$. Groups of this kind will also be denoted by $\mathcal{M}$.

The third subset is divided into two subdivisions:

(3*a*) Groups $\mathcal{M}_T$, where $\mathcal{D}$ is an equi-translational (*translationengleiche*) subgroup of $\mathcal{F}$.
(3*b*) Groups $\mathcal{M}_R$, where $\mathcal{D}$ is an equi-class (*klassengleiche*) subgroup of $\mathcal{F}$.[6]

Two magnetic groups $\mathcal{F}_1(\mathcal{D}_1)$ and $\mathcal{F}_2(\mathcal{D}_2)$ are called *equivalent* if there exists an affine transformation that maps $\mathcal{F}_1$ onto $\mathcal{F}_2$ and $\mathcal{D}_1$ onto $\mathcal{D}_2$ (Opechowski, 1986). If only non-equivalent groups $\mathcal{F}(\mathcal{D})$ are included, then the above set of groups is referred to as the reduced magnetic superfamily of $\mathcal{F}$.

Example

We consider the crystallographic point group $\mathcal{F} = 222$. The magnetic superfamily of the group $222 = \{1, 2_{100}, 2_{010}, 2_{001}\}$ consists of five groups: $\mathcal{F} = 222$, the group $\mathcal{F}1' = 2221'$, and the three groups $\mathcal{F}(\mathcal{D}) = 222(2_{100})$, $222(2_{010})$ and $222(2_{001})$. Since the latter three groups are all equivalent, the reduced magnetic superfamily of the group $\mathcal{F} = 222$ consists of only three groups: 222, $2221'$, and one of the three groups $222(2_{100})$, $222(2_{010})$ and $222(2_{001})$.

Example

In the reduced magnetic space group superfamily of $\mathcal{F} = Pnn2$ there are five groups: $\mathcal{F} = Pnn2$, $\mathcal{F}1' = Pnn21'$, and three groups $\mathcal{F}(\mathcal{D}) = Pnn2(Pc)$, $Pnn2(P2)$ and $Pnn2(Fdd2)$. The groups $Pnn2(Pc)$ and $Pnn2(P2)$ are equi-translational magnetic space groups $\mathcal{M}_T$ and $Pnn2(Fdd2)$ is an equi-class magnetic space group $\mathcal{M}_R$.

A magnetic group has been defined as a symmetry group of a spin arrangement $\mathbf{S}(r)$ (Opechowski, 1986). With this definition, since $1'\mathbf{S}(r) = -\mathbf{S}(r)$, a group $\mathcal{F}1'$ is then not a magnetic group. However, there is not universal agreement on the definition or usage of the term *magnetic group*. Two definitions (Opechowski, 1986) have magnetic groups as symmetry groups of spin arrangements, with one having only groups $\mathcal{F}(\mathcal{D})$, of the three types of groups $\mathcal{F}$, $\mathcal{F}1'$ and $\mathcal{F}(\mathcal{D})$, defined as magnetic groups, while a second having both group $\mathcal{F}$ and $\mathcal{F}(\mathcal{D})$ defined as magnetic groups. Here we shall refer to all groups in a magnetic superfamily of a group $\mathcal{F}$ as magnetic groups, while cognizant of the fact that groups $\mathcal{F}1'$ cannot be a symmetry group of a spin arrangement.

1.7.4.2.2. Survey of magnetic point groups, magnetic subperiodic groups and magnetic space groups

The survey consists of listing the reduced magnetic superfamily of one group from each type of one-, two- and three-dimensional crystallographic point groups, two- and three-dimensional crystallographic subperiodic groups, and one-, two- and three-dimensional space groups (Litvin, 1999, 2001, 2013). The numbers of types of groups $\mathcal{F}$, $\mathcal{F}1'$ and $\mathcal{F}(\mathcal{D})$ in the reduced superfamilies of these groups is given in Table 1.7.4.1. The one group from each type, called the *representative group* of that type, is specified by giving a symbol for its translational subgroup and listing a set of coset representatives, called the standard set of coset representatives, of the decomposition of the group with respect to its translational subgroup. The survey provides the following information for each magnetic group type and its associated representative group:

(1) The serial number of the magnetic group type.
(2) A Hermann–Mauguin-like symbol of the magnetic group type which serves also as the symbol of the group type's representative group.
(3) For group types $\mathcal{F}(\mathcal{D})$: The symbol of the group type of the non-primed subgroup $\mathcal{D}$ of index 2 of the representative group $\mathcal{F}(\mathcal{D})$, and the position and orientation of the coordinate system of the representative group $\mathcal{D}$ of the group type $\mathcal{D}$ in the coordinate system of the representative group $\mathcal{F}(\mathcal{D})$.
(4) The standard set of coset representatives of the decomposition of the representative group with respect to its translational subgroup.

Examples of entries in the survey of magnetic groups are given in Table 1.7.4.2. The survey of the three-dimensional magnetic

Table 1.7.4.1
Numbers of types of groups in the reduced magnetic superfamilies of one-, two- and three-dimensional crystallographic point groups, subperiodic groups and space groups

Type of group	$\mathcal{F}$	$\mathcal{F}1'$	$\mathcal{F}(\mathcal{D})$	Total
One-dimensional magnetic point groups	2	2	1	5
Two-dimensional magnetic point groups	10	10	11	31
Three-dimensional magnetic point groups	32	32	58	122
Magnetic frieze groups	7	7	17	31
Magnetic rod groups	75	75	244	394
Magnetic layer groups	80	80	368	528
One-dimensional magnetic space groups	2	2	3	7
Two-dimensional magnetic space groups	17	17	46	80
Three-dimensional magnetic space groups	230	230	1191	1651

[6] Replacing time inversion $1'$ by an operation of 'changing two colours', the two-colour groups corresponding to the types 1, 2, 3*a* and 3*b* magnetic groups are known as type I, II, III and IV Shubnikov groups, respectively (Bradley & Cracknell, 1972).

Table 1.7.4.2
Examples of the format of the survey of magnetic groups of three-dimensional magnetic space-group types

Serial No.	Symbol	Non-primed subgroup of index 2	Standard set of coset representatives			
10.1.49	$P2/m$		$\{1\|0\}$	$\{2_{010}\|0\}$	$\{\bar{1}\|0\}$	$\{m_{010}\|0\}$
10.3.51	$P2'/m$	Pm (0, 0, 0; **a**, **b**, **c**)	$\{1\|0\}$	$\{2_{010}\|0\}'$	$\{\bar{1}\|0\}'$	$\{m_{010}\|0\}$
10.9.57	$P_{2b}2'/m$	$P2_1/m$ (0, $\frac{1}{2}$, 0; **a**, 2**b**, **c**)	$\{1\|0\}$	$\{2_{010}\|0, 1, 0\}$	$\{\bar{1}\|0, 1, 0\}$	$\{m_{010}\|0\}$
50.9.385	$P_{2c}b'an$	$Pnna$ ($\frac{1}{4}$, $\frac{1}{4}$, $\frac{1}{2}$; **a**, 2$\bar{\mathbf{c}}$, **b**)	$\{1\|0\}$	$\{2_{100}\|0\}$	$\{2_{010}\|0, 0, 1\}$	$\{2_{001}\|0, 0, 1\}$
			$\{\bar{1}\|\frac{1}{2}, \frac{1}{2}, 1\}$	$\{m_{100}\|\frac{1}{2}, \frac{1}{2}, 1\}$	$\{m_{010}\|\frac{1}{2}, \frac{1}{2}, 0\}$	$\{m_{001}\|\frac{1}{2}, \frac{1}{2}, 0\}$

space groups (Litvin, 2001, 2013) was incorporated into the survey of three-dimensional magnetic space groups given by Stokes & Campbell (2009) and the coset representatives can also be found on the Bilbao Crystallographic Server (http://www.cryst.ehu.es; Aroyo *et al.*, 2006).

Magnetic group type serial number

For each set of magnetic group types, one-, two- and three-dimensional crystallographic magnetic point groups, magnetic subperiodic groups and magnetic space groups, a separate numbering system is used. A three-part composite number $N_1.N_2.N_3$ is given in the first column, see Table 1.7.4.2. N_1 is a sequential number for the group type to which $\mathcal{F}$ belongs. N_2 is a sequential numbering of the magnetic group types of the reduced magnetic superfamily of $\mathcal{F}$. Group types $\mathcal{F}$ always have the assigned number $N_1.1.N_3$, and group types $\mathcal{F}1'$ the assigned number $N_1.2.N_3$. N_3 is a global sequential numbering for each set of magnetic group types. The sequential numbering N_1 for subperiodic groups and space groups follows the numbering in *IT* E and *IT* A, respectively.

Magnetic group type symbol

A Hermann–Mauguin-like type symbol is given for each magnetic group type in the second column. This symbol denotes both the group type and the representative group of that type. For example, the symbol for the three-dimensional magnetic space-group type 25.4.158 is $Pm'm'2$. This symbol denotes both the group type, which consists of an infinite set of groups, and the representative group $Pm'_{100}m'_{010}2_{001}$. While this representative group may be referred to as 'the group $Pm'm'2$', other groups of this group type, *e.g.* $Pm'_{110}m'_{\bar{1}10}2_{001}$, will always be written with subindices. The representative group of the magnetic group type is defined by its translational subgroup, implied by the first letter in the magnetic group type symbol and defined in Table 1.1 of Litvin (2013), and a given set of coset representatives, called the *standard set of coset representatives*, of the representative group with respect to its translational subgroup.

Only the relative lengths and mutual orientations of the translation vectors of the translational subgroup are given, see Table 1.2 of Litvin (2013). The symmetry directions of symmetry operations represented by characters in the Hermann–Mauguin symbols are implied by the character's position in the symbol and are given in Table 1.3 of Litvin (2013). The standard set of coset representatives are given with respect to an implied coordinate system. The absolute lengths of translation vectors, the position in space of the origin of the coordinate system and the orientation in that space of the basis vectors of that coordinate system are not explicitly given.

Standard set of coset representatives

The standard set of coset representatives of each representative group is listed on the right-hand side of the survey of magnetic group types, see *e.g.* Table 1.7.4.2. Each coset in the standard set of coset representatives is given in Seitz notation (Seitz, 1934, 1935*a*,*b*, 1936), *i.e.* $\{\boldsymbol{R}|\boldsymbol{\tau}\}$ or $\{\boldsymbol{R}|\boldsymbol{\tau}\}'$. $\boldsymbol{R}$ denotes a proper or improper rotation (rotation-inversion), $\boldsymbol{\tau}$ a non-primitive translation with respect to the non-primed translational subgroup of the magnetic group, and the prime denotes that $\{\boldsymbol{R}|\boldsymbol{\tau}\}$ is coupled with time inversion. The subindex notation on $\boldsymbol{R}$, denoting the orientation of the proper or improper rotation, is given in Table 1.4 of Litvin (2013). [Note that the Seitz notation used in Litvin (2013) predates and is different from the IUCr standard convention for Seitz symbolism, see Section 1.4.2.2 and Glazer *et al.* (2014).]

Symbol of the subgroup $\mathcal{D}$ of index 2 of $\mathcal{F}(\mathcal{D})$

For magnetic group types $\mathcal{F}(\mathcal{D})$, the magnetic group type symbol of the subgroup $\mathcal{D}$ is given in the third column of the survey of magnetic groups, see *e.g.* Table 1.7.4.2. If $\mathcal{F}(\mathcal{D})$ is a group $\mathcal{M}_T$, then the subgroup $\mathcal{D}$ is defined by the translational group of $\mathcal{F}(\mathcal{D})$ and the unprimed coset representatives of $\mathcal{F}(\mathcal{D})$.

Example

Consider the three-dimensional magnetic space-group type 16.3.101 $P2'2'2$. The representative group $P2'2'2$ is defined by the translational subgroup $\mathcal{T}$ denoted by the letter P generated by the translations

$$\{1|1, 0, 0)\} \quad \{1|0, 1, 0\} \quad \{1|0, 0, 1\}$$

and the standard set of coset representatives

$$\{1|0\} \quad \{2_{100}|0\}' \quad \{2_{010}|0\}' \quad \{2_{001}|0\}.$$

The subgroup $\mathcal{D}$ of index 2 of the representative group $\mathcal{F}(\mathcal{D}) = P2'2'2$ is defined by the translational group $\mathcal{T}$ denoted by the letter P and the cosets $\{1|0\}$ and $\{2_{001}|0\}$, and is a group of type $P2$.

If $\mathcal{F}(\mathcal{D})$ is a group $\mathcal{M}_R$, then the subgroup $\mathcal{D}$ is defined by the non-primed translational group of $\mathcal{F}(\mathcal{D})$ and all the cosets of the standard set of coset representatives of the group $\mathcal{F}(\mathcal{D})$.

Example

Consider the three-dimensional magnetic space-group type 16.4.102 $P_{2a}222$. The representative group $P_{2a}222$ is defined by the translational group $\mathcal{T}$ denoted by the symbol P_{2a} generated by the translations

$$\{1|1, 0, 0\}' \quad \{1|0, 1, 0\} \quad \{1|0, 0, 1\}$$

and the standard set of coset representatives

$$\{1|0\} \quad \{2_{100}|0\} \quad \{2_{010}|0\} \quad \{2_{001}|0\}.$$

The subgroup $\mathcal{D}$ of index 2 of the representative group $\mathcal{F}(\mathcal{D}) = P_{2a}222$ is defined by the translational subgroup $\mathcal{T}$ denoted by the symbol P_{2a}, *i.e.* the translations generated by

$$\{1|2, 0, 0\} \quad \{1|0, 1, 0\} \quad \{1|0, 0, 1\}$$

and the standard set of cosets of $P_{2a}222$. The group $\mathcal{D}$ is a group of type $P222$.

While the group type symbol of $\mathcal{D}$ is given, the coset representatives of the subgroup $\mathcal{D}$ of $\mathcal{F}(\mathcal{D})$ derived from the standard set of coset representatives of $\mathcal{F}(\mathcal{D})$ may not be identical with the standard set of coset representatives of the representative group of type $\mathcal{D}$ found in the survey of magnetic group types. Consequently, to show the relationship between this subgroup $\mathcal{D}$ and the listed representative group of groups of type $\mathcal{D}$ additional information is provided: a new coordinate system is defined in which the coset representatives of this subgroup $\mathcal{D}$ are identical with the standard set of coset representatives listed for the representative group of groups of type $\mathcal{D}$: Let (O; **a**, **b**, **c**) be the coordinate system in which the group $\mathcal{F}(\mathcal{D})$ is defined. O is the origin of the coordinate system, and **a**, **b** and **c** are the basis vectors of the coordinate system. **a**, **b** and **c** represent a set of basis vectors of a primitive cell for primitive lattices and of a conventional cell for centred lattices. A second coordinate system, defined by (O + **p**; **a**′, **b**′, **c**′), is given in which the coset representatives of this subgroup $\mathcal{D}$ are identical with the standard set of coset representatives listed for the representative group of groups of type $\mathcal{D}$. O + **p** is referred to as the *location* of the subgroup $\mathcal{D}$ in the coordinate system of the group $\mathcal{F}(\mathcal{D})$ (Kopský, 2011). The origin is first translated from O to O + **p**. On translating the origin from O to O + **p**, a coset representative $\{\boldsymbol{R}|\boldsymbol{\tau}\}$ becomes $\{\boldsymbol{R}|\boldsymbol{\tau} + \boldsymbol{Rp} - \boldsymbol{p}\}$ (Litvin, 2005, 2008; see also Section 1.5.2.3). This is followed by changing the basis vectors **a**, **b** and **c** to **a**′, **b**′ and **c**′, respectively. The basis vectors **a**′, **b**′, **c**′ define the conventional unit cell of the non-primed subgroup $\mathcal{D}$ of $\mathcal{F}(\mathcal{D})$ in the coordinate system (O; **a**, **b**, **c**) in which $\mathcal{F}(\mathcal{D})$ is defined. (O + **p**; **a**′, **b**′, **c**′) is given immediately following the group type symbol for the subgroup $\mathcal{D}$ of $\mathcal{F}(\mathcal{D})$. [In Litvin (2013), for typographical simplicity, the symbols 'O +' are omitted.]

Example

For the three-dimensional magnetic space-group type 10.4.52, $\mathcal{F}(\mathcal{D}) = P2/m'$, one finds in Litvin (2013)[7]

Serial No.	Symbol	Non-primed subgroup of index 2	Standard set of coset representatives
10.4.52	$P2/m'$	$P2$ (0, 0, 0; **a**, **b**, **c**)	$\{1|0\}$ $\{2_{010}|0\}$ $\{\bar{1}|0\}'$ $\{m_{010}|0\}'$

The translational subgroup of the subgroup $\mathcal{D} = P2$ of $\mathcal{F}(\mathcal{D}) = P2/m'$ is generated by the translations $\{1|1, 0, 0\}$, $\{1|0, 1, 0\}$ and $\{1|0, 0, 1\}$ and the coset representatives of this group are $\{1|0\}$ and $\{2_{010}|0\}$, the unprimed coset representatives on the right. This subgroup $\mathcal{D}$ is of type $P2$. In Litvin (2013), listed for the group type 3.1.8, $P2$, one finds the identical two coset representatives. Consequently, there is no change in the coordinate system, *i.e.* $\mathbf{p} = (0, 0, 0)$ and **a**′ = **a**, **b**′ = **b** and **c**′ = **c**. In the coordinate system of the magnetic group $P2/m'$, the coset representatives of its subgroup $\mathcal{D} = P2$ are identical with the standard set of coset representatives of the group type $P2$.

Example

For the three-dimensional magnetic space-group type 16.7.105, $\mathcal{F}(\mathcal{D}) = P_{2c}22'2'$ one has

Serial No.	Symbol	Non-primed subgroup of index 2	Standard set of coset representatives
16.7.105	$P_{2c}22'2'$	$P222_1$ (0, 0, 0; **a**, **b**, 2**c**)	$\{1|0\}$ $\{2_{100}|0\}$ $\{2_{010}|0, 0, 1\}$ $\{2_{001}|0, 0, 1\}$

The translational subgroup of the subgroup $\mathcal{D} = P222_1$ of $\mathcal{F}(\mathcal{D}) = P_{2c}22'2'$ is generated by the translations $\{1|1, 0, 0\}$, $\{1|0, 1, 0\}$ and $\{1|0, 0, 2\}$, and the coset representatives of this group are all those coset representatives on the right. This subgroup $\mathcal{D}$ is of type $P222_1$. Listed for the group type 17.1.106 $P222_1$, one finds a different set of coset representatives:

$$\{1|0\} \quad \{2_{100}|0\} \quad \{2_{010}|0, 0, \tfrac{1}{2}\} \quad \{2_{001}|0, 0, \tfrac{1}{2}\}.$$

Consequently, to show the relationship between this subgroup $\mathcal{D}$ of $\mathcal{F}(\mathcal{D})$ and the listed representative group of the group type $P222_1$ we change the coordinate system in which $\mathcal{D}$ is defined to (0, 0, 0; **a**, **b**, 2**c**). In this new coordinate system the coset representatives of the subgroup $\mathcal{D}$ are identical with the coset representatives of the representative group of the group type $P222_1$.

Example

For the three-dimensional magnetic space-group type 18.4.116, $P2_12_1'2'$, one has

Serial No.	Symbol	Non-primed subgroup of index 2	Standard set of coset representatives
18.4.116	$P2_12_1'2'$	$P2_1$ (0, $\frac{1}{4}$, 0; **c**, **a**, **b**)	$\{1|0\}$ $\{2_{100}|\frac{1}{2}, \frac{1}{2}, 0\}$ $\{2_{010}|\frac{1}{2}, \frac{1}{2}, 0\}'$ $\{2_{001}|0\}'$

The translational subgroup of $\mathcal{D}$ is generated by the translations $\{1|1, 0, 0\}$, $\{1|0, 1, 0\}$ and $\{1|0, 0, 1\}$ and the coset representatives of this group are $\{1|0\}$ and $\{2_{100}|\frac{1}{2}, \frac{1}{2}, 0\}$, the unprimed coset representatives on the right. The group $\mathcal{D}$ is of type $P2_1$. For the magnetic group type 4.1.15 $P2_1$ one finds a different set of coset representatives: $\{1|0\}$ and $\{2_{010}|0, \frac{1}{2}, 0\}$. Consequently, to show the relationship between the subgroup $\mathcal{D}$ of $\mathcal{F}(\mathcal{D})$ and the listed representative group of the group type $P2_1$, we change the coordinate system in which the subgroup $\mathcal{D}$ is defined to (0, $\frac{1}{4}$, 0; **c**, **a**, **b**). The origin is first translated from O to O + **p**, where $\mathbf{p} = (0, \frac{1}{4}, 0)$, and then a new set of basis vectors, **a**′ = **c**, **b**′ = **a** and **c**′ = **b**, is defined. In this new coordinate system the coset representatives of the subgroup $\mathcal{D}$ are identical with the standard set of coset representatives of the representative group of the group type $P2_1$.

[7] In Litvin (2013) the terminology 'non-magnetic' is used in place of 'non-primed' in the column headings in these tables.

References

Alexander, L. E. (1969). *X-ray Diffraction Methods in Polymer Science*, pp. 19–22. New York: Wiley.

Aroyo, M. I., Perez-Mato, J. M., Capillas, C., Kroumova, E., Ivantchev, S., Madariaga, G., Kirov, A. & Wondratschek, H. (2006). *Bilbao Crystallographic Server I: Databases and crystallographic computing programs. Z. Kristallogr.* **221**, 15–27.

Ascher, E., Gramlich, V. & Wondratschek, H. (1969). *Korrekturen zu den Angaben 'Untergruppen' in den Raumgruppen der Internationalen Tabellen zur Bestimmung von Kristallstrukturen (1935), Band 1. Acta Cryst.* B**25**, 2154–2156.

Avitabile, G., Napolitano, R. & Pirozzi, B. (1975). *Low-temperature crystal structure of polyethylene: Results from neutron diffraction. Polymer Lett.* **73**, 351–355.

Bachmann, M., Gordon, W. L., Weinhold, S. & Lando, J. B. (1980). *The crystal structure of phase IV of poly(vinylidene fluoride). J. Appl. Phys.* **51**, 5095–5099.

Bärnighausen, H. (1980). *Group–subgroup relations between space groups: a useful tool in crystal chemistry. MATCH Commun. Math. Chem.* **9**, 139–175.

Belov, N. V. (1956*a*). *The one-dimensional infinite crystallographic groups. Kristallografiya*, **1**, 474–476.

Belov, N. V. (1956*b*). *The one-dimensional infinite crystallographic groups. Sov. Phys. Crystallogr.* **1**, 372–374.

Belov, N. V., Neronova, N. N. & Smirnova, T. S. (1957). *1651 Shubnikov groups. Sov. Phys. Crystallogr.* **1**, 487–488. See also (1955) *Trudy Inst. Krist. Acad. SSSR*, **11**, 33–67 (in Russian), English translation in Shubnikov, A. V., Belov, N. V. & others (1964). *Colored Symmetry*. London: Pergamon Press.

Bradley, C. J. & Cracknell, A. P. (1972). *The Mathematical Theory of Symmetry in Solids*. Oxford: Clarendon Press.

Brock, C. P. (2020). *Pseudosymmetric layers in high-Z′ and P1 structures of organic molecules. CrystEngComm*, https://dx.doi.org/10.1039/d0ce00302f.

Buerger, M. J. (1947). *Derivative crystal structures. J. Chem. Phys.* **15**, 1–16.

Buerger, M. J. (1951). *Phase Transformations in Solids*, ch. 6. New York: Wiley.

Bunn, C. W. (1939). *The crystal structure of long-chain normal paraffin hydrocarbons. The shape of the CH_2 group. Trans. Faraday Soc.* **35**, 482–491.

Cochran, W. (1952). *The symmetry of real periodic two-dimensional functions. Acta Cryst.* **5**, 630–633.

Davies, B. L. & Dirl, R. (1993*a*). *Space-group subgroups generated by sublattice relations: software of IBM-compatible PCs.* Anales de Física, Mongrafías, Vol. 2, edited by M. A. del Olmo, M. Santander & J. M. Mateos Guilarte, pp. 338–341. Madrid: CIEMAT/RSEF.

Davies, B. L. & Dirl, R. (1993*b*). *Space-group subgroups, coset decompositions, layer and rod symmetries: integrated software for IBM-compatible PCs.* Third Wigner Colloquium, Oxford, September 1993.

Fuksa, J., Kopský, V. & Litvin, D. B. (1993). *Spatial distribution of rod and layer symmetries in a crystal.* Anales de Física, Mongrafías, Vol. 2, edited by M. A. del Olmo, M. Santander & J. M. Mateos Guilarte, pp. 346–369. Madrid: CIEMAT/RSEF.

Galyarskii, E. I. & Zamorzaev, A. M. (1965*a*). *A complete description of crystallographic stem groups of symmetry and different types of antisymmetry. Kristallografiya*, **10**, 147–154.

Galyarskii, E. I. & Zamorzaev, A. M. (1965*b*). *A complete description of crystallographic stem groups of symmetry and different types of antisymmetry. Sov. Phys. Crystallogr.* **10**, 109–115.

Glazer, A. M., Aroyo, M. I. & Authier, A. (2014). *Acta Cryst.* A**70**, 300–302.

Groom, C. R., Bruno, I. J., Lightfoot, M. P. & Ward, S. C. (2016). *The Cambridge Structural Database. Acta Cryst.* B**72**, 171–179.

Guccione, R. (1963*a*). *Magnetic Space Groups*. PhD Thesis, University of British Columbia, Canada.

Guccione, R. (1963*b*). *On the construction of the magnetic space groups. Phys. Lett.* **5**, 105–107.

Guigas, B. (1971). *PROSEC*. Institut für Kristallographie, Universtität Karlsruhe, Germany. (Unpublished.)

Guinier, A., Bokij, G. B., Boll-Dornberger, K., Cowley, J. M., Ďurovič, S., Jagodzinski, H., Krishna, P., de Wolff, P. M., Zvyagin, B. B., Cox, D. E., Goodman, P., Hahn, Th., Kuchitsu, K. & Abrahams, S. C. (1984). *Nomenclature of polytype structures. Report of the International Union of Crystallography Ad hoc Committee on the Nomenclature of Disordered, Modulated and Polytype Structures. Acta Cryst.* A**40**, 399–404.

Hao, X., Chen, J., Cammers, A., Parkin, S. & Brock, C. P. (2005). *A helical structure with Z′ = 10. Acta Cryst.* B**61**, 218–226.

Heesch, H. (1929). *Zur strukturtheorie der ebenen symmetriegruppen. Z. Kristallogr.* **71**, 95–102.

Heesch, H. (1930). *Uber die vierdimensionalen gruppen des dreidimensionalen raumes. Z. Kristallogr.* **73**, 325–345.

Hermann, C. (1929). *Zur systematischen Strukturtheorie. IV. Untergruppen. Z. Kristallogr.* **69**, 533–555.

Holser, W. T. (1958*a*). *The relation of structure to symmetry in twinning. Z. Kristallogr.* **110**, 249–263.

Holser, W. T. (1958*b*). *Point groups and plane groups in a two-sided plane and their subgroups. Z. Kristallogr.* **110**, 266–281.

International Tables for Crystallography (2010). Vol. A1, *Symmetry Relations between Space Groups*, 2nd ed., edited by H. Wondratschek & U. Müller. Chichester: John Wiley & Sons. [Abbreviated as *IT* A1.]

International Tables for Crystallography (2010). Vol. E, *Subperiodic Groups*, 2nd ed., edited by V. Kopský & D. B. Litvin. Chichester: Wiley. [Abbreviated as *IT* E.]

International Tables for Crystallography (2016). Vol. A, *Space-Group Symmetry*, 6th ed., edited by M. I. Aroyo. Chichester: Wiley. [Abbreviated as *IT* A.]

Internationale Tabellen zur Bestimmung von Kristallstrukturen (1935). 1. Band, edited by C. Hermann. Berlin: Borntraeger.

Janovec, V. (1972). *Group analysis of domains and domain pairs. Czech. J. Phys.* **22**, 974–994.

Janovec, V. (1981). *Symmetry and structure of domain walls. Ferroelectrics*, **35**, 105–110.

Janovec, V., Grocký, M., Kopský, V. & Kluiber, Z. (2004). *On atomic displacements in 90° ferroelectric domain walls of tetragonal $BaTiO_3$ crystals. Ferroelectrics*, **303**, 65–68.

Janovec, V., Kopský, V. & Litvin, D. B. (1988). *Subperiodic subgroups of space groups. Z. Kristallogr.* **185**, 282.

Janovec, V. & Přívratská, J. (2014). *Domain structures*. In *International Tables for Crystallography*, Vol. D, *Physical Properties of Crystals*, 2nd ed., edited by A. Authier, ch. 3.4. Chichester: Wiley.

Kopský, V. (2011). Private communication.

Kopský, V. & Litvin, D. B. (1989). *Scanning of space groups*. In *Group Theoretical Methods in Physics*, edited by Y. Saint Aubin & L. Vinet, pp. 263–266. Singapore: World Scientific.

Kopský, V. & Litvin, D. B. (2004). *Space-group scanning tables. Acta Cryst.* A**60**, 637.

Koptsik, V. A. (1966). *Shubnikov Groups. Handbook on the Symmetry and Physical Properties of Crystal Structures*. Izd. MGU (in Russian). English translation of text: Kopecky, J. & Loopstra, B. O. (1971). *Fysica Memo 175*. Stichting, Reactor Centrum Nederland.

Koptsik, V. A. (1967). *A general sketch of the development of the theory of symmetry and its applications in physical crystallography over the past 50 years. Kristallografiya*, **12**, 755–774.

Koptsik, V. A. (1968). *A general sketch of the development of the theory of symmetry and its applications in physical crystallography over the past 50 years. Sov. Phys. Crystallogr.* **12**, 667–683.

Landau, L. I. & Lifschitz, E. M. (1951). *Statistical Physics*. Moscow: Gostekhizdat (in Russian); English translation (1958). Oxford: Pergamon Press.

Landau, L. I. & Lifschitz, E. M. (1957). *Electrodynamics of Continuous Media*. Moscow: Gostekhizdat (in Russian); English translation (1960). Reading, MA: Addison-Wesley.

Li, M., Wondergem, H. J., Spijkman, M.-J., Asadi, K., Katsouras, I., Blom, P. W. & de Leeuw, D. M. (2013). *Revisiting the δ-phase of poly(vinylidene fluoride). Nat. Mat.* **12**, 433–438.

Li, Y., Rao, Y., Mak, K. F., You, Y., Wang, S., Dean, C. R. & Heinz, T. F. (2013). *Probing symmetry properties of few-layer MoS_2 and h-BN by optical second-harmonic generation. Nano Lett.* **13**, 3329–3333.

Lima-de-Faria, J., Hellner, E., Liebau, F., Makovicky, E. & Parthé, E. (1990). *Nomenclature of inorganic structure types. Report of the International Union of Crystallography Commission on Crystallographic Nomenclature Subcommittee on the Nomenclature of Inorganic Structure Types. Acta Cryst.* A**46**, 1–11.

Litvin, D. B. (1999). *Magnetic subperiodic groups. Acta Cryst.* A**55**, 963–964.

Litvin, D. B. (2001). *Magnetic space-group types. Acta Cryst.* A**57**, 729–730.

Litvin, D. B. (2005). *Tables of properties of magnetic subperiodic groups. Acta Cryst.* A**61**, 382–385.

Litvin, D. B. (2008). *Tables of crystallographic properties of magnetic space groups. Acta Cryst.* A**64**, 419–424.

Litvin, D. B. (2013). *Magnetic Group Tables, 1-, 2- and 3-Dimensional Magnetic Subperiodic Groups and Space Groups*. Chester: International Union of Crystallography. Freely available from http://www.iucr.org/publ/978-0-9553602-2-0.

Locherer, K. R., Swainson, I. P. & Salje, E. K. H. (1999). *Transition to a new tetragonal phase of WO_3: crystal structure and distortion parameters. J. Phys. Condens. Matter*, **11**, 4143–4156.

Macrae, C. F., Sovago, I., Cottrell, S. J., Galek, P. T. A., McCabe, P., Pidcock, E., Platings, M., Shields, G. P., Stevens, J. S., Towler, M. & Wood, P. A. (2020). *Mercury 4.0: from visualization to analysis, design and prediction. J. Appl. Cryst.* **53**, 226–235.

Megaw, H. D. (1973). *Crystal Structures: A Working Approach.* Philadelphia: Saunders.

Müller, U. (2013). *Symmetry Relationships Between Crystal Structures.* Oxford University Press. [German: *Symmetriebeziehungen zwischen verwandten Kristallstrukturen*; Wiesbaden: Vieweg+Teubner, 2012. Spanish: *Relaciones de simetría entre estructuras cristalinas*; Madrid: Síntesis, 2013.]

Müller, U. (2017). *Die Symmetrie von Spiralketten. Acta Cryst.* **B73**, 443–452. [English translation available in the supplementary material and at https://doi.org/10.17192/es2017.0002.]

Neronova, N. N. & Belov, N. V. (1961*a*). *A single scheme for the classical and black-and-white crystallographic symmetry groups. Kristallografiya*, **6**, 3–12.

Neronova, N. N. & Belov, N. V. (1961*b*). *A single scheme for the classical and black-and-white crystallographic symmetry groups. Sov. Phys. Crystallogr.* **6**, 1–9.

Opechowski, W. (1986). *Crystallographic and Metacrystallographic Groups.* Amsterdam: North Holland.

Opechowski, W. & Guccione, R. (1965). *Magnetic symmetry.* In *Magnetism*, edited by G. T. Rado & H. Suhl, Vol. 2A, ch. 3. New York: Academic Press.

Ortiz, A. L., Sánchez-Bajo, F., Cumbrera, F. L. & Guiberteau, F. (2013). *The prolific polytypism of silicon carbide. J. Appl. Cryst.* **46**, 242–247.

Palistrant, A. F. & Zamorzaev, A. M. (1964*a*). *Groups of symmetry and different types of antisymmetry of borders and ribbons. Kristallografiya*, **9**, 155–161.

Palistrant, A. F. & Zamorzaev, A. M. (1964*b*). *Groups of symmetry and different types of antisymmetry of borders and ribbons. Sov. Phys. Crystallogr.* **9**, 123–128.

Parrish, D. A., Deschamps, J. R., Gilardi, R. D. & Butcher, R. J. (2008). *Polymorphs of picryl bromide. Cryst. Growth Des.* **8**, 57–62.

Pfister, D., Schäfer, K., Ott, C., Gerke, B., Pöttgen, R., Janka, O., Baumgartner, M., Efimova, A., Hohmann, A., Schmidt, P., Venkatachalam, S., von Wüllen, L., Schürmann, U., Kienle, L., Duppel, V., Parzinger, E., Miller, B., Becker, J., Holleitner, A., Weihrich, R. & Nilges, T. (2016). *Inorganic double helices in semiconducting SnIP. Adv. Mater.* **28**, 9783–9791. doi:10.1002/adma.201603135.

Pond, R. C. & Bollmann, W. (1979). *The symmetry and interfacial structure of bicrystals. Philos. Trans. R. Soc. London Ser. A*, **292**, 449–472.

Pond, R. C. & Vlachavas, D. S. (1983). *Bicrystallography. Proc. R. Soc. London Ser. A*, **386**, 95–143.

Preindl, J., Leitner, C., Baldauf, S. & Mulzer, J. (2014). *A short access to the skeleton of Elisabethin A and formal syntheses of Elisapterosin B and Colombiasin A. Org. Lett.* **16**, 4276–4279.

Seitz, F. Z. (1934). *A matrix-algebraic development of the crystallographic groups. I. Z. Kristallogr.* **88**, 433–459.

Seitz, F. Z. (1935*a*). *A matrix-algebraic development of the crystallographic groups. II. Z. Kristallogr.* **90**, 289–313.

Seitz, F. Z. (1935*b*). *A matrix-algebraic development of the crystallographic groups. III. Z. Kristallogr.* **91**, 336–366.

Seitz, F. Z. (1936). *A matrix-algebraic development of the crystallographic groups. IV. Z. Kristallogr.* **94**, 100–130.

Shubnikov, A. V. (1945). *New ideas in the theory of symmetry and its applications.* In *Report of the General Assembly of the Academy of Sciences of the USSR*, 14–17 October 1944. *Izd. Acad. Nauk SSSR*, 212–227.

Shubnikov, A. V. (1959*a*). *Symmetry and antisymmetry of rods and semicontinua with principal axis of infinite order and finite transfers along it. Kristallografiya*, **4**, 279–285.

Shubnikov, A. V. (1959*b*). *Symmetry and antisymmetry of rods and semicontinua with principal axis of infinite order and finite transfers along it. Sov. Phys. Crystallogr.* **4**, 262–266.

Shubnikov, A. V., Belov, N. V. & others (1964). *Colored Symmetry*, London: Pergamon Press.

Shubnikov, A. V. & Koptsik, V. A. (1974). *Symmetry in Science and Art.* New York: Plenum Press.

Spruiell, J. E. & Clark, E. S. (1980). *Methods in Experimental Physics*, Vol. 16, *Polymers*, edited by R. Fava, Part B, ch. 6, pp. 19–22. New York: Academic Press.

Stokes, H. T. & Campbell, B. J. (2009). *Table of magnetic space groups*, http://stokes.byu.edu/magneticspacegroupshelp.html.

Trucano, P. & Chen, R. (1975). *Structure of graphite by neutron diffraction. Nature*, **258**, 136–137.

Vlachavas, D. S. (1985). *Symmetry of bicrystals corresponding to a given misorientation relationship. Acta Cryst.* **A41**, 371–376.

Wondratschek, H. (1971). Institut für Kristallographie, Universtität Karlsruhe, Germany. (Unpublished.)

Wood, E. A. (1964). *Vocabulary of surface crystallography. J. Appl. Phys.* **35**, 1306–1312.

Zalkin, A. & Raymond, K. N. (1969). *Structure of di-π-cyclooctatetraeneuranium. J. Am. Chem. Soc.* **91**, 5667–5668.

Zamorzaev, A. M. (1953). Dissertation, Leningrad State University, Russia. (In Russian.)

Zamorzaev, A. M. (1957*a*). *Generalizations of Fedorov groups. Kristallografiya*, **2**, 15–20.

Zamorzaev, A. M. (1957*b*). *Generalizations of Fedorov groups. Sov. Phys. Crystallogr.* **2**, 10–15.

Zamorzaev, A. M. (1976). *The Theory of Simple and Multiple Antisymmetry.* Kishinev: Shtiintsa. (In Russian.)

Zamorzaev, A. M. & Palistrant, A. P. (1980). *Antisymmetry, its generalizations and geometrical applications. Z. Kristallogr.* **151**, 231–248.

Zikmund, Z. (1984). *Symmetry of domain pairs and domain twins. Czech. J. Phys.* **34**, 932–949.

2. CRYSTALLOGRAPHIC SYMMETRY DATA

By M. I. Aroyo, H. D. Flack, G. de la Flor, Th. Hahn, E. Kroumova, D. B. Litvin, A. Looijenga-Vos, K. Momma, U. Müller and H. Wondratschek

2.1. Guide to and examples of the space-group tables in *IT* A 108

2.1.1. Conventional descriptions of plane and space groups 108

2.1.2. Symbols of symmetry elements 110

2.1.3. Contents and arrangement of the tables 116

2.1.4. Examples of plane- and space-group tables 131

2.2. The symmetry-relations tables of *IT* A1 212

2.2.1. Guide to the subgroup tables 212

2.2.2. Examples of the subgroup tables 214

2.2.3. Guide to the tables of relations between Wyckoff positions 217

2.2.4. Examples of the tables of relations between Wyckoff positions 220

2.3. The subperiodic group tables of *IT* E 224

2.3.1. Guide to the subperiodic group tables 224

2.3.2. Examples of subperiodic group tables 225

2.4. The Symmetry Database 232

2.4.1. Space-group symmetry data 232

2.4.2. Symmetry relations between space groups 233

2.4.3. 3D Crystallographic point groups 233

2.4.4. Availability 233

2.1. Guide to and examples of the space-group tables in *IT* A

Theo Hahn, Aafje Looijenga-Vos, Mois I. Aroyo, Howard D. Flack and Koichi Momma

In this chapter, tables and diagrams of crystallographic data for the 17 types of plane groups [Chapter 2.2 of *International Tables for Crystallography* Volume A (2016), hereafter abbreviated as *IT* A] and the 230 types of space groups (Chapter 2.3 of *IT* A) are presented.

Only a minimum of theory is provided here, as the emphasis is on the practical use of the data. For the theoretical background to these data, the reader is referred to Part 1, which also includes suitable references (for further details, see Parts 1 and 3 of *IT* A). A textbook explaining space-group symmetry and the use of the crystallographic data (with exercises) is provided by Hahn & Wondratschek (1994); see also Müller (2013).

Section 2.1.1 displays, with the help of an extensive synoptic table, the classification of the 17 plane groups and 230 space groups. This is followed by an explanation of the characterization of the conventional crystallographic coordinate systems, including the symbols for the centring types of lattices and cells. Section 2.1.2 lists the alphanumeric and graphical symbols for symmetry elements and symmetry operations used throughout *IT* A. The lists are accompanied by notes and cross-references to related IUCr nomenclature reports. Section 2.1.3 explains in a systematic fashion, with many examples and figures, all the entries and diagrams in the order in which they occur in the plane-group and space-group tables of Part 2 of *IT* A. Finally, Section 2.1.4 contains a selection of plane- and space-group tables from *IT* A.

2.1.1. Conventional descriptions of plane and space groups

By Theo Hahn and Aafje Looijenga-Vos

2.1.1.1. Classification of space groups

In *IT* A, the plane groups and space groups are classified according to three criteria:

(i) According to *geometric crystal classes*, *i.e.* according to the crystallographic point group to which a particular space group belongs. There are 10 crystal classes in two dimensions and 32 in three dimensions. They are listed in column 4 of Table 2.1.1.1 (see Chapter 3.2 of *IT* A for a detailed description).

(ii) According to *crystal families*. The term crystal family designates the classification of the 17 plane groups into four categories and of the 230 space groups into *six* categories, as displayed in column 1 of Table 2.1.1.1. Here all 'hexagonal', 'trigonal' and 'rhombohedral' space groups are contained in one family, the hexagonal crystal family. The 'crystal family' thus corresponds to the term 'crystal system', as used frequently in the American and Russian literature.

The crystal families are symbolized by the lower-case letters *a*, *m*, *o*, *t*, *h*, *c*, as listed in column 2 of Table 2.1.1.1. If these letters are combined with the appropriate capital letters for the lattice-centring types (*cf.* Table 2.1.1.2), symbols for the 14 Bravais lattices result. These symbols and their occurrence in the crystal families are shown in column 8 of Table 2.1.1.1; *mS* and *oS* are the standard setting-independent symbols for the centred monoclinic and the one-face-centred orthorhombic Bravais lattices, *cf.* de Wolff *et al.* (1985); symbols between parentheses represent alternative settings of these Bravais lattices.

(iii) According to *crystal systems*. This classification collects the plane groups into four categories and the space groups into *seven* categories. The classifications according to crystal families and crystal systems are the same for two dimensions.

For three dimensions, this applies to the triclinic, monoclinic, orthorhombic, tetragonal and cubic systems. The only complication exists in the hexagonal crystal family, for which several subdivisions into systems have been proposed in the literature. In *IT* A [as well as in *International Tables for X-ray Crystallography* (1952), hereafter *IT* (1952), and the subsequent editions of *IT*], the space groups of the hexagonal crystal family are grouped into two 'crystal systems' as follows: all space groups belonging to the five crystal classes 3, $\bar{3}$, 32, 3*m* and $\bar{3}m$, *i.e.* having 3, 3_1, 3_2 or $\bar{3}$ as principal axis, form the *trigonal* crystal system, irrespective of whether the Bravais lattice is *hP* or *hR*; all space groups belonging to the seven crystal classes 6, $\bar{6}$, 6/*m*, 622, 6*mm*, $\bar{6}2m$ and 6/*mmm*, *i.e.* having 6, 6_1, 6_2, 6_3, 6_4, 6_5 or $\bar{6}$ as principal axis, form the *hexagonal* crystal system; here the lattice is always *hP*. The crystal systems, as defined above, are listed in column 3 of Table 2.1.1.1.

A different subdivision of the hexagonal crystal family is in use, mainly in the French literature. It consists of grouping all space groups based on the hexagonal Bravais lattice *hP* (lattice point symmetry 6/*mmm*) into the 'hexagonal' system and all space groups based on the rhombohedral Bravais lattice *hR* (lattice point symmetry $\bar{3}m$) into the 'rhombohedral' system. In Section 1.3.4, these systems are called 'lattice systems'. They were called 'Bravais systems' in earlier editions of *IT* A.

The theoretical background for the classification of space groups is provided in Section 1.3.4.

2.1.1.2. Conventional coordinate systems and cells

A plane group or space group usually is described by means of a *crystallographic coordinate system*, consisting of a *crystallographic basis* (basis vectors are lattice vectors) and a *crystallographic origin* (origin at a centre of symmetry or at a point of high site symmetry). The choice of such a coordinate system is not mandatory, since in principle a crystal structure can be referred to any coordinate system; *cf.* Chapters 1.3 and 1.5.

The selection of a crystallographic coordinate system is not unique. Conventionally, a right-handed set of basis vectors is taken such that the symmetry of the plane or space group is displayed best. With this convention, which is followed in *IT* A, the specific restrictions imposed on the cell parameters by each crystal family become particularly simple. They are listed in columns 6 and 7 of Table 2.1.1.1. If within these restrictions the smallest cell is chosen, a *conventional* (crystallographic) *basis* results. Together with the selection of an appropriate *conven-*

Table 2.1.1.1

Crystal families, crystal systems, conventional coordinate systems and Bravais lattices in one, two and three dimensions

Crystal family	Symbol†	Crystal system	Crystallographic point groups‡	No. of space groups	Conventional coordinate system: Restrictions on cell parameters	Conventional coordinate system: Parameters to be determined	Bravais lattices†
One dimension							
–	–	–	1, m	2	None	a	$\not{p}$
Two dimensions							
Oblique (monoclinic)	m	Oblique	1, 2	2	None	a, b γ§	mp
Rectangular (orthorhombic)	o	Rectangular	m, $2mm$	7	$\gamma = 90°$	a, b	op oc
Square (tetragonal)	t	Square	4, $4mm$	3	$a = b$ $\gamma = 90°$	a	tp
Hexagonal	h	Hexagonal	3, 6 $3m$, $6mm$	5	$a = b$ $\gamma = 120°$	a	hp
Three dimensions							
Triclinic (anorthic)	a	Triclinic	1, $\bar{1}$	2	None	a, b, c α, β, γ	aP
Monoclinic	m	Monoclinic	$2, m, 2/m$	13	b-unique setting $\alpha = \gamma = 90°$	a, b, c β §	mP mS¶ (mC, mA, mI)
					c-unique setting $\alpha = \beta = 90°$	a, b, c γ §	mP mS¶ (mA, mB, mI)
Orthorhombic	o	Orthorhombic	$222, mm2, mmm$	59	$\alpha = \beta = \gamma = 90°$	a, b, c	oP oS¶ (oC, oA, oB) oI oF
Tetragonal	t	Tetragonal	$4, \bar{4}, 4/m$ $422, 4mm, \bar{4}2m,$ $4/mmm$	68	$a = b$ $\alpha = \beta = \gamma = 90°$	a, c	tP tI
Hexagonal	h	Trigonal	$3, \bar{3}$ $32, 3m, \bar{3}m$	18	$a = b$ $\alpha = \beta = 90°,\ \gamma = 120°$	a, c	hP
				7	$a = b = c$ $\alpha = \beta = \gamma$ (rhombohedral axes, primitive cell) $a = b$ $\alpha = \beta = 90°, \gamma = 120°$ (hexagonal axes, triple obverse cell)	a, α	hR
		Hexagonal	$6, \bar{6}, 6/m$ $622, 6mm, \bar{6}2m,$ $6/mmm$	27	$a = b$ $\alpha = \beta = 90°, \gamma = 120°$	a, c	hP
Cubic	c	Cubic	$23, m\bar{3}$ $432, \bar{4}3m, m\bar{3}m$	36	$a = b = c$ $\alpha = \beta = \gamma = 90°$	a	cP cI cF

† The symbols for crystal families (column 2) and Bravais lattices (column 8) were adopted by the International Union of Crystallography in 1985; *cf.* de Wolff *et al.* (1985).
‡ Symbols surrounded by dashed or full lines indicate Laue groups; full lines indicate Laue groups which are also lattice point symmetries (holohedries).
§ These angles are conventionally taken to be non-acute, *i.e.* $\geq 90°$.
¶ For the use of the letter S as a new general, setting-independent 'centring symbol' for monoclinic and orthorhombic Bravais lattices, see de Wolff *et al.* (1985).

tional (crystallographic) *origin* (*cf.* Sections 2.1.3.2 and 2.1.3.7), such a basis defines a *conventional* (crystallographic) *coordinate system* and a *conventional cell.* The conventional cell of a point lattice or a space group, obtained in this way, turns out to be either *primitive* or to exhibit one of the *centring types* listed in Table 2.1.1.2. The centring type of a conventional cell is transferred to the lattice which is described by this cell; hence, we speak of primitive, face-centred, body-centred *etc.* lattices. Similarly, the cell parameters are often called lattice parameters; *cf.* Chapter 1.3 and Chapter 3.1 of *IT* A for further details.

In the triclinic, monoclinic and orthorhombic crystal systems, additional conventions (for instance cell reduction or metrical conventions based on the lengths of the cell edges) are needed to determine the choice and the labelling of the axes.

In *IT* A, all space groups within a crystal family are referred to the same kind of conventional coordinate system, with the exception of the hexagonal crystal family in three dimensions. Here, two kinds of coordinate systems are used, the hexagonal and the rhombohedral systems. In accordance with common crystallographic practice, all space groups based on the hexagonal Bravais lattice hP (18 trigonal and 27 hexagonal space groups) are described only with a hexagonal coordinate system (primitive cell), whereas the seven space groups based on the rhombohedral Bravais lattice hR (the so-called 'rhombohedral space groups')

Table 2.1.1.2
Symbols for the conventional centring types of one-, two- and three-dimensional cells

Symbol	Centring type of cell	Number of lattice points per cell	Coordinates of lattice points within cell
One dimension			
$\mathscr{p}$	Primitive	1	0
Two dimensions			
p	Primitive	1	0, 0
c	Centred	2	0, 0; $\frac{1}{2}, \frac{1}{2}$
h†	Hexagonally centred	3	0, 0; $\frac{2}{3}, \frac{1}{3}; \frac{1}{3}, \frac{2}{3}$
Three dimensions			
P	Primitive	1	0, 0, 0
C	*C*-face centred	2	0, 0, 0; $\frac{1}{2}, \frac{1}{2}$, 0
A	*A*-face centred	2	0, 0, 0; 0, $\frac{1}{2}, \frac{1}{2}$
B	*B*-face centred	2	0, 0, 0; $\frac{1}{2}$, 0, $\frac{1}{2}$
I	Body centred	2	0, 0, 0; $\frac{1}{2}, \frac{1}{2}, \frac{1}{2}$
F	All-face centred	4	0, 0, 0; $\frac{1}{2}, \frac{1}{2}$, 0; 0, $\frac{1}{2}, \frac{1}{2}$; $\frac{1}{2}$, 0, $\frac{1}{2}$
R‡	Rhombohedrally centred (description with 'hexagonal axes')	3	0, 0, 0; $\frac{2}{3}, \frac{1}{3}, \frac{1}{3}$; $\frac{1}{3}, \frac{2}{3}, \frac{2}{3}$ ('obverse setting') 0, 0, 0; $\frac{1}{3}, \frac{2}{3}, \frac{1}{3}$; $\frac{2}{3}, \frac{1}{3}, \frac{2}{3}$ ('reverse setting')
	Primitive (description with 'rhombohedral axes')	1	0, 0, 0
H§	Hexagonally centred	3	0, 0, 0; $\frac{2}{3}, \frac{1}{3}$, 0; $\frac{1}{3}, \frac{2}{3}$, 0

† The two-dimensional triple hexagonal cell h is an alternative description of the hexagonal plane net. It is not used for systematic plane-group description in *IT* A; it is introduced, however, in the sub- and supergroup entries of the plane-group tables of *International Tables for Crystallography*, Vol. A1 (2010), abbreviated as *IT* A1. (For a discussion of the two-dimensional triple hexagonal cell h, its illustration and the related transformation matrices, see Chapter 1.5 of *IT* A.)
‡ In the space-group tables of *IT* A, as well as in *IT* (1935) and *IT* (1952), the seven rhombohedral R space groups are presented with two descriptions, one based on *hexagonal axes* (triple cell), one on *rhombohedral axes* (primitive cell). In *IT* A, as well as in *IT* (1952) and *IT* A (2002), the *obverse* setting of the triple hexagonal cell R is used. Note that in *IT* (1935) the *reverse* setting was employed. The two settings are related by a rotation of the hexagonal cell with respect to the rhombohedral lattice around a threefold axis, involving a rotation angle of 60, 180 or 300° (*cf.* Fig. 1.5.1.6). Further details may be found in Section 1.5.4 and in Chapter 3.1 of *IT* A. Transformation matrices are contained in Table 1.5.1.1.
§ The triple hexagonal cell H is an alternative description of the hexagonal Bravais lattice. It was used for systematic space-group description in *IT* (1935), but replaced by P in *IT* (1952). It is used in the tables of maximal subgroups and minimal supergroups of the space groups in *IT* A1 (2010). (For a discussion of the triple hexagonal cell H, its illustration and the related transformation matrices, see Chapter 1.5 of *IT* A.).

are treated in two versions, one referred to 'hexagonal axes' (triple obverse cell) and one to 'rhombohedral axes' (primitive cell); *cf.* Table 2.1.1.2. In practice, hexagonal axes are preferred because they are easier to visualize.

Table 2.1.1.2 contains only those conventional centring symbols which occur in the Hermann–Mauguin space-group symbols. There exist, of course, further kinds of centred cells which are unconventional, see for example the synoptic tables of plane and space groups discussed in Chapter 1.5 of *IT* A. The centring type of a cell may change with a change of the basis vectors; in particular, a primitive cell may become a centred cell and *vice versa*. Examples of relevant transformation matrices are contained in Table 1.5.1.1.

2.1.2. Symbols of symmetry elements

By Theo Hahn and Mois I. Aroyo

As already introduced in Section 1.2.3, a 'symmetry element' (of a given structure or object) is defined as a concept with two components; it is the combination of a 'geometric element' (that allows the fixed points of a reduced symmetry operation to be located and oriented in space) with the set of symmetry operations having this geometric element in common ('element set'). The element set of a symmetry element is represented by the so-called 'defining operation', which is the simplest symmetry operation from the element set that suffices to identify the geometric element. The alphanumeric and graphical symbols of symmetry elements and the related symmetry operations used throughout the tables of plane (Chapter 2.2 of *IT* A) and space groups (Chapter 2.3 of *IT* A) are listed in Tables 2.1.2.1 to 2.1.2.7. For detailed discussion of the definition and symbols of symmetry elements, *cf.* Section 1.2.3, de Wolff *et al.* (1989, 1992) and Flack *et al.* (2000).

The alphanumeric symbols shown in Table 2.1.2.1 correspond to those symmetry elements and symmetry operations which occur in the conventional Hermann–Mauguin symbols of point groups and space groups. Further so-called 'additional symmetry elements' (see Section 1.4.2.4 for a description of these) correspond to additional symmetry operations that appear in the so-called 'extended Hermann–Mauguin symbols' (*cf.* Section 1.5.4). The symbols of symmetry elements (symmetry operations), except for glide planes (glide reflections), are independent of the choice and the labelling of the basis vectors and of the origin. The symbols of glide planes (glide reflections), however, may change with a change of the basis vectors. For this reason, the possible orientations of glide planes and the glide vectors of the corresponding operations are listed explicitly in columns 2 and 3 of Table 2.1.2.1.

In 1992, following a proposal of the Commission on Crystallographic Nomenclature (de Wolff *et al.*, 1992), the International Union of Crystallography introduced the symbol '*e*' and graphical symbols for the designation of the so-called 'double' glide planes. The double- or *e*-glide plane occurs only in centred cells and its geometric element is a plane shared by glide reflections with perpendicular glide vectors related by a centring translation (for details on *e*-glide planes, *cf.* Section 1.2.3). The introduction of the symbol *e* for the designation of double-glide planes (*cf.* de Wolff *et al.*, 1992) results in the modification of the Hermann–Mauguin symbols of five orthorhombic groups:

Table 2.1.2.1
Symbols for symmetry elements and for the corresponding symmetry operations in one, two and three dimensions

Symbol	Symmetry element and its orientation	Defining symmetry operation with glide or screw vector
m	Reflection plane, mirror plane	Reflection through the plane
	Reflection line, mirror line (two dimensions)	Reflection through the line
	Reflection point, mirror point (one dimension)	Reflection through the point
a, b or c	'Axial' glide plane	Glide reflection through the plane, with glide vector
a	$\perp$ [010] or $\perp$ [001]	$\frac{1}{2}\mathbf{a}$
b	$\perp$ [001] or $\perp$ [100]	$\frac{1}{2}\mathbf{b}$
c †	$\perp$ [100] or $\perp$ [010]	$\frac{1}{2}\mathbf{c}$
	$\perp$ [1$\bar{1}$0] or $\perp$ [110]	$\frac{1}{2}\mathbf{c}$
	$\perp$ [100] or $\perp$ [010] or $\perp$ [$\bar{1}\bar{1}$0]	$\frac{1}{2}\mathbf{c}$ } hexagonal coordinate system
	$\perp$ [1$\bar{1}$0] or $\perp$ [120] or $\perp$ [$\bar{2}\bar{1}$0]	$\frac{1}{2}\mathbf{c}$ } hexagonal coordinate system
e ‡	'Double' glide plane (in centred cells only)	*Two* glide reflections through *one* plane, with perpendicular glide vectors
	$\perp$ [001]	$\frac{1}{2}\mathbf{a}$ *and* $\frac{1}{2}\mathbf{b}$
	$\perp$ [100]	$\frac{1}{2}\mathbf{b}$ *and* $\frac{1}{2}\mathbf{c}$
	$\perp$ [010]	$\frac{1}{2}\mathbf{a}$ *and* $\frac{1}{2}\mathbf{c}$
	$\perp$ [1$\bar{1}$0]; $\perp$ [110]	$\frac{1}{2}(\mathbf{a}+\mathbf{b})$ *and* $\frac{1}{2}\mathbf{c}$; $\frac{1}{2}(\mathbf{a}-\mathbf{b})$ *and* $\frac{1}{2}\mathbf{c}$
	$\perp$ [01$\bar{1}$]; $\perp$ [011]	$\frac{1}{2}(\mathbf{b}+\mathbf{c})$ *and* $\frac{1}{2}\mathbf{a}$; $\frac{1}{2}(\mathbf{b}-\mathbf{c})$ *and* $\frac{1}{2}\mathbf{a}$
	$\perp$ [$\bar{1}$01]; $\perp$ [101]	$\frac{1}{2}(\mathbf{a}+\mathbf{c})$ *and* $\frac{1}{2}\mathbf{b}$; $\frac{1}{2}(\mathbf{a}-\mathbf{c})$ *and* $\frac{1}{2}\mathbf{b}$
n	'Diagonal' glide plane	Glide reflection through the plane, with glide vector
	$\perp$ [001]; $\perp$ [100]; $\perp$ [010]	$\frac{1}{2}(\mathbf{a}+\mathbf{b})$; $\frac{1}{2}(\mathbf{b}+\mathbf{c})$; $\frac{1}{2}(\mathbf{a}+\mathbf{c})$
	$\perp$ [1$\bar{1}$0] or $\perp$ [01$\bar{1}$] or $\perp$ [$\bar{1}$01]	$\frac{1}{2}(\mathbf{a}+\mathbf{b}+\mathbf{c})$
	$\perp$ [110]; $\perp$ [011]; $\perp$ [101]	$\frac{1}{2}(-\mathbf{a}+\mathbf{b}+\mathbf{c})$; $\frac{1}{2}(\mathbf{a}-\mathbf{b}+\mathbf{c})$; $\frac{1}{2}(\mathbf{a}+\mathbf{b}-\mathbf{c})$
d §	'Diamond' glide plane	Glide reflection through the plane, with glide vector
	$\perp$ [001]; $\perp$ [100]; $\perp$ [010]	$\frac{1}{4}(\mathbf{a}\pm\mathbf{b})$; $\frac{1}{4}(\mathbf{b}\pm\mathbf{c})$; $\frac{1}{4}(\pm\mathbf{a}+\mathbf{c})$
	$\perp$ [1$\bar{1}$0]; $\perp$ [01$\bar{1}$]; $\perp$ [$\bar{1}$01]	$\frac{1}{4}(\mathbf{a}+\mathbf{b}\pm\mathbf{c})$; $\frac{1}{4}(\pm\mathbf{a}+\mathbf{b}+\mathbf{c})$; $\frac{1}{4}(\mathbf{a}\pm\mathbf{b}+\mathbf{c})$
	$\perp$ [110]; $\perp$ [011]; $\perp$ [101]	$\frac{1}{4}(-\mathbf{a}+\mathbf{b}\pm\mathbf{c})$; $\frac{1}{4}(\pm\mathbf{a}-\mathbf{b}+\mathbf{c})$; $\frac{1}{4}(\mathbf{a}\pm\mathbf{b}-\mathbf{c})$
g	Glide line (two dimensions)	Glide reflection through the line, with glide vector
	$\perp$ [01]; $\perp$ [10]	$\frac{1}{2}\mathbf{a}$; $\frac{1}{2}\mathbf{b}$
1	None	Identity
2, 3, 4, 6	n-fold rotation axis, n	Counter-clockwise rotation of 360/n degrees around the axis
	n-fold rotation point, n (two dimensions)	Counter-clockwise rotation of 360/n degrees around the point
$\bar{1}$	Centre of symmetry, inversion centre	Inversion through the point
$\bar{2} = m$, ¶ $\bar{3}$, $\bar{4}$, $\bar{6}$	Rotoinversion axis, $\bar{n}$, and inversion point on the axis ††	Counter-clockwise rotation of 360/n degrees around the axis, followed by inversion through the point on the axis ††
2_1 3_1, 3_2 4_1, 4_2, 4_3 6_1, 6_2, 6_3, 6_4, 6_5	n-fold screw axis, n_p	Right-handed screw rotation of 360/n degrees around the axis, with screw vector (pitch) (p/n) **t**; here **t** is the shortest lattice translation vector parallel to the axis in the direction of the screw

† In the rhombohedral space-group symbols $R3c$ (161) and $R\bar{3}c$ (167), the symbol c refers to the description with 'hexagonal axes'; *i.e.* the glide vector is $\frac{1}{2}\mathbf{c}$, along [001]. In the description with 'rhombohedral axes', this glide vector is $\frac{1}{2}(\mathbf{a}+\mathbf{b}+\mathbf{c})$, along [111], *i.e.* the symbol of the glide plane would be n.
‡ Glide planes 'e' occur in orthorhombic A-, C- and F-centred space groups, tetragonal I-centred and cubic F- and I-centred space groups. The geometric element of an e-glide plane is a plane shared by glide reflections with perpendicular glide vectors, with at least one glide vector along a crystal axis [*cf.* Sections 1.2.3 and 2.1.2, and de Wolff *et al.* (1992)].
§ Glide planes d occur only in orthorhombic F space groups, in tetragonal I space groups, and in cubic I and F space groups. They always occur in pairs with alternating glide vectors, for instance $\frac{1}{4}(\mathbf{a}+\mathbf{b})$ and $\frac{1}{4}(\mathbf{a}-\mathbf{b})$. The second power of a glide reflection d is a centring vector.
¶ Only the symbol m is used in the Hermann–Mauguin symbols, for both point groups and space groups.
†† The inversion point is a centre of symmetry if n is odd.

Space group No.	39	41	64	67	68
New symbol:	$Aem2$	$Aea2$	$Cmce$	$Cmme$	$Ccce$
Former symbol:	$Abm2$	$Aba2$	$Cmca$	$Cmma$	$Ccca$

Since the introduction of its use in *IT* A (2002) the new symbol is the standard one; it is indicated in the headline of these space groups, while the former symbol is given underneath.

The graphical symbols of symmetry planes are shown in Tables 2.1.2.2 to 2.1.2.4. Like the alphanumeric symbols, the graphical symbols and their explanations (columns 2 and 3) are independent of the projection direction and the labelling of the basis vectors. They are, therefore, applicable to any projection diagram of a space group. The alphanumeric symbols of glide planes (column 4), however, may change with a change of the basis vectors. For example, the dash-dotted n glide in the hexagonal description becomes an a, b or c glide in the rhombohedral description. In monoclinic space groups, the 'parallel' vector of a glide plane may be along a lattice translation vector that is inclined to the projection plane.

The 'e'-glide graphical symbols are applied to the diagrams of seven orthorhombic A-, C- and F- centred space groups, five tetragonal I-centred space groups, and five cubic F- and I-centred space groups. The 'double-dotted-dash' symbol for e glides 'normal' and 'inclined' to the plane of projection was introduced in 1992 (de Wolff *et al.*, 1992), while the 'double-arrowed' graphical symbol for e-glide planes oriented 'parallel' to the projection plane had already been used in *IT* (1935) and *IT* (1952).

Table 2.1.2.2
Graphical symbols of symmetry planes normal to the plane of projection (three dimensions) and symmetry lines in the plane of the figure (two dimensions)

Description	Graphical symbol	Glide vector(s) of the defining operation(s) of the glide plane (in units of the shortest lattice translation vectors parallel and normal to the projection plane)	Symmetry element represented by the graphical symbol
Reflection plane, mirror plane; Reflection line, mirror line (two dimensions)		None	*m*
'Axial' glide plane; Glide line (two dimensions)		$\frac{1}{2}$ parallel to line in projection plane; $\frac{1}{2}$ parallel to line in figure plane	*a*, *b* or *c*; *g*
'Axial' glide plane		$\frac{1}{2}$ normal to projection plane	*a*, *b* or *c*
'Double' glide plane†		*Two* glide vectors: $\frac{1}{2}$ parallel to line in, and $\frac{1}{2}$ normal to projection plane	*e*
'Diagonal' glide plane		*One* glide vector with *two* components: $\frac{1}{2}$ parallel to line in, and $\frac{1}{2}$ normal to projection plane	*n*
'Diamond' glide plane‡ (pair of planes)	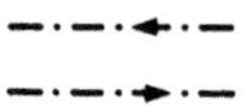	$\frac{1}{4}$ parallel to line in projection plane, combined with $\frac{1}{4}$ normal to projection plane (arrow indicates direction parallel to the projection plane for which the normal component is positive)	*d*

† The graphical symbols of the '*e*'-glide planes are applied to the diagrams of seven orthorhombic *A*-, *C*- and *F*-centred space groups, five tetragonal *I*-centred space groups, and five cubic *F*- and *I*-centred space groups.
‡ Glide planes *d* occur only in orthorhombic *F* space groups, in tetragonal *I* space groups, and in cubic *I* and *F* space groups. They always occur in pairs with alternating glide vectors, for instance $\frac{1}{4}(\mathbf{a}+\mathbf{b})$ and $\frac{1}{4}(\mathbf{a}-\mathbf{b})$. The second power of a glide reflection *d* is a centring vector.

Table 2.1.2.3
Graphical symbols of symmetry planes parallel to the plane of projection

Description	Graphical symbol†	Glide vector(s) of the defining operation(s) of the glide plane (in units of the shortest lattice translation vectors parallel to the projection plane)	Symmetry element represented by the graphical symbol
Reflection plane, mirror plane	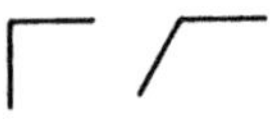	None	*m*
'Axial' glide plane		$\frac{1}{2}$ in the direction of the arrow	*a*, *b* or *c*
'Double' glide plane‡		*Two* glide vectors: $\frac{1}{2}$ in either of the directions of the two arrows	*e*
'Diagonal' glide plane		*One* glide vector with *two* components $\frac{1}{2}$ in the direction of the arrow	*n*
'Diamond' glide plane§ (pair of planes)	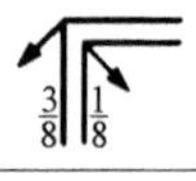	$\frac{1}{2}$ in the direction of the arrow; the glide vector is always half of a centring vector, *i.e.* one quarter of a diagonal of the conventional face-centred cell	*d*

† The symbols are given at the upper left corner of the space-group diagrams. A fraction *h* attached to a symbol indicates two symmetry planes with 'heights' *h* and $h+\frac{1}{2}$ above the plane of projection; *e.g.* $\frac{1}{8}$ stands for $h=\frac{1}{8}$ and $\frac{5}{8}$. No fraction means $h=0$ and $\frac{1}{2}$ (*cf.* Section 2.1.3.6).
‡ The graphical symbols of the '*e*'-glide planes are applied to the diagrams of seven orthorhombic *A*-, *C*- and *F*-centred space groups, five tetragonal *I*-centred space groups, and five cubic *F*- and *I*-centred space groups.
§ Glide planes *d* occur only in orthorhombic *F* space groups, in tetragonal *I* space groups, and in cubic *I* and *F* space groups. They always occur in pairs with alternating glide vectors, for instance $\frac{1}{4}(\mathbf{a}+\mathbf{b})$ and $\frac{1}{4}(\mathbf{a}-\mathbf{b})$. The second power of a glide reflection *d* is a centring vector.

The graphical symbols of symmetry axes and their descriptions are shown in Tables 2.1.2.5–2.1.2.7. The screw vectors of the defining operations of screw axes are given in units of the shortest lattice translation vectors parallel to the axes. The symbols in the last column of the tables indicate the symmetry elements that are represented by the graphical symbols in the symmetry-element diagrams of the space groups. Two main cases may be distinguished:

(i) graphical symbols of symmetry elements that in the space-group diagrams represent just one symmetry element. Thus, the graphical symbol of a fourfold rotation axis or an inversion centre represent the symmetry element 4 or $\bar{1}$. Similarly,

Table 2.1.2.4
Graphical symbols of symmetry planes inclined to the plane of projection (in cubic space groups of classes $\bar{4}3m$ and $m\bar{3}m$ only)

Description	Graphical symbol† for planes normal to [011] and [01$\bar{1}$]	Graphical symbol† for planes normal to [101] and [10$\bar{1}$]	Glide vector(s) (in units of the shortest lattice translation vectors) of the defining operation(s) of the glide plane normal to [011] and [01$\bar{1}$]	Glide vector(s) (in units of the shortest lattice translation vectors) of the defining operation(s) of the glide plane normal to [101] and [10$\bar{1}$]	Symmetry element represented by the graphical symbol
Reflection plane, mirror plane	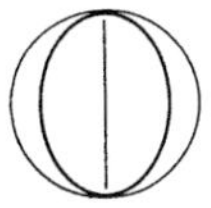	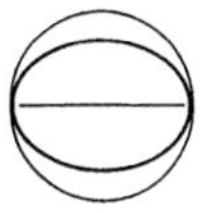	None	None	*m*
'Axial' glide plane	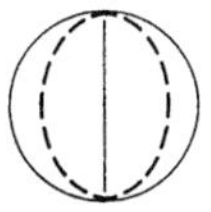	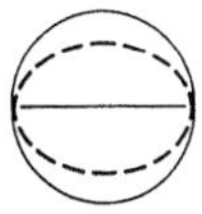	$\frac{1}{2}$ along [100]	$\frac{1}{2}$ along [010]	*a* or *b*
'Axial' glide plane	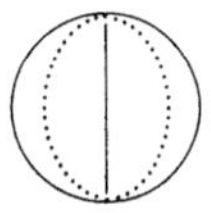	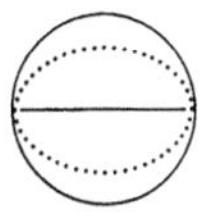	$\frac{1}{2}$ along [01$\bar{1}$] or along [011]	$\frac{1}{2}$ along [10$\bar{1}$] or along [101]	*a* or *b*
'Double' glide plane [in space groups $I\bar{4}3m$ (217) and $Im\bar{3}m$ (229) only]	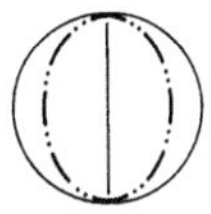	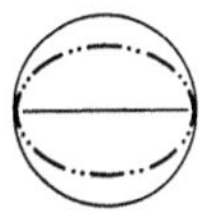	*Two* glide vectors: $\frac{1}{2}$ along [100] *and* $\frac{1}{2}$ along [01$\bar{1}$] or $\frac{1}{2}$ along [011]	*Two* glide vectors: $\frac{1}{2}$ along [010] *and* $\frac{1}{2}$ along [10$\bar{1}$] or $\frac{1}{2}$ along [101]	*e*
'Diagonal' glide plane	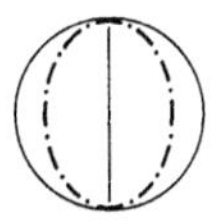	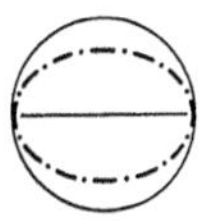	*One* glide vector: $\frac{1}{2}$ along [11$\bar{1}$] or along [111]‡	*One* glide vector: $\frac{1}{2}$ along [1$\bar{1}$1] or along [111]‡	*n*
'Diamond' glide plane§ (pair of planes)	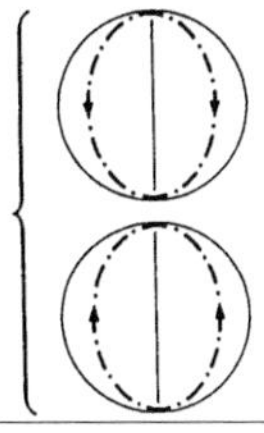	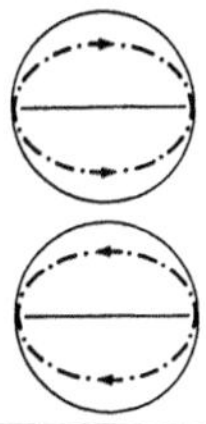	$\frac{1}{2}$ along [1$\bar{1}$1] or along [111]¶	$\frac{1}{2}$ along [$\bar{1}$11] or along [111] ¶	*d*
			$\frac{1}{2}$ along [$\bar{1}\bar{1}$1] or along [$\bar{1}$11]¶	$\frac{1}{2}$ along [$\bar{1}\bar{1}$1] or along [1$\bar{1}$1] ¶	*d*

† The symbols represent orthographic projections. In the cubic space-group diagrams, complete orthographic projections of the symmetry elements around high-symmetry points, such as 0, 0, 0; $\frac{1}{2}$, 0, 0; $\frac{1}{4}$, $\frac{1}{4}$, 0, are given as 'inserts'.
‡ In the space groups $F\bar{4}3m$ (216), $Fm\bar{3}m$ (225) and $Fd\bar{3}m$ (227), the shortest lattice translation vectors in the glide directions are $\mathbf{t}(1, \frac{1}{2}, \bar{\frac{1}{2}})$ or $\mathbf{t}(1, \frac{1}{2}, \frac{1}{2})$ and $\mathbf{t}(\frac{1}{2}, 1, \bar{\frac{1}{2}})$ or $\mathbf{t}(\frac{1}{2}, 1, \frac{1}{2})$, respectively.
§ Glide planes *d* occur only in orthorhombic *F* space groups, in tetragonal *I* space groups, and in cubic *I* and *F* space groups. They always occur in pairs with alternating glide vectors, for instance $\frac{1}{4}(\mathbf{a}+\mathbf{b})$ and $\frac{1}{4}(\mathbf{a}-\mathbf{b})$. The second power of a glide reflection *d* is a centring vector.
¶ The glide vector is half of a centring vector, *i.e.* one quarter of the diagonal of the conventional body-centred cell in space groups $I\bar{4}3d$ (220) and $Ia\bar{3}d$ (230).

the graphical symbols of symmetry planes (Tables 2.1.2.2–2.1.2.4) represent just one symmetry element (namely, mirror or glide plane) in the space-group diagrams;

(ii) graphical symbols of symmetry elements that in the space-group diagrams represent more than one symmetry element. For example, the graphical symbol described in Table 2.1.2.5 as 'Inversion axis: 3 bar' ($\bar{3}$),

represents in the diagrams the three different symmetry elements $\bar{3}$, 3, $\bar{1}$.

The last six entries of Table 2.1.2.5 are combinations of symbols of symmetry axes with that of a centre of inversion. When displayed on the space-group diagrams, the combined graphical symbols represent more than one symmetry element. For example, the symbol for a fourfold rotation axis with a centre of inversion (4/*m*),

 or 

represents the symmetry elements $\bar{4}$, 4 and $\bar{1}$.

The meaning of a graphical symbol on the space-group diagrams is often confused with the set of symmetry elements that constitute the site-symmetry group associated with the symmetry element displayed. As an example, consider the rotoinversion axis $\bar{6}$ (described as 'Inversion axis: 6 bar' in Table 2.1.2.5). The site-symmetry group $\bar{6}$ can be decomposed into three symmetry elements: $\bar{6}$, 3 and *m* (*cf.* de Wolff *et al.*, 1989). However, the graphical symbol of $\bar{6}$ in the diagrams represents the two symmetry elements $\bar{6}$ and 3, as the symmetry element '*m*' (that 'belongs' to $\bar{6}$) is represented by a separate graphical symbol.

Table 2.1.2.5
Graphical symbols of symmetry axes normal to the plane of projection and symmetry points in the plane of the figure

Description	Alphanumeric symbol	Graphical symbol†	Screw vector of the defining operation of the screw axis (in units of the shortest lattice translation vector parallel to the axis)	Symmetry elements represented by the graphical symbol
Twofold rotation axis Twofold rotation point (two dimensions)	2		None	2
Twofold screw axis: '2 sub 1'	2_1		$\frac{1}{2}$	2_1
Threefold rotation axis Threefold rotation point (two dimensions)	3		None	3
Threefold screw axis: '3 sub 1'	3_1		$\frac{1}{3}$	3_1
Threefold screw axis: '3 sub 2'	3_2		$\frac{2}{3}$	3_2
Fourfold rotation axis Fourfold rotation point (two dimensions)	4		None	4
Fourfold screw axis: '4 sub 1'	4_1	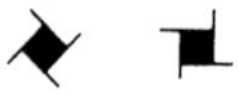	$\frac{1}{4}$	4_1
Fourfold screw axis: '4 sub 2'	4_2		$\frac{1}{2}$	4_2
Fourfold screw axis: '4 sub 3'	4_3	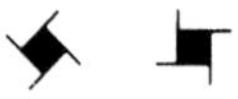	$\frac{3}{4}$	4_3
Sixfold rotation axis Sixfold rotation point (two dimensions)	6		None	6
Sixfold screw axis: '6 sub 1'	6_1		$\frac{1}{6}$	6_1
Sixfold screw axis: '6 sub 2'	6_2		$\frac{1}{3}$	6_2
Sixfold screw axis: '6 sub 3'	6_3		$\frac{1}{2}$	6_3
Sixfold screw axis: '6 sub 4'	6_4		$\frac{2}{3}$	6_4
Sixfold screw axis: '6 sub 5'	6_5		$\frac{5}{6}$	6_5
Centre of symmetry, inversion centre: '1 bar' Reflection point, mirror point (one dimension)	$\bar{1}$		None	$\bar{1}$
Inversion axis: '3 bar'	$\bar{3}$		None	$\bar{3}, \bar{1}, 3$
Inversion axis: '4 bar'	$\bar{4}$		None	$\bar{4}, 2$
Inversion axis: '6 bar'	$\bar{6}$		None	$\bar{6}, 3$
Twofold rotation axis with centre of symmetry	$2/m$		None	$2, \bar{1}$
Twofold screw axis with centre of symmetry	$2_1/m$		$\frac{1}{2}$	$2_1, \bar{1}$
Fourfold rotation axis with centre of symmetry	$4/m$		None	$4, \bar{4}, \bar{1}$
'4 sub 2' screw axis with centre of symmetry	$4_2/m$		$\frac{1}{2}$	$4_2, \bar{4}, \bar{1}$
Sixfold rotation axis with centre of symmetry	$6/m$		None	$6, \bar{6}, \bar{3}, \bar{1}$
'6 sub 3' screw axis with centre of symmetry	$6_3/m$		$\frac{1}{2}$	$6_3, \bar{6}, \bar{3}, \bar{1}$

† Notes on the 'heights' h of symmetry points $\bar{1}$, $\bar{3}$, $\bar{4}$ and $\bar{6}$:
(1) Centres of symmetry $\bar{1}$ and $\bar{3}$, as well as inversion points $\bar{4}$ and $\bar{6}$ on $\bar{4}$ and $\bar{6}$ axes parallel to [001], occur in pairs at 'heights' h and $h + \frac{1}{2}$. In the space-group diagrams, only one fraction h is given, *e.g.* $\frac{1}{4}$ stands for $h = \frac{1}{4}$ and $\frac{3}{4}$. No fraction means $h = 0$ and $\frac{1}{2}$. In *cubic* space groups, however, because of their complexity, *both* fractions are given for vertical $\bar{4}$ axes, including $h = 0$ and $\frac{1}{2}$.
(2) Symmetries $4/m$ and $6/m$ contain vertical $\bar{4}$ and $\bar{6}$ axes; their $\bar{4}$ and $\bar{6}$ inversion points coincide with the centres of symmetry. This is not indicated in the space-group diagrams.
(3) Symmetries $4_2/m$ and $6_3/m$ also contain vertical $\bar{4}$ and $\bar{6}$ axes, but their $\bar{4}$ and $\bar{6}$ inversion points alternate with the centres of symmetry; *i.e.* $\bar{1}$ points at h and $h + \frac{1}{2}$ interleave with $\bar{4}$ or $\bar{6}$ points at $h + \frac{1}{4}$ and $h + \frac{3}{4}$. In the tetragonal and hexagonal space-group diagrams, only *one* fraction for $\bar{1}$ and one for $\bar{4}$ or $\bar{6}$ is given. In the cubic diagrams, *all four* fractions are listed for $4_2/m$; *e.g.* $Pm\bar{3}n$ (223): $\bar{1}$: $0, \frac{1}{2}$; $\bar{4}$: $\frac{1}{4}, \frac{3}{4}$.

Table 2.1.2.6
Graphical symbols of symmetry axes parallel to the plane of projection

Description	Graphical symbol†	Screw vector of the defining operation of the screw axis (in units of the shortest lattice translation vector parallel to the axis)	Symmetry elements represented by the graphical symbol
Twofold rotation axis		None	2
Twofold screw axis: '2 sub 1'		$\frac{1}{2}$	2_1
Fourfold rotation axis	 (in cubic space groups only)	None	4
Fourfold screw axis: '4 sub 1'	 (in cubic space groups only)	$\frac{1}{4}$	4_1
Fourfold screw axis: '4 sub 2'	 (in cubic space groups only)	$\frac{1}{2}$	4_2
Fourfold screw axis: '4 sub 3'	 (in cubic space groups only)	$\frac{3}{4}$	4_3
Inversion axis: '4 bar'	 (in cubic space groups only)	None	$\bar{4}$, 2
Inversion point on '4 bar' axis	 (in cubic space groups only)	None	None

† The symbols for horizontal symmetry axes are given outside the unit cell of the space-group diagrams. *Twofold* axes always occur in pairs, at 'heights' h and $h+\frac{1}{2}$ above the plane of projection; here, a fraction h attached to such a symbol indicates two axes with heights h and $h+\frac{1}{2}$. No fraction stands for $h=0$ and $\frac{1}{2}$. The rule of pairwise occurrence, however, is not valid for the horizontal *fourfold* axes in cubic space groups; here, *all* heights are given, including $h=0$ and $\frac{1}{2}$. This applies also to the horizontal $\bar{4}$ axes and the $\bar{4}$ inversion points located on these axes.

Table 2.1.2.7
Graphical symbols of symmetry axes inclined to the plane of projection (in cubic space groups only)

Description	Graphical symbol†	Screw vector of the defining operation of the screw axis (in units of the shortest lattice translation vector parallel to the axis)	Symmetry elements represented by the graphical symbol
Twofold rotation axis	 (Parallel to a face diagonal of the cube)	None	2
Twofold screw axis: '2 sub 1'	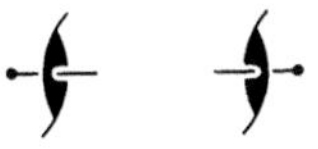 (Parallel to a face diagonal of the cube)	$\frac{1}{2}$	2_1
Threefold rotation axis	 (Parallel to a body diagonal of the cube)	None	3
Threefold screw axis: '3 sub 1'	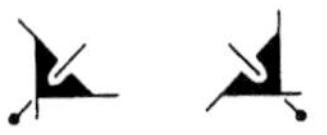 (Parallel to a body diagonal of the cube)	$\frac{1}{3}$	3_1
Threefold screw axis: '3 sub 2'	 (Parallel to a body diagonal of the cube)	$\frac{2}{3}$	3_2
Inversion axis: '3 bar'	 (Parallel to a body diagonal of the cube)	None	$\bar{3}$, 3, $\bar{1}$

† The dots mark the intersection points of axes with the plane at $h=0$. In some cases, the intersection points are obscured by symbols of symmetry elements with height $h \geq 0$; examples: $Fd\bar{3}$ (203), origin choice 2; $Pn\bar{3}n$ (222), origin choice 2; $Pm\bar{3}n$ (223); $Im\bar{3}m$ (229); $Ia\bar{3}d$ (230).

2.1.3. Contents and arrangement of the tables

By Theo Hahn and Aafje Looijenga-Vos

2.1.3.1. General layout

The presentation of the plane-group and space-group data in *IT* A follows the style of the previous editions of *International Tables*. The data for most of the space groups are displayed on one page or on two facing pages. A typical distribution of the data is shown below and is illustrated by the example of *Cccm* (66) provided inside the front and back covers.

Left-hand page:
(1) *Headline*
(2) Diagrams for the symmetry elements and the general position (for graphical symbols of symmetry elements see Section 2.1.2)
(3) *Origin*
(4) *Asymmetric unit*
(5) *Symmetry operations*

Right-hand page:
(6) *Headline* in abbreviated form
(7) *Generators selected*; this information is the basis for the order of the entries under *Symmetry operations* and *Positions*
(8) General and special *Positions*, with the following columns:
Multiplicity
Wyckoff letter
Site symmetry, given by the oriented site-symmetry symbol
Coordinates
Reflection conditions
Note: In a few space groups, two special positions with the same reflection conditions are printed on the same line
(9) *Symmetry of special projections* (not given for plane groups)

It is important to note that the symmetry data are displayed in the same sequence for all the space groups. The actual distribution of the data between pages can vary depending on the amount and nature of the data that are shown.

The symmetry data for the ten space groups of the crystal class $m\bar{3}m$ [$Pm\bar{3}m$ (221) to $Ia\bar{3}d$ (230)] are displayed on four pages. Additional general-position diagrams in tilted projection are shown on the fourth page, providing a three-dimensional-style view of these complicated general-position diagrams.

2.1.3.2. Space groups with more than one description

For several space groups, more than one description is available. Three cases occur:

(i) *Two choices of origin (cf. Section 2.1.3.7)*
For all centrosymmetric space groups, the tables contain a description with a centre of symmetry as origin. Some centrosymmetric space groups, however, contain points of high site symmetry that do not coincide with a centre of symmetry. For these 24 cases, a further description (including diagrams) with a high-symmetry point as origin is provided. Neither of the two origin choices is considered standard. Noncentrosymmetric space groups and all plane groups are described with only one choice of origin.

Examples
(1) $I4_1/amd$ (141)
Origin choice 1 at a point with site symmetry $\bar{4}m2$
Origin choice 2 at a centre with site symmetry $2/m$.
(2) $Fd\bar{3}m$ (227)
Origin choice 1 at a point with site symmetry $\bar{4}3m$
Origin choice 2 at a centre with site symmetry $\bar{3}m$.

(ii) *Monoclinic space groups*
Two complete descriptions are given for each of the 13 monoclinic space groups, one for the setting with 'unique axis *b*', followed by one for the setting with 'unique axis *c*'.
Additional descriptions in synoptic form are provided for the following eight monoclinic space groups with centred lattices or glide planes:

$C2$ (5), Pc (7), Cm (8), Cc (9), $C2/m$ (12), $P2/c$ (13), $P2_1/c$ (14), $C2/c$ (15)

These synoptic descriptions consist of abbreviated treatments for three 'cell choices', here called 'cell choices 1, 2 and 3'. Cell choice 1 corresponds to the complete treatment, mentioned above; for comparative purposes, it is repeated among the synoptic descriptions which, for each setting, are printed on two facing pages. The cell choices and their relations are introduced in Section 1.5.3.1, see also Table 1.5.1.1. (For more details, *cf.* Section 2.1.3.15 of *IT* A.)

(iii) *Rhombohedral space groups*
The seven rhombohedral space groups $R3$ (146), $R\bar{3}$ (148), $R32$ (155), $R3m$ (160), $R3c$ (161), $R\bar{3}m$ (166) and $R\bar{3}c$ (167) are described with two coordinate systems, first with *hexagonal axes* (triple hexagonal cell) and second with *rhombohedral axes* (primitive rhombohedral cell). The same space-group symbol is used for both descriptions. For convenience, the relations between the cell parameters a, c of the triple hexagonal cell and the cell parameters a' and α' of the primitive rhombohedral cell (*cf.* Table 2.1.1.1) are listed:

$$a = a'\sqrt{2}\sqrt{1-\cos\alpha'} = 2a'\sin\frac{\alpha'}{2}$$

$$c = a'\sqrt{3}\sqrt{1+2\cos\alpha'}$$

$$\frac{c}{a} = \sqrt{\frac{3}{2}}\sqrt{\frac{1+2\cos\alpha'}{1-\cos\alpha'}} = \sqrt{\frac{9}{4\sin^2(\alpha'/2)}-3}$$

$$a' = \tfrac{1}{3}\sqrt{3a^2+c^2}$$

$$\sin\frac{\alpha'}{2} = \frac{3}{2\sqrt{3+(c^2/a^2)}} \quad \text{or} \quad \cos\alpha' = \frac{(c^2/a^2)-\frac{3}{2}}{(c^2/a^2)+3}.$$

The hexagonal triple cell is given in the *obverse* setting (centring points $\frac{2}{3},\frac{1}{3},\frac{1}{3};\ \frac{1}{3},\frac{2}{3},\frac{2}{3}$). In *IT* (1935), the *reverse* setting (centring points $\frac{1}{3},\frac{2}{3},\frac{1}{3};\ \frac{2}{3},\frac{1}{3},\frac{2}{3}$) was employed; *cf.* Table 2.1.1.2.

Coordinate transformations between different space-group descriptions are treated in detail in Section 1.5.3.

2.1.3.3. Headline

The description of each plane group or space group starts with a headline consisting of two (sometimes three) lines which contain the following information, when read from left to right.

First line

(1) The *short international* (Hermann–Mauguin) *symbol* for the plane or space group. These symbols will be further referred to as Hermann–Mauguin symbols. A detailed discussion of space-group symbols is given in Section 1.4.1 and in Chapters 1.4 and 3.3 of *IT* A; for convenience, a summary is given in Section 2.1.3.4.

Note on standard monoclinic space-group symbols: In order to facilitate recognition of a monoclinic space-group type, the familiar short symbol for the *b*-axis setting (*e.g.* $P2_1/c$ for No. 14 or $C2/c$ for No. 15) has been adopted as the *standard symbol* for a space-group type. It appears in the headline of *every description of this space group* and thus does not carry any information about the setting or the cell choice of this particular description. No other short symbols for monoclinic space groups are used in *IT* A.

(2) The *Schoenflies symbol* for the space group (*cf.* Section 1.4.1).

Note: No Schoenflies symbols exist for the plane groups.

(3) The *short international* (Hermann–Mauguin) *symbol* for the point group to which the plane or space group belongs (*cf.* Section 1.4.1, and Chapter 3.3 of *IT* A).

(4) The name of the *crystal system* (*cf.* Table 2.1.1.1).

Second line

(5) The sequential *number of the plane or space group*, as introduced in *IT* (1952).

(6) The *full international* (Hermann–Mauguin) *symbol* for the plane or space group.

For monoclinic space groups, the headline of every description contains the full symbol appropriate to that description.

(7) The *Patterson symmetry* (see Section 2.1.3.5).

Third line

This line is used, where appropriate, to indicate origin choices, settings, cell choices and coordinate axes (see Section 2.1.3.2). For five orthorhombic space groups, an entry 'Former space-group symbol' is given; *cf.* Section 2.1.2.

Table 2.1.3.1
Lattice symmetry directions for two and three dimensions

Directions that belong to the same set of equivalent symmetry directions are collected between braces. The first entry in each set is taken as the representative of that set.

Lattice	Symmetry direction (position in Hermann–Mauguin symbol)		
	Primary	Secondary	Tertiary
Two dimensions			
Oblique	Rotation point in plane		
Rectangular		[10]	[01]
Square		$\{[10], [01]\}$	$\{[1\bar{1}], [11]\}$
Hexagonal		$\{[10], [01], [\bar{1}\bar{1}]\}$	$\{[1\bar{1}], [12], [\bar{2}\bar{1}]\}$
Three dimensions			
Triclinic	None		
Monoclinic†	[010] ('unique axis *b*') [001] ('unique axis *c*')		
Orthorhombic	[100]	[010]	[001]
Tetragonal	[001]	$\{[100], [010]\}$	$\{[1\bar{1}0], [110]\}$
Hexagonal	[001]	$\{[100], [010], [\bar{1}\bar{1}0]\}$	$\{[1\bar{1}0], [120], [\bar{2}\bar{1}0]\}$
Rhombohedral (hexagonal axes)	[001]	$\{[100], [010], [\bar{1}\bar{1}0]\}$	
Rhombohedral (rhombohedral axes)	[111]	$\{[1\bar{1}0], [01\bar{1}], [\bar{1}01]\}$	
Cubic	$\{[100], [010], [001]\}$	$\{[111], [1\bar{1}\bar{1}], [\bar{1}1\bar{1}], [\bar{1}\bar{1}1]\}$	$\{[1\bar{1}0]\ [110], [01\bar{1}]\ [011], [\bar{1}01]\ [101]\}$

† For the full Hermann–Mauguin symbols see Sections 1.4.1 and 2.1.3.4.

2.1.3.4. International (Hermann–Mauguin) symbols for plane groups and space groups

(For more details, *cf.* Section 1.4.1, and Chapters 1.4 and 3.3 of *IT* A.)

Current symbols. Both the short and the full Hermann–Mauguin symbols consist of two parts: (i) a letter indicating the centring type of the conventional cell, and (ii) a set of characters indicating symmetry elements of the space group (modified point-group symbol).

(i) The letters for the centring types of cells are listed in Table 2.1.1.2. Lower-case letters are used for two dimensions (nets), capital letters for three dimensions (lattices).

(ii) The one, two or three entries after the centring letter refer to the one, two or three kinds of *symmetry directions* of the lattice belonging to the space group. These symmetry directions were called *Blickrichtungen* by Heesch (1929). Symmetry directions occur either as singular directions (as in the monoclinic and orthorhombic crystal systems) or as sets of symmetry-equivalent symmetry directions (as in the higher-symmetry crystal systems). Only one representative of each set is required. The (sets of) symmetry directions and their sequence for the different lattices are summarized in Table 2.1.3.1. According to their position in this sequence, the symmetry directions are referred to as 'primary', 'secondary' and 'tertiary' directions.

This sequence of lattice symmetry directions is transferred to the sequence of positions in the corresponding Hermann–Mauguin space-group symbols. Each position contains one or two characters designating symmetry elements (axes and planes) of the space group (*cf.* Section 2.1.2) that occur for the corresponding lattice symmetry direction. Symmetry planes are represented by their normals; if a symmetry axis and a normal to a symmetry plane are parallel, the two characters (symmetry symbols) are separated by a slash, as in $P6_3/m$ or $P2/m$ ('two over *m*').

Short and *full* Hermann–Mauguin symbols differ only for the plane groups of class *m*, for the monoclinic space groups, and for the space groups of crystal classes mmm, $4/mmm$, $\bar{3}m$, $6/mmm$,

$m\bar{3}$ and $m\bar{3}m$. In the full symbols, symmetry axes *and* symmetry planes for each symmetry direction are listed; in the short symbols, symmetry axes are suppressed as much as possible. Thus, for space group No. 62, the full symbol is $P2_1/n\,2_1/m\,2_1/a$ and the short symbol is *Pnma*. For No. 194, the full symbol is $P6_3/m\,2/m\,2/c$ and the short symbol is $P6_3/mmc$. For No. 230, the full symbol is $I4_1/a\,\bar{3}\,2/d$ and the short symbol is $Ia\bar{3}d$.

Many space groups contain more kinds of symmetry elements than are indicated in the full symbol ('additional symmetry operations and elements', *cf.* Section 1.4.2.4). A listing of additional symmetry operations for the monoclinic and orthorhombic space groups is given in Table 1.5.4.1 under the heading *Extended full symbols*. Note that a centre of symmetry is never explicitly indicated (except for space group $P\bar{1}$); its presence or absence, however, can be readily inferred from the space-group symbol.

2.1.3.5. Patterson symmetry

By Howard D. Flack

The entry *Patterson symmetry* in the headline gives the symmetry of the 'vector set' generated by the action of the space group on an arbitrary set of general positions. More prosaically, it may be described as the symmetry of the set of the interatomic vectors of a crystal structure with the selected space group. The Patterson symmetry is a crystallographic space group denoted by its Hermann–Mauguin symbol. It is in fact one of the 24 centrosymmetric symmorphic space groups (see Section 1.3.3.3) in three dimensions and one of 7 in two dimensions. For each of the 230 space groups, the Patterson symmetry has the same Bravais-lattice type as the space group itself and its point group is the lowest-index centrosymmetric supergroup of the point group of the space group. The 'point-group part' of the symbol of the Patterson symmetry represents the Laue class to which the plane group or space group belongs (*cf.* Table 2.1.1.1). By way of examples: space group No. 100, *P*4*bm*, has a Bravais lattice of type *tP* and point group 4*mm*. The centrosymmetric supergroup of 4*mm* is 4/*mmm*, so the Patterson symmetry is *P*4/*mmm*; space group No. 66, *Cccm*, has a Bravais lattice of type *oC* and point group *mmm*. This point group is centrosymmetric, so the Patterson symmetry is *Cmmm*.

Note: For the four space groups *Amm*2 (38), *Aem*2 (39), *Ama*2 (40) and *Aea*2 (41), the standard symbol for their Patterson symmetry, *Cmmm*, is added (between parentheses) after the actual symbol *Ammm* in the space-group tables.

The Patterson symmetry is intimately related to the symmetry of the Patterson function (see Flack, 2015). The latter, $P_{|F|^2}(uvw)$,

Table 2.1.3.2
Patterson symmetries and symmetries of Patterson functions for space groups and plane groups

The space-group types of each row form an arithmetic crystal class. (In three instances the row is typeset on two lines.) The arithmetic crystal class is identified by its representative symmorphic space group for which both the Hermann–Mauguin symbol and the space-group-type number are shown in bold. A set of space groups with sequential numbers is indicated by the symbols of the first and last space group of the sequence separated by a dash.

The column 'Patterson symmetry' indicates the symmetry of the set of interatomic vectors of crystal structures described in the space groups given in the column 'Space-group types'. The Patterson symmetry is given in the headline of each space-group table in Chapter 2.3 of *IT* A.

The setting and origin choice of the chosen space group should also be used for the space group of the Patterson symmetry and the symmorphic space group.

Similar remarks apply to the plane groups listed in part (*b*) of the table.

(*a*) Space groups.

Space-group types: Hermann–Mauguin symbols	Nos.	Patterson symmetry
Crystal family triclinic (anorthic), Bravais-lattice type *aP*		
***P*1**	**1**	$P\bar{1}$
$\mathbf{P\bar{1}}$	**2**	$P\bar{1}$
Crystal family monoclinic, Bravais-lattice type *mP*		
***P*2**–$P2_1$	**3**–4	*P*2/*m*
Pm–*Pc*	**6**–7	*P*2/*m*
P*2/*m–$P2_1/m$,	**10**–11	*P*2/*m*
*P*2/*c*–$P2_1/c$	13–14	*P*2/*m*
Crystal family monoclinic, Bravais-lattice type *mS*		
***C*2**	**5**	*C*2/*m*
Cm–*Cc*	**8**–9	*C*2/*m*
C*2/*m, *C*2/*c*	**12**, 15	*C*2/*m*
Crystal family orthorhombic, Bravais-lattice type *oP*		
***P*222**–$P2_12_12_1$	**16**–19	*Pmmm*
***Pmm*2**–*Pnn*2	**25**–34	*Pmmm*
Pmmm–*Pnma*	**47**–62	*Pmmm*
Crystal family orthorhombic, Bravais-lattice type *oS*		
$C222_1$, ***C*222**	20, **21**	*Cmmm*
***Cmm*2**–*Ccc*2	**35**–37	*Cmmm*
***Amm*2**–*Aea*2	**38**–41	*Ammm*
Cmcm–*Cmce*, ***Cmmm***,	63–64, **65**	*Cmmm*
Cccm–*Ccce*	66–68	*Cmmm*

Space-group types: Hermann–Mauguin symbols	Nos.	Patterson symmetry
Crystal family orthorhombic, Bravais-lattice type *oF*		
***F*222**	**22**	*Fmmm*
***Fmm*2**–*Fdd*2	**42**–43	*Fmmm*
Fmmm–*Fddd*	**69**–70	*Fmmm*
Crystal family orthorhombic, Bravais-lattice type *oI*		
***I*222**–$I2_12_12_1$	**23**–24	*Immm*
***Imm*2**–*Ima*2	**44**–46	*Immm*
Immm–*Imma*	**71**–74	*Immm*
Crystal family tetragonal, Bravais-lattice type *tP*		
***P*4**–$P4_3$	**75**–78	*P*4/*m*
$\mathbf{P\bar{4}}$	**81**	*P*4/*m*
P*4/*m–$P4_2/n$	**83**–86	*P*4/*m*
***P*422**–$P4_32_12$	**89**–96	*P*4/*mmm*
P*4*mm–$P4_2bc$	**99**–106	*P*4/*mmm*
$\mathbf{P\bar{4}2m}$–$P\bar{4}2_1c$	**111**–114	*P*4/*mmm*
$\mathbf{P\bar{4}m2}$–$P\bar{4}n2$	**115**–118	*P*4/*mmm*
P*4/*mmm–$P4_2/ncm$	**123**–138	*P*4/*mmm*
Crystal family tetragonal, Bravais-lattice type *tI*		
***I*4**, $I4_1$	**79**–80	*I*4/*m*
$\mathbf{I\bar{4}}$	**82**	*I*4/*m*
I*4/*m–$I4_1/a$	**87**–88	*I*4/*m*
***I*422**–$I4_122$	**97**–98	*I*4/*mmm*

is the inverse Fourier transform of the squared structure-factor amplitudes. Patterson functions possess the crystallographic symmetry of the symmorphic space-group representative of the arithmetic crystal class (see Section 1.3.4.4.1 for the definition and discussion of arithmetic crystal classes) to which the space group belongs. Table 2.1.3.2 lists these crystallographic symmetries of the Patterson function and the Patterson symmetries for the space groups and plane groups.

Table 2.1.3.2 (continued)

Space-group types		Patterson symmetry
Hermann–Mauguin symbols	Nos.	
$\mathbf{I4mm}$–$I4_1cd$	**107**–110	$I4/mmm$
$\mathbf{I\bar{4}m2}$–$I\bar{4}c2$	**119**–120	$I4/mmm$
$\mathbf{I\bar{4}2m}$–$I\bar{4}2d$	**121**–122	$I4/mmm$
$\mathbf{I4/mmm}$–$I4_1/acd$	**139**–142	$I4/mmm$
Crystal family hexagonal, Bravais-lattice type hP		
$\mathbf{P3}$–$P3_2$	**143**–145	$P\bar{3}$
$\mathbf{P\bar{3}}$	**147**	$P\bar{3}$
$\mathbf{P312}$, $P3_112$, $P3_212$	**149**, 151, 153	$P\bar{3}1m$
$\mathbf{P321}$, $P3_121$, $P3_221$	**150**, 152, 154	$P\bar{3}m1$
$\mathbf{P3m1}$, $P3c1$	**156**, 158	$P\bar{3}m1$
$\mathbf{P31m}$, $P31c$	**157**, 159	$P\bar{3}1m$
$\mathbf{P\bar{3}1m}$–$P\bar{3}1c$	**162**–163	$P\bar{3}1m$
$\mathbf{P\bar{3}m1}$–$P\bar{3}c1$	**164**–165	$P\bar{3}m1$
$\mathbf{P6}$–$P6_3$	**168**–173	$P6/m$
$\mathbf{P\bar{6}}$	**174**	$P6/m$
$\mathbf{P6/m}$–$P6_3/m$	**175**–176	$P6/m$
$\mathbf{P622}$–$P6_322$	**177**–182	$P6/mmm$
$\mathbf{P6mm}$–$P6_3mc$	**183**–186	$P6/mmm$
$\mathbf{P\bar{6}m2}$–$P\bar{6}c2$	**187**–188	$P6/mmm$
$\mathbf{P\bar{6}2m}$–$P\bar{6}2c$	**189**–190	$P6/mmm$
$\mathbf{P6/mmm}$–$P6_3/mmc$	**191**–194	$P6/mmm$
Crystal family hexagonal, Bravais-lattice type hR		
$\mathbf{R3}$	**146**	$R\bar{3}$
$\mathbf{R\bar{3}}$	**148**	$R\bar{3}$
$\mathbf{R32}$	**155**	$R\bar{3}m$
$\mathbf{R3m}$–$R3c$	**160**–161	$R\bar{3}m$
$\mathbf{R\bar{3}m}$–$R\bar{3}c$	**166**–167	$R\bar{3}m$
Crystal family cubic, Bravais-lattice type cP		
$\mathbf{P23}$, $P2_13$	**195**, 198	$Pm\bar{3}$
$\mathbf{Pm\bar{3}}$–$Pn\bar{3}$, $Pa\bar{3}$	**200**–201, 205	$Pm\bar{3}$
$\mathbf{P432}$–$P4_232$,	**207**–208,	$Pm\bar{3}m$
$P4_332$–$P4_132$	212–213	$Pm\bar{3}m$
$\mathbf{P\bar{4}3m}$, $P\bar{4}3n$	**215**, 218	$Pm\bar{3}m$
$\mathbf{Pm\bar{3}m}$–$Pn\bar{3}m$	**221**–224	$Pm\bar{3}m$
Crystal family cubic, Bravais-lattice type cF		
$\mathbf{F23}$	**196**	$Fm\bar{3}$
$\mathbf{Fm\bar{3}}$–$Fd\bar{3}$	**202**–203	$Fm\bar{3}$
$\mathbf{F432}$–$F4_132$	**209**–210	$Fm\bar{3}m$
$\mathbf{F\bar{4}3m}$–$F\bar{4}3c$	**216**, 219	$Fm\bar{3}m$
$\mathbf{Fm\bar{3}m}$–$Fd\bar{3}c$	**225**–228	$Fm\bar{3}m$
Crystal family cubic, Bravais-lattice type cI		
$\mathbf{I23}$, $I2_13$	**197**, 199	$Im\bar{3}$
$\mathbf{Im\bar{3}}$, $Ia\bar{3}$	**204**, 206	$Im\bar{3}$
$\mathbf{I432}$, $I4_132$	**211**, 214	$Im\bar{3}m$
$\mathbf{I\bar{4}3m}$, $I\bar{4}3d$	**217**, 220	$Im\bar{3}m$
$\mathbf{Im\bar{3}m}$–$Ia\bar{3}d$	**229**–230	$Im\bar{3}m$

2.1.3.6. Space-group diagrams

(For further discussion, see Section 1.4.2.5.)

The space-group diagrams serve two purposes: (i) to show the relative locations and orientations of the symmetry elements and (ii) to illustrate the arrangement of a set of symmetry-equivalent points of the general position.

With the exception of general-position diagrams in perspective projection for some space groups (*cf.* Section 2.1.3.6.8), all of the diagrams are orthogonal projections, *i.e.* the projection direction is perpendicular to the plane of the figure. Apart from the descriptions of the rhombohedral space groups with 'rhombohedral axes' (*cf.* Section 2.1.3.6.6), the projection direction is always a cell axis. If other axes are not parallel to the plane of the figure, they are indicated by the subscript p, as a_p, b_p or c_p in the case of one or two axes for monoclinic and triclinic space groups, respectively (*cf.* Figs. 2.1.3.1 to 2.1.3.3), or by the subscript rh (as in $a_{\rm rh}$, $b_{\rm rh}$ or $c_{\rm rh}$) for the three rhombohedral axes in Fig. 2.1.3.9.

The graphical symbols for symmetry elements, as used in the drawings, are displayed in Tables 2.1.2.2 to 2.1.2.7.

In the diagrams, 'heights' h above the projection plane are indicated for symmetry planes and symmetry axes *parallel* to the projection plane, as well as for centres of symmetry. The heights are given as fractions of the shortest lattice translation normal to the projection plane and, if different from 0, are printed next to the graphical symbols. Each symmetry element at height h is accompanied by another symmetry element of the same type at height $h+\frac{1}{2}$ (this does not apply to the horizontal fourfold axes in the diagrams for the cubic space groups). In the space-group diagrams, only the symmetry element at height h is indicated (*cf.* Section 2.1.2).

Schematic representations of the diagrams, displaying the origin, the labels of the axes, and the projection direction [uvw], are given in Figs. 2.1.3.1 to 2.1.3.10 (except Fig. 2.1.3.6). The general-position diagrams are indicated by the letter G.

Table 2.1.3.2 (continued)

(b) Plane groups.

Plane-group types		Patterson symmetry
Hermann–Mauguin symbols	Nos.	
Crystal family oblique (monoclinic), Bravais-lattice type mp		
$\mathbf{p1}$	**1**	$p2$
$\mathbf{p2}$	**2**	$p2$
Crystal family rectangular (orthorhombic), Bravais-lattice type op		
$\mathbf{pm}$–pg	**3**–4	$p2mm$
$\mathbf{p2mm}$–$p2gg$	**6**–8	$p2mm$
Crystal family rectangular (orthorhombic), Bravais-lattice type oc		
$\mathbf{cm}$	**5**	$c2mm$
$\mathbf{c2mm}$	**9**	$c2mm$
Crystal family square (tetragonal), Bravais-lattice type tp		
$\mathbf{p4}$	**10**	$p4$
$\mathbf{p4mm}$–$p4gm$	**11**–12	$p4mm$
Crystal family hexagonal, Bravais-lattice type hp		
$\mathbf{p3}$	**13**	$p6$
$\mathbf{p3m1}$	**14**	$p6mm$
$\mathbf{p31m}$	**15**	$p6mm$
$\mathbf{p6}$	**16**	$p6$
$\mathbf{p6mm}$	**17**	$p6mm$

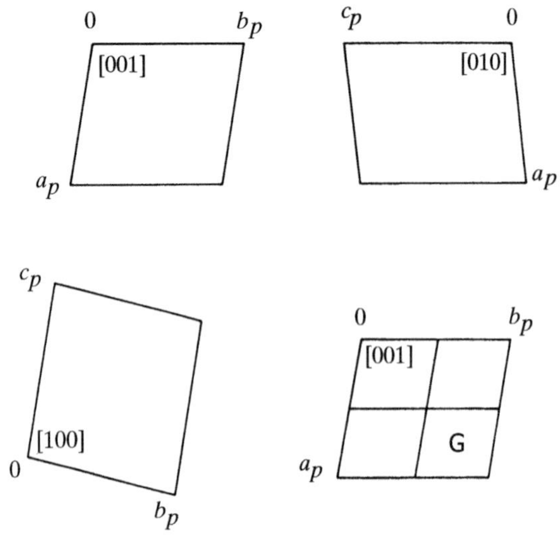

Figure 2.1.3.1
Triclinic space groups (G = general-position diagram).

2.1.3.6.1. Plane groups

Each description of a plane group contains two diagrams, one for the symmetry elements (left) and one for the general position (right). The two axes are labelled *a* and *b*, with *a* pointing downwards and *b* running from left to right.

2.1.3.6.2. Triclinic space groups

For each of the two triclinic space groups, three elevations (along *a*, *b* and *c*) are given, in addition to the general-position diagram G (projected along *c*) at the lower right of the set, as illustrated in Fig. 2.1.3.1.

The diagrams represent a reduced cell of type II for which the three interaxial angles are non-acute, *i.e.* $\alpha, \beta, \gamma \geq 90°$. For a cell of type I, all angles are acute, *i.e.* $\alpha, \beta, \gamma < 90°$. For a discussion of the two types of reduced cells, see Section 3.1.3 of *IT* A.

2.1.3.6.3. Monoclinic space groups (cf. Section 2.1.3.2, and Section 2.1.3.15 of IT A)

The 'complete treatment' of each of the two settings contains four diagrams (Figs. 2.1.3.2 and 2.1.3.3). Three of them are projections of the symmetry elements, taken along the unique axis (upper left) and along the other two axes (lower left and upper right). For the general position, only the projection along the unique axis is given (lower right).

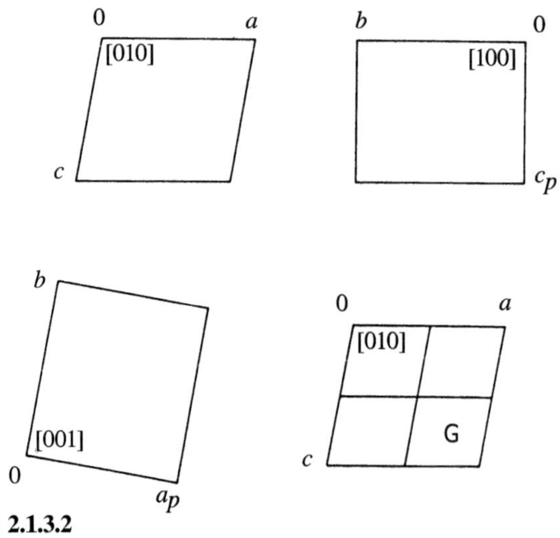

Figure 2.1.3.2
Monoclinic space groups, setting with unique axis *b* (G = general-position diagram).

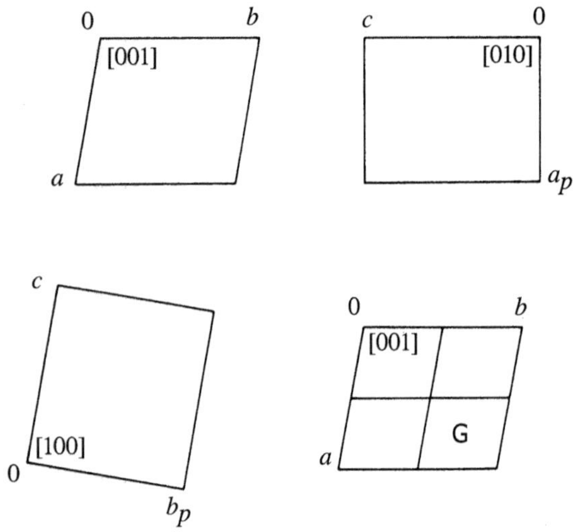

Figure 2.1.3.3
Monoclinic space groups, setting with unique axis *c* (G = general-position diagram).

The 'synoptic descriptions' of the three cell choices (for each setting) are headed by a pair of diagrams, as illustrated in Fig. 2.1.3.4. The drawings on the left display the symmetry elements and the ones on the right the general position (labelled G). Each diagram is a projection of four neighbouring unit cells along the unique axis. It contains the outlines of the three cell choices drawn as heavy lines. For the labelling of the axes, see Fig. 2.1.3.4. The headline of the description of each cell choice contains a small-scale drawing, indicating the basis vectors and the cell that apply to that description.

2.1.3.6.4. Orthorhombic space groups and orthorhombic settings

The space-group tables contain a set of four diagrams for each orthorhombic space group. The set consists of three projections of the symmetry elements [along the *c* axis (upper left), the *a* axis

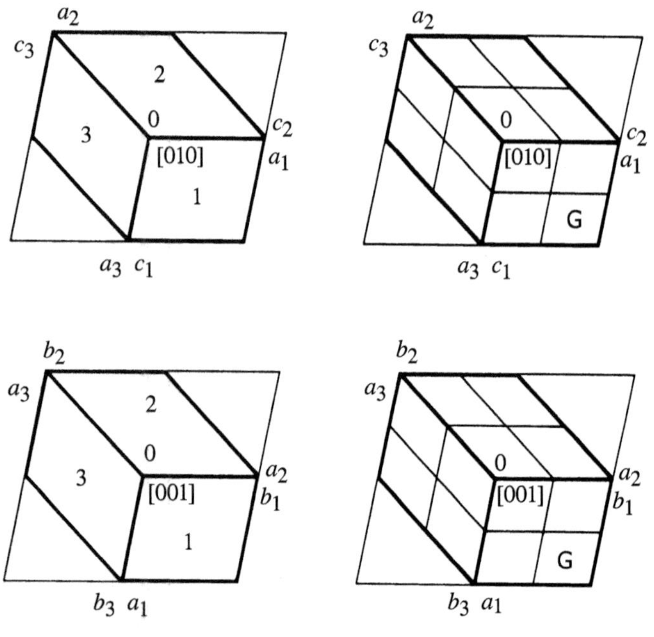

Figure 2.1.3.4
Monoclinic space groups, cell choices 1, 2, 3. Upper pair of diagrams: setting with unique axis *b*. Lower pair of diagrams: setting with unique axis *c*. The numbers 1, 2, 3 within the cells and the subscripts of the labels of the axes indicate the cell choice (*cf.* Section 1.5.3.1, and Section 2.1.3.15 of *IT* A). The unique axis points upwards from the page. G = general-position diagram.

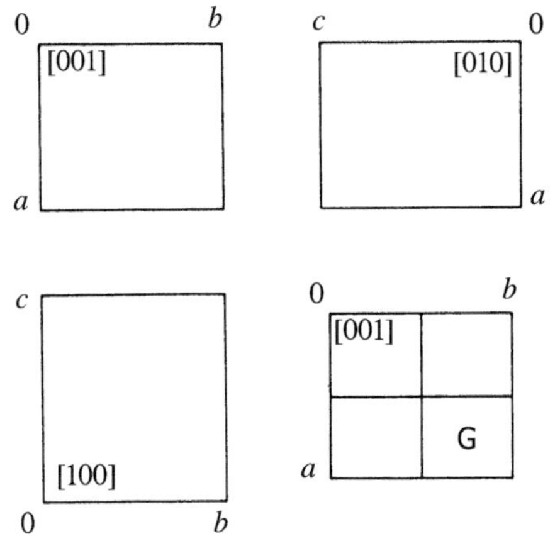

Figure 2.1.3.5
Orthorhombic space groups. Diagrams for the 'standard setting' as described in the space-group tables (G = general-position diagram).

(lower left) and the b axis (upper right)] in addition to the general-position diagram, which is given only in the projection along c (lower right). The projected axes, the origins and the projection directions of these diagrams are illustrated in Fig. 2.1.3.5. They refer to the so-called 'standard setting' of the space group, *i.e.* the setting described in the space-group tables and indicated by the 'standard Hermann–Mauguin symbol' in the headline.

For each orthorhombic space group, *six settings* exist, *i.e.* six different ways of assigning the labels a, b, c to the three orthorhombic symmetry directions; thus the shape and orientation of the cell are the same for each setting. These settings correspond to the six permutations of the labels of the axes (including the identity permutation); *cf.* Section 1.5.4:

$$\mathbf{abc} \quad \mathbf{ba\bar{c}} \quad \mathbf{cab} \quad \mathbf{\bar{c}ba} \quad \mathbf{bca} \quad \mathbf{a\bar{c}b}.$$

The symbol for each setting, here called 'setting symbol', is a shorthand notation for the (3×3) transformation matrix $\boldsymbol{P}$ of the basis vectors of the standard setting, **a**, **b**, **c**, into those of the setting considered (*cf.* Section 1.5.1 for a detailed discussion of coordinate transformations). For instance, the setting symbol **cab** stands for the cyclic permutation

$$\mathbf{a}' = \mathbf{c}, \quad \mathbf{b}' = \mathbf{a}, \quad \mathbf{c}' = \mathbf{b}$$

or

$$(\mathbf{a}', \mathbf{b}', \mathbf{c}') = (\mathbf{a}, \mathbf{b}, \mathbf{c})\boldsymbol{P} = (\mathbf{a}, \mathbf{b}, \mathbf{c})\begin{pmatrix} 0 & 1 & 0 \\ 0 & 0 & 1 \\ 1 & 0 & 0 \end{pmatrix} = (\mathbf{c}, \mathbf{a}, \mathbf{b}),$$

where $\mathbf{a}'$, $\mathbf{b}'$, $\mathbf{c}'$ is the new set of basis vectors. An interchange of two axes reverses the handedness of the coordinate system; in order to keep the system right-handed, each interchange is accompanied by the reversal of the sense of one axis, *i.e.* by an element $\bar{1}$ in the transformation matrix. Thus, $\mathbf{ba\bar{c}}$ denotes the transformation

$$(\mathbf{a}', \mathbf{b}', \mathbf{c}') = (\mathbf{a}, \mathbf{b}, \mathbf{c})\begin{pmatrix} 0 & 1 & 0 \\ 1 & 0 & 0 \\ 0 & 0 & \bar{1} \end{pmatrix} = (\mathbf{b}, \mathbf{a}, \bar{\mathbf{c}}).$$

In the earlier (1935 and 1952) editions of *International Tables*, only one setting was illustrated, in a projection along **c**, so that it was usual to consider it as the 'standard setting' and to accept its cell edges as crystal axes and its space-group symbol as the 'standard Hermann–Mauguin symbol'. In *IT* A, following *IT* A (2002), however, *all six* orthorhombic settings are illustrated, as explained below.

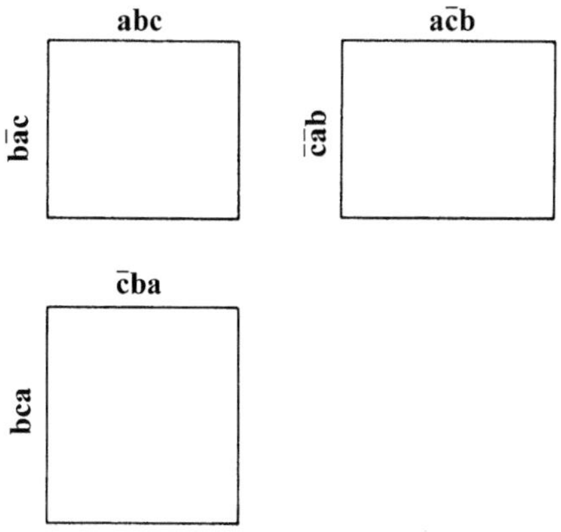

Figure 2.1.3.6
Orthorhombic space groups. The three projections of the symmetry elements with the six setting symbols (see text). For setting symbols printed vertically, the page has to be turned clockwise by 90° or viewed from the side. Note that in the actual space-group tables instead of the setting symbols the corresponding full Hermann–Mauguin space-group symbols are printed.

The three projections of the symmetry elements can be interpreted in two ways. First, in the sense indicated above, that is, as different projections of a *single* (standard) setting of the space group, with the projected basis vectors **a**, **b**, **c** labelled as in Fig. 2.1.3.5. Second, each one of the three diagrams can be considered as the projection along $\mathbf{c}'$ of either one of *two different* settings: one setting in which $\mathbf{b}'$ is horizontal and one in which $\mathbf{b}'$ is vertical ($\mathbf{a}'$, $\mathbf{b}'$, $\mathbf{c}'$ refer to the setting under consideration). This second interpretation is used to illustrate in the same figure the space-group symbols corresponding to these two settings. In order to view these projections in conventional orientation ($\mathbf{b}'$ horizontal, $\mathbf{a}'$ vertical, origin in the upper left corner, projection down the positive $\mathbf{c}'$ axis), the setting with $\mathbf{b}'$ horizontal can be inspected directly with the figure upright; hence, the corresponding space-group symbol is printed above the projection. The other setting with $\mathbf{b}'$ vertical and $\mathbf{a}'$ horizontal, however, requires turning the figure by 90°, or looking at it from the side; thus, the space-group symbol is printed at the left, and it runs upwards.

The 'setting symbols' for the six settings are attached to the three diagrams of Fig. 2.1.3.6, which correspond to those of Fig. 2.1.3.5. In the orientation of the diagram where the setting symbol is read in the usual way, $\mathbf{a}'$ is vertical pointing downwards, $\mathbf{b}'$ is horizontal pointing to the right, and $\mathbf{c}'$ is pointing upwards from the page. Each setting symbol is printed in the position that in the space-group tables is actually occupied by the corresponding full Hermann–Mauguin symbol. The changes in the space-group symbol that are associated with a particular setting symbol can easily be deduced by comparing Fig. 2.1.3.6 with the diagrams for the space group under consideration.

The six setting symbols, *i.e.* the six permutations of the labels of the axes, form the column headings of the orthorhombic entries in Table 1.5.4.1, which contains the extended Hermann–Mauguin symbols for the six settings of each orthorhombic space group. Note that some of these setting symbols exhibit different sign changes compared with those in Fig. 2.1.3.6.

2.1.3.6.5. *Tetragonal, trigonal P and hexagonal P space groups*

The pairs of diagrams for these space groups are similar to those in the previous editions of *IT*. Each pair consists of a

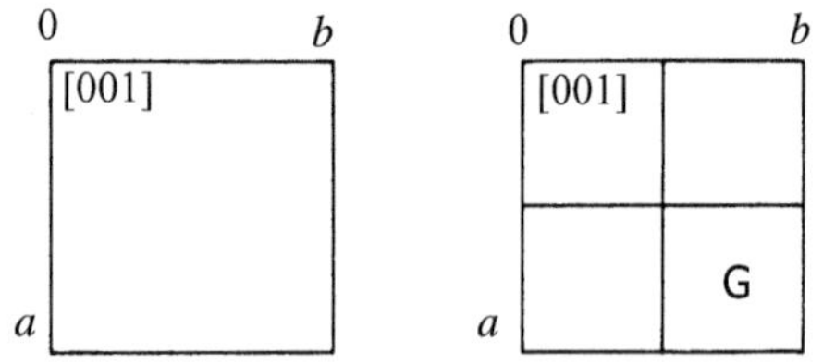

Figure 2.1.3.7
Tetragonal space groups (G = general-position diagram).

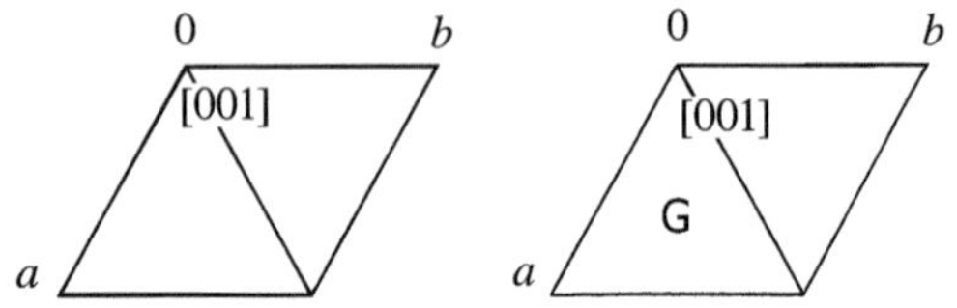

Figure 2.1.3.8
Trigonal *P* and hexagonal *P* space groups (G = general-position diagram).

general-position diagram (right) and a diagram of the symmetry elements (left), both projected along **c**, as illustrated in Figs. 2.1.3.7 and 2.1.3.8.

2.1.3.6.6. Trigonal R (rhombohedral) space groups

The seven rhombohedral space groups are treated in two versions, the first based on 'hexagonal axes' (obverse setting), the second on 'rhombohedral axes' (*cf.* Sections 2.1.1.2 and 2.1.3.2). The pairs of diagrams are similar to those in *IT* (1952) and *IT* A (2002); the left or top one displays the symmetry elements, the right or bottom one the general position. This is illustrated in Fig. 2.1.3.9, which gives the axes a and b of the triple hexagonal cell and the projections of the axes of the primitive rhombohedral cell, labelled $a_{\rm rh}$, $b_{\rm rh}$ and $c_{\rm rh}$. For convenience, all 'heights' in the space-group diagrams are fractions of the hexagonal c axis. For 'hexagonal axes', the projection direction is [001], for 'rhombohedral axes' it is [111]. In the general-position diagrams, the circles drawn in heavier lines represent positions that lie within the primitive rhombohedral cell (provided the symbol '$-$' is read as $1 - z$ rather than as $-z$).

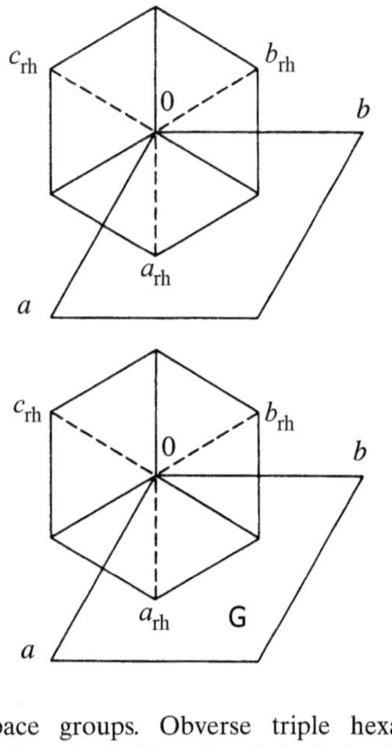

Figure 2.1.3.9
Rhombohedral space groups. Obverse triple hexagonal cell with 'hexagonal axes' a, b and primitive rhombohedral cell with projections of 'rhombohedral axes' $a_{\rm rh}$, $b_{\rm rh}$, $c_{\rm rh}$. Note: In the actual space-group diagrams the edges of the primitive rhombohedral cell (dashed lines) are only indicated in the general-position diagram of the rhombohedral-axes description (G = general-position diagram).

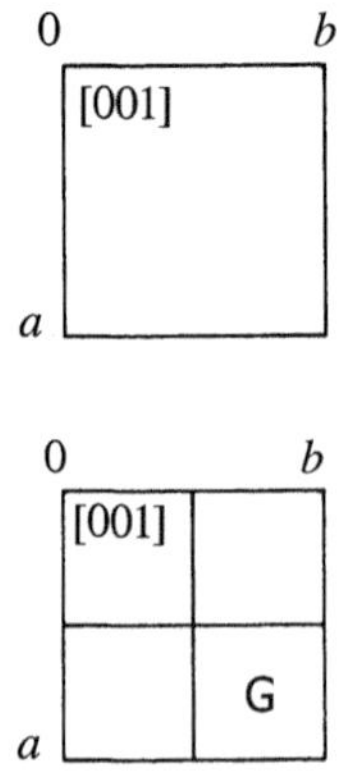

Figure 2.1.3.10
Cubic space groups. G = general-position diagram, in which the equivalent positions are shown as the vertices of polyhedra.

The symmetry-element diagrams for the hexagonal and the rhombohedral descriptions of a space group are the same. The edges of the primitive rhombohedral cell (*cf.* Fig. 2.1.3.9) are only indicated in the general-position diagram of the rhombohedral description.

2.1.3.6.7. Cubic space groups

For each cubic space group, one projection of the symmetry elements along [001] is given, Fig. 2.1.3.10; for details of the diagrams, see Section 2.1.2 and Buerger (1956). For face-centred lattices *F*, only a quarter of the unit cell is shown; this is sufficient since the projected arrangement of the symmetry elements is translation-equivalent in the four quarters of an *F* cell. It is important to note that symmetry axes inclined to the projection plane are indicated where they intersect the plane of projection. Symmetry planes inclined to the projection plane that occur in classes $\bar{4}3m$ and $m\bar{3}m$ are shown as 'inserts' around the high-symmetry points, such as 0, 0, 0; $\frac{1}{2}$, 0, 0; *etc.*

The cubic diagrams given in *IT* (1935) are different from the ones used in *IT* A. No drawings for cubic space groups were provided in *IT* (1952).

2.1.3.6.8. Diagrams of the general position (by Koichi Momma and Mois I. Aroyo)

Non-cubic space groups. In these diagrams, the 'heights' of the points are z coordinates, except for monoclinic space groups with unique axis b where they are y coordinates. For rhombohedral space groups, the heights are always fractions of the hexagonal c axis. The symbols $+$ and $-$ stand for $+z$ and $-z$ (or $+y$ and $-y$) in which z or y can assume any value. For points with symbols $+$ or $-$ preceded by a fraction, *e.g.* $\frac{1}{2}+$ or $\frac{1}{3}-$, the relative z or y coordinate is $\frac{1}{2}$ *etc.* higher than that of the point with symbol $+$ or $-$.

Where a mirror plane exists parallel to the plane of projection, the two positions superimposed in projection are indicated by the use of a ring divided through the centre. The information given on each side refers to one of the two positions related by the mirror plane, as in $-$ ⦶ $+$.

Diagrams for cubic space groups (Fig. 2.1.3.10). Following the approach of *IT* (1935), for each cubic space group a diagram showing the points of the general position as the vertices of polyhedra is given. In these diagrams, the polyhedra are transparent, but the spheres at the vertices are opaque. For most of the space groups, 'starting points' with the same coordinate values, $x = 0.048$, $y = 0.12$, $z = 0.089$, have been used. The origins of the polyhedra are chosen at special points of highest site symmetry, which for most space groups coincide with the origin (and its equivalent points in the unit cell). Polyhedra with origins at sites $(\frac{1}{8}, \frac{1}{8}, \frac{1}{8})$ have been chosen for the space groups $P4_332$ (212) and $I4_132$ (214), and $(\frac{3}{8}, \frac{3}{8}, \frac{3}{8})$ for $P4_132$ (213). The two diagrams shown for the space groups $I\bar{4}3d$ (220) and $Ia\bar{3}d$ (230) correspond to polyhedra with origins chosen at two different special sites with site-symmetry groups of equal (32 *versus* $\bar{3}$ in $Ia\bar{3}d$) or nearly equal order (3 *versus* $\bar{4}$ in $I\bar{4}3d$). The height h of the centre of each polyhedron is given on the diagram, if different from zero. For space-group Nos. 198, 199 and 220, h refers to the special point to which the polyhedron (triangle) is connected. Polyhedra with height 1 are omitted in all the diagrams. A grid of four squares is drawn to represent the four quarters of the basal plane of the cell. For space groups $F\bar{4}3c$ (219), $Fm\bar{3}c$ (226) and $Fd\bar{3}c$ (228), where the number of points is too large for one diagram, two diagrams are provided, one for the upper half and one for the lower half of the cell.

Notes:

(i) For space group $P4_132$ (213), the coordinates $\bar{x}, \bar{y}, \bar{z}$ have been chosen for the 'starting point' to show the enantiomorphism with $P4_332$ (212).

(ii) For the description of a space group with 'origin choice 2', the coordinates x, y, z of all points have been shifted with the origin to retain the same polyhedra for both origin choices.

An additional general-position diagram is shown on the fourth page for each of the ten space groups of the $m\bar{3}m$ crystal class. To provide a clearer three-dimensional-style overview of the arrangements of the polyhedra, these general-position diagrams are shown in tilted projection (in contrast to the orthogonal-projection diagrams described above).

The general-position diagrams of the cubic groups in both orthogonal and tilted projections were generated using the program *VESTA* (Momma & Izumi, 2011).

Readers who wish to compare other approaches to space-group diagrams and their history are referred to *IT* (1935), *IT* (1952), the fifth edition of *IT* A (2002) (where general-position stereodiagrams of the cubic space groups are shown) and the following publications: Astbury & Yardley (1924), Belov *et al.* (1980), Buerger (1956), Fedorov (1895; English translation, 1971), Friedel (1926), Hilton (1903), Niggli (1919) and Schiebold (1929).

2.1.3.7. Origin

The determination and description of crystal structures and particularly the application of direct methods are greatly facilitated by the choice of a suitable origin and its proper identification. This is even more important if related structures are to be compared or if 'chains' of group–subgroup relations are to be constructed. In *IT* A, as well as in *IT* (1952) and *IT* A (2002), the origin of the unit cell has been chosen according to the following conventions (*cf.* Sections 2.1.1 and 2.1.3.2):

(i) All centrosymmetric space groups are described with an inversion centre as origin. A further description is given if a centrosymmetric space group contains points of high site symmetry that do not coincide with a centre of symmetry. As an example, study the origin choice 1 and origin choice 2 descriptions of $I4_1/amd$ (141).

(ii) For noncentrosymmetric space groups, the origin is at a point of highest site symmetry, as in $F\bar{4}3m$ (216). If no site symmetry is higher than 1, except for the cases listed below under (iii), the origin is placed on a screw axis, or a glide plane, or at the intersection of several such symmetry elements, see for example space groups $P2_1$ (4) and $Pna2_1$ (33).

(iii) In space group $P2_12_12_1$ (19), the origin is chosen in such a way that it is surrounded symmetrically by three pairs of 2_1 axes. This principle is maintained in the following noncentrosymmetric cubic space groups of classes 23 and 432, which contain $P2_12_12_1$ as subgroup: $P2_13$ (198), $I2_13$ (199), $F4_132$ (210). It has been extended to other noncentrosymmetric orthorhombic and cubic space groups with $P2_12_12_1$ as subgroup, even though in these cases points of higher site symmetry *are* available: $I2_12_12_1$ (24), $P4_332$ (212), $P4_132$ (213), $I4_132$ (214).

There are several ways of determining the location and site symmetry of the origin. First, the origin can be inspected directly in the space-group diagrams (*cf.* Section 2.1.3.6). This method permits visualization of all symmetry elements that intersect the chosen origin.

Another procedure for finding the site symmetry at the origin is to look for a special position that contains the coordinate triplet 0, 0, 0 or that includes it for special values of the parameters, *e.g.* position 1*a*: $0, 0, z$ in space group $P4mm$ (99), or position 4*a*: $x, x, 0$; $\bar{x}, \bar{x}, \frac{1}{2}$; $\bar{x}+\frac{1}{2}, x+\frac{1}{2}, \frac{1}{4}$; $x+\frac{1}{2}, \bar{x}+\frac{1}{2}, \frac{3}{4}$ of space group $P4_12_12$ (92). If such a special position occurs, the symmetry at the origin is given by the oriented site-symmetry symbol (see Section 2.1.3.12) of that special position; if it does not occur, the site symmetry at the origin is 1. For most practical purposes, these two methods are sufficient for the identification of the site symmetry at the origin.

Origin statement. In the line *Origin* immediately below the diagrams, the site symmetry of the origin is stated, if different from the identity. A further symbol indicates all symmetry elements (including glide planes and screw axes) that pass through the origin, if any. For space groups with two *origin choices*, for each of the two origins the location relative to the other origin is also given. An example is space group $I4_1/amd$ (141).

2.1.3.8. Asymmetric unit

An asymmetric unit of a space group is a (simply connected) smallest closed part of space from which, by application of all symmetry operations of the space group, the whole of space is filled. This implies that mirror planes and rotation axes must form boundary planes and boundary edges of the asymmetric unit. A twofold rotation axis may bisect a boundary plane. Centres of inversion must either form vertices of the asymmetric unit or be located at the midpoints of boundary planes or boundary edges. For glide planes and screw axes, these simple restrictions do not hold. An asymmetric unit contains all the information necessary for the complete description of the crystal structure. In mathematics, an asymmetric unit is called 'fundamental region' or 'fundamental domain'.

Example

The boundary planes of the asymmetric unit in space group *Pmmm* (47) are fixed by the six mirror planes $x, y, 0$; $x, y, \frac{1}{2}$; $x, 0, z$; $x, \frac{1}{2}, z$; $0, y, z$; and $\frac{1}{2}, y, z$. For space group $P2_12_12_1$ (19), on the other hand, a large number of connected regions, each with a volume of $\frac{1}{4}V(\text{cell})$, may be chosen as asymmetric unit.

In cases where the asymmetric unit is not uniquely determined by symmetry, its choice may depend on the purpose of its application. For the description of the structures of molecular crystals, for instance, it is advantageous to select asymmetric units that contain one or more complete molecules. In the space-group tables of *IT* A, following *IT* A (2002), the asymmetric units are chosen in such a way that Fourier summations can be performed conveniently.

For all triclinic, monoclinic and orthorhombic space groups, the asymmetric unit is chosen as a parallelepiped with one vertex at the origin of the cell and with boundary planes parallel to the faces of the cell. It is given by the notation

$$0 \le x_i \le \text{upper limit of } x_i,$$

where x_i stands for x, y or z.

For space groups with higher symmetry, cases occur where the origin does not coincide with a vertex of the asymmetric unit or where not all boundary planes of the asymmetric unit are parallel to those of the cell. In all these cases, parallelepipeds

$$\text{lower limit of } x_i \le x_i \le \text{upper limit of } x_i$$

are given that are equal to or larger than the asymmetric unit. Where necessary, the boundary planes lying within these parallelepipeds are given by additional inequalities, such as $x \le y$, $y \le \frac{1}{2} - x$ *etc.*

In the trigonal, hexagonal and especially the cubic crystal systems, the asymmetric units have complicated shapes. For this reason, they are also specified by the coordinates of their vertices. Drawings of asymmetric units for cubic space groups have been published by Koch & Fischer (1974). Fig. 2.1.3.11 shows the boundary planes occurring in the tetragonal, trigonal and hexagonal systems, together with their algebraic equations.

Examples

(1) In space group *P4mm* (99), the boundary plane $y = x$ occurs in addition to planes parallel to the unit-cell faces; the asymmetric unit is given by

$$0 \le x \le \tfrac{1}{2};\quad 0 \le y \le \tfrac{1}{2};\quad 0 \le z \le 1;\quad x \le y.$$

(2) In space group *R*32 (155; hexagonal axes), the boundary planes are, among others, $x = (1+y)/2$, $y = 1-x$, $y = (1+x)/2$. The asymmetric unit is defined by

$$0 \le x \le \tfrac{2}{3};\quad 0 \le y \le \tfrac{2}{3};\quad 0 \le z \le \tfrac{1}{6};$$
$$x \le (1+y)/2;\quad y \le \min(1-x, (1+x)/2)$$

$$\text{Vertices:}\quad 0,0,0\quad \tfrac{1}{2},0,0\quad \tfrac{2}{3},\tfrac{1}{3},0\quad \tfrac{1}{3},\tfrac{2}{3},0\quad 0,\tfrac{1}{2},0$$
$$0,0,\tfrac{1}{6}\quad \tfrac{1}{2},0,\tfrac{1}{6}\quad \tfrac{2}{3},\tfrac{1}{3},\tfrac{1}{6}\quad \tfrac{1}{3},\tfrac{2}{3},\tfrac{1}{6}\quad 0,\tfrac{1}{2},\tfrac{1}{6}.$$

It is obvious that the indication of the vertices is of great help in drawing the asymmetric unit.

Fourier syntheses. For complicated space groups, the easiest way to calculate Fourier syntheses is to consider the parallelepiped listed, without taking into account the additional boundary planes of the asymmetric unit. These planes should be drawn afterwards in the Fourier synthesis. For the computation of integrated properties from Fourier syntheses, such as the number of electrons for parts of the structure, the values at the boundaries of the asymmetric unit must be applied with a reduced weight if the property is to be obtained as the product of the content of the asymmetric unit and the multiplicity.

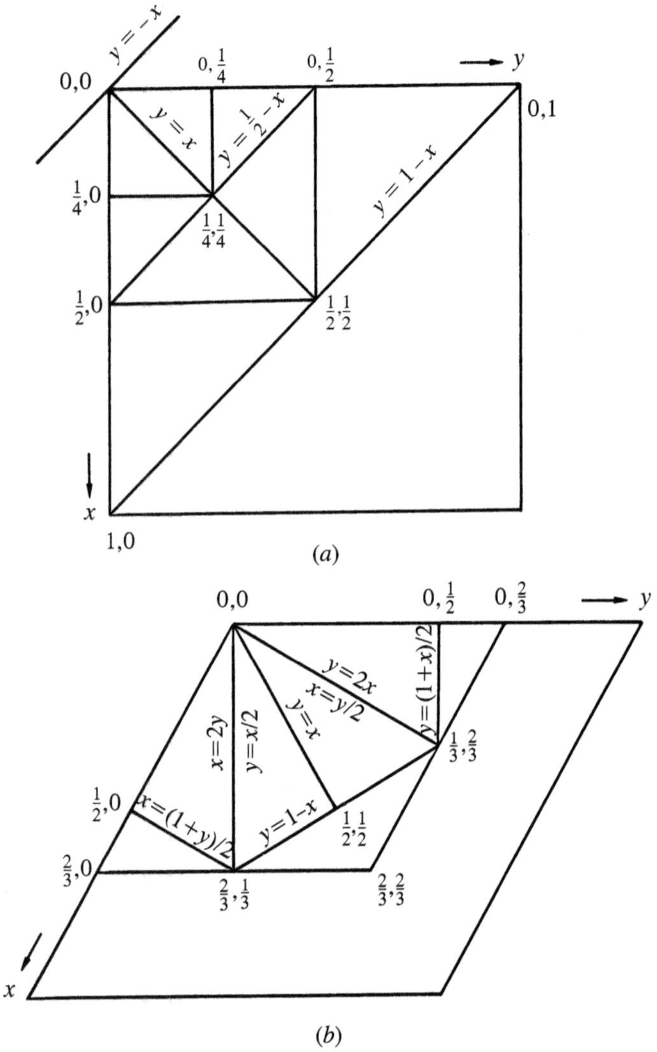

Figure 2.1.3.11

Boundary planes of asymmetric units occurring in the space-group tables. (*a*) Tetragonal system. (*b*) Trigonal and hexagonal systems. The point coordinates refer to the vertices in the plane $z = 0$.

Example

In the parallelepiped of space group *Pmmm* (47), the weights for boundary planes, edges and vertices are $\frac{1}{2}$, $\frac{1}{4}$ and $\frac{1}{8}$, respectively.

An overview of an online gallery of *exact* asymmetric units (*i.e.* taking into account the boundary conditions) for all 230 space groups can be found in Grosse-Kunstleve *et al.* (2011). Asymmetric units of the plane groups have been discussed by Buerger (1949, 1960) in connection with Fourier summations.

2.1.3.9. Symmetry operations

As explained in Sections 1.3.3.2 and 1.4.2.3, the coordinate triplets of the *General position* of a space group may be interpreted as a shorthand description of the symmetry operations in matrix notation. The geometric description of the symmetry operations is found in the space-group tables under the heading *Symmetry operations*.

Numbering scheme. The numbering $(1)\ldots(p)\ldots$ of the entries in the blocks *Symmetry operations* and *General position* (first block below *Positions*) is the same. Each listed coordinate triplet of the general position is preceded by a number between parentheses (p). The same number (p) precedes the corresponding symmetry operation. For space groups with *primitive* cells, the two lists contain the same number of entries.

For space groups with *centred* cells, several (2, 3 or 4) blocks of *Symmetry operations* correspond to the one *General position* block. The numbering scheme of the general position is applied to each one of these blocks. The number of blocks equals the multiplicity of the centred cell, *i.e.* the number of centring translations below the subheading *Coordinates*, such as $(0, 0, 0)+$, $(\frac{2}{3}, \frac{1}{3}, \frac{1}{3})+$, $(\frac{1}{3}, \frac{2}{3}, \frac{2}{3})+$.

Whereas for the *Positions* the reader is expected to add these centring translations to each printed coordinate triplet themselves (in order to obtain the complete general position), for the *Symmetry operations* the corresponding data are listed explicitly. The different blocks have the subheadings 'For $(0, 0, 0)+$ set', 'For $(\frac{1}{2}, \frac{1}{2}, \frac{1}{2})+$ set', *etc.* Thus, an obvious one-to-one correspondence exists between the analytical description of a symmetry operation in the form of its general-position coordinate triplet and the geometrical description under *Symmetry operations*. Note that the coordinates are reduced modulo 1, where applicable, as shown in the example below.

Example: Ibca (73)

The centring translation is $t(\frac{1}{2}, \frac{1}{2}, \frac{1}{2})$. Accordingly, above the general position one finds $(0, 0, 0)+$ and $(\frac{1}{2}, \frac{1}{2}, \frac{1}{2})+$. In the block *Symmetry operations*, under the subheading 'For $(0, 0, 0)+$ set', entry (2) refers to the coordinate triplet $\bar{x}+\frac{1}{2}$, $\bar{y}$, $z+\frac{1}{2}$. Under the subheading 'For $(\frac{1}{2}, \frac{1}{2}, \frac{1}{2})+$ set', however, entry (2) refers to $\bar{x}$, $\bar{y}+\frac{1}{2}$, z. The triplet $\bar{x}$, $\bar{y}+\frac{1}{2}$, z is selected rather than $\bar{x}+1$, $\bar{y}+\frac{1}{2}$, $z+1$, because the coordinates are reduced modulo 1.

The coordinate triplets of the general position represent the symmetry operations chosen as coset representatives of the decomposition of the space group with respect to its translation subgroup (*cf.* Section 1.4.2 for a detailed discussion). In space groups with two origins the origin shift may lead to the choice of symmetry operations of different types as coset representatives of the same coset, *i.e.* the coset of symmetry operations with the same rotation parts (*e.g.* mirror *versus* glide plane, rotation *versus* screw axis, see Tables 1.5.4.1 and 1.5.4.2 of *IT* A) and designated by the same number (p) in the general-position blocks of the two descriptions. Thus, in $P4/nmm$ (129), $(p) = (7)$ represents a 2 and a 2_1 axis, both in $x, x, 0$, whereas $(p) = (16)$ represents a g and an m plane, both in x, x, z.

Designation of symmetry operations. An entry in the block *Symmetry operations* is characterized as follows.

(i) A symbol denoting the *type* of the symmetry operation (*cf.* Section 2.1.2), including its glide or screw part, if present. In most cases, the glide or screw part is given explicitly by fractional coordinates between parentheses. The sense of a rotation is indicated by the superscript + or −. Abbreviated notations are used for the glide reflections $a(\frac{1}{2}, 0, 0) \equiv a$; $b(0, \frac{1}{2}, 0) \equiv b$; $c(0, 0, \frac{1}{2}) \equiv c$. Glide reflections with complicated and unconventional glide parts are designated by the letter g, followed by the glide part between parentheses.

(ii) A coordinate triplet indicating the *location* and *orientation* of the symmetry element which corresponds to the symmetry operation. For rotoinversions, the location of the inversion point is also given.

Examples

(1) $g(\frac{1}{4}, \frac{1}{4}, \frac{1}{2})$ x, x, z
Glide reflection with glide component $(\frac{1}{4}, \frac{1}{4}, \frac{1}{2})$ through the plane x, x, z, *i.e.* the plane parallel to $(1\bar{1}0)$ containing the point 0, 0, 0.

(2) $g(\frac{1}{3}, \frac{1}{6}, \frac{2}{3})$ $2x-\frac{1}{2}, x, z$ (hexagonal axes)
Glide reflection with glide component $(\frac{1}{3}, \frac{1}{6}, \frac{2}{3})$ through the plane $2x-\frac{1}{2}, x, z$, *i.e.* the plane parallel to $(1\bar{2}10)$, which intersects the a axis at $-\frac{1}{2}$ and the b axis at $\frac{1}{4}$; this operation occurs in $R\bar{3}m$ (166, hexagonal axes).

(3) Symmetry operations in *Ibca* (73)
Under the subheading 'For $(0, 0, 0)+$ set', the operation generating the coordinate triplet (2) $\bar{x}+\frac{1}{2}, \bar{y}, z+\frac{1}{2}$ from (1) x, y, z is symbolized by $2(0, 0, \frac{1}{2})$ $\frac{1}{4}, 0, z$. This indicates a twofold screw rotation with screw part $(0, 0, \frac{1}{2})$ for which the corresponding screw axis coincides with the line $\frac{1}{4}, 0, z$, *i.e.* runs parallel to [001] through the point $\frac{1}{4}, 0, 0$. Under the subheading 'For $(\frac{1}{2}, \frac{1}{2}, \frac{1}{2})+$ set', the operation generating the coordinate triplet (2) $\bar{x}, \bar{y}+\frac{1}{2}, z$ from (1) x, y, z is symbolized by $2\ 0, \frac{1}{4}, z$. It is thus a twofold rotation (without screw part) around the line $0, \frac{1}{4}, z$.

Details on the symbolism and further illustrative examples are presented in Section 1.4.2.1.

2.1.3.10. Generators

The line *Generators selected* states the symmetry operations and their sequence, selected to generate all symmetry-equivalent points of the *General position* from a point with coordinates x, y, z. Generating translations are listed as $t(1, 0, 0)$, $t(0, 1, 0)$, $t(0, 0, 1)$; likewise for additional centring translations. The other symmetry operations are given as numbers (p) that refer to the corresponding coordinate triplets of the general position and the corresponding entries under *Symmetry operations*, as explained in Section 2.1.3.9 [for centred space groups the first block 'For $(0, 0, 0)+$ set' must be used].

For all space groups, the identity operation given by (1) is selected as the first generator. It is followed by the generators $t(1, 0, 0)$, $t(0, 1, 0)$, $t(0, 0, 1)$ of the integral lattice translations and, if necessary, by those of the centring translations, *e.g.* $t(\frac{1}{2}, \frac{1}{2}, 0)$ for a C-centred lattice. In this way, point x, y, z and all its translationally equivalent points are generated. (The remark 'and its translationally equivalent points' will hereafter be omitted.) The sequence chosen for the generators following the translations depends on the crystal class of the space group and is set out in Table 1.4.3.1.

Example: $P12_1/c1$ (14, unique axis b, cell choice 1)

After the generation of (1) x, y, z, the operation (2) which stands for a twofold screw rotation around the axis $0, y, \frac{1}{4}$ generates point (2) of the general position with coordinate triplet $\bar{x}, y+\frac{1}{2}, \bar{z}+\frac{1}{2}$. Finally, the inversion (3) generates point (3) $\bar{x}, \bar{y}, \bar{z}$ from point (1), and point (4′) $x, \bar{y}-\frac{1}{2}, z-\frac{1}{2}$ from point (2). Instead of (4′), however, the coordinate triplet (4) $x, \bar{y}+\frac{1}{2}, z+\frac{1}{2}$ is listed, because the coordinates are reduced modulo 1.

The example shows that for the space group $P12_1/c1$ two operations, apart from the identity and the generating translations, are sufficient to generate all symmetry-equivalent points. Alternatively, the inversion (3) plus the glide reflection (4), or the glide reflection (4) plus the twofold screw rotation (2), might have been chosen as generators. The process of generation and the selection of the generators for the space-group tables, as well as the resulting sequence of the symmetry operations, are discussed in Section 1.4.3.

The generating operations for different descriptions of the same space group (settings, cell choices, origin choices) are chosen in such a way that the transformation relating the two coordinate systems also transforms the generators of one description into those of the other (*cf.* Section 1.5.3).

2.1.3.11. Positions

The entries under *Positions*[1] (more explicitly called *Wyckoff positions*) consist of the one *General position* (upper block) and the *Special positions* (blocks below). The columns in each block, from left to right, contain the following information for each Wyckoff position.

(i) *Multiplicity of the Wyckoff position*. This is the number of equivalent points per unit cell. For primitive cells, the multiplicity of the general position is equal to the order of the point group of the space group; for centred cells, it is the product of the order of the point group and the number (2, 3 or 4) of lattice points per cell. The multiplicity of a special position is always a divisor of the multiplicity of the general position and the quotient of the two is equal to the order of the site-symmetry group.

(ii) *Wyckoff letter*. This letter is merely a coding scheme for the Wyckoff positions, starting with *a* at the bottom position and continuing upwards in alphabetical order.

(iii) *Site symmetry*. This is explained in Section 2.1.3.12.

(iv) *Coordinates*. The sequence of coordinate triplets is produced in the same order as the symmetry operations, generated by the chosen set of generators, omitting duplicates (*cf.* Sections 1.4.3 and 2.1.3.10). For centred space groups, the centring translations, for instance $(0, 0, 0)+$ $(\frac{1}{2}, \frac{1}{2}, \frac{1}{2})+$, are listed above the coordinate triplets. The symbol '+' indicates that, in order to obtain a complete Wyckoff position, the components of these centring translations have to be added to the listed coordinate triplets.

A graphic representation of the points of the general position is provided by the general-position diagram; *cf.* Section 2.1.3.6.

(v) *Reflection conditions*. These are described in Section 2.1.3.13.

Detailed treatment of general and special Wyckoff positions, including definitions, theoretical background and examples, is given in Section 1.4.4.

The two types of positions, general and special, are characterized as follows:

(i) *General position*

A point is said to be in general position if it is left invariant only by the identity operation but by no other symmetry operation of the space group. Each space group has only one general position.

The coordinate triplets of a general position (which always start with x, y, z) can also be interpreted as a shorthand form of the matrix representation of the symmetry operations of the space group; this viewpoint is described further in Sections 1.3.3.2 and 1.4.2.3.

(ii) *Special position(s)*

A point is said to be in 'special position' if it is mapped onto itself by the identity and at least one further symmetry operation of the space group. This implies that specific constraints are imposed on the coordinates of each point of a special position; *e.g.* $x = \frac{1}{4}, y = 0$, leading to the triplet $\frac{1}{4}, 0, z$; or $y = x + \frac{1}{2}$, leading to the triplet $x, x + \frac{1}{2}, z$. The number of special positions of a space group depends on the space-group type and can vary from 0 for the so-called 'fixed-point-free' groups containing only symmetry operations *with* intrinsic translation parts (*e.g.* $Pna2_1$, No. 33) to 26 (for *Pmmm*, No. 47). For further discussion, see Section 1.4.4.2.

The set of *all* symmetry operations that map a point onto itself forms a group, known as the 'site-symmetry group' of that point. It is given in the third column by the 'oriented site-symmetry symbol' which is explained in Section 2.1.3.12. General positions always have site symmetry 1, whereas special positions have higher site symmetries, which can differ from one special position to another.

Example: Space group C12/c1 (15, unique axis b, cell choice 1)

The general position 8*f* of this space group contains eight equivalent points per cell, each with site symmetry 1. The coordinate triplets of four points, (1) to (4), are given explicitly, the coordinates of the other four points are obtained by adding the components $\frac{1}{2}, \frac{1}{2}, 0$ of the *C*-centring translation to the coordinate triplets (1) to (4).

The space group has five special positions with Wyckoff letters *a* to *e*. The positions 4*a* to 4*d* require inversion symmetry, $\bar{1}$, whereas Wyckoff position 4*e* requires twofold rotation symmetry, 2, for any object in such a position. For position 4*e*, for instance, the four equivalent points have the coordinates $0, y, \frac{1}{4};\ 0, \bar{y}, \frac{3}{4};\ \frac{1}{2}, y + \frac{1}{2}, \frac{1}{4};\ \frac{1}{2}, \bar{y} + \frac{1}{2}, \frac{3}{4}$. The values of x and z are specified, whereas y may take any value. Since each point of position 4*e* is mapped onto itself by a twofold rotation, the multiplicity of the position is reduced from 8 to 4, whereas the order of the site-symmetry group is increased from 1 to 2.

From the symmetry-element diagram of *C*2/*c*, the locations of the four twofold axes can be deduced as $0, y, \frac{1}{4};\ 0, y, \frac{3}{4};\ \frac{1}{2}, y, \frac{1}{4};\ \frac{1}{2}, y, \frac{3}{4}$.

From this example, the general rule is apparent that the product of the position multiplicity and the order of the corresponding site-symmetry group is constant for all Wyckoff positions of a given space group; it is the multiplicity of the general position.

2.1.3.12. Oriented site-symmetry symbols

The third column of each Wyckoff position gives the *Site symmetry*[2] of that position. The site-symmetry group is isomorphic to a (proper or improper) subgroup of the point group to which the space group under consideration belongs. The site-symmetry groups of the different points of the same special position are conjugate (symmetry-equivalent) subgroups of the space group. For this reason, all points of one special position are described by the same site-symmetry symbol. (See Section 1.4.4 for a detailed discussion of site-symmetry groups.)

[1] The term *Position* (singular) is defined as a *set* of symmetry-equivalent points, in agreement with *IT* (1935): Point position; *Punktlage* (German); *position* (French). Note that in *IT* (1952) the plural, equivalent positions, was used.

[2] Often called point symmetry: *Punktsymmetrie* or *Lagesymmetrie* (German): *symétrie ponctuelle* (French).

Table 2.1.3.3
Integral reflection conditions for centred cells (lattices)

Reflection condition	Centring type of cell	Centring symbol
None	Primitive	P ; R† (rhombohedral axes)
$h+k=2n$	C-face centred	C
$k+l=2n$	A-face centred	A
$h+l=2n$	B-face centred	B
$h+k+l=2n$	Body centred	I
$h+k, h+l$ and $k+l=2n$ or: h, k, l all odd or all even ('unmixed')	All-face centred	F
$-h+k+l=3n$	Rhombohedrally centred, obverse setting (standard)	R† (hexagonal axes)
$h-k+l=3n$	Rhombohedrally centred, reverse setting	R† (hexagonal axes)
$h-k=3n$	Hexagonally centred	H‡

† For further explanations see Section 2.1.1 and Table 2.1.1.2.
‡ For the use of the unconventional H cell, see Table 2.1.1.2.

Oriented site-symmetry symbols (*cf.* Fischer *et al.*, 1973) are employed to show how the symmetry elements at a site are related to the symmetry elements of the crystal lattice. The site-symmetry symbols display the same sequence of symmetry directions as the space-group symbol (*cf.* Table 2.1.3.1). Sets of equivalent symmetry directions that do not contribute any element to the site-symmetry group are represented by a dot. In this way, the orientation of the symmetry elements at the site is emphasized, as illustrated by the following examples.

Examples

(1) In the tetragonal space group $I4/mmm$ (139), Wyckoff position $16k$ has site symmetry $..2$ and position $8h$ has site symmetry $m.2m$. The easiest way to interpret the symbols is to look at the dots first. For position $16k$, the 2 is preceded by two dots and thus must belong to a tertiary symmetry direction. Only one tertiary direction is used. Consequently, the site symmetry is the monoclinic point group 2 with one of the two tetragonal tertiary directions as twofold axis.
Position $8h$ has one dot, with one symmetry symbol before and two symmetry symbols after it. The dot corresponds, therefore, to the secondary symmetry directions. The first symbol m indicates a mirror plane with a normal along the primary symmetry direction (c axis). The final symbols $2m$ indicate a twofold axis and a mirror plane with a normal along the two mutually perpendicular tertiary directions $[1\bar{1}0]$ and $[110]$. The site symmetry is thus orthorhombic, $m2m$ (isomorphic to $mm2$).

(2) In the cubic space group $Fm\bar{3}m$ (225), position $24e$ has $4m.m$ as its site-symmetry symbol. This 'cubic' site-symmetry symbol displays a tetragonal site symmetry. The position of the dot indicates that there is no symmetry along the four secondary cubic directions. The fourfold axis is connected with one of the three primary cubic symmetry directions and two equivalent mirror planes occur along the remaining two primary directions. Moreover, the group contains two mutually perpendicular (equivalent) mirror planes with normals along those two of the six tertiary cubic directions $\langle 110\rangle$ that are normal to the fourfold axis. Each pair of equivalent mirror planes is given by just one symbol m. (Note that at the six representative sites of position $24e$, the fourfold axes are twice oriented along a, twice along b and twice along c.)

The above examples show:

(i) The oriented site-symmetry symbols become identical to Hermann–Mauguin point-group symbols if the dots are omitted.

(ii) Sets of symmetry directions having more than one equivalent direction may require more than one character if the site-symmetry group belongs to a lower crystal system than the space group under consideration.

2.1.3.13. Reflection conditions

The *Reflection conditions*[3] are listed in the right-hand column of each Wyckoff position.

These conditions are formulated here, in accordance with general practice, as 'conditions of occurrence' (structure factor not systematically zero) and not as 'extinctions' or 'systematic absences' (structure factor zero). Reflection conditions are listed for those three-, two- and one-dimensional sets of reflections for which extinctions exist; those nets and rows for which no conditions apply are not listed. The theoretical background of reflection conditions and their derivation are discussed in Section 1.6.3.

There are two types of systematic reflection conditions for diffraction of radiation by crystals:

(1) *General conditions*. They are associated with systematic absences caused by the presence of lattice centrings, screw axes and glide planes. The general conditions are always obeyed, irrespective of which Wyckoff positions are occupied by atoms in a particular crystal structure.

(2) *Special conditions* ('extra' conditions). They apply only to *special* Wyckoff positions and always occur in addition to the general conditions of the space group. Note that each extra condition is valid only for the scattering contribution of those atoms that are located in the relevant special Wyckoff position. If the special position is occupied by atoms whose scattering power is high in comparison with the other atoms in the structure, reflections violating the extra condition will be weak. One should note that the special conditions apply only to isotropic and spherical atoms (*cf.* Section 1.6.3 of *IT* A).

General reflection conditions. These are due to one of three effects:

(i) *Centred cells*. The resulting conditions apply to the whole three-dimensional set of reflections hkl. Accordingly, they are called *integral reflection conditions*. They are given in Table 2.1.3.3. These conditions result from the centring vectors of centred cells. They disappear if a primitive cell is chosen instead of a centred cell. Note that the centring symbol and the corresponding integral reflection condition may change with a change of the basis vectors (*e.g.* monoclinic: $C \rightarrow A \rightarrow I$).

[3] The reflection conditions were called *Auslöschungen* (German), missing spectra (English) and *extinctions* (French) in *IT* (1935) and 'Conditions limiting possible reflections' in *IT* (1952); they are often referred to as 'Systematic or space-group absences' (*cf.* Section 3.3.3 of *IT* A).

Table 2.1.3.4
Zonal and serial reflection conditions for glide planes and screw axes (*cf.* Table 2.1.2.1)

(*a*) Glide planes

Type of reflections	Reflection condition	Glide plane: Orientation of plane	Glide vector	Symbol	Crystallographic coordinate system to which condition applies	
$0kl$	$k = 2n$	(100)	$\mathbf{b}/2$	b	Monoclinic (a unique), Tetragonal	Orthorhombic, Cubic
	$l = 2n$		$\mathbf{c}/2$	c		
	$k + l = 2n$		$\mathbf{b}/2 + \mathbf{c}/2$	n		
	$k + l = 4n$ $(k, l = 2n)$†		$\mathbf{b}/4 \pm \mathbf{c}/4$	d		
$h0l$	$l = 2n$	(010)	$\mathbf{c}/2$	c	Monoclinic (b unique), Tetragonal	Orthorhombic, Cubic
	$h = 2n$		$\mathbf{a}/2$	a		
	$l + h = 2n$		$\mathbf{c}/2 + \mathbf{a}/2$	n		
	$l + h = 4n$ $(l, h = 2n)$†		$\mathbf{c}/4 \pm \mathbf{a}/4$	d		
$hk0$	$h = 2n$	(001)	$\mathbf{a}/2$	a	Monoclinic (c unique), Tetragonal	Orthorhombic, Cubic
	$k = 2n$		$\mathbf{b}/2$	b		
	$h + k = 2n$		$\mathbf{a}/2 + \mathbf{b}/2$	n		
	$h + k = 4n$ $(h, k = 2n)$†		$\mathbf{a}/4 \pm \mathbf{b}/4$	d		
$h\bar{h}0l$ $0k\bar{k}l$ $\bar{h}0hl$	$l = 2n$	$(11\bar{2}0)$ $(\bar{2}110)$ $(1\bar{2}10)$ } $\{11\bar{2}0\}$	$\mathbf{c}/2$	c	Hexagonal	
$hh.\overline{2h}.l$ $\overline{2h}.hhl$ $h.\overline{2h}.hl$	$l = 2n$	$(1\bar{1}00)$ $(01\bar{1}0)$ $(\bar{1}010)$ } $\{1\bar{1}00\}$	$\mathbf{c}/2$	c	Hexagonal	
hhl hkk hkh	$l = 2n$ $h = 2n$ $k = 2n$	$(1\bar{1}0)$ $(01\bar{1})$ $(\bar{1}01)$ } $\{1\bar{1}0\}$	$\mathbf{c}/2$ $\mathbf{a}/2$ $\mathbf{b}/2$	c, n a, n b, n	Rhombohedral‡	
$hhl, h\bar{h}l$	$l = 2n$	$(1\bar{1}0), (110)$	$\mathbf{c}/2$	c, n	Tetragonal§	Cubic¶
	$2h + l = 4n$		$\mathbf{a}/4 \pm \mathbf{b}/4 \pm \mathbf{c}/4$	d		
$hkk, hk\bar{k}$	$h = 2n$	$(01\bar{1}), (011)$	$\mathbf{a}/2$	a, n		
	$2k + h = 4n$		$\pm\mathbf{a}/4 + \mathbf{b}/4 \pm \mathbf{c}/4$	d		
$hkh, \bar{h}kh$	$k = 2n$	$(\bar{1}01), (101)$	$\mathbf{b}/2$	b, n		
	$2h + k = 4n$		$\pm\mathbf{a}/4 \pm \mathbf{b}/4 + \mathbf{c}/4$	d		

† Glide planes d with orientations (100), (010) and (001) occur only in orthorhombic and cubic F space groups. Combination of the integral reflection condition (hkl: all odd or all even) with the zonal conditions for the d glide planes leads to the further conditions given between parentheses.
‡ For rhombohedral space groups described with 'rhombohedral axes', the three reflection conditions ($l = 2n, h = 2n, k = 2n$) imply interleaving of c and n glides, a and n glides, and b and n glides, respectively. In the Hermann–Mauguin space-group symbols, c is always used, as in $R3c$ (161) and $R\bar{3}c$ (167), because c glides also occur in the hexagonal description of these space groups.
§ For tetragonal P space groups, the two reflection conditions (hhl and $h\bar{h}l$ with $l = 2n$) imply interleaving of c and n glides. In the Hermann–Mauguin space-group symbols, c is always used, irrespective of which glide planes contain the origin: *cf.* $P4cc$ (103), $P\bar{4}2c$ (112) and $P4/nnc$ (126).
¶ For cubic space groups, the three reflection conditions ($l = 2n, h = 2n, k = 2n$) imply interleaving of c and n glides, a and n glides, and b and n glides, respectively. In the Hermann–Mauguin space-group symbols, either c or n is used, depending upon which glide plane contains the origin, *cf.* $P\bar{4}3n$ (218), $Pn\bar{3}n$ (222), $Pm\bar{3}n$ (223) *versus* $F\bar{4}3c$ (219), $Fm\bar{3}c$ (226), $Fd\bar{3}c$ (228).

(ii) *Glide planes.* The resulting conditions apply only to two-dimensional sets of reflections, *i.e.* to reciprocal-lattice nets containing the origin (such as $hk0$, $h0l$, $0kl$, hhl). For this reason, they are called *zonal reflection conditions.* The indices hkl of these 'zonal reflections' obey the relation $hu + kv + lw = 0$, where $[uvw]$, the direction of the zone axis, is normal to the reciprocal-lattice net. Note that the symbol of a glide plane and the corresponding zonal reflection condition may change with a change of the basis vectors (*e.g.* monoclinic: $c \rightarrow n \rightarrow a$).

(iii) *Screw axes.* The resulting conditions apply only to one-dimensional sets of reflections, *i.e.* reciprocal-lattice rows containing the origin (such as $h00$, $0k0$, $00l$). They are called *serial reflection conditions.* It is interesting to note that some diagonal screw axes do not give rise to systematic absences (*cf.* Section 1.6.3 of *IT* A for more details).

Reflection conditions of types (ii) and (iii) are listed in Table 2.1.3.4. They can be understood as follows: Zonal and serial reflections form two- or one-dimensional sections through the origin of reciprocal space. In direct space, they correspond to projections of a crystal structure onto a plane or onto a line. Glide planes or screw axes may reduce the translation periods in these projections (*cf.* Section 2.1.3.14) and thus decrease the size of the projected cell. As a consequence, the cells in the corresponding reciprocal-lattice sections are increased, which means that systematic absences of reflections occur.

For the two-dimensional groups, the reasoning is analogous. The reflection conditions for the plane groups are assembled in Table 2.1.3.5.

For the *interpretation of observed reflections*, the general reflection conditions must be studied in the order (i) to (iii), as conditions of type (ii) may be included in those of type (i), while

Table 2.1.3.4 (continued)

(*b*) Screw axes

Type of reflections	Reflection conditions	Screw axis: Direction of axis	Screw axis: Screw vector	Screw axis: Symbol	Crystallographic coordinate system to which condition applies
$h00$	$h = 2n$	[100]	$\mathbf{a}/2$	2_1	Monoclinic (*a* unique), Orthorhombic, Tetragonal; Cubic
				4_2	Cubic
	$h = 4n$		$\mathbf{a}/4$	$4_1, 4_3$	Cubic
$0k0$	$k = 2n$	[010]	$\mathbf{b}/2$	2_1	Monoclinic (*b* unique), Orthorhombic, Tetragonal; Cubic
				4_2	Cubic
	$k = 4n$		$\mathbf{b}/4$	$4_1, 4_3$	Cubic
$00l$	$l = 2n$	[001]	$\mathbf{c}/2$	2_1	Monoclinic (*c* unique), Orthorhombic; Cubic
				4_2	Tetragonal; Cubic
	$l = 4n$		$\mathbf{c}/4$	$4_1, 4_3$	Tetragonal; Cubic
$000l$	$l = 2n$	[001]	$\mathbf{c}/2$	6_3	Hexagonal
	$l = 3n$		$\mathbf{c}/3$	$3_1, 3_2, 6_2, 6_4$	Hexagonal
	$l = 6n$		$\mathbf{c}/6$	$6_1, 6_5$	Hexagonal

conditions of type (iii) may be included in those of types (i) or (ii). This is shown in the example below.

In the *space-group tables*, the reflection conditions are given according to the following rules:

(i) for a given space group, all reflection conditions [up to symmetry equivalence, *cf.* rule (v)] are listed; hence for those nets or rows that are *not* listed no conditions apply. No distinction is made between 'independent' and 'included' conditions, as was done in *IT* (1952), where 'included' conditions were placed in parentheses;

(ii) the integral condition, if present, is always listed first, followed by the zonal and serial conditions;

(iii) conditions that have to be satisified simultaneously are separated by a comma or by 'AND'. Thus, if two indices must be even, say h and l, the condition is written $h, l = 2n$ rather than $h = 2n$ and $l = 2n$. The same applies to sums of indices. Thus, there are several different ways to express the integral conditions for an F-centred lattice: '$h + k, h + l, k + l = 2n$' or '$h + k, h + l = 2n$ and $k + l = 2n$' or '$h + k = 2n$ and $h + l, k + l = 2n$' (*cf.* Table 2.1.3.3);

(iv) conditions separated by 'OR' are alternative conditions. For example, 'hkl: $h = 2n + 1$ or $h + k + l = 4n$' means that hkl is 'present' if either the condition $h = 2n + 1$ *or* the alternative condition $h + k + l = 4n$ is fulfilled. Obviously, hkl is also a 'present' reflection if both conditions are satisfied. Note that 'or' conditions occur only for the *special conditions* described below;

(v) in crystal systems with two or more symmetry-equivalent nets or rows (tetragonal and higher), only *one* representative set (the first one in Table 2.1.3.4) is listed; *e.g.* tetragonal: only the first members of the equivalent sets $0kl$ and $h0l$ or $h00$ and $0k0$ are listed;

(vi) for cubic space groups, it is stated that the indices hkl are 'cyclically permutable' or 'permutable'. The cyclic permutability of h, k and l in all rhombohedral space groups, described with 'rhombohedral axes', and of h and k in some tetragonal space groups are not stated;

(vii) in the 'hexagonal-axes' descriptions of trigonal and hexagonal space groups, Bravais–Miller indices $hkil$ are used. They obey two conditions:

(*a*) $h + k + i = 0$, *i.e.* $i = -(h + k)$;

(*b*) the indices h, k, i are cyclically permutable; this is not stated. Further details can be found in textbooks of crystallography.

Note that the integral reflection conditions for a rhombohedral lattice, described with 'hexagonal axes', permit the presence of only one member of the pair $hkil$ and $\bar{h}\bar{k}\bar{i}l$ for $l \neq 3n$ (*cf.* Table 2.1.3.3). This applies also to the zonal reflections $h\bar{h}0l$ and $\bar{h}h0l$, which for the rhombohedral space groups must be considered separately.

Example

For a monoclinic crystal (*b* unique), the following reflection conditions have been observed:

(1) hkl: $h + k = 2n$;

(2) $0kl$: $k = 2n$; $h0l$: $h, l = 2n$; $hk0$: $h + k = 2n$;

(3) $h00$: $h = 2n$; $0k0$: $k = 2n$; $00l$: $l = 2n$.

Table 2.1.3.5

Reflection conditions for the plane groups

Type of reflections	Reflection condition	Centring type of plane cell; or glide line with glide vector	Coordinate system to which condition applies
hk	None	Primitive p	All systems
	$h + k = 2n$	Centred c	Rectangular
	$h - k = 3n$	Hexagonally centred h†	Hexagonal
$h0$	$h = 2n$	Glide line g normal to b axis; glide vector $\frac{1}{2}\mathbf{a}$	Rectangular, Square
$0k$	$k = 2n$	Glide line g normal to a axis; glide vector $\frac{1}{2}\mathbf{b}$	Rectangular, Square

† For the use of the unconventional h cell see Table 2.1.1.2.

Line (1) states that the cell used for the description of the space group is C centred. In line (2), the conditions $0kl$ with $k = 2n$, $h0l$ with $h = 2n$ and $hk0$ with $h + k = 2n$ are a consequence of the integral condition (1), leaving only $h0l$ with $l = 2n$ as a new condition. This indicates a glide plane c. Line (3) presents no new condition, since $h00$ with $h = 2n$ and $0k0$ with $k = 2n$ follow from the integral condition (1), whereas $00l$ with $l = 2n$ is a consequence of a zonal condition (2). Accordingly, there need not be a twofold screw axis along [010]. Space groups obeying the conditions are Cc (9, b unique, cell choice 1) and $C2/c$ (15, b unique, cell choice 1). Under certain conditions, using methods based on resonant scattering, it is possible to determine whether the structure space group is centrosymmetric or not (*cf.* Section 1.6.5.1 of *IT* A).

For a different choice of the basis vectors, the reflection conditions would appear in a different form owing to the transformation of the reflection indices (for a discussion of this and an illustrative example, see Sections 1.5.2 and 1.5.3.2).

Special or 'extra' reflection conditions. These apply either to the integral reflections hkl or to particular sets of zonal or serial reflections. In the space-group tables, the minimal special conditions are listed that, on combination with the general conditions, are sufficient to generate the complete set of conditions. (For more details and illustrative examples of special reflection conditions, *cf.* Section 2.1.3.13 of *IT* A.)

Structural or non-space-group absences. Note that in addition *non-space-group absences* may occur that are not due to the symmetry of the space group (*i.e.* centred cells, glide planes or screw axes). Atoms in general or special positions may cause additional systematic absences if their coordinates assume special values [*e.g.* 'noncharacteristic orbits'; *cf.* Section 1.4.4.4 of *IT* A and Engel *et al.* (1984)]. Non-space-group absences may also occur for special arrangements of atoms ('false symmetry') in a crystal structure (*cf.* Templeton, 1956; Sadanaga *et al.*, 1978). Non-space-group absences may occur also for polytypic structures; this is briefly discussed by Durovič in Section 9.2.2.2.5 of *International Tables for Crystallography* (2004), Vol. C. Even though all these 'structural absences' are fortuitous and due to the special arrangements of atoms in a particular crystal structure, they have the appearance of space-group absences. Occurrence of structural absences thus may lead to an *incorrect assignment of the space group*. Accordingly, the reflection conditions in the space-group tables must be considered as a minimal set of conditions.

The use of reflection conditions and of the symmetry of reflection intensities for space-group determination is described in Chapter 1.6.

2.1.3.14. Symmetry of special projections

Projections of crystal structures are used by crystallographers in special cases. Use of so-called 'two-dimensional data' (zero-layer intensities) results in the projection of a crystal structure along the normal to the reciprocal-lattice net. A detailed treatment of projections of space groups, including basic definitions and illustrative examples, is given in Section 1.4.5.3.

Even though the projection of a finite object along *any* direction may be useful, the projection of a *periodic* object such as a crystal structure is only sensible along a rational lattice direction (lattice row). Projection along a nonrational direction results in a constant density in at least one direction.

Data listed in the space-group tables. Under the heading *Symmetry of special projections*, the following data are listed for three projections of each space group; no projection data are given for the plane groups.

(i) *The projection direction.* All projections are orthogonal, *i.e.* the projection is made onto a plane normal to the projection direction. This ensures that spherical atoms appear as circles in the projection. For each space group, three projections are listed. If a lattice has three kinds of symmetry directions, the three projection directions correspond to the primary, secondary and tertiary symmetry directions of the lattice (as displayed in Table 2.1.3.1). If a lattice contains fewer than three kinds of symmetry directions, as in the triclinic, monoclinic and rhombohedral cases, the additional projection direction(s) are taken along coordinate axes, *i.e.* lattice rows lacking symmetry.

(ii) *The Hermann–Mauguin symbol of the plane group* resulting from the projection of the space group. If necessary, the symbols are given in oriented form; for example, plane group pm is expressed either as $p1m1$ or as $p11m$ (*cf.* Section 1.4.1.5 for explanations of Hermann–Mauguin symbols of plane groups).

(iii) *Relations between the basis vectors* $\mathbf{a}'$, $\mathbf{b}'$ of the plane group and the basis vectors $\mathbf{a}$, $\mathbf{b}$, $\mathbf{c}$ of the space group. Each set of basis vectors refers to the conventional coordinate system of the plane group or space group, as employed in Chapters 2.2 and 2.3 of *IT* A. The basis vectors of the two-dimensional cell are always called $\mathbf{a}'$ and $\mathbf{b}'$ irrespective of which two of the basis vectors $\mathbf{a}$, $\mathbf{b}$, $\mathbf{c}$ of the three-dimensional cell are projected to form the plane cell. All relations between the basis vectors of the two cells are expressed as vector equations, *i.e.* $\mathbf{a}'$ and $\mathbf{b}'$ are given as linear combinations of $\mathbf{a}$, $\mathbf{b}$ and $\mathbf{c}$. For the triclinic or monoclinic space groups, basis vectors $\mathbf{a}$, $\mathbf{b}$ or $\mathbf{c}$ inclined to the plane of projection are replaced by the projected vectors $\mathbf{a}_p$, $\mathbf{b}_p$, $\mathbf{c}_p$.

(iv) *Location of the origin* of the plane group with respect to the unit cell of the space group. The same description is used as for the location of symmetry elements (*cf.* Section 2.1.3.9).

Example

'Origin at $x, 0, 0$' or 'Origin at $\frac{1}{4}, \frac{1}{4}, z$'.

Projections of centred cells (lattices). For centred lattices, two different cases may occur:

(i) The projection direction is parallel to a lattice-centring vector. In this case, the projected plane cell is primitive for the centring types A, B, C, I and R. For F-centred lattices, the multiplicity is reduced from 4 to 2 because c-centred plane cells result from projections along face diagonals of three-dimensional F cells.

(ii) The projection direction is not parallel to a lattice-centring vector (general projection direction). In this case, the plane cell has the same multiplicity as the three-dimensional cell. Usually, however, this centred plane cell is unconventional and a transformation is required to obtain the conventional plane cell. This transformation has been carried out for the projection data in *IT* A.

Examples

(1) Projection along [010] of a cubic I-centred cell leads to an unconventional quadratic c-centred plane cell. A simple cell transformation leads to the conventional quadratic p cell.

Table 2.1.3.6
Projections of crystallographic symmetry elements

Symmetry element in three dimensions	Symmetry element in projection
Arbitrary orientation	
Symmetry centre $\bar{1}$ Rotoinversion axis $\bar{3} \equiv 3 \times \bar{1}$	Rotation point 2 (at projection of centre)
Parallel to projection direction	
Rotation axis 2; 3; 4; 6	Rotation point 2; 3; 4; 6
Screw axis 2_1	Rotation point 2
$3_1, 3_2$	3
$4_1, 4_2, 4_3$	4
$6_1, 6_2, 6_3, 6_4, 6_5$	6
Rotoinversion axis $\bar{4}$	Rotation point 4
$\bar{6} \equiv 3/m$	3, with overlap of atoms
$\bar{3} \equiv 3 \times \bar{1}$	6
Reflection plane m	Reflection line m
Glide plane with $\perp$ component†	Glide line g
Glide plane without $\perp$ component†	Reflection line m
Normal to projection direction	
Rotation axis 2;4;6	Reflection line m
3	None
Screw axis 4_2; $6_2, 6_4$	Reflection line m
2_1; $4_1, 4_3$; $6_1, 6_3, 6_5$	Glide line g
$3_1, 3_2$	None
Rotoinversion axis $\bar{4}$	Reflection line m parallel to axis
$\bar{6} \equiv 3/m$	Reflection line m perpendicular to axis (through projection of inversion point)
$\bar{3} \equiv 3 \times \bar{1}$	Rotation point 2 (at projection of centre)
Reflection plane m	None, but overlap of atoms
Glide plane with glide vector **t**	Translation with translation vector **t**

† The term 'with $\perp$ component' refers to the component of the glide vector normal to the projection direction.

(2) Projection along [010] of an orthorhombic I-centred cell leads to a rectangular c-centred plane cell, which is conventional.

(3) Projection along [001] of an R-centred cell (both in obverse and reverse setting) results in a triple hexagonal plane cell h (the two-dimensional analogue of the H cell, *cf.* Table 2.1.1.2). A simple cell transformation leads to the conventional hexagonal p cell.

Projections of symmetry elements. A symmetry element of a space group does not project as a symmetry element unless its orientation bears a special relation to the projection direction; all translation components of a symmetry operation along the projection direction vanish, whereas those perpendicular to the projection direction (*i.e.* parallel to the plane of projection) may be retained. This is summarized in Table 2.1.3.6 for the various crystallographic symmetry elements.

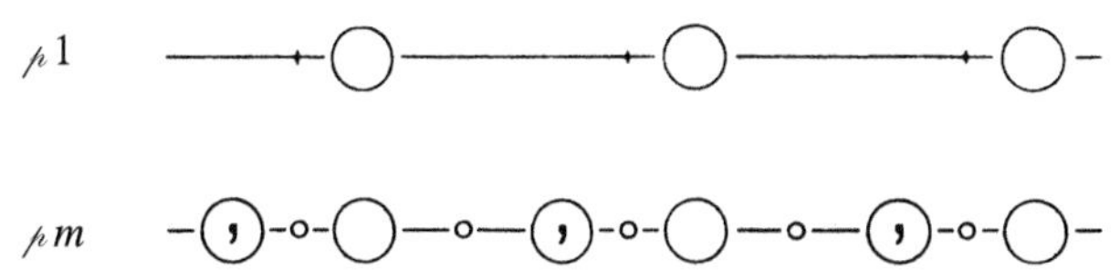

Figure 2.1.3.12
The two line groups (one-dimensional space groups). Small circles are reflection points; large circles represent the general position; in line group $\mathscr{p}1$, the vertical bars are the origins of the unit cells.

A detailed discussion of the correspondence between the symmetry elements and their projections is given in Section 1.4.5.3.

*Example: C*12/*c*1 *(15, b unique, cell choice 1)*
The C-centred cell has lattice points at 0, 0, 0 and $\frac{1}{2}, \frac{1}{2}$, 0. In all projections, the centre $\bar{1}$ projects as a twofold rotation point.
Projection along [001]: The plane cell is centred; 2 ∥ [010] projects as m; the glide component $(0, 0, \frac{1}{2})$ of glide plane c vanishes and thus c projects as m.
Result: Plane group $c2mm$ (9), $\mathbf{a}' = \mathbf{a}_p$, $\mathbf{b}' = \mathbf{b}$.
Projection along [100]: The periodicity along b is halved because of the C centring; 2 ∥ [010] projects as m; the glide component $(0, 0, \frac{1}{2})$ of glide plane c is retained and thus c projects as g.
Result: Plane group $p2gm$ (7), $\mathbf{a}' = \mathbf{b}/2$, $\mathbf{b}' = \mathbf{c}_p$.
Projection along [010]: The periodicity along a is halved because of the C centring; that along c is halved owing to the glide component $(0, 0, \frac{1}{2})$ of glide plane c; 2 ∥ [010] projects as 2.
Result: Plane group $p2$ (2), $\mathbf{a}' = \mathbf{c}/2$, $\mathbf{b}' = \mathbf{a}/2$.

Further details about the geometry of projections can be found in publications by Buerger (1965) and Biedl (1966).

2.1.3.15. Crystallographic groups in one dimension

In one dimension, only one crystal family, one crystal system and one Bravais lattice exist. No name or common symbol is required for any of them. All one-dimensional lattices are primitive, which is symbolized by the script letter $\mathscr{p}$; *cf.* Table 2.1.1.1.

There occur two types of one-dimensional point groups, 1 and $m \equiv \bar{1}$. The latter contains reflections through a point (reflection point or mirror point). This operation can also be described as inversion through a point, thus $m \equiv \bar{1}$ for one dimension; *cf.* Section 2.1.2.

Two types of line groups (one-dimensional space groups) exist, with Hermann–Mauguin symbols $\mathscr{p}1$ and $\mathscr{p}m \equiv \mathscr{p}\bar{1}$, which are illustrated in Fig. 2.1.3.12. Line group $\mathscr{p}1$, which consists of one-dimensional translations only, has merely one (general) position with coordinate x. Line group $\mathscr{p}m$ consists of one-dimensional translations and reflections through points. It has one general and two special positions. The coordinates of the general position are x and $\bar{x}$; the coordinate of one special position is 0, that of the other $\frac{1}{2}$. The site symmetries of both special positions are $m \equiv \bar{1}$. For $\mathscr{p}1$, the origin is arbitrary, for $\mathscr{p}m$ it is at a reflection point.

The one-dimensional *point groups* are of interest as 'edge symmetries' of two-dimensional 'edge forms'; they are listed in Table 3.2.3.1 of *IT* A. The one-dimensional *space groups* occur as projection and section symmetries of crystal structures.

2.1.4. Examples of plane- and space-group tables

A representative set of plane- and space-group tables from *IT* A, as listed in Table 2.1.4.1, are shown at the end of this chapter. The 7 plane-group and 35 space-group tables have been selected in accordance with three criteria: (i) to illustrate the wide variety of space groups (according to different classification schemes) and of their descriptions found in *IT* A; (ii) to include most of the frequently occurring space groups for both inorganic and organic crystals; and finally (in order to make this Teaching Edition as

Table 2.1.4.1
Examples of plane- and space-group tables from *IT* A shown on

(*a*) Plane-group tables.

*p*2 (2)	*p*2*mg* (7)	*p*2*gg* (8)	*c*2*mm* (9)	*p*4*gm* (12)	*p*6 (16)	*p*6*mm* (17)

(*b*) Space-group tables.

$P1$ (1)	$P\bar{1}$ (2)	$P12_11$ (4)	$C1c1$ (9)	$C12/m1$ (12)	$P12_1/c1$ (14)
$P112_1/a$ (14)	$C12/c1$ (15)	$P2_12_12$ (18)	$P2_12_12_1$ (19)	$Pmm2$ (25)	$Pna2_1$ (33)
$Aem2$ (39)	$Pmmm$ (47)	$Pnma$ (62)	$Cmcm$ (63)	$Fddd$ (70), origin choice 1	$Ibca$ (73)
$I4/m$ (87)	$P4_12_12$ (92)	$P4_32_12$ (96)	$P4mm$ (99)	$P4bm$ (100)	$I4/mmm$ (139)
$I4_1/amd$ (141)	$P3m1$ (156)	$P31m$ (157)	$R3m$ (160)	$P6_3/m$ (176)	$P6/mmm$ (191)
$P6_3/mmc$ (194)	$Pa\bar{3}$ (205)	$F\bar{4}3m$ (216)	$Fm\bar{3}m$ (225)	$Fd\bar{3}m$ (227), origin choice 2	$Ia\bar{3}d$ (230)

self-contained as possible) (iii) to list most of the space groups used in the illustrative examples in this book.

In addition, the space-group table for *Cccm* (66) is shown on the inside of the front and back covers.

References

Astbury, W. T. & Yardley, K. (1924). *Tabulated data for the examination of the 230 space groups by homogeneous X-rays. Philos. Trans. R. Soc. London Ser. A*, **224**, 221–257.

Belov, N. V., Zagal'skaja, Ju. G., Litvinskaja, G. P. & Egorov-Tismenko, Ju. K. (1980). *Atlas of the Space Groups of the Cubic System.* Moscow: Nauka. (In Russian.)

Biedl, A. W. (1966). *The projection of a crystal structure. Z. Kristallogr.* **123**, 21–26.

Buerger, M. J. (1949). *Fourier summations for symmetrical crystals. Am. Mineral.* **34**, 771–788.

Buerger, M. J. (1956). *Elementary Crystallography*. New York: Wiley.

Buerger, M. J. (1960). *Crystal-Structure Analysis*, ch. 17. New York: Wiley.

Buerger, M. J. (1965). *The geometry of projections. Tschermaks Mineral. Petrogr. Mitt.* **10**, 595–607.

Engel, P., Matsumoto, T., Steinmann, G. & Wondratschek, H. (1984). *The non-characteristic orbits of the space groups. Z. Kristallogr.*, Supplement Issue No. 1.

Fedorov, E. S. (1895). *Theorie der Kristallstruktur. Einleitung. Regelmässige Punktsysteme (mit übersichtlicher graphischer Darstellung). Z. Kristallogr.* **24**, 209–252, Tafel V, VI. [English translation by D. & K. Harker (1971). *Symmetry of Crystals*, esp. pp. 206–213. Am. Crystallogr. Assoc., ACA Monograph No. 7.]

Fischer, W., Burzlaff, H., Hellner, E. & Donnay, J. D. H. (1973). *Space Groups and Lattice Complexes*. NBS Monograph No. 134. Washington, DC: National Bureau of Standards.

Flack, H. D. (2015). *Patterson functions. Z. Kristallogr.* **230**, 743–748.

Flack, H. D., Wondratschek, H., Hahn, Th. & Abrahams, S. C. (2000). *Symmetry elements in space groups and point groups. Addenda to two IUCr Reports on the Nomenclature of Symmetry. Acta Cryst.* **A56**, 96–98.

Friedel, G. (1926). *Leçons de Cristallographie*. Nancy/Paris/Strasbourg: Berger-Levrault. [Reprinted: Paris: Blanchard (1964).]

Grosse-Kunstleve, R. W., Wong, B., Mustyakimov, M. & Adams, P. D. (2011). *Exact direct-space asymmetric units for the 230 crystallographic space groups. Acta Cryst.* **A67**, 269–275.

Hahn, Th. & Wondratschek, H. (1994). *Symmetry of Crystals*. Sofia: Heron Press.

Heesch, H. (1929). *Zur systematischen Strukturtheorie. II. Z. Kristallogr.* **72**, 177–201.

Hilton, H. (1903). *Mathematical Crystallography*. Oxford: Clarendon Press. [Reprint: New York: Dover (1963).]

International Tables for Crystallography (2002). Volume A, 5th ed., edited by Th. Hahn. Dordrecht: Kluwer Academic Publishers. [Abbreviated as *IT* A (2002).]

International Tables for Crystallography (2004). Volume C, 3rd ed., edited by E. Prince. Dordrecht: Kluwer Academic Publishers.

International Tables for Crystallography (2010). Volume A1, 2nd ed., edited by H. Wondratschek & U. Müller. Chichester: Wiley. [Abbreviated as *IT* A1.]

International Tables for Crystallography (2016). Volume A, 6th ed., edited by M. I. Aroyo. Chichester: Wiley. [Abbreviated as *IT* A.]

International Tables for X-ray Crystallography (1952). Volume I, edited by N. F. M. Henry & K. Lonsdale. Birmingham: Kynoch Press. [Revised editions: 1965, 1969 and 1977. Abbreviated as *IT* (1952).]

Internationale Tabellen zur Bestimmung von Kristallstrukturen (1935). 1. Band, edited by C. Hermann. Berlin: Borntraeger. [Revised edition: Ann Arbor: Edwards (1944). Abbreviated as *IT* (1935).]

Koch, E. & Fischer, W. (1974). *Zur Bestimmung asymmetrischer Einheiten kubischer Raumgruppen mit Hilfe von Wirkungsbereichen. Acta Cryst.* A**30**, 490–496.

Momma, K. & Izumi, F. (2011). *VESTA 3 for three-dimensional visualization of crystal, volumetric and morphology data. J. Appl. Cryst.* **44**, 1272–1276.

Müller, U. (2013). *Symmetry Relationships between Crystal Structures.* Oxford: IUCr/Oxford University Press.

Niggli, P. (1919). *Geometrische Kristallographie des Diskontinuums.* Leipzig: Borntraeger. [Reprint: Wiesbaden: Sändig (1973).]

Sadanaga, R., Takeuchi, Y. & Morimoto, N. (1978). *Complex structures of minerals. Recent Prog. Nat. Sci. Jpn*, **3**, 141–206, esp. pp. 149–151.

Schiebold, E. (1929). *Über eine neue Herleitung und Nomenklatur der 230 kristallographischen Raumgruppen mit Atlas der 230 Raumgruppen-Projektionen.* Text, Atlas. In *Abhandlungen der Mathematisch-Physikalischen Klasse der Sächsischen Akademie der Wissenschaften*, Band 40, Heft 5. Leipzig: Hirzel.

Templeton, D. H. (1956). *Systematic absences corresponding to false symmetry. Acta Cryst.* **9**, 199–200.

Wilson, A. J. C. (1993). *Laue and Patterson symmetry in the complex case. Z. Kristallogr.* **208**, 199–206.

Wolff, P. M. de, Belov, N. V., Bertaut, E. F., Buerger, M. J., Donnay, J. D. H., Fischer, W., Hahn, Th., Koptsik, V. A., Mackay, A. L., Wondratschek, H., Wilson, A. J. C. & Abrahams, S. C. (1985). *Nomenclature for crystal families, Bravais-lattice types and arithmetic classes. Report of the International Union of Crystallography Ad-hoc Committee on the Nomenclature of Symmetry. Acta Cryst.* **A41**, 278–280.

Wolff, P. M. de, Billiet, Y., Donnay, J. D. H., Fischer, W., Galiulin, R. B., Glazer, A. M., Hahn, Th., Senechal, M., Shoemaker, D. P., Wondratschek, H., Wilson, A. J. C. & Abrahams, S. C. (1992). *Symbols for symmetry elements and symmetry operations. Final Report of the International Union of Crystallography Ad-hoc Committee on the Nomenclature of Symmetry. Acta Cryst.* **A48**, 727–732.

Wolff, P. M. de, Billiet, Y., Donnay, J. D. H., Fischer, W., Galiulin, R. B., Glazer, A. M., Senechal, M., Shoemaker, D. P., Wondratschek, H., Hahn, Th., Wilson, A. J. C. & Abrahams, S. C. (1989). *Definition of symmetry elements in space groups and point groups. Report of the International Union of Crystallography Ad-hoc Committee on the Nomenclature of Symmetry. Acta Cryst.* **A45**, 494–499.

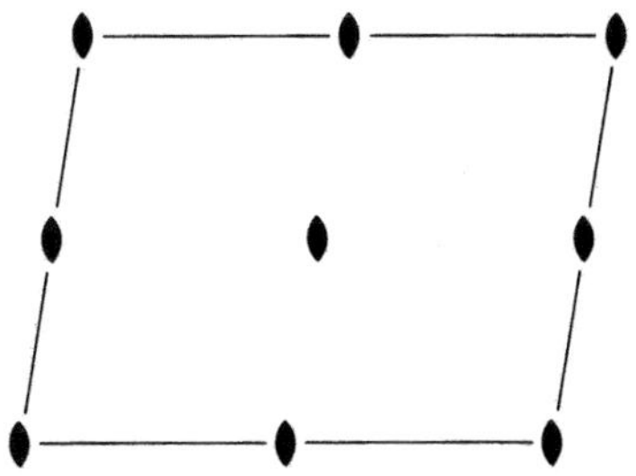

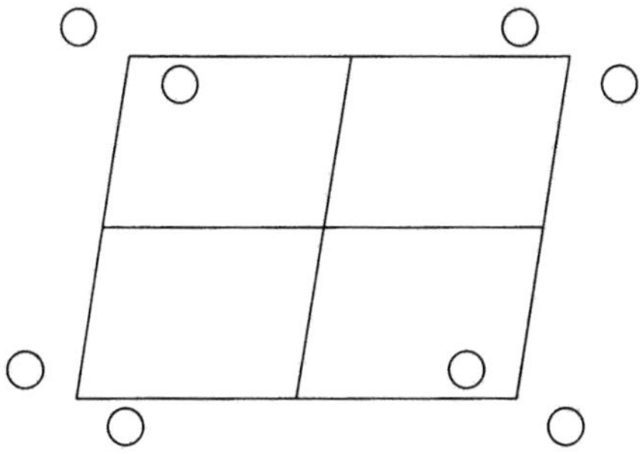

Origin at 2

Asymmetric unit $0 \le x \le \frac{1}{2}$; $0 \le y \le 1$

Symmetry operations

(1) 1 (2) 2 0,0

Generators selected (1); $t(1,0)$; $t(0,1)$; (2)

Positions

Multiplicity, Wyckoff letter, Site symmetry			Coordinates		Reflection conditions
					General:
2	e	1	(1) x,y	(2) $\bar{x},\bar{y}$	no conditions
					Special: no extra conditions
1	d	2	$\frac{1}{2},\frac{1}{2}$		
1	c	2	$\frac{1}{2},0$		
1	b	2	$0,\frac{1}{2}$		
1	a	2	$0,0$		

$p2mg$ $2mm$ Rectangular

No. 7 $p2mg$ Patterson symmetry $p2mm$

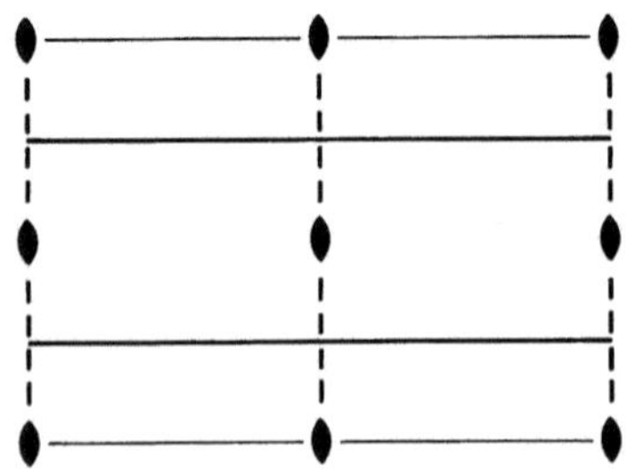

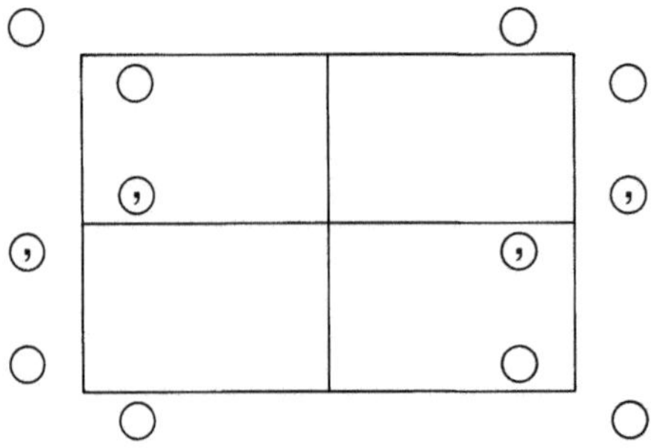

Origin at $21g$

Asymmetric unit $0 \le x \le \frac{1}{4}$; $0 \le y \le 1$

Symmetry operations

(1) 1 (2) 2 0,0 (3) m $\frac{1}{4}, y$ (4) a $x, 0$

Generators selected (1); $t(1,0)$; $t(0,1)$; (2); (3)

Positions

Multiplicity, Wyckoff letter, Site symmetry			Coordinates				Reflection conditions
							General:
4	d	1	(1) x, y	(2) $\bar{x}, \bar{y}$	(3) $\bar{x}+\frac{1}{2}, y$	(4) $x+\frac{1}{2}, \bar{y}$	$h0$: $h = 2n$
							Special: as above, plus
2	c	$.m.$	$\frac{1}{4}, y$	$\frac{3}{4}, \bar{y}$			no extra conditions
2	b	$2..$	$0, \frac{1}{2}$	$\frac{1}{2}, \frac{1}{2}$			hk: $h = 2n$
2	a	$2..$	$0, 0$	$\frac{1}{2}, 0$			hk: $h = 2n$

Rectangular $2mm$ $p2gg$

Patterson symmetry $p2mm$ $p2gg$ No. 8

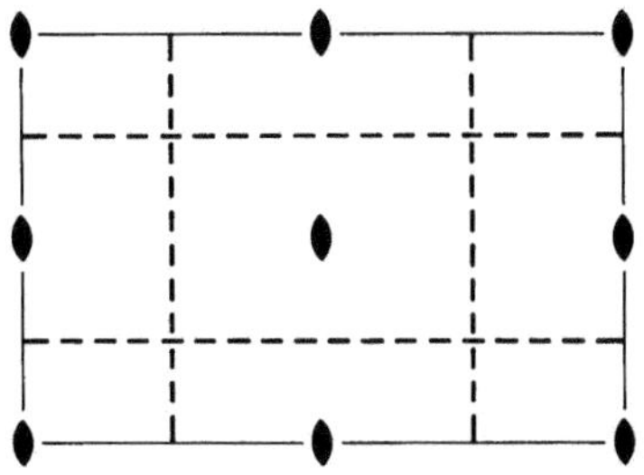

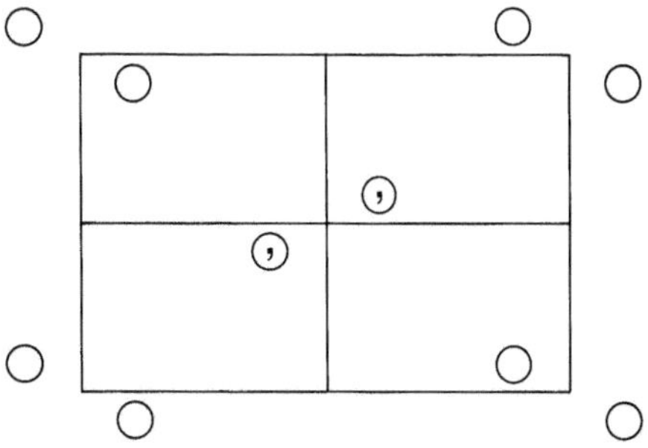

Origin at 2

Asymmetric unit $0 \le x \le \frac{1}{2}$; $0 \le y \le \frac{1}{2}$

Symmetry operations

(1) 1 (2) 2 0,0 (3) b $\frac{1}{4},y$ (4) a $x,\frac{1}{4}$

Generators selected (1); $t(1,0)$; $t(0,1)$; (2); (3)

Positions

Multiplicity, Wyckoff letter, Site symmetry			Coordinates				Reflection conditions
							General:
4	c	1	(1) x,y	(2) $\bar{x},\bar{y}$	(3) $\bar{x}+\frac{1}{2},y+\frac{1}{2}$	(4) $x+\frac{1}{2},\bar{y}+\frac{1}{2}$	$h0$: $h=2n$ $0k$: $k=2n$
							Special: as above, plus
2	b	$2\,.\,.$	$\frac{1}{2},0$	$0,\frac{1}{2}$			hk: $h+k=2n$
2	a	$2\,.\,.$	$0,0$	$\frac{1}{2},\frac{1}{2}$			hk: $h+k=2n$

$c2mm$ $2mm$ Rectangular

No. 9 $c2mm$ Patterson symmetry $c2mm$

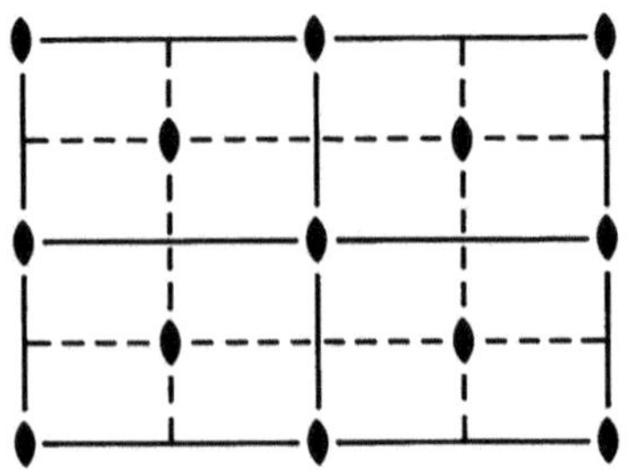

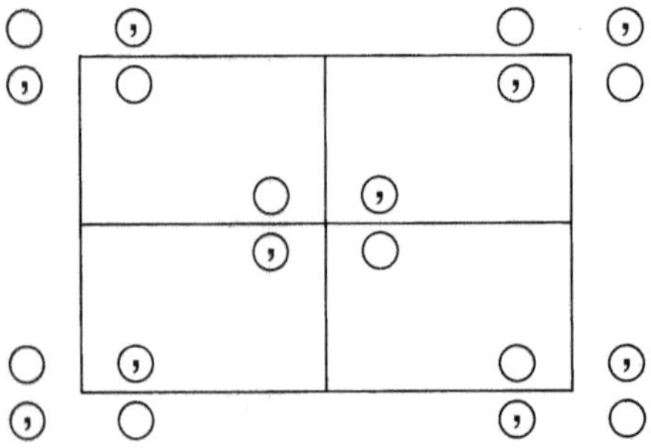

Origin at $2mm$

Asymmetric unit $0 \le x \le \frac{1}{4}$; $0 \le y \le \frac{1}{2}$

Symmetry operations

For $(0,0)+$ set

(1) 1 (2) $2\ \ 0,0$ (3) $m\ \ 0,y$ (4) $m\ \ x,0$

For $(\frac{1}{2},\frac{1}{2})+$ set

(1) $t(\frac{1}{2},\frac{1}{2})$ (2) $2\ \ \frac{1}{4},\frac{1}{4}$ (3) $b\ \ \frac{1}{4},y$ (4) $a\ \ x,\frac{1}{4}$

Generators selected (1); $t(1,0)$; $t(0,1)$; $t(\frac{1}{2},\frac{1}{2})$; (2); (3)

Positions

Multiplicity, Wyckoff letter, Site symmetry			Coordinates $(0,0)+$ $(\frac{1}{2},\frac{1}{2})+$				Reflection conditions
							General:
8	f	1	(1) x,y	(2) $\bar{x},\bar{y}$	(3) $\bar{x},y$	(4) $x,\bar{y}$	hk: $h+k=2n$ $h0$: $h=2n$ $0k$: $k=2n$
							Special: as above, plus
4	e	$.m.$	$0,y$	$0,\bar{y}$			no extra conditions
4	d	$..m$	$x,0$	$\bar{x},0$			no extra conditions
4	c	$2..$	$\frac{1}{4},\frac{1}{4}$	$\frac{3}{4},\frac{1}{4}$			hk: $h=2n$
2	b	$2mm$	$0,\frac{1}{2}$				no extra conditions
2	a	$2mm$	$0,0$				no extra conditions

Square $4mm$ $p4gm$

Patterson symmetry $p4mm$ $p4gm$ No. 12

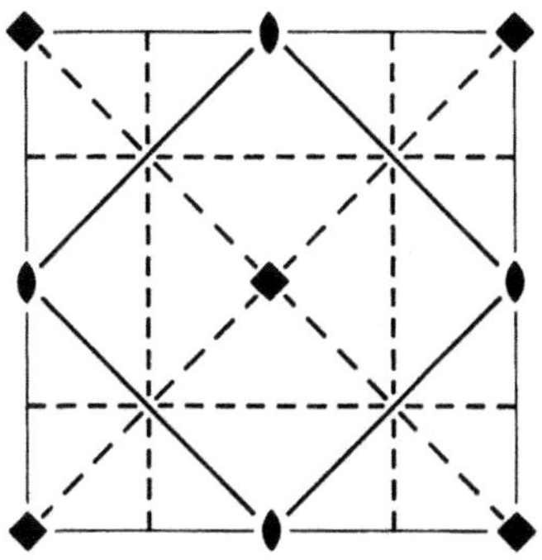

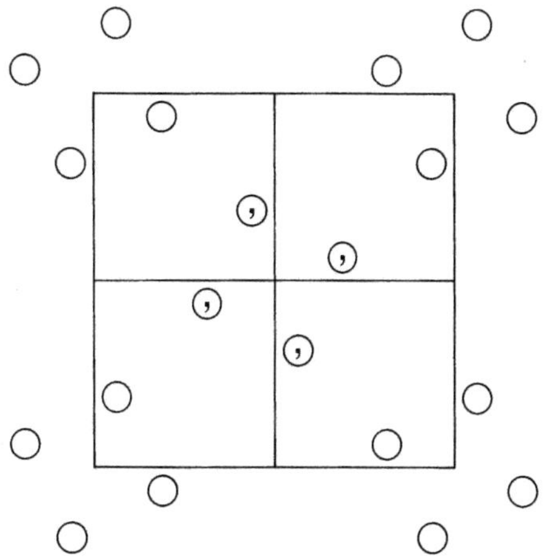

Origin at $4\,1\,g$

Asymmetric unit $0 \le x \le \frac{1}{2}$; $0 \le y \le \frac{1}{2}$; $y \le \frac{1}{2} - x$

Symmetry operations

(1) 1	(2) 2 $0,0$	(3) 4^{+} $0,0$	(4) 4^{-} $0,0$
(5) b $\frac{1}{4},y$	(6) a $x,\frac{1}{4}$	(7) $g(\frac{1}{2},\frac{1}{2})$ x,x	(8) m $x+\frac{1}{2},\bar{x}$

Generators selected (1); $t(1,0)$; $t(0,1)$; (2); (3); (5)

Positions

Multiplicity, Wyckoff letter, Site symmetry			Coordinates				Reflection conditions
							General:
8	d	1	(1) x,y	(2) $\bar{x},\bar{y}$	(3) $\bar{y},x$	(4) $y,\bar{x}$	$h0$: $h=2n$
			(5) $\bar{x}+\frac{1}{2},y+\frac{1}{2}$	(6) $x+\frac{1}{2},\bar{y}+\frac{1}{2}$	(7) $y+\frac{1}{2},x+\frac{1}{2}$	(8) $\bar{y}+\frac{1}{2},\bar{x}+\frac{1}{2}$	$0k$: $k=2n$
							Special: as above, plus
4	c	$..m$	$x,x+\frac{1}{2}$	$\bar{x},\bar{x}+\frac{1}{2}$	$\bar{x}+\frac{1}{2},x$	$x+\frac{1}{2},\bar{x}$	no extra conditions
2	b	$2.mm$	$\frac{1}{2},0$	$0,\frac{1}{2}$			hk: $h+k=2n$
2	a	$4..$	$0,0$	$\frac{1}{2},\frac{1}{2}$			hk: $h+k=2n$

$p6$ 6 Hexagonal

No. 16 $p6$ Patterson symmetry $p6$

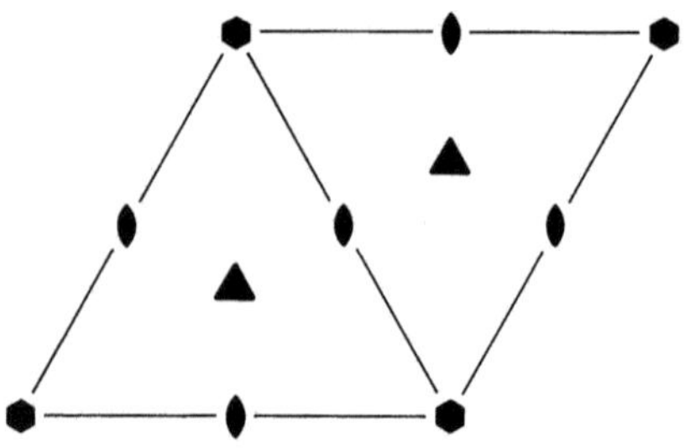
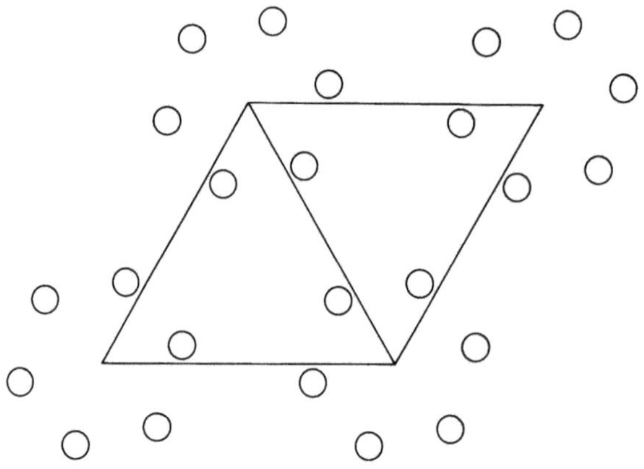

Origin at 6

Asymmetric unit $0 \le x \le \frac{2}{3}$; $0 \le y \le \frac{1}{2}$; $x \le (1+y)/2$; $y \le \min(1-x, x)$

Vertices $0,0$ $\frac{1}{2},0$ $\frac{2}{3},\frac{1}{3}$ $\frac{1}{2},\frac{1}{2}$

Symmetry operations

(1) 1	(2) 3^{+} 0,0	(3) 3^{-} 0,0
(4) 2 0,0	(5) 6^{-} 0,0	(6) 6^{+} 0,0

Generators selected (1); $t(1,0)$; $t(0,1)$; (2); (4)

Positions

Multiplicity, Wyckoff letter, Site symmetry			Coordinates			Reflection conditions
						General:
6	*d*	1	(1) x,y	(2) $\bar{y},x-y$	(3) $\bar{x}+y,\bar{x}$	no conditions
			(4) $\bar{x},\bar{y}$	(5) $y,\bar{x}+y$	(6) $x-y,x$	
						Special: no extra conditions
3	*c*	2 . .	$\frac{1}{2},0$	$0,\frac{1}{2}$	$\frac{1}{2},\frac{1}{2}$	
2	*b*	3 . .	$\frac{1}{3},\frac{2}{3}$	$\frac{2}{3},\frac{1}{3}$		
1	*a*	6 . .	$0,0$			

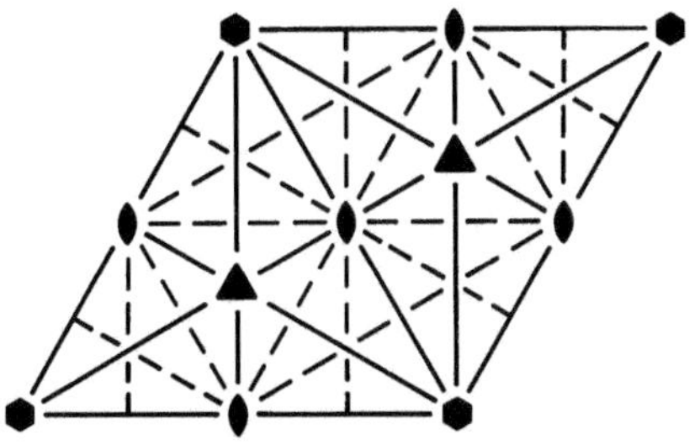

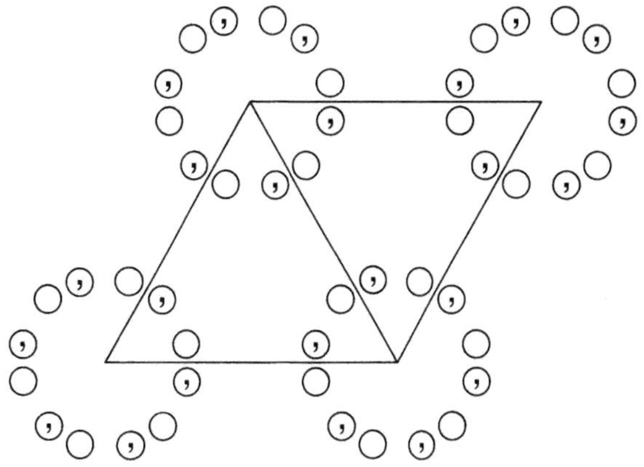

Origin at $6mm$

Asymmetric unit $0 \le x \le \frac{2}{3}$; $0 \le y \le \frac{1}{3}$; $x \le (1+y)/2$; $y \le x/2$

Vertices $0,0$ $\frac{1}{2},0$ $\frac{2}{3},\frac{1}{3}$

Symmetry operations

(1) 1	(2) 3^{+} $0,0$	(3) 3^{-} $0,0$
(4) 2 $0,0$	(5) 6^{-} $0,0$	(6) 6^{+} $0,0$
(7) m $x,\bar{x}$	(8) m $x,2x$	(9) m $2x,x$
(10) m x,x	(11) m $x,0$	(12) m $0,y$

Generators selected (1); $t(1,0)$; $t(0,1)$; (2); (4); (7)

Positions

Multiplicity, Wyckoff letter, Site symmetry			Coordinates						Reflection conditions
									General:
12	f	1	(1) x,y	(2) $\bar{y},x-y$	(3) $\bar{x}+y,\bar{x}$				no conditions
			(4) $\bar{x},\bar{y}$	(5) $y,\bar{x}+y$	(6) $x-y,x$				
			(7) $\bar{y},\bar{x}$	(8) $\bar{x}+y,y$	(9) $x,x-y$				
			(10) y,x	(11) $x-y,\bar{y}$	(12) $\bar{x},\bar{x}+y$				
									Special: no extra conditions
6	e	$.m.$	$x,\bar{x}$	$x,2x$	$2\bar{x},\bar{x}$	$\bar{x},x$	$\bar{x},2\bar{x}$	$2x,x$	
6	d	$..m$	$x,0$	$0,x$	$\bar{x},\bar{x}$	$\bar{x},0$	$0,\bar{x}$	x,x	
3	c	$2mm$	$\frac{1}{2},0$	$0,\frac{1}{2}$	$\frac{1}{2},\frac{1}{2}$				
2	b	$3m.$	$\frac{1}{3},\frac{2}{3}$	$\frac{2}{3},\frac{1}{3}$					
1	a	$6mm$	$0,0$						

$P1$ C_1^1 1 Triclinic

No. 1 $P1$ Patterson symmetry $P\bar{1}$

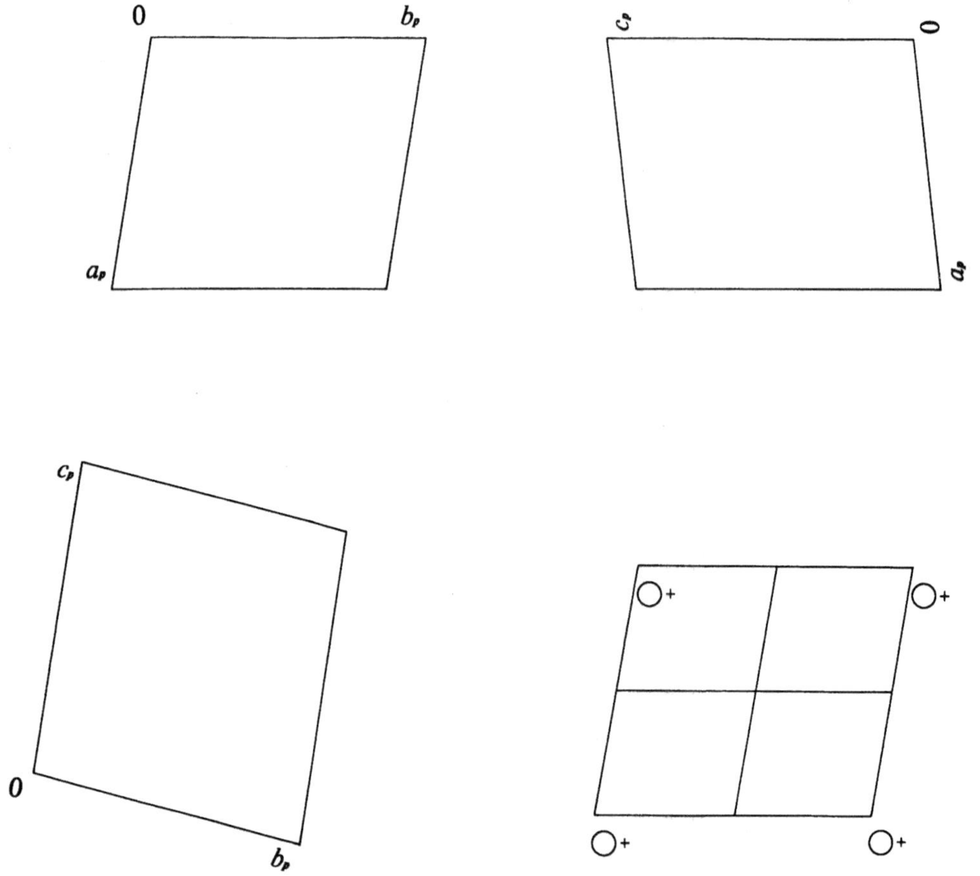

Drawings for type II cell. Proper cell reduction gives either
a type I (α, β, γ acute) or a type II (α, β, γ non-acute) cell.

Origin arbitrary

Asymmetric unit $0 \le x \le 1$; $0 \le y \le 1$; $0 \le z \le 1$

Symmetry operations

(1) 1

Generators selected (1); $t(1,0,0)$; $t(0,1,0)$; $t(0,0,1)$

Positions

Multiplicity, Wyckoff letter, Site symmetry	Coordinates	Reflection conditions
		General:
1 *a* 1	(1) x,y,z	no conditions

Symmetry of special projections

Along [001] $p1$
$\mathbf{a}' = \mathbf{a}_p$ $\mathbf{b}' = \mathbf{b}_p$
Origin at $0,0,z$

Along [100] $p1$
$\mathbf{a}' = \mathbf{b}_p$ $\mathbf{b}' = \mathbf{c}_p$
Origin at $x,0,0$

Along [010] $p1$
$\mathbf{a}' = \mathbf{c}_p$ $\mathbf{b}' = \mathbf{a}_p$
Origin at $0,y,0$

Triclinic $\bar{1}$ C_i^1 $P\bar{1}$

Patterson symmetry $P\bar{1}$ $P\bar{1}$ **No. 2**

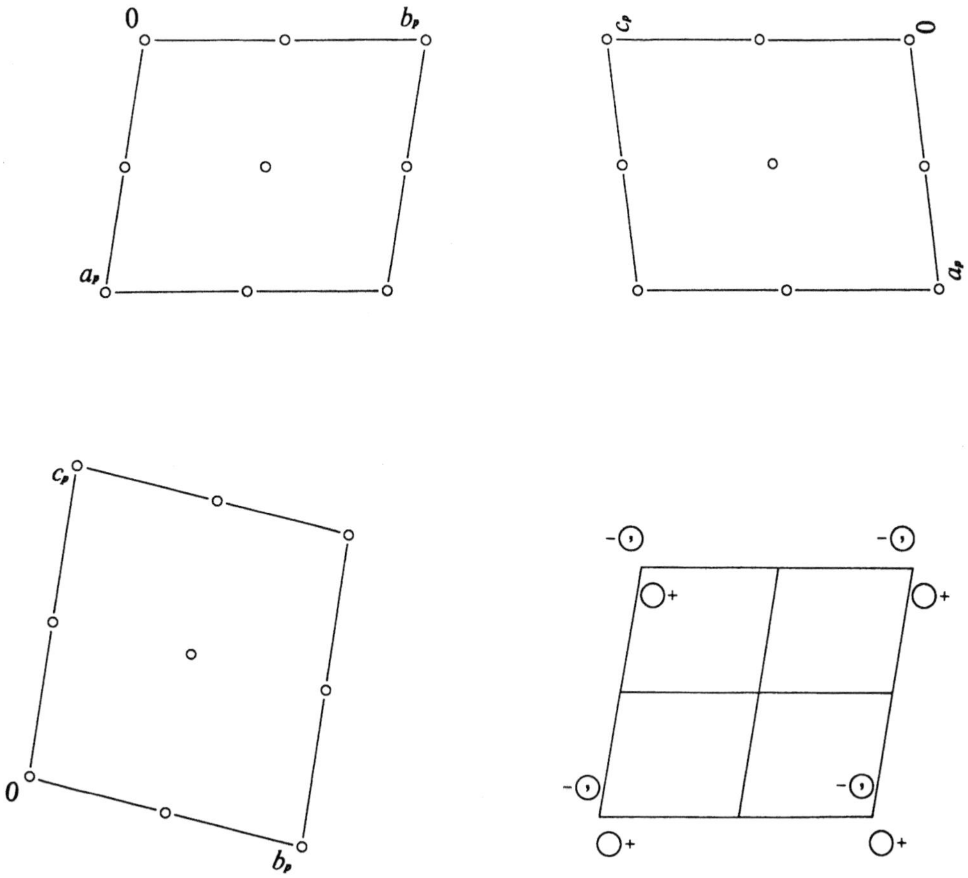

Drawings for type II cell. Proper cell reduction gives either a type I (α, β, γ acute) or a type II (α, β, γ non-acute) cell.

Origin at $\bar{1}$

Asymmetric unit $0 \leq x \leq \frac{1}{2}$; $0 \leq y \leq 1$; $0 \leq z \leq 1$

Symmetry operations

(1) 1 (2) $\bar{1}$ 0,0,0

Generators selected (1); $t(1,0,0)$; $t(0,1,0)$; $t(0,0,1)$; (2)

Positions

Multiplicity, Wyckoff letter, Site symmetry			Coordinates		Reflection conditions
					General:
2	*i*	1	(1) x,y,z	(2) $\bar{x},\bar{y},\bar{z}$	no conditions
					Special: no extra conditions

1	*h*	$\bar{1}$	$\frac{1}{2},\frac{1}{2},\frac{1}{2}$	1	*d*	$\bar{1}$	$\frac{1}{2},0,0$
1	*g*	$\bar{1}$	$0,\frac{1}{2},\frac{1}{2}$	1	*c*	$\bar{1}$	$0,\frac{1}{2},0$
1	*f*	$\bar{1}$	$\frac{1}{2},0,\frac{1}{2}$	1	*b*	$\bar{1}$	$0,0,\frac{1}{2}$
1	*e*	$\bar{1}$	$\frac{1}{2},\frac{1}{2},0$	1	*a*	$\bar{1}$	$0,0,0$

Symmetry of special projections

Along [001] $p2$
$\mathbf{a}' = \mathbf{a}_p$ $\mathbf{b}' = \mathbf{b}_p$
Origin at $0,0,z$

Along [100] $p2$
$\mathbf{a}' = \mathbf{b}_p$ $\mathbf{b}' = \mathbf{c}_p$
Origin at $x,0,0$

Along [010] $p2$
$\mathbf{a}' = \mathbf{c}_p$ $\mathbf{b}' = \mathbf{a}_p$
Origin at $0,y,0$

$P2_1$ C_2^2 2 Monoclinic

No. 4 $P12_11$ Patterson symmetry $P12/m1$

UNIQUE AXIS b

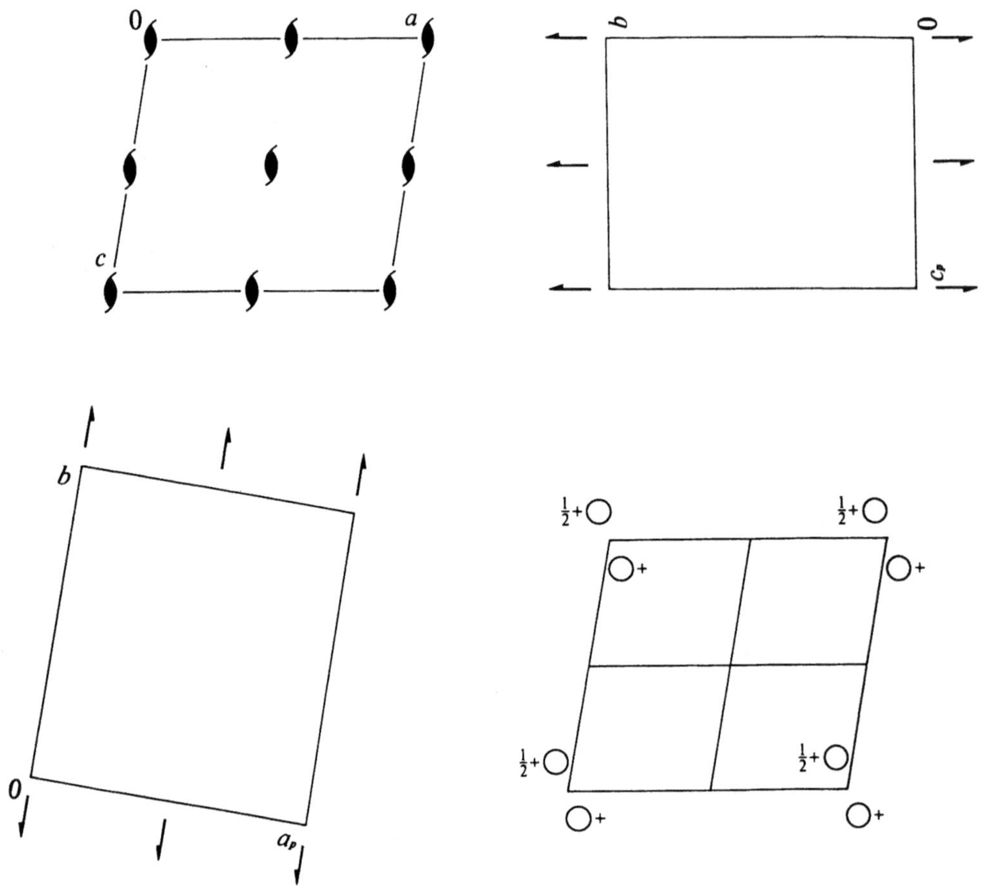

Origin on 2_1

Asymmetric unit $0 \le x \le 1;\quad 0 \le y \le 1;\quad 0 \le z \le \frac{1}{2}$

Symmetry operations

(1) 1 (2) $2(0,\frac{1}{2},0)$ $0,y,0$

Generators selected (1); $t(1,0,0)$; $t(0,1,0)$; $t(0,0,1)$; (2)

Positions

Multiplicity, Wyckoff letter, Site symmetry	Coordinates	Reflection conditions
		General:
2 a 1	(1) x,y,z (2) $\bar{x},y+\frac{1}{2},\bar{z}$	$0k0$: $k=2n$

Symmetry of special projections

Along [001] $p1g1$
$\mathbf{a}'=\mathbf{a}_p$ $\mathbf{b}'=\mathbf{b}$
Origin at $0,0,z$

Along [100] $p11g$
$\mathbf{a}'=\mathbf{b}$ $\mathbf{b}'=\mathbf{c}_p$
Origin at $x,0,0$

Along [010] $p2$
$\mathbf{a}'=\mathbf{c}$ $\mathbf{b}'=\mathbf{a}$
Origin at $0,y,0$

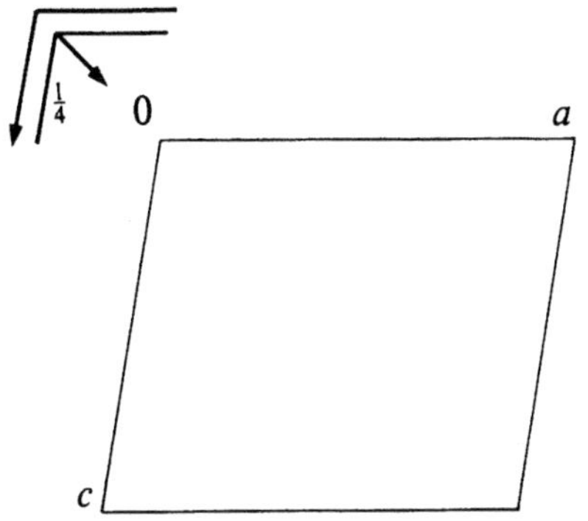

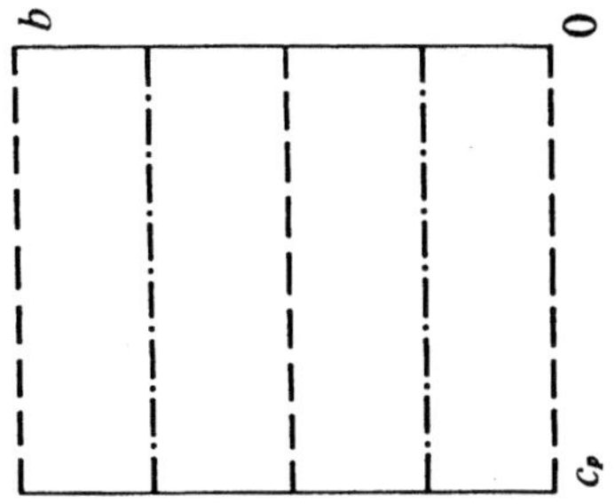

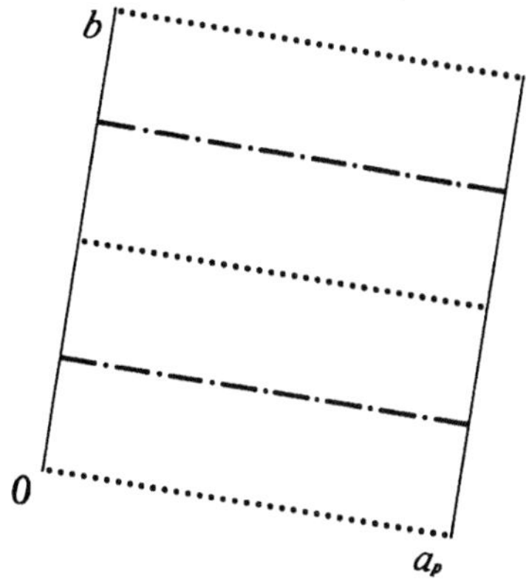

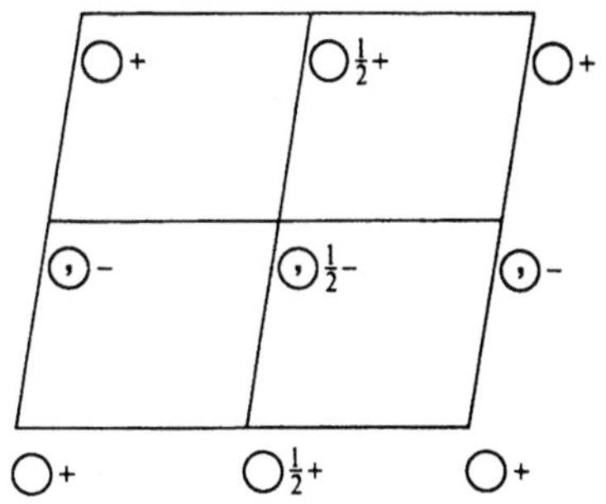

Origin on glide plane c

Asymmetric unit $0 \le x \le 1$; $0 \le y \le \frac{1}{4}$; $0 \le z \le 1$

Symmetry operations

For $(0,0,0)+$ set

(1) 1 (2) c $x,0,z$

For $(\frac{1}{2},\frac{1}{2},0)+$ set

(1) $t(\frac{1}{2},\frac{1}{2},0)$ (2) $n(\frac{1}{2},0,\frac{1}{2})$ $x,\frac{1}{4},z$

Generators selected (1); $t(1,0,0)$; $t(0,1,0)$; $t(0,0,1)$; $t(\frac{1}{2},\frac{1}{2},0)$; (2)

Positions

Multiplicity, Wyckoff letter, Site symmetry	Coordinates		Reflection conditions
	$(0,0,0)+$	$(\frac{1}{2},\frac{1}{2},0)+$	General:
4 a 1	(1) x,y,z	(2) $x,\bar{y},z+\frac{1}{2}$	hkl: $h+k=2n$ $h0l$: $h,l=2n$ $0kl$: $k=2n$ $hk0$: $h+k=2n$ $0k0$: $k=2n$ $h00$: $h=2n$ $00l$: $l=2n$

Symmetry of special projections

Along [001] $c11m$
$\mathbf{a}'=\mathbf{a}_p$ $\mathbf{b}'=\mathbf{b}$
Origin at $0,0,z$

Along [100] $p1g1$
$\mathbf{a}'=\frac{1}{2}\mathbf{b}$ $\mathbf{b}'=\mathbf{c}_p$
Origin at $x,0,0$

Along [010] $p1$
$\mathbf{a}'=\frac{1}{2}\mathbf{c}$ $\mathbf{b}'=\frac{1}{2}\mathbf{a}$
Origin at $0,y,0$

$C2/m$ $\quad C_{2h}^{3}$ $\quad 2/m$ $\quad$ Monoclinic

No. 12 $\quad C12/m1$ $\quad$ Patterson symmetry $C12/m1$

UNIQUE AXIS b, CELL CHOICE 1

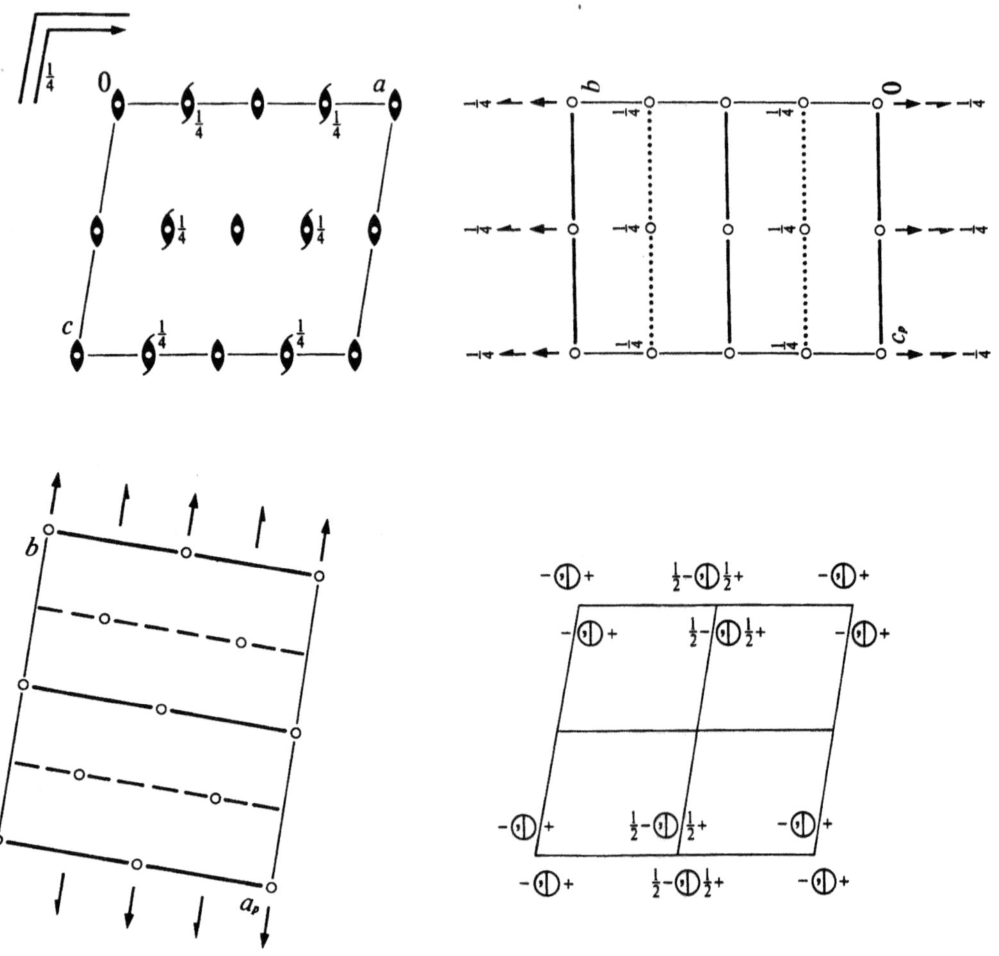

Origin at centre $(2/m)$

Asymmetric unit $\quad 0 \leq x \leq \frac{1}{2}; \quad 0 \leq y \leq \frac{1}{4}; \quad 0 \leq z \leq 1$

Symmetry operations

For $(0,0,0)+$ set

(1) 1 $\quad$ (2) $2\;\; 0,y,0$ $\quad$ (3) $\bar{1}\;\; 0,0,0$ $\quad$ (4) $m\;\; x,0,z$

For $(\frac{1}{2},\frac{1}{2},0)+$ set

(1) $t(\frac{1}{2},\frac{1}{2},0)$ $\quad$ (2) $2(0,\frac{1}{2},0)\;\; \frac{1}{4},y,0$ $\quad$ (3) $\bar{1}\;\; \frac{1}{4},\frac{1}{4},0$ $\quad$ (4) $a\;\; x,\frac{1}{4},z$

Generators selected (1); $t(1,0,0)$; $t(0,1,0)$; $t(0,0,1)$; $t(\frac{1}{2},\frac{1}{2},0)$; (2); (3)

Positions

Multiplicity, Wyckoff letter, Site symmetry			Coordinates				Reflection conditions
			$(0,0,0)+$	$(\frac{1}{2},\frac{1}{2},0)+$			General:
8	j	1	(1) x,y,z	(2) $\bar{x},y,\bar{z}$	(3) $\bar{x},\bar{y},\bar{z}$	(4) $x,\bar{y},z$	hkl: $h+k=2n$ $h0l$: $h=2n$ $0kl$: $k=2n$ $hk0$: $h+k=2n$ $0k0$: $k=2n$ $h00$: $h=2n$
							Special: as above, plus
4	i	m	$x,0,z$	$\bar{x},0,\bar{z}$			no extra conditions
4	h	2	$0,y,\frac{1}{2}$	$0,\bar{y},\frac{1}{2}$			no extra conditions
4	g	2	$0,y,0$	$0,\bar{y},0$			no extra conditions
4	f	$\bar{1}$	$\frac{1}{4},\frac{1}{4},\frac{1}{2}$	$\frac{3}{4},\frac{1}{4},\frac{1}{2}$			hkl: $h=2n$
4	e	$\bar{1}$	$\frac{1}{4},\frac{1}{4},0$	$\frac{3}{4},\frac{1}{4},0$			hkl: $h=2n$
2	d	$2/m$	$0,\frac{1}{2},\frac{1}{2}$				no extra conditions
2	c	$2/m$	$0,0,\frac{1}{2}$				no extra conditions
2	b	$2/m$	$0,\frac{1}{2},0$				no extra conditions
2	a	$2/m$	$0,0,0$				no extra conditions

Symmetry of special projections

Along [001] $c2mm$
$\mathbf{a}'=\mathbf{a}_p$ $\mathbf{b}'=\mathbf{b}$
Origin at $0,0,z$

Along [100] $p2mm$
$\mathbf{a}'=\frac{1}{2}\mathbf{b}$ $\mathbf{b}'=\mathbf{c}_p$
Origin at $x,0,0$

Along [010] $p2$
$\mathbf{a}'=\mathbf{c}$ $\mathbf{b}'=\frac{1}{2}\mathbf{a}$
Origin at $0,y,0$

$P2_1/c$ $\quad C_{2h}^5$ $\quad 2/m$ $\quad$ Monoclinic

No. 14 $\quad P1\,2_1/c\,1$ $\quad$ Patterson symmetry $P1\,2/m\,1$

UNIQUE AXIS b, CELL CHOICE 1

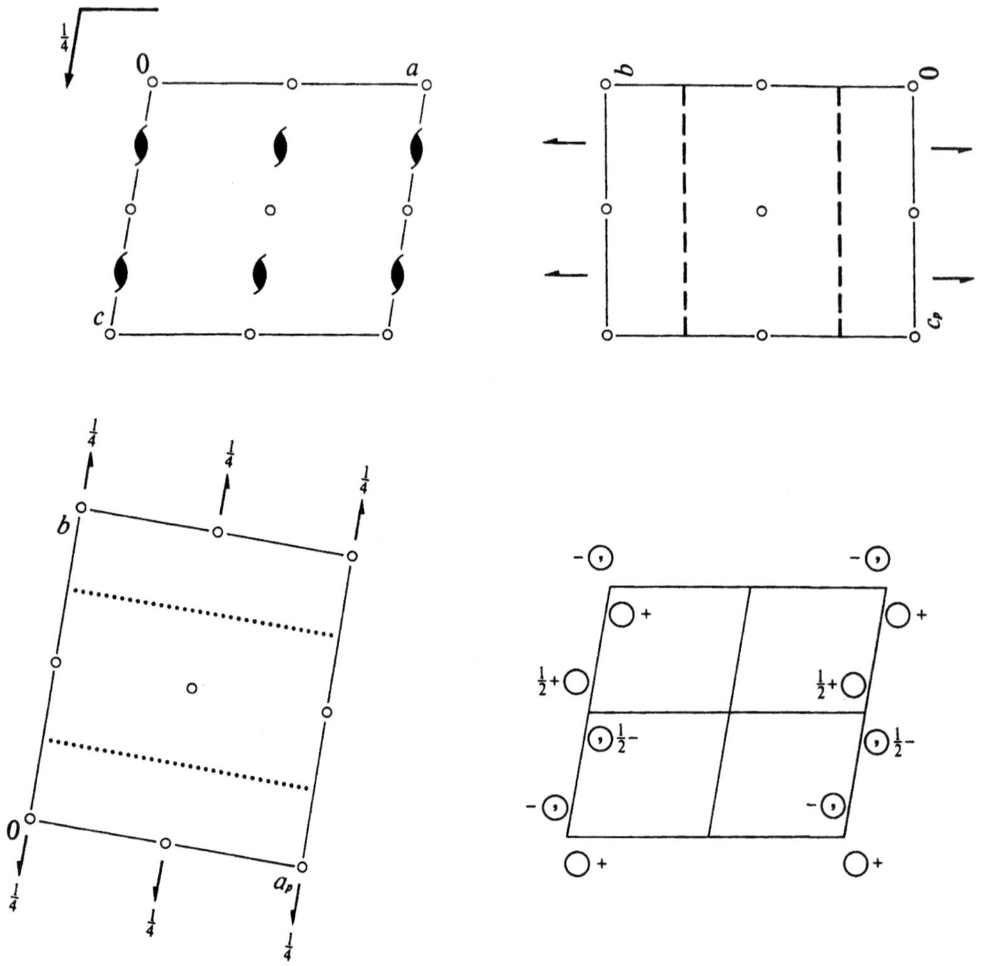

Origin at $\bar{1}$

Asymmetric unit $\quad 0 \le x \le 1; \quad 0 \le y \le \frac{1}{4}; \quad 0 \le z \le 1$

Symmetry operations

(1) 1 $\quad$ (2) $2(0,\frac{1}{2},0)\;\; 0,y,\frac{1}{4}$ $\quad$ (3) $\bar{1}\;\; 0,0,0$ $\quad$ (4) $c\;\; x,\frac{1}{4},z$

Generators selected (1); $t(1,0,0)$; $t(0,1,0)$; $t(0,0,1)$; (2); (3)

Positions

Multiplicity, Wyckoff letter, Site symmetry			Coordinates				Reflection conditions
							General:
4	*e*	1	(1) x,y,z	(2) $\bar{x},y+\frac{1}{2},\bar{z}+\frac{1}{2}$	(3) $\bar{x},\bar{y},\bar{z}$	(4) $x,\bar{y}+\frac{1}{2},z+\frac{1}{2}$	$h0l$: $l=2n$ $0k0$: $k=2n$ $00l$: $l=2n$
							Special: as above, plus
2	*d*	$\bar{1}$	$\frac{1}{2},0,\frac{1}{2}$	$\frac{1}{2},\frac{1}{2},0$			hkl: $k+l=2n$
2	*c*	$\bar{1}$	$0,0,\frac{1}{2}$	$0,\frac{1}{2},0$			hkl: $k+l=2n$
2	*b*	$\bar{1}$	$\frac{1}{2},0,0$	$\frac{1}{2},\frac{1}{2},\frac{1}{2}$			hkl: $k+l=2n$
2	*a*	$\bar{1}$	$0,0,0$	$0,\frac{1}{2},\frac{1}{2}$			hkl: $k+l=2n$

Symmetry of special projections

Along [001] $p2gm$
$\mathbf{a}'=\mathbf{a}_p$ $\mathbf{b}'=\mathbf{b}$
Origin at $0,0,z$

Along [100] $p2gg$
$\mathbf{a}'=\mathbf{b}$ $\mathbf{b}'=\mathbf{c}_p$
Origin at $x,0,0$

Along [010] $p2$
$\mathbf{a}'=\frac{1}{2}\mathbf{c}$ $\mathbf{b}'=\mathbf{a}$
Origin at $0,y,0$

No. 14

UNIQUE AXIS b, DIFFERENT CELL CHOICES

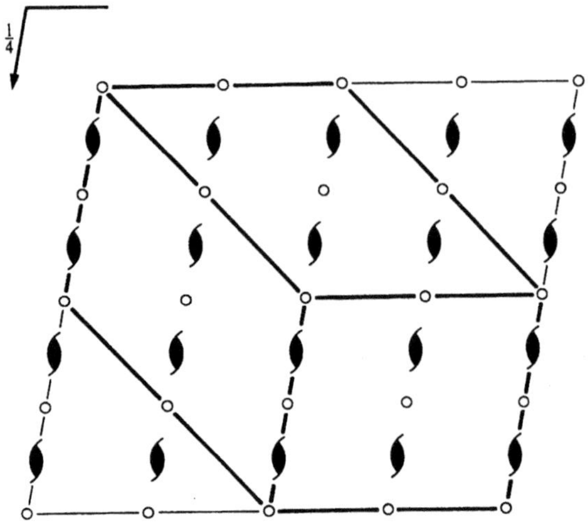

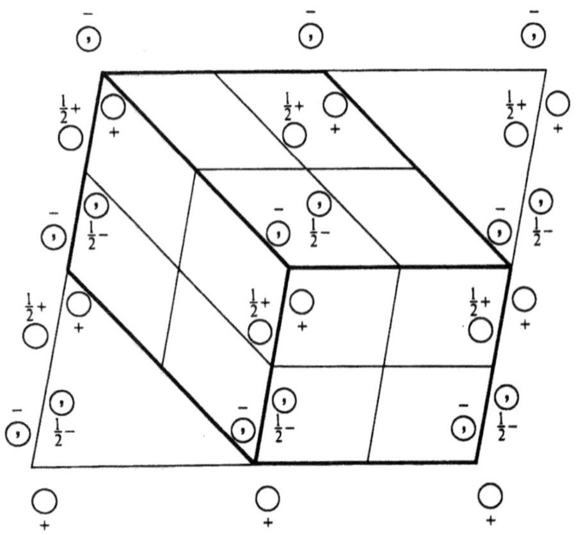

$P12_1/c1$

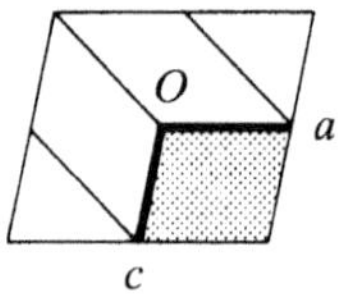

UNIQUE AXIS b, CELL CHOICE 1

Origin at $\bar{1}$

Asymmetric unit $0 \le x \le 1;\quad 0 \le y \le \frac{1}{4};\quad 0 \le z \le 1$

Generators selected (1); $t(1,0,0)$; $t(0,1,0)$; $t(0,0,1)$; (2); (3)

Positions

Multiplicity, Wyckoff letter, Site symmetry			Coordinates				Reflection conditions
							General:
4	e	1	(1) x,y,z	(2) $\bar{x},y+\frac{1}{2},\bar{z}+\frac{1}{2}$	(3) $\bar{x},\bar{y},\bar{z}$	(4) $x,\bar{y}+\frac{1}{2},z+\frac{1}{2}$	$h0l$: $l=2n$ $0k0$: $k=2n$ $00l$: $l=2n$
							Special: as above, plus
2	d	$\bar{1}$	$\frac{1}{2},0,\frac{1}{2}$	$\frac{1}{2},\frac{1}{2},0$			hkl: $k+l=2n$
2	c	$\bar{1}$	$0,0,\frac{1}{2}$	$0,\frac{1}{2},0$			hkl: $k+l=2n$
2	b	$\bar{1}$	$\frac{1}{2},0,0$	$\frac{1}{2},\frac{1}{2},\frac{1}{2}$			hkl: $k+l=2n$
2	a	$\bar{1}$	$0,0,0$	$0,\frac{1}{2},\frac{1}{2}$			hkl: $k+l=2n$

$P\,1\,2_1/n\,1$

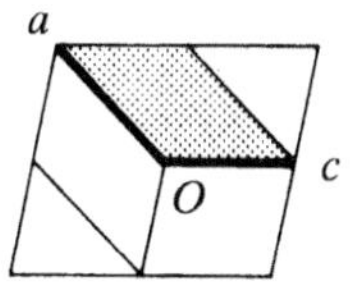

UNIQUE AXIS b, CELL CHOICE 2

Origin at $\bar{1}$

Asymmetric unit $0 \le x \le 1;\quad 0 \le y \le \frac{1}{4};\quad 0 \le z \le 1$

Generators selected (1); $t(1,0,0)$; $t(0,1,0)$; $t(0,0,1)$; (2); (3)

Positions

Multiplicity, Wyckoff letter, Site symmetry			Coordinates				Reflection conditions
							General:
4	e	1	(1) x,y,z	(2) $\bar{x}+\frac{1}{2},y+\frac{1}{2},\bar{z}+\frac{1}{2}$	(3) $\bar{x},\bar{y},\bar{z}$	(4) $x+\frac{1}{2},\bar{y}+\frac{1}{2},z+\frac{1}{2}$	$h0l$: $h+l=2n$ $0k0$: $k=2n$ $h00$: $h=2n$ $00l$: $l=2n$
							Special: as above, plus
2	d	$\bar{1}$	$\frac{1}{2},0,0$	$0,\frac{1}{2},\frac{1}{2}$			hkl: $h+k+l=2n$
2	c	$\bar{1}$	$\frac{1}{2},0,\frac{1}{2}$	$0,\frac{1}{2},0$			hkl: $h+k+l=2n$
2	b	$\bar{1}$	$0,0,\frac{1}{2}$	$\frac{1}{2},\frac{1}{2},0$			hkl: $h+k+l=2n$
2	a	$\bar{1}$	$0,0,0$	$\frac{1}{2},\frac{1}{2},\frac{1}{2}$			hkl: $h+k+l=2n$

$P\,1\,2_1/a\,1$

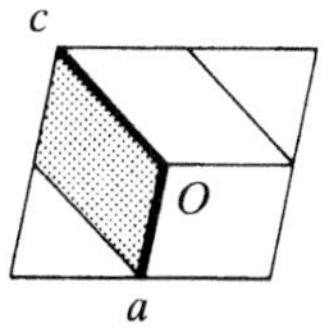

UNIQUE AXIS b, CELL CHOICE 3

Origin at $\bar{1}$

Asymmetric unit $0 \le x \le 1;\quad 0 \le y \le \frac{1}{4};\quad 0 \le z \le 1$

Generators selected (1); $t(1,0,0)$; $t(0,1,0)$; $t(0,0,1)$; (2); (3)

Positions

Multiplicity, Wyckoff letter, Site symmetry			Coordinates				Reflection conditions
							General:
4	e	1	(1) x,y,z	(2) $\bar{x}+\frac{1}{2},y+\frac{1}{2},\bar{z}$	(3) $\bar{x},\bar{y},\bar{z}$	(4) $x+\frac{1}{2},\bar{y}+\frac{1}{2},z$	$h0l$: $h=2n$ $0k0$: $k=2n$ $h00$: $h=2n$
							Special: as above, plus
2	d	$\bar{1}$	$0,0,\frac{1}{2}$	$\frac{1}{2},\frac{1}{2},\frac{1}{2}$			hkl: $h+k=2n$
2	c	$\bar{1}$	$\frac{1}{2},0,0$	$0,\frac{1}{2},0$			hkl: $h+k=2n$
2	b	$\bar{1}$	$\frac{1}{2},0,\frac{1}{2}$	$0,\frac{1}{2},\frac{1}{2}$			hkl: $h+k=2n$
2	a	$\bar{1}$	$0,0,0$	$\frac{1}{2},\frac{1}{2},0$			hkl: $h+k=2n$

$P2_1/c$ C_{2h}^5 $2/m$ Monoclinic

No. 14 $P112_1/a$ Patterson symmetry $P112/m$

UNIQUE AXIS c, CELL CHOICE 1

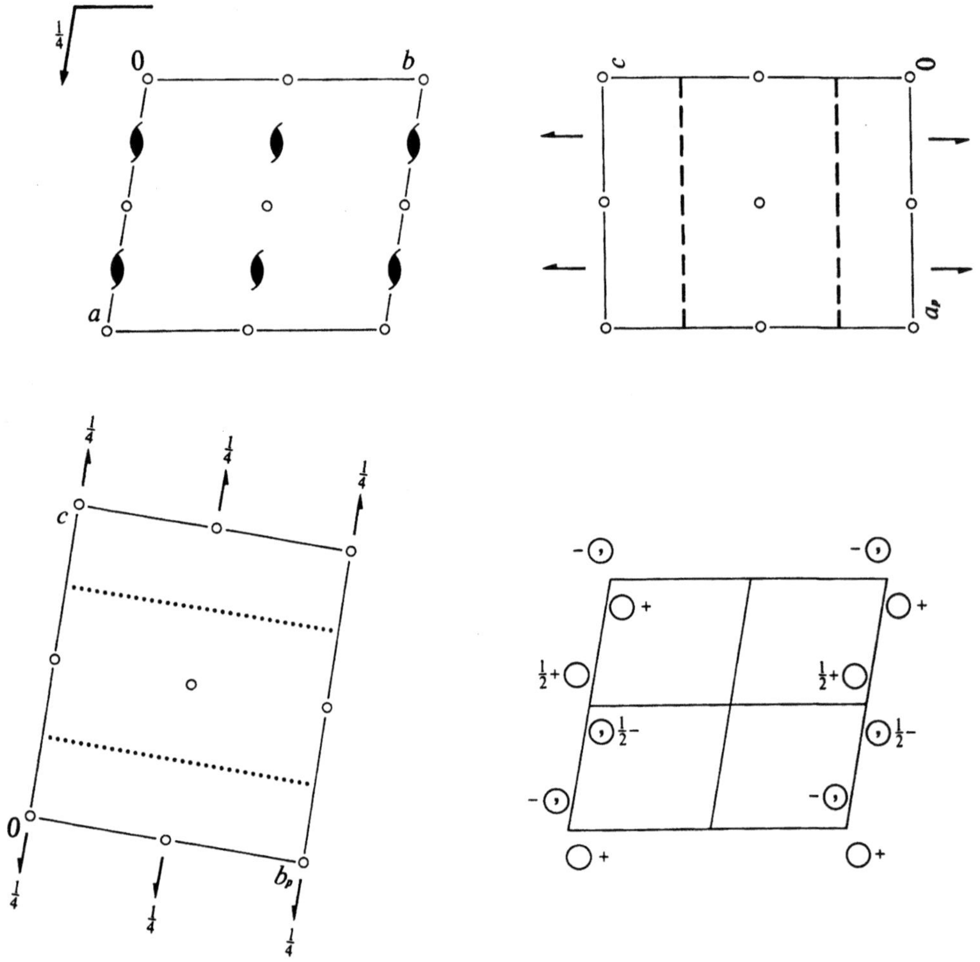

Origin at $\bar{1}$

Asymmetric unit $0 \leq x \leq 1;\quad 0 \leq y \leq 1;\quad 0 \leq z \leq \frac{1}{4}$

Symmetry operations

(1) 1 (2) $2(0,0,\frac{1}{2})\ \ \frac{1}{4},0,z$ (3) $\bar{1}\ \ 0,0,0$ (4) $a\ \ x,y,\frac{1}{4}$

Generators selected (1); $t(1,0,0)$; $t(0,1,0)$; $t(0,0,1)$; (2); (3)

Positions

Multiplicity, Wyckoff letter, Site symmetry			Coordinates				Reflection conditions
							General:
4	*e*	1	(1) x,y,z	(2) $\bar{x}+\frac{1}{2},\bar{y},z+\frac{1}{2}$	(3) $\bar{x},\bar{y},\bar{z}$	(4) $x+\frac{1}{2},y,\bar{z}+\frac{1}{2}$	$hk0$: $h=2n$ $00l$: $l=2n$ $h00$: $h=2n$
							Special: as above, plus
2	*d*	$\bar{1}$	$\frac{1}{2},\frac{1}{2},0$	$0,\frac{1}{2},\frac{1}{2}$			hkl: $h+l=2n$
2	*c*	$\bar{1}$	$\frac{1}{2},0,0$	$0,0,\frac{1}{2}$			hkl: $h+l=2n$
2	*b*	$\bar{1}$	$0,\frac{1}{2},0$	$\frac{1}{2},\frac{1}{2},\frac{1}{2}$			hkl: $h+l=2n$
2	*a*	$\bar{1}$	$0,0,0$	$\frac{1}{2},0,\frac{1}{2}$			hkl: $h+l=2n$

Symmetry of special projections

Along [001] $p2$
$\mathbf{a}'=\frac{1}{2}\mathbf{a}$ $\mathbf{b}'=\mathbf{b}$
Origin at $0,0,z$

Along [100] $p2gm$
$\mathbf{a}'=\mathbf{b}_p$ $\mathbf{b}'=\mathbf{c}$
Origin at $x,0,0$

Along [010] $p2gg$
$\mathbf{a}'=\mathbf{c}$ $\mathbf{b}'=\mathbf{a}_p$
Origin at $0,y,0$

$P2_1/c$ $\quad C_{2h}^5$ $\quad 2/m$ $\quad$ Monoclinic

No. 14

UNIQUE AXIS c, DIFFERENT CELL CHOICES

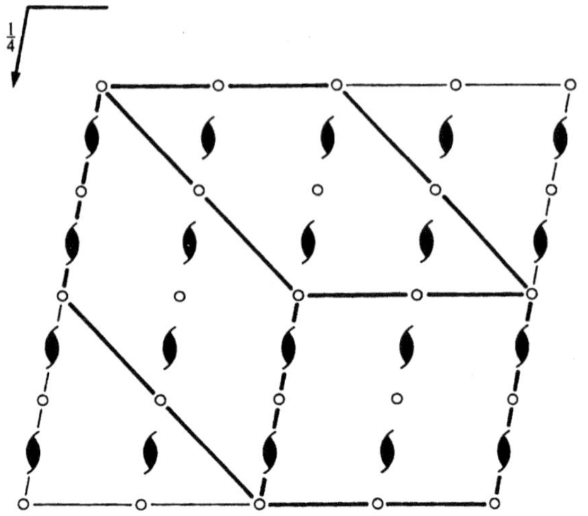

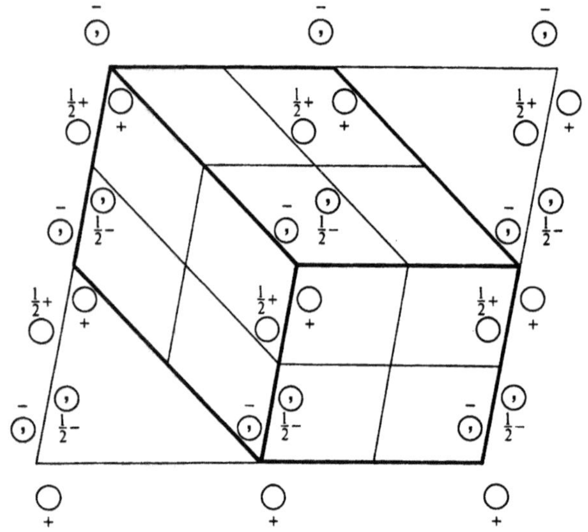

$P112_1/a$

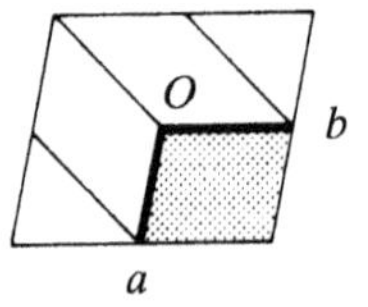

UNIQUE AXIS c, CELL CHOICE 1

Origin at $\bar{1}$

Asymmetric unit $\quad 0 \le x \le 1; \quad 0 \le y \le 1; \quad 0 \le z \le \frac{1}{4}$

Generators selected $\quad$ (1); $t(1,0,0)$; $t(0,1,0)$; $t(0,0,1)$; (2); (3)

Positions

Multiplicity, Wyckoff letter, Site symmetry			Coordinates				Reflection conditions
							General:
4	*e*	1	(1) x,y,z	(2) $\bar{x}+\frac{1}{2},\bar{y},z+\frac{1}{2}$	(3) $\bar{x},\bar{y},\bar{z}$	(4) $x+\frac{1}{2},y,\bar{z}+\frac{1}{2}$	$hk0$: $h=2n$ $00l$: $l=2n$ $h00$: $h=2n$
							Special: as above, plus
2	*d*	$\bar{1}$	$\frac{1}{2},\frac{1}{2},0$	$0,\frac{1}{2},\frac{1}{2}$			hkl: $h+l=2n$
2	*c*	$\bar{1}$	$\frac{1}{2},0,0$	$0,0,\frac{1}{2}$			hkl: $h+l=2n$
2	*b*	$\bar{1}$	$0,\frac{1}{2},0$	$\frac{1}{2},\frac{1}{2},\frac{1}{2}$			hkl: $h+l=2n$
2	*a*	$\bar{1}$	$0,0,0$	$\frac{1}{2},0,\frac{1}{2}$			hkl: $h+l=2n$

$P112_1/n$

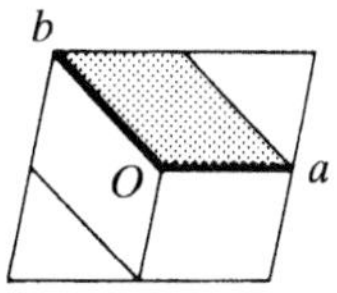

UNIQUE AXIS c, CELL CHOICE 2

Origin at $\bar{1}$

Asymmetric unit $0 \le x \le 1;\quad 0 \le y \le 1;\quad 0 \le z \le \frac{1}{4}$

Generators selected (1); $t(1,0,0)$; $t(0,1,0)$; $t(0,0,1)$; (2); (3)

Positions

Multiplicity, Wyckoff letter, Site symmetry			Coordinates				Reflection conditions
							General:
4	e	1	(1) x,y,z	(2) $\bar{x}+\frac{1}{2},\bar{y}+\frac{1}{2},z+\frac{1}{2}$	(3) $\bar{x},\bar{y},\bar{z}$	(4) $x+\frac{1}{2},y+\frac{1}{2},\bar{z}+\frac{1}{2}$	$hk0$: $h+k=2n$ $00l$: $l=2n$ $h00$: $h=2n$ $0k0$: $k=2n$
							Special: as above, plus
2	d	$\bar{1}$	$0,\frac{1}{2},0$	$\frac{1}{2},0,\frac{1}{2}$			hkl: $h+k+l=2n$
2	c	$\bar{1}$	$\frac{1}{2},\frac{1}{2},0$	$0,0,\frac{1}{2}$			hkl: $h+k+l=2n$
2	b	$\bar{1}$	$\frac{1}{2},0,0$	$0,\frac{1}{2},\frac{1}{2}$			hkl: $h+k+l=2n$
2	a	$\bar{1}$	$0,0,0$	$\frac{1}{2},\frac{1}{2},\frac{1}{2}$			hkl: $h+k+l=2n$

$P112_1/b$

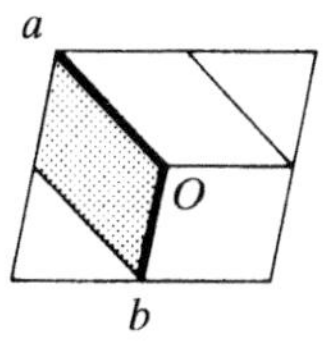

UNIQUE AXIS c, CELL CHOICE 3

Origin at $\bar{1}$

Asymmetric unit $0 \le x \le 1;\quad 0 \le y \le 1;\quad 0 \le z \le \frac{1}{4}$

Generators selected (1); $t(1,0,0)$; $t(0,1,0)$; $t(0,0,1)$; (2); (3)

Positions

Multiplicity, Wyckoff letter, Site symmetry			Coordinates				Reflection conditions
							General:
4	e	1	(1) x,y,z	(2) $\bar{x},\bar{y}+\frac{1}{2},z+\frac{1}{2}$	(3) $\bar{x},\bar{y},\bar{z}$	(4) $x,y+\frac{1}{2},\bar{z}+\frac{1}{2}$	$hk0$: $k=2n$ $00l$: $l=2n$ $0k0$: $k=2n$
							Special: as above, plus
2	d	$\bar{1}$	$\frac{1}{2},0,0$	$\frac{1}{2},\frac{1}{2},\frac{1}{2}$			hkl: $k+l=2n$
2	c	$\bar{1}$	$0,\frac{1}{2},0$	$0,0,\frac{1}{2}$			hkl: $k+l=2n$
2	b	$\bar{1}$	$\frac{1}{2},\frac{1}{2},0$	$\frac{1}{2},0,\frac{1}{2}$			hkl: $k+l=2n$
2	a	$\bar{1}$	$0,0,0$	$0,\frac{1}{2},\frac{1}{2}$			hkl: $k+l=2n$

$C2/c$ $\quad$ C_{2h}^6 $\quad$ $2/m$ $\quad$ Monoclinic

No. 15 $\quad$ $C12/c1$ $\quad$ Patterson symmetry $C12/m1$

UNIQUE AXIS b, CELL CHOICE 1

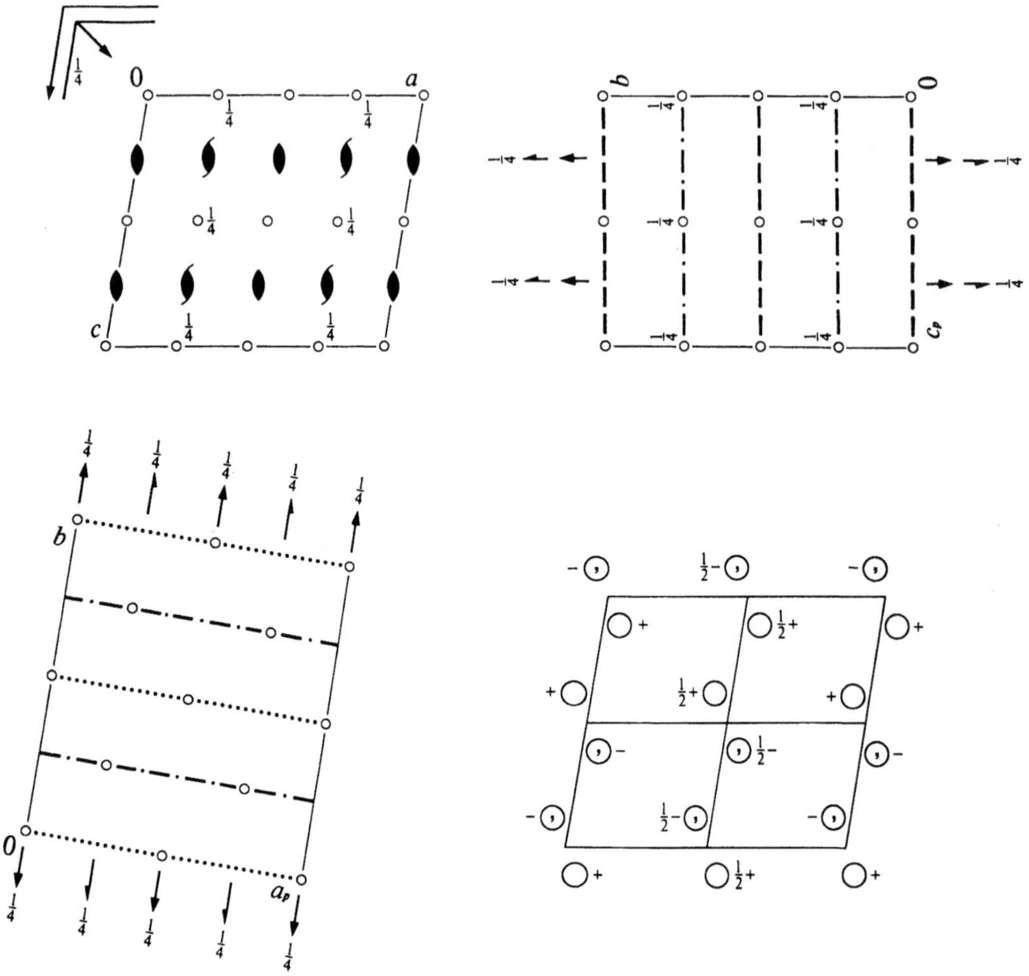

Origin at $\bar{1}$ on glide plane c

Asymmetric unit $\quad 0 \leq x \leq \frac{1}{2}; \quad 0 \leq y \leq \frac{1}{2}; \quad 0 \leq z \leq \frac{1}{2}$

Symmetry operations

For $(0,0,0)+$ set

(1) 1 $\quad$ (2) $2 \quad 0,y,\frac{1}{4}$ $\quad$ (3) $\bar{1} \quad 0,0,0$ $\quad$ (4) $c \quad x,0,z$

For $(\frac{1}{2},\frac{1}{2},0)+$ set

(1) $t(\frac{1}{2},\frac{1}{2},0)$ $\quad$ (2) $2(0,\frac{1}{2},0) \quad \frac{1}{4},y,\frac{1}{4}$ $\quad$ (3) $\bar{1} \quad \frac{1}{4},\frac{1}{4},0$ $\quad$ (4) $n(\frac{1}{2},0,\frac{1}{2}) \quad x,\frac{1}{4},z$

Generators selected (1); $t(1,0,0)$; $t(0,1,0)$; $t(0,0,1)$; $t(\frac{1}{2},\frac{1}{2},0)$; (2); (3)

Positions

Multiplicity, Wyckoff letter, Site symmetry			Coordinates				Reflection conditions
			$(0,0,0)+$	$(\frac{1}{2},\frac{1}{2},0)+$			General:
8	f	1	(1) x,y,z	(2) $\bar{x},y,\bar{z}+\frac{1}{2}$	(3) $\bar{x},\bar{y},\bar{z}$	(4) $x,\bar{y},z+\frac{1}{2}$	hkl: $h+k=2n$ $h0l$: $h,l=2n$ $0kl$: $k=2n$ $hk0$: $h+k=2n$ $0k0$: $k=2n$ $h00$: $h=2n$ $00l$: $l=2n$
							Special: as above, plus
4	e	2	$0,y,\frac{1}{4}$	$0,\bar{y},\frac{3}{4}$			no extra conditions
4	d	$\bar{1}$	$\frac{1}{4},\frac{1}{4},\frac{1}{2}$	$\frac{3}{4},\frac{1}{4},0$			hkl: $k+l=2n$
4	c	$\bar{1}$	$\frac{1}{4},\frac{1}{4},0$	$\frac{3}{4},\frac{1}{4},\frac{1}{2}$			hkl: $k+l=2n$
4	b	$\bar{1}$	$0,\frac{1}{2},0$	$0,\frac{1}{2},\frac{1}{2}$			hkl: $l=2n$
4	a	$\bar{1}$	$0,0,0$	$0,0,\frac{1}{2}$			hkl: $l=2n$

Symmetry of special projections

Along [001] $c2mm$
$\mathbf{a}'=\mathbf{a}_p$ $\mathbf{b}'=\mathbf{b}$
Origin at $0,0,z$

Along [100] $p2gm$
$\mathbf{a}'=\frac{1}{2}\mathbf{b}$ $\mathbf{b}'=\mathbf{c}_p$
Origin at $x,0,0$

Along [010] $p2$
$\mathbf{a}'=\frac{1}{2}\mathbf{c}$ $\mathbf{b}'=\frac{1}{2}\mathbf{a}$
Origin at $0,y,0$

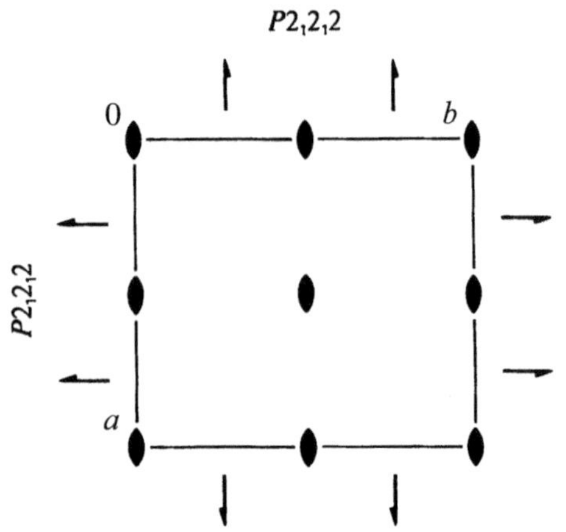

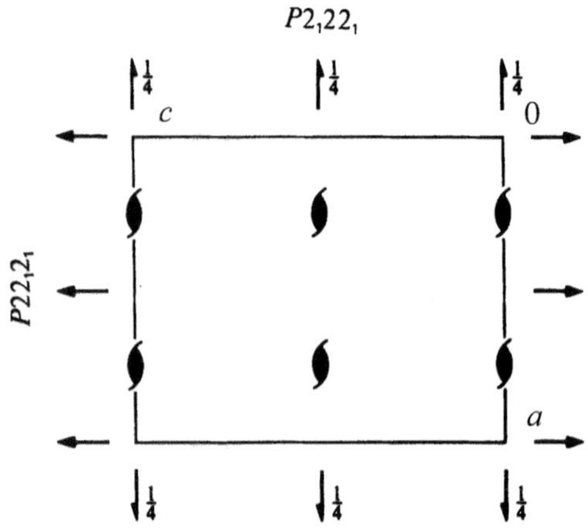

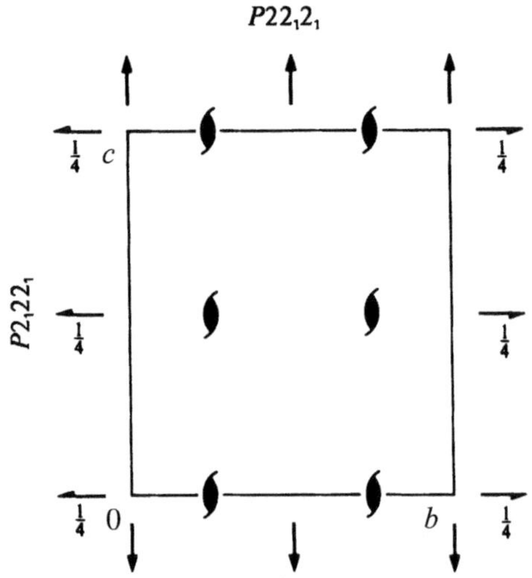

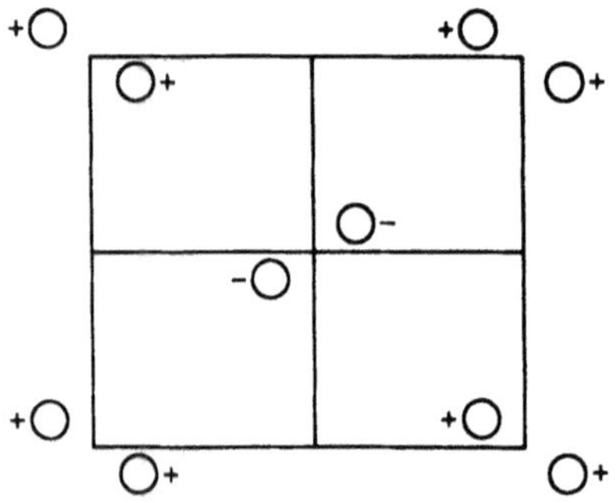

Origin at intersection of 2 with perpendicular plane containing 2_1 axes

Asymmetric unit $0 \le x \le \frac{1}{2}$; $0 \le y \le \frac{1}{2}$; $0 \le z \le 1$

Symmetry operations

(1) 1 (2) 2 $0,0,z$ (3) $2(0,\frac{1}{2},0)$ $\frac{1}{4},y,0$ (4) $2(\frac{1}{2},0,0)$ $x,\frac{1}{4},0$

Generators selected (1); $t(1,0,0)$; $t(0,1,0)$; $t(0,0,1)$; (2); (3)

Positions

Multiplicity, Wyckoff letter, Site symmetry	Coordinates	Reflection conditions
		General:
4 c 1	(1) x,y,z (2) $\bar{x},\bar{y},z$ (3) $\bar{x}+\frac{1}{2},y+\frac{1}{2},\bar{z}$ (4) $x+\frac{1}{2},\bar{y}+\frac{1}{2},\bar{z}$	$h00$: $h=2n$ $0k0$: $k=2n$
		Special: as above, plus
2 b ..2	$0,\frac{1}{2},z$ $\frac{1}{2},0,\bar{z}$	$hk0$: $h+k=2n$
2 a ..2	$0,0,z$ $\frac{1}{2},\frac{1}{2},\bar{z}$	$hk0$: $h+k=2n$

Symmetry of special projections

Along [001] $p2gg$
a′ = **a** **b**′ = **b**
Origin at $0,0,z$

Along [100] $p2mg$
a′ = **b** **b**′ = **c**
Origin at $x,\frac{1}{4},0$

Along [010] $p2gm$
a′ = **c** **b**′ = **a**
Origin at $\frac{1}{4},y,0$

Orthorhombic 222 D_2^4 $P2_12_12_1$

Patterson symmetry $Pmmm$ $P2_12_12_1$ No. 19

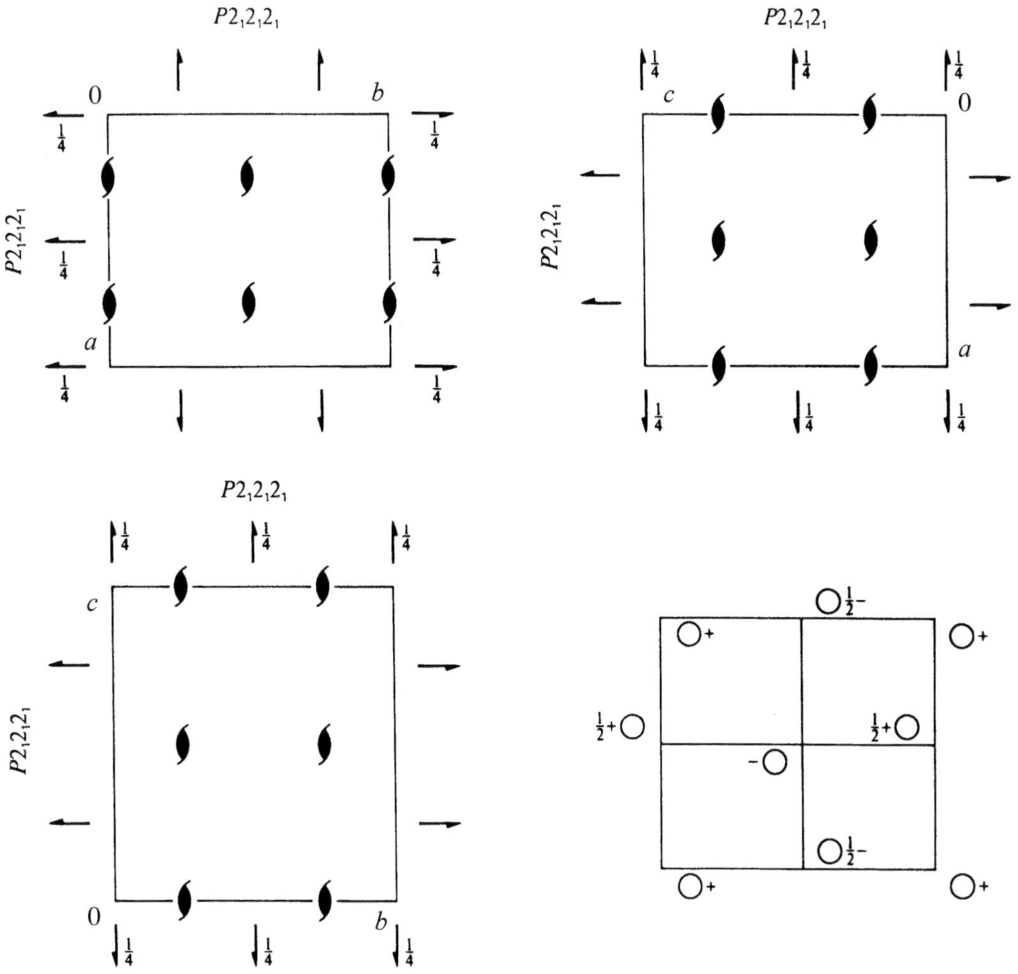

Origin at midpoint of three non-intersecting pairs of parallel 2_1 axes

Asymmetric unit $0 \le x \le \frac{1}{2}$; $0 \le y \le \frac{1}{2}$; $0 \le z \le 1$

Symmetry operations

(1) 1 (2) $2(0,0,\frac{1}{2})$ $\frac{1}{4},0,z$ (3) $2(0,\frac{1}{2},0)$ $0,y,\frac{1}{4}$ (4) $2(\frac{1}{2},0,0)$ $x,\frac{1}{4},0$

Generators selected (1); $t(1,0,0)$; $t(0,1,0)$; $t(0,0,1)$; (2); (3)

Positions

Multiplicity, Wyckoff letter, Site symmetry	Coordinates				Reflection conditions
					General:
4 *a* 1	(1) x,y,z	(2) $\bar{x}+\frac{1}{2},\bar{y},z+\frac{1}{2}$	(3) $\bar{x},y+\frac{1}{2},\bar{z}+\frac{1}{2}$	(4) $x+\frac{1}{2},\bar{y}+\frac{1}{2},\bar{z}$	$h00$: $h=2n$ $0k0$: $k=2n$ $00l$: $l=2n$

Symmetry of special projections

Along [001] $p2gg$
$\mathbf{a}'=\mathbf{a}$ $\mathbf{b}'=\mathbf{b}$
Origin at $\frac{1}{4},0,z$

Along [100] $p2gg$
$\mathbf{a}'=\mathbf{b}$ $\mathbf{b}'=\mathbf{c}$
Origin at $x,\frac{1}{4},0$

Along [010] $p2gg$
$\mathbf{a}'=\mathbf{c}$ $\mathbf{b}'=\mathbf{a}$
Origin at $0,y,\frac{1}{4}$

$Pmm2$ C_{2v}^1 $mm2$ Orthorhombic

No. 25 $Pmm2$ Patterson symmetry $Pmmm$

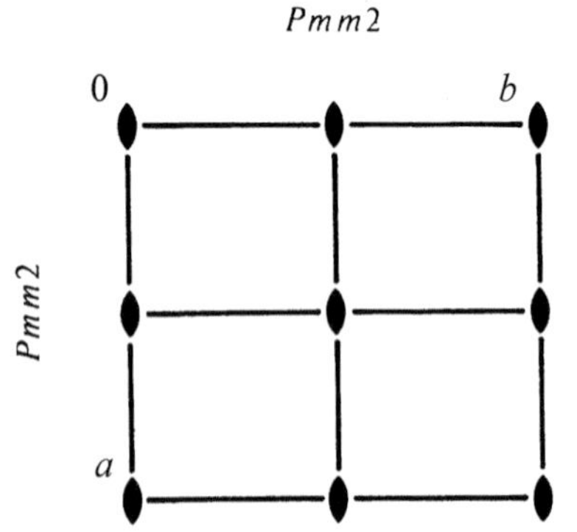

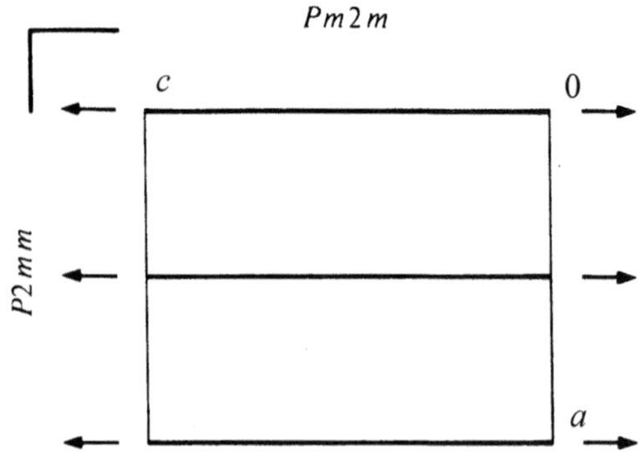

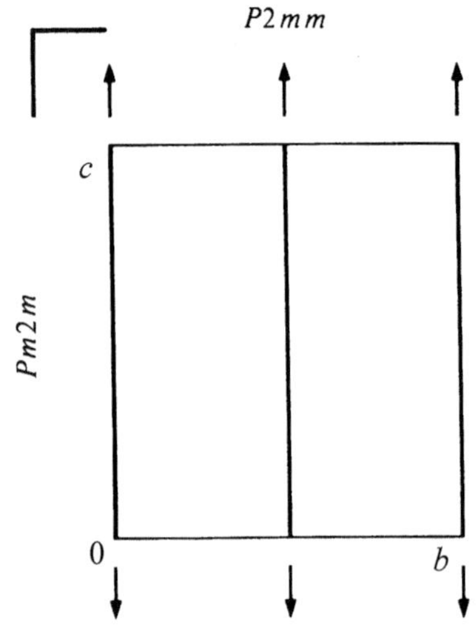

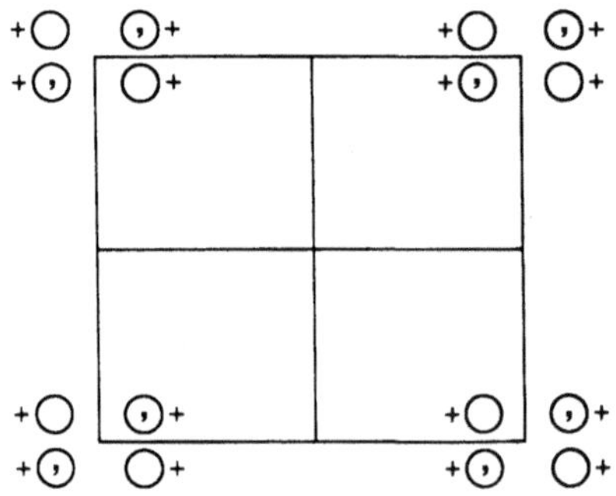

Origin on $mm2$

Asymmetric unit $0 \le x \le \frac{1}{2}$; $0 \le y \le \frac{1}{2}$; $0 \le z \le 1$

Symmetry operations

(1) 1   (2) 2 $0,0,z$ (3) m $x,0,z$ (4) m $0,y,z$

Generators selected   (1); $t(1,0,0)$; $t(0,1,0)$; $t(0,0,1)$; (2); (3)

Positions

Multiplicity, Wyckoff letter, Site symmetry		Coordinates			Reflection conditions
					General:
4	i	1	(1) x,y,z (2) $\bar{x},\bar{y},z$ (3) $x,\bar{y},z$ (4) $\bar{x},y,z$		no conditions
					Special: no extra conditions

2	h	$m..$	$\frac{1}{2},y,z$	$\frac{1}{2},\bar{y},z$	1	d	$mm2$	$\frac{1}{2},\frac{1}{2},z$
2	g	$m..$	$0,y,z$	$0,\bar{y},z$	1	c	$mm2$	$\frac{1}{2},0,z$
2	f	$.m.$	$x,\frac{1}{2},z$	$\bar{x},\frac{1}{2},z$	1	b	$mm2$	$0,\frac{1}{2},z$
2	e	$.m.$	$x,0,z$	$\bar{x},0,z$	1	a	$mm2$	$0,0,z$

Symmetry of special projections

Along [001] $p2mm$ Along [100] $p1m1$ Along [010] $p11m$
$\mathbf{a}' = \mathbf{a}$ $\mathbf{b}' = \mathbf{b}$ $\mathbf{a}' = \mathbf{b}$ $\mathbf{b}' = \mathbf{c}$ $\mathbf{a}' = \mathbf{c}$ $\mathbf{b}' = \mathbf{a}$
Origin at $0,0,z$ Origin at $x,0,0$ Origin at $0,y,0$

Orthorhombic $mm2$ C_{2v}^{9} $Pna2_1$

Patterson symmetry $Pmmm$ $Pna2_1$ No. 33

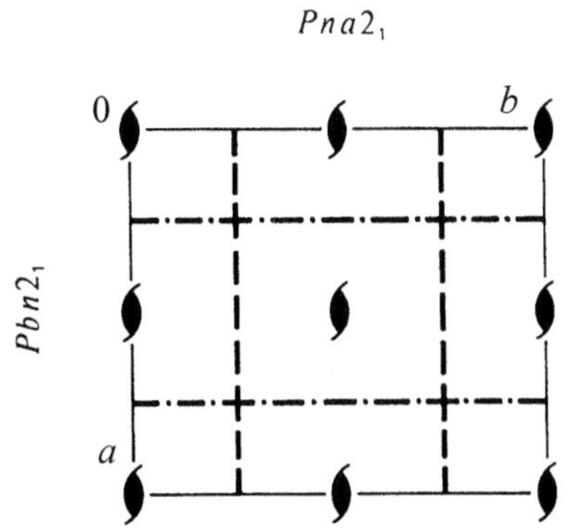

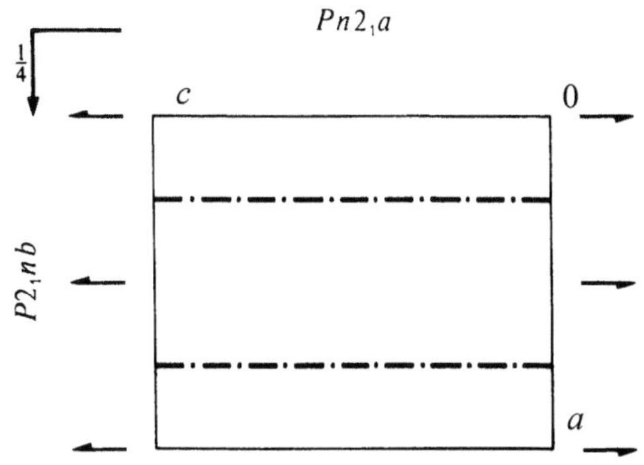

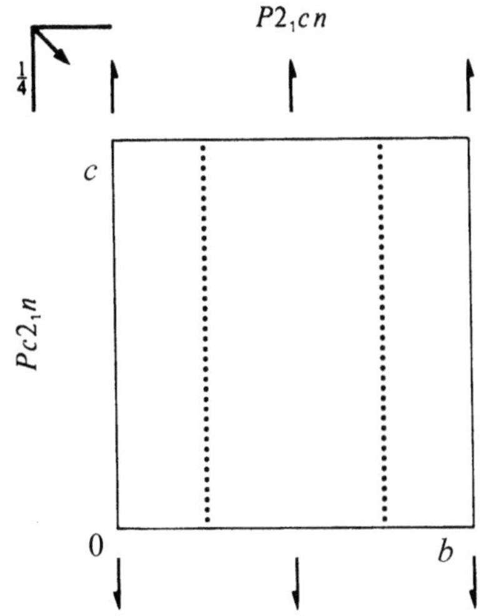

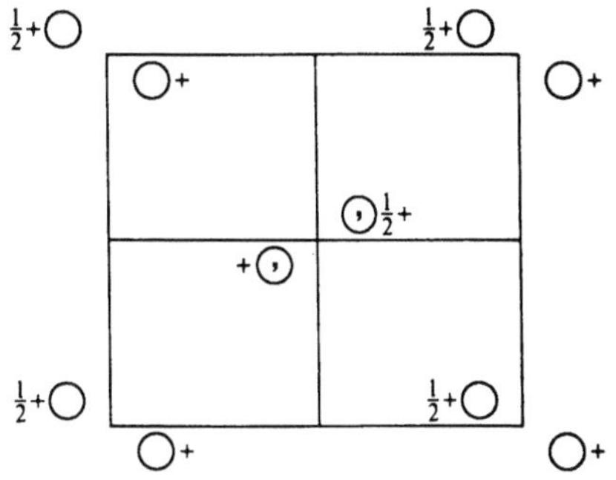

Origin on 112_1

Asymmetric unit $0 \le x \le \frac{1}{2};\quad 0 \le y \le \frac{1}{2};\quad 0 \le z \le 1$

Symmetry operations

(1) 1 (2) $2(0,0,\frac{1}{2})\ \ 0,0,z$ (3) $a\ \ x,\frac{1}{4},z$ (4) $n(0,\frac{1}{2},\frac{1}{2})\ \ \frac{1}{4},y,z$

Generators selected (1); $t(1,0,0)$; $t(0,1,0)$; $t(0,0,1)$; (2); (3)

Positions

Multiplicity, Wyckoff letter, Site symmetry	Coordinates				Reflection conditions
					General:
4 a 1	(1) x,y,z	(2) $\bar{x},\bar{y},z+\frac{1}{2}$	(3) $x+\frac{1}{2},\bar{y}+\frac{1}{2},z$	(4) $\bar{x}+\frac{1}{2},y+\frac{1}{2},z+\frac{1}{2}$	$0kl$: $k+l=2n$ $h0l$: $h=2n$ $h00$: $h=2n$ $0k0$: $k=2n$ $00l$: $l=2n$

Symmetry of special projections

Along [001] $p2gg$
$\mathbf{a}'=\mathbf{a}$ $\mathbf{b}'=\mathbf{b}$
Origin at $0,0,z$

Along [100] $c1m1$
$\mathbf{a}'=\mathbf{b}$ $\mathbf{b}'=\mathbf{c}$
Origin at $x,\frac{1}{4},0$

Along [010] $p11g$
$\mathbf{a}'=\mathbf{c}$ $\mathbf{b}'=\frac{1}{2}\mathbf{a}$
Origin at $0,y,0$

No. 39 $Aem2$ Patterson symmetry $Ammm$ ($Cmmm$)

Former space-group symbol $Abm2$; *cf.* Section 2.1.2

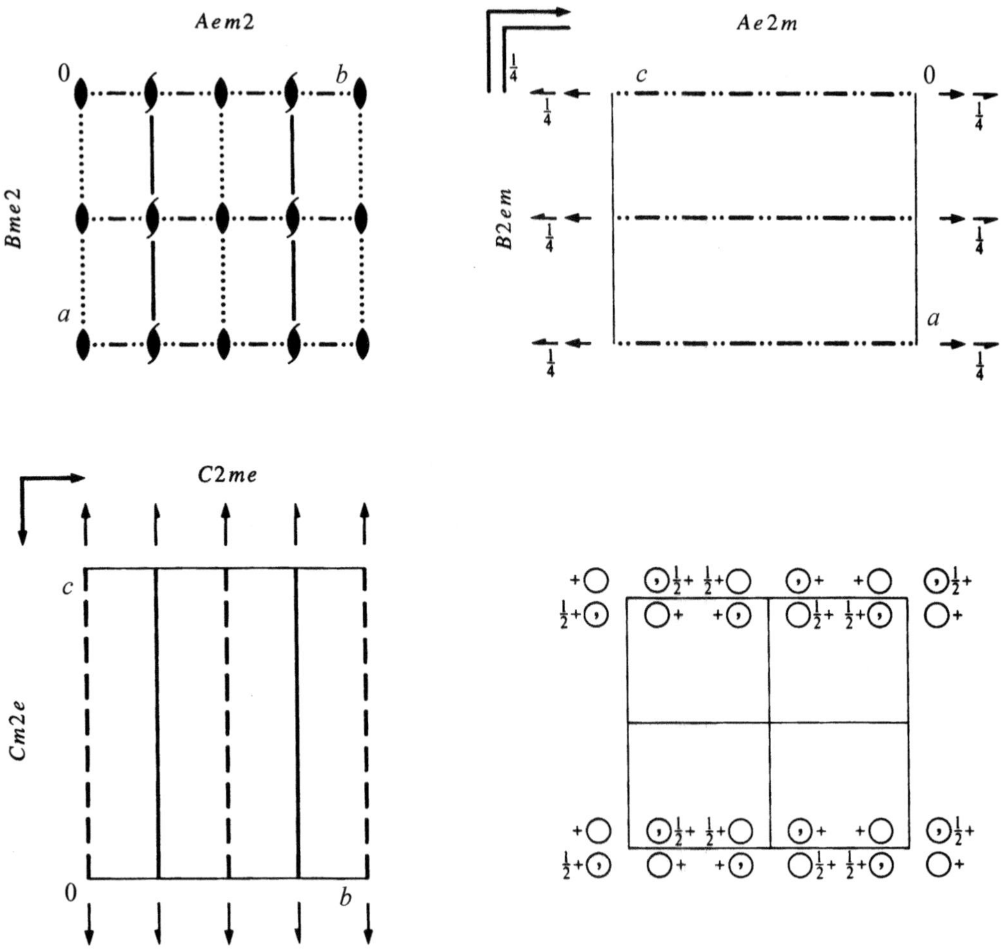

Origin on $ec2$

Asymmetric unit $0 \le x \le \frac{1}{2}$; $0 \le y \le \frac{1}{4}$; $0 \le z \le 1$

Symmetry operations

For $(0,0,0)+$ set

(1) 1 (2) 2 $0,0,z$ (3) m $x,\frac{1}{4},z$ (4) b $0,y,z$

For $(0,\frac{1}{2},\frac{1}{2})+$ set

(1) $t(0,\frac{1}{2},\frac{1}{2})$ (2) $2(0,0,\frac{1}{2})$ $0,\frac{1}{4},z$ (3) c $x,0,z$ (4) c $0,y,z$

Generators selected (1); $t(1,0,0)$; $t(0,1,0)$; $t(0,0,1)$; $t(0,\frac{1}{2},\frac{1}{2})$; (2); (3)

Positions

Multiplicity, Wyckoff letter, Site symmetry			Coordinates				Reflection conditions
			$(0,0,0)+$ $(0,\frac{1}{2},\frac{1}{2})+$				General:
8	d	1	(1) x,y,z	(2) $\bar{x},\bar{y},z$	(3) $x,\bar{y}+\frac{1}{2},z$	(4) $\bar{x},y+\frac{1}{2},z$	hkl: $k+l=2n$ $0kl$: $k,l=2n$ $h0l$: $l=2n$ $hk0$: $k=2n$ $0k0$: $k=2n$ $00l$: $l=2n$
							Special: as above, plus
4	c	$.m.$	$x,\frac{1}{4},z$	$\bar{x},\frac{3}{4},z$			no extra conditions
4	b	$..2$	$\frac{1}{2},0,z$	$\frac{1}{2},\frac{1}{2},z$			hkl: $k=2n$
4	a	$..2$	$0,0,z$	$0,\frac{1}{2},z$			hkl: $k=2n$

Symmetry of special projections

Along [001] $p2mm$
$\mathbf{a}'=\mathbf{a}$ $\mathbf{b}'=\frac{1}{2}\mathbf{b}$
Origin at $0,0,z$

Along [100] $p1m1$
$\mathbf{a}'=\frac{1}{2}\mathbf{b}$ $\mathbf{b}'=\frac{1}{2}\mathbf{c}$
Origin at $x,0,0$

Along [010] $p11m$
$\mathbf{a}'=\frac{1}{2}\mathbf{c}$ $\mathbf{b}'=\mathbf{a}$
Origin at $0,y,0$

Pmmm	D_{2h}^{1}	*mmm*	Orthorhombic
No. 47	$P\ 2/m\ 2/m\ 2/m$		Patterson symmetry *Pmmm*

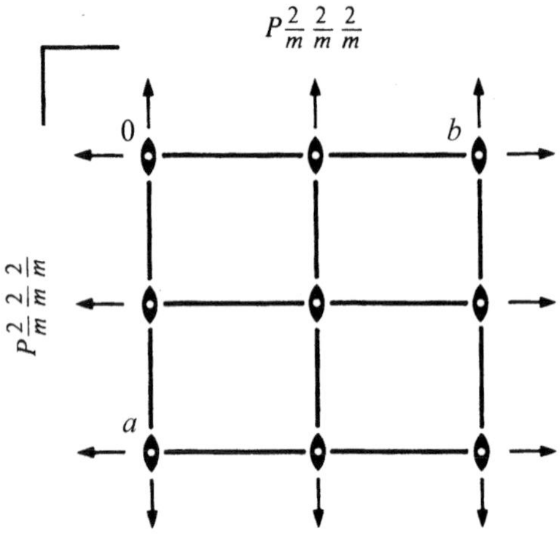

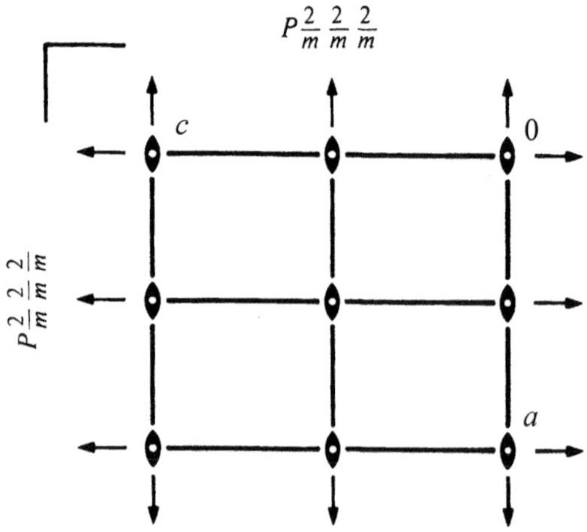

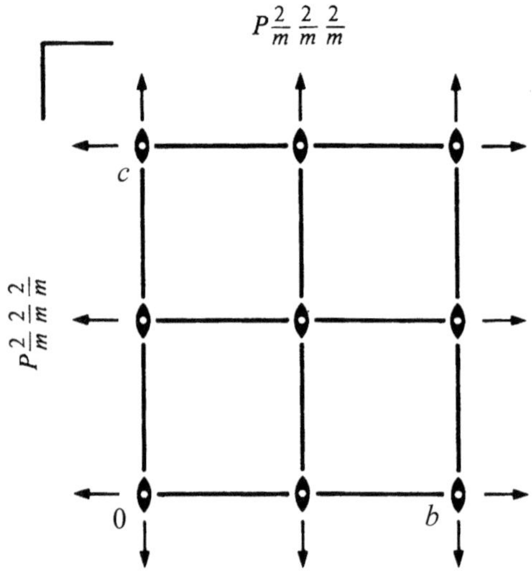

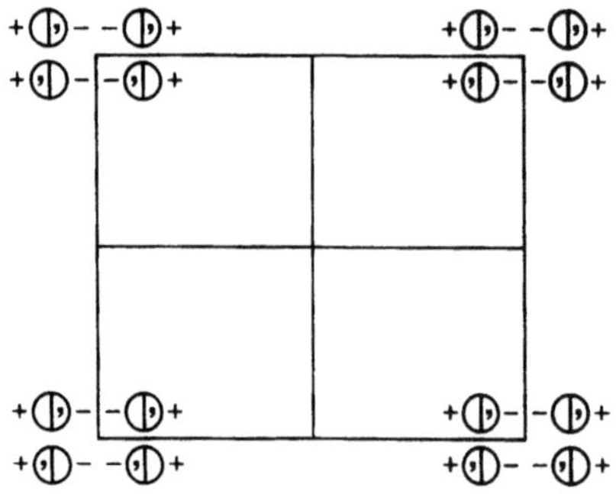

Origin at centre (*mmm*)

Asymmetric unit $0 \le x \le \frac{1}{2}$; $0 \le y \le \frac{1}{2}$; $0 \le z \le \frac{1}{2}$

Symmetry operations

(1) 1	(2) 2 $0,0,z$	(3) 2 $0,y,0$	(4) 2 $x,0,0$
(5) $\bar{1}$ $0,0,0$	(6) m $x,y,0$	(7) m $x,0,z$	(8) m $0,y,z$

Generators selected (1); $t(1,0,0)$; $t(0,1,0)$; $t(0,0,1)$; (2); (3); (5)

Positions

Multiplicity	Wyckoff letter	Site symmetry	Coordinates				Reflection conditions
							General:
8	α	1	(1) x,y,z	(2) $\bar{x},\bar{y},z$	(3) $\bar{x},y,\bar{z}$	(4) $x,\bar{y},\bar{z}$	no conditions
			(5) $\bar{x},\bar{y},\bar{z}$	(6) $x,y,\bar{z}$	(7) $x,\bar{y},z$	(8) $\bar{x},y,z$	
							Special: no extra conditions
4	z	$..m$	$x,y,\frac{1}{2}$	$\bar{x},\bar{y},\frac{1}{2}$	$\bar{x},y,\frac{1}{2}$	$x,\bar{y},\frac{1}{2}$	
4	y	$..m$	$x,y,0$	$\bar{x},\bar{y},0$	$\bar{x},y,0$	$x,\bar{y},0$	
4	x	$.m.$	$x,\frac{1}{2},z$	$\bar{x},\frac{1}{2},z$	$\bar{x},\frac{1}{2},\bar{z}$	$x,\frac{1}{2},\bar{z}$	
4	w	$.m.$	$x,0,z$	$\bar{x},0,z$	$\bar{x},0,\bar{z}$	$x,0,\bar{z}$	
4	v	$m..$	$\frac{1}{2},y,z$	$\frac{1}{2},\bar{y},z$	$\frac{1}{2},y,\bar{z}$	$\frac{1}{2},\bar{y},\bar{z}$	
4	u	$m..$	$0,y,z$	$0,\bar{y},z$	$0,y,\bar{z}$	$0,\bar{y},\bar{z}$	
2	t	$mm2$	$\frac{1}{2},\frac{1}{2},z$	$\frac{1}{2},\frac{1}{2},\bar{z}$			
2	s	$mm2$	$\frac{1}{2},0,z$	$\frac{1}{2},0,\bar{z}$			
2	r	$mm2$	$0,\frac{1}{2},z$	$0,\frac{1}{2},\bar{z}$			
2	q	$mm2$	$0,0,z$	$0,0,\bar{z}$			
2	p	$m2m$	$\frac{1}{2},y,\frac{1}{2}$	$\frac{1}{2},\bar{y},\frac{1}{2}$			
2	o	$m2m$	$\frac{1}{2},y,0$	$\frac{1}{2},\bar{y},0$			
2	n	$m2m$	$0,y,\frac{1}{2}$	$0,\bar{y},\frac{1}{2}$			
2	m	$m2m$	$0,y,0$	$0,\bar{y},0$			
2	l	$2mm$	$x,\frac{1}{2},\frac{1}{2}$	$\bar{x},\frac{1}{2},\frac{1}{2}$			
2	k	$2mm$	$x,\frac{1}{2},0$	$\bar{x},\frac{1}{2},0$			
2	j	$2mm$	$x,0,\frac{1}{2}$	$\bar{x},0,\frac{1}{2}$			
2	i	$2mm$	$x,0,0$	$\bar{x},0,0$			
1	h	mmm	$\frac{1}{2},\frac{1}{2},\frac{1}{2}$				
1	g	mmm	$0,\frac{1}{2},\frac{1}{2}$				
1	f	mmm	$\frac{1}{2},\frac{1}{2},0$				
1	e	mmm	$0,\frac{1}{2},0$				
1	d	mmm	$\frac{1}{2},0,\frac{1}{2}$				
1	c	mmm	$0,0,\frac{1}{2}$				
1	b	mmm	$\frac{1}{2},0,0$				
1	a	mmm	$0,0,0$				

Symmetry of special projections

Along [001] $p2mm$
$\mathbf{a}' = \mathbf{a}$ $\mathbf{b}' = \mathbf{b}$
Origin at $0,0,z$

Along [100] $p2mm$
$\mathbf{a}' = \mathbf{b}$ $\mathbf{b}' = \mathbf{c}$
Origin at $x,0,0$

Along [010] $p2mm$
$\mathbf{a}' = \mathbf{c}$ $\mathbf{b}' = \mathbf{a}$
Origin at $0,y,0$

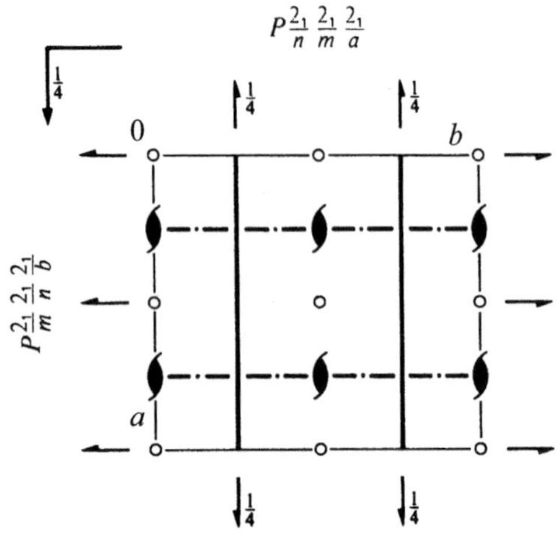

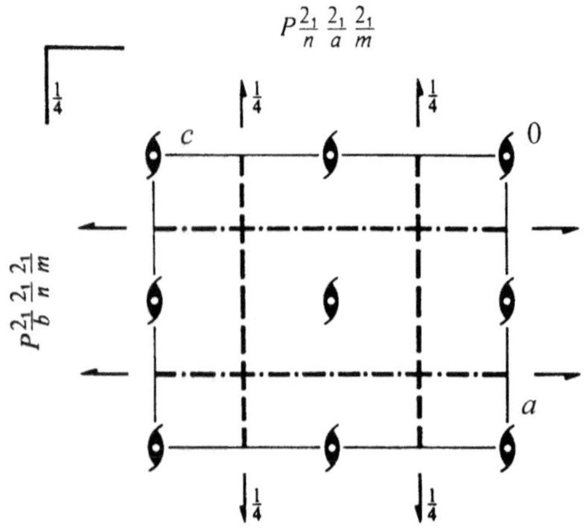

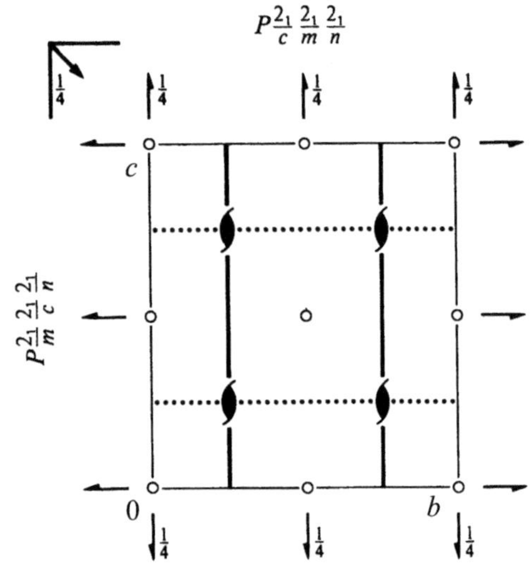

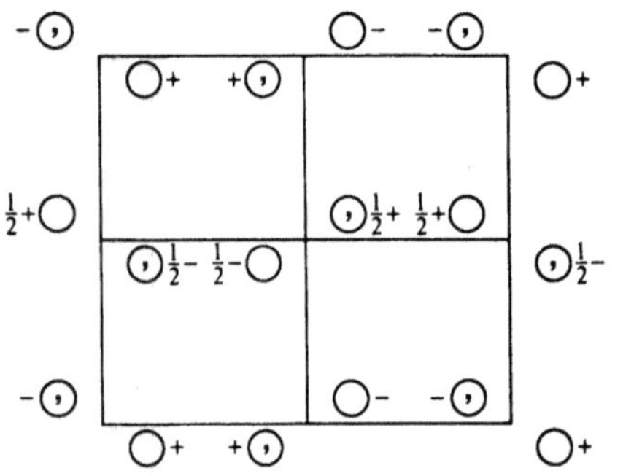

Origin at $\bar{1}$ on 12_11

Asymmetric unit $0 \le x \le \frac{1}{2};\quad 0 \le y \le \frac{1}{4};\quad 0 \le z \le 1$

Symmetry operations

(1) 1	(2) $2(0,0,\frac{1}{2})\ \ \frac{1}{4},0,z$	(3) $2(0,\frac{1}{2},0)\ \ 0,y,0$	(4) $2(\frac{1}{2},0,0)\ \ x,\frac{1}{4},\frac{1}{4}$
(5) $\bar{1}\ \ 0,0,0$	(6) $a\ \ x,y,\frac{1}{4}$	(7) $m\ \ x,\frac{1}{4},z$	(8) $n(0,\frac{1}{2},\frac{1}{2})\ \ \frac{1}{4},y,z$

Generators selected (1); $t(1,0,0)$; $t(0,1,0)$; $t(0,0,1)$; (2); (3); (5)

Positions

Multiplicity, Wyckoff letter, Site symmetry			Coordinates				Reflection conditions
							General:
8	*d*	1	(1) x,y,z	(2) $\bar{x}+\frac{1}{2},\bar{y},z+\frac{1}{2}$	(3) $\bar{x},y+\frac{1}{2},\bar{z}$	(4) $x+\frac{1}{2},\bar{y}+\frac{1}{2},\bar{z}+\frac{1}{2}$	$0kl$: $k+l=2n$
			(5) $\bar{x},\bar{y},\bar{z}$	(6) $x+\frac{1}{2},y,\bar{z}+\frac{1}{2}$	(7) $x,\bar{y}+\frac{1}{2},z$	(8) $\bar{x}+\frac{1}{2},y+\frac{1}{2},z+\frac{1}{2}$	$hk0$: $h=2n$
							$h00$: $h=2n$
							$0k0$: $k=2n$
							$00l$: $l=2n$
							Special: as above, plus
4	*c*	$.m.$	$x,\frac{1}{4},z$	$\bar{x}+\frac{1}{2},\frac{3}{4},z+\frac{1}{2}$	$\bar{x},\frac{3}{4},\bar{z}$	$x+\frac{1}{2},\frac{1}{4},\bar{z}+\frac{1}{2}$	no extra conditions
4	*b*	$\bar{1}$	$0,0,\frac{1}{2}$	$\frac{1}{2},0,0$	$0,\frac{1}{2},\frac{1}{2}$	$\frac{1}{2},\frac{1}{2},0$	hkl: $h+l,k=2n$
4	*a*	$\bar{1}$	$0,0,0$	$\frac{1}{2},0,\frac{1}{2}$	$0,\frac{1}{2},0$	$\frac{1}{2},\frac{1}{2},\frac{1}{2}$	hkl: $h+l,k=2n$

Symmetry of special projections

Along [001] $p2gm$
$\mathbf{a}'=\frac{1}{2}\mathbf{a}$ $\mathbf{b}'=\mathbf{b}$
Origin at $0,0,z$

Along [100] $c2mm$
$\mathbf{a}'=\mathbf{b}$ $\mathbf{b}'=\mathbf{c}$
Origin at $x,\frac{1}{4},\frac{1}{4}$

Along [010] $p2gg$
$\mathbf{a}'=\mathbf{c}$ $\mathbf{b}'=\mathbf{a}$
Origin at $0,y,0$

$Cmcm$ D_{2h}^{17} mmm Orthorhombic

No. 63 $C\ 2/m\ 2/c\ 2_1/m$ Patterson symmetry $Cmmm$

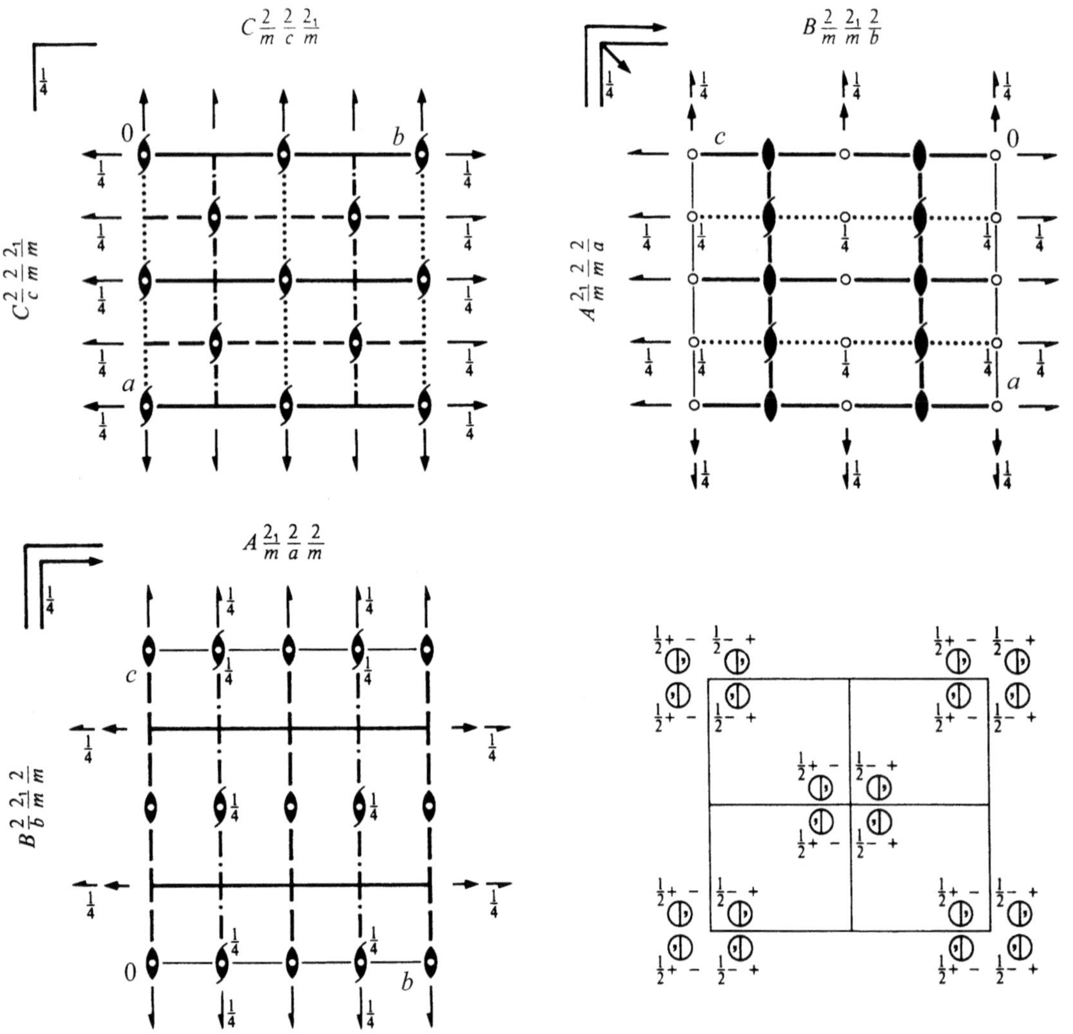

Origin at centre ($2/m$) at $2/mc2_1$

Asymmetric unit $0 \le x \le \frac{1}{2};\quad 0 \le y \le \frac{1}{2};\quad 0 \le z \le \frac{1}{4}$

Symmetry operations

For $(0,0,0)+$ set

(1) 1	(2) $2(0,0,\frac{1}{2})\ \ 0,0,z$	(3) $2\ \ 0,y,\frac{1}{4}$	(4) $2\ \ x,0,0$
(5) $\bar{1}\ \ 0,0,0$	(6) $m\ \ x,y,\frac{1}{4}$	(7) $c\ \ x,0,z$	(8) $m\ \ 0,y,z$

For $(\frac{1}{2},\frac{1}{2},0)+$ set

(1) $t(\frac{1}{2},\frac{1}{2},0)$	(2) $2(0,0,\frac{1}{2})\ \ \frac{1}{4},\frac{1}{4},z$	(3) $2(0,\frac{1}{2},0)\ \ \frac{1}{4},y,\frac{1}{4}$	(4) $2(\frac{1}{2},0,0)\ \ x,\frac{1}{4},0$
(5) $\bar{1}\ \ \frac{1}{4},\frac{1}{4},0$	(6) $n(\frac{1}{2},\frac{1}{2},0)\ \ x,y,\frac{1}{4}$	(7) $n(\frac{1}{2},0,\frac{1}{2})\ \ x,\frac{1}{4},z$	(8) $b\ \ \frac{1}{4},y,z$

Generators selected (1); $t(1,0,0)$; $t(0,1,0)$; $t(0,0,1)$; $t(\frac{1}{2},\frac{1}{2},0)$; (2); (3); (5)

Positions

Multiplicity, Wyckoff letter, Site symmetry			Coordinates				Reflection conditions
			$(0,0,0)+$	$(\frac{1}{2},\frac{1}{2},0)+$			General:
16	*h*	1	(1) x,y,z (5) $\bar{x},\bar{y},\bar{z}$	(2) $\bar{x},\bar{y},z+\frac{1}{2}$ (6) $x,y,\bar{z}+\frac{1}{2}$	(3) $\bar{x},y,\bar{z}+\frac{1}{2}$ (7) $x,\bar{y},z+\frac{1}{2}$	(4) $x,\bar{y},\bar{z}$ (8) $\bar{x},y,z$	hkl: $h+k=2n$ $0kl$: $k=2n$ $h0l$: $h,l=2n$ $hk0$: $h+k=2n$ $h00$: $h=2n$ $0k0$: $k=2n$ $00l$: $l=2n$
							Special: as above, plus
8	*g*	$..m$	$x,y,\frac{1}{4}$	$\bar{x},\bar{y},\frac{3}{4}$	$\bar{x},y,\frac{1}{4}$	$x,\bar{y},\frac{3}{4}$	no extra conditions
8	*f*	$m..$	$0,y,z$	$0,\bar{y},z+\frac{1}{2}$	$0,y,\bar{z}+\frac{1}{2}$	$0,\bar{y},\bar{z}$	no extra conditions
8	*e*	$2..$	$x,0,0$	$\bar{x},0,\frac{1}{2}$	$\bar{x},0,0$	$x,0,\frac{1}{2}$	hkl: $l=2n$
8	*d*	$\bar{1}$	$\frac{1}{4},\frac{1}{4},0$	$\frac{3}{4},\frac{3}{4},\frac{1}{2}$	$\frac{3}{4},\frac{1}{4},\frac{1}{2}$	$\frac{1}{4},\frac{3}{4},0$	hkl: $k,l=2n$
4	*c*	$m2m$	$0,y,\frac{1}{4}$	$0,\bar{y},\frac{3}{4}$			no extra conditions
4	*b*	$2/m..$	$0,\frac{1}{2},0$	$0,\frac{1}{2},\frac{1}{2}$			hkl: $l=2n$
4	*a*	$2/m..$	$0,0,0$	$0,0,\frac{1}{2}$			hkl: $l=2n$

Symmetry of special projections

Along [001] $c2mm$
$\mathbf{a}'=\mathbf{a}$ $\mathbf{b}'=\mathbf{b}$
Origin at $0,0,z$

Along [100] $p2gm$
$\mathbf{a}'=\frac{1}{2}\mathbf{b}$ $\mathbf{b}'=\mathbf{c}$
Origin at $x,0,0$

Along [010] $p2mm$
$\mathbf{a}'=\frac{1}{2}\mathbf{c}$ $\mathbf{b}'=\frac{1}{2}\mathbf{a}$
Origin at $0,y,0$

$Fddd$ D_{2h}^{24} mmm Orthorhombic

No. 70 $F\,2/d\,2/d\,2/d$ Patterson symmetry $Fmmm$

ORIGIN CHOICE 1

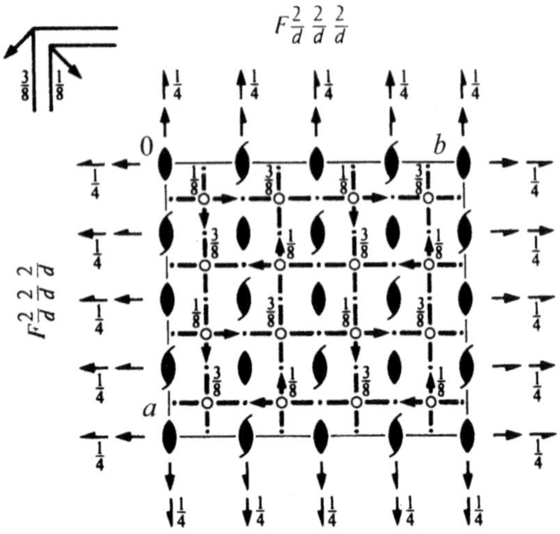

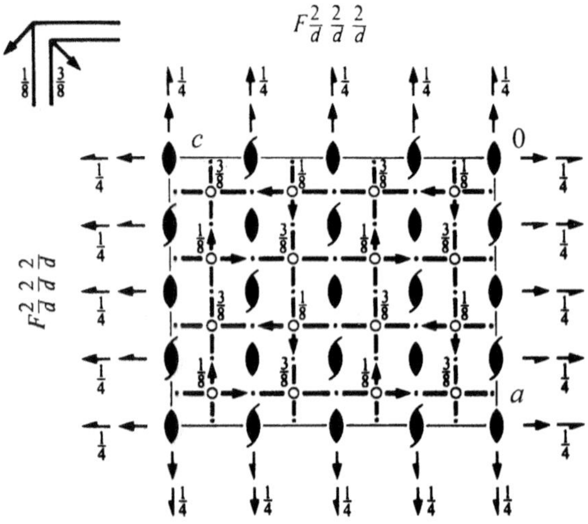

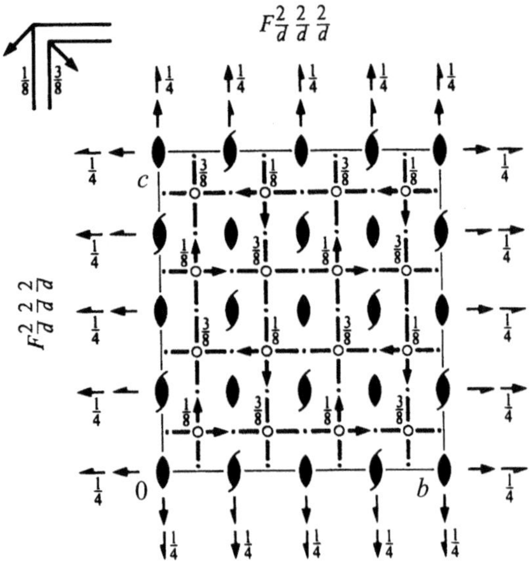

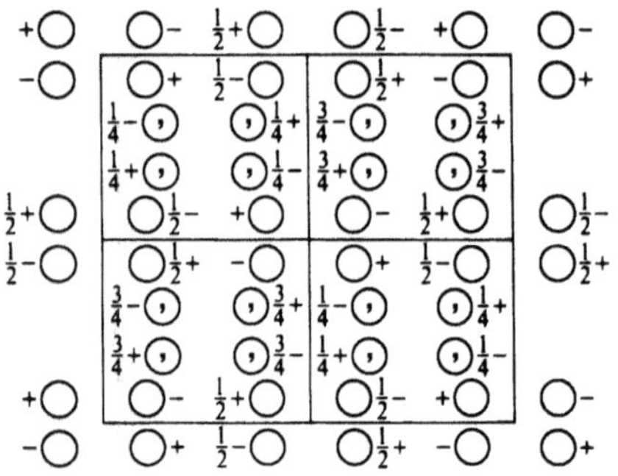

Origin at 222, at $-\frac{1}{8},-\frac{1}{8},-\frac{1}{8}$ from $\bar{1}$

Asymmetric unit $0 \le x \le \frac{1}{8};\quad 0 \le y \le \frac{1}{4};\quad 0 \le z \le 1$

Symmetry operations

For $(0,0,0)+$ set

(1) 1	(2) $2\quad 0,0,z$	(3) $2\quad 0,y,0$	(4) $2\quad x,0,0$
(5) $\bar{1}\quad \frac{1}{8},\frac{1}{8},\frac{1}{8}$	(6) $d(\frac{1}{4},\frac{1}{4},0)\quad x,y,\frac{1}{8}$	(7) $d(\frac{1}{4},0,\frac{1}{4})\quad x,\frac{1}{8},z$	(8) $d(0,\frac{1}{4},\frac{1}{4})\quad \frac{1}{8},y,z$

For $(0,\frac{1}{2},\frac{1}{2})+$ set

(1) $t(0,\frac{1}{2},\frac{1}{2})$	(2) $2(0,0,\frac{1}{2})\quad 0,\frac{1}{4},z$	(3) $2(0,\frac{1}{2},0)\quad 0,y,\frac{1}{4}$	(4) $2\quad x,\frac{1}{4},\frac{1}{4}$
(5) $\bar{1}\quad \frac{1}{8},\frac{3}{8},\frac{3}{8}$	(6) $d(\frac{1}{4},\frac{3}{4},0)\quad x,y,\frac{3}{8}$	(7) $d(\frac{1}{4},0,\frac{3}{4})\quad x,\frac{3}{8},z$	(8) $d(0,\frac{3}{4},\frac{3}{4})\quad \frac{1}{8},y,z$

For $(\frac{1}{2},0,\frac{1}{2})+$ set

(1) $t(\frac{1}{2},0,\frac{1}{2})$	(2) $2(0,0,\frac{1}{2})\quad \frac{1}{4},0,z$	(3) $2\quad \frac{1}{4},y,\frac{1}{4}$	(4) $2(\frac{1}{2},0,0)\quad x,0,\frac{1}{4}$
(5) $\bar{1}\quad \frac{3}{8},\frac{1}{8},\frac{3}{8}$	(6) $d(\frac{3}{4},\frac{1}{4},0)\quad x,y,\frac{3}{8}$	(7) $d(\frac{3}{4},0,\frac{3}{4})\quad x,\frac{1}{8},z$	(8) $d(0,\frac{1}{4},\frac{3}{4})\quad \frac{3}{8},y,z$

For $(\frac{1}{2},\frac{1}{2},0)+$ set

(1) $t(\frac{1}{2},\frac{1}{2},0)$	(2) $2\quad \frac{1}{4},\frac{1}{4},z$	(3) $2(0,\frac{1}{2},0)\quad \frac{1}{4},y,0$	(4) $2(\frac{1}{2},0,0)\quad x,\frac{1}{4},0$
(5) $\bar{1}\quad \frac{3}{8},\frac{3}{8},\frac{1}{8}$	(6) $d(\frac{3}{4},\frac{3}{4},0)\quad x,y,\frac{1}{8}$	(7) $d(\frac{3}{4},0,\frac{1}{4})\quad x,\frac{3}{8},z$	(8) $d(0,\frac{3}{4},\frac{1}{4})\quad \frac{3}{8},y,z$

Generators selected (1); $t(1,0,0)$; $t(0,1,0)$; $t(0,0,1)$; $t(0,\frac{1}{2},\frac{1}{2})$; $t(\frac{1}{2},0,\frac{1}{2})$; (2); (3); (5)

Positions

Multiplicity, Wyckoff letter, Site symmetry			Coordinates				Reflection conditions
			$(0,0,0)+$	$(0,\frac{1}{2},\frac{1}{2})+$	$(\frac{1}{2},0,\frac{1}{2})+$	$(\frac{1}{2},\frac{1}{2},0)+$	General:
32	*h*	1	(1) x,y,z (5) $\bar{x}+\frac{1}{4},\bar{y}+\frac{1}{4},\bar{z}+\frac{1}{4}$	(2) $\bar{x},\bar{y},z$ (6) $x+\frac{1}{4},y+\frac{1}{4},\bar{z}+\frac{1}{4}$	(3) $\bar{x},y,\bar{z}$ (7) $x+\frac{1}{4},\bar{y}+\frac{1}{4},z+\frac{1}{4}$	(4) $x,\bar{y},\bar{z}$ (8) $\bar{x}+\frac{1}{4},y+\frac{1}{4},z+\frac{1}{4}$	hkl: $h+k=2n$ and $h+l,k+l=2n$ $0kl$: $k+l=4n$ and $k,l=2n$ $h0l$: $h+l=4n$ and $h,l=2n$ $hk0$: $h+k=4n$ and $h,k=2n$ $h00$: $h=4n$ $0k0$: $k=4n$ $00l$: $l=4n$
							Special: as above, plus
16	*g*	..2	$0,0,z$	$0,0,\bar{z}$	$\frac{1}{4},\frac{1}{4},\bar{z}+\frac{1}{4}$	$\frac{1}{4},\frac{1}{4},z+\frac{1}{4}$	hkl: $h=2n+1$ or $h+k+l=4n$
16	*f*	.2.	$0,y,0$	$0,\bar{y},0$	$\frac{1}{4},\bar{y}+\frac{1}{4},\frac{1}{4}$	$\frac{1}{4},y+\frac{1}{4},\frac{1}{4}$	
16	*e*	2..	$x,0,0$	$\bar{x},0,0$	$\bar{x}+\frac{1}{4},\frac{1}{4},\frac{1}{4}$	$x+\frac{1}{4},\frac{1}{4},\frac{1}{4}$	
16	*d*	$\bar{1}$	$\frac{5}{8},\frac{5}{8},\frac{5}{8}$	$\frac{3}{8},\frac{3}{8},\frac{5}{8}$	$\frac{3}{8},\frac{5}{8},\frac{3}{8}$	$\frac{5}{8},\frac{3}{8},\frac{3}{8}$	hkl: $h=2n+1$ or $h,k,l=4n+2$ or $h,k,l=4n$
16	*c*	$\bar{1}$	$\frac{1}{8},\frac{1}{8},\frac{1}{8}$	$\frac{7}{8},\frac{7}{8},\frac{1}{8}$	$\frac{7}{8},\frac{1}{8},\frac{7}{8}$	$\frac{1}{8},\frac{7}{8},\frac{7}{8}$	
8	*b*	222	$0,0,\frac{1}{2}$	$\frac{1}{4},\frac{1}{4},\frac{3}{4}$			hkl: $h=2n+1$ or $h+k+l=4n$
8	*a*	222	$0,0,0$	$\frac{1}{4},\frac{1}{4},\frac{1}{4}$			

Symmetry of special projections

Along [001] $c2mm$
$\mathbf{a}'=\frac{1}{2}\mathbf{a}$ $\mathbf{b}'=\frac{1}{2}\mathbf{b}$
Origin at $0,0,z$

Along [100] $c2mm$
$\mathbf{a}'=\frac{1}{2}\mathbf{b}$ $\mathbf{b}'=\frac{1}{2}\mathbf{c}$
Origin at $x,0,0$

Along [010] $c2mm$
$\mathbf{a}'=\frac{1}{2}\mathbf{c}$ $\mathbf{b}'=\frac{1}{2}\mathbf{a}$
Origin at $0,y,0$

$Ibca$ $\quad D_{2h}^{27}$ $\quad mmm$ $\quad$ Orthorhombic

No. 73 $\quad I\ 2_1/b\ 2_1/c\ 2_1/a$

Patterson symmetry $Immm$

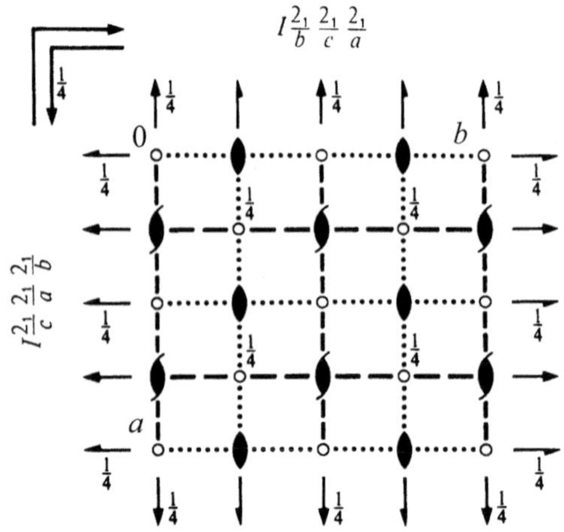

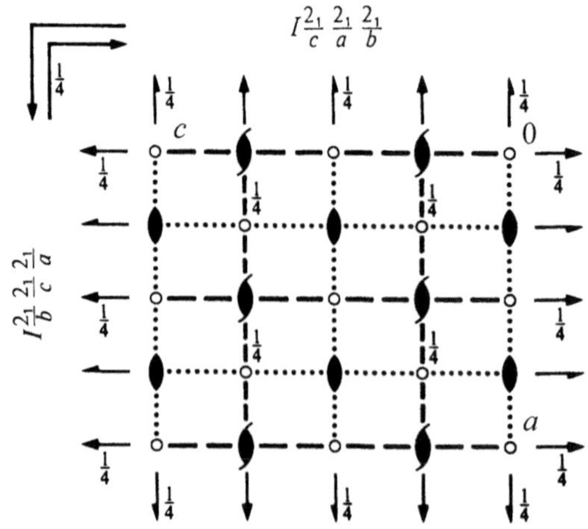

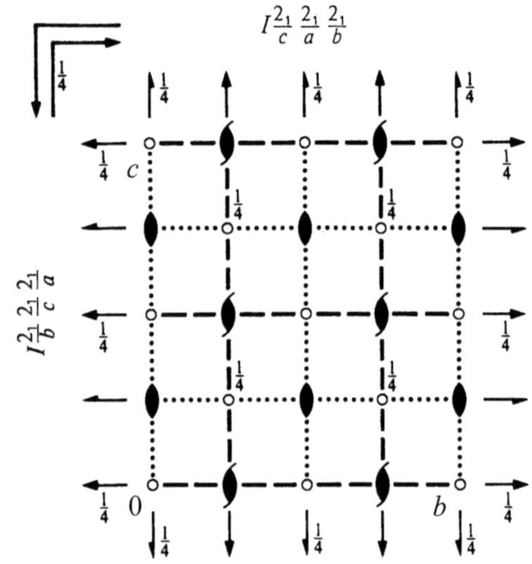

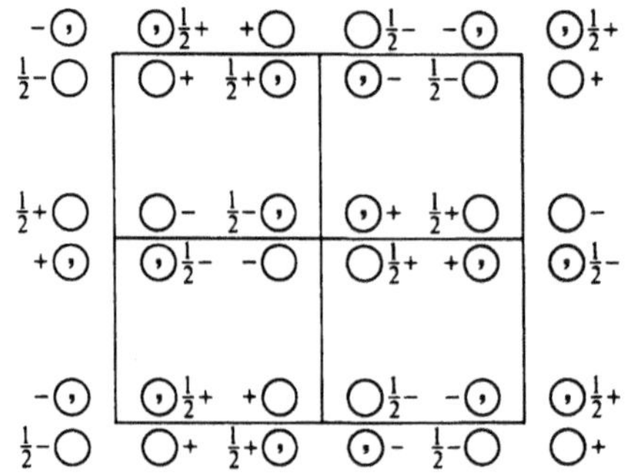

Origin at $\bar{1}$ at cab

Asymmetric unit $\quad 0 \le x \le \frac{1}{4}; \quad 0 \le y \le \frac{1}{2}; \quad 0 \le z \le \frac{1}{2}$

Symmetry operations

For $(0,0,0)+$ set

(1) 1	(2) $2(0,0,\frac{1}{2})\ \ \frac{1}{4},0,z$	(3) $2(0,\frac{1}{2},0)\ \ 0,y,\frac{1}{4}$	(4) $2(\frac{1}{2},0,0)\ \ x,\frac{1}{4},0$
(5) $\bar{1}\ \ 0,0,0$	(6) $a\ \ x,y,\frac{1}{4}$	(7) $c\ \ x,\frac{1}{4},z$	(8) $b\ \ \frac{1}{4},y,z$

For $(\frac{1}{2},\frac{1}{2},\frac{1}{2})+$ set

(1) $t(\frac{1}{2},\frac{1}{2},\frac{1}{2})$	(2) $2\ \ 0,\frac{1}{4},z$	(3) $2\ \ \frac{1}{4},y,0$	(4) $2\ \ x,0,\frac{1}{4}$
(5) $\bar{1}\ \ \frac{1}{4},\frac{1}{4},\frac{1}{4}$	(6) $b\ \ x,y,0$	(7) $a\ \ x,0,z$	(8) $c\ \ 0,y,z$

Generators selected (1); $t(1,0,0)$; $t(0,1,0)$; $t(0,0,1)$; $t(\frac{1}{2},\frac{1}{2},\frac{1}{2})$; (2); (3); (5)

Positions

Multiplicity, Wyckoff letter, Site symmetry			Coordinates				Reflection conditions
			$(0,0,0)+$ $(\frac{1}{2},\frac{1}{2},\frac{1}{2})+$				General:
16	*f*	1	(1) x,y,z (5) $\bar{x},\bar{y},\bar{z}$	(2) $\bar{x}+\frac{1}{2},\bar{y},z+\frac{1}{2}$ (6) $x+\frac{1}{2},y,\bar{z}+\frac{1}{2}$	(3) $\bar{x},y+\frac{1}{2},\bar{z}+\frac{1}{2}$ (7) $x,\bar{y}+\frac{1}{2},z+\frac{1}{2}$	(4) $x+\frac{1}{2},\bar{y}+\frac{1}{2},\bar{z}$ (8) $\bar{x}+\frac{1}{2},y+\frac{1}{2},z$	hkl: $h+k+l=2n$ $0kl$: $k,l=2n$ $h0l$: $h,l=2n$ $hk0$: $h,k=2n$ $h00$: $h=2n$ $0k0$: $k=2n$ $00l$: $l=2n$
							Special: as above, plus
8	*e*	. . 2	$0,\frac{1}{4},z$	$0,\frac{3}{4},\bar{z}+\frac{1}{2}$	$0,\frac{3}{4},\bar{z}$	$0,\frac{1}{4},z+\frac{1}{2}$	hkl: $l=2n$
8	*d*	. 2 .	$\frac{1}{4},y,0$	$\frac{1}{4},\bar{y},\frac{1}{2}$	$\frac{3}{4},\bar{y},0$	$\frac{3}{4},y,\frac{1}{2}$	hkl: $k=2n$
8	*c*	2 . .	$x,0,\frac{1}{4}$	$\bar{x}+\frac{1}{2},0,\frac{3}{4}$	$\bar{x},0,\frac{3}{4}$	$x+\frac{1}{2},0,\frac{1}{4}$	hkl: $h=2n$
8	*b*	$\bar{1}$	$\frac{1}{4},\frac{1}{4},\frac{1}{4}$	$\frac{1}{4},\frac{3}{4},\frac{3}{4}$	$\frac{3}{4},\frac{3}{4},\frac{1}{4}$	$\frac{3}{4},\frac{1}{4},\frac{3}{4}$	hkl: $k,l=2n$
8	*a*	$\bar{1}$	$0,0,0$	$\frac{1}{2},0,\frac{1}{2}$	$0,\frac{1}{2},\frac{1}{2}$	$\frac{1}{2},\frac{1}{2},0$	hkl: $k,l=2n$

Symmetry of special projections

Along [001] $p2mm$
$\mathbf{a}'=\frac{1}{2}\mathbf{a}$ $\mathbf{b}'=\frac{1}{2}\mathbf{b}$
Origin at $0,0,z$

Along [100] $p2mm$
$\mathbf{a}'=\frac{1}{2}\mathbf{b}$ $\mathbf{b}'=\frac{1}{2}\mathbf{c}$
Origin at $x,0,0$

Along [010] $p2mm$
$\mathbf{a}'=\frac{1}{2}\mathbf{c}$ $\mathbf{b}'=\frac{1}{2}\mathbf{a}$
Origin at $0,y,0$

$I4/m$ $\quad C_{4h}^5$ $\quad 4/m$ $\quad$ Tetragonal

No. 87 $\quad I4/m$ $\quad$ Patterson symmetry $I4/m$

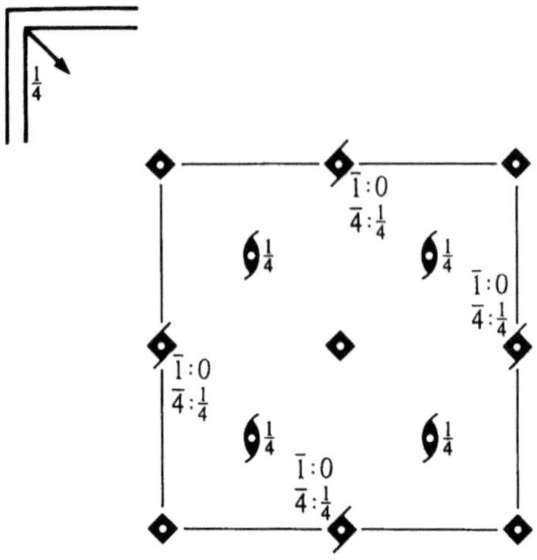

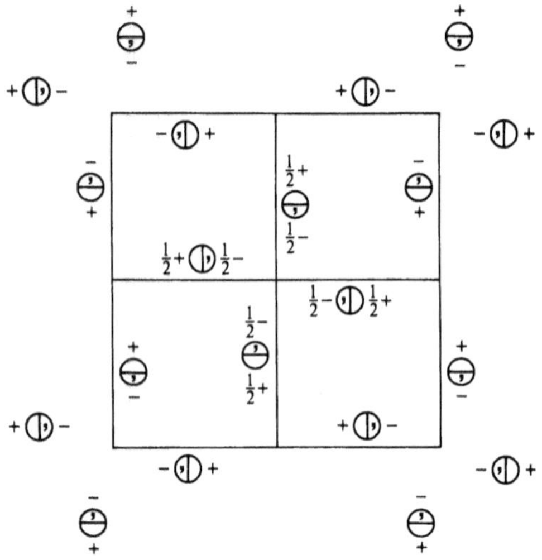

Origin at centre ($4/m$)

Asymmetric unit $\quad 0 \le x \le \frac{1}{2}; \quad 0 \le y \le \frac{1}{2}; \quad 0 \le z \le \frac{1}{4}$

Symmetry operations

For $(0,0,0)+$ set

(1) 1	(2) $2 \quad 0,0,z$	(3) $4^+ \quad 0,0,z$	(4) $4^- \quad 0,0,z$
(5) $\bar{1} \quad 0,0,0$	(6) $m \quad x,y,0$	(7) $\bar{4}^+ \quad 0,0,z; \; 0,0,0$	(8) $\bar{4}^- \quad 0,0,z; \; 0,0,0$

For $(\frac{1}{2},\frac{1}{2},\frac{1}{2})+$ set

(1) $t(\frac{1}{2},\frac{1}{2},\frac{1}{2})$	(2) $2(0,0,\frac{1}{2}) \quad \frac{1}{4},\frac{1}{4},z$	(3) $4^+(0,0,\frac{1}{2}) \quad 0,\frac{1}{2},z$	(4) $4^-(0,0,\frac{1}{2}) \quad \frac{1}{2},0,z$
(5) $\bar{1} \quad \frac{1}{4},\frac{1}{4},\frac{1}{4}$	(6) $n(\frac{1}{2},\frac{1}{2},0) \quad x,y,\frac{1}{4}$	(7) $\bar{4}^+ \quad \frac{1}{2},0,z; \; \frac{1}{2},0,\frac{1}{4}$	(8) $\bar{4}^- \quad 0,\frac{1}{2},z; \; 0,\frac{1}{2},\frac{1}{4}$

Generators selected (1); $t(1,0,0)$; $t(0,1,0)$; $t(0,0,1)$; $t(\frac{1}{2},\frac{1}{2},\frac{1}{2})$; (2); (3); (5)

Positions

Multiplicity, Wyckoff letter, Site symmetry			Coordinates $(0,0,0)+ \quad (\frac{1}{2},\frac{1}{2},\frac{1}{2})+$				Reflection conditions
							General:
16	i	1	(1) x,y,z (5) $\bar{x},\bar{y},\bar{z}$	(2) $\bar{x},\bar{y},z$ (6) $x,y,\bar{z}$	(3) $\bar{y},x,z$ (7) $y,\bar{x},\bar{z}$	(4) $y,\bar{x},z$ (8) $\bar{y},x,\bar{z}$	hkl: $h+k+l=2n$ $hk0$: $h+k=2n$ $0kl$: $k+l=2n$ hhl: $l=2n$ $00l$: $l=2n$ $h00$: $h=2n$
							Special: as above, plus
8	h	$m\,.\,.$	$x,y,0$	$\bar{x},\bar{y},0$	$\bar{y},x,0$	$y,\bar{x},0$	no extra conditions
8	g	$2\,.\,.$	$0,\frac{1}{2},z$	$\frac{1}{2},0,z$	$0,\frac{1}{2},\bar{z}$	$\frac{1}{2},0,\bar{z}$	hkl: $l=2n$
8	f	$\bar{1}$	$\frac{1}{4},\frac{1}{4},\frac{1}{4}$	$\frac{3}{4},\frac{3}{4},\frac{1}{4}$	$\frac{3}{4},\frac{1}{4},\frac{1}{4}$	$\frac{1}{4},\frac{3}{4},\frac{1}{4}$	hkl: $k,l=2n$
4	e	$4\,.\,.$	$0,0,z$	$0,0,\bar{z}$			no extra conditions
4	d	$\bar{4}\,.\,.$	$0,\frac{1}{2},\frac{1}{4}$	$\frac{1}{2},0,\frac{1}{4}$			hkl: $l=2n$
4	c	$2/m\,.\,.$	$0,\frac{1}{2},0$	$\frac{1}{2},0,0$			hkl: $l=2n$
2	b	$4/m\,.\,.$	$0,0,\frac{1}{2}$				no extra conditions
2	a	$4/m\,.\,.$	$0,0,0$				no extra conditions

Symmetry of special projections

Along [001] $p4$
$\mathbf{a}'=\frac{1}{2}(\mathbf{a}-\mathbf{b}) \quad \mathbf{b}'=\frac{1}{2}(\mathbf{a}+\mathbf{b})$
Origin at $0,0,z$

Along [100] $c2mm$
$\mathbf{a}'=\mathbf{b} \quad \mathbf{b}'=\mathbf{c}$
Origin at $x,0,0$

Along [110] $p2mm$
$\mathbf{a}'=\frac{1}{2}(-\mathbf{a}+\mathbf{b}) \quad \mathbf{b}'=\frac{1}{2}\mathbf{c}$
Origin at $x,x,0$

$P4_12_12$ D_4^4 422 Tetragonal

No. 92 $P4_12_12$ Patterson symmetry $P4/mmm$

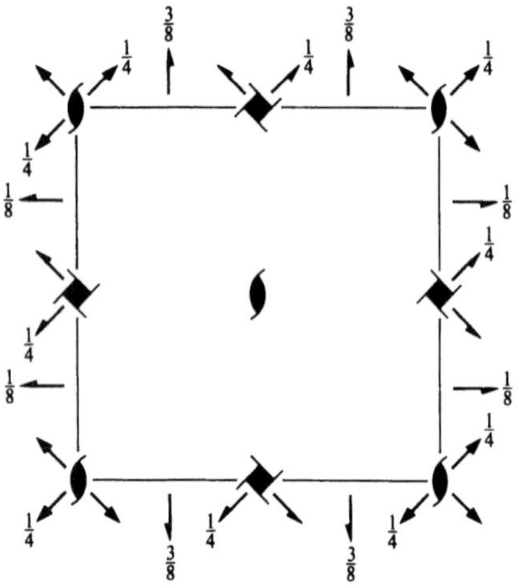

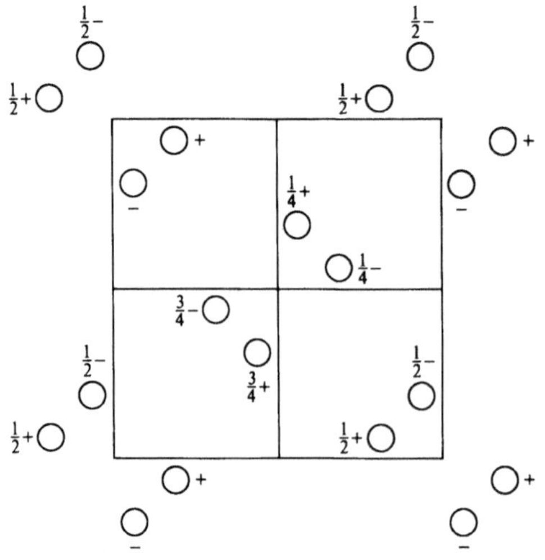

Origin on $2[1\,1\,0]$ at $2_1\,1\,(1,2)$

Asymmetric unit $0 \le x \le 1;\quad 0 \le y \le 1;\quad 0 \le z \le \frac{1}{8}$

Symmetry operations

(1) 1 (2) $2(0,0,\frac{1}{2})\ \ 0,0,z$ (3) $4^+(0,0,\frac{1}{4})\ \ 0,\frac{1}{2},z$ (4) $4^-(0,0,\frac{3}{4})\ \ \frac{1}{2},0,z$
(5) $2(0,\frac{1}{2},0)\ \ \frac{1}{4},y,\frac{1}{8}$ (6) $2(\frac{1}{2},0,0)\ \ x,\frac{1}{4},\frac{3}{8}$ (7) $2\ \ x,x,0$ (8) $2\ \ x,\bar{x},\frac{1}{4}$

Generators selected (1); $t(1,0,0)$; $t(0,1,0)$; $t(0,0,1)$; (2); (3); (5)

Positions

Multiplicity, Wyckoff letter, Site symmetry	Coordinates				Reflection conditions
					General:
8 b 1	(1) x,y,z	(2) $\bar{x},\bar{y},z+\frac{1}{2}$	(3) $\bar{y}+\frac{1}{2},x+\frac{1}{2},z+\frac{1}{4}$	(4) $y+\frac{1}{2},\bar{x}+\frac{1}{2},z+\frac{3}{4}$	$00l$: $l=4n$
	(5) $\bar{x}+\frac{1}{2},y+\frac{1}{2},\bar{z}+\frac{1}{4}$	(6) $x+\frac{1}{2},\bar{y}+\frac{1}{2},\bar{z}+\frac{3}{4}$	(7) $y,x,\bar{z}$	(8) $\bar{y},\bar{x},\bar{z}+\frac{1}{2}$	$h00$: $h=2n$
					Special: as above, plus
4 a ..2	$x,x,0$	$\bar{x},\bar{x},\frac{1}{2}$	$\bar{x}+\frac{1}{2},x+\frac{1}{2},\frac{1}{4}$	$x+\frac{1}{2},\bar{x}+\frac{1}{2},\frac{3}{4}$	$0kl$: $l=2n+1$ or $2k+l=4n$

Symmetry of special projections

Along [001] $p4gm$
$\mathbf{a}'=\mathbf{a}$ $\mathbf{b}'=\mathbf{b}$
Origin at $0,\frac{1}{2},z$

Along [100] $p2gg$
$\mathbf{a}'=\mathbf{b}$ $\mathbf{b}'=\mathbf{c}$
Origin at $x,\frac{1}{4},\frac{3}{8}$

Along [110] $p2gm$
$\mathbf{a}'=\frac{1}{2}(-\mathbf{a}+\mathbf{b})$ $\mathbf{b}'=\mathbf{c}$
Origin at $x,x,0$

Patterson symmetry $P4/mmm$ $P4_32_12$ No. 96

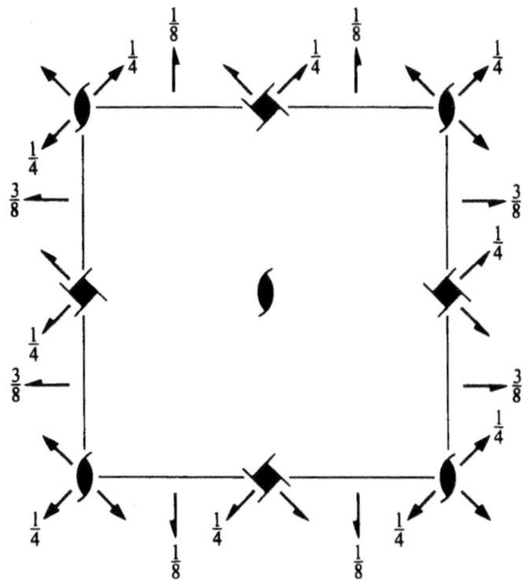

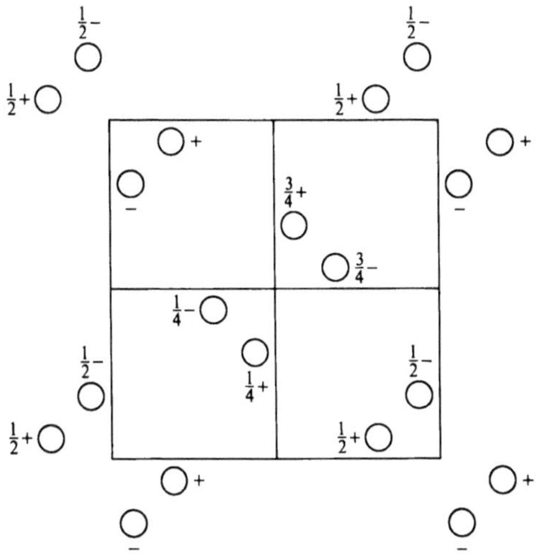

Origin on 2[110] at $2_1$1(1,2)

Asymmetric unit $0 \le x \le 1;\quad 0 \le y \le 1;\quad 0 \le z \le \frac{1}{8}$

Symmetry operations

(1) 1 (2) $2(0,0,\frac{1}{2})$ $0,0,z$ (3) $4^+(0,0,\frac{3}{4})$ $0,\frac{1}{2},z$ (4) $4^-(0,0,\frac{1}{4})$ $\frac{1}{2},0,z$
(5) $2(0,\frac{1}{2},0)$ $\frac{1}{4},y,\frac{3}{8}$ (6) $2(\frac{1}{2},0,0)$ $x,\frac{1}{4},\frac{1}{8}$ (7) 2 $x,x,0$ (8) 2 $x,\bar{x},\frac{1}{4}$

Generators selected (1); $t(1,0,0)$; $t(0,1,0)$; $t(0,0,1)$; (2); (3); (5)

Positions

Multiplicity, Wyckoff letter, Site symmetry	Coordinates				Reflection conditions
					General:
8 *b* 1	(1) x,y,z	(2) $\bar{x},\bar{y},z+\frac{1}{2}$	(3) $\bar{y}+\frac{1}{2},x+\frac{1}{2},z+\frac{3}{4}$	(4) $y+\frac{1}{2},\bar{x}+\frac{1}{2},z+\frac{1}{4}$	$00l$: $l=4n$
	(5) $\bar{x}+\frac{1}{2},y+\frac{1}{2},\bar{z}+\frac{3}{4}$	(6) $x+\frac{1}{2},\bar{y}+\frac{1}{2},\bar{z}+\frac{1}{4}$	(7) $y,x,\bar{z}$	(8) $\bar{y},\bar{x},\bar{z}+\frac{1}{2}$	$h00$: $h=2n$
					Special: as above, plus
4 *a* ..2	$x,x,0$	$\bar{x},\bar{x},\frac{1}{2}$	$\bar{x}+\frac{1}{2},x+\frac{1}{2},\frac{3}{4}$	$x+\frac{1}{2},\bar{x}+\frac{1}{2},\frac{1}{4}$	$0kl$: $l=2n+1$ or $2k+l=4n$

Symmetry of special projections

Along [001] $p4gm$
$\mathbf{a}'=\mathbf{a}$ $\mathbf{b}'=\mathbf{b}$
Origin at $0,\frac{1}{2},z$

Along [100] $p2gg$
$\mathbf{a}'=\mathbf{b}$ $\mathbf{b}'=\mathbf{c}$
Origin at $x,\frac{1}{4},\frac{1}{8}$

Along [110] $p2gm$
$\mathbf{a}'=\frac{1}{2}(-\mathbf{a}+\mathbf{b})$ $\mathbf{b}'=\mathbf{c}$
Origin at $x,x,0$

$P4mm$ C_{4v}^1 $4mm$ Tetragonal

No. 99 $P4mm$ Patterson symmetry $P4/mmm$

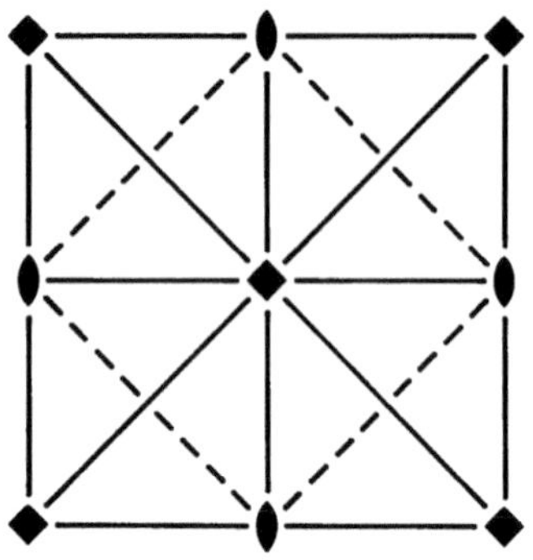

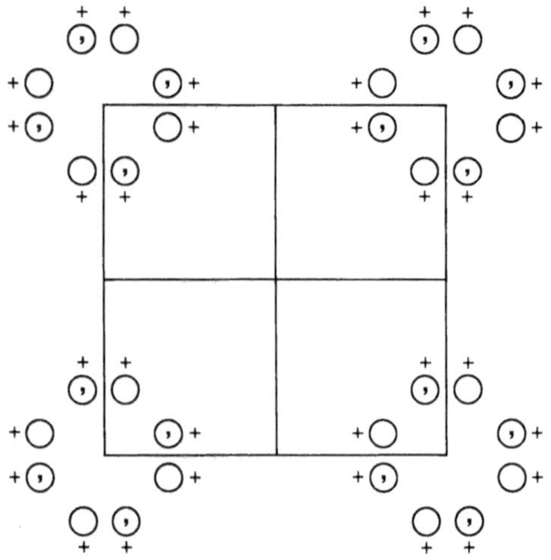

Origin on $4mm$

Asymmetric unit $0 \leq x \leq \frac{1}{2}$; $0 \leq y \leq \frac{1}{2}$; $0 \leq z \leq 1$; $x \leq y$

Symmetry operations

(1) 1	(2) 2 $0,0,z$	(3) 4^+ $0,0,z$	(4) 4^- $0,0,z$
(5) m $x,0,z$	(6) m $0,y,z$	(7) m $x,\bar{x},z$	(8) m x,x,z

Generators selected (1); $t(1,0,0)$; $t(0,1,0)$; $t(0,0,1)$; (2); (3); (5)

Positions

Multiplicity, Wyckoff letter, Site symmetry			Coordinates				Reflection conditions
							General:
8	g	1	(1) x,y,z (5) $x,\bar{y},z$	(2) $\bar{x},\bar{y},z$ (6) $\bar{x},y,z$	(3) $\bar{y},x,z$ (7) $\bar{y},\bar{x},z$	(4) $y,\bar{x},z$ (8) y,x,z	no conditions
							Special:
4	f	$.m.$	$x,\frac{1}{2},z$	$\bar{x},\frac{1}{2},z$	$\frac{1}{2},x,z$	$\frac{1}{2},\bar{x},z$	no extra conditions
4	e	$.m.$	$x,0,z$	$\bar{x},0,z$	$0,x,z$	$0,\bar{x},z$	no extra conditions
4	d	$..m$	x,x,z	$\bar{x},\bar{x},z$	$\bar{x},x,z$	$x,\bar{x},z$	no extra conditions
2	c	$2mm.$	$\frac{1}{2},0,z$	$0,\frac{1}{2},z$			hkl: $h+k=2n$
1	b	$4mm$	$\frac{1}{2},\frac{1}{2},z$				no extra conditions
1	a	$4mm$	$0,0,z$				no extra conditions

Symmetry of special projections

Along [001] $p4mm$
$\mathbf{a}' = \mathbf{a}$ $\mathbf{b}' = \mathbf{b}$
Origin at $0,0,z$

Along [100] $p1m1$
$\mathbf{a}' = \mathbf{b}$ $\mathbf{b}' = \mathbf{c}$
Origin at $x,0,0$

Along [110] $p1m1$
$\mathbf{a}' = \frac{1}{2}(-\mathbf{a}+\mathbf{b})$ $\mathbf{b}' = \mathbf{c}$
Origin at $x,x,0$

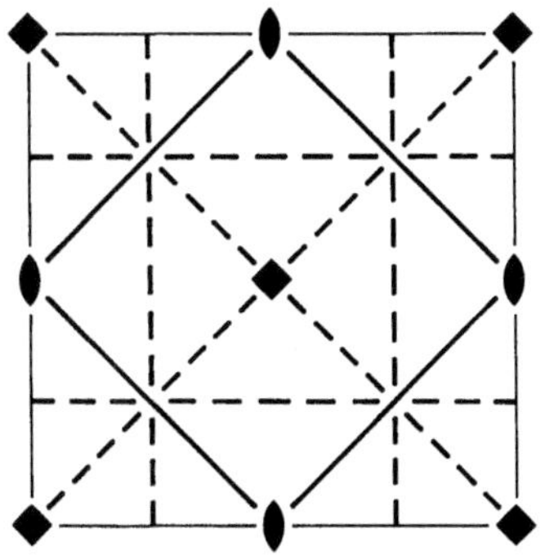

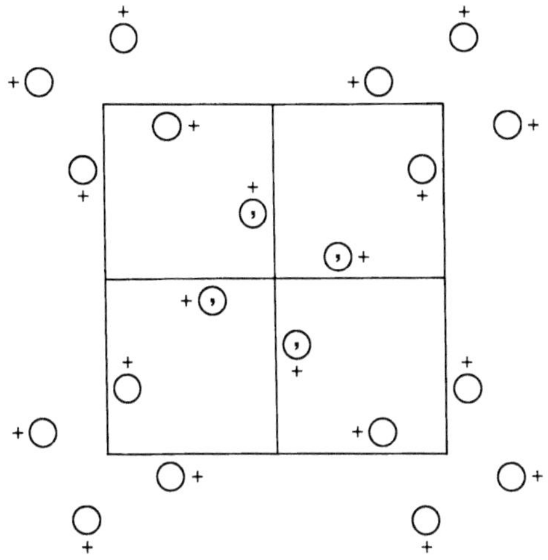

Origin on $41g$

Asymmetric unit $0 \le x \le \frac{1}{2}$; $0 \le y \le \frac{1}{2}$; $0 \le z \le 1$; $y \le \frac{1}{2} - x$

Symmetry operations

(1) 1	(2) 2 $0,0,z$	(3) 4^+ $0,0,z$	(4) 4^- $0,0,z$
(5) a $x,\frac{1}{4},z$	(6) b $\frac{1}{4},y,z$	(7) m $x+\frac{1}{2},\bar{x},z$	(8) $g(\frac{1}{2},\frac{1}{2},0)$ x,x,z

Generators selected (1); $t(1,0,0)$; $t(0,1,0)$; $t(0,0,1)$; (2); (3); (5)

Positions

Multiplicity, Wyckoff letter, Site symmetry	Coordinates				Reflection conditions
					General:
8 d 1	(1) x,y,z	(2) $\bar{x},\bar{y},z$	(3) $\bar{y},x,z$	(4) $y,\bar{x},z$	$0kl$: $k=2n$
	(5) $x+\frac{1}{2},\bar{y}+\frac{1}{2},z$	(6) $\bar{x}+\frac{1}{2},y+\frac{1}{2},z$	(7) $\bar{y}+\frac{1}{2},\bar{x}+\frac{1}{2},z$	(8) $y+\frac{1}{2},x+\frac{1}{2},z$	$h00$: $h=2n$
					Special: as above, plus
4 c $..m$	$x,x+\frac{1}{2},z$	$\bar{x},\bar{x}+\frac{1}{2},z$	$\bar{x}+\frac{1}{2},x,z$	$x+\frac{1}{2},\bar{x},z$	no extra conditions
2 b $2.mm$	$\frac{1}{2},0,z$	$0,\frac{1}{2},z$			hkl: $h+k=2n$
2 a $4..$	$0,0,z$	$\frac{1}{2},\frac{1}{2},z$			hkl: $h+k=2n$

Symmetry of special projections

Along [001] $p4gm$
$\mathbf{a}'=\mathbf{a}$ $\mathbf{b}'=\mathbf{b}$
Origin at $0,0,z$

Along [100] $p1m1$
$\mathbf{a}'=\frac{1}{2}\mathbf{b}$ $\mathbf{b}'=\mathbf{c}$
Origin at $x,0,0$

Along [110] $p1m1$
$\mathbf{a}'=\frac{1}{2}(-\mathbf{a}+\mathbf{b})$ $\mathbf{b}'=\mathbf{c}$
Origin at $x,x,0$

$I4/mmm$ $\quad D_{4h}^{17}$ $\quad 4/mmm$ $\quad$ Tetragonal

No. 139 $\quad I\ 4/m\ 2/m\ 2/m$ $\quad$ Patterson symmetry $I4/mmm$

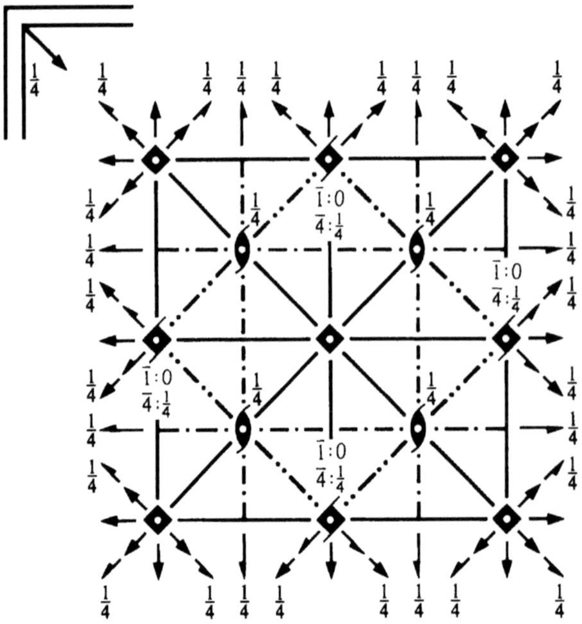

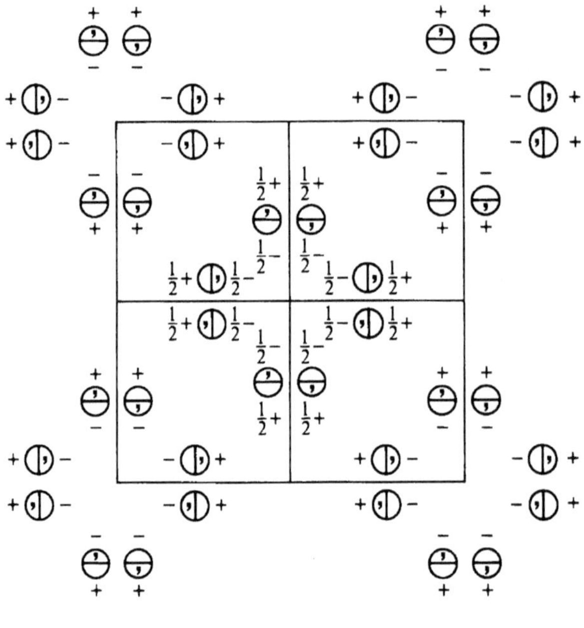

Origin at centre ($4/mmm$)

Asymmetric unit $\quad 0 \le x \le \frac{1}{2};\quad 0 \le y \le \frac{1}{2};\quad 0 \le z \le \frac{1}{4};\quad x \le y$

Symmetry operations

For $(0,0,0)+$ set

(1) 1	(2) $2\quad 0,0,z$	(3) $4^{+}\quad 0,0,z$	(4) $4^{-}\quad 0,0,z$
(5) $2\quad 0,y,0$	(6) $2\quad x,0,0$	(7) $2\quad x,x,0$	(8) $2\quad x,\bar{x},0$
(9) $\bar{1}\quad 0,0,0$	(10) $m\quad x,y,0$	(11) $\bar{4}^{+}\quad 0,0,z;\ 0,0,0$	(12) $\bar{4}^{-}\quad 0,0,z;\ 0,0,0$
(13) $m\quad x,0,z$	(14) $m\quad 0,y,z$	(15) $m\quad x,\bar{x},z$	(16) $m\quad x,x,z$

For $(\frac{1}{2},\frac{1}{2},\frac{1}{2})+$ set

(1) $t(\frac{1}{2},\frac{1}{2},\frac{1}{2})$	(2) $2(0,0,\frac{1}{2})\quad \frac{1}{4},\frac{1}{4},z$	(3) $4^{+}(0,0,\frac{1}{2})\quad 0,\frac{1}{2},z$	(4) $4^{-}(0,0,\frac{1}{2})\quad \frac{1}{2},0,z$
(5) $2(0,\frac{1}{2},0)\quad \frac{1}{4},y,\frac{1}{4}$	(6) $2(\frac{1}{2},0,0)\quad x,\frac{1}{4},\frac{1}{4}$	(7) $2(\frac{1}{2},\frac{1}{2},0)\quad x,x,\frac{1}{4}$	(8) $2\quad x,\bar{x}+\frac{1}{2},\frac{1}{4}$
(9) $\bar{1}\quad \frac{1}{4},\frac{1}{4},\frac{1}{4}$	(10) $n(\frac{1}{2},\frac{1}{2},0)\quad x,y,\frac{1}{4}$	(11) $\bar{4}^{+}\quad \frac{1}{2},0,z;\ \frac{1}{2},0,\frac{1}{4}$	(12) $\bar{4}^{-}\quad 0,\frac{1}{2},z;\ 0,\frac{1}{2},\frac{1}{4}$
(13) $n(\frac{1}{2},0,\frac{1}{2})\quad x,\frac{1}{4},z$	(14) $n(0,\frac{1}{2},\frac{1}{2})\quad \frac{1}{4},y,z$	(15) $c\quad x+\frac{1}{2},\bar{x},z$	(16) $n(\frac{1}{2},\frac{1}{2},\frac{1}{2})\quad x,x,z$

Generators selected (1); $t(1,0,0)$; $t(0,1,0)$; $t(0,0,1)$; $t(\frac{1}{2},\frac{1}{2},\frac{1}{2})$; (2); (3); (5); (9)

Positions

Multiplicity, Wyckoff letter, Site symmetry			Coordinates $(0,0,0)+$ $(\frac{1}{2},\frac{1}{2},\frac{1}{2})+$				Reflection conditions
							General:
32	o	1	(1) x,y,z	(2) $\bar{x},\bar{y},z$	(3) $\bar{y},x,z$	(4) $y,\bar{x},z$	hkl: $h+k+l=2n$
			(5) $\bar{x},y,\bar{z}$	(6) $x,\bar{y},\bar{z}$	(7) $y,x,\bar{z}$	(8) $\bar{y},\bar{x},\bar{z}$	$hk0$: $h+k=2n$
			(9) $\bar{x},\bar{y},\bar{z}$	(10) $x,y,\bar{z}$	(11) $y,\bar{x},\bar{z}$	(12) $\bar{y},x,\bar{z}$	$0kl$: $k+l=2n$
			(13) $x,\bar{y},z$	(14) $\bar{x},y,z$	(15) $\bar{y},\bar{x},z$	(16) y,x,z	hhl: $l=2n$
							$00l$: $l=2n$
							$h00$: $h=2n$
							Special: as above, plus
16	n	$.m.$	$0,y,z$	$0,\bar{y},z$	$\bar{y},0,z$	$y,0,z$	no extra conditions
			$0,y,\bar{z}$	$0,\bar{y},\bar{z}$	$y,0,\bar{z}$	$\bar{y},0,\bar{z}$	
16	m	$..m$	x,x,z	$\bar{x},\bar{x},z$	$\bar{x},x,z$	$x,\bar{x},z$	no extra conditions
			$\bar{x},x,\bar{z}$	$x,\bar{x},\bar{z}$	$x,x,\bar{z}$	$\bar{x},\bar{x},\bar{z}$	
16	l	$m..$	$x,y,0$	$\bar{x},\bar{y},0$	$\bar{y},x,0$	$y,\bar{x},0$	no extra conditions
			$\bar{x},y,0$	$x,\bar{y},0$	$y,x,0$	$\bar{y},\bar{x},0$	
16	k	$..2$	$x,x+\frac{1}{2},\frac{1}{4}$	$\bar{x},\bar{x}+\frac{1}{2},\frac{1}{4}$	$\bar{x}+\frac{1}{2},x,\frac{1}{4}$	$x+\frac{1}{2},\bar{x},\frac{1}{4}$	hkl: $l=2n$
			$\bar{x},\bar{x}+\frac{1}{2},\frac{3}{4}$	$x,x+\frac{1}{2},\frac{3}{4}$	$x+\frac{1}{2},\bar{x},\frac{3}{4}$	$\bar{x}+\frac{1}{2},x,\frac{3}{4}$	
8	j	$m2m.$	$x,\frac{1}{2},0$	$\bar{x},\frac{1}{2},0$	$\frac{1}{2},x,0$	$\frac{1}{2},\bar{x},0$	no extra conditions
8	i	$m2m.$	$x,0,0$	$\bar{x},0,0$	$0,x,0$	$0,\bar{x},0$	no extra conditions
8	h	$m.2m$	$x,x,0$	$\bar{x},\bar{x},0$	$\bar{x},x,0$	$x,\bar{x},0$	no extra conditions
8	g	$2mm.$	$0,\frac{1}{2},z$	$\frac{1}{2},0,z$	$0,\frac{1}{2},\bar{z}$	$\frac{1}{2},0,\bar{z}$	hkl: $l=2n$
8	f	$..2/m$	$\frac{1}{4},\frac{1}{4},\frac{1}{4}$	$\frac{3}{4},\frac{3}{4},\frac{1}{4}$	$\frac{3}{4},\frac{1}{4},\frac{1}{4}$	$\frac{1}{4},\frac{3}{4},\frac{1}{4}$	hkl: $k,l=2n$
4	e	$4mm$	$0,0,z$	$0,0,\bar{z}$			no extra conditions
4	d	$\bar{4}m2$	$0,\frac{1}{2},\frac{1}{4}$	$\frac{1}{2},0,\frac{1}{4}$			hkl: $l=2n$
4	c	$mmm.$	$0,\frac{1}{2},0$	$\frac{1}{2},0,0$			hkl: $l=2n$
2	b	$4/mmm$	$0,0,\frac{1}{2}$				no extra conditions
2	a	$4/mmm$	$0,0,0$				no extra conditions

Symmetry of special projections

Along [001] $p4mm$
$\mathbf{a}'=\frac{1}{2}(\mathbf{a}-\mathbf{b})$ $\mathbf{b}'=\frac{1}{2}(\mathbf{a}+\mathbf{b})$
Origin at $0,0,z$

Along [100] $c2mm$
$\mathbf{a}'=\mathbf{b}$ $\mathbf{b}'=\mathbf{c}$
Origin at $x,0,0$

Along [110] $p2mm$
$\mathbf{a}'=\frac{1}{2}(-\mathbf{a}+\mathbf{b})$ $\mathbf{b}'=\frac{1}{2}\mathbf{c}$
Origin at $x,x,0$

$I4_1/amd$ $\quad D_{4h}^{19}$ $\quad 4/mmm$ $\quad$ Tetragonal

No. 141 $\quad I\ 4_1/a\ 2/m\ 2/d$ $\quad$ Patterson symmetry $I4/mmm$

ORIGIN CHOICE 1

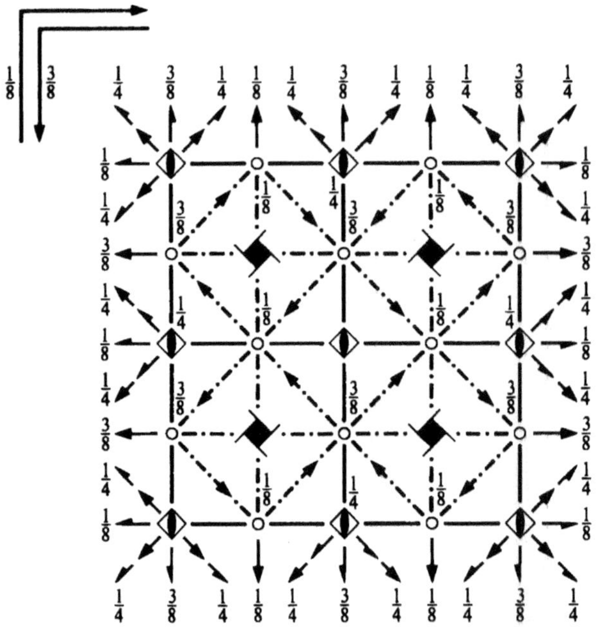

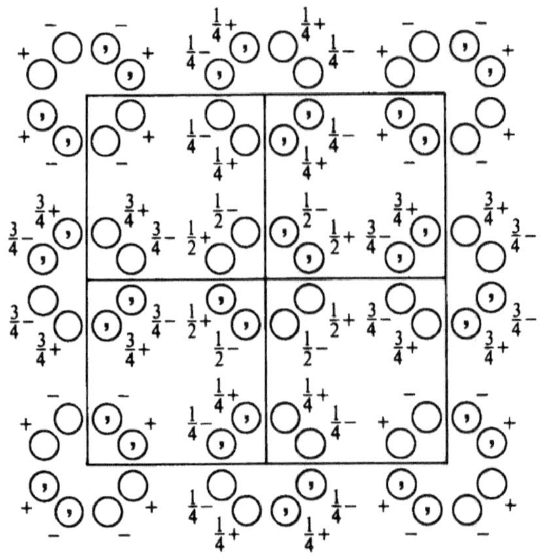

Origin at $\bar{4}m2$, at $0,\frac{1}{4},-\frac{1}{8}$ from centre $(2/m)$

Asymmetric unit $\quad 0\le x\le \frac{1}{2};\quad 0\le y\le \frac{1}{2};\quad 0\le z\le \frac{1}{8}$

Symmetry operations

For $(0,0,0)+$ set

(1) 1 $\quad$ (2) $2(0,0,\frac{1}{2})\ \ \frac{1}{4},\frac{1}{4},z$ $\quad$ (3) $4^+(0,0,\frac{1}{4})\ \ -\frac{1}{4},\frac{1}{4},z$ $\quad$ (4) $4^-(0,0,\frac{3}{4})\ \ \frac{1}{4},-\frac{1}{4},z$
(5) $2\ \ \frac{1}{4},y,\frac{3}{8}$ $\quad$ (6) $2\ \ x,\frac{1}{4},\frac{1}{8}$ $\quad$ (7) $2(\frac{1}{2},\frac{1}{2},0)\ \ x,x,\frac{1}{4}$ $\quad$ (8) $2\ \ x,\bar{x},0$
(9) $\bar{1}\ \ 0,\frac{1}{4},\frac{1}{8}$ $\quad$ (10) $a\ \ x,y,\frac{3}{8}$ $\quad$ (11) $\bar{4}^+\ \ 0,0,z;\ 0,0,0$ $\quad$ (12) $\bar{4}^-\ \ 0,\frac{1}{2},z;\ 0,\frac{1}{2},\frac{1}{4}$
(13) $n(\frac{1}{2},0,\frac{1}{2})\ \ x,\frac{1}{4},z$ $\quad$ (14) $m\ \ 0,y,z$ $\quad$ (15) $d(\frac{1}{4},-\frac{1}{4},\frac{3}{4})\ \ x+\frac{1}{4},\bar{x},z$ $\quad$ (16) $d(\frac{1}{4},\frac{1}{4},\frac{1}{4})\ \ x-\frac{1}{4},x,z$

For $(\frac{1}{2},\frac{1}{2},\frac{1}{2})+$ set

(1) $t(\frac{1}{2},\frac{1}{2},\frac{1}{2})$ $\quad$ (2) $2\ \ 0,0,z$ $\quad$ (3) $4^+(0,0,\frac{3}{4})\ \ \frac{1}{4},\frac{1}{4},z$ $\quad$ (4) $4^-(0,0,\frac{1}{4})\ \ \frac{1}{4},\frac{1}{4},z$
(5) $2(0,\frac{1}{2},0)\ \ 0,y,\frac{1}{8}$ $\quad$ (6) $2(\frac{1}{2},0,0)\ \ x,0,\frac{3}{8}$ $\quad$ (7) $2\ \ x,x,0$ $\quad$ (8) $2\ \ x,\bar{x}+\frac{1}{2},\frac{1}{4}$
(9) $\bar{1}\ \ \frac{1}{4},0,\frac{3}{8}$ $\quad$ (10) $b\ \ x,y,\frac{1}{8}$ $\quad$ (11) $\bar{4}^+\ \ \frac{1}{2},0,z;\ \frac{1}{2},0,\frac{1}{4}$ $\quad$ (12) $\bar{4}^-\ \ 0,0,z;\ 0,0,0$
(13) $m\ \ x,0,z$ $\quad$ (14) $n(0,\frac{1}{2},\frac{1}{2})\ \ \frac{1}{4},y,z$ $\quad$ (15) $d(-\frac{1}{4},\frac{1}{4},\frac{1}{4})\ \ x+\frac{1}{4},\bar{x},z$ $\quad$ (16) $d(\frac{1}{4},\frac{1}{4},\frac{3}{4})\ \ x+\frac{1}{4},x,z$

Generators selected (1); $t(1,0,0)$; $t(0,1,0)$; $t(0,0,1)$; $t(\frac{1}{2},\frac{1}{2},\frac{1}{2})$; (2); (3); (5); (9)

Positions

Multiplicity, Wyckoff letter, Site symmetry			Coordinates $(0,0,0)+\ \ (\frac{1}{2},\frac{1}{2},\frac{1}{2})+$				Reflection conditions
							General:
32	i	1	(1) x,y,z	(2) $\bar{x}+\frac{1}{2},\bar{y}+\frac{1}{2},z+\frac{1}{2}$	(3) $\bar{y},x+\frac{1}{2},z+\frac{1}{4}$	(4) $y+\frac{1}{2},\bar{x},z+\frac{3}{4}$	hkl: $h+k+l=2n$
			(5) $\bar{x}+\frac{1}{2},y,\bar{z}+\frac{3}{4}$	(6) $x,\bar{y}+\frac{1}{2},\bar{z}+\frac{1}{4}$	(7) $y+\frac{1}{2},x+\frac{1}{2},\bar{z}+\frac{1}{2}$	(8) $\bar{y},\bar{x},\bar{z}$	$hk0$: $h,k=2n$
			(9) $\bar{x},\bar{y}+\frac{1}{2},\bar{z}+\frac{1}{4}$	(10) $x+\frac{1}{2},y,\bar{z}+\frac{3}{4}$	(11) $y,\bar{x},\bar{z}$	(12) $\bar{y}+\frac{1}{2},x+\frac{1}{2},\bar{z}+\frac{1}{2}$	$0kl$: $k+l=2n$
			(13) $x+\frac{1}{2},\bar{y}+\frac{1}{2},z+\frac{1}{2}$	(14) $\bar{x},y,z$	(15) $\bar{y}+\frac{1}{2},\bar{x},z+\frac{3}{4}$	(16) $y,x+\frac{1}{2},z+\frac{1}{4}$	hhl: $2h+l=4n$ $00l$: $l=4n$ $h00$: $h=2n$ $h\bar{h}0$: $h=2n$
							Special: as above, plus
16	h	$.m.$	$0,y,z$	$\frac{1}{2},\bar{y}+\frac{1}{2},z+\frac{1}{2}$	$\bar{y},\frac{1}{2},z+\frac{1}{4}$	$y+\frac{1}{2},0,z+\frac{3}{4}$	no extra conditions
			$\frac{1}{2},y,\bar{z}+\frac{3}{4}$	$0,\bar{y}+\frac{1}{2},\bar{z}+\frac{1}{4}$	$y+\frac{1}{2},\frac{1}{2},\bar{z}+\frac{1}{2}$	$\bar{y},0,\bar{z}$	
16	g	$..2$	$x,x,0$	$\bar{x}+\frac{1}{2},\bar{x}+\frac{1}{2},\frac{1}{2}$	$\bar{x},x+\frac{1}{2},\frac{1}{4}$	$x+\frac{1}{2},\bar{x},\frac{3}{4}$	hkl: $l=2n+1$
			$\bar{x},\bar{x}+\frac{1}{2},\frac{1}{4}$	$x+\frac{1}{2},x,\frac{3}{4}$	$x,\bar{x},0$	$\bar{x}+\frac{1}{2},x+\frac{1}{2},\frac{1}{2}$	or $2h+l=4n$
16	f	$.2.$	$x,\frac{1}{4},\frac{1}{8}$	$\bar{x}+\frac{1}{2},\frac{1}{4},\frac{5}{8}$	$\frac{3}{4},x+\frac{1}{2},\frac{3}{8}$	$\frac{3}{4},\bar{x},\frac{7}{8}$	hkl: $l=2n+1$
			$\bar{x},\frac{1}{4},\frac{1}{8}$	$x+\frac{1}{2},\frac{1}{4},\frac{5}{8}$	$\frac{1}{4},\bar{x},\frac{7}{8}$	$\frac{1}{4},x+\frac{1}{2},\frac{3}{8}$	or $h=2n$
8	e	$2mm.$	$0,0,z$	$0,\frac{1}{2},z+\frac{1}{4}$	$\frac{1}{2},0,\bar{z}+\frac{3}{4}$	$\frac{1}{2},\frac{1}{2},\bar{z}+\frac{1}{2}$	hkl: $l=2n+1$ or $2h+l=4n$
8	d	$.2/m.$	$0,\frac{1}{4},\frac{5}{8}$	$\frac{1}{2},\frac{1}{4},\frac{1}{8}$	$\frac{3}{4},\frac{1}{2},\frac{7}{8}$	$\frac{3}{4},0,\frac{3}{8}$ }	hkl: $l=2n+1$ or $h,k=2n,\ h+k+l=4n$
8	c	$.2/m.$	$0,\frac{1}{4},\frac{1}{8}$	$\frac{1}{2},\frac{1}{4},\frac{5}{8}$	$\frac{3}{4},\frac{1}{2},\frac{3}{8}$	$\frac{3}{4},0,\frac{7}{8}$ }	
4	b	$\bar{4}m2$	$0,0,\frac{1}{2}$	$0,\frac{1}{2},\frac{3}{4}$ }			hkl: $l=2n+1$ or $2h+l=4n$
4	a	$\bar{4}m2$	$0,0,0$	$0,\frac{1}{2},\frac{1}{4}$ }			

Symmetry of special projections

Along [001] $p4mm$
$\mathbf{a}'=\frac{1}{2}\mathbf{a}$ $\mathbf{b}'=\frac{1}{2}\mathbf{b}$
Origin at $0,0,z$

Along [100] $c2mm$
$\mathbf{a}'=\mathbf{b}$ $\mathbf{b}'=\mathbf{c}$
Origin at $x,0,\frac{3}{8}$

Along [110] $c2mm$
$\mathbf{a}'=\frac{1}{2}(-\mathbf{a}+\mathbf{b})$ $\mathbf{b}'=\frac{1}{2}\mathbf{c}$
Origin at $x,x,0$

$I4_1/amd$ D_{4h}^{19} $4/mmm$ Tetragonal

No. 141 $I\ 4_1/a\ 2/m\ 2/d$ Patterson symmetry $I4/mmm$

ORIGIN CHOICE 2

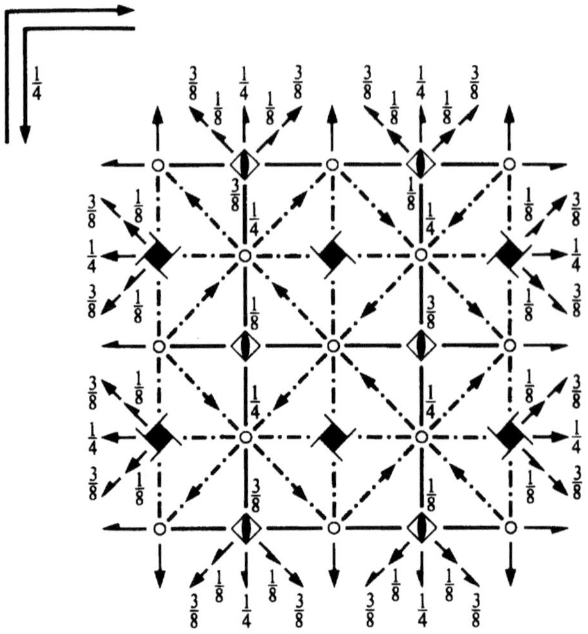

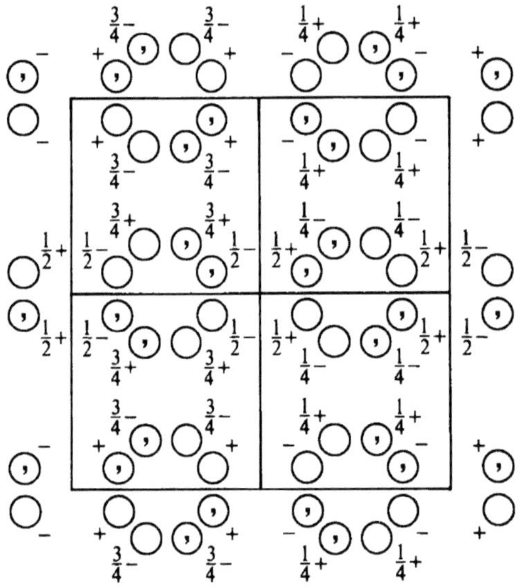

Origin at centre $(2/m)$ at $b\,(2/m, 2_1/n)d$, at $0, -\frac{1}{4}, \frac{1}{8}$ from $\bar{4}m2$

Asymmetric unit $0 \le x \le \frac{1}{2};\quad -\frac{1}{4} \le y \le \frac{1}{4};\quad 0 \le z \le \frac{1}{8}$

Symmetry operations

For $(0,0,0)+$ set

(1) 1	(2) $2(0,0,\frac{1}{2})\ \ \frac{1}{4},0,z$	(3) $4^+(0,0,\frac{1}{4})\ \ -\frac{1}{4},\frac{1}{2},z$	(4) $4^-(0,0,\frac{3}{4})\ \ \frac{1}{4},0,z$
(5) $2\ \ \frac{1}{4},y,\frac{1}{4}$	(6) $2\ \ x,0,0$	(7) $2(\frac{1}{2},\frac{1}{2},0)\ \ x,x+\frac{1}{4},\frac{1}{8}$	(8) $2\ \ x,\bar{x}+\frac{1}{4},\frac{3}{8}$
(9) $\bar{1}\ \ 0,0,0$	(10) $a\ \ x,y,\frac{1}{4}$	(11) $\bar{4}^+\ \ \frac{1}{2},-\frac{1}{4},z;\ \ \frac{1}{2},-\frac{1}{4},\frac{3}{8}$	(12) $\bar{4}^-\ \ 0,\frac{3}{4},z;\ \ 0,\frac{3}{4},\frac{1}{8}$
(13) $n(\frac{1}{2},0,\frac{1}{2})\ \ x,0,z$	(14) $m\ \ 0,y,z$	(15) $d(\frac{1}{4},-\frac{1}{4},\frac{3}{4})\ \ x+\frac{1}{2},\bar{x},z$	(16) $d(\frac{3}{4},\frac{3}{4},\frac{1}{4})\ \ x,x,z$

For $(\frac{1}{2},\frac{1}{2},\frac{1}{2})+$ set

(1) $t(\frac{1}{2},\frac{1}{2},\frac{1}{2})$	(2) $2\ \ 0,\frac{1}{4},z$	(3) $4^+(0,0,\frac{3}{4})\ \ \frac{1}{4},\frac{1}{2},z$	(4) $4^-(0,0,\frac{1}{4})\ \ \frac{3}{4},0,z$
(5) $2(0,\frac{1}{2},0)\ \ 0,y,0$	(6) $2(\frac{1}{2},0,0)\ \ x,\frac{1}{4},\frac{1}{4}$	(7) $2(\frac{1}{2},\frac{1}{2},0)\ \ x,x-\frac{1}{4},\frac{3}{8}$	(8) $2\ \ x,\bar{x}+\frac{3}{4},\frac{1}{8}$
(9) $\bar{1}\ \ \frac{1}{4},\frac{1}{4},\frac{1}{4}$	(10) $b\ \ x,y,0$	(11) $\bar{4}^+\ \ \frac{1}{2},\frac{1}{4},z;\ \ \frac{1}{2},\frac{1}{4},\frac{1}{8}$	(12) $\bar{4}^-\ \ 0,\frac{1}{4},z;\ \ 0,\frac{1}{4},\frac{3}{8}$
(13) $m\ \ x,\frac{1}{4},z$	(14) $n(0,\frac{1}{2},\frac{1}{2})\ \ \frac{1}{4},y,z$	(15) $d(-\frac{1}{4},\frac{1}{4},\frac{1}{4})\ \ x+\frac{1}{2},\bar{x},z$	(16) $d(\frac{1}{4},\frac{1}{4},\frac{3}{4})\ \ x,x,z$

Generators selected (1); $t(1,0,0)$; $t(0,1,0)$; $t(0,0,1)$; $t(\frac{1}{2},\frac{1}{2},\frac{1}{2})$; (2); (3); (5); (9)

Positions

Multiplicity, Wyckoff letter, Site symmetry — Coordinates $(0,0,0)+\quad(\frac{1}{2},\frac{1}{2},\frac{1}{2})+$ — Reflection conditions

Multiplicity	Wyckoff letter	Site symmetry	Coordinates				Reflection conditions
							General:
32	i	1	(1) x,y,z	(2) $\bar{x}+\frac{1}{2},\bar{y},z+\frac{1}{2}$	(3) $\bar{y}+\frac{1}{4},x+\frac{3}{4},z+\frac{1}{4}$	(4) $y+\frac{1}{4},\bar{x}+\frac{1}{4},z+\frac{3}{4}$	hkl: $h+k+l=2n$
			(5) $\bar{x}+\frac{1}{2},y,\bar{z}+\frac{1}{2}$	(6) $x,\bar{y},\bar{z}$	(7) $y+\frac{1}{4},x+\frac{3}{4},\bar{z}+\frac{1}{4}$	(8) $\bar{y}+\frac{1}{4},\bar{x}+\frac{1}{4},\bar{z}+\frac{3}{4}$	$hk0$: $h,k=2n$
			(9) $\bar{x},\bar{y},\bar{z}$	(10) $x+\frac{1}{2},y,\bar{z}+\frac{1}{2}$	(11) $y+\frac{3}{4},\bar{x}+\frac{1}{4},\bar{z}+\frac{3}{4}$	(12) $\bar{y}+\frac{3}{4},x+\frac{3}{4},\bar{z}+\frac{1}{4}$	$0kl$: $k+l=2n$
			(13) $x+\frac{1}{2},\bar{y},z+\frac{1}{2}$	(14) $\bar{x},y,z$	(15) $\bar{y}+\frac{3}{4},\bar{x}+\frac{1}{4},z+\frac{3}{4}$	(16) $y+\frac{3}{4},x+\frac{3}{4},z+\frac{1}{4}$	hhl: $2h+l=4n$
							$00l$: $l=4n$
							$h00$: $h=2n$
							$h\bar{h}0$: $h=2n$
							Special: as above, plus
16	h	$.m.$	$0,y,z$	$\frac{1}{2},\bar{y},z+\frac{1}{2}$	$\bar{y}+\frac{1}{4},\frac{3}{4},z+\frac{1}{4}$	$y+\frac{1}{4},\frac{1}{4},z+\frac{3}{4}$	no extra conditions
			$\frac{1}{2},y,\bar{z}+\frac{1}{2}$	$0,\bar{y},\bar{z}$	$y+\frac{1}{4},\frac{3}{4},\bar{z}+\frac{1}{4}$	$\bar{y}+\frac{1}{4},\frac{1}{4},\bar{z}+\frac{3}{4}$	
16	g	$..2$	$x,x+\frac{1}{4},\frac{7}{8}$	$\bar{x}+\frac{1}{2},\bar{x}+\frac{3}{4},\frac{3}{8}$	$\bar{x},x+\frac{3}{4},\frac{1}{8}$	$x+\frac{1}{2},\bar{x}+\frac{1}{4},\frac{5}{8}$	hkl: $l=2n+1$
			$\bar{x},\bar{x}+\frac{3}{4},\frac{1}{8}$	$x+\frac{1}{2},x+\frac{1}{4},\frac{5}{8}$	$x,\bar{x}+\frac{1}{4},\frac{7}{8}$	$\bar{x}+\frac{1}{2},x+\frac{3}{4},\frac{3}{8}$	or $2h+l=4n$
16	f	$.2.$	$x,0,0$	$\bar{x}+\frac{1}{2},0,\frac{1}{2}$	$\frac{1}{4},x+\frac{3}{4},\frac{1}{4}$	$\frac{1}{4},\bar{x}+\frac{1}{4},\frac{3}{4}$	hkl: $l=2n+1$
			$\bar{x},0,0$	$x+\frac{1}{2},0,\frac{1}{2}$	$\frac{3}{4},\bar{x}+\frac{1}{4},\frac{3}{4}$	$\frac{3}{4},x+\frac{3}{4},\frac{1}{4}$	or $h=2n$
8	e	$2mm.$	$0,\frac{1}{4},z$	$0,\frac{3}{4},z+\frac{1}{4}$	$\frac{1}{2},\frac{1}{4},\bar{z}+\frac{1}{2}$	$\frac{1}{2},\frac{3}{4},\bar{z}+\frac{1}{4}$	hkl: $l=2n+1$ or $2h+l=4n$
8	d	$.2/m.$	$0,0,\frac{1}{2}$	$\frac{1}{2},0,0$	$\frac{1}{4},\frac{3}{4},\frac{3}{4}$	$\frac{1}{4},\frac{1}{4},\frac{1}{4}$	} hkl: $l=2n+1$ or $h,k=2n,\ h+k+l=4n$
8	c	$.2/m.$	$0,0,0$	$\frac{1}{2},0,\frac{1}{2}$	$\frac{1}{4},\frac{3}{4},\frac{1}{4}$	$\frac{1}{4},\frac{1}{4},\frac{3}{4}$	}
4	b	$\bar{4}m2$	$0,\frac{1}{4},\frac{3}{8}$	$0,\frac{3}{4},\frac{5}{8}$			} hkl: $l=2n+1$ or $2h+l=4n$
4	a	$\bar{4}m2$	$0,\frac{3}{4},\frac{1}{8}$	$\frac{1}{2},\frac{3}{4},\frac{3}{8}$			}

Symmetry of special projections

Along [001] $p4mm$
$\mathbf{a}'=\frac{1}{2}\mathbf{a}$ $\quad\mathbf{b}'=\frac{1}{2}\mathbf{b}$
Origin at $\frac{1}{4},0,z$

Along [100] $c2mm$
$\mathbf{a}'=\mathbf{b}$ $\quad\mathbf{b}'=\mathbf{c}$
Origin at $x,\frac{1}{4},\frac{1}{4}$

Along [110] $c2mm$
$\mathbf{a}'=\frac{1}{2}(-\mathbf{a}+\mathbf{b})$ $\quad\mathbf{b}'=\frac{1}{2}\mathbf{c}$
Origin at $x,x+\frac{1}{4},\frac{1}{8}$

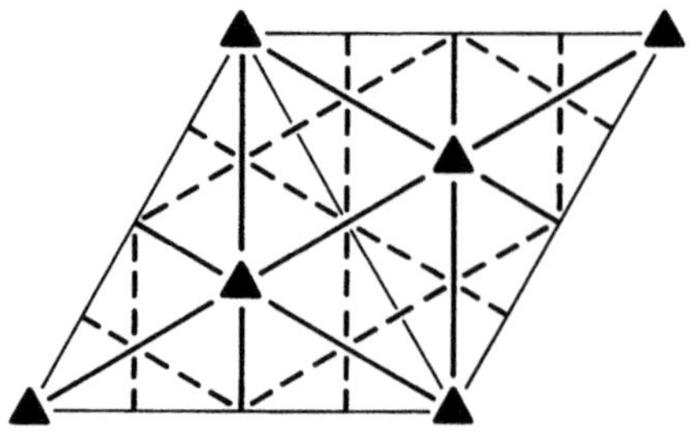

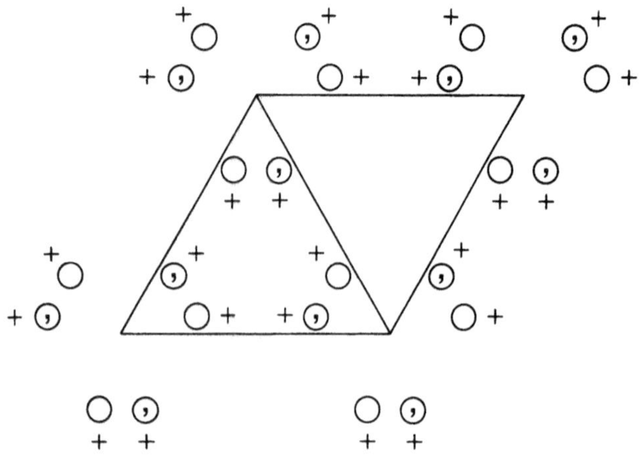

Origin on $3m1$

Asymmetric unit $0 \leq x \leq \frac{2}{3}$; $0 \leq y \leq \frac{2}{3}$; $0 \leq z \leq 1$; $x \leq 2y$; $y \leq \min(1-x, 2x)$

Vertices $0,0,0$ $\frac{2}{3},\frac{1}{3},0$ $\frac{1}{3},\frac{2}{3},0$

$0,0,1$ $\frac{2}{3},\frac{1}{3},1$ $\frac{1}{3},\frac{2}{3},1$

Symmetry operations

(1) 1 (2) 3^+ $0,0,z$ (3) 3^- $0,0,z$

(4) m $x,\bar{x},z$ (5) m $x,2x,z$ (6) m $2x,x,z$

Generators selected (1); $t(1,0,0)$; $t(0,1,0)$; $t(0,0,1)$; (2); (4)

Positions

Multiplicity, Wyckoff letter, Site symmetry			Coordinates			Reflection conditions
						General:
6	e	1	(1) x,y,z	(2) $\bar{y},x-y,z$	(3) $\bar{x}+y,\bar{x},z$	no conditions
			(4) $\bar{y},\bar{x},z$	(5) $\bar{x}+y,y,z$	(6) $x,x-y,z$	
						Special: no extra conditions
3	d	$.m.$	$x,\bar{x},z$	$x,2x,z$	$2\bar{x},\bar{x},z$	
1	c	$3m.$	$\frac{2}{3},\frac{1}{3},z$			
1	b	$3m.$	$\frac{1}{3},\frac{2}{3},z$			
1	a	$3m.$	$0,0,z$			

Symmetry of special projections

Along [001] $p3m1$
$\mathbf{a}' = \mathbf{a}$ $\mathbf{b}' = \mathbf{b}$
Origin at $0,0,z$

Along [100] $p1$
$\mathbf{a}' = \frac{1}{2}(\mathbf{a}+2\mathbf{b})$ $\mathbf{b}' = \mathbf{c}$
Origin at $x,0,0$

Along [210] $p1m1$
$\mathbf{a}' = \frac{1}{2}\mathbf{b}$ $\mathbf{b}' = \mathbf{c}$
Origin at $x,\frac{1}{2}x,0$

Patterson symmetry $P\bar{3}\,1\,m$ $P\,3\,1\,m$ No. 157

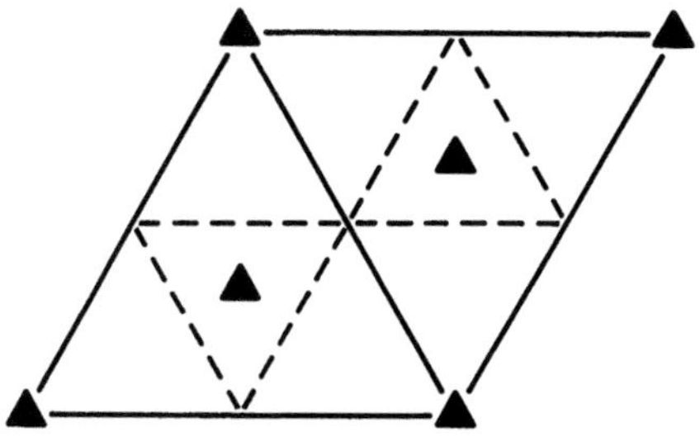

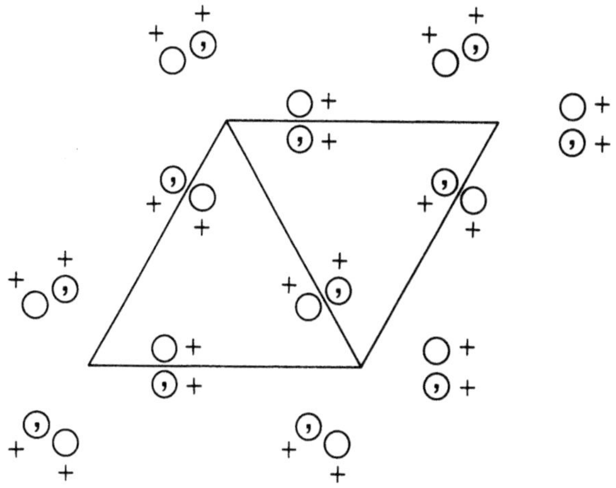

Origin on $3\,1\,m$

Asymmetric unit $0 \le x \le \frac{2}{3};\quad 0 \le y \le \frac{1}{2};\quad 0 \le z \le 1;\quad x \le (y+1)/2;\quad y \le \min(1-x,x)$

Vertices $0,0,0\quad \frac{1}{2},0,0\quad \frac{2}{3},\frac{1}{3},0\quad \frac{1}{2},\frac{1}{2},0$

$0,0,1\quad \frac{1}{2},0,1\quad \frac{2}{3},\frac{1}{3},1\quad \frac{1}{2},\frac{1}{2},1$

Symmetry operations

(1) 1	(2) 3^+ $0,0,z$	(3) 3^- $0,0,z$
(4) m x,x,z	(5) m $x,0,z$	(6) m $0,y,z$

Generators selected (1); $t(1,0,0)$; $t(0,1,0)$; $t(0,0,1)$; (2); (4)

Positions

Multiplicity, Wyckoff letter, Site symmetry			Coordinates			Reflection conditions
						General:
6	d	1	(1) x,y,z (4) y,x,z	(2) $\bar{y},x-y,z$ (5) $x-y,\bar{y},z$	(3) $\bar{x}+y,\bar{x},z$ (6) $\bar{x},\bar{x}+y,z$	no conditions
						Special: no extra conditions
3	c	$..m$	$x,0,z$	$0,x,z$	$\bar{x},\bar{x},z$	
2	b	$3..$	$\frac{1}{3},\frac{2}{3},z$	$\frac{2}{3},\frac{1}{3},z$		
1	a	$3.m$	$0,0,z$			

Symmetry of special projections

Along [001] $p\,3\,1\,m$
$\mathbf{a}' = \mathbf{a}$ $\mathbf{b}' = \mathbf{b}$
Origin at $0,0,z$

Along [100] $p\,1\,m\,1$
$\mathbf{a}' = \frac{1}{2}(\mathbf{a}+2\mathbf{b})$ $\mathbf{b}' = \mathbf{c}$
Origin at $x,0,0$

Along [210] $p\,1$
$\mathbf{a}' = \frac{1}{2}\mathbf{b}$ $\mathbf{b}' = \mathbf{c}$
Origin at $x,\frac{1}{2}x,0$

$R3m$ $\quad C_{3v}^5$ $\quad 3m$ $\quad$ Trigonal

No. 160 $\quad R3m$ $\quad$ Patterson symmetry $R\bar{3}m$

HEXAGONAL AXES

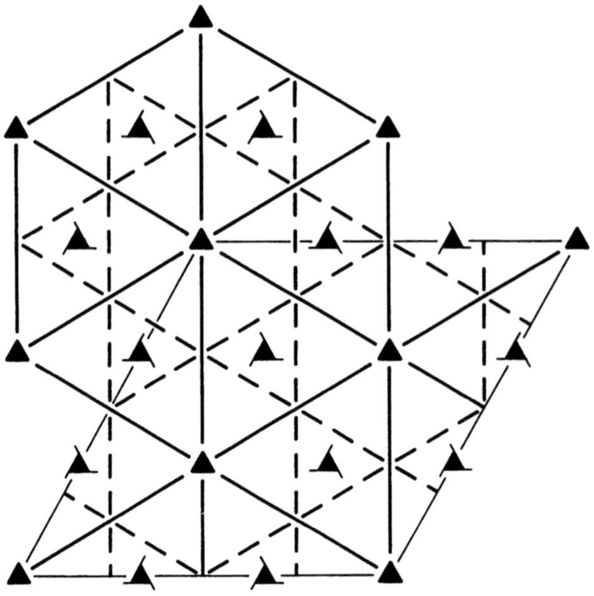

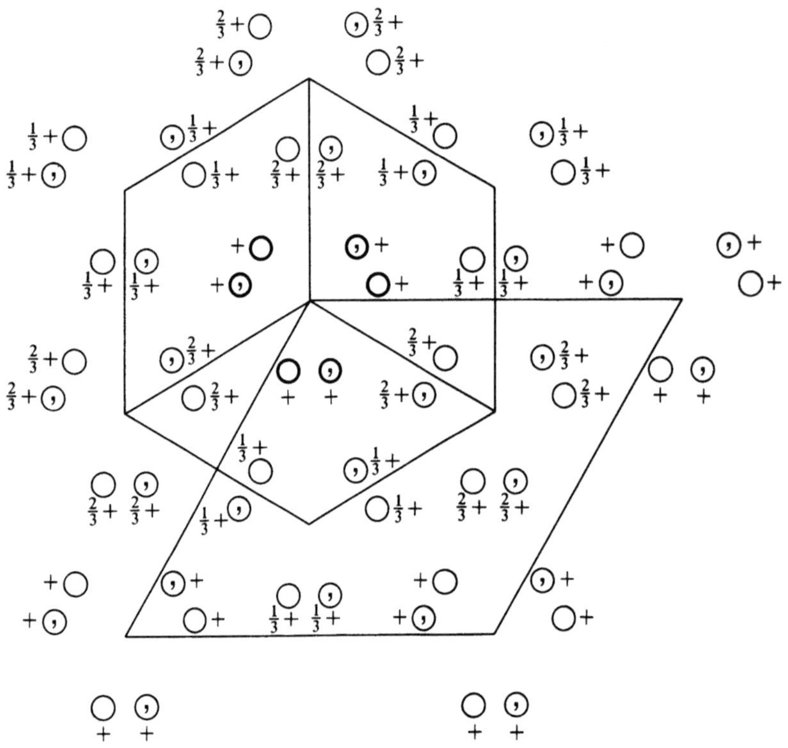

Origin on $3m$

Asymmetric unit $\quad 0 \le x \le \frac{2}{3};\quad 0 \le y \le \frac{2}{3};\quad 0 \le z \le \frac{1}{3};\quad x \le 2y;\quad y \le \min(1-x, 2x)$

Vertices $\quad 0,0,0 \quad \frac{2}{3},\frac{1}{3},0 \quad \frac{1}{3},\frac{2}{3},0$

$0,0,\frac{1}{3} \quad \frac{2}{3},\frac{1}{3},\frac{1}{3} \quad \frac{1}{3},\frac{2}{3},\frac{1}{3}$

Symmetry operations

For $(0,0,0)+$ set

(1) 1 (2) $3^{+}\ 0,0,z$ (3) $3^{-}\ 0,0,z$
(4) $m\ \ x,\bar{x},z$ (5) $m\ \ x,2x,z$ (6) $m\ \ 2x,x,z$

For $(\frac{2}{3},\frac{1}{3},\frac{1}{3})+$ set

(1) $t(\frac{2}{3},\frac{1}{3},\frac{1}{3})$ (2) $3^{+}(0,0,\frac{1}{3})\ \ \frac{1}{3},\frac{1}{3},z$ (3) $3^{-}(0,0,\frac{1}{3})\ \ \frac{1}{3},0,z$
(4) $g(\frac{1}{6},-\frac{1}{6},\frac{1}{3})\ \ x+\frac{1}{2},\bar{x},z$ (5) $g(\frac{1}{6},\frac{1}{3},\frac{1}{3})\ \ x+\frac{1}{4},2x,z$ (6) $g(\frac{2}{3},\frac{1}{3},\frac{1}{3})\ \ 2x,x,z$

For $(\frac{1}{3},\frac{2}{3},\frac{2}{3})+$ set

(1) $t(\frac{1}{3},\frac{2}{3},\frac{2}{3})$ (2) $3^{+}(0,0,\frac{2}{3})\ \ 0,\frac{1}{3},z$ (3) $3^{-}(0,0,\frac{2}{3})\ \ \frac{1}{3},\frac{1}{3},z$
(4) $g(-\frac{1}{6},\frac{1}{6},\frac{2}{3})\ \ x+\frac{1}{2},\bar{x},z$ (5) $g(\frac{1}{3},\frac{2}{3},\frac{2}{3})\ \ x,2x,z$ (6) $g(\frac{1}{3},\frac{1}{6},\frac{2}{3})\ \ 2x-\frac{1}{2},x,z$

Generators selected (1); $t(1,0,0)$; $t(0,1,0)$; $t(0,0,1)$; $t(\frac{2}{3},\frac{1}{3},\frac{1}{3})$; (2); (4)

Positions

Multiplicity, Wyckoff letter, Site symmetry			Coordinates $(0,0,0)+\ \ (\frac{2}{3},\frac{1}{3},\frac{1}{3})+\ \ (\frac{1}{3},\frac{2}{3},\frac{2}{3})+$			Reflection conditions
						General:
18	c	1	(1) x,y,z	(2) $\bar{y},x-y,z$	(3) $\bar{x}+y,\bar{x},z$	$hkil$: $-h+k+l=3n$
			(4) $\bar{y},\bar{x},z$	(5) $\bar{x}+y,y,z$	(6) $x,x-y,z$	$hki0$: $-h+k=3n$
						$hh\overline{2h}l$: $l=3n$
						$h\bar{h}0l$: $h+l=3n$
						$000l$: $l=3n$
						$h\bar{h}00$: $h=3n$
						Special: no extra conditions
9	b	$.m$	$x,\bar{x},z$	$x,2x,z$	$2\bar{x},\bar{x},z$	
3	a	$3m$	$0,0,z$			

Symmetry of special projections

Along [001] $p31m$
$\mathbf{a}'=\frac{1}{3}(2\mathbf{a}+\mathbf{b})$ $\mathbf{b}'=\frac{1}{3}(-\mathbf{a}+\mathbf{b})$
Origin at $0,0,z$

Along [100] $p1$
$\mathbf{a}'=\frac{1}{2}(\mathbf{a}+2\mathbf{b})$ $\mathbf{b}'=\frac{1}{3}(-\mathbf{a}-2\mathbf{b}+\mathbf{c})$
Origin at $x,0,0$

Along [210] $p1m1$
$\mathbf{a}'=\frac{1}{2}\mathbf{b}$ $\mathbf{b}'=\frac{1}{3}\mathbf{c}$
Origin at $x,\frac{1}{2}x,0$

$R3m$	C_{3v}^{5}	$3m$	Trigonal
No. 160	$R3m$		Patterson symmetry $R\bar{3}m$

RHOMBOHEDRAL AXES

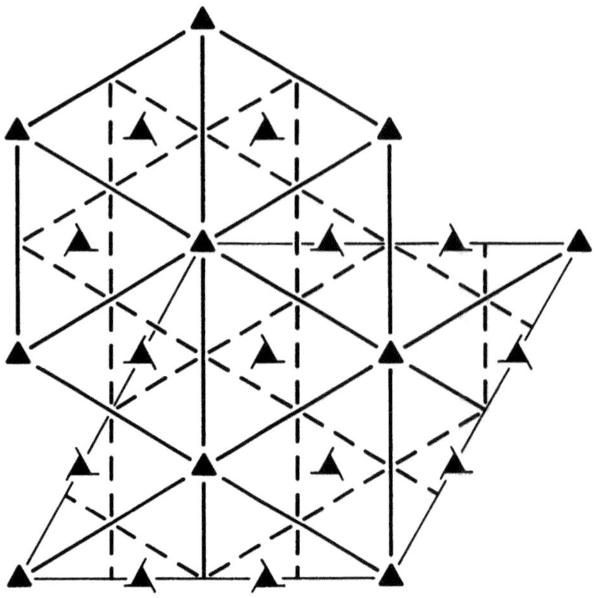

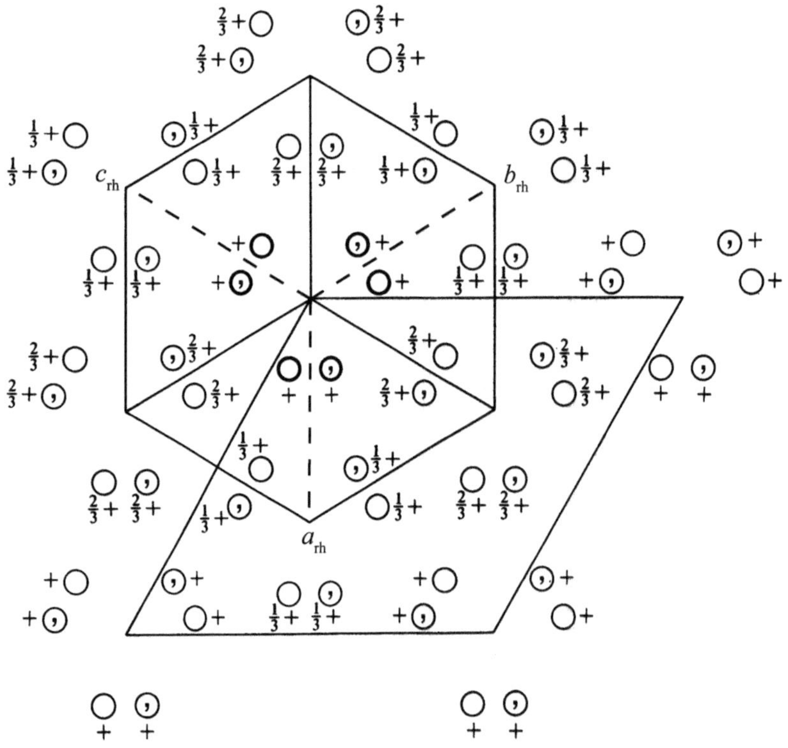

Heights refer to hexagonal axes

Origin on $3m$

Asymmetric unit $0 \le x \le 1$; $0 \le y \le 1$; $0 \le z \le 1$; $y \le x$; $z \le y$

Vertices 0,0,0 1,0,0 1,1,0 1,1,1

Symmetry operations

(1) 1	(2) 3^{+} x,x,x	(3) 3^{-} x,x,x
(4) m x,y,x	(5) m x,x,z	(6) m x,y,y

Generators selected (1); $t(1,0,0)$; $t(0,1,0)$; $t(0,0,1)$; (2); (4)

Positions

Multiplicity, Wyckoff letter, Site symmetry			Coordinates			Reflection conditions
						General:
6	c	1	(1) x,y,z (4) z,y,x	(2) z,x,y (5) y,x,z	(3) y,z,x (6) x,z,y	no conditions
						Special: no extra conditions
3	b	$.m$	x,y,x	x,x,y	y,x,x	
1	a	$3m$	x,x,x			

Symmetry of special projections

Along [111] $p31m$
$\mathbf{a}' = \frac{1}{3}(2\mathbf{a}-\mathbf{b}-\mathbf{c})$ $\mathbf{b}' = \frac{1}{3}(-\mathbf{a}+2\mathbf{b}-\mathbf{c})$
Origin at x,x,x

Along [$1\bar{1}0$] $p1$
$\mathbf{a}' = \frac{1}{2}(\mathbf{a}+\mathbf{b}-2\mathbf{c})$ $\mathbf{b}' = \mathbf{c}$
Origin at $x,\bar{x},0$

Along [$2\bar{1}\bar{1}$] $p1m1$
$\mathbf{a}' = \frac{1}{2}(\mathbf{b}-\mathbf{c})$ $\mathbf{b}' = \frac{1}{3}(\mathbf{a}+\mathbf{b}+\mathbf{c})$
Origin at $2x,\bar{x},\bar{x}$

$P6_3/m$ C_{6h}^2 $6/m$ Hexagonal

No. 176 $P6_3/m$ Patterson symmetry $P6/m$

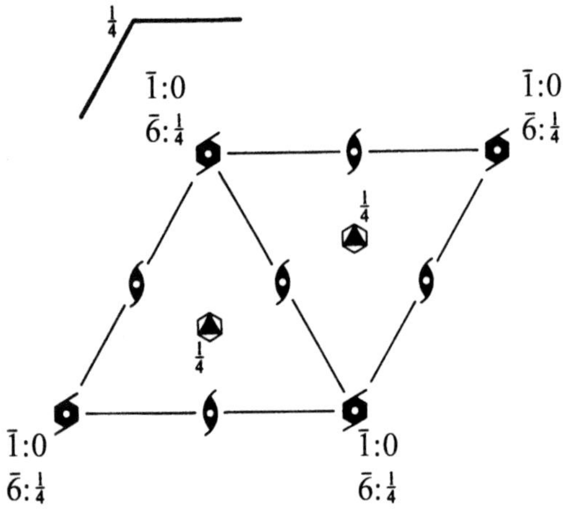

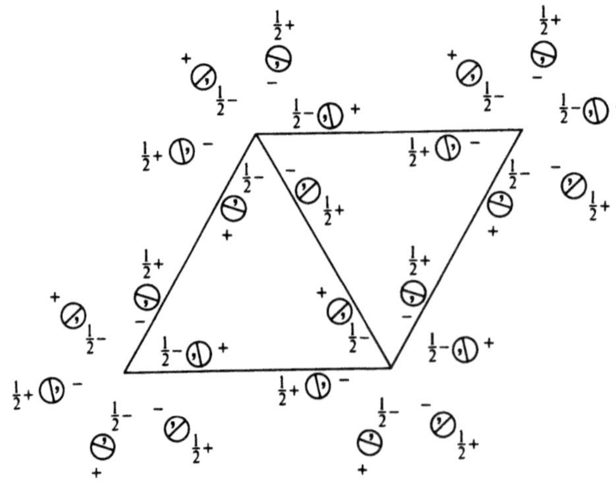

Origin at centre ($\bar{3}$) on 6_3

Asymmetric unit $0 \le x \le \frac{2}{3}$; $0 \le y \le \frac{2}{3}$; $0 \le z \le \frac{1}{4}$; $x \le (1+y)/2$; $y \le \min(1-x,(1+x)/2)$

Vertices $0,0,0$ $\frac{1}{2},0,0$ $\frac{2}{3},\frac{1}{3},0$ $\frac{1}{3},\frac{2}{3},0$ $0,\frac{1}{2},0$

$0,0,\frac{1}{4}$ $\frac{1}{2},0,\frac{1}{4}$ $\frac{2}{3},\frac{1}{3},\frac{1}{4}$ $\frac{1}{3},\frac{2}{3},\frac{1}{4}$ $0,\frac{1}{2},\frac{1}{4}$

Symmetry operations

(1) 1	(2) 3^+ $0,0,z$	(3) 3^- $0,0,z$
(4) $2(0,0,\frac{1}{2})$ $0,0,z$	(5) $6^-(0,0,\frac{1}{2})$ $0,0,z$	(6) $6^+(0,0,\frac{1}{2})$ $0,0,z$
(7) $\bar{1}$ $0,0,0$	(8) $\bar{3}^+$ $0,0,z$; $0,0,0$	(9) $\bar{3}^-$ $0,0,z$; $0,0,0$
(10) m $x,y,\frac{1}{4}$	(11) $\bar{6}^-$ $0,0,z$; $0,0,\frac{1}{4}$	(12) $\bar{6}^+$ $0,0,z$; $0,0,\frac{1}{4}$

Generators selected (1); $t(1,0,0)$; $t(0,1,0)$; $t(0,0,1)$; (2); (4); (7)

Positions

Multiplicity, Wyckoff letter, Site symmetry			Coordinates						Reflection conditions
									General:
12	i	1	(1) x,y,z (4) $\bar{x},\bar{y},z+\frac{1}{2}$ (7) $\bar{x},\bar{y},\bar{z}$ (10) $x,y,\bar{z}+\frac{1}{2}$	(2) $\bar{y},x-y,z$ (5) $y,\bar{x}+y,z+\frac{1}{2}$ (8) $y,\bar{x}+y,\bar{z}$ (11) $\bar{y},x-y,\bar{z}+\frac{1}{2}$	(3) $\bar{x}+y,\bar{x},z$ (6) $x-y,x,z+\frac{1}{2}$ (9) $x-y,x,\bar{z}$ (12) $\bar{x}+y,\bar{x},\bar{z}+\frac{1}{2}$				$000l$: $l=2n$
									Special: as above, plus
6	h	$m\,.\,.$	$x,y,\frac{1}{4}$	$\bar{y},x-y,\frac{1}{4}$	$\bar{x}+y,\bar{x},\frac{1}{4}$	$\bar{x},\bar{y},\frac{3}{4}$	$y,\bar{x}+y,\frac{3}{4}$	$x-y,x,\frac{3}{4}$	no extra conditions
6	g	$\bar{1}$	$\frac{1}{2},0,0$	$0,\frac{1}{2},0$	$\frac{1}{2},\frac{1}{2},0$	$\frac{1}{2},0,\frac{1}{2}$	$0,\frac{1}{2},\frac{1}{2}$	$\frac{1}{2},\frac{1}{2},\frac{1}{2}$	$hkil$: $l=2n$
4	f	$3\,.\,.$	$\frac{1}{3},\frac{2}{3},z$	$\frac{2}{3},\frac{1}{3},z+\frac{1}{2}$	$\frac{2}{3},\frac{1}{3},\bar{z}$	$\frac{1}{3},\frac{2}{3},\bar{z}+\frac{1}{2}$			$hkil$: $l=2n$ or $h-k=3n+1$ or $h-k=3n+2$
4	e	$3\,.\,.$	$0,0,z$	$0,0,z+\frac{1}{2}$	$0,0,\bar{z}$	$0,0,\bar{z}+\frac{1}{2}$			$hkil$: $l=2n$
2	d	$\bar{6}\,.\,.$	$\frac{2}{3},\frac{1}{3},\frac{1}{4}$	$\frac{1}{3},\frac{2}{3},\frac{3}{4}$					$hkil$: $l=2n$ or $h-k=3n+1$ or $h-k=3n+2$
2	c	$\bar{6}\,.\,.$	$\frac{1}{3},\frac{2}{3},\frac{1}{4}$	$\frac{2}{3},\frac{1}{3},\frac{3}{4}$					$hkil$: $l=2n$ or $h-k=3n+1$ or $h-k=3n+2$
2	b	$\bar{3}\,.\,.$	$0,0,0$	$0,0,\frac{1}{2}$					$hkil$: $l=2n$
2	a	$\bar{6}\,.\,.$	$0,0,\frac{1}{4}$	$0,0,\frac{3}{4}$					$hkil$: $l=2n$

Symmetry of special projections

Along [001] $p6$
$\mathbf{a}'=\mathbf{a}$ $\mathbf{b}'=\mathbf{b}$
Origin at $0,0,z$

Along [100] $p2gm$
$\mathbf{a}'=\frac{1}{2}(\mathbf{a}+2\mathbf{b})$ $\mathbf{b}'=\mathbf{c}$
Origin at $x,0,0$

Along [210] $p2gm$
$\mathbf{a}'=\frac{1}{2}\mathbf{b}$ $\mathbf{b}'=\mathbf{c}$
Origin at $x,\frac{1}{2}x,0$

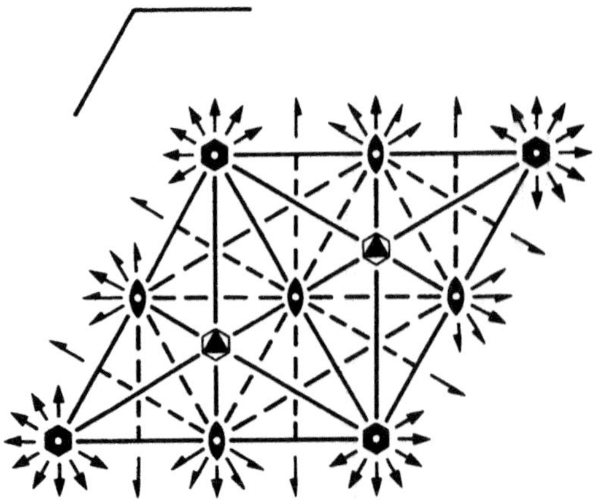

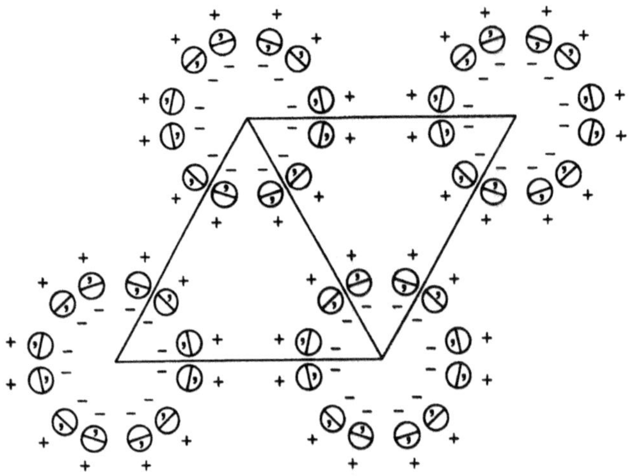

Origin at centre ($6/mmm$)

Asymmetric unit $0 \le x \le \frac{2}{3}$; $0 \le y \le \frac{1}{3}$; $0 \le z \le \frac{1}{2}$; $x \le (1+y)/2$; $y \le x/2$

Vertices $0,0,0$ $\frac{1}{2},0,0$ $\frac{2}{3},\frac{1}{3},0$

$0,0,\frac{1}{2}$ $\frac{1}{2},0,\frac{1}{2}$ $\frac{2}{3},\frac{1}{3},\frac{1}{2}$

Symmetry operations

(1) 1	(2) 3^+ $0,0,z$	(3) 3^- $0,0,z$
(4) 2 $0,0,z$	(5) 6^- $0,0,z$	(6) 6^+ $0,0,z$
(7) 2 $x,x,0$	(8) 2 $x,0,0$	(9) 2 $0,y,0$
(10) 2 $x,\bar{x},0$	(11) 2 $x,2x,0$	(12) 2 $2x,x,0$
(13) $\bar{1}$ $0,0,0$	(14) $\bar{3}^+$ $0,0,z$; $0,0,0$	(15) $\bar{3}^-$ $0,0,z$; $0,0,0$
(16) m $x,y,0$	(17) $\bar{6}^-$ $0,0,z$; $0,0,0$	(18) $\bar{6}^+$ $0,0,z$; $0,0,0$
(19) m $x,\bar{x},z$	(20) m $x,2x,z$	(21) m $2x,x,z$
(22) m x,x,z	(23) m $x,0,z$	(24) m $0,y,z$

Generators selected (1); $t(1,0,0)$; $t(0,1,0)$; $t(0,0,1)$; (2); (4); (7); (13)

Positions

Multiplicity, Wyckoff letter, Site symmetry			Coordinates	Reflection conditions
				General:
24	r	1	(1) x,y,z (2) $\bar{y},x-y,z$ (3) $\bar{x}+y,\bar{x},z$ (4) $\bar{x},\bar{y},z$ (5) $y,\bar{x}+y,z$ (6) $x-y,x,z$ (7) $y,x,\bar{z}$ (8) $x-y,\bar{y},\bar{z}$ (9) $\bar{x},\bar{x}+y,\bar{z}$ (10) $\bar{y},\bar{x},\bar{z}$ (11) $\bar{x}+y,y,\bar{z}$ (12) $x,x-y,\bar{z}$ (13) $\bar{x},\bar{y},\bar{z}$ (14) $y,\bar{x}+y,\bar{z}$ (15) $x-y,x,\bar{z}$ (16) $x,y,\bar{z}$ (17) $\bar{y},x-y,\bar{z}$ (18) $\bar{x}+y,\bar{x},\bar{z}$ (19) $\bar{y},\bar{x},z$ (20) $\bar{x}+y,y,z$ (21) $x,x-y,z$ (22) y,x,z (23) $x-y,\bar{y},z$ (24) $\bar{x},\bar{x}+y,z$	no conditions
				Special: no extra conditions
12	q	$m\,.\,.$	$x,y,\frac{1}{2}$ $\bar{y},x-y,\frac{1}{2}$ $\bar{x}+y,\bar{x},\frac{1}{2}$ $\bar{x},\bar{y},\frac{1}{2}$ $y,\bar{x}+y,\frac{1}{2}$ $x-y,x,\frac{1}{2}$ $y,x,\frac{1}{2}$ $x-y,\bar{y},\frac{1}{2}$ $\bar{x},\bar{x}+y,\frac{1}{2}$ $\bar{y},\bar{x},\frac{1}{2}$ $\bar{x}+y,y,\frac{1}{2}$ $x,x-y,\frac{1}{2}$	
12	p	$m\,.\,.$	$x,y,0$ $\bar{y},x-y,0$ $\bar{x}+y,\bar{x},0$ $\bar{x},\bar{y},0$ $y,\bar{x}+y,0$ $x-y,x,0$ $y,x,0$ $x-y,\bar{y},0$ $\bar{x},\bar{x}+y,0$ $\bar{y},\bar{x},0$ $\bar{x}+y,y,0$ $x,x-y,0$	
12	o	$.\,m\,.$	$x,2x,z$ $2\bar{x},\bar{x},z$ $x,\bar{x},z$ $\bar{x},2\bar{x},z$ $2x,x,z$ $\bar{x},x,z$ $2x,x,\bar{z}$ $\bar{x},2\bar{x},\bar{z}$ $\bar{x},x,\bar{z}$ $2\bar{x},\bar{x},\bar{z}$ $x,2x,\bar{z}$ $x,\bar{x},\bar{z}$	
12	n	$.\,.\,m$	$x,0,z$ $0,x,z$ $\bar{x},\bar{x},z$ $\bar{x},0,z$ $0,\bar{x},z$ x,x,z $0,x,\bar{z}$ $x,0,\bar{z}$ $\bar{x},\bar{x},\bar{z}$ $0,\bar{x},\bar{z}$ $\bar{x},0,\bar{z}$ $x,x,\bar{z}$	
6	m	$m\,m\,2$	$x,2x,\frac{1}{2}$ $2\bar{x},\bar{x},\frac{1}{2}$ $x,\bar{x},\frac{1}{2}$ $\bar{x},2\bar{x},\frac{1}{2}$ $2x,x,\frac{1}{2}$ $\bar{x},x,\frac{1}{2}$	
6	l	$m\,m\,2$	$x,2x,0$ $2\bar{x},\bar{x},0$ $x,\bar{x},0$ $\bar{x},2\bar{x},0$ $2x,x,0$ $\bar{x},x,0$	
6	k	$m\,2\,m$	$x,0,\frac{1}{2}$ $0,x,\frac{1}{2}$ $\bar{x},\bar{x},\frac{1}{2}$ $\bar{x},0,\frac{1}{2}$ $0,\bar{x},\frac{1}{2}$ $x,x,\frac{1}{2}$	
6	j	$m\,2\,m$	$x,0,0$ $0,x,0$ $\bar{x},\bar{x},0$ $\bar{x},0,0$ $0,\bar{x},0$ $x,x,0$	
6	i	$2\,m\,m$	$\frac{1}{2},0,z$ $0,\frac{1}{2},z$ $\frac{1}{2},\frac{1}{2},z$ $0,\frac{1}{2},\bar{z}$ $\frac{1}{2},0,\bar{z}$ $\frac{1}{2},\frac{1}{2},\bar{z}$	
4	h	$3\,m\,.$	$\frac{1}{3},\frac{2}{3},z$ $\frac{2}{3},\frac{1}{3},z$ $\frac{2}{3},\frac{1}{3},\bar{z}$ $\frac{1}{3},\frac{2}{3},\bar{z}$	
3	g	$m\,m\,m$	$\frac{1}{2},0,\frac{1}{2}$ $0,\frac{1}{2},\frac{1}{2}$ $\frac{1}{2},\frac{1}{2},\frac{1}{2}$	
3	f	$m\,m\,m$	$\frac{1}{2},0,0$ $0,\frac{1}{2},0$ $\frac{1}{2},\frac{1}{2},0$	
2	e	$6\,m\,m$	$0,0,z$ $0,0,\bar{z}$	
2	d	$\bar{6}\,m\,2$	$\frac{1}{3},\frac{2}{3},\frac{1}{2}$ $\frac{2}{3},\frac{1}{3},\frac{1}{2}$	
2	c	$\bar{6}\,m\,2$	$\frac{1}{3},\frac{2}{3},0$ $\frac{2}{3},\frac{1}{3},0$	
1	b	$6/m\,m\,m$	$0,0,\frac{1}{2}$	
1	a	$6/m\,m\,m$	$0,0,0$	

Symmetry of special projections

Along [001] $p6mm$
$\mathbf{a}'=\mathbf{a}$ $\mathbf{b}'=\mathbf{b}$
Origin at $0,0,z$

Along [100] $p2mm$
$\mathbf{a}'=\frac{1}{2}(\mathbf{a}+2\mathbf{b})$ $\mathbf{b}'=\mathbf{c}$
Origin at $x,0,0$

Along [210] $p2mm$
$\mathbf{a}'=\frac{1}{2}\mathbf{b}$ $\mathbf{b}'=\mathbf{c}$
Origin at $x,\frac{1}{2}x,0$

$P6_3/mmc$ $\quad D_{6h}^4$ $\quad 6/mmm$ $\quad$ Hexagonal

No. 194 $\quad P\ 6_3/m\ 2/m\ 2/c$ $\quad$ Patterson symmetry $P6/mmm$

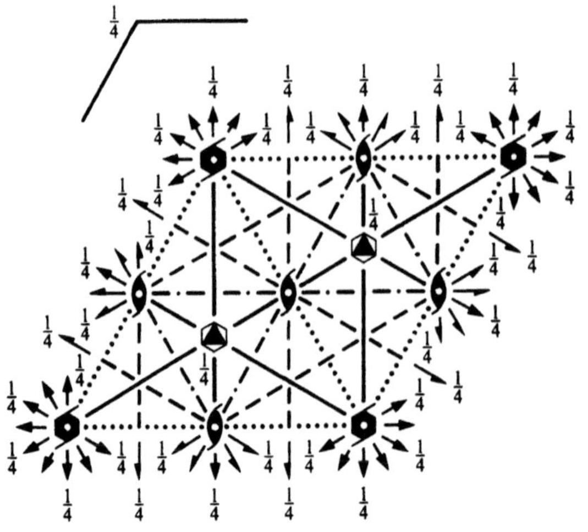

For inversion-point heights of $\bar{1}$ and $\bar{6}$ see $P6_3/m$ (No. 176)

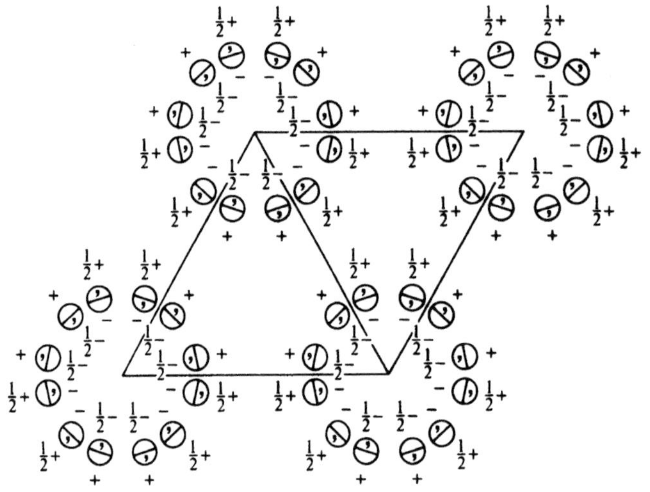

Origin at centre ($\bar{3}m1$) at $\bar{3}2/mc$

Asymmetric unit $\quad 0 \le x \le \frac{2}{3};\quad 0 \le y \le \frac{2}{3};\quad 0 \le z \le \frac{1}{4};\quad x \le 2y;\quad y \le \min(1-x, 2x)$

Vertices $\quad 0,0,0 \quad \frac{2}{3},\frac{1}{3},0 \quad \frac{1}{3},\frac{2}{3},0$

$0,0,\frac{1}{4} \quad \frac{2}{3},\frac{1}{3},\frac{1}{4} \quad \frac{1}{3},\frac{2}{3},\frac{1}{4}$

Symmetry operations

(1) 1	(2) 3^+ $0,0,z$	(3) 3^- $0,0,z$
(4) $2(0,0,\frac{1}{2})$ $0,0,z$	(5) $6^-(0,0,\frac{1}{2})$ $0,0,z$	(6) $6^+(0,0,\frac{1}{2})$ $0,0,z$
(7) 2 $x,x,0$	(8) 2 $x,0,0$	(9) 2 $0,y,0$
(10) 2 $x,\bar{x},\frac{1}{4}$	(11) 2 $x,2x,\frac{1}{4}$	(12) 2 $2x,x,\frac{1}{4}$
(13) $\bar{1}$ $0,0,0$	(14) $\bar{3}^+$ $0,0,z$; $0,0,0$	(15) $\bar{3}^-$ $0,0,z$; $0,0,0$
(16) m $x,y,\frac{1}{4}$	(17) $\bar{6}^-$ $0,0,z$; $0,0,\frac{1}{4}$	(18) $\bar{6}^+$ $0,0,z$; $0,0,\frac{1}{4}$
(19) m $x,\bar{x},z$	(20) m $x,2x,z$	(21) m $2x,x,z$
(22) c x,x,z	(23) c $x,0,z$	(24) c $0,y,z$

Generators selected (1); $t(1,0,0)$; $t(0,1,0)$; $t(0,0,1)$; (2); (4); (7); (13)

Positions

Multiplicity, Wyckoff letter, Site symmetry			Coordinates						Reflection conditions
									General:
24	l	1	(1) x,y,z	(2) $\bar{y},x-y,z$	(3) $\bar{x}+y,\bar{x},z$				$hh\overline{2h}l$: $l=2n$
			(4) $\bar{x},\bar{y},z+\frac{1}{2}$	(5) $y,\bar{x}+y,z+\frac{1}{2}$	(6) $x-y,x,z+\frac{1}{2}$				$000l$: $l=2n$
			(7) $y,x,\bar{z}$	(8) $x-y,\bar{y},\bar{z}$	(9) $\bar{x},\bar{x}+y,\bar{z}$				
			(10) $\bar{y},\bar{x},\bar{z}+\frac{1}{2}$	(11) $\bar{x}+y,y,\bar{z}+\frac{1}{2}$	(12) $x,x-y,\bar{z}+\frac{1}{2}$				
			(13) $\bar{x},\bar{y},\bar{z}$	(14) $y,\bar{x}+y,\bar{z}$	(15) $x-y,x,\bar{z}$				
			(16) $x,y,\bar{z}+\frac{1}{2}$	(17) $\bar{y},x-y,\bar{z}+\frac{1}{2}$	(18) $\bar{x}+y,\bar{x},\bar{z}+\frac{1}{2}$				
			(19) $\bar{y},\bar{x},z$	(20) $\bar{x}+y,y,z$	(21) $x,x-y,z$				
			(22) $y,x,z+\frac{1}{2}$	(23) $x-y,\bar{y},z+\frac{1}{2}$	(24) $\bar{x},\bar{x}+y,z+\frac{1}{2}$				
									Special: as above, plus
12	k	$.m.$	$x,2x,z$	$2\bar{x},\bar{x},z$	$x,\bar{x},z$	$\bar{x},2\bar{x},z+\frac{1}{2}$			no extra conditions
			$2x,x,z+\frac{1}{2}$	$\bar{x},x,z+\frac{1}{2}$	$2x,x,\bar{z}$	$\bar{x},2\bar{x},\bar{z}$			
			$\bar{x},x,\bar{z}$	$2\bar{x},\bar{x},\bar{z}+\frac{1}{2}$	$x,2x,\bar{z}+\frac{1}{2}$	$x,\bar{x},\bar{z}+\frac{1}{2}$			
12	j	$m..$	$x,y,\frac{1}{4}$	$\bar{y},x-y,\frac{1}{4}$	$\bar{x}+y,\bar{x},\frac{1}{4}$	$\bar{x},\bar{y},\frac{3}{4}$	$y,\bar{x}+y,\frac{3}{4}$	$x-y,x,\frac{3}{4}$	no extra conditions
			$y,x,\frac{3}{4}$	$x-y,\bar{y},\frac{3}{4}$	$\bar{x},\bar{x}+y,\frac{3}{4}$	$\bar{y},\bar{x},\frac{1}{4}$	$\bar{x}+y,y,\frac{1}{4}$	$x,x-y,\frac{1}{4}$	
12	i	$.2.$	$x,0,0$	$0,x,0$	$\bar{x},\bar{x},0$	$\bar{x},0,\frac{1}{2}$	$0,\bar{x},\frac{1}{2}$	$x,x,\frac{1}{2}$	$hkil$: $l=2n$
			$\bar{x},0,0$	$0,\bar{x},0$	$x,x,0$	$x,0,\frac{1}{2}$	$0,x,\frac{1}{2}$	$\bar{x},\bar{x},\frac{1}{2}$	
6	h	$mm2$	$x,2x,\frac{1}{4}$	$2\bar{x},\bar{x},\frac{1}{4}$	$x,\bar{x},\frac{1}{4}$	$\bar{x},2\bar{x},\frac{3}{4}$	$2x,x,\frac{3}{4}$	$\bar{x},x,\frac{3}{4}$	no extra conditions
6	g	$.2/m.$	$\frac{1}{2},0,0$	$0,\frac{1}{2},0$	$\frac{1}{2},\frac{1}{2},0$	$\frac{1}{2},0,\frac{1}{2}$	$0,\frac{1}{2},\frac{1}{2}$	$\frac{1}{2},\frac{1}{2},\frac{1}{2}$	$hkil$: $l=2n$
4	f	$3m.$	$\frac{1}{3},\frac{2}{3},z$	$\frac{2}{3},\frac{1}{3},z+\frac{1}{2}$	$\frac{2}{3},\frac{1}{3},\bar{z}$	$\frac{1}{3},\frac{2}{3},\bar{z}+\frac{1}{2}$			$hkil$: $l=2n$ or $h-k=3n+1$ or $h-k=3n+2$
4	e	$3m.$	$0,0,z$	$0,0,z+\frac{1}{2}$	$0,0,\bar{z}$	$0,0,\bar{z}+\frac{1}{2}$			$hkil$: $l=2n$
2	d	$\bar{6}m2$	$\frac{1}{3},\frac{2}{3},\frac{3}{4}$	$\frac{2}{3},\frac{1}{3},\frac{1}{4}$					$hkil$: $l=2n$ or $h-k=3n+1$ or $h-k=3n+2$ (for d and c)
2	c	$\bar{6}m2$	$\frac{1}{3},\frac{2}{3},\frac{1}{4}$	$\frac{2}{3},\frac{1}{3},\frac{3}{4}$					
2	b	$\bar{6}m2$	$0,0,\frac{1}{4}$	$0,0,\frac{3}{4}$					$hkil$: $l=2n$
2	a	$\bar{3}m.$	$0,0,0$	$0,0,\frac{1}{2}$					$hkil$: $l=2n$

Symmetry of special projections

Along [001] $p6mm$
$\mathbf{a}'=\mathbf{a}$ $\mathbf{b}'=\mathbf{b}$
Origin at $0,0,z$

Along [100] $p2gm$
$\mathbf{a}'=\frac{1}{2}(\mathbf{a}+2\mathbf{b})$ $\mathbf{b}'=\mathbf{c}$
Origin at $x,0,0$

Along [210] $p2mm$
$\mathbf{a}'=\frac{1}{2}\mathbf{b}$ $\mathbf{b}'=\frac{1}{2}\mathbf{c}$
Origin at $x,\frac{1}{2}x,0$

$Pa\bar{3}$ T_h^6 $m\bar{3}$ Cubic

No. 205 $P2_1/a\bar{3}$ Patterson symmetry $Pm\bar{3}$

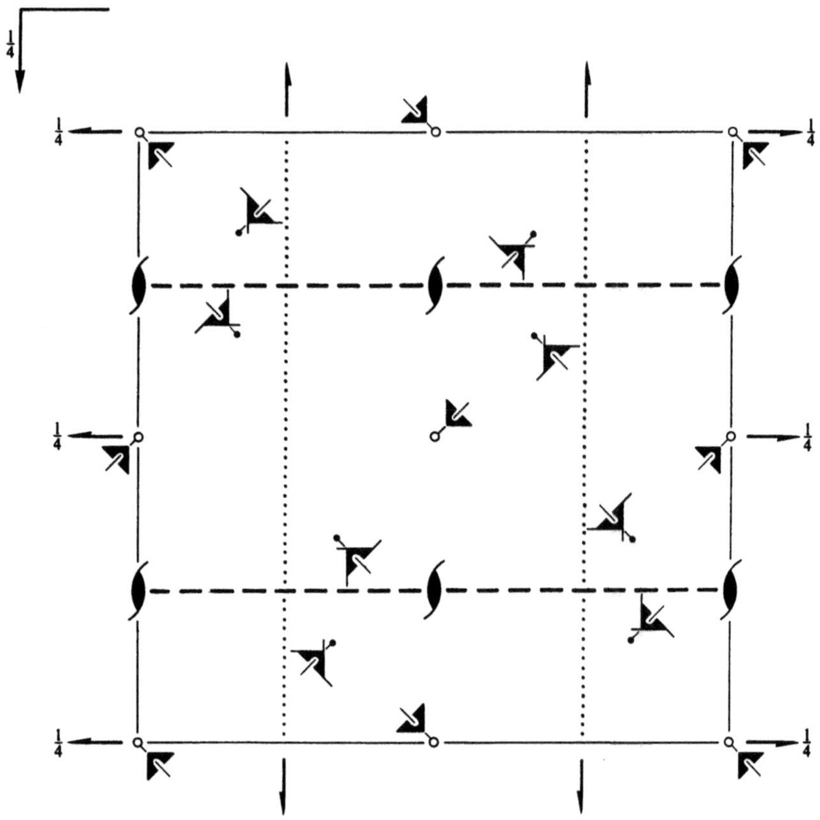

Origin at centre ($\bar{3}$)

Asymmetric unit $0 \leq x \leq \frac{1}{2}$; $0 \leq y \leq \frac{1}{2}$; $0 \leq z \leq \frac{1}{2}$; $z \leq \min(x,y)$

Vertices $0,0,0$ $\frac{1}{2},0,0$ $\frac{1}{2},\frac{1}{2},0$ $0,\frac{1}{2},0$ $\frac{1}{2},\frac{1}{2},\frac{1}{2}$

Symmetry operations

(1) 1	(2) $2(0,0,\frac{1}{2})\ \ \frac{1}{4},0,z$	(3) $2(0,\frac{1}{2},0)\ \ 0,y,\frac{1}{4}$	(4) $2(\frac{1}{2},0,0)\ \ x,\frac{1}{4},0$
(5) $3^{+}\ \ x,x,x$	(6) $3^{+}\ \ \bar{x}+\frac{1}{2},x,\bar{x}$	(7) $3^{+}\ \ x+\frac{1}{2},\bar{x}-\frac{1}{2},\bar{x}$	(8) $3^{+}\ \ \bar{x},\bar{x}+\frac{1}{2},x$
(9) $3^{-}\ \ x,x,x$	(10) $3^{-}(-\frac{1}{3},\frac{1}{3},\frac{1}{3})\ \ x+\frac{1}{6},\bar{x}+\frac{1}{6},\bar{x}$	(11) $3^{-}(\frac{1}{3},\frac{1}{3},-\frac{1}{3})\ \ \bar{x}+\frac{1}{3},\bar{x}+\frac{1}{6},x$	(12) $3^{-}(\frac{1}{3},-\frac{1}{3},\frac{1}{3})\ \ \bar{x}-\frac{1}{6},x+\frac{1}{3},\bar{x}$
(13) $\bar{1}\ \ 0,0,0$	(14) $a\ \ x,y,\frac{1}{4}$	(15) $c\ \ x,\frac{1}{4},z$	(16) $b\ \ \frac{1}{4},y,z$
(17) $\bar{3}^{+}\ \ x,x,x;\ 0,0,0$	(18) $\bar{3}^{+}\ \ \bar{x}-\frac{1}{2},x+1,\bar{x};\ 0,\frac{1}{2},\frac{1}{2}$	(19) $\bar{3}^{+}\ \ x+\frac{1}{2},\bar{x}+\frac{1}{2},\bar{x};\ \frac{1}{2},\frac{1}{2},0$	(20) $\bar{3}^{+}\ \ \bar{x}+1,\bar{x}+\frac{1}{2},x;\ \frac{1}{2},0,\frac{1}{2}$
(21) $\bar{3}^{-}\ \ x,x,x;\ 0,0,0$	(22) $\bar{3}^{-}\ \ x+\frac{1}{2},\bar{x}-\frac{1}{2},\bar{x};\ 0,0,\frac{1}{2}$	(23) $\bar{3}^{-}\ \ \bar{x},\bar{x}+\frac{1}{2},x;\ 0,\frac{1}{2},0$	(24) $\bar{3}^{-}\ \ \bar{x}+\frac{1}{2},x,\bar{x};\ \frac{1}{2},0,0$

Generators selected (1); $t(1,0,0)$; $t(0,1,0)$; $t(0,0,1)$; (2); (3); (5); (13)

Positions

Multiplicity, Wyckoff letter, Site symmetry	Coordinates				Reflection conditions
					h,k,l cyclically permutable General:
24 d 1	(1) x,y,z	(2) $\bar{x}+\frac{1}{2},\bar{y},z+\frac{1}{2}$	(3) $\bar{x},y+\frac{1}{2},\bar{z}+\frac{1}{2}$	(4) $x+\frac{1}{2},\bar{y}+\frac{1}{2},\bar{z}$	$0kl$: $k=2n$
	(5) z,x,y	(6) $z+\frac{1}{2},\bar{x}+\frac{1}{2},\bar{y}$	(7) $\bar{z}+\frac{1}{2},\bar{x},y+\frac{1}{2}$	(8) $\bar{z},x+\frac{1}{2},\bar{y}+\frac{1}{2}$	$h00$: $h=2n$
	(9) y,z,x	(10) $\bar{y},z+\frac{1}{2},\bar{x}+\frac{1}{2}$	(11) $y+\frac{1}{2},\bar{z}+\frac{1}{2},\bar{x}$	(12) $\bar{y}+\frac{1}{2},\bar{z},x+\frac{1}{2}$	
	(13) $\bar{x},\bar{y},\bar{z}$	(14) $x+\frac{1}{2},y,\bar{z}+\frac{1}{2}$	(15) $x,\bar{y}+\frac{1}{2},z+\frac{1}{2}$	(16) $\bar{x}+\frac{1}{2},y+\frac{1}{2},z$	
	(17) $\bar{z},\bar{x},\bar{y}$	(18) $\bar{z}+\frac{1}{2},x+\frac{1}{2},y$	(19) $z+\frac{1}{2},x,\bar{y}+\frac{1}{2}$	(20) $z,\bar{x}+\frac{1}{2},y+\frac{1}{2}$	
	(21) $\bar{y},\bar{z},\bar{x}$	(22) $y,\bar{z}+\frac{1}{2},x+\frac{1}{2}$	(23) $\bar{y}+\frac{1}{2},z+\frac{1}{2},x$	(24) $y+\frac{1}{2},z,\bar{x}+\frac{1}{2}$	
					Special: as above, plus
8 c $.3.$	x,x,x	$\bar{x}+\frac{1}{2},\bar{x},x+\frac{1}{2}$	$\bar{x},x+\frac{1}{2},\bar{x}+\frac{1}{2}$	$x+\frac{1}{2},\bar{x}+\frac{1}{2},\bar{x}$	no extra conditions
	$\bar{x},\bar{x},\bar{x}$	$x+\frac{1}{2},x,\bar{x}+\frac{1}{2}$	$x,\bar{x}+\frac{1}{2},x+\frac{1}{2}$	$\bar{x}+\frac{1}{2},x+\frac{1}{2},x$	
4 b $.\bar{3}.$	$\frac{1}{2},\frac{1}{2},\frac{1}{2}$	$0,\frac{1}{2},0$	$\frac{1}{2},0,0$	$0,0,\frac{1}{2}$	hkl: $h+k,h+l,k+l=2n$
4 a $.\bar{3}.$	$0,0,0$	$\frac{1}{2},0,\frac{1}{2}$	$0,\frac{1}{2},\frac{1}{2}$	$\frac{1}{2},\frac{1}{2},0$	hkl: $h+k,h+l,k+l=2n$

Symmetry of special projections

Along [001] $p2gm$	Along [111] $p6$	Along [110] $p2gg$
$\mathbf{a}'=\frac{1}{2}\mathbf{a}$ $\quad\mathbf{b}'=\mathbf{b}$	$\mathbf{a}'=\frac{1}{3}(2\mathbf{a}-\mathbf{b}-\mathbf{c})$ $\quad\mathbf{b}'=\frac{1}{3}(-\mathbf{a}+2\mathbf{b}-\mathbf{c})$	$\mathbf{a}'=\frac{1}{2}(-\mathbf{a}+\mathbf{b})$ $\quad\mathbf{b}'=\mathbf{c}$
Origin at $0,0,z$	Origin at x,x,x	Origin at $x,x,0$

$F\bar{4}3m$ | T_d^2 | $\bar{4}3m$ | Cubic

No. 216 | $F\bar{4}3m$ | Patterson symmetry $Fm\bar{3}m$

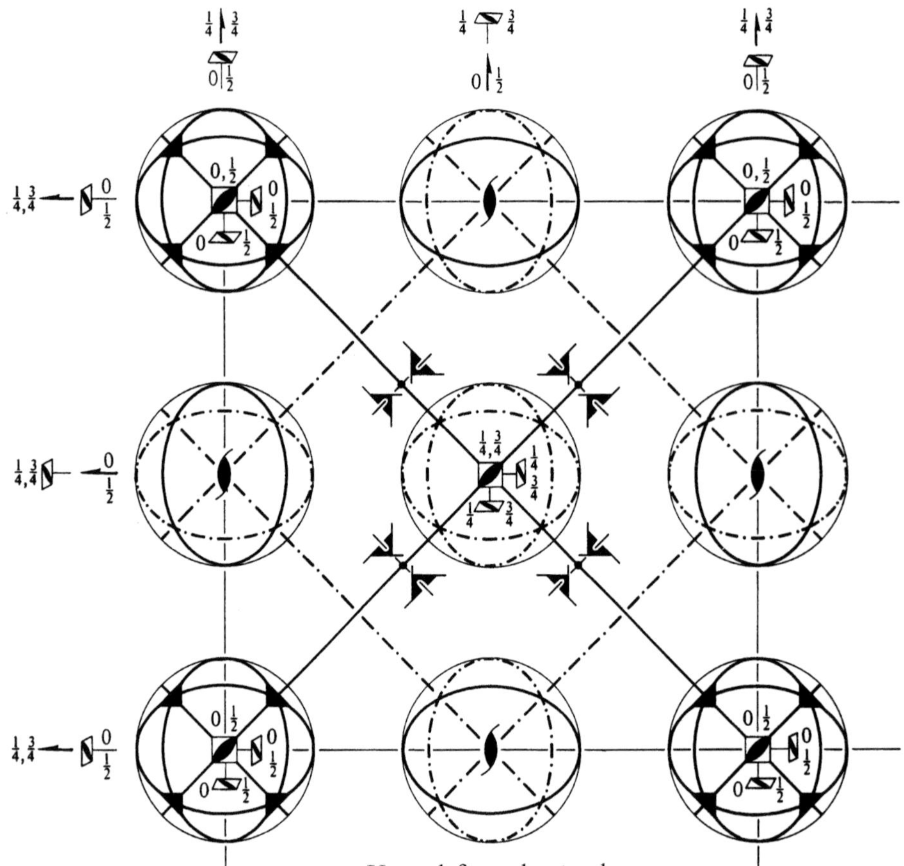

Upper left quadrant only

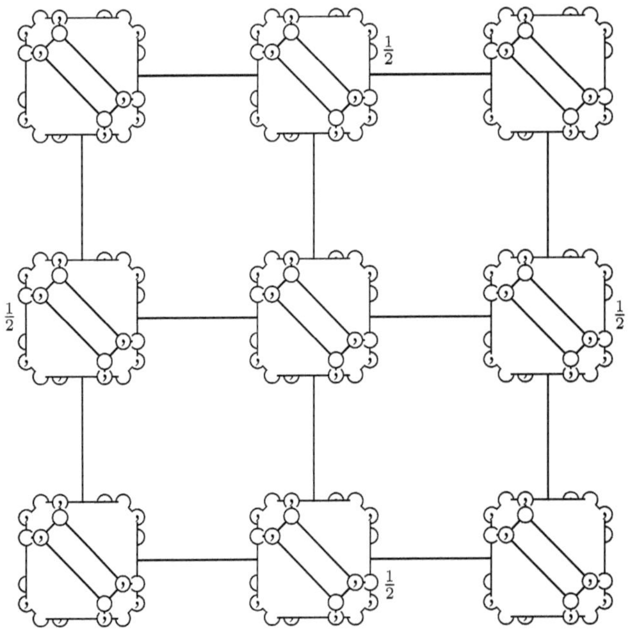

Origin at $\bar{4}3m$

Asymmetric unit $0 \leq x \leq \frac{1}{2}$; $0 \leq y \leq \frac{1}{4}$; $-\frac{1}{4} \leq z \leq \frac{1}{4}$; $y \leq \min(x, \frac{1}{2} - x)$; $-y \leq z \leq y$

Vertices $0,0,0$ $\frac{1}{2},0,0$ $\frac{1}{4},\frac{1}{4},\frac{1}{4}$ $\frac{1}{4},\frac{1}{4},-\frac{1}{4}$

Symmetry operations

For (0,0,0)+ set

(1) 1	(2) 2 $0,0,z$	(3) 2 $0,y,0$	(4) 2 $x,0,0$
(5) 3^+ x,x,x	(6) 3^+ $\bar{x},x,\bar{x}$	(7) 3^+ $x,\bar{x},\bar{x}$	(8) 3^+ $\bar{x},\bar{x},x$
(9) 3^- x,x,x	(10) 3^- $x,\bar{x},\bar{x}$	(11) 3^- $\bar{x},\bar{x},x$	(12) 3^- $\bar{x},x,\bar{x}$
(13) m x,x,z	(14) m $x,\bar{x},z$	(15) $\bar{4}^+$ $0,0,z$; $0,0,0$	(16) $\bar{4}^-$ $0,0,z$; $0,0,0$
(17) m x,y,y	(18) $\bar{4}^+$ $x,0,0$; $0,0,0$	(19) $\bar{4}^-$ $x,0,0$; $0,0,0$	(20) m $x,y,\bar{y}$
(21) m x,y,x	(22) $\bar{4}^-$ $0,y,0$; $0,0,0$	(23) m $\bar{x},y,x$	(24) $\bar{4}^+$ $0,y,0$; $0,0,0$

Symmetry operations *(continued)*

For $(0,\frac{1}{2},\frac{1}{2})+$ set

(1) $t(0,\frac{1}{2},\frac{1}{2})$	(2) $2(0,0,\frac{1}{2})\ \ 0,\frac{1}{4},z$	(3) $2(0,\frac{1}{2},0)\ \ 0,y,\frac{1}{4}$	(4) $2\ \ x,\frac{1}{4},\frac{1}{4}$
(5) $3^{+}(\frac{1}{3},\frac{1}{3},\frac{1}{3})\ \ x-\frac{1}{3},x-\frac{1}{6},x$	(6) $3^{+}\ \ \bar{x},x+\frac{1}{2},\bar{x}$	(7) $3^{+}(-\frac{1}{3},\frac{1}{3},\frac{1}{3})\ \ x+\frac{1}{3},\bar{x}-\frac{1}{6},\bar{x}$	(8) $3^{+}\ \ \bar{x},\bar{x}+\frac{1}{2},x$
(9) $3^{-}(\frac{1}{3},\frac{1}{3},\frac{1}{3})\ \ x-\frac{1}{6},x+\frac{1}{6},x$	(10) $3^{-}(-\frac{1}{3},\frac{1}{3},\frac{1}{3})\ \ x+\frac{1}{6},\bar{x}+\frac{1}{6},\bar{x}$	(11) $3^{-}\ \ \bar{x}+\frac{1}{2},\bar{x}+\frac{1}{2},x$	(12) $3^{-}\ \ \bar{x}-\frac{1}{2},x+\frac{1}{2},\bar{x}$
(13) $g(\frac{1}{4},\frac{1}{4},\frac{1}{2})\ \ x-\frac{1}{4},x,z$	(14) $g(-\frac{1}{4},\frac{1}{4},\frac{1}{2})\ \ x+\frac{1}{4},\bar{x},z$	(15) $\bar{4}^{+}\ \ \frac{1}{4},\frac{1}{4},z;\ \frac{1}{4},\frac{1}{4},\frac{1}{4}$	(16) $\bar{4}^{-}\ \ -\frac{1}{4},\frac{1}{4},z;\ -\frac{1}{4},\frac{1}{4},\frac{1}{4}$
(17) $g(0,\frac{1}{2},\frac{1}{2})\ \ x,y,y$	(18) $\bar{4}^{+}\ \ x,\frac{1}{2},0;\ 0,\frac{1}{2},0$	(19) $\bar{4}^{-}\ \ x,0,\frac{1}{2};\ 0,0,\frac{1}{2}$	(20) $m\ \ x,y+\frac{1}{2},\bar{y}$
(21) $g(\frac{1}{4},\frac{1}{2},\frac{1}{4})\ \ x-\frac{1}{4},y,x$	(22) $\bar{4}^{-}\ \ \frac{1}{4},y,\frac{1}{4};\ \frac{1}{4},\frac{1}{4},\frac{1}{4}$	(23) $g(-\frac{1}{4},\frac{1}{2},\frac{1}{4})\ \ \bar{x}+\frac{1}{4},y,x$	(24) $\bar{4}^{+}\ \ -\frac{1}{4},y,\frac{1}{4};\ -\frac{1}{4},\frac{1}{4},\frac{1}{4}$

For $(\frac{1}{2},0,\frac{1}{2})+$ set

(1) $t(\frac{1}{2},0,\frac{1}{2})$	(2) $2(0,0,\frac{1}{2})\ \ \frac{1}{4},0,z$	(3) $2\ \ \frac{1}{4},y,\frac{1}{4}$	(4) $2(\frac{1}{2},0,0)\ \ x,0,\frac{1}{4}$
(5) $3^{+}(\frac{1}{3},\frac{1}{3},\frac{1}{3})\ \ x+\frac{1}{6},x-\frac{1}{6},x$	(6) $3^{+}(\frac{1}{3},-\frac{1}{3},\frac{1}{3})\ \ \bar{x}+\frac{1}{6},x+\frac{1}{6},\bar{x}$	(7) $3^{+}\ \ x+\frac{1}{2},\bar{x}-\frac{1}{2},\bar{x}$	(8) $3^{+}\ \ \bar{x}+\frac{1}{2},\bar{x}+\frac{1}{2},x$
(9) $3^{-}(\frac{1}{3},\frac{1}{3},\frac{1}{3})\ \ x-\frac{1}{6},x-\frac{1}{3},x$	(10) $3^{-}\ \ x+\frac{1}{2},\bar{x},\bar{x}$	(11) $3^{-}\ \ \bar{x}+\frac{1}{2},\bar{x},x$	(12) $3^{-}(\frac{1}{3},-\frac{1}{3},\frac{1}{3})\ \ \bar{x}-\frac{1}{6},x+\frac{1}{3},\bar{x}$
(13) $g(\frac{1}{4},\frac{1}{4},\frac{1}{2})\ \ x+\frac{1}{4},x,z$	(14) $g(\frac{1}{4},-\frac{1}{4},\frac{1}{2})\ \ x+\frac{1}{4},\bar{x},z$	(15) $\bar{4}^{+}\ \ \frac{1}{4},-\frac{1}{4},z;\ \frac{1}{4},-\frac{1}{4},\frac{1}{4}$	(16) $\bar{4}^{-}\ \ \frac{1}{4},\frac{1}{4},z;\ \frac{1}{4},\frac{1}{4},\frac{1}{4}$
(17) $g(\frac{1}{2},\frac{1}{4},\frac{1}{4})\ \ x,y-\frac{1}{4},y$	(18) $\bar{4}^{+}\ \ x,\frac{1}{4},\frac{1}{4};\ \frac{1}{4},\frac{1}{4},\frac{1}{4}$	(19) $\bar{4}^{-}\ \ x,-\frac{1}{4},\frac{1}{4};\ \frac{1}{4},-\frac{1}{4},\frac{1}{4}$	(20) $g(\frac{1}{2},-\frac{1}{4},\frac{1}{4})\ \ x,y+\frac{1}{4},\bar{y}$
(21) $g(\frac{1}{2},0,\frac{1}{2})\ \ x,y,x$	(22) $\bar{4}^{-}\ \ \frac{1}{2},y,0;\ \frac{1}{2},0,0$	(23) $m\ \ \bar{x}+\frac{1}{2},y,x$	(24) $\bar{4}^{+}\ \ 0,y,\frac{1}{2};\ 0,0,\frac{1}{2}$

For $(\frac{1}{2},\frac{1}{2},0)+$ set

(1) $t(\frac{1}{2},\frac{1}{2},0)$	(2) $2\ \ \frac{1}{4},\frac{1}{4},z$	(3) $2(0,\frac{1}{2},0)\ \ \frac{1}{4},y,0$	(4) $2(\frac{1}{2},0,0)\ \ x,\frac{1}{4},0$
(5) $3^{+}(\frac{1}{3},\frac{1}{3},\frac{1}{3})\ \ x+\frac{1}{6},x+\frac{1}{3},x$	(6) $3^{+}\ \ \bar{x}+\frac{1}{2},x,\bar{x}$	(7) $3^{+}\ \ x+\frac{1}{2},\bar{x},\bar{x}$	(8) $3^{+}(\frac{1}{3},\frac{1}{3},-\frac{1}{3})\ \ \bar{x}+\frac{1}{6},\bar{x}+\frac{1}{3},x$
(9) $3^{-}(\frac{1}{3},\frac{1}{3},\frac{1}{3})\ \ x+\frac{1}{3},x+\frac{1}{6},x$	(10) $3^{-}\ \ x,\bar{x}+\frac{1}{2},\bar{x}$	(11) $3^{-}(\frac{1}{3},\frac{1}{3},-\frac{1}{3})\ \ \bar{x}+\frac{1}{3},\bar{x}+\frac{1}{6},x$	(12) $3^{-}\ \ \bar{x},x+\frac{1}{2},\bar{x}$
(13) $g(\frac{1}{2},\frac{1}{2},0)\ \ x,x,z$	(14) $m\ \ x+\frac{1}{2},\bar{x},z$	(15) $\bar{4}^{+}\ \ \frac{1}{2},0,z;\ \frac{1}{2},0,0$	(16) $\bar{4}^{-}\ \ 0,\frac{1}{2},z;\ 0,\frac{1}{2},0$
(17) $g(\frac{1}{2},\frac{1}{4},\frac{1}{4})\ \ x,y+\frac{1}{4},y$	(18) $\bar{4}^{+}\ \ x,\frac{1}{4},-\frac{1}{4};\ \frac{1}{4},\frac{1}{4},-\frac{1}{4}$	(19) $\bar{4}^{-}\ \ x,\frac{1}{4},\frac{1}{4};\ \frac{1}{4},\frac{1}{4},\frac{1}{4}$	(20) $g(\frac{1}{2},\frac{1}{4},-\frac{1}{4})\ \ x,y+\frac{1}{4},\bar{y}$
(21) $g(\frac{1}{4},\frac{1}{2},\frac{1}{4})\ \ x+\frac{1}{4},y,x$	(22) $\bar{4}^{-}\ \ \frac{1}{4},y,-\frac{1}{4};\ \frac{1}{4},\frac{1}{4},-\frac{1}{4}$	(23) $g(\frac{1}{4},\frac{1}{2},-\frac{1}{4})\ \ \bar{x}+\frac{1}{4},y,x$	(24) $\bar{4}^{+}\ \ \frac{1}{4},y,\frac{1}{4};\ \frac{1}{4},\frac{1}{4},\frac{1}{4}$

Generators selected (1); $t(1,0,0)$; $t(0,1,0)$; $t(0,0,1)$; $t(0,\frac{1}{2},\frac{1}{2})$; $t(\frac{1}{2},0,\frac{1}{2})$; (2); (3); (5); (13)

Positions

Multiplicity, Wyckoff letter, Site symmetry			Coordinates						Reflection conditions
			$(0,0,0)+$	$(0,\frac{1}{2},\frac{1}{2})+$	$(\frac{1}{2},0,\frac{1}{2})+$	$(\frac{1}{2},\frac{1}{2},0)+$			h,k,l permutable General:
96	i	1	(1) x,y,z	(2) $\bar{x},\bar{y},z$	(3) $\bar{x},y,\bar{z}$	(4) $x,\bar{y},\bar{z}$			hkl: $h+k,h+l,k+l=2n$
			(5) z,x,y	(6) $z,\bar{x},\bar{y}$	(7) $\bar{z},\bar{x},y$	(8) $\bar{z},x,\bar{y}$			$0kl$: $k,l=2n$
			(9) y,z,x	(10) $\bar{y},z,\bar{x}$	(11) $y,\bar{z},\bar{x}$	(12) $\bar{y},\bar{z},x$			hhl: $h+l=2n$
			(13) y,x,z	(14) $\bar{y},\bar{x},z$	(15) $y,\bar{x},\bar{z}$	(16) $\bar{y},x,\bar{z}$			$h00$: $h=2n$
			(17) x,z,y	(18) $\bar{x},z,\bar{y}$	(19) $\bar{x},\bar{z},y$	(20) $x,\bar{z},\bar{y}$			
			(21) z,y,x	(22) $z,\bar{y},\bar{x}$	(23) $\bar{z},y,\bar{x}$	(24) $\bar{z},\bar{y},x$			
									Special: no extra conditions
48	h	$..m$	x,x,z	$\bar{x},\bar{x},z$	$\bar{x},x,\bar{z}$	$x,\bar{x},\bar{z}$	z,x,x	$z,\bar{x},\bar{x}$	
			$\bar{z},\bar{x},x$	$\bar{z},x,\bar{x}$	x,z,x	$\bar{x},z,\bar{x}$	$x,\bar{z},\bar{x}$	$\bar{x},\bar{z},x$	
24	g	$2.mm$	$x,\frac{1}{4},\frac{1}{4}$	$\bar{x},\frac{3}{4},\frac{1}{4}$	$\frac{1}{4},x,\frac{1}{4}$	$\frac{1}{4},\bar{x},\frac{3}{4}$	$\frac{1}{4},\frac{1}{4},x$	$\frac{3}{4},\frac{1}{4},\bar{x}$	
24	f	$2.mm$	$x,0,0$	$\bar{x},0,0$	$0,x,0$	$0,\bar{x},0$	$0,0,x$	$0,0,\bar{x}$	
16	e	$.3m$	x,x,x	$\bar{x},\bar{x},x$	$\bar{x},x,\bar{x}$	$x,\bar{x},\bar{x}$			
4	d	$\bar{4}3m$	$\frac{3}{4},\frac{3}{4},\frac{3}{4}$						
4	c	$\bar{4}3m$	$\frac{1}{4},\frac{1}{4},\frac{1}{4}$						
4	b	$\bar{4}\,3\,m$	$\frac{1}{2},\frac{1}{2},\frac{1}{2}$						
4	a	$\bar{4}3m$	$0,0,0$						

Symmetry of special projections

Along [001] $p4mm$
$\mathbf{a}'=\frac{1}{2}\mathbf{a}$ $\mathbf{b}'=\frac{1}{2}\mathbf{b}$
Origin at $0,0,z$

Along [111] $p31m$
$\mathbf{a}'=\frac{1}{6}(2\mathbf{a}-\mathbf{b}-\mathbf{c})$ $\mathbf{b}'=\frac{1}{6}(-\mathbf{a}+2\mathbf{b}-\mathbf{c})$
Origin at x,x,x

Along [110] $c1m1$
$\mathbf{a}'=\frac{1}{2}(-\mathbf{a}+\mathbf{b})$ $\mathbf{b}'=\mathbf{c}$
Origin at $x,x,0$

$Fm\bar{3}m$ O_h^5 $m\bar{3}m$ Cubic

No. 225 $F\ 4/m\ \bar{3}\ 2/m$ Patterson symmetry $Fm\bar{3}m$

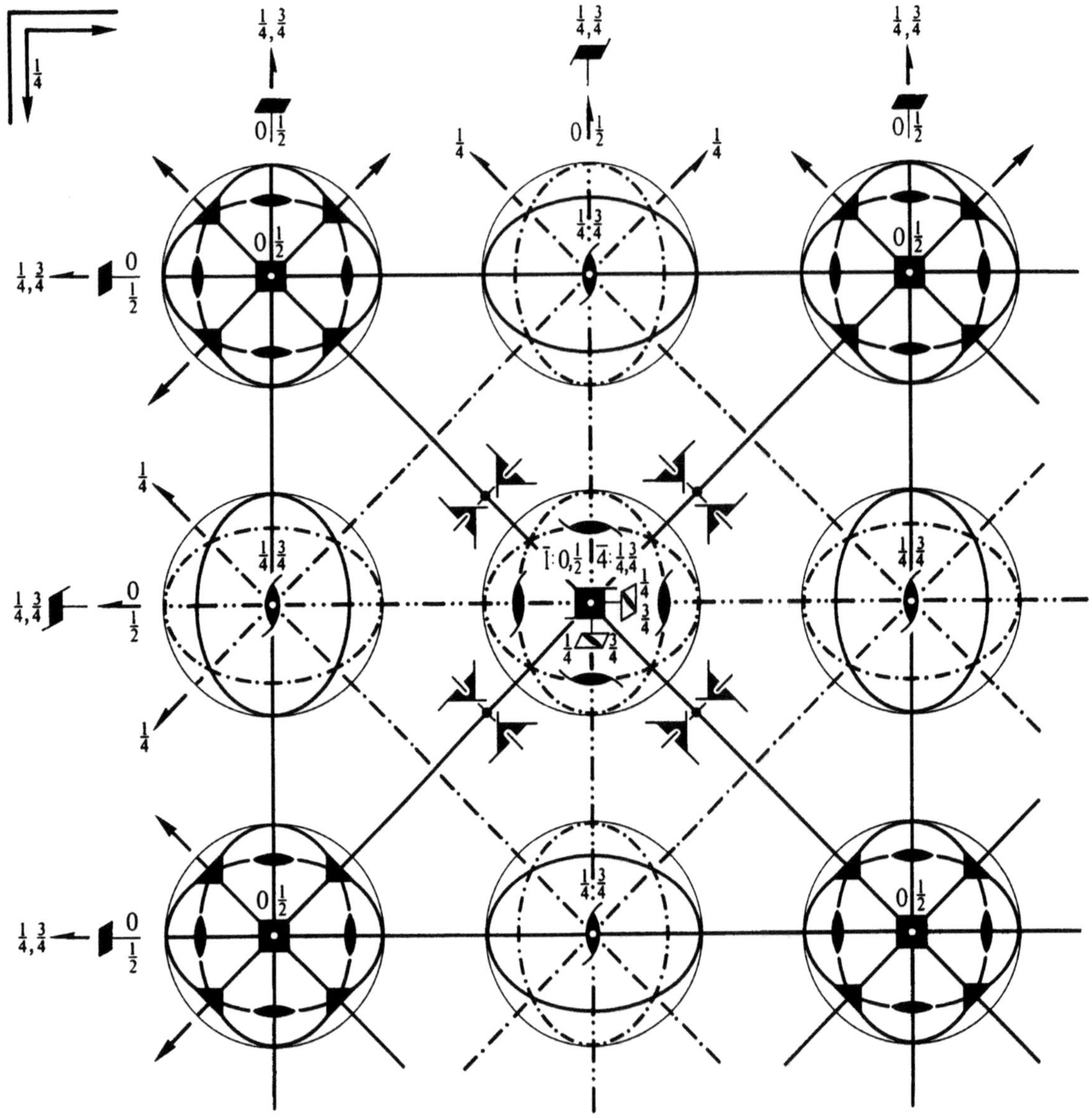

Upper left quadrant only

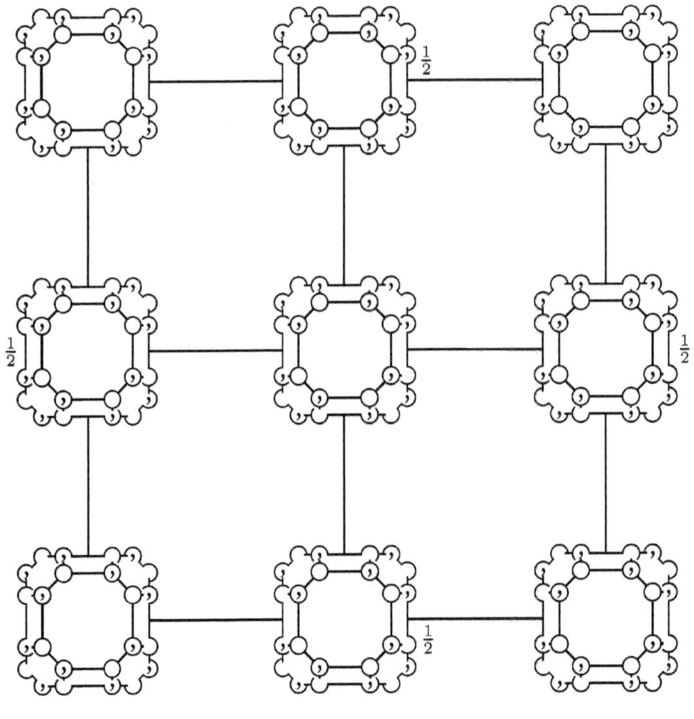

Origin at centre ($m\bar{3}m$)

Asymmetric unit $0 \le x \le \frac{1}{2}$; $0 \le y \le \frac{1}{4}$; $0 \le z \le \frac{1}{4}$; $y \le \min(x, \frac{1}{2} - x)$; $z \le y$

Vertices $0,0,0$ $\frac{1}{2},0,0$ $\frac{1}{4},\frac{1}{4},0$ $\frac{1}{4},\frac{1}{4},\frac{1}{4}$

Symmetry operations

For $(0,0,0)+$ set

(1) 1	(2) $2\ 0,0,z$	(3) $2\ 0,y,0$	(4) $2\ x,0,0$
(5) $3^{+}\ x,x,x$	(6) $3^{+}\ \bar{x},x,\bar{x}$	(7) $3^{+}\ x,\bar{x},\bar{x}$	(8) $3^{+}\ \bar{x},\bar{x},x$
(9) $3^{-}\ x,x,x$	(10) $3^{-}\ x,\bar{x},\bar{x}$	(11) $3^{-}\ \bar{x},\bar{x},x$	(12) $3^{-}\ \bar{x},x,\bar{x}$
(13) $2\ x,x,0$	(14) $2\ x,\bar{x},0$	(15) $4^{-}\ 0,0,z$	(16) $4^{+}\ 0,0,z$
(17) $4^{-}\ x,0,0$	(18) $2\ 0,y,y$	(19) $2\ 0,y,\bar{y}$	(20) $4^{+}\ x,0,0$
(21) $4^{+}\ 0,y,0$	(22) $2\ x,0,x$	(23) $4^{-}\ 0,y,0$	(24) $2\ \bar{x},0,x$
(25) $\bar{1}\ 0,0,0$	(26) $m\ x,y,0$	(27) $m\ x,0,z$	(28) $m\ 0,y,z$
(29) $\bar{3}^{+}\ x,x,x;\ 0,0,0$	(30) $\bar{3}^{+}\ \bar{x},x,\bar{x};\ 0,0,0$	(31) $\bar{3}^{+}\ x,\bar{x},\bar{x};\ 0,0,0$	(32) $\bar{3}^{+}\ \bar{x},\bar{x},x;\ 0,0,0$
(33) $\bar{3}^{-}\ x,x,x;\ 0,0,0$	(34) $\bar{3}^{-}\ x,\bar{x},\bar{x};\ 0,0,0$	(35) $\bar{3}^{-}\ \bar{x},\bar{x},x;\ 0,0,0$	(36) $\bar{3}^{-}\ \bar{x},x,\bar{x};\ 0,0,0$
(37) $m\ x,\bar{x},z$	(38) $m\ x,x,z$	(39) $\bar{4}^{-}\ 0,0,z;\ 0,0,0$	(40) $\bar{4}^{+}\ 0,0,z;\ 0,0,0$
(41) $\bar{4}^{-}\ x,0,0;\ 0,0,0$	(42) $m\ x,y,\bar{y}$	(43) $m\ x,y,y$	(44) $\bar{4}^{+}\ x,0,0;\ 0,0,0$
(45) $\bar{4}^{+}\ 0,y,0;\ 0,0,0$	(46) $m\ \bar{x},y,x$	(47) $\bar{4}^{-}\ 0,y,0;\ 0,0,0$	(48) $m\ x,y,x$

For $(0,\frac{1}{2},\frac{1}{2})+$ set

(1) $t(0,\frac{1}{2},\frac{1}{2})$	(2) $2(0,0,\frac{1}{2})\ 0,\frac{1}{4},z$	(3) $2(0,\frac{1}{2},0)\ 0,y,\frac{1}{4}$	(4) $2\ x,\frac{1}{4},\frac{1}{4}$
(5) $3^{+}(\frac{1}{3},\frac{1}{3},\frac{1}{3})\ x-\frac{1}{3},x-\frac{1}{6},x$	(6) $3^{+}\ \bar{x},x+\frac{1}{2},\bar{x}$	(7) $3^{+}(-\frac{1}{3},\frac{1}{3},\frac{1}{3})\ x+\frac{1}{3},\bar{x}-\frac{1}{6},\bar{x}$	(8) $3^{+}\ \bar{x},\bar{x}+\frac{1}{2},x$
(9) $3^{-}(\frac{1}{3},\frac{1}{3},\frac{1}{3})\ x-\frac{1}{6},x+\frac{1}{6},x$	(10) $3^{-}(-\frac{1}{3},\frac{1}{3},\frac{1}{3})\ x+\frac{1}{6},\bar{x}+\frac{1}{6},\bar{x}$	(11) $3^{-}\ \bar{x}+\frac{1}{2},\bar{x}+\frac{1}{2},x$	(12) $3^{-}\ \bar{x}-\frac{1}{2},x+\frac{1}{2},\bar{x}$
(13) $2(\frac{1}{4},\frac{1}{4},0)\ x,x+\frac{1}{4},\frac{1}{4}$	(14) $2(-\frac{1}{4},\frac{1}{4},0)\ x,\bar{x}+\frac{1}{4},\frac{1}{4}$	(15) $4^{-}(0,0,\frac{1}{2})\ \frac{1}{4},\frac{1}{4},z$	(16) $4^{+}(0,0,\frac{1}{2})\ -\frac{1}{4},\frac{1}{4},z$
(17) $4^{-}\ x,\frac{1}{2},0$	(18) $2(0,\frac{1}{2},\frac{1}{2})\ 0,y,y$	(19) $2\ 0,y+\frac{1}{2},\bar{y}$	(20) $4^{+}\ x,0,\frac{1}{2}$
(21) $4^{+}(0,\frac{1}{2},0)\ \frac{1}{4},y,\frac{1}{4}$	(22) $2(\frac{1}{4},0,\frac{1}{4})\ x-\frac{1}{4},\frac{1}{4},x$	(23) $4^{-}(0,\frac{1}{2},0)\ -\frac{1}{4},y,\frac{1}{4}$	(24) $2(-\frac{1}{4},0,\frac{1}{4})\ \bar{x}+\frac{1}{4},\frac{1}{4},x$
(25) $\bar{1}\ 0,\frac{1}{4},\frac{1}{4}$	(26) $b\ x,y,\frac{1}{4}$	(27) $c\ x,\frac{1}{4},z$	(28) $n(0,\frac{1}{2},\frac{1}{2})\ 0,y,z$
(29) $\bar{3}^{+}\ x,x+\frac{1}{2},x;\ 0,\frac{1}{2},0$	(30) $\bar{3}^{+}\ \bar{x}-1,x+\frac{1}{2},\bar{x};\ -\frac{1}{2},0,\frac{1}{2}$	(31) $\bar{3}^{+}\ x,\bar{x}+\frac{1}{2},\bar{x};\ 0,\frac{1}{2},0$	(32) $\bar{3}^{+}\ \bar{x}+1,\bar{x}+\frac{1}{2},x;\ \frac{1}{2},0,\frac{1}{2}$
(33) $\bar{3}^{-}\ x-\frac{1}{2},x-\frac{1}{2},x;\ 0,0,\frac{1}{2}$	(34) $\bar{3}^{-}\ x+\frac{1}{2},\bar{x}-\frac{1}{2},\bar{x};\ 0,0,\frac{1}{2}$	(35) $\bar{3}^{-}\ \bar{x}-\frac{1}{2},\bar{x}+\frac{1}{2},x;\ -\frac{1}{2},\frac{1}{2},0$	(36) $\bar{3}^{-}\ \bar{x}+\frac{1}{2},x+\frac{1}{2},\bar{x};\ \frac{1}{2},\frac{1}{2},0$
(37) $g(-\frac{1}{4},\frac{1}{4},\frac{1}{2})\ x+\frac{1}{4},\bar{x},z$	(38) $g(\frac{1}{4},\frac{1}{4},\frac{1}{2})\ x-\frac{1}{4},x,z$	(39) $\bar{4}^{-}\ -\frac{1}{4},\frac{1}{4},z;\ -\frac{1}{4},\frac{1}{4},\frac{1}{4}$	(40) $\bar{4}^{+}\ \frac{1}{4},\frac{1}{4},z;\ \frac{1}{4},\frac{1}{4},\frac{1}{4}$
(41) $\bar{4}^{-}\ x,0,\frac{1}{2};\ 0,0,\frac{1}{2}$	(42) $m\ x,y+\frac{1}{2},\bar{y}$	(43) $g(0,\frac{1}{2},\frac{1}{2})\ x,y,y$	(44) $\bar{4}^{+}\ x,\frac{1}{2},0;\ 0,\frac{1}{2},0$
(45) $\bar{4}^{+}\ -\frac{1}{4},y,\frac{1}{4};\ -\frac{1}{4},\frac{1}{4},\frac{1}{4}$	(46) $g(-\frac{1}{4},\frac{1}{2},\frac{1}{4})\ \bar{x}+\frac{1}{4},y,x$	(47) $\bar{4}^{-}\ \frac{1}{4},y,\frac{1}{4};\ \frac{1}{4},\frac{1}{4},\frac{1}{4}$	(48) $g(\frac{1}{4},\frac{1}{2},\frac{1}{4})\ x-\frac{1}{4},y,x$

For $(\frac{1}{2},0,\frac{1}{2})+$ set

(1) $t(\frac{1}{2},0,\frac{1}{2})$	(2) $2(0,0,\frac{1}{2})\ \frac{1}{4},0,z$	(3) $2\ \frac{1}{4},y,\frac{1}{4}$	(4) $2(\frac{1}{2},0,0)\ x,0,\frac{1}{4}$
(5) $3^{+}(\frac{1}{3},\frac{1}{3},\frac{1}{3})\ x+\frac{1}{6},x-\frac{1}{6},x$	(6) $3^{+}(\frac{1}{3},-\frac{1}{3},\frac{1}{3})\ \bar{x}+\frac{1}{6},x+\frac{1}{6},\bar{x}$	(7) $3^{+}\ x+\frac{1}{2},\bar{x}-\frac{1}{2},\bar{x}$	(8) $3^{+}\ \bar{x}+\frac{1}{2},\bar{x}+\frac{1}{2},x$
(9) $3^{-}(\frac{1}{3},\frac{1}{3},\frac{1}{3})\ x-\frac{1}{6},x-\frac{1}{3},x$	(10) $3^{-}\ x+\frac{1}{2},\bar{x},\bar{x}$	(11) $3^{-}\ \bar{x}+\frac{1}{2},\bar{x},x$	(12) $3^{-}(\frac{1}{3},-\frac{1}{3},\frac{1}{3})\ \bar{x}-\frac{1}{6},x+\frac{1}{3},\bar{x}$
(13) $2(\frac{1}{4},\frac{1}{4},0)\ x,x-\frac{1}{4},\frac{1}{4}$	(14) $2(\frac{1}{4},-\frac{1}{4},0)\ x,\bar{x}+\frac{1}{4},\frac{1}{4}$	(15) $4^{-}(0,0,\frac{1}{2})\ \frac{1}{4},-\frac{1}{4},z$	(16) $4^{+}(0,0,\frac{1}{2})\ \frac{1}{4},\frac{1}{4},z$
(17) $4^{-}(\frac{1}{2},0,0)\ x,\frac{1}{4},\frac{1}{4}$	(18) $2(0,\frac{1}{4},\frac{1}{4})\ \frac{1}{4},y-\frac{1}{4},y$	(19) $2(0,-\frac{1}{4},\frac{1}{4})\ \frac{1}{4},y+\frac{1}{4},\bar{y}$	(20) $4^{+}(\frac{1}{2},0,0)\ x,-\frac{1}{4},\frac{1}{4}$
(21) $4^{+}\ \frac{1}{2},y,0$	(22) $2(\frac{1}{2},0,\frac{1}{2})\ x,0,x$	(23) $4^{-}\ 0,y,\frac{1}{2}$	(24) $2\ \bar{x}+\frac{1}{2},0,x$
(25) $\bar{1}\ \frac{1}{4},0,\frac{1}{4}$	(26) $a\ x,y,\frac{1}{4}$	(27) $n(\frac{1}{2},0,\frac{1}{2})\ x,0,z$	(28) $c\ \frac{1}{4},y,z$
(29) $\bar{3}^{+}\ x-\frac{1}{2},x-\frac{1}{2},x;\ 0,0,\frac{1}{2}$	(30) $\bar{3}^{+}\ \bar{x}-\frac{1}{2},x+\frac{1}{2},\bar{x};\ 0,0,\frac{1}{2}$	(31) $\bar{3}^{+}\ x+\frac{1}{2},\bar{x}+\frac{1}{2},\bar{x};\ \frac{1}{2},\frac{1}{2},0$	(32) $\bar{3}^{+}\ \bar{x}+\frac{1}{2},\bar{x}-\frac{1}{2},x;\ \frac{1}{2},-\frac{1}{2},0$
(33) $\bar{3}^{-}\ x+\frac{1}{2},x,x;\ \frac{1}{2},0,0$	(34) $\bar{3}^{-}\ x+\frac{1}{2},\bar{x}-1,\bar{x};\ 0,-\frac{1}{2},\frac{1}{2}$	(35) $\bar{3}^{-}\ \bar{x}+\frac{1}{2},\bar{x}+1,x;\ 0,\frac{1}{2},\frac{1}{2}$	(36) $\bar{3}^{-}\ \bar{x}+\frac{1}{2},x,\bar{x};\ \frac{1}{2},0,0$
(37) $g(\frac{1}{4},-\frac{1}{4},\frac{1}{2})\ x+\frac{1}{4},\bar{x},z$	(38) $g(\frac{1}{4},\frac{1}{4},\frac{1}{2})\ x+\frac{1}{4},x,z$	(39) $\bar{4}^{-}\ \frac{1}{4},\frac{1}{4},z;\ \frac{1}{4},\frac{1}{4},\frac{1}{4}$	(40) $\bar{4}^{+}\ \frac{1}{4},-\frac{1}{4},z;\ \frac{1}{4},-\frac{1}{4},\frac{1}{4}$
(41) $\bar{4}^{-}\ x,-\frac{1}{4},\frac{1}{4};\ \frac{1}{4},-\frac{1}{4},\frac{1}{4}$	(42) $g(\frac{1}{2},-\frac{1}{4},\frac{1}{4})\ x,y+\frac{1}{4},\bar{y}$	(43) $g(\frac{1}{2},\frac{1}{4},\frac{1}{4})\ x,y-\frac{1}{4},y$	(44) $\bar{4}^{+}\ x,\frac{1}{4},\frac{1}{4};\ \frac{1}{4},\frac{1}{4},\frac{1}{4}$
(45) $\bar{4}^{+}\ 0,y,\frac{1}{2};\ 0,0,\frac{1}{2}$	(46) $m\ \bar{x}+\frac{1}{2},y,x$	(47) $\bar{4}^{-}\ \frac{1}{2},y,0;\ \frac{1}{2},0,0$	(48) $g(\frac{1}{2},0,\frac{1}{2})\ x,y,x$

For $(\frac{1}{2},\frac{1}{2},0)+$ set

(1) $t(\frac{1}{2},\frac{1}{2},0)$	(2) $2\ \frac{1}{4},\frac{1}{4},z$	(3) $2(0,\frac{1}{2},0)\ \frac{1}{4},y,0$	(4) $2(\frac{1}{2},0,0)\ x,\frac{1}{4},0$
(5) $3^{+}(\frac{1}{3},\frac{1}{3},\frac{1}{3})\ x+\frac{1}{6},x+\frac{1}{3},x$	(6) $3^{+}\ \bar{x}+\frac{1}{2},x,\bar{x}$	(7) $3^{+}\ x+\frac{1}{2},\bar{x},\bar{x}$	(8) $3^{+}(\frac{1}{3},\frac{1}{3},-\frac{1}{3})\ \bar{x}+\frac{1}{6},\bar{x}+\frac{1}{3},x$
(9) $3^{-}(\frac{1}{3},\frac{1}{3},\frac{1}{3})\ x+\frac{1}{3},x+\frac{1}{6},x$	(10) $3^{-}\ x,\bar{x}+\frac{1}{2},\bar{x}$	(11) $3^{-}(\frac{1}{3},\frac{1}{3},-\frac{1}{3})\ \bar{x}+\frac{1}{3},\bar{x}+\frac{1}{6},x$	(12) $3^{-}\ \bar{x},x+\frac{1}{2},\bar{x}$
(13) $2(\frac{1}{2},\frac{1}{2},0)\ x,x,0$	(14) $2\ x,\bar{x}+\frac{1}{2},0$	(15) $4^{-}\ \frac{1}{2},0,z$	(16) $4^{+}\ 0,\frac{1}{2},z$
(17) $4^{-}(\frac{1}{2},0,0)\ x,\frac{1}{4},-\frac{1}{4}$	(18) $2(0,\frac{1}{4},\frac{1}{4})\ \frac{1}{4},y+\frac{1}{4},y$	(19) $2(0,\frac{1}{4},-\frac{1}{4})\ \frac{1}{4},y+\frac{1}{4},\bar{y}$	(20) $4^{+}(\frac{1}{2},0,0)\ x,\frac{1}{4},\frac{1}{4}$
(21) $4^{+}(0,\frac{1}{2},0)\ \frac{1}{4},y,-\frac{1}{4}$	(22) $2(\frac{1}{4},0,\frac{1}{4})\ x+\frac{1}{4},\frac{1}{4},x$	(23) $4^{-}(0,\frac{1}{2},0)\ \frac{1}{4},y,\frac{1}{4}$	(24) $2(\frac{1}{4},0,-\frac{1}{4})\ \bar{x}+\frac{1}{4},\frac{1}{4},x$
(25) $\bar{1}\ \frac{1}{4},\frac{1}{4},0$	(26) $n(\frac{1}{2},\frac{1}{2},0)\ x,y,0$	(27) $a\ x,\frac{1}{4},z$	(28) $b\ \frac{1}{4},y,z$
(29) $\bar{3}^{+}\ x+\frac{1}{2},x,x;\ \frac{1}{2},0,0$	(30) $\bar{3}^{+}\ \bar{x}-\frac{1}{2},x+1,\bar{x};\ 0,\frac{1}{2},\frac{1}{2}$	(31) $\bar{3}^{+}\ x-\frac{1}{2},\bar{x}+1,\bar{x};\ 0,\frac{1}{2},-\frac{1}{2}$	(32) $\bar{3}^{+}\ \bar{x}+\frac{1}{2},\bar{x},x;\ \frac{1}{2},0,0$
(33) $\bar{3}^{-}\ x,x+\frac{1}{2},x;\ 0,\frac{1}{2},0$	(34) $\bar{3}^{-}\ x+1,\bar{x}-\frac{1}{2},\bar{x};\ \frac{1}{2},0,\frac{1}{2}$	(35) $\bar{3}^{-}\ \bar{x},\bar{x}+\frac{1}{2},x;\ 0,\frac{1}{2},0$	(36) $\bar{3}^{-}\ \bar{x}+1,x-\frac{1}{2},\bar{x};\ \frac{1}{2},0,-\frac{1}{2}$
(37) $m\ x+\frac{1}{2},\bar{x},z$	(38) $g(\frac{1}{2},\frac{1}{2},0)\ x,x,z$	(39) $\bar{4}^{-}\ 0,\frac{1}{2},z;\ 0,\frac{1}{2},0$	(40) $\bar{4}^{+}\ \frac{1}{2},0,z;\ \frac{1}{2},0,0$
(41) $\bar{4}^{-}\ x,\frac{1}{4},\frac{1}{4};\ \frac{1}{4},\frac{1}{4},\frac{1}{4}$	(42) $g(\frac{1}{2},\frac{1}{4},-\frac{1}{4})\ x,y+\frac{1}{4},\bar{y}$	(43) $g(\frac{1}{2},\frac{1}{4},\frac{1}{4})\ x,y+\frac{1}{4},y$	(44) $\bar{4}^{+}\ x,\frac{1}{4},-\frac{1}{4};\ \frac{1}{4},\frac{1}{4},-\frac{1}{4}$
(45) $\bar{4}^{+}\ \frac{1}{4},y,\frac{1}{4};\ \frac{1}{4},\frac{1}{4},\frac{1}{4}$	(46) $g(\frac{1}{4},\frac{1}{2},-\frac{1}{4})\ \bar{x}+\frac{1}{4},y,x$	(47) $\bar{4}^{-}\ \frac{1}{4},y,-\frac{1}{4};\ \frac{1}{4},\frac{1}{4},-\frac{1}{4}$	(48) $g(\frac{1}{4},\frac{1}{2},\frac{1}{4})\ x+\frac{1}{4},y,x$

Generators selected (1); $t(1,0,0)$; $t(0,1,0)$; $t(0,0,1)$; $t(0,\frac{1}{2},\frac{1}{2})$; $t(\frac{1}{2},0,\frac{1}{2})$; (2); (3); (5); (13); (25)

Positions

Multiplicity, Wyckoff letter, Site symmetry

Coordinates

$(0,0,0)+$ $(0,\frac{1}{2},\frac{1}{2})+$ $(\frac{1}{2},0,\frac{1}{2})+$ $(\frac{1}{2},\frac{1}{2},0)+$

Reflection conditions

h,k,l permutable

General:

192	l	1	(1) x,y,z	(2) $\bar{x},\bar{y},z$	(3) $\bar{x},y,\bar{z}$	(4) $x,\bar{y},\bar{z}$	hkl: $h+k,h+l,k+l=2n$
			(5) z,x,y	(6) $z,\bar{x},\bar{y}$	(7) $\bar{z},\bar{x},y$	(8) $\bar{z},x,\bar{y}$	$0kl$: $k,l=2n$
			(9) y,z,x	(10) $\bar{y},z,\bar{x}$	(11) $y,\bar{z},\bar{x}$	(12) $\bar{y},\bar{z},x$	hhl: $h+l=2n$
			(13) $y,x,\bar{z}$	(14) $\bar{y},\bar{x},\bar{z}$	(15) $y,\bar{x},z$	(16) $\bar{y},x,z$	$h00$: $h=2n$
			(17) $x,z,\bar{y}$	(18) $\bar{x},z,y$	(19) $\bar{x},\bar{z},\bar{y}$	(20) $x,\bar{z},y$	
			(21) $z,y,\bar{x}$	(22) $z,\bar{y},x$	(23) $\bar{z},y,x$	(24) $\bar{z},\bar{y},\bar{x}$	
			(25) $\bar{x},\bar{y},\bar{z}$	(26) $x,y,\bar{z}$	(27) $x,\bar{y},z$	(28) $\bar{x},y,z$	
			(29) $\bar{z},\bar{x},\bar{y}$	(30) $\bar{z},x,y$	(31) $z,x,\bar{y}$	(32) $z,\bar{x},y$	
			(33) $\bar{y},\bar{z},\bar{x}$	(34) $y,\bar{z},x$	(35) $\bar{y},z,x$	(36) $y,z,\bar{x}$	
			(37) $\bar{y},\bar{x},z$	(38) y,x,z	(39) $\bar{y},x,\bar{z}$	(40) $y,\bar{x},\bar{z}$	
			(41) $\bar{x},\bar{z},y$	(42) $x,\bar{z},\bar{y}$	(43) x,z,y	(44) $\bar{x},z,\bar{y}$	
			(45) $\bar{z},\bar{y},x$	(46) $\bar{z},y,\bar{x}$	(47) $z,\bar{y},\bar{x}$	(48) z,y,x	

Special: as above, plus

96	k	$..m$	x,x,z	$\bar{x},\bar{x},z$	$\bar{x},x,\bar{z}$	$x,\bar{x},\bar{z}$	z,x,x	$z,\bar{x},\bar{x}$	no extra conditions
			$\bar{z},\bar{x},x$	$\bar{z},x,\bar{x}$	x,z,x	$\bar{x},z,\bar{x}$	$x,\bar{z},\bar{x}$	$\bar{x},\bar{z},x$	
			$x,x,\bar{z}$	$\bar{x},\bar{x},\bar{z}$	$x,\bar{x},z$	$\bar{x},x,z$	$x,z,\bar{x}$	$\bar{x},z,x$	
			$\bar{x},\bar{z},\bar{x}$	$x,\bar{z},x$	$z,x,\bar{x}$	$z,\bar{x},x$	$\bar{z},x,x$	$\bar{z},\bar{x},\bar{x}$	
96	j	$m..$	$0,y,z$	$0,\bar{y},z$	$0,y,\bar{z}$	$0,\bar{y},\bar{z}$	$z,0,y$	$z,0,\bar{y}$	no extra conditions
			$\bar{z},0,y$	$\bar{z},0,\bar{y}$	$y,z,0$	$\bar{y},z,0$	$y,\bar{z},0$	$\bar{y},\bar{z},0$	
			$y,0,\bar{z}$	$\bar{y},0,\bar{z}$	$y,0,z$	$\bar{y},0,z$	$0,z,\bar{y}$	$0,z,y$	
			$0,\bar{z},\bar{y}$	$0,\bar{z},y$	$z,y,0$	$z,\bar{y},0$	$\bar{z},y,0$	$\bar{z},\bar{y},0$	
48	i	$m.m2$	$\frac{1}{2},y,y$	$\frac{1}{2},\bar{y},y$	$\frac{1}{2},y,\bar{y}$	$\frac{1}{2},\bar{y},\bar{y}$	$y,\frac{1}{2},y$	$y,\frac{1}{2},\bar{y}$	no extra conditions
			$\bar{y},\frac{1}{2},y$	$\bar{y},\frac{1}{2},\bar{y}$	$y,y,\frac{1}{2}$	$\bar{y},y,\frac{1}{2}$	$y,\bar{y},\frac{1}{2}$	$\bar{y},\bar{y},\frac{1}{2}$	
48	h	$m.m2$	$0,y,y$	$0,\bar{y},y$	$0,y,\bar{y}$	$0,\bar{y},\bar{y}$	$y,0,y$	$y,0,\bar{y}$	no extra conditions
			$\bar{y},0,y$	$\bar{y},0,\bar{y}$	$y,y,0$	$\bar{y},y,0$	$y,\bar{y},0$	$\bar{y},\bar{y},0$	
48	g	$2.mm$	$x,\frac{1}{4},\frac{1}{4}$	$\bar{x},\frac{3}{4},\frac{1}{4}$	$\frac{1}{4},x,\frac{1}{4}$	$\frac{1}{4},\bar{x},\frac{3}{4}$	$\frac{1}{4},\frac{1}{4},x$	$\frac{3}{4},\frac{1}{4},\bar{x}$	hkl: $h=2n$
			$\frac{1}{4},x,\frac{3}{4}$	$\frac{3}{4},\bar{x},\frac{3}{4}$	$x,\frac{1}{4},\frac{3}{4}$	$\bar{x},\frac{1}{4},\frac{1}{4}$	$\frac{1}{4},\frac{1}{4},\bar{x}$	$\frac{1}{4},\frac{3}{4},x$	
32	f	$.3m$	x,x,x	$\bar{x},\bar{x},x$	$\bar{x},x,\bar{x}$	$x,\bar{x},\bar{x}$			no extra conditions
			$x,x,\bar{x}$	$\bar{x},\bar{x},\bar{x}$	$x,\bar{x},x$	$\bar{x},x,x$			
24	e	$4m.m$	$x,0,0$	$\bar{x},0,0$	$0,x,0$	$0,\bar{x},0$	$0,0,x$	$0,0,\bar{x}$	no extra conditions
24	d	$m.mm$	$0,\frac{1}{4},\frac{1}{4}$	$0,\frac{3}{4},\frac{1}{4}$	$\frac{1}{4},0,\frac{1}{4}$	$\frac{1}{4},0,\frac{3}{4}$	$\frac{1}{4},\frac{1}{4},0$	$\frac{3}{4},\frac{1}{4},0$	hkl: $h=2n$
8	c	$\bar{4}3m$	$\frac{1}{4},\frac{1}{4},\frac{1}{4}$	$\frac{1}{4},\frac{1}{4},\frac{3}{4}$					hkl: $h=2n$
4	b	$m\bar{3}m$	$\frac{1}{2},\frac{1}{2},\frac{1}{2}$						no extra conditions
4	a	$m\bar{3}m$	$0,0,0$						no extra conditions

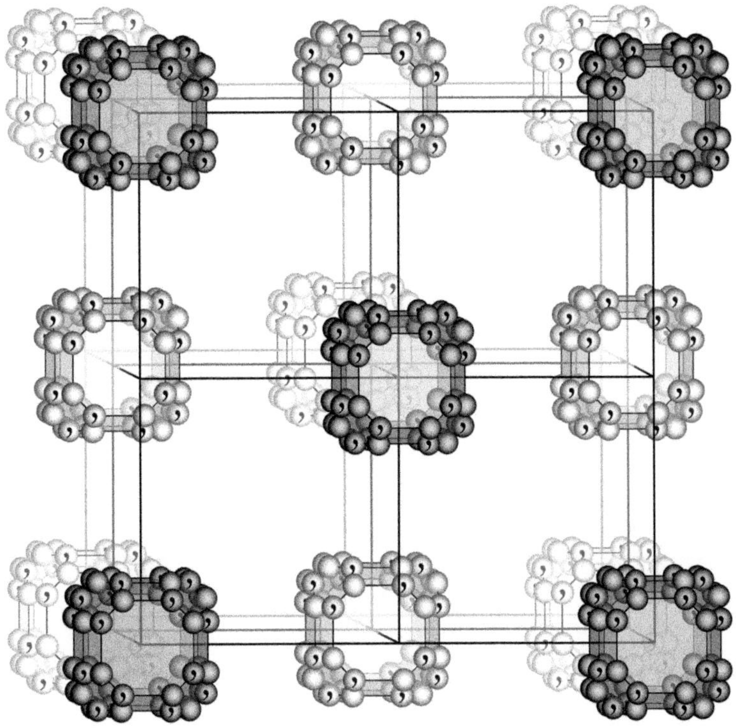

Symmetry of special projections

Along [001] $p4mm$
$\mathbf{a}' = \frac{1}{2}\mathbf{a}$ $\mathbf{b}' = \frac{1}{2}\mathbf{b}$
Origin at $0,0,z$

Along [111] $p6mm$
$\mathbf{a}' = \frac{1}{6}(2\mathbf{a} - \mathbf{b} - \mathbf{c})$ $\mathbf{b}' = \frac{1}{6}(-\mathbf{a} + 2\mathbf{b} - \mathbf{c})$
Origin at x,x,x

Along [110] $c2mm$
$\mathbf{a}' = \frac{1}{2}(-\mathbf{a} + \mathbf{b})$ $\mathbf{b}' = \mathbf{c}$
Origin at $x,x,0$

$Fd\bar{3}m$ O_h^7 $m\bar{3}m$ Cubic

No. 227 $F\ 4_1/d\ \bar{3}\ 2/m$ Patterson symmetry $Fm\bar{3}m$

ORIGIN CHOICE 2

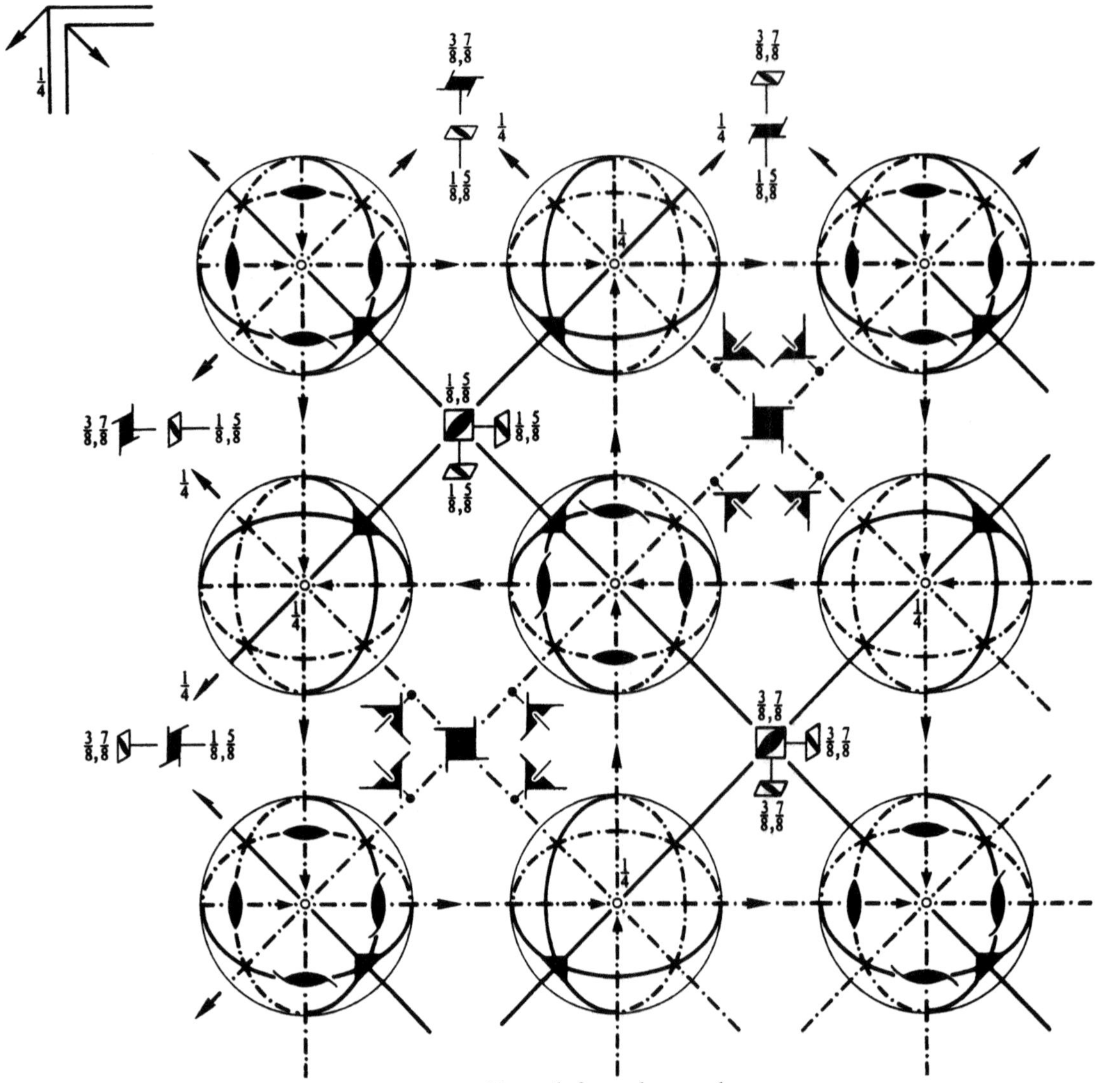

Upper left quadrant only

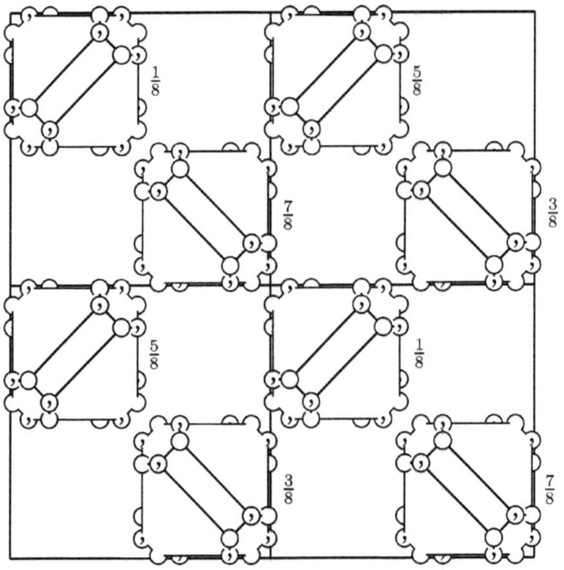

Origin at centre ($\bar{3}m$), at $\frac{1}{8},\frac{1}{8},\frac{1}{8}$ from $\bar{4}3m$

Asymmetric unit $-\frac{1}{8}\le x\le\frac{3}{8};\quad -\frac{1}{8}\le y\le 0;\quad -\frac{1}{4}\le z\le 0;\quad y\le\min(\frac{1}{4}-x,x);\quad -y-\frac{1}{4}\le z\le y$

Vertices $-\frac{1}{8},-\frac{1}{8},-\frac{1}{8}\quad \frac{3}{8},-\frac{1}{8},-\frac{1}{8}\quad \frac{1}{4},0,0\quad 0,0,0\quad \frac{1}{4},0,-\frac{1}{4}\quad 0,0,-\frac{1}{4}$

Symmetry operations

For (0,0,0)+ set

(1) 1
(2) $2(0,0,\frac{1}{2})\quad \frac{3}{8},\frac{1}{8},z$
(3) $2(0,\frac{1}{2},0)\quad \frac{1}{8},y,\frac{3}{8}$
(4) $2(\frac{1}{2},0,0)\quad x,\frac{3}{8},\frac{1}{8}$
(5) $3^{+}\quad x,x,x$
(6) $3^{+}\quad \bar{x}+\frac{1}{2},x+\frac{1}{4},\bar{x}$
(7) $3^{+}\quad x+\frac{3}{4},\bar{x}-\frac{1}{2},\bar{x}$
(8) $3^{+}\quad \bar{x}+\frac{1}{4},\bar{x}+\frac{3}{4},x$
(9) $3^{-}\quad x,x,x$
(10) $3^{-}(-\frac{1}{3},\frac{1}{3},\frac{1}{3})\quad x+\frac{5}{12},\bar{x}+\frac{1}{6},\bar{x}$
(11) $3^{-}(\frac{1}{3},\frac{1}{3},-\frac{1}{3})\quad \bar{x}+\frac{7}{12},\bar{x}+\frac{5}{12},x$
(12) $3^{-}(\frac{1}{3},-\frac{1}{3},\frac{1}{3})\quad \bar{x}-\frac{1}{6},x+\frac{7}{12},\bar{x}$
(13) $2(\frac{1}{2},\frac{1}{2},0)\quad x,x-\frac{1}{4},\frac{1}{4}$
(14) $2\quad x,\bar{x},0$
(15) $4^{-}(0,0,\frac{3}{4})\quad \frac{3}{8},\frac{1}{8},z$
(16) $4^{+}(0,0,\frac{1}{4})\quad -\frac{1}{8},\frac{5}{8},z$
(17) $4^{-}(\frac{3}{4},0,0)\quad x,\frac{3}{8},\frac{1}{8}$
(18) $2(0,\frac{1}{2},\frac{1}{2})\quad \frac{1}{4},y+\frac{1}{4},y$
(19) $2\quad 0,y,\bar{y}$
(20) $4^{+}(\frac{1}{4},0,0)\quad x,-\frac{1}{8},\frac{5}{8}$
(21) $4^{+}(0,\frac{1}{4},0)\quad \frac{5}{8},y,-\frac{1}{8}$
(22) $2(\frac{1}{2},0,\frac{1}{2})\quad x-\frac{1}{4},\frac{1}{4},x$
(23) $4^{-}(0,\frac{3}{4},0)\quad \frac{1}{8},y,\frac{3}{8}$
(24) $2\quad \bar{x},0,x$
(25) $\bar{1}\quad 0,0,0$
(26) $d(\frac{1}{4},\frac{3}{4},0)\quad x,y,\frac{1}{4}$
(27) $d(\frac{3}{4},0,\frac{1}{4})\quad x,\frac{1}{4},z$
(28) $d(0,\frac{1}{4},\frac{3}{4})\quad \frac{1}{4},y,z$
(29) $\bar{3}^{+}\quad x,x,x;\ 0,0,0$
(30) $\bar{3}^{+}\quad \bar{x}-1,x+\frac{3}{4},\bar{x};\ -\frac{1}{4},0,\frac{3}{4}$
(31) $\bar{3}^{+}\quad x-\frac{1}{4},\bar{x}+1,\bar{x};\ 0,\frac{3}{4},-\frac{1}{4}$
(32) $\bar{3}^{+}\quad \bar{x}+\frac{3}{4},\bar{x}-\frac{1}{4},x;\ \frac{3}{4},-\frac{1}{4},0$
(33) $\bar{3}^{-}\quad x,x,x;\ 0,0,0$
(34) $\bar{3}^{-}\quad x+\frac{5}{4},\bar{x}-1,\bar{x};\ \frac{1}{2},-\frac{1}{4},\frac{3}{4}$
(35) $\bar{3}^{-}\quad \bar{x}+\frac{1}{4},\bar{x}+\frac{5}{4},x;\ -\frac{1}{4},\frac{3}{4},\frac{1}{2}$
(36) $\bar{3}^{-}\quad \bar{x}+1,x+\frac{1}{4},\bar{x};\ \frac{3}{4},\frac{1}{2},-\frac{1}{4}$
(37) $g(-\frac{1}{4},\frac{1}{4},\frac{1}{2})\quad x+\frac{1}{2},\bar{x},z$
(38) $m\quad x,x,z$
(39) $\bar{4}^{-}\quad \frac{1}{8},\frac{5}{8},z;\ \frac{1}{8},\frac{5}{8},\frac{1}{8}$
(40) $\bar{4}^{+}\quad \frac{3}{8},-\frac{1}{8},z;\ \frac{3}{8},-\frac{1}{8},\frac{3}{8}$
(41) $\bar{4}^{-}\quad x,\frac{1}{8},\frac{5}{8};\ \frac{1}{8},\frac{1}{8},\frac{5}{8}$
(42) $g(\frac{1}{2},-\frac{1}{4},\frac{1}{4})\quad x,y+\frac{1}{2},\bar{y}$
(43) $m\quad x,y,y$
(44) $\bar{4}^{+}\quad x,\frac{3}{8},-\frac{1}{8};\ \frac{3}{8},\frac{3}{8},-\frac{1}{8}$
(45) $\bar{4}^{+}\quad -\frac{1}{8},y,\frac{3}{8};\ -\frac{1}{8},\frac{3}{8},\frac{3}{8}$
(46) $g(\frac{1}{4},\frac{1}{2},-\frac{1}{4})\quad \bar{x}+\frac{1}{2},y,x$
(47) $\bar{4}^{-}\quad \frac{5}{8},y,\frac{1}{8};\ \frac{5}{8},\frac{1}{8},\frac{1}{8}$
(48) $m\quad x,y,x$

For $(0,\frac{1}{2},\frac{1}{2})$+ set

(1) $t(0,\frac{1}{2},\frac{1}{2})$
(2) $2\quad \frac{3}{8},\frac{3}{8},z$
(3) $2\quad \frac{1}{8},y,\frac{1}{8}$
(4) $2(\frac{1}{2},0,0)\quad x,\frac{1}{8},\frac{3}{8}$
(5) $3^{+}(\frac{1}{3},\frac{1}{3},\frac{1}{3})\quad x-\frac{1}{3},x-\frac{1}{6},x$
(6) $3^{+}(\frac{1}{3},-\frac{1}{3},\frac{1}{3})\quad \bar{x}+\frac{1}{6},x+\frac{5}{12},\bar{x}$
(7) $3^{+}\quad x+\frac{3}{4},\bar{x},\bar{x}$
(8) $3^{+}\quad \bar{x}+\frac{1}{4},\bar{x}+\frac{1}{4},x$
(9) $3^{-}(\frac{1}{3},\frac{1}{3},\frac{1}{3})\quad x-\frac{1}{6},x+\frac{1}{6},x$
(10) $3^{-}\quad x+\frac{1}{4},\bar{x},\bar{x}$
(11) $3^{-}\quad \bar{x}+\frac{3}{4},\bar{x}+\frac{1}{4},x$
(12) $3^{-}\quad \bar{x},x+\frac{3}{4},\bar{x}$
(13) $2(\frac{3}{4},\frac{3}{4},0)\quad x,x,0$
(14) $2(-\frac{1}{4},\frac{1}{4},0)\quad x,\bar{x}+\frac{1}{4},\frac{1}{4}$
(15) $4^{-}(0,0,\frac{1}{4})\quad \frac{1}{8},-\frac{1}{8},z$
(16) $4^{+}(0,0,\frac{3}{4})\quad \frac{1}{8},\frac{3}{8},z$
(17) $4^{-}(\frac{3}{4},0,0)\quad x,\frac{3}{8},-\frac{3}{8}$
(18) $2(0,\frac{1}{2},\frac{1}{2})\quad \frac{1}{4},y-\frac{1}{4},y$
(19) $2\quad 0,y+\frac{1}{2},\bar{y}$
(20) $4^{+}(\frac{1}{4},0,0)\quad x,-\frac{1}{8},\frac{1}{8}$
(21) $4^{+}(0,\frac{3}{4},0)\quad \frac{3}{8},y,-\frac{3}{8}$
(22) $2(\frac{1}{4},0,\frac{1}{4})\quad x,0,x$
(23) $4^{-}(0,\frac{1}{4},0)\quad -\frac{1}{8},y,\frac{5}{8}$
(24) $2(-\frac{1}{4},0,\frac{1}{4})\quad \bar{x}+\frac{1}{4},\frac{1}{4},x$
(25) $\bar{1}\quad 0,\frac{1}{4},\frac{1}{4}$
(26) $d(\frac{1}{4},\frac{1}{4},0)\quad x,y,0$
(27) $d(\frac{3}{4},0,\frac{3}{4})\quad x,0,z$
(28) $d(0,\frac{3}{4},\frac{1}{4})\quad \frac{1}{4},y,z$
(29) $\bar{3}^{+}\quad x,x+\frac{1}{2},x;\ 0,\frac{1}{2},0$
(30) $\bar{3}^{+}\quad \bar{x}-1,x+\frac{5}{4},\bar{x};\ -\frac{1}{4},\frac{1}{2},\frac{3}{4}$
(31) $\bar{3}^{+}\quad x-\frac{1}{4},\bar{x}+\frac{1}{2},\bar{x};\ 0,\frac{1}{4},-\frac{1}{4}$
(32) $\bar{3}^{+}\quad \bar{x}+\frac{3}{4},\bar{x}-\frac{3}{4},x;\ \frac{3}{4},-\frac{3}{4},0$
(33) $\bar{3}^{-}\quad x-\frac{1}{2},x-\frac{1}{2},x;\ 0,0,\frac{1}{2}$
(34) $\bar{3}^{-}\quad x+\frac{3}{4},\bar{x}-\frac{3}{2},\bar{x};\ 0,-\frac{3}{4},\frac{3}{4}$
(35) $\bar{3}^{-}\quad \bar{x}-\frac{1}{4},\bar{x}+\frac{3}{4},x;\ -\frac{1}{4},\frac{3}{4},0$
(36) $\bar{3}^{-}\quad \bar{x}+\frac{1}{2},x-\frac{1}{4},\bar{x};\ \frac{1}{4},0,-\frac{1}{4}$
(37) $m\quad x+\frac{1}{4},\bar{x},z$
(38) $g(\frac{1}{4},\frac{1}{4},\frac{1}{2})\quad x-\frac{1}{4},x,z$
(39) $\bar{4}^{-}\quad \frac{3}{8},\frac{3}{8},z;\ \frac{3}{8},\frac{3}{8},\frac{3}{8}$
(40) $\bar{4}^{+}\quad \frac{5}{8},\frac{1}{8},z;\ \frac{5}{8},\frac{1}{8},\frac{1}{8}$
(41) $\bar{4}^{-}\quad x,\frac{1}{8},\frac{1}{8};\ \frac{1}{8},\frac{1}{8},\frac{1}{8}$
(42) $g(\frac{1}{2},\frac{1}{4},-\frac{1}{4})\quad x,y+\frac{1}{2},\bar{y}$
(43) $g(0,\frac{1}{2},\frac{1}{2})\quad x,y,y$
(44) $\bar{4}^{+}\quad x,\frac{3}{8},\frac{3}{8};\ \frac{3}{8},\frac{3}{8},\frac{3}{8}$
(45) $\bar{4}^{+}\quad \frac{1}{8},y,\frac{1}{8};\ \frac{1}{8},\frac{1}{8},\frac{1}{8}$
(46) $m\quad \bar{x}+\frac{3}{4},y,x$
(47) $\bar{4}^{-}\quad \frac{3}{8},y,-\frac{1}{8};\ \frac{3}{8},\frac{3}{8},-\frac{1}{8}$
(48) $g(\frac{1}{4},\frac{1}{2},\frac{1}{4})\quad x-\frac{1}{4},y,x$

For $(\frac{1}{2},0,\frac{1}{2})$+ set

(1) $t(\frac{1}{2},0,\frac{1}{2})$
(2) $2\quad \frac{1}{8},\frac{1}{8},z$
(3) $2(0,\frac{1}{2},0)\quad \frac{3}{8},y,\frac{1}{8}$
(4) $2\quad x,\frac{3}{8},\frac{3}{8}$
(5) $3^{+}(\frac{1}{3},\frac{1}{3},\frac{1}{3})\quad x+\frac{1}{6},x-\frac{1}{6},x$
(6) $3^{+}\quad \bar{x},x+\frac{3}{4},\bar{x}$
(7) $3^{+}\quad x+\frac{1}{4},\bar{x},\bar{x}$
(8) $3^{+}(\frac{1}{3},\frac{1}{3},-\frac{1}{3})\quad \bar{x}+\frac{5}{12},\bar{x}+\frac{7}{12},x$
(9) $3^{-}(\frac{1}{3},\frac{1}{3},\frac{1}{3})\quad x-\frac{1}{6},x-\frac{1}{3},x$
(10) $3^{-}\quad x+\frac{1}{4},\bar{x}+\frac{1}{2},\bar{x}$
(11) $3^{-}\quad \bar{x}+\frac{3}{4},\bar{x}+\frac{3}{4},x$
(12) $3^{-}\quad \bar{x},x+\frac{1}{4},\bar{x}$
(13) $2(\frac{1}{4},\frac{1}{4},0)\quad x,x,0$
(14) $2(\frac{1}{4},-\frac{1}{4},0)\quad x,\bar{x}+\frac{1}{4},\frac{1}{4}$
(15) $4^{-}(0,0,\frac{1}{4})\quad \frac{5}{8},-\frac{1}{8},z$
(16) $4^{+}(0,0,\frac{3}{4})\quad -\frac{3}{8},\frac{3}{8},z$
(17) $4^{-}(\frac{1}{4},0,0)\quad x,\frac{1}{8},-\frac{1}{8}$
(18) $2(0,\frac{3}{4},\frac{3}{4})\quad 0,y,y$
(19) $2(0,-\frac{1}{4},\frac{1}{4})\quad \frac{1}{4},y+\frac{1}{4},\bar{y}$
(20) $4^{+}(\frac{3}{4},0,0)\quad x,\frac{1}{8},\frac{3}{8}$
(21) $4^{+}(0,\frac{1}{4},0)\quad \frac{1}{8},y,-\frac{1}{8}$
(22) $2(\frac{1}{2},0,\frac{1}{2})\quad x+\frac{1}{4},\frac{1}{4},x$
(23) $4^{-}(0,\frac{3}{4},0)\quad -\frac{3}{8},y,\frac{3}{8}$
(24) $2\quad \bar{x}+\frac{1}{2},0,x$
(25) $\bar{1}\quad \frac{1}{4},0,\frac{1}{4}$
(26) $d(\frac{3}{4},\frac{3}{4},0)\quad x,y,0$
(27) $d(\frac{1}{4},0,\frac{3}{4})\quad x,\frac{1}{4},z$
(28) $d(0,\frac{1}{4},\frac{1}{4})\quad 0,y,z$
(29) $\bar{3}^{+}\quad x-\frac{1}{2},x-\frac{1}{2},x;\ 0,0,\frac{1}{2}$
(30) $\bar{3}^{+}\quad \bar{x}-\frac{1}{2},x+\frac{1}{4},\bar{x};\ -\frac{1}{4},0,\frac{1}{4}$
(31) $\bar{3}^{+}\quad x-\frac{3}{4},\bar{x}+\frac{3}{2},\bar{x};\ 0,\frac{3}{4},-\frac{3}{4}$
(32) $\bar{3}^{+}\quad \bar{x}+\frac{5}{4},\bar{x}+\frac{1}{4},x;\ \frac{3}{4},-\frac{1}{4},\frac{1}{2}$
(33) $\bar{3}^{-}\quad x+\frac{1}{2},x,x;\ \frac{1}{2},0,0$
(34) $\bar{3}^{-}\quad x+\frac{3}{4},\bar{x}-1,\bar{x};\ 0,-\frac{1}{4},\frac{3}{4}$
(35) $\bar{3}^{-}\quad \bar{x}-\frac{1}{4},\bar{x}+\frac{1}{4},x;\ -\frac{1}{4},\frac{1}{4},0$
(36) $\bar{3}^{-}\quad \bar{x}+\frac{3}{2},x-\frac{3}{4},\bar{x};\ \frac{3}{4},0,-\frac{3}{4}$
(37) $m\quad x+\frac{3}{4},\bar{x},z$
(38) $g(\frac{1}{4},\frac{1}{4},\frac{1}{2})\quad x+\frac{1}{4},x,z$
(39) $\bar{4}^{-}\quad -\frac{1}{8},\frac{3}{8},z;\ -\frac{1}{8},\frac{3}{8},\frac{3}{8}$
(40) $\bar{4}^{+}\quad \frac{1}{8},\frac{1}{8},z;\ \frac{1}{8},\frac{1}{8},\frac{1}{8}$
(41) $\bar{4}^{-}\quad x,\frac{3}{8},\frac{3}{8};\ \frac{3}{8},\frac{3}{8},\frac{3}{8}$
(42) $m\quad x,y+\frac{1}{4},\bar{y}$
(43) $g(\frac{1}{2},\frac{1}{4},\frac{1}{4})\quad x,y-\frac{1}{4},y$
(44) $\bar{4}^{+}\quad x,\frac{5}{8},\frac{1}{8};\ \frac{1}{8},\frac{5}{8},\frac{1}{8}$
(45) $\bar{4}^{+}\quad \frac{3}{8},y,\frac{3}{8};\ \frac{3}{8},\frac{3}{8},\frac{3}{8}$
(46) $g(-\frac{1}{4},\frac{1}{2},\frac{1}{4})\quad \bar{x}+\frac{1}{2},y,x$
(47) $\bar{4}^{-}\quad \frac{1}{8},y,\frac{1}{8};\ \frac{1}{8},\frac{1}{8},\frac{1}{8}$
(48) $g(\frac{1}{2},0,\frac{1}{2})\quad x,y,x$

For $(\frac{1}{2},\frac{1}{2},0)$+ set

(1) $t(\frac{1}{2},\frac{1}{2},0)$
(2) $2(0,0,\frac{1}{2})\quad \frac{1}{8},\frac{3}{8},z$
(3) $2\quad \frac{3}{8},y,\frac{3}{8}$
(4) $2\quad x,\frac{1}{8},\frac{1}{8}$
(5) $3^{+}(\frac{1}{3},\frac{1}{3},\frac{1}{3})\quad x+\frac{1}{6},x+\frac{1}{3},x$
(6) $3^{+}\quad \bar{x},x+\frac{1}{4},\bar{x}$
(7) $3^{+}(-\frac{1}{3},\frac{1}{3},\frac{1}{3})\quad x+\frac{7}{12},\bar{x}-\frac{1}{6},\bar{x}$
(8) $3^{+}\quad \bar{x}+\frac{3}{4},\bar{x}+\frac{3}{4},x$
(9) $3^{-}(\frac{1}{3},\frac{1}{3},\frac{1}{3})\quad x+\frac{1}{3},x+\frac{1}{6},x$
(10) $3^{-}\quad x+\frac{3}{4},\bar{x},\bar{x}$
(11) $3^{-}\quad \bar{x}+\frac{1}{4},\bar{x}+\frac{1}{4},x$
(12) $3^{-}\quad \bar{x}-\frac{1}{2},x+\frac{3}{4},\bar{x}$
(13) $2(\frac{1}{2},\frac{1}{2},0)\quad x,x+\frac{1}{4},\frac{1}{4}$
(14) $2\quad x,\bar{x}+\frac{1}{2},0$
(15) $4^{-}(0,0,\frac{3}{4})\quad \frac{3}{8},-\frac{3}{8},z$
(16) $4^{+}(0,0,\frac{1}{4})\quad -\frac{1}{8},\frac{1}{8},z$
(17) $4^{-}(\frac{1}{4},0,0)\quad x,\frac{5}{8},-\frac{1}{8}$
(18) $2(0,\frac{1}{4},\frac{1}{4})\quad 0,y,y$
(19) $2(0,\frac{1}{4},-\frac{1}{4})\quad \frac{1}{4},y+\frac{1}{4},\bar{y}$
(20) $4^{+}(\frac{3}{4},0,0)\quad x,-\frac{3}{8},\frac{3}{8}$
(21) $4^{+}(0,\frac{3}{4},0)\quad \frac{3}{8},y,\frac{1}{8}$
(22) $2(\frac{3}{4},0,\frac{3}{4})\quad x,0,x$
(23) $4^{-}(0,\frac{1}{4},0)\quad -\frac{1}{8},y,\frac{1}{8}$
(24) $2(\frac{1}{4},0,-\frac{1}{4})\quad \bar{x}+\frac{1}{4},\frac{1}{4},x$
(25) $\bar{1}\quad \frac{1}{4},\frac{1}{4},0$
(26) $d(\frac{3}{4},\frac{1}{4},0)\quad x,y,\frac{1}{4}$
(27) $d(\frac{1}{4},0,\frac{1}{4})\quad x,0,z$
(28) $d(0,\frac{3}{4},\frac{3}{4})\quad 0,y,z$
(29) $\bar{3}^{+}\quad x+\frac{1}{2},x,x;\ \frac{1}{2},0,0$
(30) $\bar{3}^{+}\quad \bar{x}-\frac{3}{2},x+\frac{3}{4},\bar{x};\ -\frac{3}{4},0,\frac{3}{4}$
(31) $\bar{3}^{+}\quad x+\frac{1}{4},\bar{x}+1,\bar{x};\ \frac{1}{2},\frac{3}{4},-\frac{1}{4}$
(32) $\bar{3}^{+}\quad \bar{x}+\frac{1}{4},\bar{x}-\frac{1}{4},x;\ \frac{1}{4},-\frac{1}{4},0$
(33) $\bar{3}^{-}\quad x,x+\frac{1}{2},x;\ 0,\frac{1}{2},0$
(34) $\bar{3}^{-}\quad x+\frac{1}{4},\bar{x}-\frac{1}{2},\bar{x};\ 0,-\frac{1}{4},\frac{1}{4}$
(35) $\bar{3}^{-}\quad \bar{x}-\frac{3}{4},\bar{x}+\frac{3}{4},x;\ -\frac{3}{4},\frac{3}{4},0$
(36) $\bar{3}^{-}\quad \bar{x}+1,x-\frac{1}{4},\bar{x};\ \frac{3}{4},0,-\frac{1}{4}$
(37) $g(\frac{1}{4},-\frac{1}{4},\frac{1}{2})\quad x+\frac{1}{2},\bar{x},z$
(38) $g(\frac{1}{2},\frac{1}{2},0)\quad x,x,z$
(39) $\bar{4}^{-}\quad \frac{1}{8},\frac{1}{8},z;\ \frac{1}{8},\frac{1}{8},\frac{1}{8}$
(40) $\bar{4}^{+}\quad \frac{3}{8},\frac{3}{8},z;\ \frac{3}{8},\frac{3}{8},\frac{3}{8}$
(41) $\bar{4}^{-}\quad x,-\frac{1}{8},\frac{3}{8};\ \frac{3}{8},-\frac{1}{8},\frac{3}{8}$
(42) $m\quad x,y+\frac{3}{4},\bar{y}$
(43) $g(\frac{1}{2},\frac{1}{4},\frac{1}{4})\quad x,y+\frac{1}{4},y$
(44) $\bar{4}^{+}\quad x,\frac{1}{8},\frac{1}{8};\ \frac{1}{8},\frac{1}{8},\frac{1}{8}$
(45) $\bar{4}^{+}\quad \frac{1}{8},y,\frac{5}{8};\ \frac{1}{8},\frac{1}{8},\frac{5}{8}$
(46) $m\quad \bar{x}+\frac{1}{4},y,x$
(47) $\bar{4}^{-}\quad \frac{3}{8},y,\frac{3}{8};\ \frac{3}{8},\frac{3}{8},\frac{3}{8}$
(48) $g(\frac{1}{4},\frac{1}{2},\frac{1}{4})\quad x+\frac{1}{4},y,x$

ORIGIN CHOICE 2

Generators selected (1); $t(1,0,0)$; $t(0,1,0)$; $t(0,0,1)$; $t(0,\frac{1}{2},\frac{1}{2})$; $t(\frac{1}{2},0,\frac{1}{2})$; (2); (3); (5); (13); (25)

Positions

Multiplicity, Wyckoff letter, Site symmetry			Coordinates				Reflection conditions
			$(0,0,0)+$	$(0,\frac{1}{2},\frac{1}{2})+$	$(\frac{1}{2},0,\frac{1}{2})+$	$(\frac{1}{2},\frac{1}{2},0)+$	h,k,l permutable General:
192	i	1	(1) x,y,z	(2) $\bar{x}+\frac{3}{4},\bar{y}+\frac{1}{4},z+\frac{1}{2}$	(3) $\bar{x}+\frac{1}{4},y+\frac{1}{2},\bar{z}+\frac{3}{4}$	(4) $x+\frac{1}{2},\bar{y}+\frac{3}{4},\bar{z}+\frac{1}{4}$	hkl: $h+k=2n$ and $h+l,k+l=2n$
			(5) z,x,y	(6) $z+\frac{1}{2},\bar{x}+\frac{3}{4},\bar{y}+\frac{1}{4}$	(7) $\bar{z}+\frac{3}{4},\bar{x}+\frac{1}{4},y+\frac{1}{2}$	(8) $\bar{z}+\frac{1}{4},x+\frac{1}{2},\bar{y}+\frac{3}{4}$	
			(9) y,z,x	(10) $\bar{y}+\frac{1}{4},z+\frac{1}{2},\bar{x}+\frac{3}{4}$	(11) $y+\frac{1}{2},\bar{z}+\frac{3}{4},\bar{x}+\frac{1}{4}$	(12) $\bar{y}+\frac{3}{4},\bar{z}+\frac{1}{4},x+\frac{1}{2}$	$0kl$: $k+l=4n$ and $k,l=2n$
			(13) $y+\frac{3}{4},x+\frac{1}{4},\bar{z}+\frac{1}{2}$	(14) $\bar{y},\bar{x},\bar{z}$	(15) $y+\frac{1}{4},\bar{x}+\frac{1}{2},z+\frac{3}{4}$	(16) $\bar{y}+\frac{1}{2},x+\frac{3}{4},z+\frac{1}{4}$	
			(17) $x+\frac{3}{4},z+\frac{1}{4},\bar{y}+\frac{1}{2}$	(18) $\bar{x}+\frac{1}{2},z+\frac{3}{4},y+\frac{1}{4}$	(19) $\bar{x},\bar{z},\bar{y}$	(20) $x+\frac{1}{4},\bar{z}+\frac{1}{2},y+\frac{3}{4}$	hhl: $h+l=2n$
			(21) $z+\frac{3}{4},y+\frac{1}{4},\bar{x}+\frac{1}{2}$	(22) $z+\frac{1}{4},\bar{y}+\frac{1}{2},x+\frac{3}{4}$	(23) $\bar{z}+\frac{1}{2},y+\frac{3}{4},x+\frac{1}{4}$	(24) $\bar{z},\bar{y},\bar{x}$	$h00$: $h=4n$
			(25) $\bar{x},\bar{y},\bar{z}$	(26) $x+\frac{1}{4},y+\frac{3}{4},\bar{z}+\frac{1}{2}$	(27) $x+\frac{3}{4},\bar{y}+\frac{1}{2},z+\frac{1}{4}$	(28) $\bar{x}+\frac{1}{2},y+\frac{1}{4},z+\frac{3}{4}$	
			(29) $\bar{z},\bar{x},\bar{y}$	(30) $\bar{z}+\frac{1}{2},x+\frac{1}{4},y+\frac{3}{4}$	(31) $z+\frac{1}{4},x+\frac{3}{4},\bar{y}+\frac{1}{2}$	(32) $z+\frac{3}{4},\bar{x}+\frac{1}{2},y+\frac{1}{4}$	
			(33) $\bar{y},\bar{z},\bar{x}$	(34) $y+\frac{3}{4},\bar{z}+\frac{1}{2},x+\frac{1}{4}$	(35) $\bar{y}+\frac{1}{2},z+\frac{1}{4},x+\frac{3}{4}$	(36) $y+\frac{1}{4},z+\frac{3}{4},\bar{x}+\frac{1}{2}$	
			(37) $\bar{y}+\frac{1}{4},\bar{x}+\frac{3}{4},z+\frac{1}{2}$	(38) y,x,z	(39) $\bar{y}+\frac{3}{4},x+\frac{1}{2},\bar{z}+\frac{1}{4}$	(40) $y+\frac{1}{2},\bar{x}+\frac{1}{4},\bar{z}+\frac{3}{4}$	
			(41) $\bar{x}+\frac{1}{4},\bar{z}+\frac{3}{4},y+\frac{1}{2}$	(42) $x+\frac{1}{2},\bar{z}+\frac{1}{4},\bar{y}+\frac{3}{4}$	(43) x,z,y	(44) $\bar{x}+\frac{3}{4},z+\frac{1}{2},\bar{y}+\frac{1}{4}$	
			(45) $\bar{z}+\frac{1}{4},\bar{y}+\frac{3}{4},x+\frac{1}{2}$	(46) $\bar{z}+\frac{3}{4},y+\frac{1}{2},\bar{x}+\frac{1}{4}$	(47) $z+\frac{1}{2},\bar{y}+\frac{1}{4},\bar{x}+\frac{3}{4}$	(48) z,y,x	

Special: as above, plus

							Reflection conditions
96	h	$..2$	$0,y,\bar{y}$	$\frac{3}{4},\bar{y}+\frac{1}{4},\bar{y}+\frac{1}{2}$	$\frac{1}{4},y+\frac{1}{2},y+\frac{3}{4}$	$\frac{1}{2},\bar{y}+\frac{3}{4},y+\frac{1}{4}$	no extra conditions
			$\bar{y},0,y$	$\bar{y}+\frac{1}{2},\frac{3}{4},\bar{y}+\frac{1}{4}$	$y+\frac{3}{4},\frac{1}{4},y+\frac{1}{2}$	$y+\frac{1}{4},\frac{1}{2},\bar{y}+\frac{3}{4}$	
			$y,\bar{y},0$	$\bar{y}+\frac{1}{4},\bar{y}+\frac{1}{2},\frac{3}{4}$	$y+\frac{1}{2},y+\frac{3}{4},\frac{1}{4}$	$\bar{y}+\frac{3}{4},y+\frac{1}{4},\frac{1}{2}$	
			$0,\bar{y},y$	$\frac{1}{4},y+\frac{3}{4},y+\frac{1}{2}$	$\frac{3}{4},\bar{y}+\frac{1}{2},\bar{y}+\frac{1}{4}$	$\frac{1}{2},y+\frac{1}{4},\bar{y}+\frac{3}{4}$	
			$y,0,\bar{y}$	$y+\frac{1}{2},\frac{1}{4},y+\frac{3}{4}$	$\bar{y}+\frac{1}{4},\frac{3}{4},\bar{y}+\frac{1}{2}$	$\bar{y}+\frac{3}{4},\frac{1}{2},y+\frac{1}{4}$	
			$\bar{y},y,0$	$y+\frac{3}{4},y+\frac{1}{2},\frac{1}{4}$	$\bar{y}+\frac{1}{2},\bar{y}+\frac{1}{4},\frac{3}{4}$	$y+\frac{1}{4},\bar{y}+\frac{3}{4},\frac{1}{2}$	
96	g	$..m$	x,x,z	$\bar{x}+\frac{3}{4},\bar{x}+\frac{1}{4},z+\frac{1}{2}$	$\bar{x}+\frac{1}{4},x+\frac{1}{2},\bar{z}+\frac{3}{4}$	$x+\frac{1}{2},\bar{x}+\frac{3}{4},\bar{z}+\frac{1}{4}$	no extra conditions
			z,x,x	$z+\frac{1}{2},\bar{x}+\frac{3}{4},\bar{x}+\frac{1}{4}$	$\bar{z}+\frac{3}{4},\bar{x}+\frac{1}{4},x+\frac{1}{2}$	$\bar{z}+\frac{1}{4},x+\frac{1}{2},\bar{x}+\frac{3}{4}$	
			x,z,x	$\bar{x}+\frac{1}{4},z+\frac{1}{2},\bar{x}+\frac{3}{4}$	$x+\frac{1}{2},\bar{z}+\frac{3}{4},\bar{x}+\frac{1}{4}$	$\bar{x}+\frac{3}{4},\bar{z}+\frac{1}{4},x+\frac{1}{2}$	
			$x+\frac{3}{4},x+\frac{1}{4},\bar{z}+\frac{1}{2}$	$\bar{x},\bar{x},\bar{z}$	$x+\frac{1}{4},\bar{x}+\frac{1}{2},z+\frac{3}{4}$	$\bar{x}+\frac{1}{2},x+\frac{3}{4},z+\frac{1}{4}$	
			$x+\frac{3}{4},z+\frac{1}{4},\bar{x}+\frac{1}{2}$	$\bar{x}+\frac{1}{2},z+\frac{3}{4},x+\frac{1}{4}$	$\bar{x},\bar{z},\bar{x}$	$x+\frac{1}{4},\bar{z}+\frac{1}{2},x+\frac{3}{4}$	
			$z+\frac{3}{4},x+\frac{1}{4},\bar{x}+\frac{1}{2}$	$z+\frac{1}{4},\bar{x}+\frac{1}{2},x+\frac{3}{4}$	$\bar{z}+\frac{1}{2},x+\frac{3}{4},x+\frac{1}{4}$	$\bar{z},\bar{x},\bar{x}$	

									Reflection conditions
48	f	$2.mm$	$x,\frac{1}{8},\frac{1}{8}$	$\bar{x}+\frac{3}{4},\frac{1}{8},\frac{5}{8}$	$\frac{1}{8},x,\frac{1}{8}$	$\frac{5}{8},\bar{x}+\frac{3}{4},\frac{1}{8}$	$\frac{1}{8},\frac{1}{8},x$	$\frac{1}{8},\frac{5}{8},\bar{x}+\frac{3}{4}$	hkl: $h=2n+1$ or $h+k+l=4n$
			$\frac{7}{8},x+\frac{1}{4},\frac{3}{8}$	$\frac{7}{8},\bar{x},\frac{7}{8}$	$x+\frac{3}{4},\frac{3}{8},\frac{3}{8}$	$\bar{x}+\frac{1}{2},\frac{7}{8},\frac{3}{8}$	$\frac{7}{8},\frac{3}{8},\bar{x}+\frac{1}{2}$	$\frac{3}{8},\frac{3}{8},x+\frac{3}{4}$	

							Reflection conditions
32	e	$.3m$	x,x,x	$\bar{x}+\frac{3}{4},\bar{x}+\frac{1}{4},x+\frac{1}{2}$			no extra conditions
			$\bar{x}+\frac{1}{4},x+\frac{1}{2},\bar{x}+\frac{3}{4}$	$x+\frac{1}{2},\bar{x}+\frac{3}{4},\bar{x}+\frac{1}{4}$			
			$x+\frac{3}{4},x+\frac{1}{4},\bar{x}+\frac{1}{2}$	$\bar{x},\bar{x},\bar{x}$			
			$x+\frac{1}{4},\bar{x}+\frac{1}{2},x+\frac{3}{4}$	$\bar{x}+\frac{1}{2},x+\frac{3}{4},x+\frac{1}{4}$			
16	d	$.\bar{3}m$	$\frac{1}{2},\frac{1}{2},\frac{1}{2}$	$\frac{1}{4},\frac{3}{4},0$	$\frac{3}{4},0,\frac{1}{4}$	$0,\frac{1}{4},\frac{3}{4}$	hkl: $h=2n+1$ or $h,k,l=4n+2$ or $h,k,l=4n$ (16 d and 16 c)
16	c	$.\bar{3}m$	$0,0,0$	$\frac{3}{4},\frac{1}{4},\frac{1}{2}$	$\frac{1}{4},\frac{1}{2},\frac{3}{4}$	$\frac{1}{2},\frac{3}{4},\frac{1}{4}$	
8	b	$\bar{4}3m$	$\frac{3}{8},\frac{3}{8},\frac{3}{8}$	$\frac{1}{8},\frac{5}{8},\frac{1}{8}$			hkl: $h=2n+1$ or $h+k+l=4n$ (8 b and 8 a)
8	a	$\bar{4}3m$	$\frac{1}{8},\frac{1}{8},\frac{1}{8}$	$\frac{7}{8},\frac{3}{8},\frac{3}{8}$			

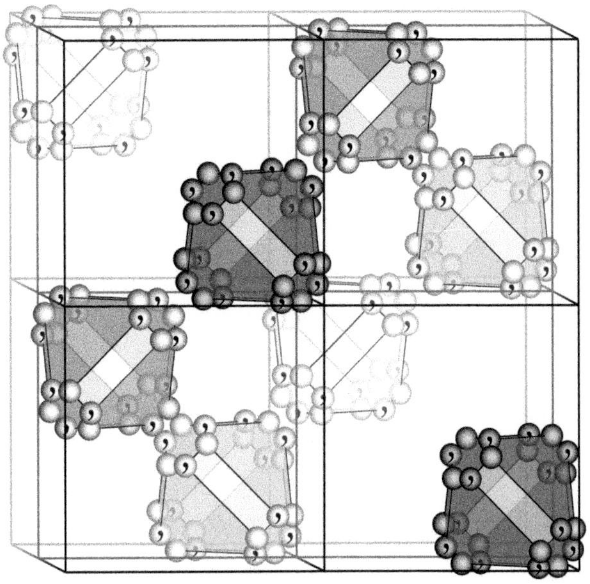

Symmetry of special projections

Along [001] $p4mm$
$\mathbf{a}' = \frac{1}{4}(\mathbf{a}-\mathbf{b})$ $\mathbf{b}' = \frac{1}{4}(\mathbf{a}+\mathbf{b})$
Origin at $\frac{1}{8}, \frac{3}{8}, z$

Along [111] $p6mm$
$\mathbf{a}' = \frac{1}{6}(2\mathbf{a}-\mathbf{b}-\mathbf{c})$ $\mathbf{b}' = \frac{1}{6}(-\mathbf{a}+2\mathbf{b}-\mathbf{c})$
Origin at x, x, x

Along [110] $c2mm$
$\mathbf{a}' = \frac{1}{2}(-\mathbf{a}+\mathbf{b})$ $\mathbf{b}' = \mathbf{c}$
Origin at $x, x, 0$

$I\,a\bar{3}d$ $\qquad$ O_h^{10} $\qquad$ $m\bar{3}m$ $\qquad$ Cubic

No. 230 $\qquad$ $I\ 4_1/a\ \bar{3}\ 2/d$ $\qquad$ Patterson symmetry $Im\bar{3}m$

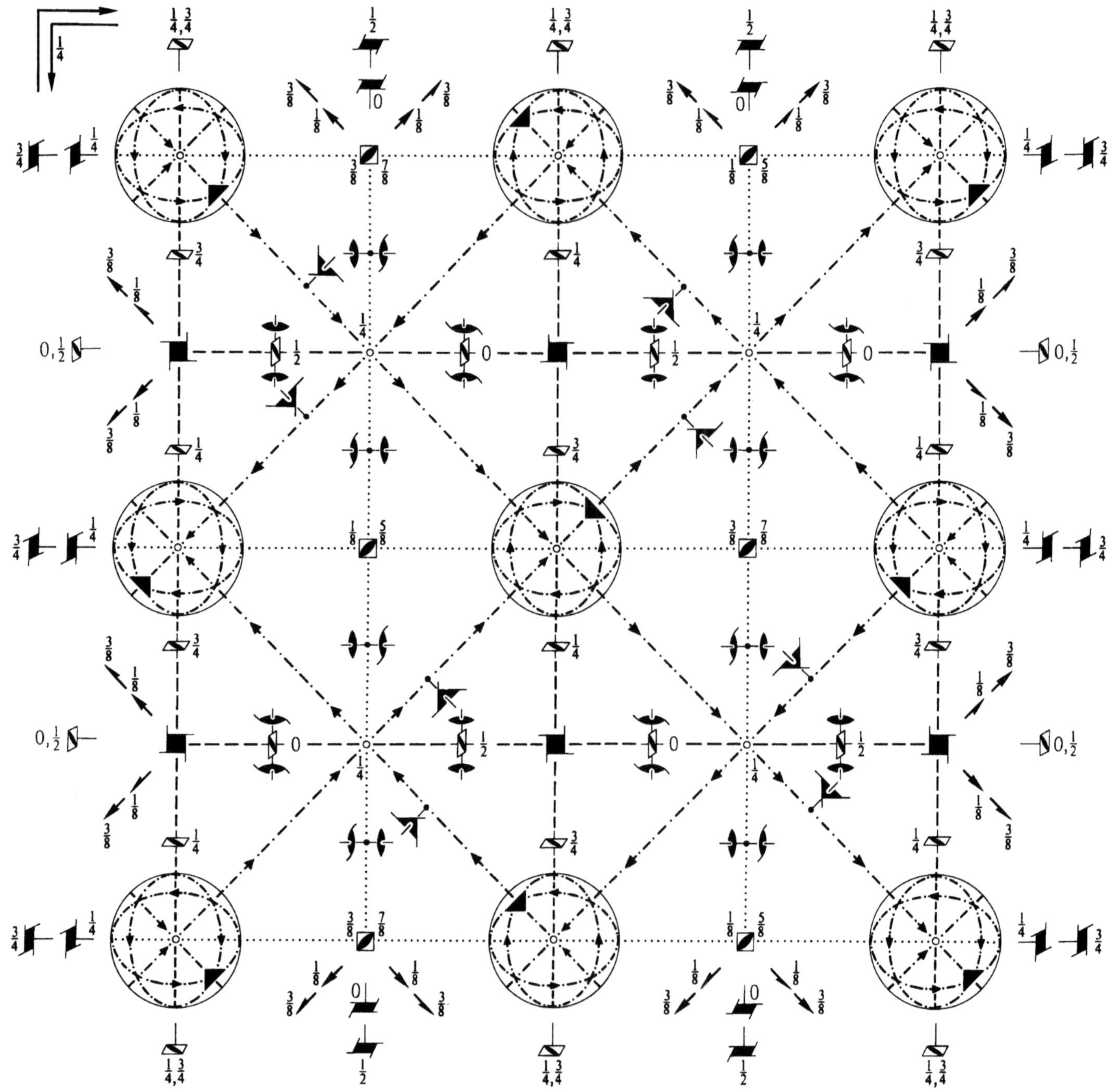

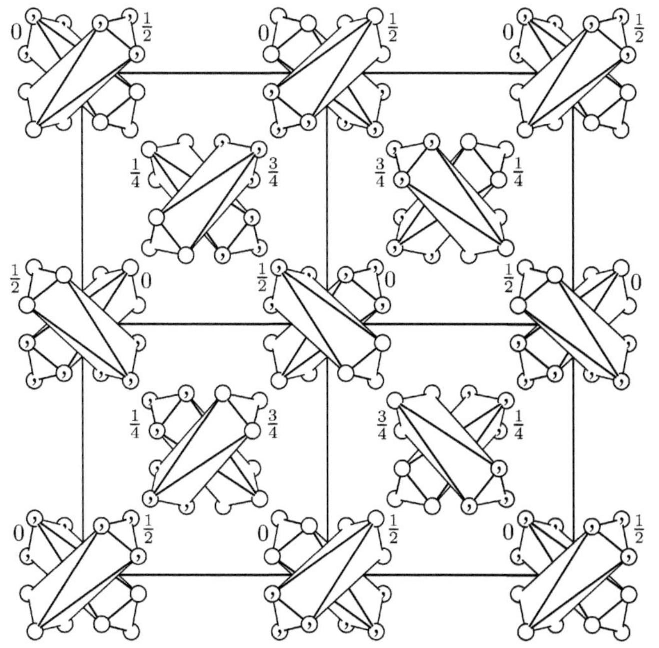

Polyhedron centre at 0, 0, 0

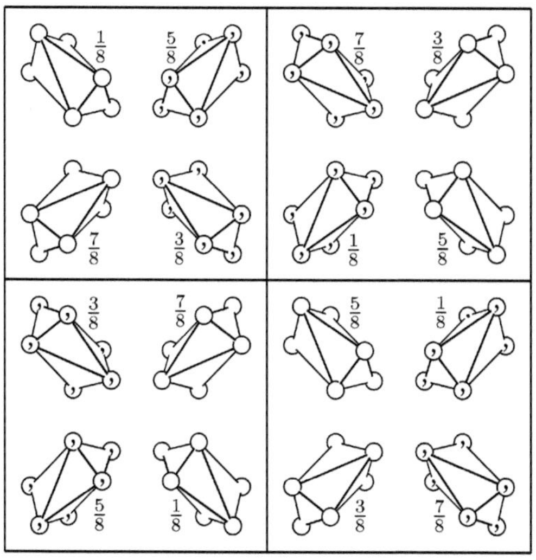

Polyhedron centre at $\frac{1}{8},\frac{1}{8},\frac{1}{8}$

Origin at centre ($\bar{3}$)

Asymmetric unit $-\frac{1}{8}\le x\le\frac{1}{8};\quad -\frac{1}{8}\le y\le\frac{1}{8};\quad 0\le z\le\frac{1}{4};\quad \max(x,-x,y,-y)\le z$

Vertices $0,0,0\quad \frac{1}{8},\frac{1}{8},\frac{1}{8}\quad -\frac{1}{8},\frac{1}{8},\frac{1}{8}\quad -\frac{1}{8},-\frac{1}{8},\frac{1}{8}\quad \frac{1}{8},-\frac{1}{8},\frac{1}{8}$

$\frac{1}{8},\frac{1}{8},\frac{1}{4}\quad -\frac{1}{8},\frac{1}{8},\frac{1}{4}\quad -\frac{1}{8},-\frac{1}{8},\frac{1}{4}\quad \frac{1}{8},-\frac{1}{8},\frac{1}{4}$

Symmetry operations

For (0,0,0)+ set

(1) 1	(2) $2(0,0,\frac{1}{2})\ \ \frac{1}{4},0,z$	(3) $2(0,\frac{1}{2},0)\ \ 0,y,\frac{1}{4}$	(4) $2(\frac{1}{2},0,0)\ \ x,\frac{1}{4},0$
(5) $3^{+}\ \ x,x,x$	(6) $3^{+}\ \ \bar{x}+\frac{1}{2},x,\bar{x}$	(7) $3^{+}\ \ x+\frac{1}{2},\bar{x}-\frac{1}{2},\bar{x}$	(8) $3^{+}\ \ \bar{x},\bar{x}+\frac{1}{2},x$
(9) $3^{-}\ \ x,x,x$	(10) $3^{-}(-\frac{1}{3},\frac{1}{3},\frac{1}{3})\ \ x+\frac{1}{6},\bar{x}+\frac{1}{6},\bar{x}$	(11) $3^{-}(\frac{1}{3},\frac{1}{3},-\frac{1}{3})\ \ \bar{x}+\frac{1}{3},\bar{x}+\frac{1}{6},x$	(12) $3^{-}(\frac{1}{3},-\frac{1}{3},\frac{1}{3})\ \ \bar{x}-\frac{1}{6},x+\frac{1}{3},\bar{x}$
(13) $2(\frac{1}{2},\frac{1}{2},0)\ \ x,x-\frac{1}{4},\frac{1}{8}$	(14) $2\ \ x,\bar{x}+\frac{3}{4},\frac{3}{8}$	(15) $4^{-}(0,0,\frac{3}{4})\ \ \frac{1}{4},0,z$	(16) $4^{+}(0,0,\frac{1}{4})\ \ -\frac{1}{4},\frac{1}{2},z$
(17) $4^{-}(\frac{3}{4},0,0)\ \ x,\frac{1}{4},0$	(18) $2(0,\frac{1}{2},\frac{1}{2})\ \ \frac{1}{8},y+\frac{1}{4},y$	(19) $2\ \ \frac{3}{8},y+\frac{3}{4},\bar{y}$	(20) $4^{+}(\frac{1}{4},0,0)\ \ x,-\frac{1}{4},\frac{1}{2}$
(21) $4^{+}(0,\frac{1}{4},0)\ \ \frac{1}{2},y,-\frac{1}{4}$	(22) $2(\frac{1}{2},0,\frac{1}{2})\ \ x-\frac{1}{4},\frac{1}{8},x$	(23) $4^{-}(0,\frac{3}{4},0)\ \ 0,y,\frac{1}{4}$	(24) $2\ \ \bar{x}+\frac{3}{4},\frac{3}{8},x$
(25) $\bar{1}\ \ 0,0,0$	(26) $a\ \ x,y,\frac{1}{4}$	(27) $c\ \ x,\frac{1}{4},z$	(28) $b\ \ \frac{1}{4},y,z$
(29) $\bar{3}^{+}\ \ x,x,x;\ 0,0,0$	(30) $\bar{3}^{+}\ \ \bar{x}-\frac{1}{2},x+1,\bar{x};\ 0,\frac{1}{2},\frac{1}{2}$	(31) $\bar{3}^{+}\ \ x+\frac{1}{2},\bar{x}+\frac{1}{2},\bar{x};\ \frac{1}{2},\frac{1}{2},0$	(32) $\bar{3}^{+}\ \ \bar{x}+1,\bar{x}+\frac{1}{2},x;\ \frac{1}{2},0,\frac{1}{2}$
(33) $\bar{3}^{-}\ \ x,x,x;\ 0,0,0$	(34) $\bar{3}^{-}\ \ x+\frac{1}{2},\bar{x}-\frac{1}{2},\bar{x};\ 0,0,\frac{1}{2}$	(35) $\bar{3}^{-}\ \ \bar{x},\bar{x}+\frac{1}{2},x;\ 0,\frac{1}{2},0$	(36) $\bar{3}^{-}\ \ \bar{x}+\frac{1}{2},x,\bar{x};\ \frac{1}{2},0,0$
(37) $d(-\frac{1}{4},\frac{1}{4},\frac{3}{4})\ \ x+\frac{1}{2},\bar{x},z$	(38) $d(\frac{1}{4},\frac{1}{4},\frac{1}{4})\ \ x,x,z$	(39) $\bar{4}^{-}\ \ 0,\frac{3}{4},z;\ 0,\frac{3}{4},\frac{1}{8}$	(40) $\bar{4}^{+}\ \ \frac{1}{2},-\frac{1}{4},z;\ \frac{1}{2},-\frac{1}{4},\frac{3}{8}$
(41) $\bar{4}^{-}\ \ x,0,\frac{3}{4};\ \frac{1}{8},0,\frac{3}{4}$	(42) $d(\frac{3}{4},-\frac{1}{4},\frac{1}{4})\ \ x,y+\frac{1}{2},\bar{y}$	(43) $d(\frac{1}{4},\frac{1}{4},\frac{1}{4})\ \ x,y,y$	(44) $\bar{4}^{+}\ \ x,\frac{1}{2},-\frac{1}{4};\ \frac{3}{8},\frac{1}{2},-\frac{1}{4}$
(45) $\bar{4}^{+}\ \ -\frac{1}{4},y,\frac{1}{2};\ -\frac{1}{4},\frac{3}{8},\frac{1}{2}$	(46) $d(\frac{1}{4},\frac{3}{4},-\frac{1}{4})\ \ \bar{x}+\frac{1}{2},y,x$	(47) $\bar{4}^{-}\ \ \frac{3}{4},y,0;\ \frac{3}{4},\frac{1}{8},0$	(48) $d(\frac{1}{4},\frac{1}{4},\frac{1}{4})\ \ x,y,x$

For $(\frac{1}{2},\frac{1}{2},\frac{1}{2})$+ set

(1) $t(\frac{1}{2},\frac{1}{2},\frac{1}{2})$	(2) $2\ \ 0,\frac{1}{4},z$	(3) $2\ \ \frac{1}{4},y,0$	(4) $2\ \ x,0,\frac{1}{4}$
(5) $3^{+}(\frac{1}{2},\frac{1}{2},\frac{1}{2})\ \ x,x,x$	(6) $3^{+}(\frac{1}{6},-\frac{1}{6},\frac{1}{6})\ \ \bar{x}-\frac{1}{6},x+\frac{1}{3},\bar{x}$	(7) $3^{+}(-\frac{1}{6},\frac{1}{6},\frac{1}{6})\ \ x+\frac{1}{6},\bar{x}+\frac{1}{6},\bar{x}$	(8) $3^{+}(\frac{1}{6},\frac{1}{6},-\frac{1}{6})\ \ \bar{x}+\frac{1}{3},\bar{x}+\frac{1}{6},x$
(9) $3^{-}(\frac{1}{2},\frac{1}{2},\frac{1}{2})\ \ x,x,x$	(10) $3^{-}(\frac{1}{6},-\frac{1}{6},-\frac{1}{6})\ \ x+\frac{1}{6},\bar{x}+\frac{1}{6},\bar{x}$	(11) $3^{-}(-\frac{1}{6},-\frac{1}{6},\frac{1}{6})\ \ \bar{x}+\frac{1}{3},\bar{x}+\frac{1}{6},x$	(12) $3^{-}(-\frac{1}{6},\frac{1}{6},-\frac{1}{6})\ \ \bar{x}-\frac{1}{6},x+\frac{1}{3},\bar{x}$
(13) $2(\frac{1}{2},\frac{1}{2},0)\ \ x,x+\frac{1}{4},\frac{3}{8}$	(14) $2\ \ x,\bar{x}+\frac{1}{4},\frac{1}{8}$	(15) $4^{-}(0,0,\frac{1}{4})\ \ \frac{3}{4},0,z$	(16) $4^{+}(0,0,\frac{3}{4})\ \ \frac{1}{4},\frac{1}{2},z$
(17) $4^{-}(\frac{1}{4},0,0)\ \ x,\frac{3}{4},0$	(18) $2(0,\frac{1}{2},\frac{1}{2})\ \ \frac{3}{8},y-\frac{1}{4},y$	(19) $2\ \ \frac{1}{8},y+\frac{1}{4},\bar{y}$	(20) $4^{+}(\frac{3}{4},0,0)\ \ x,\frac{1}{4},\frac{1}{2}$
(21) $4^{+}(0,\frac{3}{4},0)\ \ \frac{1}{2},y,\frac{1}{4}$	(22) $2(\frac{1}{2},0,\frac{1}{2})\ \ x+\frac{1}{4},\frac{3}{8},x$	(23) $4^{-}(0,\frac{1}{4},0)\ \ 0,y,\frac{3}{4}$	(24) $2\ \ \bar{x}+\frac{1}{4},\frac{1}{8},x$
(25) $\bar{1}\ \ \frac{1}{4},\frac{1}{4},\frac{1}{4}$	(26) $b\ \ x,y,0$	(27) $a\ \ x,0,z$	(28) $c\ \ 0,y,z$
(29) $\bar{3}^{+}\ \ x,x,x;\ \frac{1}{4},\frac{1}{4},\frac{1}{4}$	(30) $\bar{3}^{+}\ \ \bar{x}-\frac{1}{2},x,\bar{x};\ -\frac{1}{4},-\frac{1}{4},\frac{1}{4}$	(31) $\bar{3}^{+}\ \ x-\frac{1}{2},\bar{x}+\frac{1}{2},\bar{x};\ -\frac{1}{4},\frac{1}{4},-\frac{1}{4}$	(32) $\bar{3}^{+}\ \ \bar{x},\bar{x}-\frac{1}{2},x;\ \frac{1}{4},-\frac{1}{4},-\frac{1}{4}$
(33) $\bar{3}^{-}\ \ x,x,x;\ \frac{1}{4},\frac{1}{4},\frac{1}{4}$	(34) $\bar{3}^{-}\ \ x+\frac{1}{2},\bar{x}-\frac{1}{2},\bar{x};\ \frac{1}{4},-\frac{1}{4},\frac{1}{4}$	(35) $\bar{3}^{-}\ \ \bar{x},\bar{x}+\frac{1}{2},x;\ -\frac{1}{4},\frac{1}{4},\frac{1}{4}$	(36) $\bar{3}^{-}\ \ \bar{x}+\frac{1}{2},x,\bar{x};\ \frac{1}{4},\frac{1}{4},-\frac{1}{4}$
(37) $d(\frac{1}{4},-\frac{1}{4},\frac{1}{4})\ \ x+\frac{1}{2},\bar{x},z$	(38) $d(\frac{3}{4},\frac{3}{4},\frac{3}{4})\ \ x,x,z$	(39) $\bar{4}^{-}\ \ 0,\frac{1}{4},z;\ 0,\frac{1}{4},\frac{3}{8}$	(40) $\bar{4}^{+}\ \ \frac{1}{2},\frac{1}{4},z;\ \frac{1}{2},\frac{1}{4},\frac{1}{8}$
(41) $\bar{4}^{-}\ \ x,0,\frac{1}{4};\ \frac{3}{8},0,\frac{1}{4}$	(42) $d(\frac{1}{4},\frac{1}{4},-\frac{1}{4})\ \ x,y+\frac{1}{2},\bar{y}$	(43) $d(\frac{3}{4},\frac{3}{4},\frac{3}{4})\ \ x,y,y$	(44) $\bar{4}^{+}\ \ x,\frac{1}{2},\frac{1}{4};\ \frac{1}{8},\frac{1}{2},\frac{1}{4}$
(45) $\bar{4}^{+}\ \ \frac{1}{4},y,\frac{1}{2};\ \frac{1}{4},\frac{1}{8},\frac{1}{2}$	(46) $d(-\frac{1}{4},\frac{1}{4},\frac{1}{4})\ \ \bar{x}+\frac{1}{2},y,x$	(47) $\bar{4}^{-}\ \ \frac{1}{4},y,0;\ \frac{1}{4},\frac{3}{8},0$	(48) $d(\frac{3}{4},\frac{3}{4},\frac{3}{4})\ \ x,y,x$

Generators selected (1); $t(1,0,0)$; $t(0,1,0)$; $t(0,0,1)$; $t(\frac{1}{2},\frac{1}{2},\frac{1}{2})$; (2); (3); (5); (13); (25)

Positions

Multiplicity, Wyckoff letter, Site symmetry	Coordinates	Reflection conditions
	$(0,0,0)+\quad(\frac{1}{2},\frac{1}{2},\frac{1}{2})+$	h,k,l permutable General:

96 h 1

				Reflection conditions
(1) x,y,z	(2) $\bar{x}+\frac{1}{2},\bar{y},z+\frac{1}{2}$	(3) $\bar{x},y+\frac{1}{2},\bar{z}+\frac{1}{2}$	(4) $x+\frac{1}{2},\bar{y}+\frac{1}{2},\bar{z}$	hkl: $h+k+l=2n$
(5) z,x,y	(6) $z+\frac{1}{2},\bar{x}+\frac{1}{2},\bar{y}$	(7) $\bar{z}+\frac{1}{2},\bar{x},y+\frac{1}{2}$	(8) $\bar{z},x+\frac{1}{2},\bar{y}+\frac{1}{2}$	$0kl$: $k,l=2n$
(9) y,z,x	(10) $\bar{y},z+\frac{1}{2},\bar{x}+\frac{1}{2}$	(11) $y+\frac{1}{2},\bar{z}+\frac{1}{2},\bar{x}$	(12) $\bar{y}+\frac{1}{2},\bar{z},x+\frac{1}{2}$	hhl: $2h+l=4n$
(13) $y+\frac{3}{4},x+\frac{1}{4},\bar{z}+\frac{1}{4}$	(14) $\bar{y}+\frac{3}{4},\bar{x}+\frac{3}{4},\bar{z}+\frac{3}{4}$	(15) $y+\frac{1}{4},\bar{x}+\frac{1}{4},z+\frac{3}{4}$	(16) $\bar{y}+\frac{1}{4},x+\frac{3}{4},z+\frac{1}{4}$	$h00$: $h=4n$
(17) $x+\frac{3}{4},z+\frac{1}{4},\bar{y}+\frac{1}{4}$	(18) $\bar{x}+\frac{1}{4},z+\frac{3}{4},y+\frac{1}{4}$	(19) $\bar{x}+\frac{3}{4},\bar{z}+\frac{3}{4},\bar{y}+\frac{3}{4}$	(20) $x+\frac{1}{4},\bar{z}+\frac{1}{4},y+\frac{3}{4}$	
(21) $z+\frac{3}{4},y+\frac{1}{4},\bar{x}+\frac{1}{4}$	(22) $z+\frac{1}{4},\bar{y}+\frac{1}{4},x+\frac{3}{4}$	(23) $\bar{z}+\frac{1}{4},y+\frac{3}{4},x+\frac{1}{4}$	(24) $\bar{z}+\frac{3}{4},\bar{y}+\frac{3}{4},\bar{x}+\frac{3}{4}$	
(25) $\bar{x},\bar{y},\bar{z}$	(26) $x+\frac{1}{2},y,\bar{z}+\frac{1}{2}$	(27) $x,\bar{y}+\frac{1}{2},z+\frac{1}{2}$	(28) $\bar{x}+\frac{1}{2},y+\frac{1}{2},z$	
(29) $\bar{z},\bar{x},\bar{y}$	(30) $\bar{z}+\frac{1}{2},x+\frac{1}{2},y$	(31) $z+\frac{1}{2},x,\bar{y}+\frac{1}{2}$	(32) $z,\bar{x}+\frac{1}{2},y+\frac{1}{2}$	
(33) $\bar{y},\bar{z},\bar{x}$	(34) $y,\bar{z}+\frac{1}{2},x+\frac{1}{2}$	(35) $\bar{y}+\frac{1}{2},z+\frac{1}{2},x$	(36) $y+\frac{1}{2},z,\bar{x}+\frac{1}{2}$	
(37) $\bar{y}+\frac{1}{4},\bar{x}+\frac{3}{4},z+\frac{3}{4}$	(38) $y+\frac{1}{4},x+\frac{1}{4},z+\frac{1}{4}$	(39) $\bar{y}+\frac{3}{4},x+\frac{3}{4},\bar{z}+\frac{1}{4}$	(40) $y+\frac{3}{4},\bar{x}+\frac{1}{4},\bar{z}+\frac{3}{4}$	
(41) $\bar{x}+\frac{1}{4},\bar{z}+\frac{3}{4},y+\frac{3}{4}$	(42) $x+\frac{3}{4},\bar{z}+\frac{1}{4},\bar{y}+\frac{3}{4}$	(43) $x+\frac{1}{4},z+\frac{1}{4},y+\frac{1}{4}$	(44) $\bar{x}+\frac{3}{4},z+\frac{3}{4},\bar{y}+\frac{1}{4}$	
(45) $\bar{z}+\frac{1}{4},\bar{y}+\frac{3}{4},x+\frac{3}{4}$	(46) $\bar{z}+\frac{3}{4},y+\frac{3}{4},\bar{x}+\frac{1}{4}$	(47) $z+\frac{3}{4},\bar{y}+\frac{1}{4},\bar{x}+\frac{3}{4}$	(48) $z+\frac{1}{4},y+\frac{1}{4},x+\frac{1}{4}$	

Special: as above, plus

48 g ..2

				Reflection conditions
$\frac{1}{8},y,\bar{y}+\frac{1}{4}$	$\frac{3}{8},\bar{y},\bar{y}+\frac{3}{4}$	$\frac{7}{8},y+\frac{1}{2},y+\frac{1}{4}$	$\frac{5}{8},\bar{y}+\frac{1}{2},y+\frac{3}{4}$	hkl: $h=2n+1$
$\bar{y}+\frac{1}{4},\frac{1}{8},y$	$\bar{y}+\frac{3}{4},\frac{3}{8},\bar{y}$	$y+\frac{1}{4},\frac{7}{8},y+\frac{1}{2}$	$y+\frac{3}{4},\frac{5}{8},\bar{y}+\frac{1}{2}$	or $h=4n$
$y,\bar{y}+\frac{1}{4},\frac{1}{8}$	$\bar{y},\bar{y}+\frac{3}{4},\frac{3}{8}$	$y+\frac{1}{2},y+\frac{1}{4},\frac{7}{8}$	$\bar{y}+\frac{1}{2},y+\frac{3}{4},\frac{5}{8}$	
$\frac{7}{8},\bar{y},y+\frac{3}{4}$	$\frac{5}{8},y,y+\frac{1}{4}$	$\frac{1}{8},\bar{y}+\frac{1}{2},\bar{y}+\frac{3}{4}$	$\frac{3}{8},y+\frac{1}{2},\bar{y}+\frac{1}{4}$	
$y+\frac{3}{4},\frac{7}{8},\bar{y}$	$y+\frac{1}{4},\frac{5}{8},y$	$\bar{y}+\frac{3}{4},\frac{1}{8},\bar{y}+\frac{1}{2}$	$\bar{y}+\frac{1}{4},\frac{3}{8},y+\frac{1}{2}$	
$\bar{y},y+\frac{3}{4},\frac{7}{8}$	$y,y+\frac{1}{4},\frac{5}{8}$	$\bar{y}+\frac{1}{2},\bar{y}+\frac{3}{4},\frac{1}{8}$	$y+\frac{1}{2},\bar{y}+\frac{1}{4},\frac{3}{8}$	

48 f 2..

						Reflection conditions
$x,0,\frac{1}{4}$	$\bar{x}+\frac{1}{2},0,\frac{3}{4}$	$\frac{1}{4},x,0$	$\frac{3}{4},\bar{x}+\frac{1}{2},0$	$0,\frac{1}{4},x$	$0,\frac{3}{4},\bar{x}+\frac{1}{2}$	hkl: $2h+l=4n$
$\frac{3}{4},x+\frac{1}{4},0$	$\frac{3}{4},\bar{x}+\frac{3}{4},\frac{1}{2}$	$x+\frac{3}{4},\frac{1}{2},\frac{1}{4}$	$\bar{x}+\frac{1}{4},0,\frac{1}{4}$	$0,\frac{1}{4},\bar{x}+\frac{1}{4}$	$\frac{1}{2},\frac{1}{4},x+\frac{3}{4}$	
$\bar{x},0,\frac{3}{4}$	$x+\frac{1}{2},0,\frac{1}{4}$	$\frac{3}{4},\bar{x},0$	$\frac{1}{4},x+\frac{1}{2},0$	$0,\frac{3}{4},\bar{x}$	$0,\frac{1}{4},x+\frac{1}{2}$	
$\frac{1}{4},\bar{x}+\frac{3}{4},0$	$\frac{1}{4},x+\frac{1}{4},\frac{1}{2}$	$\bar{x}+\frac{1}{4},\frac{1}{2},\frac{3}{4}$	$x+\frac{3}{4},0,\frac{3}{4}$	$0,\frac{3}{4},x+\frac{3}{4}$	$\frac{1}{2},\frac{3}{4},\bar{x}+\frac{1}{4}$	

32 e .3.

				Reflection conditions
x,x,x	$\bar{x}+\frac{1}{2},\bar{x},x+\frac{1}{2}$	$\bar{x},x+\frac{1}{2},\bar{x}+\frac{1}{2}$	$x+\frac{1}{2},\bar{x}+\frac{1}{2},\bar{x}$	hkl: $h=2n+1$
$x+\frac{3}{4},x+\frac{1}{4},\bar{x}+\frac{1}{4}$	$\bar{x}+\frac{3}{4},\bar{x}+\frac{3}{4},\bar{x}+\frac{3}{4}$	$x+\frac{1}{4},\bar{x}+\frac{1}{4},x+\frac{3}{4}$	$\bar{x}+\frac{1}{4},x+\frac{3}{4},x+\frac{1}{4}$	or $h+k+l=4n$
$\bar{x},\bar{x},\bar{x}$	$x+\frac{1}{2},x,\bar{x}+\frac{1}{2}$	$x,\bar{x}+\frac{1}{2},x+\frac{1}{2}$	$\bar{x}+\frac{1}{2},x+\frac{1}{2},x$	
$\bar{x}+\frac{1}{4},\bar{x}+\frac{3}{4},x+\frac{3}{4}$	$x+\frac{1}{4},x+\frac{1}{4},x+\frac{1}{4}$	$\bar{x}+\frac{3}{4},x+\frac{3}{4},\bar{x}+\frac{1}{4}$	$x+\frac{3}{4},\bar{x}+\frac{1}{4},\bar{x}+\frac{3}{4}$	

24 d $\bar{4}$.. and **24 c 2.22**

Position							Reflection conditions (d and c)
24 d $\bar{4}$..	$\frac{3}{8},0,\frac{1}{4}$	$\frac{1}{8},0,\frac{3}{4}$	$\frac{1}{4},\frac{3}{8},0$	$\frac{3}{4},\frac{1}{8},0$	$0,\frac{1}{4},\frac{3}{8}$	$0,\frac{3}{4},\frac{1}{8}$	hkl: $h,k=2n$, $h+k+l=4n$
	$\frac{3}{4},\frac{5}{8},0$	$\frac{3}{4},\frac{3}{8},\frac{1}{2}$	$\frac{1}{8},\frac{1}{2},\frac{1}{4}$	$\frac{7}{8},0,\frac{1}{4}$	$0,\frac{1}{4},\frac{7}{8}$	$\frac{1}{2},\frac{1}{4},\frac{1}{8}$	or $h,k=2n+1$, $l=4n+2$
24 c 2.22	$\frac{1}{8},0,\frac{1}{4}$	$\frac{3}{8},0,\frac{3}{4}$	$\frac{1}{4},\frac{1}{8},0$	$\frac{3}{4},\frac{3}{8},0$	$0,\frac{1}{4},\frac{1}{8}$	$0,\frac{3}{4},\frac{3}{8}$	or $h=8n$, $k=8n+4$ and
	$\frac{7}{8},0,\frac{3}{4}$	$\frac{5}{8},0,\frac{1}{4}$	$\frac{3}{4},\frac{7}{8},0$	$\frac{1}{4},\frac{5}{8},0$	$0,\frac{3}{4},\frac{7}{8}$	$0,\frac{1}{4},\frac{5}{8}$	$h+k+l=4n+2$

16 b .32

								Reflection conditions
$\frac{1}{8},\frac{1}{8},\frac{1}{8}$	$\frac{3}{8},\frac{7}{8},\frac{5}{8}$	$\frac{7}{8},\frac{5}{8},\frac{3}{8}$	$\frac{5}{8},\frac{3}{8},\frac{7}{8}$	$\frac{7}{8},\frac{7}{8},\frac{7}{8}$	$\frac{5}{8},\frac{1}{8},\frac{3}{8}$	$\frac{1}{8},\frac{3}{8},\frac{5}{8}$	$\frac{3}{8},\frac{5}{8},\frac{1}{8}$	hkl: $h,k=2n+1$, $l=4n+2$ or $h,k,l=4n$

16 a .$\bar{3}$.

								Reflection conditions
$0,0,0$	$\frac{1}{2},0,\frac{1}{2}$	$0,\frac{1}{2},\frac{1}{2}$	$\frac{1}{2},\frac{1}{2},0$	$\frac{3}{4},\frac{1}{4},\frac{1}{4}$	$\frac{3}{4},\frac{3}{4},\frac{3}{4}$	$\frac{1}{4},\frac{1}{4},\frac{3}{4}$	$\frac{1}{4},\frac{3}{4},\frac{1}{4}$	hkl: $h,k=2n$, $h+k+l=4n$

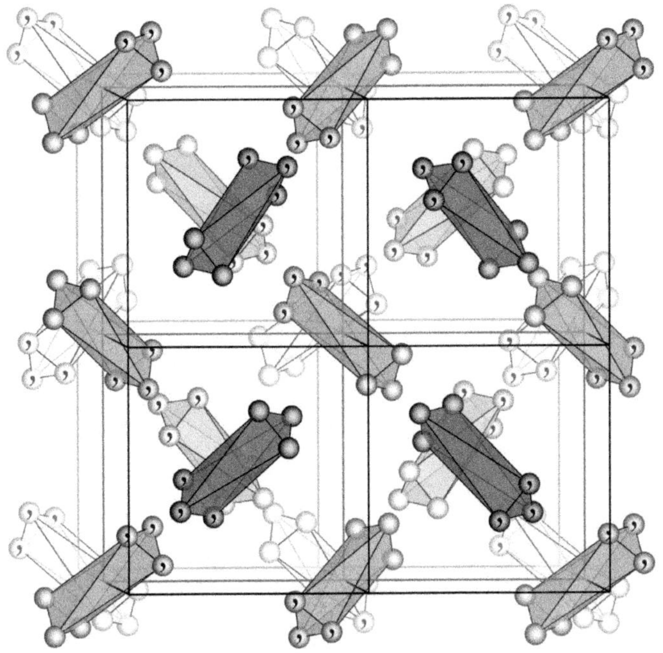

Polyhedron centre at 0, 0, 0

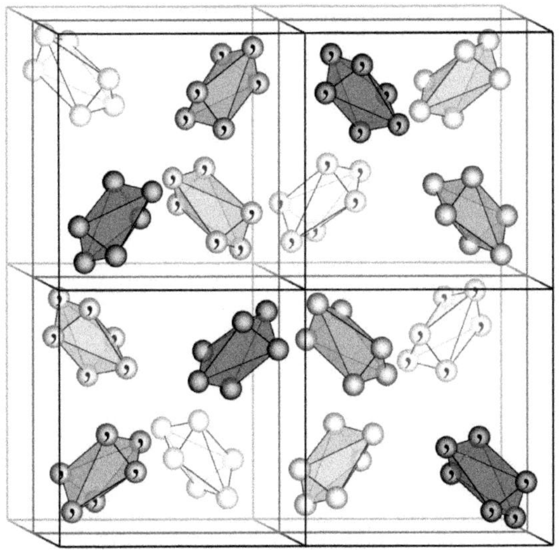

Polyhedron centre at $\frac{1}{8}, \frac{1}{8}, \frac{1}{8}$

Symmetry of special projections

Along [001] $p4mm$
$\mathbf{a}' = \frac{1}{2}\mathbf{a}$ $\mathbf{b}' = \frac{1}{2}\mathbf{b}$
Origin at $\frac{1}{4}, 0, z$

Along [111] $p6mm$
$\mathbf{a}' = \frac{1}{3}(2\mathbf{a} - \mathbf{b} - \mathbf{c})$ $\mathbf{b}' = \frac{1}{3}(-\mathbf{a} + 2\mathbf{b} - \mathbf{c})$
Origin at x, x, x

Along [110] $c2mm$
$\mathbf{a}' = \frac{1}{2}(-\mathbf{a} + \mathbf{b})$ $\mathbf{b}' = \frac{1}{2}\mathbf{c}$
Origin at $x, x + \frac{1}{4}, \frac{1}{8}$

2.2. The symmetry-relations tables of *IT* A1

Hans Wondratschek, Mois I. Aroyo and Ulrich Müller

Volume A1 of *International Tables for Crystallography* (2010), referred to as *IT* A1, presents a systematic treatment of the symmetry relations of space groups. For each plane and space group *IT* A1 provides a complete listing of all maximal subgroups and minimal supergroups. There are diagrams of *translationengleiche* and *klassengleiche* subgroups which contain for each space group all kinds of subgroups, not only the maximal ones. These data are complemented by the relations of the Wyckoff positions of the group–maximal subgroup pairs. A short introduction to group–subgroup relations of space groups is given in Chapter 1.7. Here we provide a brief guide to the subgroup tables of *IT* A1 and the related Wyckoff-position relationships; a few representative examples of these tables are included in Sections 2.2.2 and 2.2.4, respectively. For an introduction to and a detailed description of the subgroup data and the Wyckoff-position data, the reader is referred to the corresponding chapters of *IT* A1.

2.2.1. Guide to the subgroup tables

By Hans Wondratschek and Mois I. Aroyo

2.2.1.1. Contents and arrangement of the subgroup tables

In the subgroup tables of *IT* A1 for every space (or plane) group there is a separate table of its maximal subgroups and minimal supergroups. The sequence of the plane groups and space groups $\mathcal{G}$ follows exactly that of the tables of Volume A of *International Tables for Crystallography* (2016), referred to as *IT* A. The listed data consist of the following blocks:

Headline
Generators selected
General position
I Maximal *translationengleiche* subgroups
II Maximal *klassengleiche* subgroups
I Minimal *translationengleiche* supergroups
II Minimal non-isomorphic *klassengleiche* supergroups.

In the *Headline* block the space group whose subgroups are listed is specified by: (i) the *short (international) Hermann–Mauguin* (HM) *symbol* for the space (or plane) group (HM symbols are discussed in Section 1.4.1); (ii) the space- (or plane-) group number as introduced in Vol. I of *International Tables for X-ray Crystallography* (1952); (iii) the *full (international) Hermann–Mauguin symbol* for the space (or plane) group, abbreviated as 'full HM symbol'; and (iv) the *Schoenflies symbol* for the space group (there are no Schoenflies symbols for the plane groups), *cf.* Section 1.4.1.3.

Many of the subgroups $\mathcal{H}$ in these tables are characterized by the elements of their *general position* specified by numbers which refer to the coordinate triplets of the general position of $\mathcal{G}$. Other subgroups are defined by their *generators* specified by numbers which again refer to the general position of $\mathcal{G}$. To ensure the independent use of the tables, the subgroup data for each group $\mathcal{G}$ are preceded by the listings of its general position and the set of generators copied from *IT* A (for a detailed discussion of the general position see Section 1.4.4; for the generators see Section 1.4.3).

In the subgroup tables, depending on their kind, the maximal subgroups are collected under the blocks of *translationengleiche* and *klassengleiche* subgroups.

2.2.1.2. I Maximal *translationengleiche* subgroups (*t*-subgroups)

In this block, all maximal *t*-subgroups $\mathcal{H}$ of the plane and space groups $\mathcal{G}$ are listed individually. The sequence of the subgroups is determined by the rising value of the index and by the decreasing space-group number. For each subgroup $\mathcal{H}$ of index [*i*] in $\mathcal{G}$, the following information is presented in one line:

[*i*] HMS1 (No., HMS2)   sequence   matrix   shift

The symbol HMS1 is the HM symbol of $\mathcal{H}$ referred to the coordinate system and setting of $\mathcal{G}$. The space-group number (No.) of $\mathcal{H}$ and the conventional HM symbol HMS2 of $\mathcal{H}$ if HMS1 is not a conventional one are given in brackets. Each *t*-subgroup $\mathcal{H} < \mathcal{G}$ is defined by its general-position representatives, listed under 'sequence' and specified by numbers, each of which designates an element of $\mathcal{G}$. They are taken from the general position of $\mathcal{G}$ and, therefore, are referred to the coordinate system of $\mathcal{G}$. The matrix–column pair $(\boldsymbol{P}, \boldsymbol{p})$ for the transformation from the conventional basis of $\mathcal{G}$ to that of $\mathcal{H}$ in concise notation can be read off the columns 'matrix' and 'shift' of the subgroup tables [for details of the concise notation of $(\boldsymbol{P}, \boldsymbol{p})$, see Section 1.5.1.3]. The column 'matrix' is empty if there is no change of basis, *i.e.* if $\boldsymbol{P}$ is the unit matrix $\boldsymbol{I}$. The column 'shift' is empty if there is no origin shift, *i.e.* if $\boldsymbol{p}$ is the column $\boldsymbol{o}$ consisting of zeroes only.

Conjugate subgroups are listed together and are connected by a left brace.

The way in which the subgroups are described is illustrated by the example of the *t*-subgroups of $P3_112$ (151), *cf.* Section 2.2.2.

I Maximal *translationengleiche* subgroups

	[2] $P3_111$ (144, $P3_1$)	1; 2; 3		
{	[3] $P112$ (5, $C121$)	1; 6	$\mathbf{b}, -2\mathbf{a}-\mathbf{b}, \mathbf{c}$	
{	[3] $P112$ (5, $C121$)	1; 4	$-\mathbf{a}-\mathbf{b}, \mathbf{a}-\mathbf{b}, \mathbf{c}$	0, 0, 1/3
{	[3] $P112$ (5, $C121$)	1; 5	$\mathbf{a}, \mathbf{a}+2\mathbf{b}, \mathbf{c}$	0, 0, 2/3

There are four maximal *t*-subgroups: one trigonal and three conjugate monoclinic subgroups connected by a brace on the left. Note that $P112$ is not the conventional HM symbol for the monoclinic unique axis *c* setting; here, the constituent '2' in the symbols of each of the three subgroups refers to the directions $-2\mathbf{a}-\mathbf{b}$, $\mathbf{a}-\mathbf{b}$ and $\mathbf{a}+2\mathbf{b}$, respectively, in the hexagonal basis. Unique axis *b* is the conventional setting of the monoclinic groups, as expressed by the standard HM symbol $C121$.

The entries in the columns 'matrix' and 'shift' are used for the transformation of matrix–column pairs $(\boldsymbol{W}, \boldsymbol{w})$ of the elements of $\mathcal{H}$ to their conventional form. In the case of the trigonal subgroup, these columns are empty because the subgroup is already in the conventional setting, *i.e.* $(\boldsymbol{P}, \boldsymbol{p}) = (\boldsymbol{I}, \boldsymbol{o})$. The monoclinic subgroups are not in the conventional setting: for example, the relation between the conventional basis $\mathbf{a}', \mathbf{b}', \mathbf{c}'$ of the last monoclinic subgroup and the hexagonal basis $\mathbf{a}, \mathbf{b}, \mathbf{c}$ is given by $\mathbf{a}' = \mathbf{a}$, $\mathbf{b}' = \mathbf{a} + 2\mathbf{b}$, $\mathbf{c}' = \mathbf{c}$, *i.e.*

$$\boldsymbol{P} = \begin{pmatrix} 1 & 1 & 0 \\ 0 & 2 & 0 \\ 0 & 0 & 1 \end{pmatrix}, \text{ and } \boldsymbol{p} = \begin{pmatrix} 0 \\ 0 \\ \frac{2}{3} \end{pmatrix}$$

as the corresponding origin shift.

Note that the conventional cell of the monoclinic subgroups is C-centred. In fact, the three conventional bases of the subgroups span ortho-hexagonal cells with twice the volume of the original hexagonal cell (see *e.g.* Fig. 1.5.1.7 of *IT* A). As we are dealing with t-subgroups, all translation vectors of $\mathcal{G}$ are retained in $\mathcal{H}$. On the other hand, the volume of the cells is doubled: therefore, there must be centring-translation vectors in the new cells. For example, the transformation of the translation $t(1,1,0)$ of the hexagonal lattice results in the C-centring translation $t'(\frac{1}{2},\frac{1}{2},0)$ of the ortho-hexagonal cell of the third monoclinic subgroup.

2.2.1.3. II Maximal *klassengleiche* subgroups (*k*-subgroups)

For practical reasons, the listing of the maximal *klassengleiche* subgroups (maximal k-subgroups) $\mathcal{H}$ of the space group $\mathcal{G}$ is divided into three blocks:

The block **Loss of centring translations** exists only for space groups $\mathcal{G}$ with centred lattices, and it contains those maximal subgroups of $\mathcal{G}$ which have fully or partly lost their centring translations. Maximal subgroups $\mathcal{H}$ of this block have the same conventional unit cell as the original space group $\mathcal{G}$. They are always non-isomorphic and have index 2 for plane groups and index 2, 3 or 4 for space groups.

All these subgroups are listed individually and each subgroup is characterized by the same type of data as that for t-subgroups (*cf.* Section 2.2.1.2). The sequence of the subgroups in this block is determined by the decreasing space-group number of the subgroups.

Under the heading **Enlarged unit cell** are listed those maximal k-subgroups $\mathcal{H}$ for which the conventional unit cell is enlarged relative to the unit cell of the original space group $\mathcal{G}$. *All* maximal k-subgroups with enlarged unit cell of index 2, 3 or 4 of the plane groups and of the space groups are listed *individually*. The listing is restricted to these indices because 4 is the highest index of a maximal *non-isomorphic* subgroup, and the number of these subgroups is finite. Maximal subgroups of higher indices are always isomorphic to $\mathcal{G}$, and their number is infinite.

Subgroups with the same lattice are collected in sub-blocks. The heading of each sub-block consists of the index of the subgroup and the lattice relations of the sublattice relative to the original lattice. Basis vectors that are not mentioned are not changed. The sequence of the subgroups is determined by the index of the subgroup, the kind of cell enlargement and by the number of the subgroup, so that the subgroup of highest space-group number is given first.

Each of the subgroups is represented by the following information:

[*i*] HMS1 (No., HMS2) generators matrix shift

The index [*i*], the HM symbol of $\mathcal{H}$ referred to the setting of $\mathcal{G}$ (often nonconventional) (HMS1), the space-group number and conventional symbol of $\mathcal{H}$ (HMS2) are given in the first two columns. The description of the subgroup is completed by a set of space-group generators and the transformation from the setting of the space group $\mathcal{G}$ to the conventional setting of the subgroup $\mathcal{H}$.

As an example, consider one of the conjugacy classes of maximal k-subgroups of index 3 of $P3_112$ (151) as shown in Fig. 2.2.1.1 (for the rest of the maximal k-subgroups, see the subgroup table of $P3_112$ in Section 2.2.2). In this example, the nonconventional symbols $H3_112$ of the subgroups of index 3 are followed by the space-group number (152) and the conventional HM symbol $P3_121$. The brace on the left indicates that the three subgroups $P3_121$ are conjugated in $\mathcal{G}$. The generators of the k-subgroups are obtained by adding translations to the generators of $\mathcal{G}$. For example, the shorthand description of the pair of generators $\langle 2 + (2,1,0);\ 4 + (2,2,0)\rangle$ of the last of the $P3_121$ subgroups is given by $\bar{y}+2, x-y+1, z+\frac{1}{3}$; $\bar{y}+2, \bar{x}+2, \bar{z}+\frac{2}{3}$ (for the generators of $P3_112$, see its subgroup table in Section 2.2.2). The last two columns display the transformations to the conventional basis of $\mathcal{H}$ and the corresponding origin shifts (referred to the basis of $\mathcal{G}$) in the same concise form as in the rest of the subgroup tables, *cf.* Sections 2.2.1.2 and 1.5.1.3.

Maximal subgroups of index higher than 4 have index p, p^2 or p^3, p prime, are necessarily isomorphic subgroups and are infinite in number. They cannot be listed individually but are listed in *IT* A1 as members of series under the heading **Series of maximal isomorphic subgroups**. In most of the series the HM symbol for each isomorphic subgroup $\mathcal{H} < \mathcal{G}$ is the same as that of $\mathcal{G}$. However, if $\mathcal{G}$ is an enantiomorphic space group, the HM symbol of $\mathcal{H}$ is either that of $\mathcal{G}$ or that of its enantiomorphic partner. The sequence of the isomorphic subgroups is determined by the index, the kind of cell enlargement and by the number of the subgroup.

The series of isomorphic subgroups of the space group $P3_112$ can be described by three series distributed in two blocks of different cell enlargement (*cf.* the subgroup table of $P3_112$ in Section 2.2.2). Two of the series correspond to a cell enlargement of index [p] and the subgroups of each of the series belong to one of the enantiomorphic pair of space groups $P3_112$ and $P3_212$. The conventional bases $\mathbf{a}', \mathbf{b}', \mathbf{c}'$ of the isomorphic subgroups in a series are defined relative to the basis $\mathbf{a}, \mathbf{b}, \mathbf{c}$ of the original space

II Maximal *klassengleiche* subgroups

- **Enlarged unit cell**

...

[3] $\mathbf{a}' = 3\mathbf{a}$, $\mathbf{b}' = 3\mathbf{b}$

{ $H3_112$ (152, $P3_121$)	$\langle 2; 4\rangle$	$\mathbf{a}-\mathbf{b}, \mathbf{a}+2\mathbf{b}, \mathbf{c}$	
$H3_112$ (152, $P3_121$)	$\langle 2+(1,-1,0);\ 4+(1,1,0)\rangle$	$\mathbf{a}-\mathbf{b}, \mathbf{a}+2\mathbf{b}, \mathbf{c}$	1,0,0
$H3_112$ (152, $P3_121$)	$\langle 2+(2,1,0);\ 4+(2,2,0)\rangle$	$\mathbf{a}-\mathbf{b}, \mathbf{a}+2\mathbf{b}, \mathbf{c}$	1,1,0

Figure 2.2.1.1
Part of the maximal k-subgroups of $P3_112$ (151) with enlarged unit cells (for the rest of the maximal k-subgroups see the subgroup table of $P3_112$ in Section 2.2.2).

$[p^2]$ $\mathbf{a}' = p\mathbf{a}$, $\mathbf{b}' = p\mathbf{b}$			
$P3_112$ (151)	$\langle 2+(u+v,-u+2v,0);\ 4+(u+v,u+v,0)\rangle$ prime $p \neq 3$; $0 \leq u < p$; $0 \leq v < p$ p^2 conjugate subgroups	$p\mathbf{a}, p\mathbf{b}, \mathbf{c}$	$u, v, 0$

Figure 2.2.1.2
Part of the series of isomorphic subgroups of the space group $P3_112$ (151) (for the rest of the series of isomorphic subgroups see the subgroup table of $P3_112$ in Section 2.2.2).

group with $\mathbf{c}' = p\mathbf{c}$. The p subgroups in a conjugacy class differ by the positions of their conventional origins relative to the origin of the space group $\mathcal{G}$, which are defined by the p permitted values of u. For the third of the series, shown in Fig. 2.2.1.2, the index is $[p^2]$ and the lattice relations are $\mathbf{a}' = p\mathbf{a}$, $\mathbf{b}' = p\mathbf{b}$, listed as $p\mathbf{a}, p\mathbf{b}, \mathbf{c}$ for the transformation matrix. For each value of p there exist exactly p^2 conjugate subgroups with origins at the points u, v, 0, where the parameters u and v run independently: $0 \leq u < p$ and $0 \leq v < p$.

The generators of the p (or p^2) conjugate isomorphic subgroups $\mathcal{H}$ are obtained from the generators (2) and (4) of $P3_112$ by adding translational components. These components are determined by the parameters p and u (or p, and u and v, if relevant).

As an example, consider the $[p^2]$ isomorphic subgroups of $P3_112$ with $p = 2$ (*cf.* Fig. 2.2.1.2). There are four conjugate subgroups whose generators and transformation matrices are obtained by assigning the values (0, 0), (1, 0), (0, 1) and (1, 1) to the pair of parameters (u, v). As expected, the subgroup data obtained coincide with the data of the k-subgroups of index 4 shown in the subgroup table of $P3_112$ in Section 2.2.2. For example, substituting $p = 2$, $u = 1$, $v = 1$ in the expressions for the generators and transformation matrices of the $[p^2]$ series, we get

$P3_112$ (151) $\langle 2 + (2, 1, 0); 4 + (2, 2, 0)\rangle$ $2\mathbf{a}, 2\mathbf{b}, \mathbf{c}$ 1, 1, 0

which corresponds to the last subgroup of index 4 of $P3_112$ (*cf.* Section 2.2.2).

2.2.1.4. Minimal supergroups

According to Hermann's theorem, a minimal supergroup $\mathcal{G}$ of a space group $\mathcal{H}$ is either a *translationengleiche* supergroup (t-supergroup) or a *klassengleiche* supergroup (k-supergroup). A proper minimal t-supergroup always has an index i, $1 < i < 5$, and is never isomorphic. A minimal k-supergroup with index i, $1 < i < 5$, may be isomorphic or non-isomorphic; for indices $i > 4$ a minimal k-supergroup can only be an isomorphic k-supergroup. The infinite in number isomorphic supergroups are not listed in the tables of *IT* A1 because they are implicitly shown among the subgroup data (for the derivation of the isomorphic supergroups *cf.* Section 2.1.7.2 of *IT* A1).

The supergroup data of *IT* A1 are partitioned into two main blocks: minimal t-supergroups and minimal non-isomorphic k-supergroups.

For each space group $\mathcal{H}$, under the heading **I Minimal translationengleiche supergroups** are listed those space-group types $\mathcal{G}$ for which $\mathcal{H}$ appears as an entry under the heading **I Minimal translationengleiche subgroups**. Note that the minimal t-supergroups $\mathcal{G}$ of $\mathcal{H}$ are not listed individually; the entries correspond to the space-group types of $\mathcal{G}$, specified by the index [...], the conventional HM symbol and the space-group number (in parentheses). Not listed is the number of supergroups belonging to one entry. The supergroups are ordered according to their rising indices and space-group numbers.

It is important to mention that the supergroups listed in this block represent space groups only if the lattice conditions of $\mathcal{H}$ fulfil the lattice conditions for $\mathcal{G}$. This requirement is always satisfied if the group $\mathcal{H}$ and the supergroup $\mathcal{G}$ belong to the same crystal family (for the term crystal family, see Section 1.3.4.4). If $\mathcal{G}$ is a k-supergroup of $\mathcal{H}$, $\mathcal{G}$ and $\mathcal{H}$ always belong to the same crystal family and there are no lattice restrictions on $\mathcal{H}$.

The block **II Minimal non-isomorphic *klassengleiche* supergroups** is divided into two sub-blocks:

Under the heading **Additional centring translations** the supergroups are listed by their indices and either by their unconventional HM symbols, followed by the space-group numbers and the standard HM symbols in parentheses, or by their conventional HM symbols and the space-group numbers in parentheses.

Under the heading **Decreased unit cell** each supergroup is listed by its index and by its lattice relations, where the basis vectors $\mathbf{a}'$, $\mathbf{b}'$ and $\mathbf{c}'$ refer to the supergroup $\mathcal{G}$ and the basis vectors $\mathbf{a}$, $\mathbf{b}$ and $\mathbf{c}$ to the original group $\mathcal{H}$. The supergroups are specified either by the unconventional HM symbol, followed by the space-group number and the conventional HM symbol in parentheses, or by the conventional HM symbol with the space-group number in parentheses.

All minimal non-isomorphic k-supergroups are listed individually and their sequence is determined by the rising indices and space-group numbers.

2.2.2. Examples of the subgroup tables

The listing of the maximal subgroup data and minimal supergroup data in *IT* A1 is illustrated by the subgroup tables of the space groups $P3_112$ (151) and $P3_121$ (152). These two space groups have been selected in order to demonstrate the impact of certain space-group properties on the (maximal) subgroup and (minimal) supergroup data. For example, the property that the point groups of the space groups are of the same point-group type determines the similarity of the sets of maximal t-subgroups and minimal t-supergroups of the two space groups. (In fact, similar sets of maximal t-subgroups and minimal t-supergroups are observed in all seven space groups that belong to the geometric crystal class 32.) On the other hand, $P3_112$ and $P3_121$ are distinguished by different orientations of some of their symmetry elements with respect to the lattice symmetry directions. This feature gives rise to a significant difference in their maximal k-subgroups and minimal k-supergroups, expressed by the lack of reciprocity of their group–subgroup relations. For example, $P3_112$ has nine $P3_121$ subgroups of index 3, distributed into three conjugacy classes, while $P3_121$ has just one normal $P3_112$ of index 3. Differences in the sets of maximal k-subgroups and minimal k-supergroups of similar nature are observed for other pairs of space groups such as $P321$ (149) and $P312$ (150), and $P3_212$ (153) and $P3_221$ (154) *etc.*

Generators selected (1); $t(1,0,0)$; $t(0,1,0)$; $t(0,0,1)$; (2); (4)

General position

Multiplicity, Wyckoff letter, Site symmetry			Coordinates		
6	c	1	(1) x,y,z	(2) $\bar{y},x-y,z+\frac{1}{3}$	(3) $\bar{x}+y,\bar{x},z+\frac{2}{3}$
			(4) $\bar{y},\bar{x},\bar{z}+\frac{2}{3}$	(5) $\bar{x}+y,y,\bar{z}+\frac{1}{3}$	(6) $x,x-y,\bar{z}$

I Maximal *translationengleiche* subgroups

[2] $P3_111$ (144, $P3_1$)	1; 2; 3		
{ [3] $P112$ (5, $C121$)	1; 6	$\mathbf{b},-2\mathbf{a}-\mathbf{b},\mathbf{c}$	
[3] $P112$ (5, $C121$)	1; 4	$-\mathbf{a}-\mathbf{b},\mathbf{a}-\mathbf{b},\mathbf{c}$	0,0,1/3
[3] $P112$ (5, $C121$) }	1; 5	$\mathbf{a},\mathbf{a}+2\mathbf{b},\mathbf{c}$	0,0,2/3

II Maximal *klassengleiche* subgroups

- **Enlarged unit cell**

[2] $\mathbf{c}'=2\mathbf{c}$			
$P3_212$ (153)	$\langle 4;\ 2+(0,0,1)\rangle$	$\mathbf{a},\mathbf{b},2\mathbf{c}$	
$P3_212$ (153)	$\langle (2;\ 4)+(0,0,1)\rangle$	$\mathbf{a},\mathbf{b},2\mathbf{c}$	0,0,1/2
[3] $\mathbf{a}'=3\mathbf{a}$, $\mathbf{b}'=3\mathbf{b}$			
{ $H3_112$ (152, $P3_121$)	$\langle 2;\ 4\rangle$	$\mathbf{a}-\mathbf{b},\mathbf{a}+2\mathbf{b},\mathbf{c}$	
$H3_112$ (152, $P3_121$)	$\langle 2+(1,-1,0);\ 4+(1,1,0)\rangle$	$\mathbf{a}-\mathbf{b},\mathbf{a}+2\mathbf{b},\mathbf{c}$	1,0,0
$H3_112$ (152, $P3_121$) }	$\langle 2+(2,1,0);\ 4+(2,2,0)\rangle$	$\mathbf{a}-\mathbf{b},\mathbf{a}+2\mathbf{b},\mathbf{c}$	1,1,0
{ $H3_112$ (152, $P3_121$)	$\langle 4;\ 2+(1,0,0)\rangle$	$\mathbf{a}-\mathbf{b},\mathbf{a}+2\mathbf{b},\mathbf{c}$	2/3,−2/3,0
$H3_112$ (152, $P3_121$)	$\langle 2+(2,2,0);\ 4+(1,1,0)\rangle$	$\mathbf{a}-\mathbf{b},\mathbf{a}+2\mathbf{b},\mathbf{c}$	2/3,1/3,0
$H3_112$ (152, $P3_121$) }	$\langle 2+(3,4,0);\ 4+(2,2,0)\rangle$	$\mathbf{a}-\mathbf{b},\mathbf{a}+2\mathbf{b},\mathbf{c}$	2/3,4/3,0
{ $H3_112$ (152, $P3_121$)	$\langle 4;\ 2+(1,1,0)\rangle$	$\mathbf{a}-\mathbf{b},\mathbf{a}+2\mathbf{b},\mathbf{c}$	1/3,−1/3,0
$H3_112$ (152, $P3_121$)	$\langle 2+(2,3,0);\ 4+(1,1,0)\rangle$	$\mathbf{a}-\mathbf{b},\mathbf{a}+2\mathbf{b},\mathbf{c}$	1/3,2/3,0
$H3_112$ (152, $P3_121$) }	$\langle 2+(3,2,0);\ 4+(2,2,0)\rangle$	$\mathbf{a}-\mathbf{b},\mathbf{a}+2\mathbf{b},\mathbf{c}$	4/3,2/3,0
[4] $\mathbf{a}'=2\mathbf{a}$, $\mathbf{b}'=2\mathbf{b}$			
{ $P3_112$ (151)	$\langle 2;\ 4\rangle$	$2\mathbf{a},2\mathbf{b},\mathbf{c}$	
$P3_112$ (151)	$\langle 2+(1,-1,0);\ 4+(1,1,0)\rangle$	$2\mathbf{a},2\mathbf{b},\mathbf{c}$	1,0,0
$P3_112$ (151)	$\langle 2+(1,2,0);\ 4+(1,1,0)\rangle$	$2\mathbf{a},2\mathbf{b},\mathbf{c}$	0,1,0
$P3_112$ (151) }	$\langle 2+(2,1,0);\ 4+(2,2,0)\rangle$	$2\mathbf{a},2\mathbf{b},\mathbf{c}$	1,1,0

- **Series of maximal isomorphic subgroups**

$[p]$ $\mathbf{c}'=p\mathbf{c}$			
$P3_212$ (153)	$\langle 2+(0,0,\frac{2p}{3}-\frac{1}{3});\ 4+(0,0,\frac{p}{3}-\frac{2}{3}+2u)\rangle$ prime $p>4$; $0\le u<p$ p conjugate subgroups for $p=6n-1$	$\mathbf{a},\mathbf{b},p\mathbf{c}$	$0,0,u$
$P3_112$ (151)	$\langle 2+(0,0,\frac{p}{3}-\frac{1}{3});\ 4+(0,0,\frac{2p}{3}-\frac{2}{3}+2u)\rangle$ prime $p>6$; $0\le u<p$ p conjugate subgroups for $p=6n+1$	$\mathbf{a},\mathbf{b},p\mathbf{c}$	$0,0,u$
$[p^2]$ $\mathbf{a}'=p\mathbf{a}$, $\mathbf{b}'=p\mathbf{b}$			
$P3_112$ (151)	$\langle 2+(u+v,-u+2v,0);\ 4+(u+v,u+v,0)\rangle$ prime $p\neq 3$; $0\le u<p$; $0\le v<p$ p^2 conjugate subgroups	$p\mathbf{a},p\mathbf{b},\mathbf{c}$	$u,v,0$

I Minimal *translationengleiche* supergroups

[2] $P6_122$ (178); [2] $P6_422$ (181)

II Minimal non-isomorphic *klassengleiche* supergroups

- **Additional centring translations**

[3] $H3_112$ (152, $P3_121$)

- **Decreased unit cell**

[3] $\mathbf{c}'=\frac{1}{3}\mathbf{c}$ $P312$ (149)

$P3_121$ No. 152 $P3_121$ D_3^4

Generators selected (1); $t(1,0,0)$; $t(0,1,0)$; $t(0,0,1)$; (2); (4)

General position

Multiplicity, Wyckoff letter, Site symmetry			Coordinates		
6	c	1	(1) x,y,z	(2) $\bar{y},x-y,z+\frac{1}{3}$	(3) $\bar{x}+y,\bar{x},z+\frac{2}{3}$
			(4) $y,x,\bar{z}$	(5) $x-y,\bar{y},\bar{z}+\frac{2}{3}$	(6) $\bar{x},\bar{x}+y,\bar{z}+\frac{1}{3}$

I Maximal *translationengleiche* subgroups

[2] $P3_111$ (144, $P3_1$)	1; 2; 3		
{ [3] $P121$ (5, $C121$)	1; 4	$-\mathbf{a}+\mathbf{b},-\mathbf{a}-\mathbf{b},\mathbf{c}$	
{ [3] $P121$ (5, $C121$)	1; 5	$-\mathbf{a}-2\mathbf{b},\mathbf{a},\mathbf{c}$	0,0,1/3
{ [3] $P121$ (5, $C121$)	1; 6	$2\mathbf{a}+\mathbf{b},\mathbf{b},\mathbf{c}$	0,0,2/3

II Maximal *klassengleiche* subgroups

- **Enlarged unit cell**

[2] $\mathbf{c}'=2\mathbf{c}$			
$P3_221$ (154)	$\langle 4;\ 2+(0,0,1)\rangle$	$\mathbf{a},\mathbf{b},2\mathbf{c}$	
$P3_221$ (154)	$\langle (2;\ 4)+(0,0,1)\rangle$	$\mathbf{a},\mathbf{b},2\mathbf{c}$	0,0,1/2
[3] $\mathbf{a}'=3\mathbf{a},\ \mathbf{b}'=3\mathbf{b}$			
$H3_121$ (151, $P3_112$)	$\langle 2;\ 4\rangle$	$\mathbf{a}-\mathbf{b},\mathbf{a}+2\mathbf{b},\mathbf{c}$	0,0,1/3
[4] $\mathbf{a}'=2\mathbf{a},\ \mathbf{b}'=2\mathbf{b}$			
{ $P3_121$ (152)	$\langle 2;\ 4\rangle$	$2\mathbf{a},2\mathbf{b},\mathbf{c}$	
{ $P3_121$ (152)	$\langle (2;\ 4)+(1,-1,0)\rangle$	$2\mathbf{a},2\mathbf{b},\mathbf{c}$	1,0,0
{ $P3_121$ (152)	$\langle 2+(1,2,0);\ 4+(-1,1,0)\rangle$	$2\mathbf{a},2\mathbf{b},\mathbf{c}$	0,1,0
{ $P3_121$ (152)	$\langle 4;\ 2+(2,1,0)\rangle$	$2\mathbf{a},2\mathbf{b},\mathbf{c}$	1,1,0

- **Series of maximal isomorphic subgroups**

$[p]$ $\mathbf{c}'=p\mathbf{c}$			
$P3_221$ (154)	$\langle 2+(0,0,\frac{2p}{3}-\frac{1}{3});\ 4+(0,0,2u)\rangle$ prime $p>4$; $0\le u<p$ p conjugate subgroups for $p=6n-1$	$\mathbf{a},\mathbf{b},p\mathbf{c}$	$0,0,u$
$P3_121$ (152)	$\langle 2+(0,0,\frac{p}{3}-\frac{1}{3});\ 4+(0,0,2u)\rangle$ prime $p>6$; $0\le u<p$ p conjugate subgroups for $p=6n+1$	$\mathbf{a},\mathbf{b},p\mathbf{c}$	$0,0,u$
$[p^2]$ $\mathbf{a}'=p\mathbf{a},\ \mathbf{b}'=p\mathbf{b}$			
$P3_121$ (152)	$\langle 2+(u+v,-u+2v,0);\ 4+(u-v,-u+v,0)\rangle$ prime $p\neq 3$; $0\le u<p$; $0\le v<p$ p^2 conjugate subgroups	$p\mathbf{a},p\mathbf{b},\mathbf{c}$	$u,v,0$

I Minimal *translationengleiche* supergroups

[2] $P6_122$ (178); [2] $P6_422$ (181)

II Minimal non-isomorphic *klassengleiche* supergroups

- **Additional centring translations**

[3] $H3_121$ (151, $P3_112$); [3] $R_{obv}32$ (155, $R32$); [3] $R_{rev}32$ (155, $R32$)

- **Decreased unit cell**

[3] $\mathbf{c}'=\frac{1}{3}\mathbf{c}$ $P321$ (150)

2.2.3. Guide to the tables of relations between Wyckoff positions

BY ULRICH MÜLLER

2.2.3.1. Guide to the use of the tables

For an introduction to Wyckoff positions and crystallographic orbits see Section 1.4.4. In Part 3 of *IT* A1, all maximal subgroups of the space groups are listed once again. For all Wyckoff positions of a space group the relations to the Wyckoff positions of the subgroups are given.

The Wyckoff positions are always labelled by their multiplicities and their Wyckoff letters, with the same labels as in the tables of *IT* A and in Section 2.1.4. Reference to *IT* A or to Section 2.1.4 therefore is always necessary, especially when the corresponding coordinate triplets or site symmetries are needed.

The tables in Parts 2 (tables of maximal subgroups) and 3 (relations of the Wyckoff positions) of *IT* A1 and, correspondingly, in Sections 2.2.2 and 2.2.4 deal with different aspects of the reduction of crystal symmetry. Whereas the subject of Part 2 of *IT* A1 is *symmetry*, the subject of Part 3 is the *implications of symmetry changes for crystal structures*. As the subjects of Parts 2 and 3 are different, the presentation of the data in the two parts differs in order to make the listed data as convenient as possible.

Every space group begins on a new page (with the exception of $P4_3$, $P3_2$, $P6_4$ and $P6_5$, which are listed together with $P4_1$, $P3_1$, $P6_2$ and $P6_1$, respectively). If necessary, continuation occurs on the following page(s). The different settings for monoclinic space groups are continued on the same or the following page(s).

The tables are arranged in the following way:

2.2.3.1.1. *Headline*

The headline lists from the outer margin inwards:

(1) The *short Hermann–Mauguin symbol*;
(2) The number of the space group according to *IT* A;
(3) The *full Hermann–Mauguin symbol* if it differs from the short symbol;
(4) The *Schoenflies symbol*.

In the case of monoclinic space groups, the headline can have one or two additional entries with the full Hermann–Mauguin symbols for different settings.

2.2.3.1.2. *Specification of the settings*

Each of the monoclinic space groups is listed several times, namely with unique axis *b* and with unique axis *c*, and, if applicable, with the three cell choices 1, 2 and 3 according to *IT* A (see also Section 1.5.4). Space permitting, the entries for the different settings have been combined on one page or on facing pages, since in most cases the Wyckoff-position relations do not depend on the choice of setting. In the few cases where there is a dependence, arrows (⇒) in the corresponding lines show to which settings they refer. Otherwise, the Wyckoff positions of the subgroups correspond to all of the settings listed on the same page or on facing pages.

Rhombohedral space groups are listed only in the setting with hexagonal axes with a rhombohedrally centred obverse cell [*i.e.* $\pm(\frac{2}{3},\frac{1}{3},\frac{1}{3})$]. However, for cubic space groups, the rhombohedral subgroups are also given with rhombohedral axes.

Settings with different origin choices are taken account of by two separate columns 'Coordinates' with the headings 'origin 1' and 'origin 2'.

2.2.3.1.3. *List of Wyckoff positions*

Under the column heading 'Wyckoff positions', the complete sequence of the Wyckoff positions of the space group is given by their multiplicities and Wyckoff letters. If necessary, the sequence runs over two or more lines.

2.2.3.1.4. *Subgroup data*

The subgroups are divided into two sections: **I Maximal *translationengleiche* subgroups** and **II Maximal *klassengleiche* subgroups**. The latter are further subdivided into three blocks:

Loss of centring translations. This block appears only if the space group has a conventionally centred lattice. The centring has been fully or partly lost in the subgroups listed. The size of the conventional unit cell is not changed.

Enlarged unit cell, non-isomorphic. The *klassengleiche* subgroups listed in this block are non-isomorphic and have conventional unit cells that are enlarged compared with the unit cell of the space group.

Enlarged unit cell, isomorphic. This listing includes the isomorphic subgroups with the smallest possible indices for every kind of cell enlargement. If they exist, index values of 2, 3 and 4 are always given (except for $P\bar{1}$, which is restricted to index 2). If the indices 2, 3 or 4 are not possible, the smallest possible index for the kind of cell enlargement considered is listed. In addition, the infinite series of isomorphic subgroups are given for all possible kinds of cell enlargements. The factor of the cell enlargement corresponds to the index, which is p, p^2 or p^3 (p = prime number; *cf.* Section 2.2.3.1.6). If $p > 2$, the specifically listed subgroups with small index values also always belong to the infinite series, so that the corresponding information is given twice in these cases. For $p = 2$ this applies only to certain special cases.

2.2.3.1.5. *Sequence of the listed subgroups*

Within each of the aforementioned blocks, the subgroups are listed in the following order. First priority is given to the index, with smallest values first. Subgroups with the same index follow decreasing space-group numbers (according to *IT* A). Exception: the *translationengleiche* subgroup of a tetragonal space group listed last is always the one with the axes transformation to a diagonally oriented cell.

Translationengleiche subgroups of cubic space groups are in the order cubic, rhombohedral, tetragonal, orthorhombic.

In the case of the isomorphic subgroups, there is a subdivision according to the kind of cell enlargement. For monoclinic, tetragonal, trigonal and hexagonal space groups, cell enlargements in the direction of the unique axis are given first. For orthorhombic space groups, the isomorphic subgroups with increased **a** are given first, followed by increased **b** and **c**.

The sequence differs somewhat from that in Part 2 of *IT* A1 and Section 2.2.2, where the *klassengleiche* subgroups have been subdivided in more detail according to the different kinds of cell enlargements and the isomorphic subgroups with specific index values have been listed together with the *klassengleiche* subgroups, *i.e.* separately from the infinite series of isomorphic subgroups.

2.2.3.1.6. *Information for every subgroup*

Index of symmetry reduction

The entry for every subgroup begins with the index of the symmetry reduction in brackets, for example [2] or [p] or [p^2] (p = prime number).

The index for any of the infinite number of maximal isomorphic subgroups must be either a prime number p, or, in certain cases of tetragonal, trigonal and hexagonal space groups, a square of a prime number p^2; for isomorphic subgroups of cubic space groups the index may only be the cube of a prime number p^3. In many instances only certain prime numbers are allowed (Bertaut & Billiet, 1979; Billiet & Bertaut, 2005; Müller & Brelle, 1995). If restrictions exist, the prime numbers allowed are given under the axes transformations by formulae such as 'p = prime = $6n + 1$'.

Subgroup symbol

The index is followed by the short Hermann–Mauguin symbol and the space-group number of the subgroup. If a nonconventional setting has been chosen, then the space-group symbol of the conventional setting is also mentioned in the following line after the symbol $\widehat{=}$.

In some cases of nonconventional settings, the space-group symbol does not show uniquely in which manner it deviates from the conventional setting. For example, the nonconventional setting $P22_12$ of the space group $P222_1$ can result from cyclic exchange of the axes (**b**, **c**, **a**) or by interchange of **b** with **c** (**a**, −**c**, **b**). As a consequence, the relations between the Wyckoff positions can be different. In such cases, cyclic exchange has always been chosen.

Basis vectors

The column 'Axes' shows how the basis vectors of the unit cell of a subgroup result from the basis vectors **a**, **b** and **c** of the space group considered. This information is omitted if there is no change of basis vectors.

A formula such as '$q\mathbf{a} - r\mathbf{b}, r\mathbf{a} + q\mathbf{b}, \mathbf{c}$' together with the restrictions '$p = q^2 + r^2$ = prime = $4n + 1$' and '$q = 2n + 1 \geq 1$; $r = \pm 2n' \neq 0$' means that for a given index p there exist several subgroups with different lattices depending on the values of the integer parameters q (odd) and r (even) within the limits of the restriction. In this example, the prime number p must be $p \equiv 1$ modulo 4 (*i.e.* 5, 13, 17, ...); if it is, say, $p = 13 = 3^2 + (\pm 2)^2$, the values of q and r may be $q = 3, r = 2$ and $q = 3, r = -2$.

Coordinates

The column 'Coordinates' shows how the atomic coordinates of the subgroups are calculated from the coordinates x, y and z of the initial unit cell. This includes coordinate shifts whenever a shift of the origin is required. If the cell of the subgroup is enlarged, the coordinate triplet is followed by a semicolon; then follow fractional numbers in parentheses. This means that in addition to the coordinates given before the semicolon, further coordinates have to be considered; they result from adding the numbers in the parentheses. However, if the subgroup has a centring, the values to be added due to this centring are not mentioned. If no transformation of coordinates is necessary, the entry is omitted.

Example

The entry of an I-centred subgroup

$$\tfrac{1}{2}x,\ \tfrac{1}{2}y,\ \tfrac{1}{2}z;\ +(\tfrac{1}{2},0,0);\ +(0,\tfrac{1}{2},0);\ +(0,0,\tfrac{1}{2})$$

means: starting from the coordinates of, say, 0.08, 0.14, 0.20, sites with the following coordinates result in the subgroup:

0.04, 0.07, 0.10; 0.54, 0.07, 0.10;
0.04, 0.57, 0.10; 0.04, 0.07, 0.60;

in addition, there are all coordinates with $+(\tfrac{1}{2},\tfrac{1}{2},\tfrac{1}{2})$ due to the I-centring:

0.54, 0.57, 0.60; 0.04, 0.57, 0.60;
0.54, 0.07, 0.60; 0.54, 0.57, 0.10.

For the infinite series of isomorphic subgroups, coordinate formulae are, for example, in the form $x,\ y,\ \tfrac{1}{p}z;\ +(0,0,\tfrac{u}{p})$ with $u = 1, \ldots, p-1$. Then there are p coordinate values running from $x,\ y,\ \tfrac{1}{p}z$ to $x,\ y,\ \tfrac{1}{p}z + \tfrac{p-1}{p}$.

If *IT* A allows two choices for the origin, coordinate transformations for both are listed in separate columns with the headings 'origin 1' and 'origin 2'. If two origin choices are allowed for both the group as well as the subgroup, then it is understood that the origin choices of the group and the subgroup are the same (either origin choice 1 for both groups or origin choice 2 for both). If the space group has only one origin choice, but the subgroup has two choices, the coordinate transformations are given for both choices on separate lines.

Wyckoff positions

The columns under the heading 'Wyckoff positions' contain the Wyckoff symbols of all sites of the subgroups that result therefrom. They are given in the same sequence as in the top line(s). If the symbols at the top run over more than one line, then the symbols for the subgroups take a corresponding number of lines.

When an orbit splits into several independent orbits, the corresponding Wyckoff symbols are separated by semicolons, *i.e.* 1*b*; 4*h*; 4*k*. An entry such as 3 × 8*j* means that a splitting into three orbits takes place, all of which are of the same kind 8*j*; they differ in the values of their free parameters.

For the infinite series of isomorphic subgroups general formulae are given. They allow the calculation of the Wyckoff-position relations for any index in a simple manner.

Example

The entry $\frac{p(p-1)}{2} \times 24k$ means that for a given prime number p, say $p = 5$, there are $\frac{5(5-1)}{2} = 10$ orbits of the kind 24*k*.

In some cases of splittings there is not enough space to enter all Wyckoff symbols on one line; this requires them to be listed one below the other over two or more lines. Whenever a Wyckoff symbol is followed by a semicolon, another symbol follows.

Sometimes a Wyckoff label is followed by another Wyckoff label in parentheses together with a footnote marker. In this case, the Wyckoff label in parentheses is to be taken for the cases specified in the footnote.

Example

The entry 2*c*(*d**) together with the footnote *$p = 4n - 1$ means that the Wyckoff position is 2*c*, but it is 2*d* if the index is $p \equiv 3$ modulo 4 (*i.e.* $p = 3, 7, 11, \ldots$).

The Wyckoff positions of an isomorphic subgroup of a space group with two choices for the origin are only identical for the two choices if certain origin shifts are taken into account. Since origin shifts have been avoided as far as possible, in some cases some Wyckoff positions differ for the two origin choices.

Warning: *The listed Wyckoff positions of the subgroups apply only to the transformations given in the column 'Coordinates'.* If other cell transformations or origin shifts are used, this may result in an interchange of Wyckoff positions.

2.2.3.2. Cell transformations

It is generally assumed that the initial crystal structure and its space group $\mathcal{G}$ are referred to a conventional coordinate system (conventional setting). If this is not the case, a transformation to a conventional coordinate system has to precede the use of the data. However, for orthorhombic space groups nonconventional settings are listed at the bottom of the tables (see Section 2.2.3.4).

In general, the conventional coordinate system of a subgroup $\mathcal{H}$ is not identical to that of $\mathcal{G}$ and may involve a change of the basis and an origin shift. This coordinate transformation is not uniquely determined and may be chosen within certain limits. For the formalism of the transformation see Chapter 1.5.

A basis transformation is listed in Parts 2 and 3 of *IT* A1 by linear combinations such as $\mathbf{a}-\mathbf{b}, \mathbf{a}+\mathbf{b}, \mathbf{c}$. In Part 3, in addition, the transformation of the atomic coordinates is given in the column 'coordinates', *e.g.* $\frac{1}{2}(x-y), \frac{1}{2}(x+y), z$. Note that, in general, the basis transformation (and the origin shift) listed in Parts 2 and 3 of *IT* A1 for a given group–subgroup pair are different.

When comparing related crystal structures, unit-cell transformations are troublesome. They result in differing sets of atomic coordinates for corresponding atoms; this can make comparisons more complicated and structural relations may be obscured. Frequently, it is more convenient not to interchange axes and to avoid transformations if possible. The use of a nonconventional setting of a space group may be preferable if this reduces cell transformations. For this reason, in the tables in *IT* A1 the settings of the subgroups were preferentially chosen in such a way that the directions of the basis vectors of a space group and its subgroup deviate as little as possible. If this results in a nonconventional setting of the subgroup, then the way to transform the basis vectors and coordinates to those of the conventional cell is also given.

Subgroups listed in nonconventional settings concern orthorhombic and monoclinic space groups. Their transformations to conventional settings frequently only involve an interchange of axes. For more details see Section 2.2.3.4.

The settings listed for monoclinic and triclinic subgroups were chosen in such a way that axes transformations are avoided or kept to a minimum. Depending on the cell metrics, this may result in cells that do not have the shortest possible basis vectors. Unfortunately, transformation of a monoclinic or triclinic cell setting to another one may cause an interchange of Wyckoff labels within a Wyckoff set of the subgroup; a Wyckoff set comprises different Wyckoff positions that have the same kind of site symmetry (for a more exact definition *cf.* Sections 1.4.4.3, 3.4.1 and 3.4.2 in *IT* A). Frequently, several possible cell settings of the same monoclinic subgroup have been listed; the entry for the subgroup then is followed by the word 'or' or 'alternative', plus another entry.

Example

Space group *Cmcm*, No. 63, has the subgroup $P112_1/m$, No. 11. It requires a cell transformation which is given as $\mathbf{a}, \frac{1}{2}(-\mathbf{a}+\mathbf{b}), \mathbf{c}$. The following two lines list two other possible cell transformations for the *same* subgroup after the words 'or': $\frac{1}{2}(\mathbf{a}-\mathbf{b}), \mathbf{b}, \mathbf{c}$ and $\frac{1}{2}(\mathbf{a}-\mathbf{b}), \frac{1}{2}(\mathbf{a}+\mathbf{b}), \mathbf{c}$. These three options cause different relations for the Wyckoff positions 4*b* and 8*d* of *Cmcm*.

2.2.3.3. Origin shifts

In a group–subgroup relation, an origin shift may be necessary to conform to the conventional origin setting of the subgroup $\mathcal{H}$. Origin shifts can be specified in terms of the coordinate system of the initial space group $\mathcal{G}$ or of the coordinate system of the subgroup. In Part 2 of *IT* A1 and in Section 2.2.4, all origin shifts refer to the initial space group $\mathcal{G}$. They are stated by the coefficients of the column $\boldsymbol{p}$, *e.g.* 0, 1/2, 1/4. That is also the origin shift marked in the the group–subgroup arrow of a Bärnighausen tree (*cf.* Fig. 1.7.2.2).

Note: In Part 3 of *IT* A1 and in Section 2.2.4, the origin shifts are contained only in the column 'Coordinates' as additive fractional numbers, *e.g.* $\frac{1}{2}(x-y)-\frac{1}{4}, \frac{1}{2}(x+y)+\frac{1}{4}, z$. The numbers $-\frac{1}{4}, +\frac{1}{4}, 0$ are the coefficients of the shift vector $\boldsymbol{p}'$ *in the coordinate system of the subgroup* and thus are different from the coefficients of $\boldsymbol{p}$.

The origin shifts listed in the column 'Coordinates' can be converted to origin shifts that refer to the coordinate system of the initial space group in the following way:

Take:

$\mathbf{a}, \mathbf{b}, \mathbf{c}$	basis vectors of the initial space group $\mathcal{G}$;
O	origin of $\mathcal{G}$;
$\mathbf{a}', \mathbf{b}', \mathbf{c}'$	basis vectors of the subgroup $\mathcal{H}$;
O'	origin of $\mathcal{H}$;
$x_{o'}, y_{o'}, z_{o'}$	coordinates of O' expressed in the coordinate system of $\mathcal{G}$;
x'_o, y'_o, z'_o	coordinates of O expressed in the coordinate system of $\mathcal{H}$.

The basis vectors are related according to $(\mathbf{a}', \mathbf{b}', \mathbf{c}') = (\mathbf{a}, \mathbf{b}, \mathbf{c})\boldsymbol{P}$ (*cf.* Chapter 1.5). $\boldsymbol{P}$ is the 3×3 transformation matrix of the basis change. The origin shift $O \to O'$ then corresponds to the vector

$$\boldsymbol{p} = \begin{pmatrix} x_{o'} \\ y_{o'} \\ z_{o'} \end{pmatrix} = -\boldsymbol{P}\boldsymbol{p}' = -\boldsymbol{P}\begin{pmatrix} x'_o \\ y'_o \\ z'_o \end{pmatrix}.$$

If there is only an origin shift and no basis transformation (*i.e.* $\boldsymbol{P} = \boldsymbol{I}$), the two origin shifts $x_{o'}, y_{o'}, z_{o'}$ and x'_o, y'_o, z'_o only differ in their signs, $\boldsymbol{p} = -\boldsymbol{p}'$ (they have opposite directions).

For a given group–subgroup pair $\mathcal{G} > \mathcal{H}$, in general, the basis transformation and the origin shift listed in Parts 2 and 3 of *IT* A1 are different. If it is needed, the origin shift $\boldsymbol{p}$ *has to be calculated*; it cannot be taken from Part 2.

When space groups with two origin choices are involved, the origin shifts can be different depending on whether origin choice 1 or 2 has been selected. Therefore, all space groups with two origin choices have two columns 'Coordinates', one for each origin choice. The coordinate conversion formulae for a specific subgroup in the two columns only differ in the additive fractional numbers that specify the origin shift.

2.2.3.4. Nonconventional settings of orthorhombic space groups

Orthorhombic space groups can have as many as six different settings, as explained in Sections 1.5.4 and 2.1.3.6.4, and listed in Table 1.5.4.1. They result from the interchange of the axes.

The exchange has two consequences for a Hermann–Mauguin symbol:

(1) the symmetry operations given in the symbol interchange their positions in the symbol;

(2) the labels of the glide directions and of the centrings are interchanged.

In the same way, the sequences and the values of the coordinate triplets have to be interchanged.

Example

Take space group *Pbcm*, No. 57 (full symbol $P2/b\,2_1/c\,2_1/m$), and its Wyckoff position 4*c* ($x, \frac{1}{4}, 0$). The positions in the symbol change as given by the arrows, and simultaneously the labels change:

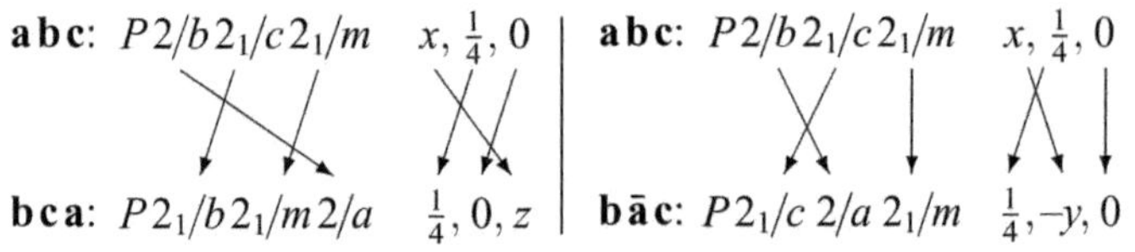

The notation **b c a** means: the former *b* axis is now in the position of the *a* axis *etc.* or: convert *b* to *a*, *c* to *b*, and *a* to *c*.

The corresponding interchanges of positions and labels for all possible nonconventional settings of a space group are listed at the end of the table of each orthorhombic space group in *IT* A1. They have to be applied to all subgroups.

Example

Consider the nonconventional setting *Bmmb* of *Cmcm*. The entry at the bottom of the page for space group *Cmcm*, No. 63, shows the necessary interchanges for the setting *Bmmb*: $C \rightleftarrows B$, $b \rightleftarrows c$, $\mathbf{b} \rightleftarrows -\mathbf{c}$ and $y \rightleftarrows -z$.

For the subgroup *Pmcn* (second entry in the block of *klassengleiche* subgroups) this means: *Pmcn* has to be replaced by *Pmnb*, the axes have to be converted to **a**, −**c**, **b** and the coordinate transformation $x+\frac{1}{4}, y+\frac{1}{4}, z$ has to be replaced by $x+\frac{1}{4}, -z, y+\frac{1}{4}$.

Pmcn and *Pmnb* are nonconventional settings of *Pnma*, No. 62.

The interchange of the axes does not affect the Wyckoff labels, just the corresponding coordinates. However, this is not generally true in the case of space groups where the Hermann–Mauguin symbol does not uniquely show how the setting differs from the conventional setting; this requires special care.

2.2.4. Examples of the tables of relations between Wyckoff positions

Two examples of tables of Wyckoff relations from *IT* A1 are given on the following pages. The first is for the space group $P6_3/mmc$ (194), a space group with a primitive lattice which, apart from maximal *translationengleiche* subgroups, has only maximal *klassengleiche* subgroups with enlarged unit cells. The second is for the space group *Cmcm* (63). This is a subgroup of $P6_3/mmc$ and has *klassengleiche* subgroups that have lost the centring translations and have unit cells that are unchanged in size.

References

Bertaut, E. F. & Billiet, Y. (1979). *On equivalent subgroups and supergroups of the space groups. Acta Cryst.* A**35**, 733–745.

Billiet, Y. & Bertaut, E. F. (2005). *Isomorphic subgroups of space groups. International Tables for Crystallography*, Vol. A, *Space-Group Symmetry*, edited by Th. Hahn, Part 13. Heidelberg: Springer.

International Tables for Crystallography (2010). Volume A1, *Symmetry Relations between Space Groups*, 2nd ed., edited by H. Wondratschek & U. Müller. Chichester: Wiley. (Abbreviated as *IT* A1.)

International Tables for Crystallography (2016). Volume A, *Space-Group Symmetry*, 6th ed., edited by M. I. Aroyo. Chichester: Wiley. (Abbreviated as *IT* A.)

International Tables for X-ray Crystallography (1952). Vol. I, edited by N. F. M. Henry & K. Lonsdale. Birmingham: Kynoch Press.

Müller, U. & Brelle, A. (1995). *Über isomorphe Untergruppen von Raumgruppen der Kristallklassen* 4, $\bar{4}$, 4/*m*, 3, $\bar{3}$, 6, $\bar{6}$ *und* 6/*m*. *Acta Cryst.* A**51**, 300–304.

D_{6h}^4 $P6_3/m2/m2/c$ No. 194 $P6_3/mmc$

Subgroup	Axes	Coordinates	Wyckoff positions					
			$2a$ $6g$	$2b$ $6h$	$2c$ $12i$	$2d$ $12j$	$4e$ $12k$	$4f$ $24l$

I Maximal *translationengleiche* subgroups

Subgroup	Axes	Coordinates	$2a$ / $6g$	$2b$ / $6h$	$2c$ / $12i$	$2d$ / $12j$	$4e$ / $12k$	$4f$ / $24l$
[2] $P\bar{6}2c$ (190)			$2a$ $6g$	$2b$ $6h$	$2c$ $2\times 6g$	$2d$ $2\times 6h$	$4e$ $12i$	$4f$ $2\times 12i$
[2] $P\bar{6}m2$ (187)		$x,y,z+\frac{1}{4}$	$2g$ $6n$	$1a$; $1b$ $3j$; $3k$	$1d$; $1e$ $12o$	$1c$; $1f$ $6l$; $6m$	$2\times 2g$ $2\times 6n$	$2h$; $2i$ $2\times 12o$
[2] $P6_3mc$ (186)			$2a$ $6c$	$2a$ $6c$	$2b$ $12d$	$2b$ $12d$	$2\times 2a$ $2\times 6c$	$2\times 2b$ $2\times 12d$
[2] $P6_322$ (182)			$2a$ $6g$	$2b$ $6h$	$2c$ $2\times 6g$	$2d$ $12i$	$4e$ $12i$	$4f$ $2\times 12i$
[2] $P6_3/m$ (176)			$2b$ $6g$	$2a$ $6h$	$2c$ $12i$	$2d$ $2\times 6h$	$4e$ $12i$	$4f$ $2\times 12i$
[2] $P\bar{3}m1$ (164)			$1a$; $1b$ $3e$; $3f$	$2c$ $6i$	$2d$ $6g$; $6h$	$2d$ $12j$	$2\times 2c$ $2\times 6i$	$2\times 2d$ $2\times 12j$
[2] $P\bar{3}1c$ (163)			$2b$ $6g$	$2a$ $6h$	$2c$ $12i$	$2d$ $12i$	$4e$ $12i$	$4f$ $2\times 12i$
[3] $Cmcm$ (63)	**a**, **a**+2**b**, **c**	$x-\frac{1}{2}y, \frac{1}{2}y, z$	$4a$ $4b$; $8d$	$4c$ $4c$; $8g$	$4c$ $8e$; $16h$	$4c$ $3\times 8g$	$8f$ $8f$; $16h$	$8f$ $3\times 16h$
conjugate:	**b**, −2**a**−**b**, **c**	$-\frac{1}{2}x+y, -\frac{1}{2}x, z$						
conjugate:	−**a**−**b**, **a**−**b**, **c**	$-\frac{1}{2}(x+y), \frac{1}{2}(x-y), z$						

II Maximal *klassengleiche* subgroups

Enlarged unit cell, non-isomorphic

Subgroup	Axes	Coordinates	$2a$ / $6g$	$2b$ / $6h$	$2c$ / $12i$	$2d$ / $12j$	$4e$ / $12k$	$4f$ / $24l$
[3] $P6_3/mcm$ (193)	2**a**+**b**, −**a**+**b**, **c**	$\frac{1}{3}(x+y), \frac{1}{3}(-x+2y)$, z; $\pm(\frac{1}{3},\frac{2}{3},0)$	$2b$; $4d$ $6f$; $12i$	$2a$; $4c$ $6g$; $12j$	$6g$ $3\times 12i$	$6g$ $3\times 12j$	$4e$; $8h$ $12k$; $24l$	$12k$ $3\times 24l$

Enlarged unit cell, isomorphic

Subgroup	Axes	Coordinates	$2a$ / $6g$	$2b$ / $6h$	$2c$ / $12i$	$2d$ / $12j$	$4e$ / $12k$	$4f$ / $24l$
[3] $P6_3/mmc$	**a**, **b**, 3**c**	$x, y, \frac{1}{3}z$; $\pm(0,0,\frac{1}{3})$	$2a$; $4e$ $6g$; $12k$	$2b$; $4e$ $6h$; $12k$	$2d$; $4f$ $12i$; $24l$	$2c$; $4f$ $12j$; $24l$	$3\times 4e$ $3\times 12k$	$3\times 4f$ $3\times 24l$
[p] $P6_3/mmc$	**a**, **b**, p**c** p = prime > 2;	$x, y, \frac{1}{p}z$; $+(0,0,\frac{u}{p})$ $u = 1\dots p-1$	$2a$; $\frac{p-1}{2}\times 4e$ $6g$; $\frac{p-1}{2}\times 12k$	$2b$; $\frac{p-1}{2}\times 4e$ $6h$; $\frac{p-1}{2}\times 12k$	$2c(d^*)$; $\frac{p-1}{2}\times 4f$ $12i$; $\frac{p-1}{2}\times 24l$	$2d(c^*)$; $\frac{p-1}{2}\times 4f$ $12j$; $\frac{p-1}{2}\times 24l$	$p\times 4e$ $p\times 12k$	$p\times 4f$ $p\times 24l$
[4] $P6_3/mmc$	2**a**, 2**b**, **c**	$\frac{1}{2}x, \frac{1}{2}y, z$; $+(\frac{1}{2},0,0)$; $+(0,\frac{1}{2},0)$; $+(\frac{1}{2},\frac{1}{2},0)$	$2a$; $6g$ $12i$; $12k$	$2b$; $6h$ $2\times 6h$; $12j$	$2d$; $6h$ $2\times 12i$; $24l$	$2c$; $6h$ $4\times 12j$	$4e$; $12k$ $2\times 12k$; $24l$	$4f$; $12k$ $4\times 24l$
[p^2] $P6_3/mmc$	p**a**, p**b**, **c** p = prime > 4;	$\frac{1}{p}x, \frac{1}{p}y, z$; $+(\frac{u}{p},\frac{v}{p},0)$ $u, v = 1,\dots,p-1$	$2a$; $\frac{p-1}{2}\times 12i$; $\frac{p-1}{2}\times 12k$; $\frac{(p-1)(p-5)}{12}\times 24l$ $6g$; $\frac{p-1}{2}\times 12i$; $\frac{p-1}{2}\times 12k$; $\frac{(p-1)^2}{4}\times 24l$	$2b$; $(p-1)\times 6h$; $\frac{(p-1)(p-2)}{6}\times 12j$ $p\times 6h$; $\frac{p(p-1)}{2}\times 12j$	$2c(d^\dagger)$; $(p-1)\times 6h$; $\frac{(p-1)(p-2)}{6}\times 12j$ $p\times 12i$; $\frac{p(p-1)}{2}\times 24l$	$2d(c^\dagger)$; $(p-1)\times 6h$; $\frac{(p-1)(p-2)}{6}\times 12j$ $p^2\times 12j$	$4e$; $(p-1)\times 12k$; $\frac{(p-1)(p-2)}{6}\times 24l$ $p\times 12k$; $\frac{p(p-1)}{2}\times 24l$	$4f$; $(p-1)\times 12k$; $\frac{(p-1)(p-2)}{6}\times 24l$ $p^2\times 24l$

$^*\,p = 4n-1$ $\quad$ $^\dagger\,p = 6n-1$

$Cmcm$ No. 63 $C2/m2/c2_1/m$ D_{2h}^{17}

	Axes	Coordinates	Wyckoff positions 4a	4b	4c	8d	8e	8f	8g	16h
I Maximal *translationengleiche* subgroups										
[2] *C2cm* (40)			4a	4a	4b	8c	2×4a	8c	2×4b	2×8c
≙ *Ama2*	**c**, **b**, −**a**	$z,y,-x$								
[2] *Cm2m* (38)		$x,y,z+\frac{1}{4}$	4c	4c	2a; 2b	8f	8f	2×4c	4d; 4e	2×8f
≙ *Amm2*	**c**, **a**, **b**	$z+\frac{1}{4},x,y$								
[2] $Cmc2_1$ (36)			4a	4a	4a	8b	8b	2×4a	8b	2×8b
[2] $C222_1$ (20)			4a	4a	4b	8c	2×4a	8c	8c	2×8c
[2] *C*12/*c*1 (15)			4a	4b	4e	4c; 4d	8f	8f	8f	2×8f
[2] *C*2/*m*11 (12)			2a; 2c	2b; 2d	4i	4e; 4f	4g; 4h	2×4i	8j	2×8j
≙ *C*12/*m*1	**b**, −**a**, **c**	$y,-x,z$								
[2] $P112_1/m$	**a**, $\frac{1}{2}$(−**a**+**b**), **c**	$x+y,2y,z$	2a	2c	2e	2b; 2d	4f	4f	2×2e	2×4f
(11)	or: $\frac{1}{2}$(**a**−**b**), **b**, **c**	$2x,x+y,z$	2a	2b	2e	2c; 2d	4f	4f	2×2e	2×4f
	or: $\frac{1}{2}$(**a**−**b**), $\frac{1}{2}$(**a**+**b**), **c**	$x-y,x+y,z$	2a	2d	2e	2b; 2c	4f	4f	2×2e	2×4f
II Maximal *klassengleiche* subgroups										
Loss of centring translations										
[2] *Pbnm* (62)			4a	4b	4c	8d	8d	8d	2×4c	2×8d
≙ *Pnma*	**b**, **c**, **a**	y,z,x								
[2] *Pmcn* (62)		$x+\frac{1}{4},y+\frac{1}{4},z$	4c	4c	4c	4a; 4b	8d	2×4c	8d	2×8d
≙ *Pnma*	**c**, **a**, **b**	$z,x+\frac{1}{4},y+\frac{1}{4}$								
[2] *Pbcn* (60)			4a	4b	4c	8d	8d	8d	8d	2×8d
[2] *Pmnm* (59)		origin 1: $x,y+\frac{1}{4},z+\frac{1}{4}$ origin 2: $x-\frac{1}{4},y+\frac{1}{4},z$	4f	4f	2a; 2b	4c; 4d	8g	2×4f	2×4e	2×8g
≙ *Pmmn*	**c**, **a**, **b**	$z+\frac{1}{4},x,y+\frac{1}{4}$ (origin 1) $z,x-\frac{1}{4},y+\frac{1}{4}$ (origin 2)								
[2] *Pmnn* (58)			2a; 2c	2b; 2d	4g	8h	4e; 4f	2×4g	8h	2×8h
≙ *Pnnm*	**b**, **c**, **a**	y,z,x								
[2] *Pbcm* (57)		$x+\frac{1}{4},y+\frac{1}{4},z$	4c	4c	4d	4a; 4b	2×4c	8e	2×4d	2×8e
[2] *Pbnn* (52)		$x+\frac{1}{4},y+\frac{1}{4},z$	4c	4c	4d	4a; 4b	2×4c	8e	8e	2×8e
≙ *Pnna*	**b**, **c**, **a**	$y+\frac{1}{4},z,x+\frac{1}{4}$								
[2] *Pmcm* (51)			2a; 2d	2b; 2c	2e; 2f	8l	4g; 4h	4i; 4j	2×4k	2×8l
≙ *Pmma*	**c**, **a**, **b**	z,x,y								
Enlarged unit cell, isomorphic										
[3] *Cmcm*	3**a**, **b**, **c**	$\frac{1}{3}x,y,z$; $\pm(\frac{1}{3},0,0)$	4a; 8e	4b; 8e	4c; 8g	8d; 16h	3×8e	8f; 16h	3×8g	3×16h
[*p*] *Cmcm*	*p***a**, **b**, **c** p = prime > 2; $u=1,\ldots,p-1$	$\frac{1}{3}x,y,z$; $+(\frac{u}{p},0,0)$	4a; $\frac{p-1}{2}$×8e	4b; $\frac{p-1}{2}$×8e	4c; $\frac{p-1}{2}$×8g	8d; $\frac{p-1}{2}$×16h	p×8e	8f; $\frac{p-1}{2}$×16h	p×8g	p×16h
[3] *Cmcm*	**a**, 3**b**, **c**	$x,\frac{1}{3}y,z$; $\pm(0,\frac{1}{3},0)$	4a; 8f	4b; 8f	3×4c	8d; 16h	8e; 16h	3×8f	3×8g	3×16h
[*p*] *Cmcm*	**a**, *p***b**, **c** p = prime > 2; $u=1,\ldots,p-1$	$x,\frac{1}{p}y,z$; $+(0,\frac{u}{p},0)$	4a; $\frac{p-1}{2}$×8f	4b; $\frac{p-1}{2}$×8f	p×4c	8d; $\frac{p-1}{2}$×16h	8e; $\frac{p-1}{2}$×16h	p×8f	p×8g	p×16h
[3] *Cmcm*	**a**, **b**, 3**c**	$x,y,\frac{1}{3}z$; $\pm(0,0,\frac{1}{3})$	4a; 8f	4b; 8f	4c; 8f	8d; 16h	8e; 16h	3×8f	8g; 16h	3×16h
[*p*] *Cmcm*	**a**, **b**, *p***c** p = prime > 2; $u=1,\ldots,p-1$	$x,y,\frac{1}{p}z$; $+(0,0,\frac{u}{p})$	4a; $\frac{p-1}{2}$×8f	4b; $\frac{p-1}{2}$×8f	4c; $\frac{p-1}{2}$×8f	8d; $\frac{p-1}{2}$×16h	8e; $\frac{p-1}{2}$×16h	p×8f	8g; $\frac{p-1}{2}$×16h	p×16h

Nonconventional settings

Interchange letters and sequences in Hermann–Mauguin symbols, axes and coordinates:

$Amma$	$C \to A$; $A \to B$	$a \to b \to c \to a$	$\mathbf{a} \to \mathbf{b} \to \mathbf{c} \to \mathbf{a}$	$x \to y \to z \to x$
$Bbmm$	$C \to B$; $A \to C$	$a \leftarrow b \leftarrow c \leftarrow a$	$\mathbf{a} \leftarrow \mathbf{b} \leftarrow \mathbf{c} \leftarrow \mathbf{a}$	$x \leftarrow y \leftarrow z \leftarrow x$
$Ccmm$	$A \to B$	$a \rightleftarrows b$	$\mathbf{a} \rightleftarrows -\mathbf{b}$	$x \rightleftarrows -y$
$Amam$	$C \rightleftarrows A$	$a \rightleftarrows c$	$\mathbf{a} \rightleftarrows -\mathbf{c}$	$x \rightleftarrows -z$
$Bmmb$	$C \to B$	$b \rightleftarrows c$	$\mathbf{b} \rightleftarrows -\mathbf{c}$	$y \rightleftarrows -z$

2.3. The subperiodic group tables of *IT* E

Daniel B. Litvin

2.3.1. Guide to the subperiodic group tables

The subperiodic group tables of *International Tables for Crystallography* Volume E (*IT* E, 2010) are an extension of the tables in Volume A, *Space-Group Symmetry* (*IT* A, 2005, 2016). The symmetry tables in *IT* A (2005, 2016) are for the 230 three-dimensional crystallographic space-group types (space groups) and the 17 two-dimensional crystallographic space-group types (plane groups). The analogous symmetry tables in *IT* E (2010) are for the two-dimensional and three-dimensional subperiodic group types: the 7 crystallographic *frieze-group* types (frieze groups, two-dimensional groups with one-dimensional translations), the 75 crystallographic *rod-group* types (rod groups, three-dimensional groups with one-dimensional translations) and the 80 crystallographic *layer-group* types (layer groups, three-dimensional groups with two-dimensional translations).

2.3.1.1. Content and arrangement of the tables

The presentation of the subperiodic group tables follows the form and content of *IT* A (2005). The entries for a subperiodic group are printed on two facing pages or continuously on a single page, where space permits, in the following order:

Left-hand page:

(1) *Headline*;
(2) *Diagrams* for the symmetry elements and the general position;
(3) *Origin*;
(4) *Asymmetric unit*;
(5) *Symmetry operations*.

Right-hand page:

(6) *Headline* in abbreviated form;
(7) *Generators selected*;
(8) General and special *Positions*, with the following columns: *Multiplicity*, *Wyckoff letter*, *Site symmetry*, *Coordinates*, *Reflection conditions*;
(9) *Symmetry of special projections*;
(10) *Maximal non-isotypic non-enantiomorphic subgroups*;
(11) *Maximal isotypic subgroups and enantiomorphic subgroups of lowest index*;
(12) *Minimal non-isotypic non-enantiomorphic supergroups*.

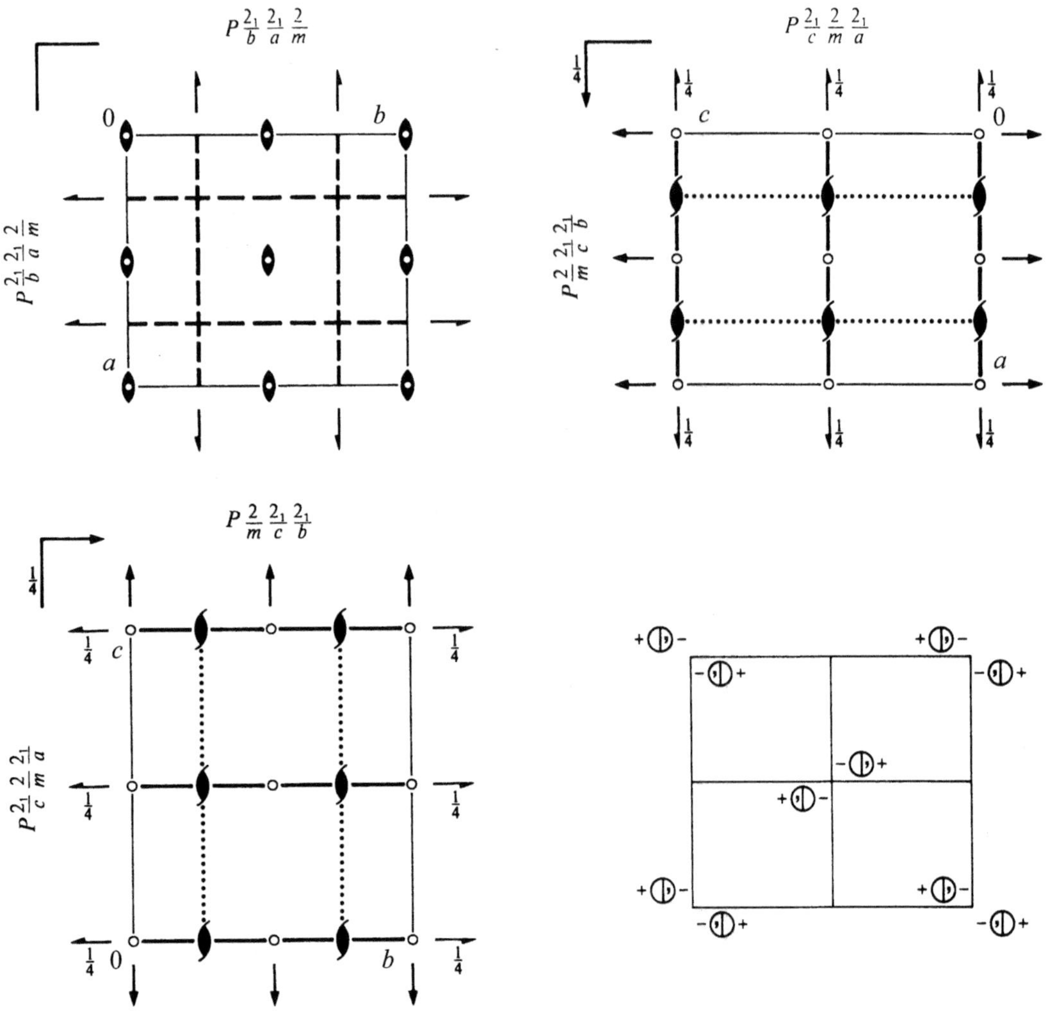

Figure 2.3.1.1
The symmetry-element and general-position diagrams of space group No. 55, *Pbam*.

(1) 1 $\{1\|0\}$	(2) $2\ \ 0,0,z$ $\{2_{001}\|0\}$	(3) $2(0,\frac{1}{2},0)\ \ \frac{1}{4},y,0$ $\{2_{010}\|\frac{1}{2},\frac{1}{2},0\}$	(4) $2(\frac{1}{2},0,0)\ \ x,\frac{1}{4},0$ $\{2_{100}\|\frac{1}{2},\frac{1}{2},0\}$
(5) $\bar{1}\ \ 0,0,0$ $\{\bar{1}\|0\}$	(6) $m\ \ x,y,0$ $\{m_{001}\|0\}$	(7) $a\ \ x,\frac{1}{4},z$ $\{m_{010}\|\frac{1}{2},\frac{1}{2},0\}$	(8) $b\ \ \frac{1}{4},y,z$ $\{m_{100}\|\frac{1}{2},\frac{1}{2},0\}$

Figure 2.3.1.2
Symmetry operations of layer group No. 44, *pbam*. The Seitz notation of each operation following the IUCr conventions (Glazer *et al.*, 2013) is given below the corresponding geometric description.

Almost all of the form and content of the subperiodic symmetry tables in *IT* E (2010) is identical with or analogous to the form and content of the symmetry tables in *IT* A (2016). That of *IT* A (2016) has been presented in Chapter 2.1. Consequently, we shall indicate below only the major differences in the form and content of the subperiodic symmetry tables in *IT* E (2010) from those of the symmetry tables in *IT* A (2016). These are the entries on diagrams, symmetry operations, and subgroups and supergroups.

2.3.1.2. Diagrams for the symmetry elements and the general position

The graphical symbols used in the symmetry-element and general-position diagrams in the symmetry tables in *IT* E (2010) are the same as those in *IT* A (2016). The format of the diagrams may be different owing to the dimensionality of the groups considered and their underlying translational periodicity. Space groups and rod groups are both three-dimensional groups, but because of the one-dimensionality of the rod-group translations, the rod-group symmetry-element and general-position diagrams are given as projections along their translations onto circular diagrams, see the rod-group table for $\not{p}mc2_1$ (No. 17) in Section 2.3.2. For triclinic, monoclinic and orthorhombic rod groups, the symmetry-element and general-position diagrams have three projections along their three coordinate axes (see the rod-group table for $\not{p}mc2_1$).

The symmetry-element and general-position diagrams of a layer group are projections along a direction perpendicular to the plane containing its two-dimensional translations. These diagrams are identical with the diagrams of a related space group. If one were to add a group of translations perpendicular to the layer group's plane of translations, one would have a space group. If one then found the diagrams of this space group by projecting along the direction of the perpendicular translations, one would have identical diagrams to the layer-group diagrams. Therefore, the symmetry-element and general-position diagrams of a layer group can be found among the diagrams of the space groups. For example, the diagrams of layer group *pbam* (No. 44) (see Section 2.3.2) can also be recognised among the diagrams of space group *Pbam* (No. 55), Fig. 2.3.1.1.

2.3.1.3. Symmetry operations

As in *IT* A (2016), for each subperiodic group in *IT* E (2010) a block of symmetry operations is given. For each symmetry operation, in *IT* E (2010) this block contains not only the symmetry operation's geometric description but also its Seitz notation (Burns & Glazer, 2013). The symmetry operations of layer group *pbam*, as given in *IT* E (2010), are shown in its table in Section 2.3.2. Subsequent to the publication of *IT* E (2010), the Commission on Crystallographic Nomenclature of the International Union of Crystallography (IUCr) adopted standard IUCr conventions for Seitz notation (Glazer *et al.*, 2014; see also Section 1.4.2.2). The Seitz notation used in *IT* E (2010) differs from these conventions. The blocks of symmetry operations for the subperiodic groups which do follow IUCr conventions can be found in Litvin & Kopský (2014). An example (for layer group *pbam*) of the block of symmetry operations following these conventions is given in Fig. 2.3.1.2.

2.3.1.4. Subgroups and supergroups

The form and content of the sections considering subgroups and supergroups of the subperiodic groups in *IT* E (2010) follow those of the fifth edition of *IT* A (2005). Those sections of *IT* A (2005) do not appear in the sixth edition of *IT* A (2016), but have been replaced by the more detailed and comprehensive survey in *International Tables for Crystallography* Volume A1, *Symmetry Relations between Space Groups* (*IT* A1, 2010). The examples of the subperiodic group tables in Section 2.3.2 show how the subgroup data in *IT* E (2010) are presented. Type-I subgroups are maximal non-isotypic equi-translational (*translationengleiche*) subgroups. Types IIa and IIb are maximal non-isotypic equi-class (*klassengleiche*) subgroups having, respectively, the same or larger conventional unit cell as the original group. There are no such subgroups for the layer group *pbam*. Maximal isotypic subgroups of lowest index are type-IIc subgroups. Minimal non-isotypic supergroups are listed as type I if equi-translational (*translationengleiche*) or type II if equi-class (*klassengleiche*).

2.3.2. Examples of subperiodic group tables

The subperiodic group tables for rod group $\not{p}mc2_1$ (No. 17), layer group $p2_1/b11$ (No. 17) and layer group *pbam* (No. 44) from *IT* E (2010) are shown on the following pages.

References

Burns, G. & Glazer, A. M. (2013). *Space Groups for Solid State Scientists*, 3rd ed. New York: Academic Press.

Glazer, A. M., Aroyo, M. I. & Authier, A. (2014). *Seitz symbols for crystallographic symmetry operations. Acta Cryst.* A**70**, 300–302.

International Tables for Crystallography (2005). Vol. A, *Space-Group Symmetry*, 5th ed., edited by Th. Hahn, Dordrecht: Kluwer. [Abbreviated as *IT* A (2005).]

International Tables for Crystallography (2010). Vol. A1, *Symmetry Relations between Space Groups*, 2nd ed., edited by H. Wondratschek & U. Müller. Chichester: John Wiley & Sons. [Abbreviated as *IT* A1.]

International Tables for Crystallography (2010). Vol. E, *Subperiodic Groups*, 2nd ed., edited by V. Kopský & D. B. Litvin. Chichester: Wiley. [Abbreviated as *IT* E.]

International Tables for Crystallography (2016). Vol. A, *Space-Group Symmetry*, 6th ed., edited by M. I. Aroyo. Chichester: Wiley. [Abbreviated as *IT* A (2016).]

Litvin, D. B. & Kopský, V. (2014). *Seitz symbols for symmetry operations of subperiodic groups. Acta Cryst.* A**70**, 677–678.

No. 17 $\mathscr{p}mc2_1$ Patterson symmetry $\mathscr{p}mmm$

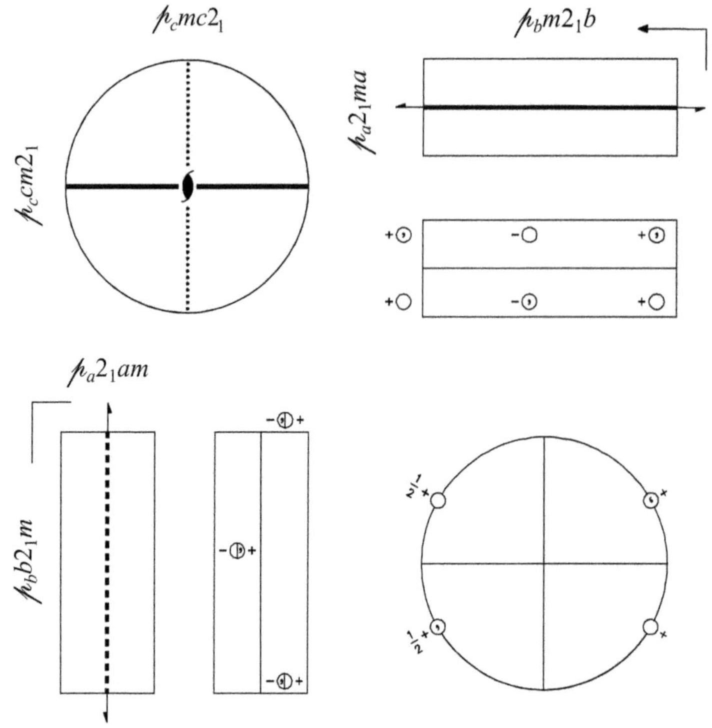

Origin on $mc2_1$

Asymmetric unit $0 \le x$; $0 \le y$; $0 \le z \le 1$

Symmetry operations

(1) 1 $(1|0,0,0)$

(2) $2(\frac{1}{2})$ $0,0,z$ $(2_z|0,0,\frac{1}{2})$

(3) c $x,0,z$ $(m_y|0,0,\frac{1}{2})$

(4) m $0,y,z$ $(m_x|0,0,0)$

Generators selected (1); $t(0,0,1)$; (2); (3)

Positions

Multiplicity, Wyckoff letter, Site symmetry			Coordinates				Reflection conditions
							General:
4	b	1	(1) x,y,z	(2) $\bar{x},\bar{y},z+\frac{1}{2}$	(3) $x,\bar{y},z+\frac{1}{2}$	(4) $\bar{x},y,z$	l : $l=2n$
							Special: no extra conditions
2	a	$m..$	$0,y,z$	$0,\bar{y},z+\frac{1}{2}$			

Symmetry of special projections

Along [001] $2mm$
Origin at $0,0,z$

Along [100] $p11g$
$\mathbf{a}'=\mathbf{c}$
Origin at $x,0,0$

Along [010] $p11m$
$\mathbf{a}'=\frac{1}{2}\mathbf{c}$
Origin at $0,y,0$

Maximal non-isotypic non-enantiomorphic subgroups

I	[2] $p112_1$ (9)	1; 2
	[2] $p1c1$ ($pc11$, 5)	1; 3
	[2] $pm11$ (4)	1; 4
IIa	none	
IIb	none	

Maximal isotypic subgroups and enantiomorphic subgroups of lowest index

IIc [3] $pmc2_1$ ($\mathbf{c}'=3\mathbf{c}$) (17)

Minimal non-isotypic non-enantiomorphic supergroups

I [2] $pmcm$ (22); [3] $p6_3mc$ (70)

II [2] $pmm2$ ($\mathbf{c}'=\frac{1}{2}\mathbf{c}$) (15)

No. 17 $p2_1/b11$ Patterson symmetry $p2/m11$

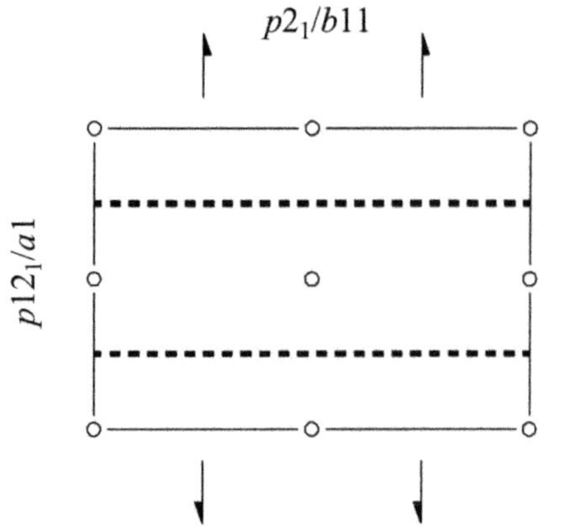

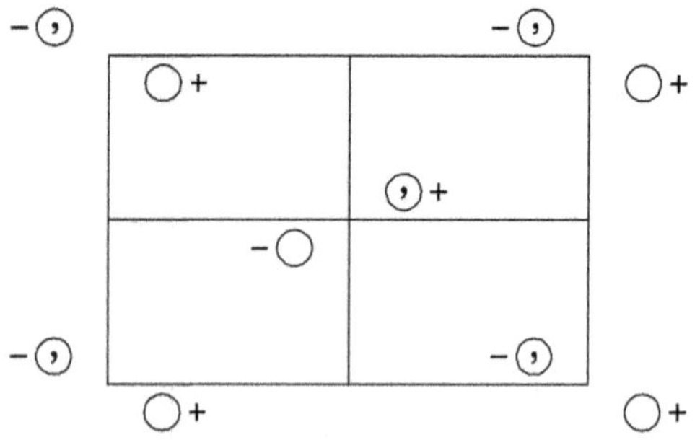

Origin at $\bar{1}$

Asymmetric unit $0 \leq x \leq \frac{1}{4}$; $0 \leq y \leq 1$

Symmetry operations

(1) 1 $(1|0,0,0)$

(2) $2(\frac{1}{2},0,0)$ $x,\frac{1}{4},0$ $(2_x|\frac{1}{2},\frac{1}{2},0)$

(3) $\bar{1}$ $0,0,0$ $(\bar{1}|0,0,0)$

(4) b $\frac{1}{4},y,z$ $(m_x|\frac{1}{2},\frac{1}{2},0)$

Generators selected (1); $t(1,0,0)$; $t(0,1,0)$; (2); (3)

Positions

Multiplicity, Wyckoff letter, Site symmetry			Coordinates				Reflection conditions
							General:
4	c	1	(1) x,y,z	(2) $x+\frac{1}{2},\bar{y}+\frac{1}{2},\bar{z}$	(3) $\bar{x},\bar{y},\bar{z}$	(4) $\bar{x}+\frac{1}{2},y+\frac{1}{2},z$	$h0$: $h=2n$ $0k$: $k=2n$
							Special: as above, plus
2	b	$\bar{1}$	$0,\frac{1}{2},0$	$\frac{1}{2},0,0$			hk: $h+k=2n$
2	a	$\bar{1}$	$0,0,0$	$\frac{1}{2},\frac{1}{2},0$			hk: $h+k=2n$

Symmetry of special projections

Along [001] $p2gg$
$\mathbf{a}'=\mathbf{a}$ $\mathbf{b}'=\mathbf{b}_p$
Origin at $0,0,z$

Along [100] $p211$
$\mathbf{a}'=\frac{1}{2}\mathbf{b}$
Origin at $x,0,0$

Along [010] $p2mg$
$\mathbf{a}'=\mathbf{a}$
Origin at $0,y,0$

Maximal non-isotypic subgroups

I [2] $pb11$ (12) 1; 4
[2] $p2_111$ (9) 1; 2
[2] $p\bar{1}$ (2) 1; 3

IIa none

IIb none

Maximal isotypic subgroups of lowest index

IIc [3] $p2_1/b11$ ($\mathbf{a}'=3\mathbf{a}$) (17)

Minimal non-isotypic supergroups

I [2] $pman$ (42); [2] $pbaa$ (43); [2] $pbam$ (44); [2] $pbma$ (45)

II [2] $c2/m11$ (18); [2] $p2_1/m11$ ($\mathbf{b}'=\frac{1}{2}\mathbf{b}$) (15); [2] $p2/b11$ ($\mathbf{a}'=\frac{1}{2}\mathbf{a}$) (16)

No. 44 $p2_1/b2_1/a2/m$

Patterson symmetry *pmmm*

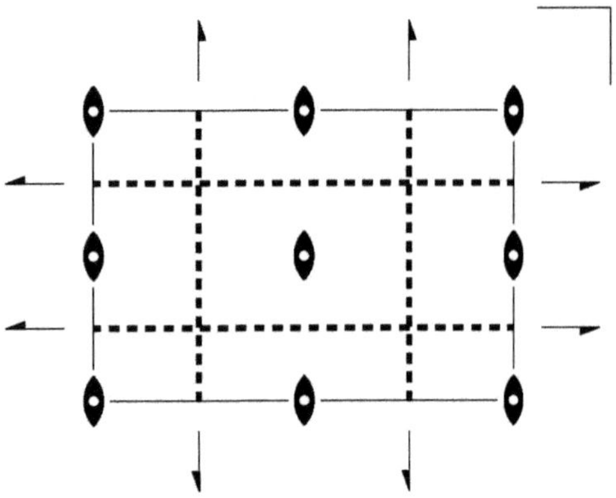

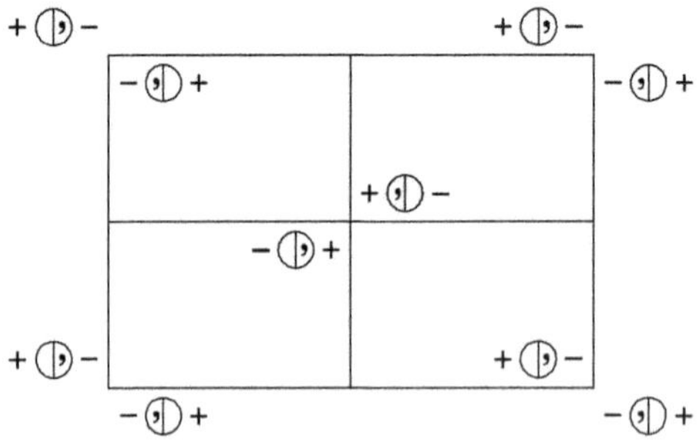

Origin at centre $(2/m)$

Asymmetric unit $0 \le x \le \frac{1}{2};\quad 0 \le y \le \frac{1}{2};\quad 0 \le z$

Symmetry operations

(1) 1 $(1\|0,0,0)$	(2) $2\quad 0,0,z$ $(2_z\|0,0,0)$	(3) $2(0,\frac{1}{2},0)\quad \frac{1}{4},y,0$ $(2_y\|\frac{1}{2},\frac{1}{2},0)$	(4) $2(\frac{1}{2},0,0)\quad x,\frac{1}{4},0$ $(2_x\|\frac{1}{2},\frac{1}{2},0)$
(5) $\bar{1}\quad 0,0,0$ $(\bar{1}\|0,0,0)$	(6) $m\quad x,y,0$ $(m_z\|0,0,0)$	(7) $a\quad x,\frac{1}{4},z$ $(m_y\|\frac{1}{2},\frac{1}{2},0)$	(8) $b\quad \frac{1}{4},y,z$ $(m_x\|\frac{1}{2},\frac{1}{2},0)$

Generators selected (1); $t(1,0,0)$; $t(0,1,0)$; (2); (3); (5)

Positions

Multiplicity, Wyckoff letter, Site symmetry			Coordinates				Reflection conditions
							General:
8	*f*	1	(1) x,y,z (5) $\bar{x},\bar{y},\bar{z}$	(2) $\bar{x},\bar{y},z$ (6) $x,y,\bar{z}$	(3) $\bar{x}+\frac{1}{2},y+\frac{1}{2},\bar{z}$ (7) $x+\frac{1}{2},\bar{y}+\frac{1}{2},z$	(4) $x+\frac{1}{2},\bar{y}+\frac{1}{2},\bar{z}$ (8) $\bar{x}+\frac{1}{2},y+\frac{1}{2},z$	$0k$: $k=2n$ $h0$: $h=2n$
							Special: as above, plus
4	*e*	$..m$	$x,y,0$	$\bar{x},\bar{y},0$	$\bar{x}+\frac{1}{2},y+\frac{1}{2},0$	$x+\frac{1}{2},\bar{y}+\frac{1}{2},0$	no extra conditions
4	*d*	$..2$	$0,\frac{1}{2},z$	$\frac{1}{2},0,\bar{z}$	$0,\frac{1}{2},\bar{z}$	$\frac{1}{2},0,z$	hk: $h+k=2n$
4	*c*	$..2$	$0,0,z$	$\frac{1}{2},\frac{1}{2},\bar{z}$	$0,0,\bar{z}$	$\frac{1}{2},\frac{1}{2},z$	hk: $h+k=2n$
2	*b*	$..2/m$	$0,\frac{1}{2},0$	$\frac{1}{2},0,0$			hk: $h+k=2n$
2	*a*	$..2/m$	$0,0,0$	$\frac{1}{2},\frac{1}{2},0$			hk: $h+k=2n$

Symmetry of special projections

Along [001] $p2gg$
$\mathbf{a}'=\mathbf{a}$ $\mathbf{b}'=\mathbf{b}$
Origin at $0,0,z$

Along [100] $p2mm$
$\mathbf{a}'=\frac{1}{2}\mathbf{b}$
Origin at $x,0,0$

Along [010] $p2mm$
$\mathbf{a}'=\frac{1}{2}\mathbf{a}$
Origin at $0,y,0$

Maximal non-isotypic subgroups

I	[2] $pb2_1m$ (29)	1; 3; 6; 8
	[2] $p2_1am$ ($pb2_1m$, 29)	1; 4; 6; 7
	[2] $pba2$ (25)	1; 2; 7; 8
	[2] $p2_12_12$ (21)	1; 2; 3; 4
	[2] $p12_1/a1$ ($p2_1/b11$, 17)	1; 3; 5; 7
	[2] $p2_1/b11$ (17)	1; 4; 5; 8
	[2] $p112/m$ (6)	1; 2; 5; 6

IIa none

IIb none

Maximal isotypic subgroups of lowest index

IIc [3] $pbam$ ($\mathbf{a}'=3\mathbf{a}$ or $\mathbf{b}'=3\mathbf{b}$) (44)

Minimal non-isotypic supergroups

I [2] $p4/mbm$ (63)

II [2] $cmmm$ (47); [2] $pmam$ ($\mathbf{b}'=\frac{1}{2}\mathbf{b}$) (40)

2.4. The Symmetry Database

Eli Kroumova, Gemma de la Flor and Mois I. Aroyo

The online edition of *International Tables for Crystallography* at https://it.iucr.org/ includes all the volumes in the series along with the Symmetry Database (https://symmdb.iucr.org/). This database contains more extensive symmetry information for the crystallographic point and space groups than the print editions of *International Tables for Crystallography* Volume A, *Space-Group Symmetry* (2016) (hereafter referred to as *IT* A) or Volume A1, *Symmetry Relations between Space Groups* (2010) (hereafter referred to as *IT* A1). The information in the database can either be retrieved directly from it or generated 'on-the fly' using a range of programs. The different types of data can be accessed through user-friendly menus, and help pages briefly explain the crystallographic data and the functionality of the programs. The data are also presented in different ways, which in some cases includes interactive visualization.

The data and programs that are currently available in the Symmetry Database are arranged into three sections: (i) space-group symmetry data, (ii) symmetry relations between space groups and (iii) 3D crystallographic point groups (*cf.* Fig. 2.4.1.1).

2.4.1. Space-group symmetry data

This section hosts crystallographic data for all 230 space groups. Generators (Section 1.4.3), general and special Wyckoff positions (Section 1.4.4), and normalizers (Section 1.1.7) of space groups can be accessed using the appropriately named programs. For all these the user only needs to choose the space group.

The symmetry operations provided by the programs `Generators` and `General position` are presented in four different ways (see Fig. 2.4.1.2): as coordinate triplets, as (3×4) matrix–column pairs, using the geometric descriptions that are used in *IT* A (which indicate the type and order of the operations, the location and the orientation of the corresponding geometric elements, and the screw or glide components if relevant) and as Seitz symbols (Section 1.4.2; see also Glazer *et al.*, 2014).

The program `Wyckoff positions` lists the Wyckoff positions for the space group along with their Wyckoff letters and multiplicities, and the oriented symbols of their site-symmetry groups (see Sections 1.4.4 and 2.1.3.12). The coordinate triplets of the points of the Wyckoff position within the unit cell are linked to more detailed information about the corresponding site-symmetry groups. There is also an option for calculating the symmetry operations of the site-symmetry group of any point specified by its coordinates.

These three programs can also provide data for different space-group settings, either by specifying the matrix for the transformation to a new basis (*cf.* Section 1.5.3) or by selecting directly one of the 530 settings listed in Table 1.5.4.4 of *IT* A, all of which are available options.

The program `Normalizers` provides data on the Euclidean, chirality-preserving Euclidean and affine normalizers, which are useful, for example, for comparing descriptions of crystal structures (for more details see Chapter 3.5 of *IT* A). They are described using sets of additional symmetry operations that generate the normalizers successively from the space groups. For triclinic and monoclinic groups whose affine normalizers are not isomorphic to groups of motions, parametric representations of the affine normalizers are shown.

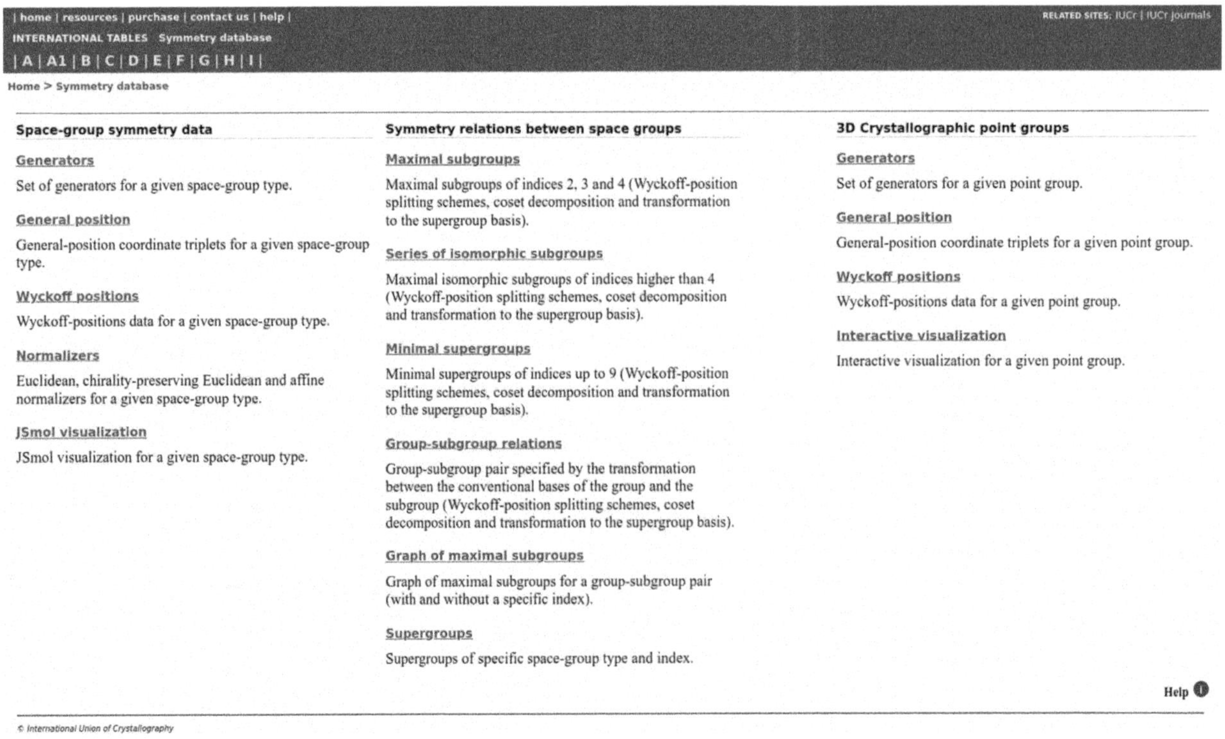
| home | resources | purchase | contact us | help |
INTERNATIONAL TABLES Symmetry database
| A | A1 | B | C | D | E | F | G | H | I |
Home > Symmetry database
Space-group symmetry data
Generators
Set of generators for a given space-group type.
General position
General-position coordinate triplets for a given space-group type.
Wyckoff positions
Wyckoff-positions data for a given space-group type.
Normalizers
Euclidean, chirality-preserving Euclidean and affine normalizers for a given space-group type.
JSmol visualization
JSmol visualization for a given space-group type.
Symmetry relations between space groups
Maximal subgroups
Maximal subgroups of indices 2, 3 and 4 (Wyckoff-position splitting schemes, coset decomposition and transformation to the supergroup basis).
Series of isomorphic subgroups
Maximal isomorphic subgroups of indices higher than 4 (Wyckoff-position splitting schemes, coset decomposition and transformation to the supergroup basis).
Minimal supergroups
Minimal supergroups of indices up to 9 (Wyckoff-position splitting schemes, coset decomposition and transformation to the supergroup basis).
Group-subgroup relations
Group-subgroup pair specified by the transformation between the conventional bases of the group and the subgroup (Wyckoff-position splitting schemes, coset decomposition and transformation to the supergroup basis).
Graph of maximal subgroups
Graph of maximal subgroups for a group-subgroup pair (with and without a specific index).
Supergroups
Supergroups of specific space-group type and index.
3D Crystallographic point groups
Generators
Set of generators for a given point group.
General position
General-position coordinate triplets for a given point group.
Wyckoff positions
Wyckoff-positions data for a given point group.
Interactive visualization
Interactive visualization for a given point group.
Help
© International Union of Crystallography

Figure 2.4.1.1
Main page of the Symmetry Database for *International Tables for Crystallography*.

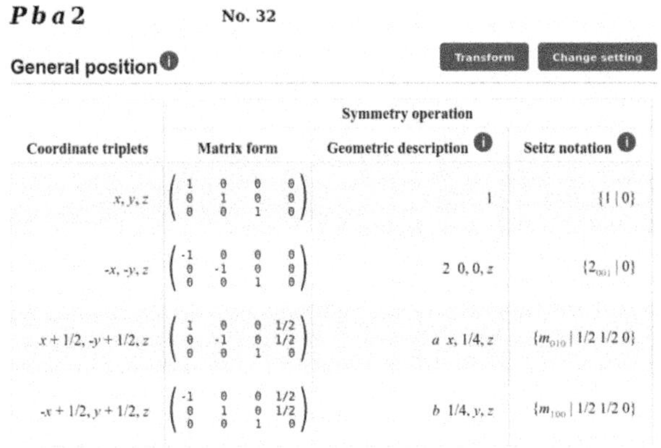

***Pba*2** No. 32

General position

Transform | Change setting

Coordinate triplets	Matrix form	Symmetry operation: Geometric description	Symmetry operation: Seitz notation
x, y, z	$\begin{pmatrix} 1 & 0 & 0 & 0 \\ 0 & 1 & 0 & 0 \\ 0 & 0 & 1 & 0 \end{pmatrix}$	1	$\{1 \mid 0\}$
$-x, -y, z$	$\begin{pmatrix} -1 & 0 & 0 & 0 \\ 0 & -1 & 0 & 0 \\ 0 & 0 & 1 & 0 \end{pmatrix}$	$2\ 0, 0, z$	$\{2_{001} \mid 0\}$
$x+1/2, -y+1/2, z$	$\begin{pmatrix} 1 & 0 & 0 & 1/2 \\ 0 & -1 & 0 & 1/2 \\ 0 & 0 & 1 & 0 \end{pmatrix}$	$a\ x, 1/4, z$	$\{m_{010} \mid 1/2\ 1/2\ 0\}$
$-x+1/2, y+1/2, z$	$\begin{pmatrix} -1 & 0 & 0 & 1/2 \\ 0 & 1 & 0 & 1/2 \\ 0 & 0 & 1 & 0 \end{pmatrix}$	$b\ 1/4, y, z$	$\{m_{100} \mid 1/2\ 1/2\ 0\}$

Figure 2.4.1.2
General positions of the space group *Pba*2 (No. 32) given by the program `General position`.

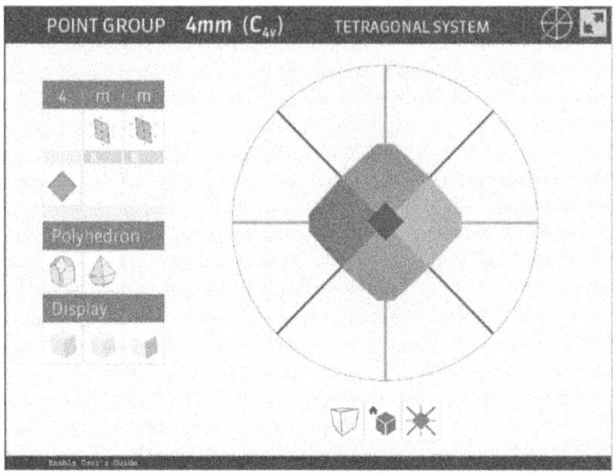

Figure 2.4.3.1
Output of the program `Interactive visualization` for the point group 4*mm*.

2.4.2. Symmetry relations between space groups

The data and programs in this section provide information on group–subgroup relations between space groups, which are essential for studying crystal-structure relationships or phase transitions (see, for example, Section 1.5.2.4). The data in Volume A1 on maximal subgroups of space groups are limited to those with indices 2, 3 and 4 (see Section 2.2.1). The Symmetry Database, however, includes data for series of isomorphic subgroups for indices up to 50 for orthorhombic, tetragonal, trigonal and hexagonal space groups, and for indices up to 27 and 125 for cubic space groups. The maximal-subgroup information is presented in blocks of *translationengleiche* and *klassengleiche* subgroups, as defined in Sections 1.7.1 and 2.2.1. Additional programs allow the data to be transformed to arbitrary settings, and also allow left- and right-coset decomposition calculations (see Section 1.1.4), indispensable for domain-structure analysis. Information is also available on how individual Wyckoff positions split under symmetry reduction, and the relations between the corresponding coordinate triplets of the group and its maximal subgroups are specified.

The program `Group-subgroup relations` enables the study of symmetry relations between space groups for an arbitrary group–subgroup index, including group–subgroup coset decompositions and the corresponding splitting of the Wyckoff positions. If the transformation matrix defining the group–subgroup pair is not known, a step-by-step procedure can be used to find it; this can also be used to calculate the shortest maximal-subgroup paths between the group and the subgroup (Aljazzar & Leue, 2011). Group–subgroup symmetry relations can be visualized as graphs of chains of maximal subgroups using the program `Graph of maximal subgroups`. This program is interactive, so one can browse the possible chains of maximal subgroups and obtain specific information on any node and edge of the group–subgroup graph generated from the maximal-subgroup data.

Data for the minimal supergroups of space groups are also provided (Section 1.7.1.4). In contrast to *IT* A1, where only the space-group types of supergroups are indicated (*cf.* Section 2.2.1.4), in the Symmetry Database each minimal supergroup is listed individually and specified by the transformation matrix that relates the conventional bases of the group and the supergroup. The program `Supergroups` allows a detailed study of supergroups of space groups of any index (not just minimal supergroups). Coset decompositions can be performed, Wyckoff-position splittings can be examined and transformations to the basis of the group or supergroup can also be carried out.

2.4.3. 3D Crystallographic point groups

The types of data that are provided for the point groups are similar to the space-group data and are presented in an analogous way. The point group is simply chosen from a table of Hermann–Mauguin and Schönflies symbols. The crystallographic data for each point group include generators, and general and special Wyckoff positions, each specified by its Wyckoff letter, multiplicity, coordinate triplets and site-symmetry group. Different types of notation are used for the symmetry operations: they are presented as coordinate triplets, in matrix form, using geometric symbols (indicating the type and order of the operations, and the orientation of the corresponding symmetry elements) and as Seitz symbols. The data can be transformed to different settings, thus enhancing and extending the data tabulated in Chapter 3.2 of *IT* A.

Clear and instructive visualization of the symmetry elements of the crystallographic point groups and their stereographic projections is provided by the program `Interactive visualization` (Arribas *et al.*, 2014). Interactive 3D polyhedra are used to represent idealized crystals and their corresponding symmetry elements, *cf.* Fig. 2.4.3.1. The user can explore different shapes of polyhedra and view them from different angles, and can also selectively enable or disable the representations of the symmetry elements.

2.4.4. Availability

The Symmetry Database is available to all subscribers to the online version of *International Tables for Crystallography*. A teaching edition of the Symmetry Database, which can be used to obtain and explore the data for several of the space groups that have been used as examples in this book, is also available online.

References

Aljazzar, H. & Leue, S. (2011). *K*: A heuristic search algorithm for finding the k shortest paths. Artif. Intell.* **175**, 2129–2154.

Arribas, V., Casas, L., Estop, E. & Labrador, M. (2014). *Interactive PDF files with embedded 3D designs as support material to study the 32 crystallographic point groups. Comput. Geosci.* **62**, 53–61.

Glazer, A. M., Aroyo, M. I. & Authier, A. (2014). *Seitz symbols for crystallographic symmetry operations. Acta Cryst.* A**70**, 300–302.

International Tables for Crystallography (2010). Volume A1, *Symmetry Relations between Space Groups*, 2nd ed., edited by H. Wondratschek & U. Müller. Chichester: John Wiley & Sons. (Abbreviated as *IT* A1.)

International Tables for Crystallography (2016). Volume A, *Space-Group Symmetry*, 6th ed., edited by M. I. Aroyo. Chichester: Wiley. (Abbreviated as *IT* A.)

Subject index

Abelian groups 2
Absences
  structural (non-space-group) 130
  systematic 76, 80, 127
Additional symmetry elements 40
Additional symmetry operations 40
Affine mapping 10
Affine space 17
Angle of rotation 12
Anorthic (triclinic) system 109
Antiphase domains 93
Aristotype 92
Arithmetic crystal class 30, 118–119
Asymmetric unit 123
Atomic heterogeneity, effect on structure-factor statistics 78
Auslöschungen (reflection conditions) 127
Automorphism group 23–24
Axis of a rotation 10, 12, 111
Axis of a rotoinversion 11, 12, 111
Axis of a screw rotation 10, 111

Bärnighausen tree 92
Basic structure 92
Basis 17, 57, 108
  change of 18, 58, 90
  conventional 19, 32
  dual 63
  primitive 19, 34
  reciprocal 21, 63
Basis vectors 32, 59, 218
Bicrystal 95
Bieberbach groups 47
Bilayers 96
Bilbao Crystallographic Server 89
Blickrichtung (symmetry direction) 117
Bravais class 28
Bravais group 23, 28
Bravais–Miller indices 129
Bravais system 108
Bravais type of lattice 28, 30, 75, 108–109, 118
Brillouin zone 19

C cell 59
Cell
  centred 20, 109–110, 125, 127
  conventional 20, 108
  multiple 20
  primitive 20, 109, 125, 127
  reduced 75, 81, 120
Cell choice, monoclinic 66, 109, 116, 120, 217
Cell parameters 18, 108–109
Centre 15
Centre of a rotoinversion 11
Centred cell 20, 109–110, 125, 127
Centred lattice 19, 34, 127
Centring vector 19, 34
Chirality 10, 78
Classification of crystal structures 89
Coaxial equivalents of a reflection or glide reflection 15
Commutative groups 2
Comparison of crystal structures 89
Composition of symmetry operations 2
Composition series 45
Conjugacy 6
Conjugacy classes 8–9
Conjugate subgroups 6, 8–9, 126, 212
Conjugation 8
Conventional basis 19, 32
Conventional cell 20, 108
Conventional coordinate system 31, 108–109
Convergent-beam electron microdiffraction (CBED) 83
Coordinate system 11, 18, 31, 108–110
  transformation of 57–63, 218–219, 232–233
Coordinates and coordinate triplets 39, 48, 126
Coplanar equivalents of a reflection or glide reflection 15
Coset decomposition 23–24, 39–40
Coset representatives 5, 23, 52, 55, 102
Cosets 5
Crystal class 30, 45, 108, 118–119
Crystal family 31, 108–109
Crystal pattern 2, 10
Crystal system 31, 36, 108–109
Cubic point groups 109
Cumulative distributions of structure-factor magnitudes 77
Cyclic groups 4, 32

D cell (hexagonal lattice) 59
Data-collection strategy 76
Daughter phase 92
Defining operation 15, 110–111
Derivative structure 92
Determinant 12
Dihedral groups 32
Dirichlet domain 19
Distorted structure 92
Domains 93–96
Dual basis 63

e glide 69, 73
e-glide plane 16, 110, 112–113
Eigensymmetry 43
Electron density 79
Element set 15, 34, 40, 110
Enantiomeric purity 78
Enantiomorphic space groups, subgroups of 91
Enantiomorphism 27
Enantioselective synthesis 81
Equivalence classes 8
Explicit space-group symbols 32
Extended Hermann–Mauguin space-group symbols 33, 36, 69–74, 110, 118
Extinction symbols 81
Extinctions 127

Factor groups 6–7
Family of structures 92
Fixed point 10, 13
Fixed-point-free space groups 47
Fourier series expansion of probability density functions 77
Fourier synthesis 124
Friedel opposites 75, 79
Frieze groups 94, 224–225
Full Hermann–Mauguin space-group symbols 33, 36, 66, 69, 117–118
Fundamental domain 19

General linear group 2
General position 39, 43, 47, 122, 125–126, 225
General reflection conditions 68, 127
General subgroups 90
Generation of a space group 44, 46
Generators 4, 45, 125
Geometric crystal class 27, 32
Geometric element 3, 15, 32, 41
Glide part of a symmetry operation 13
Glide plane 11, 128
Glide reflection 11, 13, 36–37, 43, 111
Glide vector 11, 13, 111–113, 129
Graphical symbols 112–115
Group actions 7
Group–subgroup relations 89–94
Groups 2

H and *h* cell (hexagonal lattice) 60–61, 110, 127, 131
Hall symbols 32
Handedness 10
Hermann, theorem of 90, 214
Hermann–Mauguin symbols 33–36, 99, 100, 102, 110–111, 117–118
Hettotype 92
Hexagonal axes, cell and coordinate system 60–61, 109–110
Hexagonal Bravais lattice 109
Hexagonal crystal family 108–109
Hexagonal crystal system 36, 109
Hexagonal point groups 109
Hexagonal space groups, diagrams for 121
Holohedry 29–30, 78, 109
Homeotypic structures 92

Ideal crystal 10
Identity operation 10
Image point 10
Improper isometry 10
Index of a subgroup 5, 90, 212, 217
Integral reflection conditions 127
International space-group symbols 117
Intrinsic translation part of a symmetry operation 13, 41
Inversion 11, 37
Inversion centre 11, 43, 111, 114
Isometry 10, 17
Isomorphic subgroups 91, 94, 213–214, 217
Isotypic structures 92

Jones' faithful representation symbol 38

Klassengleiche (*k*-) subgroups 90–92, 213–214

Lagesymmetrie (site symmetry) 126
Lattice 17, 19, 21, 34, 75, 127
Lattice basis 17
Lattice parameters 109
Lattice point group 29
Lattice point symmetry 109
Lattice system 30, 108
Laue class 76, 118
Laue group 75, 109
Layer groups 50–51, 94–100, 224–225
Layered structures 96–98
Left coset 5
Line groups 109, 111, 131
Linear orbit 94
Linear part of a symmetry operation 11, 17
Liquid crystals 94, 96
Location of a subgroup 103
Location part of a symmetry operation 13, 41

Macromolecular crystallography, space-group determination in 83
Macroscopic crystal 10
Magnetic groups 100–102
Magnetic superfamily 101
Mappings 10
Matrices for unit-cell (basis) transformation 59
Matrix–column pair 11, 17, 64, 212
Matrix part of a symmetry operation 11
Matrix representation of a symmetry operation 11, 124, 126
Maximal subgroups 4, 89–91, 212–213
Metric tensor 18, 64, 75
Miller indices 64
Minimal supergroups 91, 212–214, 225
Mirror plane 11, 111
Mirror point 131
Misordered (disordered) structure 92
Missing spectra 127
Monoclinic crystal system 109
Monoclinic point groups 109
Monoclinic settings and cell choices 66, 109, 116, 120, 217
Monoclinic space groups 70, 116–117, 120
Motion 10
Multi-domain crystals 79
Multiple cell 20
Multiplication table 2
Multiplicity 40, 48, 126

n-fold rotation 10
N(*z*) test for centrosymmetry 77
Nanocrystals, space-group determination 83
Negative sense of rotation 10
Non-characteristic orbit 44, 130

Non-symmorphic space groups 24
Normal subgroups 6
Normalizers 8–9

Oblique crystal system 36
Obverse setting of *R* cell 59, 67, 110, 116, 122, 127
One-dimensional groups, *see* line groups
Orbit 8, 44, 46, 94, 130
Order of a group 2
Order of an isometry 10
Oriented site-symmetry symbol 48, 126
Origin 108, 116, 123
Origin choice 25, 66, 68, 217
Origin shift 57, 90, 218–219
Orthohexagonal *C* cell 59
Orthorhombic crystal system 109
Orthorhombic point groups 109
Orthorhombic space groups 59, 71, 73, 120, 219–221

Parent structure 92
Patterson function 118
Patterson symmetry 118
Penetration line 51
Penetration rod groups 99
Phase transitions 64, 89, 92, 96
Physical properties of crystals 78
Plane groups 32, 53, 97, 117, 128
  diagrams for 120
  examples of tables from Volume A 131–139
  reflection conditions 129
  symbols for 36, 109, 117
  symmetry directions 117
  symmetry elements and operations of 111, 114
Point groups 17, 22, 75, 109, 131, 233
  point-group types 31–32
  point-group visualization 233
  symmetry elements and operations of 111, 114
Point symmetry 109, 126
Polymers 99–100
Polymorphs 98
Polytypes 98
Positive sense of rotation 10
Possible space groups 76
Preimage 10
Primitive basis 19, 34
Primitive cell 20, 109, 125, 127
Primitive lattice 19, 34
Priority rule 34
Probability density functions (p.d.f.'s) 76–77
Projection 130–131
Projection–slice theorem 50
Projection symmetry 130
Projections 36, 50, 53, 130
Proper isometry 10
Proper subgroups 4
Prototype 92
Punktlage (position) 126
Punktsymmetrie (site symmetry) 126

Quotient groups 7

R cell (rhombohedral lattice) 59, 67, 110, 116, 122, 127
R_{merge} 76, 83
Racemate 79
Rational sections 50–51
Real crystal 10
Reciprocal basis 21, 63
Reciprocal lattice 21
Rectangular crystal system 36
Reduced cell 75, 81, 120
Reduced magnetic superfamily 101
Reduced symmetry operation 10, 13
Reflection (mirror reflection) 11, 13, 37, 111–114
Reflection conditions 68, 76, 79, 127–130
Reflection plane 11
Reflection point 131
Reflections, symmetry related 80
Reflections, systematically absent 76, 80, 127
Representative group 101
Representative symmetry directions 34–36
Resonant scattering 75
Reverse setting of *R* cell 59, 67, 110, 116, 127
Reversible mapping 10
Rhombohedral axes, cell and coordinate system 59, 67, 109–110, 122, 127, 217
Rhombohedral Bravais lattice 108–109
Rhombohedral lattice 21
Rhombohedral space groups 66, 108, 110, 116, 122, 127
Right coset 5
Rigid motion 10
Rod groups 50–51, 94–100, 224–225
Rotation 10, 12–13, 36
Rotoinversion 11–12, 37, 111

S centring 109
Scanning direction 94
Scanning group 53
Scanning tables 94, 96, 99
Schoenflies symbols 32–33, 70, 117
Screw part of a symmetry operation 13
Screw rotations 10–11, 13, 37, 42, 98
  reflection conditions for 129
  symbols for 111, 114–115
Section plane 51
Sectional layer groups 94
Sections 36, 50–51
Seitz symbols 11, 37–38
Serial reflection conditions 128
Settings 66, 69, 219
  monoclinic 59, 70, 120
  orthorhombic 59, 71, 120, 219–220
  rhombohedral 59, 110, 116, 127
Shift vector 63
Short Hermann–Mauguin symbols 33, 36
Side-preserving or side-reversing operations 51
Simplest symmetry-operation rule 34
Site symmetry 48, 126
Site-symmetry group 8, 25, 46–47
Sohncke space groups 78
Space-group determination 75, 127
Space-group numbers 32
Space-group operations 17
Space-group symbols 32–36, 69–74, 117–118
  changes of 110
  Hermann–Mauguin symbols 33–36, 99, 100, 102, 110–111, 117–118
  Schoenflies symbols 32–33, 70–73
Space-group types 26
Space groups 17, 32
  classification of 25, 108
  diagrams for 42, 119
  enantiomorphic, subgroups of 91
  examples of tables from Volume A 131–132, 140–211
  fixed-point-free 47
  generation of 44, 46
  incorrect assignment of 130
  non-symmorphic 24
  one-dimensional (line groups) 109, 111, 131
  Sohncke space groups 78
  symmorphic 24, 118
  transformations between different descriptions of 66
Special position 47, 126
Special projections 130
Special reflection conditions 127, 130
Specialized metrics 29
Square crystal system 36
Stabilizers 8, 80
Structural (non-space-group) absences 130
Structure factors 76–79
Structure validation 78
Subgroups 4–6, 89–94, 126, 212, 217
  examples of tables from Volume A1 215–216
  group–subgroup relations 89–94
  of magnetic groups 101
  of subperiodic groups 225
Subperiodic groups 50, 94–100, 224–225
  examples of tables from Volume E 225–231
Supergroups 4, 89, 91, 212–214, 225
Symbols
  for Bravais types of lattice 108–109
  for centring types 110
  for crystal classes 109
  for crystal families 108–109
Symbols
  for helices 99
  for line groups 109, 131
  for magnetic group types 102
  for plane groups 36, 109, 117
  for rod groups 99
  for site symmetries 48, 126
  for space groups 32–36, 69–74, 117–118
  for symmetry elements and operations 36–37, 110–115
  Jones' faithful representation symbol 38
  Seitz symbols 11, 37–38
Symétrie ponctuelle (site symmetry) 126
Symmetry axes 114
Symmetry Database 232
Symmetry directions 34–36, 117
Symmetry elements 3, 14, 110–115
  additional 40
  diagrams of 42, 119, 225
  projection of 131
  symbols for 110–115
Symmetry of special projections 130
Symmetry operations 2, 10, 124, 126
  additional 40
  composition of 2
  defining symmetry operation 15, 110–111
  for subperiodic groups 225
  intrinsic translation part 13, 41
  location part 13, 41
  matrix representation of 124, 126
  reduced 10, 13
  relationship to general position 39
  symbols for 36–37, 110–115
  transformation of 64, 68
  type of 12, 40
Symmetry-related reflections 80
Symmorphic space groups 24, 118
Systematic absences 76, 80, 127

Tetragonal crystal system 109
Tetragonal point groups 109
Tetragonal space groups, diagrams for 121
Thin films 94
Topotactic reaction 89, 93
Trace of a matrix 12
Transformation of coordinate system 57–63, 218–219, 232–233
Transformation of reflection conditions 68
Transformation of symmetry operations 64, 68
Transformations between different space-group descriptions 66
Translation 10, 36
Translation part of a symmetry operation 11, 13, 17
Translation subgroup 17, 23, 39–40, 52, 55
Translationengleiche (*t*-) subgroups 90, 92, 212–213
Tree of group–subgroup relations 92
Triclinic crystal system 109
Triclinic space groups, diagrams for 120
Trigonal crystal system 108–109
Trigonal point groups 109
Trigonal space groups 121–122
Twinning 79, 93–94
Two-dimensional space groups, *see* plane groups

Unique monoclinic axis 66–67, 70, 116, 120
Unit cell 19–20
Unit (or identity) element 2

Vector space 17
Voronoï domain 19

Wallpaper groups 36, 53
Wigner–Seitz cell 19
Wirkungsbereich (domain of influence) 19
Wyckoff letters 48, 126
Wyckoff positions 8, 47, 126
  examples of tables from Volume A1 220–223
  interchange of 219
  relations between 89, 92, 94, 217–220
  splitting of 92, 218

Zonal reflection conditions 128

① Headline in abbreviated form:

The *sequential number of the plane or space group*, as introduced in *International Tables for X-ray Crystallography* Vol. I (1952) (*cf.* Section 1.4.1.2).

Short international (Hermann–Mauguin) symbol for the plane or space group: Sections 1.4.1.4 and 2.1.3.4.

② *Generators selected*: Sections 1.4.3 and 2.1.3.10. The line *Generators selected* states the symmetry operations and their sequence, selected to generate all symmetry-equivalent points of the *General position* from a point with coordinates x, y, z. The generating translations are listed explicitly, while the non-translational generators are given as numbers (p) that refer to the coordinate triplets of the general position and the corresponding entries under *Symmetry operations*.

③ *Positions*: Sections 1.4.4 and 2.1.3.11. The entries under *Positions* (called also *Wyckoff positions*) consist of the block *General position* given at the top followed downwards by the blocks of various special Wyckoff positions with decreasing multiplicity and increasing site symmetry. For each general and special position its multiplicity, Wyckoff letter, oriented site-symmetry symbol, the appropriate coordinate triplets and the reflection conditions are listed:

Multiplicity: The Wyckoff-position multiplicity is the number of symmetry-equivalent points that lie in the conventional unit cell. The quotient of the multiplicity for the general position by that of a special position gives the order of the site-symmetry group of the special position.

Wyckoff letter: Each Wyckoff position is labelled by a letter in alphabetical order, starting with '*a*' for a position with site-symmetry group of maximal order and ending with the highest letter for the general position.

Site symmetry: The site symmetry at the points of a special position is described by the site-symmetry group, which is isomorphic to a subgroup of the point group of the space group. The site-symmetry groups are indicated by *oriented site-symmetry symbols* (third column) that show how the symmetry elements at a site are related to the symmetry elements of the crystal lattice (*cf.* Section 2.1.3.12).

Coordinates: The coordinate triplets of a position represent the coordinates of the symmetry-equivalent points in the unit cell. The sequence of coordinate triplets is produced in the same order as the symmetry operations, generated by the chosen set of generators, omitting duplicates. For centred space groups, the centring translations have to be added to the listed coordinate triplets in order to obtain a complete set of triplets for the Wyckoff position.
The coordinate triplets of the general position can also be interpreted as a shorthand notation of the matrix–column presentation of a representative set of the symmetry operations of the space group (*cf.* Sections 1.3.3.2 and 1.4.2.3).

Reflection conditions (last column): Sections 1.6.3 and 2.1.3.13. The listed systematic reflection conditions for diffraction of radiation by crystals are formulated as 'conditions of occurrence' (structure factor not systematically zero). The *General conditions* are associated with systematic absences caused by the presence of lattice centrings, screw axes and glide planes. The *Special conditions* (or 'extra' conditions) apply only to special Wyckoff positions and always occur in addition to the general conditions of the space group.

④ *Symmetry of special projections*: Sections 1.4.5.3 and 2.1.3.14. For each space group, orthogonal projections along three (symmetry) directions are listed. Given are the projection direction, the Hermann–Mauguin symbol of the plane group of the projection, the relations between the basis vectors of the plane group and the basis vectors of the space group, and the location of the origin of the plane group with respect to the unit cell of the space group.

EXPLANATION OF THE SPACE-GROUP DATA